Lecture Notes of the Institute for Computer Sciences, Social Informatics and Telecommunications Engineering 75

Rahul Jain Rajgopal Kannan (Eds.)

Game Theory for Networks

Second International ICST Conference
GAMENETS 2011
Shanghai, China, April 16–18, 2011
Revised Selected Papers

Springer

Volume Editors

Rahul Jain
University of Southern California
Los Angeles, CA 90089-2562, USA
E-mail: rahul.jain@usc.edu

Rajgopal Kannan
Louisiana State University
Baton Rouge, LA 70803, USA
E-mail: rkannan@csc.lsu.edu

ISSN 1867-8211 e-ISSN 1867-822X
ISBN 978-3-642-30372-2 e-ISBN 978-3-642-30373-9
DOI 10.1007/978-3-642-30373-9

Springer Heidelberg Dordrecht London New York

Library of Congress Control Number: 2012937594

CR Subject Classification (1998): C.2, H.4, D.2, H.3, I.2, K.6.5

Typesetting: Camera-ready by author, data conversion by Scientific Publishing Services, Chennai, India

Printed on acid-free paper

Springer is part of Springer Science+Business Media (www.springer.com)

Preface

The second edition of the International ICST Conference on Game Theory in Networks (GAMENETS 2011) was held during April 16–18, 2011, at the Crowne Plaza Century Park in Shanghai, China, a beautiful venue internationally recognized for business, trading and rapidly becoming a conspolitan international city. This edition continued the scientific event previously held in Istanbul, Turkey (2009), with both events being dedicated to the latest advances on game-theoretic methods and techniques for solving communication and networks problems. GAMENETS 2011 brought together researchers who apply game theory to analyze, design and assess the performance of networks, including such problems as cyber and network security, spectrum sharing, social network analysis, algorithmic game theory, etc.

GAMENETS 2011 was organized as a 3-day 2-track event with 3 keynote speeches and 14 technical sessions. A total of 42 paper submissions were received from all over the world. All papers received rigorous peer reviews from the Technical Program Committee (TPC). After carefully examining all the review reports, the TPC finally selected 24 regular papers for presentations at the conference with an acceptance rate of 57%. In addition, 23 papers were invited from well-known researchers in the area. The program included three outstanding keynote speakers from academia and industry: Nimrod Meggido (IBM Almaden Research), Sergiu Hart (Hebrew University of Jerusalem), and John Chuang (UC Berkeley). A social dinner was also scheduled on the second day of the conference.

The conference had close to 60 participants from industry and academia from a wide range of countries such as the USA, Canada, France, Hong Kong, Singapore, China, India, Japan, UK, Italy, and Greece. We thank the invited keynote and other speakers for accepting our invitation, and taking out time to make it to the conference in Shanghai. We also thank all members of the TPC for their valuable input in ensuring an excellent technical program. We also thank the General Co-chairs Thanos Vasilakos and Xinbing Wang, and other members of the Organizing Committee for putting in so much hard work to make this event a great success.

Our gratitude also extends to the representatives of ICST, EAI, and CREATE-NET for sponsoring this event and for their help in arranging all the logistics. A special acknowledgment is due to Elena J. Fezzardi from EAI for running the event smoothly. Last but not least, we the General Chairs would like to thank all the reviewers, whose effort reflects their commitment to the success of this yearly event.

We eagerly look forward to another successful GAMENETS edition in 2012 and forthcoming years.

Rahul Jain
Rajgopal Kannan

Organization

Steering Committee Members

Athanasios V. Vasilakos	National Technical University of Athens, Greece
Imrich Chlamtac	Create-Net, Italy

General Chair

Athanasios V. Vasilakos	National Technical University of Athens, Greece
Xinbing Wang	Shanghai Jiaotong University, China

TPC Chairs

Rajgopal Kannan	LSU, USA
Rahul Jain	USC, USA

Publication Chair

Foad Dabiri	UCLA, USA

Workshops Chairs

Hung-Yu Wei	National Taiwan University
Jang-Won Lee	Yonsei University, Korea
Hongyang Chen	The University of Tokyo, Japan

Web Chair

Shen Gu	Shanghai Jiaotong University, China

Conference Coordinators

Elena J. Fezzardi	EAI, Italy
Richard Heffernan	EAI, Italy

Table of Contents

Session A1: Mechanism Design

Session A2: Physical Layer Games

Session B1: Network Neutrality and Regulation

Session B2: Network Mechanisms

Session C1: Stochastic and Dynamic Games

Session C2: Security Games I

Session D1: Algorithmic Game Theory and Learning

Session D2: Game-Theoretic Network Models

Session E1: Cooperative Games in Networks

Session E2: Security Games II

Session F1: Incentives for Relaying

Session F2: Spectrum Sharing Games

Session G1: P2P and Social Networks

Session G2: Economics of Network QoS

Mechanism Design with Limited Information: The Case of Nonlinear Pricing*

Dirk Bergemann[1], Ji Shen[1], Yun Xu[2], and Edmund M. Yeh[2]

[1] Department of Economics, Yale University, New Haven, CT 06520, U.S.A.
[2] Department of Electrical Engineering, Yale University, New Haven, CT 06520
{dirk.bergemann,ji.shen,yun.xu,edmund.yeh}@yale.edu

Abstract. We analyze the canonical nonlinear pricing model with limited information. A seller offers a menu with a finite number of choices to a continuum of buyers with a continuum of possible valuations. By revealing an underlying connection to quantization theory, we derive the optimal finite menu for the socially efficient and the revenue-maximizing mechanism. In both cases, we provide an estimate of the loss resulting from the usage of a finite n-class menu. We show that the losses converge to zero at a rate proportional to $1/n^2$ as n becomes large.

Keywords: Mechanism Design, Limited Information, Nonlinear Pricing, Quantization, Lloyd-Max Optimality.

1 Introduction

The theory of mechanism design addresses a wide set of questions, ranging from the design of markets and exchanges to the design of constitutions and political institutions. A central result in the theory of mechanism design is the "revelation principle" which establishes that if an allocation can be implemented incentive compatible in any mechanism, then it can be truthfully implemented in the direct revelation mechanism, where every agent reports his private information, his type, truthfully. Yet, when the private information (the type space) of the agents is large, then the direct revelation mechanism requires the agents to have abundant capacity to communicate with the principal, and the principal to have abundant capacity to process information. By contrast, the objective of this paper is to study the performance of optimal mechanisms, when the agents can communicate only limited information or equivalently when the principal can process only limited information. We pursue the analysis in the context of a representative, but suitably tractable, mechanism design environment, namely the canonical problem of nonlinear pricing. Here the principal, the seller, is offering a variety of choices to the agent, the buyer, who has private information about his willingness-to-pay for the product.

The distinct point of view, relative to the seminal analysis by [11] and [8], resides in the fact that the information conveyed by the agents, and subsequently

* The first author acknowledges financial support through NSF Grant SES 0851200.

R. Jain and R. Kannan (Eds.): GameNets 2011, LNICST 75, pp. 1–10, 2012.

the menu of possible choices offered by the seller, is finite, rather than uncountable as in the earlier analysis. The limits to information may arise for various, direct or indirect, reasons. On the demand side, it may be too difficult or too complex for the buyer to communicate his exact preferences and resulting willingness to pay to the seller. On the supply side, it may be too time-consuming for the seller to process the fine detail of the consumer's preferences, or to identify the consumer's preferences across many goods with close attributes and only subtle differences.

Our analysis adopts a linear-quadratic specification (analogous to that of [11]) in which the consumer's gross utility is the product of his willingness-to-pay (his type θ) and the consumed quantity q of the product, whereas the cost of production cost is quadratic in the quantity. For this important case, we reveal an interesting connection between the problem of optimal nonlinear pricing with limited information to the problem of optimally quantizing a source signal by using a finite number of representation levels in information theory. In our setting, the socially efficient quantity q for a customer should be equated to his valuation θ if a continuum of choices were available. In the case where a finite number of choices are accessible q can take on only some values. If we see θ as the source signal and q as the representation level, then the total social welfare can be written as the mean square error between the source signal and the representation signal. Given this, the welfare maximization problem can be characterized by the Lloyd-Max optimality conditions, a well-established result in the theory of quantization. Furthermore, we can extend the analysis to the revenue maximization problem, after replacing the customer's true valuation by the corresponding virtual valuation, as defined by [12]. We estimate the welfare and revenue loss resulting from the use of a finite n-class contract (relative to the continuum contract). In particular, we characterize the rate of convergence for the welfare and revenue loss as a function of n. We examine this problem first for a given distribution on the customer's type, and then over all possible type distributions with finite support. We establish that the maximum welfare loss shrinks towards zero at the rate proportional to $1/n^2$.

The role of limited information in mechanism design has recently attracted increased attention. In a seminal paper, [13] considers the impact of a finite number of priority classes on the efficient rationing of services. His analysis is less concerned with the optimal priority ranking for a given finite class, and more with the approximation properties of the finite priority classes. [10] rephrases the priority rationing problem as a two-sided matching problem (between consumer and services) and shows that already binary priority contract ("coarse matching") can achieve at least half of the social welfare that could be generated by a continuum of priorities. [4] extend the matching analysis and explicitly considers monetary transfers between the agents. In particular, they present lower bounds on the revenue which can be achieved with specific, not necessarily optimal, binary contracts. By contrast, [7] suggest a specific allocation, the "profit-participation" mechanism to establish approximation results, rather than finite optimality results, in the nonlinear pricing environment. While the above contributions are concerned with single agent environments, there have been a number of contributions to multi-agent mechanisms, specifically single-item auctions among many bidders. [3] consider the effect of restricted communication in auctions with either two agents or binary messages for every agent. [5] generalizes the analysis by allowing for a finite number of messages and agents. In turn, their equilibrium characterization in terms of partitions shares features with the optimal information structures in auctions as derived by [1].

2 Model

We consider a monopolist facing a continuum of heterogeneous consumers. Each consumer is characterized by a quasi-linear utility function: $u(\theta, q, t) = \theta q - t$, where q is the quantity of his consumption purchased from the monopolist, θ describes his willingness-to-pay for the good (his "type"), and t is the transfer paid by the agent. The monopoly seller offers q units of products at a cost $c(q) = \frac{1}{2}q^2$. Consequently, the net utility of the buyer and seller are given by $\theta q - t(q)$ and $t(q) - \frac{1}{2}q^2$ respectively, where $t(q)$ is the transfer price that the buyer has to pay the seller for a quantity q of the product. The specific parameterization of the utility function and the cost function is referred to as the "linear-quadratic model" and has been extensively studied in the literature beginning with [11]. The prior distribution of θ is given by F and has compact support on $\mathbb{R}$. Without loss of generality we normalize it to the unit interval $[0, 1]$. We denote the set of all distribution on the unit interval by $\Delta \equiv \Delta[0, 1]$.

3 Welfare Maximization

We first consider the social welfare maximization problem in the absence of private information by the agent. That is, the willingness-to-pay of the buyer, his type, is publicly known. Moreover, as the transfer t does not determine the level of the social surplus, but rather its distribution between buyer and seller, it does not enter the social welfare problem. In the absence of communication constraints, $n = \infty$, the social surplus, denoted by SW_∞ is then determined as the solution to the allocation problem:

$$SW_\infty \triangleq \max_{q(\theta)} \mathbb{E}\left[\theta q(\theta) - \frac{1}{2}q^2(\theta)\right]. \tag{1}$$

In the absence of private information, the optimal solution for every type θ can be obtained pointwise, and is given by $q^*(\theta) = \theta$. In other words, the socially optimal menu $M_\infty^* = \{q^*(\theta) = \theta\}$ offers a continuum of choices and assigns each consumer the quantity of the good which is equal to his willingness-to-pay. The resulting social surplus is given by $SW_\infty = \frac{1}{2}\mathbb{E}\left[\theta^2\right]$. Importantly, given its linear-quadratic structure, the welfare maximizing problem is equivalent to minimizing the mean square error (MSE), $\mathbb{E}_\theta[(\theta - q)^2]$. We shall use this equivalent representation of the problem as we now consider the problem with communication constraints.

By contrast, we seek to determine the optimal menu when we can offer only a finite number of choices, and we denote by $\mathcal{M}_n$ the set of contracts which offer at most a finite number n of quantity choices. Henceforth, such a discretized contract $M_n = \{q_k\}_{k=1}^n$ is called an $n-$class contract or $n-$class menu. The socially optimal assignment rule then seeks to assign to each buyer with type θ a specific quantity $q(\theta)$ with the property that the quantity $q(\theta)$ represents an element in the $n-$class contract. For a given number of choices n, the social welfare problem is:

$$SW_n = \max_{q(\theta)} \mathbb{E}_\theta \left[\theta q(\theta) - \frac{1}{2} (q(\theta))^2 \right] \quad \text{subject to } \{q(\theta)\}_{\theta=0}^1 \in \mathcal{M}_n. \tag{2}$$

Given that the valuation of the buyer is supermodular, i.e. $\partial^2 u(\theta, q) / \partial\theta\partial q > 0$, it follows that the optimal assignment of types to quantities has a partitional structure. Let $\{A_k = [\theta_{k-1}, \theta_k)\}_{k=1}^n$ represent a partition of the set of consumer types where $0 = \theta_0 < \cdots < \theta_{k-1} < \theta_k < \cdots < \theta_n = 1$. A consumer with type $\theta \in A_k$ will be assigned $q^*(\theta) = q_k^*$, and the socially optimal menu $M_n^* = \{q_k^*\}_{k=0}^n$ is increasing in k, so that $q_1^* < q_2^* < \cdots < q_k^*$. Now, given the relationship to the mean square error problem discussed above, if we view θ as the source signal and q_k as the representation points of θ on the quantization intervals $A_k = [\theta_{k-1}, \theta_k)$, then the solution to the social welfare maximizing contract is given by the n-level quantization problem, where both the quantization intervals A_k and the corresponding representation points q_k are chosen to minimize the mean square error (MSE):

$$MSE_n \equiv \min_{q(\theta)} \mathbb{E}_\theta \left[(\theta - q)^2 \right], \quad \text{subject to } \{q(\theta)\}_{\theta=0}^1 \in \mathcal{M}_n. \tag{3}$$

Hence, the optimal solution must satisfy the Lloyd-Max optimality conditions, see [6] and [9].

Proposition 1 (Lloyd-Max-Conditions). *The optimal menu M_n^* of the social welfare problem (2) satisfies:*

$$\theta_k^* = \frac{1}{2} \left(q_k^* + q_{k+1}^* \right), \quad q_k^* = \mathbb{E}_\theta \left[\theta | \theta \in \left[\theta_{k-1}^*, \theta_k^* \right) \right], \quad k = 0, \ldots, n. \tag{4}$$

That is, q_k^*, the production level for the interval $A_k^* = \left[\theta_{k-1}^*, \theta_k^*\right)$, must be the conditional mean for θ given that θ falls in the interval A_k^* and θ_k^*, which separates two neighboring intervals A_k^* and A_{k+1}^*, must be the arithmetic average between q_k^* and q_{k+1}^*. One can observe immediately that q_k^* is actually determined by the first-order condition with respect to (3) because MSE_n in (3) is convex in q_k when taking θ_k and θ_{k+1} as given. Similarly, θ_k^* is determined by the first-order condition when q_k and q_{k+1} are given because MSE_n in (3) is convex in θ_k when taking q_k and q_{k+1} as given. For certain family of distributions (e.g., uniform distribution and some discrete distributions) we can obtain closed-form solutions from the Lloyd-Max optimality conditions. We are interested in the relative performance of finite contracts and evaluate the difference between SW_∞^* and SW_n^*.

Definition 1. *Given any $F \in \Delta$, the welfare loss of an n-class contract compared with the optimal continuous contract is defined by $L(F;n) \equiv SW_{\infty}^{*} - SW_{n}^{*}$.*

It is easy to see that the lower bound over all densities is zero, i.e. $\inf_{F\in\Delta} L(F;n) = 0$. This can be achieved by a categorical distribution, i.e., $\Pr\left(\theta = \frac{k}{n}\right) = \frac{1}{n}$ for $k = 1, ..., n$. Our main task is to provide an upper bound over all distributions, i.e., the worst-case scenario from the point of view of total social welfare.

Definition 2. *The maximum welfare loss of an n-class contract over all $F \in \Delta$ is given by $L(n) \equiv \sup_{F\in\Delta} L(F;n)$.*

We first consider a simple example, and show in detail how to use the Lloyd-Max conditions to obtain the optimal discretized contract and measure the resulting welfare loss.

Example 1. Suppose that θ is uniformly distributed over $[0,1]$. The optimization problem (2) has a unique optimal solution given by $\theta_k^* = \frac{k}{n}$, $q_k^* = \frac{(k-1)/2}{n}$, $k = 0, 1, \ldots, n$. The expected social welfare is $SW_n^* = \frac{1}{6} - \frac{1}{24n^2}$ and the welfare loss is $SW_{\infty}^* - SW_n^* = \frac{1}{24n^2}$.

In this example, the cutoff points are uniformly distributed, which is due to the fact that the underlying distribution of θ is uniform. In addition, the convergence rate of the welfare loss induced by discretized contracts is of the order $1/n^2$. Next, we provide a general estimate of the convergence rate of the welfare loss induced by discretized contracts as the number of classes tends to infinity. A direct approach to calculate the welfare loss for general distributions would require the explicit form of the optimal quantizer, determined by the Lloyd-Max conditions. But an explicit characterization of the optimal quantizer is not known, and thus we pursue an indirect approach to obtain a bound through a series of suboptimal quantizers. For any given $F \in \Delta$, we have:

$$SW_n = \mathbb{E}_{\theta}\left[\theta q - \frac{1}{2}q^2\right] = \frac{1}{2}\mathbb{E}\left(\theta^2\right) - \frac{1}{2}MSE_n,$$

and since the social welfare with the continuous contract is $SW_{\infty} = \frac{1}{2}\mathbb{E}\left(\theta^2\right)$, we obtain

$$SW_{\infty} - SW_n = \frac{1}{2}MSE_n. \tag{5}$$

Given the necessary conditions of Proposition 1, it will suffice to confine our attention to the set of finite menus $\mathcal{M}_n^*$ with the property that, given a distribution $F \in \Delta$, the menu $M_n = \{q_k\}_{k=1}^n$ can be generated by a finite partition A_k through $q_k = \mathbb{E}(\theta|\theta \in A_k)$, $k = 1, \ldots, n$, so that $\mathcal{M}_n^*$ is the feasible set of menus $\mathcal{M}_n$ consistent with the optimality condition (4). For any $M_n \in \mathcal{M}_n^*$,

$$MSE_n = \mathbb{E}_{\theta}\left[(q-\theta)^2\right] = \sum_{k=1}^{n}\left(F(\theta_k) - F(\theta_{k-1})\right) var\left(\theta|\theta \in A_k\right). \tag{6}$$

We can write $L(F;n)$ and $L(n)$, using (5) and (6) as follows:

$$L(F;n) = \inf_{M_n \in \mathcal{M}_n^*} [SW_\infty^* - SW_n] = \inf_{M_n \in \mathcal{M}_n^*} \frac{1}{2} \sum_{k=1}^{n} (F(\theta_k) - F(\theta_{k-1}))\, var(\theta | \theta \in A_k), \tag{7}$$

and consequently:

$$L(n) = \sup_{F \in \Delta} \inf_{M_n \in \mathcal{M}_n^*} \frac{1}{2} \sum_{k=1}^{n} (F(\theta_k) - F(\theta_{k-1}))\, var(\theta | \theta \in A_k). \tag{8}$$

It is then central to estimate the variance of θ conditional on the interval A_k to provide an upper bound on $L(n)$.

Proposition 2. *For $F \in \Delta$, and any $n \geq 1$, $L(F;n) \leq \frac{1}{8n^2}$.*

Proof. For any given $F \in \Delta$, let M_n be defined by $\theta'_k = k/n$, $q'_k = \mathbb{E}[(\theta | \theta \in [\theta_{k-1}, \theta_k))]$, $k = 0, 1, \ldots, n$. Now, we have $L(F;n)$ given by:

$$\inf_{M_n \in \mathcal{M}_n^*} \frac{1}{2} \sum_{k=1}^{n} (F(\theta_k) - F(\theta_{k-1}))\, var(\theta | \theta \in [\theta_{k-1}, \theta_k))$$
$$\leq \frac{1}{2} \sum_{k=1}^{n} \left(F(\theta'_k) - F(\theta'_{k-1})\right) var\left(\theta | \theta \in \left[\theta'_{k-1}, \theta'_k\right)\right).$$

But the variance in any interval is bounded by the following elementary inequality:

$$var\left(\theta | \theta \in \left[\theta'_{k-1}, \theta'_k\right)\right) \leq \frac{1}{4}\left(\theta'_k - \theta'_{k-1}\right)^2 = \frac{1}{4n^2}.$$

It then follows that:

$$L(F;n) \leq \frac{1}{8n^2} \sum_{k=1}^{n} \left(F(\theta'_k) - F(\theta'_{k-1})\right) = \frac{1}{8n^2},$$

which concludes the proof.

By considering the uniform distribution of Example 1, we can in fact show that the maximum welfare loss is bounded both above and below by $1/n^2$ (up to some constant).

Proposition 3. *For any $n \geq 1$, $\frac{1}{24n^2} \leq L(n) \leq \frac{1}{8n^2}$, i.e. $L(n) = \Theta\left(\frac{1}{n^2}\right)$.*

Similar to us, [13] establishes that a finite priority ranking of order n induces a welfare loss of order $1/n^2$. His method of proof is different from ours, in that it does not use quantization explicitly, and in that for the limit results he proposes uniform quantization of the relevant distribution.

4 Revenue Maximization

We now analyze the problem of revenue maximization with limited information. In contrast to the social welfare maximizing problem, the seller wishes to maximizes his expected net revenue. The expected net revenue is the difference between the gross revenue that he receives from the buyer minus the cost of providing the demanded quantity. The contract offered by the principal now has to satisfy two sets of constraints, namely the participation constraint, $\theta q(\theta) - t(\theta) \geq 0$, for all $\theta \in [0,1]$, and the incentive constraints: $\theta q(\theta) - t(\theta) \geq \theta q(\theta') - t(\theta')$, of the buyer for all $\theta, \theta' \in [0,1]$. The participation constraint guarantees that the buyer receives a nonnegative net utility from his choice, and the incentive constraints account for the fact that the type θ is private information to the buyer, and hence the revelation of the information is required to be incentive compatible. The current problem is then identical to the seminal analysis by [11] and [8] with one important exception: the buyer can only access a finite number of choices due to the limited communication with the seller. Now, a menu of quantity-price bundles is designed by the monopolistic seller to extract as much profit as possible.

The revenue maximization problem, finding the optimal solution for the allocation $q(\theta)$ and the transfer $t(\theta)$ simultaneously, then appears to be rather distinct from the welfare maximization problem, which only involved the allocation $q(\theta)$. However, we can use the above incentive constraints to eliminate the transfers and rewrite the problem in terms of the allocation alone. This insight appeared prominently in the analysis of revenue maximizing auction in [12]. He showed that the revenue maximization problem can be transformed into a welfare maximization problem (without incentive constraints) as long as we replace the *true valuation* θ of the buyer with the corresponding *virtual valuation*:

$$\hat{\theta} \equiv \psi(\theta) = \theta - \frac{1 - F(\theta)}{f(\theta)}. \tag{9}$$

The virtual valuation is always below the true valuation, and the inverse of the hazard rate $(1 - F(\theta))/f(\theta)$ accounts for the information rent, the cost of the private information, as perceived by the principal in the optimal mechanism. We shall follow [12] and impose the regularity condition that $\psi(\theta)$ is strictly increasing in θ. With this standard transformation of the problem, the expected profit of the seller (without information constraints) is:

$$\Pi^*_\infty = \max_{q(\theta)} \mathbb{E}_\theta \left[q(\theta)\psi(\theta) - \frac{1}{2} q^2(\theta) \right]. \tag{10}$$

The resulting optimal contract exhibits $q^*(\theta) = \max\{\psi(\theta), 0\}$. Now, in the world with limited information, the seller can only offer a finite menu $\{(q_k, t_k), k = 1, \ldots, n\}$ to the buyer. After rewriting the revenue maximizing problem in terms of the virtual utility, we can omit the dependence on the transfers and rewrite the problem in terms of a choice over a finite set of allocations $\mathcal{M}_n$:

$$\Pi^*_n = \max_{q(\theta) \in \mathcal{M}_n} \mathbb{E}_\theta \left[q\psi(\theta) - \frac{1}{2} q^2 \right]. \tag{11}$$

We denote the distribution function and density function of $\hat{\theta}$ by G and g, respectively. We have $F(x) = \Pr(\theta \leq x) = \Pr(\hat{\theta} \leq \psi(x)) = G(\psi(x))$, and thus $f(x) = g(\psi(x))\psi'(x)$. Using the insights of the previous section, we observe that maximizing the seller's revenue is equivalent to minimizing the mean square error $\mathbb{E}_{\hat{\theta}}[(\hat{\theta}-q)^2]$, where the expectation is taking with respect to the new random variable $\widehat{\theta}$. We then appeal to the appropriately modified Lloyd-Max optimality conditions to characterize the revenue maximizing contract in the presence of information constraints:

Proposition 4. *The revenue maximizing solution (11) satisfies:*

$$\theta_k^* - \frac{1 - F(\theta_k^*)}{f(\theta_k^*)} = \frac{1}{2}\left(q_k^* + q_{k+1}^*\right) \qquad k = 0, \ldots, n-1, \tag{12}$$

and

$$q_k^* = \frac{\theta_{k-1}^*\left(1 - F\left(\theta_{k-1}^*\right)\right) - \theta_k^*\left(1 - F\left(\theta_k^*\right)\right)}{F\left(\theta_k^*\right) - F\left(\theta_{k-1}^*\right)} \qquad k = 1, \ldots, n. \tag{13}$$

Similar to the social welfare problem, we wish to evaluate the upper bound of $\Pi_\infty^* - \Pi_n^*$ across all possible distribution functions $F \in \Delta$. To this end, we define the revenue loss induced by an n-class contract compared with the optimal continuous contract, given a distribution $F \in \Delta$, as $\Lambda(F;n) \equiv \Pi_\infty^* - \Pi_n^*$ and the maximum revenue loss induced by an n-class contract across all $F \in \Delta$ as $\Lambda(n) \equiv \sup_{F \in \Delta} \Lambda(F;n)$. The example of the uniform distribution is again illustrative before turning to the general analysis.

Example 2. Suppose that θ is uniformly distributed over $[0,1]$. The optimization problem (11) has a unique solution: $\theta_k^* = \frac{n+k+1}{2n+1}$, $\quad q_k^* = \frac{2k}{2n+1}$, $\quad k = 0, \ldots, n$. The maximum expected revenue is $\Pi_n^* = \frac{n(n+1)}{3(2n+1)^2}$ and revenue loss is $\Pi_\infty^* - \Pi_n^* = \frac{1}{12(2n+1)^2}$.

It follows that the convergence rate of the revenue loss induced by discretized contracts is also of the order $1/n^2$. We find that the seller tends to serve fewer buyers as compared to the case when a continuous contract is used. This property holds for general distributions as the seller's ability of extracting revenue is limited. To compensate, the seller reduces the service coverage to pursue higher marginal revenues. We now provide the convergence rate of the revenue loss induced by discretized contracts as the number of intervals (classes) tends to infinity. Thus,

$$\Pi_\infty^* - \Pi_n = \frac{1}{2}\left[\int_0^{\hat{\theta}_0} \hat{\theta}^2 g\left(\hat{\theta}\right) d\hat{\theta} + \int_{\hat{\theta}_0}^1 \left(\hat{\theta} - q\right)^2 g\left(\hat{\theta}\right) d\hat{\theta}\right]. \tag{14}$$

The first term in the square bracket captures the revenue loss by reducing the service coverage. The second term in the square bracket and $L(F;n)$ in (7) are very much alike. One can immediately get this term by replacing θ by $\hat{\theta}$ and F by G in $L(F;n)$. We can then adapt Proposition 2 to the current environment.

Proposition 5. *For any $F \in \Delta$, and any $n \geq 1$, $\Lambda(F; n) \leq 1/8n$.*

The approximation result of the revenue maximizing problem is similar to the one of the social welfare program. Likewise, we can use the above uniform example to establish a lower bound for the revenue losses.

Proposition 6. *For any $n \geq 1$, $1/12\,(2n+1)^2 \leq \Lambda(n) \leq 1/8n^2$, and hence $\Lambda(n) = \Theta(1/n^2)$.*

5 Conclusion

We analyzed the role of limited information (or communication) in the context of the canonical nonlinear pricing environment. By focusing on the simple linear-quadratic specification of the utility and cost function, we were able to relate the limited information problem directly to the quantization problem in information theory. This allowed us to explicitly derive the optimal mechanism, both from a social efficiency as well as from a revenue-maximizing point of view. In either case, our analysis established that the worst welfare loss due to the limits of information, imposed by an n-class contract, is of the order of $1/n^2$. In Bergemann et al. (2012), we extend the present analysis for the efficient allocation to a multi-dimensional type space [2].

While the nonlinear pricing environment is of interest by itself, it also represents an elementary instance of the general mechanism design environment. The simplicity of the nonlinear pricing problem arises from the fact that it can viewed as a relationship between the principal, here the seller, and a single agent, here the buyer, even in the presence of many buyers. The reason for the simplicity is that the principal does not have to solve allocative externalities. By contrast, in auctions, and other multi-agent allocation problems, the allocation (and hence the relevant information) with respect to a given agent constrains and is constrained by the allocation to the other agents. For an information-theoretic point of view, the ensuing multi-dimensionality would suggest that the methods of vector quantization rather than the scalar quantization employed here, would become relevant.

Finally, the current analysis focused on limited information, and the ensuing problem of efficient source coding. But clearly, from an information-theoretic as well as economic viewpoint, it is natural to augment the analysis to reliable communication between agent and principal over noisy channels, the problem of channel coding, which we plan to address in future work.

References

1. Bergemann, D., Pesendorfer, M.: Information Structures in Optimal Auctions. Journal of Economic Theory 137, 580–609 (2007)
2. Bergemann, D., Shen, J., Yeh, E., Yun, X.: Multi-Dimensional Mechanism Design with Limited Information. Cowles Foundation for Research in Economics. Yale University (2012)

3. Blumrosen, L., Nisan, N., Segal, I.: Auctions with Severly Bounded Communication. Journal of Artificial Intelligence Research 28, 233–266 (2007)
4. Hoppe, H., Moldovanu, B., Ozdenoren, E.: Coarse Matching with Incomplete Information. Economic Theory 47, 75–104 (2011)
5. Kos, N.: Communication and Efficiency in Auctions. Discussion paper, Universita Bocconi. Games and Economic Behavior (forthcoming)
6. Lloyd, S.: Least Square Quantization in PCM. IEEE Transactions in Information Theory 28, 127–135 (1982)
7. Madarasz, K., Prat, A.: Screening with An Approximate Type Space. Discussion paper, London School of Economics
8. Maskin, E., Riley, J.: Monopoly with Incomplete Information. Rand Journal of Economics 15, 171–196 (1984)
9. Max, J.: Quantizing for Minimum Distortion. IEEE Transactions in Information Theory 6, 7–12 (1960)
10. McAfee, P.: Coarse Matching. Econometrica 70, 2025–2034 (2002)
11. Mussa, M., Rosen, S.: Monopoly and Product Quality. Journal of Economic Theory 18, 301–317 (1978)
12. Myerson, R.: Incentive Compatibility and the Bargaining Problem. Econometrica 47, 61–73 (1979)
13. Wilson, R.: Efficient and Competitive Rationing. Econometrica 57, 1–40 (1989)

Hierarchical Auctions for Network Resource Allocation

Wenyuan Tang and Rahul Jain*

Department of Electrical Engineering
University of Southern California
{wenyuan,rahul.jain}@usc.edu

Abstract. Motivated by allocation of cloud computing services, bandwidth and wireless spectrum in secondary network markets, we introduce a hierarchical auction model for network resource allocation. The Tier 1 provider owns all of the resource, who holds an auction in which the Tier 2 providers participate. Each of the Tier 2 providers then holds an auction to allocate the acquired resource among the Tier 3 users. The Tier 2 providers play the role of middlemen, since their utility for the resource depends entirely on the payment that they receive by selling it. We first consider the case of indivisible resource. We study a class of mechanisms where each sub-mechanism is either a first-price or a second-price auction, and show that incentive compatibility and efficiency cannot be simultaneously achieved. We then consider the resource to be divisible and propose the hierarchical network second-price mechanism in which there exists an efficient Nash equilibrium with endogenous strong budget balance.

Keywords: Network economics, mechanism design, auctions, hierarchical models.

1 Introduction

As networks have become increasingly complex, so has the ownership structure. This means that the traditional models and resource allocation mechanisms that are used for resource exchange between the primary owners and the end-users are no longer always relevant. Increasingly, there are middlemen, operators who buy network resources from the primary owners and then sell them to the end-users. This potentially causes inefficiencies in network resource allocation.

Consider the case of bandwidth allocation. The network bandwidth is primarily owned by a Tier 1 ISP (Internet Service Provider) or carrier, who then sells it to various Tier 2 ISPs. The Tier 2 ISPs then sell it further either to corporate customers or to the Tier 3 ISPs, who provide service directly to consumers. The presence of the Tier 2 ISPs can potentially skew the network resource allocation, and cause it to be inefficient from a social welfare point of view. Another case

* The second author's research on this project is supported by the NSF CAREER award CNS-0954116 and an IBM Faculty Award.

R. Jain and R. Kannan (Eds.): GameNets 2011, LNICST 75, pp. 11–26, 2012.

in point, is the emerging market of cloud computing services. Providers such as IBM, Google, Amazon and others are providing cloud computing services which end-users (e.g., enterprises having small computational or data center needs) can buy. Of course, the distribution channel for these services is likely to involve middlemen. This raises the key question what hierarchical mechanisms can be used in the presence of middlemen that are incentive compatible and/or efficient.

Auctions as mechanisms for network resource allocation have received lots of attention recently. Following-up on the network utility model proposed by Kelly [12], Johari and Tsitsiklis showed that the Kelly mechanism can have up to 25% efficiency loss [10]. This led to a flurry of activity in designing efficient network resource allocation mechanisms, including the work of Maheshwaran and Basar [15], Johari and Tsitsiklis [11], Yang and Hajek [20], Jain and Walrand [5], Jia and Caines [8] among others [2,16]. Most of the work focused on one-sided auctions for divisible resources, and is related to the approach of Lazar and Semret [13]. Double-sided network auctions for divisible resources were developed in [5]. The only work to focus on indivisible network resources is Jain and Varaiya [6] which proposed a Nash implementation combinatorial double auction. This is also the only work known so far that presents incomplete information analysis of combinatorial market mechanisms [7].

All those mechanisms either involve network resource allocation by an auctioneer among multiple buyers, or network resource exchange among multiple buyers and sellers. Most of the proposed mechanisms are Nash implementations, i.e., truth telling is a Nash equilibrium but not necessarily a dominant strategy equilibrium, and have either unique Nash equilibria which are efficient, or at least one that is. In reality, however, markets for network resources often have middlemen operators, and efficiency can be rather hard, if not impossible to achieve with their presence. Unfortunately, models with middlemen have not been studied at all, primarily due to the difficulty of designing appropriate mechanisms. Even in the economic and game theory literature, the closest related auction models are those that involve a resale after an auction. That is, there is only a single tier auction, and the winners can then resell the resources acquired in the auction [3].

There is indeed some game-theoretic work on network pricing in a more general topology. Johari, Mannor and Tsitsiklis [9] studied a network game where the nodes of the network wish to form a graph to route traffic between themselves, and they characterized connected link stable equilibria. Shakkottai and Srikant [18] examined how transit and customer prices and quality of service are set in a network consisting of multiple ISPs, where a 3-tier hierarchical model is proposed. However, such work only focused on the pricing equilibrium, and issues like mechanism design and auctions were never studied.

In this paper, we consider a multi-tier setting. We consider a homogeneous network resource. This could be bandwidth, wireless spectrum or cloud computing service, all owned by a single entity, the Tier 1 provider. He conducts an auction to allocate the resource among the Tier 2 operators. The Tier 2 operators then further allocate the network resource they have acquired in the auction among the Tier 3 entities, who may be the end-users. Each Tier 3 user has a valuation for the resource, which is strictly increasing and concave with respect to the capacity. On the other hand, the Tier 2 entities are more like middlemen. They do not have any intrinsic valuation for the network resource but a quasi-valuation which depends on the revenue that they will acquire by selling it off in an auction. Our goal is to design a *hierarchical auction mechanism* that specifies one sub-mechanism for each tier.

We develop a general hierarchical mechanism design framework and consider the setting where all auctions are conducted simultaneously. Admittedly, this does not fully meet the reality (where auctions in different tiers may take place one after another), but provides insights into the problem from a theoretical point of view. We first consider the resource to be indivisible. We investigate a class of mechanisms where each sub-mechanism is either a first-price or a second-price auction. We show that the all-tier second-price auction mechanism is incentive compatible but not efficient, i.e., social-welfare maximizing. This is a surprising observation and the only known instance of its type involving the VCG/second-price mechanism [19]. We then show that the hierarchical mechanism with a first-price or a second-price sub-mechanism at Tier 1, and first-price sub-mechanisms at all other tiers is indeed efficient but not incentive compatible.

When the resource is divisible, it is impossible for bidders to report their arbitrary real-valued valuation functions. They are thus asked to report a two-dimensional bid signal, i.e., a per-unit bid price and the maximum quantity that they want to buy/sell. We note that while the Tier 1 sub-mechanism is a single-sided auction, the sub-mechanisms at all lower tiers are double-sided auctions. In this framework, we propose a hierarchical mechanism with a VCG-type auction at each tier. We show that in this hierarchical mechanism, there exists an efficient Nash equilibrium that is strongly budget-balanced at all tiers except the top tier, where a single-sided auction is conducted.

2 Model and Problem Statement

2.1 The Hierarchical Model

Consider a Tier 1 provider who owns a homogeneous network resource, say bandwidth. Such an entity could be either a carrier (e.g., AT&T), the FCC, or a cloud service provider such as IBM. The Tier 1 provider auctions C units of the resource among Tier 2 operators via a single-sided auction. We will refer to this as the *Tier 1 auction*. Each of the Tier 2 operators then auctions off the resource acquired in the Tier 1 auction to the Tier 3 operators. These will be referred to as the *Tier 2 auctions*, and in general at Tier k as the *Tier k auctions*. We will assume that there are K tiers. The only Tier 1 provider will be considered as the social planner, and the Tier K entities are the end-users or the consumers, while operators at all other tiers will be considered as middlemen. Tier k $(1 < k \leq K)$ operators can participate in and acquire the resource only from their Tier $k-1$ parent. That is, each middleman has exclusive access to his children, and there is only one seller in each auction. This simplifies the network topology. Extension (e.g., allowing competition among sellers) is possible but would be much more complicated.

An example of the network topology is shown in Fig. 1. We now introduce the notation to ease further discussion. Let the nodes in the tree network be numbered $i = 0, 1, \ldots, N$ with M terminal nodes, where node 0 at the root of the tree is the social planner. Let $\mathcal{T}(i)$ be a function that specifies the tier to which node i belongs. The tier of node 0 is considered as Tier 1. By $\texttt{parent}(i)$, we shall denote the parent of node i in the tree network, and by $\texttt{children}(i)$, we shall denote the set of the children of node i. Each node represents a player. To avoid cumbersomeness, we shall use the redundant notation $P_i^{(k)}$ for node i that is at Tier k. Let the capacity acquired by $P_i^{(k)}$ denoted by $x_i^{(k)}$ (in the Tier $k-1$ auction), and the capacity that node i offers denoted by $y_i^{(k)}$ (in the Tier k auction). Note that $\sum_{j\in\texttt{children}(i)} x_j^{(k+1)} \leq y_i^{(k)} \leq x_i^{(k)}$, for $k = 1, ..., K-1$.

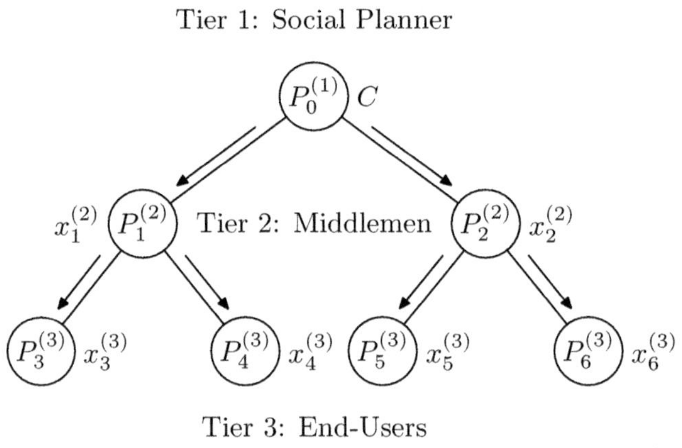

Fig. 1. An example of a 3-tier network

2.2 The Mechanism Design Framework

We now describe the mechanism design framework. We assume that each player $P_i^{(k)}$ has a quasi-linear utility function $u_i^{(k)}(x, \overline{w}_i) = v_i^{(k)}(x) - \overline{w}_i$, where $v_i^{(k)}(x)$ is the valuation of player $P_i^{(k)}$ when he is allocated a capacity x, and $\overline{w}_i$ is the payment made to his parent. Typically, for the middlemen $P_i^{(k)}$ ($k = 2, \ldots, K-1$), $v_i^{(k)}(x) = \underline{w}_i - c_i^{(k)}(x)$, where $\underline{w}_i$ is the revenue from reselling and $c_i^{(k)}(x)$ is the cost function, since they do not derive any utility from the allocation but may incur a transaction cost.

We define the *social welfare* to be the total utility derived by the end-users minus the total cost incurred by the middlemen, i.e.,

$$S(\mathbf{x}) = \sum_{i:\mathcal{T}(i)=K} v_i^{(K)}(x_i^{(K)}) - \sum_{2\leq k\leq K-1} \sum_{j:\mathcal{T}(j)=k} c_j^{(k)}(x_j^{(k)})$$

where $\mathbf{x} = (x_1, \ldots, x_N)$. The social planner's objective is to realize an *(allocatively) efficient* allocation $\mathbf{x}^{**}$ that maximizes the social welfare, and solves

$$\begin{aligned}
\textbf{HN-OPT}: \quad & \max \quad S(\mathbf{x}) && && (1)\\
\text{s.t.} \quad & \sum_{i \in \texttt{children}(0)} x_i^{(2)} && \leq C, \\
& \sum_{i \in \texttt{children}(j)} x_i^{(k+1)} && \leq x_j^{(k)}, \quad \forall (j,k) : 2 \leq k \leq K-1, \\
& x_i^{(k)} && \geq 0, \quad \forall (i,k) : 2 \leq k \leq K.
\end{aligned}$$

The first constraint follows because in the Tier 1 auction, the auctioneer (player 0) allocates the total capacity C among the Tier 2 players. The second constraint follows from the fact that the total allocation among the buyers in the Tier k auction cannot be greater than the allocation received by $P_j^{(k)}$ from the Tier $k-1$ auction. The third constraint is required to ensure non-negative allocations. Furthermore, we could consider the resource to be indivisible and let the x_i's to be integral, or consider it to be divisible and allow the x_i's to be real.

The social planner cannot achieve the objective (i.e., social-welfare maximizing) by himself as he does not know the valuation and cost functions of the end-users and the middlemen respectively. Thus, a decentralized implementation is necessary. However, the strategic players are selfish and may misreport their information. Furthermore, in the hierarchical model, the mechanism is distributed with multiple auctions at each tier. This makes the achievement of the social welfare maximization even more difficult.

Our goal thus is an incentive *mechanism* Γ that is composed of various sub-mechanisms $(\Gamma_i^{(k)}, i = 0, 1, \ldots, N-M, k = \mathcal{T}(i))$. Each sub-mechanism (i.e., auction) $\Gamma_i^{(k)}$ is conducted at each node i of the tree, except the M leaf (terminal) nodes. Note that the auction $\Gamma_i^{(k)}$ involves player $P_i^{(k)}$ as a seller, and the players $\texttt{children}(i)$ as the buyers. Note that node 0 acts only as an seller and the terminal nodes only act as buyers, whereas the middlemen $P_i^{(k)}$ $(2 \leq k \leq K-1)$ act as buyers in the Tier $k-1$ auction, and as sellers in the Tier k auction. Generally, each $\Gamma_i^{(k)}$ can be different, though we consider the setting for which $\Gamma_i^{(k)} = \Gamma^{(k)}$, i.e., a common sub-mechanism is used at each Tier k. Still, this is a simplification but subject to extension.

Since the middlemen have no intrinsic valuation for the resource itself, we define the notion of quasi-valuation functions for the middlemen. Let $\mathbb{X}$ denote the allocation space, which is $\mathbb{Z}_+$ when the resource is indivisible and $\mathbb{R}_+$ when the resource is divisible.

Definition 1. *A quasi-valuation function of player* $P_i^{(k)}$ *is a function* $\bar{v}_i^{(k)} : \mathbb{X} \to \mathbb{R}_+$ *that specifies the revenue he receives in the auction* $\Gamma_i^{(k)}$ *from his children for each possible allocation, when all the players* $\texttt{children}(i)$ *report their valuation functions (for end-users) or quasi-valuation functions (for middlemen) truthfully.*

Note the backward-recursiveness in the definition of quasi-valuation functions. They can be easily computed by the players in complete information settings. We now see the role of such functions in defining hierarchical incentive compatible and Nash implementation mechanisms.

Definition 2. *The (direct) hierarchical mechanism* $\Gamma = (\Gamma^{(1)}, \ldots, \Gamma^{(K-1)})$ *is incentive compatible (or strategy-proof) if there is a dominant strategy equilibrium wherein all the end-users report their valuation functions and all the middlemen report their quasi-valuation functions, truthfully.*

Such equilibrium strategies will be referred to as "truth telling" as a counterpart to standard notions of "truth-telling" in non-hierarchical mechanisms [17]. We now define the notion of efficiency in hierarchical mechanisms.

Definition 3. *The (direct) hierarchical mechanism* $\Gamma = (\Gamma^{(1)}, \ldots, \Gamma^{(K-1)})$ *is* efficient *if there is an equilibrium that maximizes the social welfare in the optimization problem* ***HN-OPT*** *(1).*

We study *simultaneous hierarchical mechanisms*, in which all sub-mechanism auctions take place simultaneously (which are modeled as a normal form game). Thus, usual notions of *Nash equilibrium* shall be studied [1,4].

3 Hierarchical Auctions for Indivisible Resources

When the resource is indivisible, we present a class of hierarchical mechanisms $\Gamma = (\Gamma^{(1)}, \ldots, \Gamma^{(K-1)})$ wherein a common sub-mechanism is used at each tier, and each such sub-mechanism is either a first-price auction (denoted by $\mathcal{F}$) or a second-price auction (denoted by $\mathcal{S}$), i.e., $\Gamma^{(k)} \in \{\mathcal{F}, \mathcal{S}\}$. We investigate the efficiency and incentive compatibility of such hierarchical mechanism designs.

We first consider the case where there is only a single unit to be allocated, i.e., $C = 1$. Here, we assume that the middlemen have no transaction costs, i.e., $c_i^{(k)}(x) = 0$. We note that the introduction of transaction costs would be trivial in the case of indivisible resources, and it can be easily extended if desired.

Let $b_i^{(k)}$ denote the buy-bid of player i who is at Tier k, and $x_i^{(k)}$ the unit he acquires (in the Tier $k-1$ auction as defined). Recall that there are $N - M + 1$ auctions that are conducted simultaneously, though some auction outcomes cannot be fulfilled since there is only a single indivisible unit. This, however, is not unreasonable since there is really a single winner among the end-users. The middlemen that connect this end-user to the root will also be purported to be winners.

Theorem 1. *Assume each player except the end-users has at least two children. Suppose a single indivisible unit is to be allocated through a hierarchical auction mechanism* $\hat{\Gamma}$ *such that* $\hat{\Gamma}^{(1)} \in \{\mathcal{F}, \mathcal{S}\}, \hat{\Gamma}^{(2)} = \cdots = \hat{\Gamma}^{(K-1)} = \mathcal{F}$. *Then, there exists an* ϵ*-Nash equilibrium which is efficient.*

Proof. We prove by construction. Consider a Tier $K-1$ auction $\hat{\Gamma}_i^{(K-1)}$. Find the player that has the highest valuation in that auction, i.e., $j^* \in \arg\max_{j \in \texttt{children}(i)} v_j^{(K)}$. Define the bids of the player as follows

$$b_{j^*}^{(K)} = v_{j^*}^{(K)},$$

$$b_j^{(K)} = v_{j^*}^{(K)} - \epsilon, \forall j \in \texttt{children}(i), j \neq j^*,$$

i.e., player j^* bids truthfully, while all others in that auction bid just a bit below.

Consider a Tier k auction $\Gamma_i^{(k)}$ $(1 < k < K-1)$. As before, find a player $j^* \in \arg\max_{j \in \texttt{children}(i)} \bar{v}_j^{(k+1)}$, and define the bids of the players in this auction as

$$b_{j^*}^{(k+1)} = \bar{v}_{j^*}^{(k+1)},$$

$$b_j^{(k+1)} = \bar{v}_{j^*}^{(k+1)} - \epsilon, \forall j \in \texttt{children}(i), j \neq j^*.$$

Now, consider the Tier 1 auction $\hat{\Gamma}^{(1)}$.Find a player $j^* \in \arg\max_{j \in \texttt{children}(0)} \bar{v}_j^{(2)}$. If $\hat{\Gamma}^{(1)} = \mathcal{F}$, define the bids of players in this auction as

$$b_{j^*}^{(2)} = \bar{v}_{j^*}^{(2)},$$

$$b_j^{(2)} = \bar{v}_{j^*}^{(2)} - \epsilon, \forall j \in \texttt{children}(0), j \neq j^*.$$

Otherwise, if $\hat{\Gamma}^{(1)} = \mathcal{S}$, define the bids of players in this auction as

$$b_j^{(2)} = \bar{v}_j^{(2)}, \forall j \in \texttt{children}(0).$$

It is obvious that such bids induce the efficient allocation. We argue that these bids constitute an ϵ-Nash equilibrium. Note that every player gets a non-negative payoff in such a bid profile.

Consider a player $P_i^{(K)}$. If he is a winner, he has no incentive to increase his bid since $b_i^{(K)} = v_i^{(K)}$, and he has no incentive to decrease his bid since there exists a player $P_{i'}^{(K)}$ with $\texttt{parent}(i) = \texttt{parent}(i')$ whose bid is $b_i^{(K)} - \epsilon$. If he is a loser, we have $b_i^{(K)} > v_i^{(K)} - \epsilon$. Clearly, he has no incentive to either increase or decrease his bid.

Consider a player $P_i^{(k)}$ $(2 < k < K)$. If he is a winner, he has no incentive to increase his bid since $b_i^{(k)} = \bar{v}_i^{(k)}$, and he has no incentive to decrease his bid since there exists a player $P_{i'}^{(k)}$ with $\texttt{parent}(i) = \texttt{parent}(i')$ whose bid is $b_i^{(k)} - \epsilon$. If he is a loser, we have $b_i^{(k)} > \bar{v}_i^{(k)} - \epsilon$. Clearly, he has no incentive to either increase or decrease his bid.

It is also easy to verify that such bids are the best responses of the Tier 2 players for $\hat{\Gamma}^{(1)} = \mathcal{F}$ and $\hat{\Gamma}^{(1)} = \mathcal{S}$ respectively. This proves the claim.

The following example shows that the above mechanism $\hat{\Gamma}$ achieves efficiency but is not incentive compatible.

Proposition 1. *The hierarchical mechanism $\hat{\Gamma}$ is efficient but not incentive compatible.*

Proof. We prove by providing a counter example. Assume the network topology is as in Fig. 1, i.e., there are two Tier 2 players $P_1^{(2)}$ (with his Tier 3 children $P_3^{(3)}$, $P_4^{(3)}$) and $P_2^{(2)}$ (with his Tier 3 children $P_5^{(3)}$, $P_6^{(3)}$). Let the valuation functions of the Tier 3 players be

$$v_3^{(3)} = 2, v_4^{(3)} = 3, v_5^{(3)} = 1, v_6^{(3)} = 4.$$

Since $\hat{\Gamma}^{(2)} = \mathcal{F}$, the quasi-valuation functions of the Tier 2 players can be easily computed to be

$$\bar{v}_1^{(2)} = 3, \bar{v}_2^{(2)} = 4.$$

However, truth telling is not an equilibrium in this auction. Rather, it is easy to verify that an ϵ-Nash equilibrium is

$$(b_1^{(2)}, b_2^{(2)}) = (4 - \epsilon, 4),$$

$$(b_3^{(3)}, b_4^{(3)}, b_5^{(3)}, b_6^{(3)}) = (3 - \epsilon, 3, 4 - \epsilon, 4).$$

The corresponding equilibrium allocation is

$$(x_1^{(2)}, x_2^{(2)}) = (0, 1),$$

$$(x_3^{(3)}, x_4^{(3)}, x_5^{(3)}, x_6^{(3)}) = (0, 0, 0, 1),$$

which is exactly the efficient allocation.

Thus, this mechanism is efficient but not incentive compatible.

We now introduce a natural hierarchical extension of the second-price or VCG auction mechanism.

Theorem 2. *Suppose multiple units of an indivisible resource are to be allocated through a hierarchical auction mechanism $\tilde{\Gamma}$ such that $\tilde{\Gamma}^{(1)} = \cdots = \tilde{\Gamma}^{(K-1)} = \mathcal{S}$ (which we shall call the second-price hierarchical auction). Then, the mechanism is incentive-compatible.*

Proof. We argue by backward induction that truth telling is a dominant strategy equilibrium. Consider the Tier $K-1$ auction, which is a second-price sub-mechanism. The end-user $P_i^{(K)}$ ($\mathcal{T}(i) = K$) will bid truthfully, no matter how the other players bid and what capacity his parent is allocated, since that is his dominant strategy in a second-price auction. This is the fundamental property of the VCG mechanism.

Given that all the $P_i^{(K)}$'s report truthfully, the quasi-valuation functions of the players $P_i^{(K-1)}$'s are true. Furthermore, the Tier $K-2$ auction is again a VCG mechanism in which truth-telling is a dominant strategy equilibrium.

Now, we can argue by backward induction. Assuming the Tier $k+1$ players in the Tier k auctions have true quasi-valuation functions, they will bid truthfully. So, the quasi-valuation functions of the Tier k players will be true as well. Since this is true for $k = K-1$, it is true for all $k = K-1, \ldots, 1$. Hence, all the players bid truthfully, and the hierarchical mechanism is incentive-compatible.

The second-price hierarchical auction mechanism, as can be expected, has truth telling by each player as a dominant strategy equilibrium. The surprise is that unlike non-hierarchical settings, efficiency may not be achieved.

Proposition 2. *The second-price hierarchical mechanism $\tilde{\Gamma}$ is not necessarily efficient.*

Proof. We prove by providing a counter example. Let $C = 5$ be allocated by the second-price hierarchical mechanism in a 3-tier network as in Fig. 1. Let the valuation functions of the Tier 3 players be

$$(v_3^{(3)}(x), x = 1,2,3,4,5) = (10, 18, 24, 28, 30),$$

$$(v_4^{(3)}(x), x = 1,2,3,4,5) = (20, 25, 29, 32, 34),$$

$$(v_5^{(3)}(x), x = 1,2,3,4,5) = (15, 24, 32, 39, 45),$$

$$(v_6^{(3)}(x), x = 1,2,3,4,5) = (16, 20, 24, 27, 29).$$

According to (1), the efficient allocation is

$$(x_1^{(2)**}, x_2^{(2)**}) = (2, 3),$$

$$(x_3^{(3)**}, x_4^{(3)**}, x_5^{(3)**}, x_6^{(3)**}) = (1, 1, 2, 1).$$

Since $\tilde{\Gamma}^{(2)} = \mathcal{S}$, the quasi-valuation functions of the Tier 2 players can be easily computed to be

$$(\bar{v}_1^{(2)}(x), x = 1,2,3,4,5) = (10, 13, 15, 16, 15),$$

$$(\bar{v}_2^{(2)}(x), x = 1,2,3,4,5) = (15, 13, 16, 18, 19).$$

In the mechanism $\tilde{\Gamma}$, truth telling is a Nash equilibrium as we have already proved. Thus, the corresponding equilibrium allocation is

$$(x_1^{(2)}, x_2^{(2)}) = (4, 1),$$

$$(x_3^{(3)}, x_4^{(3)}, x_5^{(3)}, x_6^{(3)}) = (3, 1, 0, 1),$$

which however, is different from the efficient allocation.

Thus, in the case of multiple units of an indivisible resource, this hierarchical mechanism is incentive-compatible but not efficient.

An even greater surprise is the following impossibility result if we restrict our attention to first-price and second-price sub-mechanisms.

Theorem 3 (Hierarchical Impossibility). *Suppose we allocate a single unit of the indivisible resource through a hierarchical auction mechanism Γ such that $\Gamma^{(k)} \in \{\mathcal{F}, \mathcal{S}\}$ (for all $k = 1, \ldots, K-1$ and $K \geq 3$). Then, there exists no such hierarchical mechanism which is both incentive-compatible and efficient.*

Proof. As we have already seen in Proposition 1 that incentive compatibility is not guaranteed if there exists a k such that $\Gamma^{(k)} = \mathcal{F}$. We have also seen in Proposition 2 that efficiency is not guaranteed if there exists a k such that $\Gamma^{(k)} = \mathcal{S}$. Thus, if the choices of the $\Gamma^{(k)}$'s are restricted to the two alternatives ($\mathcal{F}$ or $\mathcal{S}$), incentive compatibility and efficiency cannot be simultaneously achieved.

Our conjecture is that this "limited" impossibility theorem foretells a more general impossibility result for hierarchical mechanism design.

4 Hierarchical Auctions for Divisible Resources

We now consider the resource to be divisible, and propose a hierarchical auction mechanism. We will now consider the setting where the middlemen have transaction costs as well. While the Tier 1 auction will remain single-sided, Tier 2 through Tier $K-1$ auctions will be double-sided, i.e., in such auctions buyers will make buy-bids, and sellers will make sell-bids.

For simplicity of exposition, we will only consider a 3-tier network as in Fig. 1. Also, we drop the superscripts and adopt a more concise notation here, i.e., denote the ith Tier 2 player by P_i and the jth child of P_i by P_{ij} (Tier 3 player). The notations of valuation functions, bids, etc. are changed correspondingly.

We will assume that the valuation functions of the end-users, $v_{ij}(x_{ij})$ are strictly increasing and concave, and smooth, with $v_{ij}(0) = 0$. The cost functions of the middlemen, $c_i(x_i)$ are assumed to be strictly increasing and convex, and smooth, with $c_i(0) = 0$.

The end-user's payoff is $u_{ij} = v_{ij}(x_{ij}) - \overline{w}_{ij}$, where $\overline{w}_{ij}$ is the payment made by player P_{ij} to player P_i. A middleman P_i has a utility $u_i = \underline{w}_i - \overline{w}_i - c_i(x_i)$, where $\underline{w}_i$ is P_i's revenue from reselling and $\overline{w}_i$ is the payment made by P_i to player 0.

In this setting the social welfare optimization problem is as following

$$\textbf{DIV-OPT}: \quad \max \sum_{i,j} v_{ij}(x_{ij}) - \sum_i c_i(x_i) \tag{2}$$

$$\text{s.t.} \quad \sum_i x_i \leq C, \quad [\lambda_0]$$

$$\sum_j x_{ij} \leq x_i, \quad \forall i, \quad [\lambda_i]$$

$$x_i \geq 0, \quad \forall i,$$

$$x_{ij} \geq 0, \quad \forall (i,j).$$

Here, λ_0 and λ_i's are the Lagrange multipliers of the corresponding constraints above. The above is a convex optimization problem, and a solution exists, which is characterized by the KKT conditions [14]

$$\begin{aligned}
(c_i'(x_i) + \lambda_0 - \lambda_i)x_i = 0, &\quad \forall i,\\
c_i'(x_i) + \lambda_0 - \lambda_i \geq 0, &\quad \forall i,\\
(v_{ij}'(x_{ij}) - \lambda_i)x_{ij} = 0, &\quad \forall (i,j),\\
v_{ij}'(x_{ij}) - \lambda_i \leq 0, &\quad \forall (i,j),\\
\lambda_0(\sum_i x_i - C) = 0, &\\
\lambda_i(\sum_j x_{ij} - x_i) = 0, &\quad \forall i.
\end{aligned} \tag{3}$$

Our objective is to design a hierarchical mechanism to allocate the divisible resource that achieves the social welfare optimum despite the strategic behavior of the players. An important issue in the context of divisible resources is that it is impossible for a bidder to communicate a complete arbitrary real-valued valuation function. Thus, the bidders must communicate an approximation to it from a finite-dimensional bid space.

Hierarchical Network Second-Price Mechanism

We now propose the hierarchical network second-price (**HNSP**) mechanism $\bar{\Gamma}$ that can be used to allocate a divisible resource in a multi-tier network. We take a 3-tier network as an example. The mechanism $\bar{\Gamma} = (\bar{\Gamma}^{(1)}, \bar{\Gamma}^{(2)})$ is composed of two sub-mechanisms $\bar{\Gamma}^{(1)}$ and $\bar{\Gamma}^{(2)}$. The sub-mechanism $\bar{\Gamma}^{(1)}$ employed at Tier 1 is a single-sided VCG-type auction mechanism in which Tier 2 players (the middlemen) report bids $b_i = (\beta_i, d_i)$ where β_i is interpreted to be the per-unit bid price, and d_i as the maximum quantity wanted. The sub-mechanism $\bar{\Gamma}^{(2)}$ employed at Tier 2 is a double-sided VCG-type auction mechanism where Tier 2 players report sell-bids $a_i = (\alpha_i, q_i)$ where α_i is the per-unit sell-bid price and q_i is the maximum quantity offered for sale, while the Tier 3 players (end-users) report buy-bids $b_{ij} = (\beta_{ij}, d_{ij})$ where β_{ij} is the per-unit buy-bid price and d_{ij} is the maximum quantity wanted.

Once the bids are received in all the auctions, the auction outcomes are determined as follows. In the Tier 1 auction $\bar{\Gamma}^{(1)}$, the allocation $\hat{\mathbf{x}}$ is a solution of the optimization problem

$$\begin{aligned}
\textbf{HNSP-1}: \quad \max \quad & \textstyle\sum_i \beta_i x_i \\
\text{s.t.} \quad \sum_i x_i \quad & \leq \quad C, \quad [\lambda_0'] \\
x_i \quad & \leq \quad d_i, \quad \forall i, \quad [\mu_i] \\
x_i \quad & \geq \quad 0, \quad \forall i.
\end{aligned} \tag{4}$$

Let $\hat{\mathbf{x}}^{-i}$ denote the allocation as a solution of the above with $d_i = 0$, i.e., when the player P_i (a middleman) is not present. Then, the payment made by P_i is

$$\overline{w}_i = \sum_{j \neq i} \beta_j (\hat{x}_j^{-i} - \hat{x}_j), \tag{5}$$

which is the "externality" that P_i imposes on all the other players (in the Tier 1 auction) by his participation. Let λ_0' and μ_i's be the Lagrange multipliers of the corresponding constraints. Then, the solution of **HNSP-1** is characterized by the KKT conditions

$$\begin{aligned} (\beta_i - \lambda_0' - \mu_i) x_i &= 0, \quad \forall i, \\ \beta_i - \lambda_0' - \mu_i &\leq 0, \quad \forall i, \\ \lambda_0' (\sum_i x_i - C) &= 0, \\ \mu_i (x_i - d_i) &= 0, \quad \forall i. \end{aligned} \tag{6}$$

In the Tier 2 auction $\bar{\Gamma}^{(2)}$, the middleman is the seller and his children (the end-users) are the buyers. The sub-mechanism $\bar{\Gamma}^{(2)}$ is a VCG-type double-sided auction, i.e., both the seller and the buyers place bids, and the allocation $(\tilde{\mathbf{x}}, \tilde{\mathbf{y}})$ is a solution of the optimization problem

$$\begin{aligned} \textbf{HNSP-2}: \quad \max \quad & \sum_j \beta_{ij} x_{ij} - \alpha_i y_i \\ \text{s.t.} \quad & \sum_j x_{ij} \leq y_i, \quad [\lambda_i'] \\ & x_{ij} \leq d_{ij}, \quad \forall j, \quad [\mu_{ij}] \\ & y_i \leq q_i, \quad [\nu_i] \\ & x_{ij} \geq 0, \quad \forall j, \\ & y_i \geq 0. \end{aligned} \tag{7}$$

Let $(\tilde{\mathbf{x}}^{-j}, \tilde{\mathbf{y}}^{-j})$ denote the allocation as a solution of the above with $d_{ij} = 0$. Then, the payment made by player P_{ij} is

$$\overline{w}_{ij} = \sum_{k \neq j} \beta_{ik} (\tilde{x}_{ik}^{-j} - \tilde{x}_{ik}) - \alpha_i (\tilde{y}_i^{-j} - \tilde{y}_i), \tag{8}$$

and the payment received by player P_i is

$$\underline{w}_i = \sum_j \beta_{ij} \tilde{x}_{ij}. \tag{9}$$

These transfers are the externalities that the players impose on the others through their participation. Let λ_i''s, μ_{ij}'s and ν_i's be the Lagrange multipliers corresponding to the constraints in the **HNSP-2** above. Then, the solution is characterized by the following KKT conditions

$$\begin{aligned}(\beta_{ij} - \lambda_i' - \mu_{ij})x_{ij} &= 0, \quad \forall j,\\ \beta_{ij} - \lambda_i' - \mu_{ij} &\leq 0, \quad \forall j,\\ (\alpha_i - \lambda_i' + \nu_i)y_i &= 0,\\ \alpha_i - \lambda_i' + \nu_i &\geq 0,\\ \lambda_i'(\sum_j x_{ij} - y_i) &= 0,\\ \mu_{ij}(x_{ij} - d_{ij}) &= 0, \quad \forall j\\ \nu_i(y_i - q_i) &= 0.\end{aligned} \tag{10}$$

This completes the definition of the **HNSP** mechanism.

We now show the existence of an efficient Nash equilibrium in the simultaneous hierarchical network second-price mechanism by construction. Moreover, we show that the Tier 2 sub-mechanism $\bar{\Gamma}^{(2)}$ achieves *endogenous strong budget balance* at this equilibrium, i.e., the payment received by each middleman equals the total payments made by his children.

Theorem 4. *In the **HNSP** mechanism $\bar{\Gamma}$, there exists an efficient Nash equilibrium with endogenous strong budget balance.*

Proof. Let $\mathbf{x}^{**}$ be an efficient allocation corresponding to the problem **DIV-OPT** in (2). Then, there exist Lagrange multipliers λ_0 and λ_i's that satisfy the KKT conditions (3). Consider the bid profile $d_i = q_i = x_i^{**}$, $d_{ij} = x_{ij}^{**}$, $\beta_i = \sum_j v_{ij}'(x_{ij}^{**}) - c_i'(x_i^{**})$, $\alpha_i = c_i'(x_i^{**}) + \lambda_0$, and $\beta_{ij} = v_{ij}'(x_{ij}^{**})$.

First, we prove that the bid profile induces the efficient allocation. Let $\lambda_0' = \lambda_0$, $\lambda_i' = \lambda_i$, $\mu_i = \sum_j \beta_{ij} - \lambda_i$ and $\mu_{ij} = \nu_i = 0$. Then the KKT conditions (6) and (10) are equivalent to the KKT conditions (3). This implies $\mathbf{x}^{**}$ is also a solution of the problem **HNSP-1** in (4) and the problem **HNSP-2** in (7) with these bids.

Now, we prove that the strategy profile is a Nash equilibrium. Consider an end-user P_{ij} with bid (β_{ij}, d_{ij}). His payoff at the efficient allocation is $u_{ij} = v_{ij}(x_{ij}^{**}) - \overline{w}_{ij} = v_{ij}(x_{ij}^{**}) - \alpha_i x_{ij}^{**}$. Then, given the bids of others, if he changes his bid to decrease his allocation x_{ij}^{**} by a $\delta > 0$ (when $x_{ij}^{**} > 0$), then note that the allocations of buyers P_{ik} $(k \neq j)$ do not change but seller P_i sells less. His new payoff is $u_{ij}' = v_{ij}(x_{ij}^{'*}) - \alpha_i x_{ij}^{'*} = v_{ij}(x_{ij}^{**} - \delta) - \alpha_i(x_{ij}^{**} - \delta)$. Thus, his payoff changes by

$$\begin{aligned}&u_{ij}' - u_{ij}\\ &= \alpha_i\delta + v_{ij}(x_{ij}^{**} - \delta) - v_{ij}(x_{ij}^{**})\\ &= (c_i'(x_i^{**}) + \lambda_0)\delta + v_{ij}(x_{ij}^{**} - \delta) - v_{ij}(x_{ij}^{**})\\ &= \lambda_i\delta + v_{ij}(x_{ij}^{**} - \delta) - v_{ij}(x_{ij}^{**})\\ &= v_{ij}'(x_{ij}^{**})\delta + v_{ij}(x_{ij}^{**} - \delta) - v_{ij}(x_{ij}^{**})\\ &< 0.\end{aligned}$$

Thus, his payoff will decrease. Now suppose he changes his bid to increase his allocation x_{ij}^{**} by a $\delta > 0$, then note that the allocation of player P_i does not change but that of some players P_{ik} $(k \neq j)$ decrease. His new payoff is $u'_{ij} = v_{ij}(x_{ij}^{**} + \delta) - \sum_{k \neq j} \beta_{ik}(x_{ik}^{**} - x_{ik}'^{*}) - \alpha_i x_{ij}^{**}$. Thus,

$$\begin{aligned}
& u'_{ij} - u_{ij} \\
&= -\sum_{k \neq j} \beta_{ik}(x_{ik}^{**} - x_{ik}'^{*}) + v_{ij}(x_{ij}^{**} + \delta) - v_{ij}(x_{ij}^{**}) \\
&\leq -\sum_{k \neq j} \lambda'_i(x_{ik}^{**} - x_{ik}'^{*}) + v_{ij}(x_{ij}^{**} + \delta) - v_{ij}(x_{ij}^{**}) \\
&= -\lambda'_i \delta + v_{ij}(x_{ij}^{**} + \delta) - v_{ij}(x_{ij}^{**}) \\
&= -\lambda_i \delta + v_{ij}(x_{ij}^{**} + \delta) - v_{ij}(x_{ij}^{**}) \\
&\leq -v'_{ij}(x_{ij}^{**})\delta + v_{ij}(x_{ij}^{**} + \delta) - v_{ij}(x_{ij}^{**}) \\
&< 0.
\end{aligned}$$

Thus, his payoff will decrease again. Therefore, the best response of an end-user P_{ij} is to bid (β_{ij}, d_{ij}), and he has no incentive to deviate.

Consider a middleman P_i with bid (β_i, d_i) in Tier 1 auction and bid (α_i, q_i) in Tier 2 auction. His payoff at the efficient allocation is $u_i = \sum_j \beta_{ij} x_{ij}^{**} - c_i(x_i^{**})$. Then, given the bids of others, if he changes his bid to increase his allocation x_i^{**} by a $\delta > 0$, his payoff will be $u'_i = \sum_j \beta_{ij} x_{ij}^{**} - c_i(x_i^{**} + \delta) - \overline{w}'_i < u_i$. That is, his revenue remains the same, while his cost and his payment to player 0 increase. Thus, he has no incentive to increase his allocation.

Now, suppose he changes his bid to decrease his allocation x_i^{**} by a $\delta > 0$ (when $x_i^{**} > 0$). His payment to player 0 does not change but the payment he receives changes. His new payoff is $u'_i = \sum_j \beta_{ij} x_{ij}'^{*} - c_i(x_i^{**} - \delta)$. Thus,

$$\begin{aligned}
& u'_i - u_i \\
&= \sum_j \beta_{ij} x_{ij}'^{*} - \sum_j \beta_{ij} x_{ij}^{**} - c_i(x_i^{**} - \delta) + c_i(x_i^{**}) \\
&= \sum_j \beta_{ij} x_{ij}'^{*} - \sum_j \lambda'_i x_{ij}^{**} - c_i(x_i^{**} - \delta) + c_i(x_i^{**}) \\
&\leq \sum_j \lambda'_i x_{ij}'^{*} - \sum_j \lambda'_i x_{ij}^{**} - c_i(x_i^{**} - \delta) + c_i(x_i^{**}) \\
&= \lambda'_i \sum_j (x_{ij}'^{*} - x_{ij}^{**}) - c_i(x_i^{**} - \delta) + c_i(x_i^{**}) \\
&= -\lambda'_i \delta - c_i(x_i^{**} - \delta) + c_i(x_i^{**}) \\
&= -\lambda_i \delta - c_i(x_i^{**} - \delta) + c_i(x_i^{**}) \\
&= -(c'_i(x_i) + \lambda_0)\delta - c_i(x_i^{**} - \delta) + c_i(x_i^{**}) \\
&\leq -c'_i(x_i)\delta - c_i(x_i^{**} - \delta) + c_i(x_i^{**}) \\
&< 0.
\end{aligned}$$

Thus, his payoff will decrease. Since his payoff will decrease by deviation in either direction, bidding (β_i, d_i) and (α_i, q_i) is his best response. This implies that the constructed bid profile is a Nash equilibrium in the **HNSP** mechanism and yields an efficient outcome.

We now prove that there is endogenous strong budget balance at this Nash equilibrium. Note that $\overline{w}_{ij} = \alpha_i x_{ij}^{**} = \lambda_i' x_{ij}^{**} = \beta_{ij} x_{ij}^{**}$. So $\sum_j \overline{w}_{ij} = \sum_j \beta_{ij} x_{ij}^{**} = \underline{w}_i$ (for all i), which is what we wanted to prove.

Remark 1. We can easily check that each end-user and each middleman has a non-negative payoff at the Nash equilibrium constructed above.

Remark 2. We also note that the **HNSP** mechanism can be easily extended to the general multi-tier model wherein the Tier 1 sub-mechanism $\bar{\Gamma}^{(1)}$ is a VCG-type single-sided mechanism, while sub-mechanisms at all lower tiers, $\bar{\Gamma}^{(2)}, \ldots, \bar{\Gamma}^{(K-1)}$ are VCG-type double-sided mechanisms. Likewise, we can then establish the existence of an efficient Nash equilibrium with endogenous strong budget balance.

5 Conclusion

In this paper, we introduced a hierarchical auction model for network settings with multi-tier structures. We developed a general hierarchical mechanism design framework. Such a model is innovative and this paper is the first work on multi-tier auctions to our knowledge.

When the resource is indivisible, we investigated a class of mechanisms where each sub-mechanism is either a first- or a second-price auction. We showed that the hierarchical mechanism with a first- or a second-price sub-mechanism at Tier 1, and first-price sub-mechanisms at all other tiers is efficient but not incentive-compatible and surprisingly, the all-tier second-price auction mechanism is incentive-compatible but not efficient. This seems to fortell a more general impossibility of achieving incentive compatibility and efficiency at the same time in a hierarchical setting.

When the resource is divisible, we proposed the hierarchical network second-price mechanism, where the Tier 1 sub-mechanism is a single-sided VCG-type auction and the sub-mechanism at all lower tiers is a VCG-type double-sided auction. We showed that in this hierarchical mechanism, there exists an efficient Nash equilibrium with endogenous strong budget balance.

As part of future work, we intend to study more general classes of mechanisms than those where the sub-mechanisms are either first- or second-price auctions. We will also consider the *Stackelberg auction* setting, wherein the auctions at various tiers are conducted one after another. We will also consider more general network topologies wherein there may be more than one resource (e.g., bandwidth on multiple links, or bandwidth, storage and computation), and also allow for sub-mechanism auctions with multiple sellers.

References

1. Basar, T., Olsder, G.J.: Dynamic Non-cooperative Game Theory, 2nd edn. SIAM classics in applied mathematics (1999)
2. Bitsaki, M., Stamoulis, G., Courcoubetis, C.: A new strategy for bidding in the network-wide progressive second price auction for bandwidth. In: Proc. CoNEXT (2005)
3. Cheng, H.H., Tan, G.: Assymetric common-value auctions with applications to private-value auctions with resale. Economic Theory (2009)
4. Fudenberg, D., Tirole, J.: Game Theory, ch. 8. MIT Press (1991)
5. Jain, R., Walrand, J.: An efficient Nash-implementation mechanism for network resource allocation. Automatica 46, 1276–1283 (2010)
6. Jain, R., Varaiya, P.: A design for an asymptotically efficient combinatorial Bayesian market: Generalizing the Satterthwaite-Williams mechanism. In: International Conf. on Game Theory, Stony Brook (July 2007)
7. Jain, R., Varaiya, P.: The combinatorial sellers bid double auction: An asymptotically efficient market mechanism, working paper (2010)
8. Jia, P., Caines, P.: Auctions on networks: Efficiency, consensus, passivity, rates of convergence. In: Proc. IEEE Control and Decision Conf. (CDC) (December 2009)
9. Johari, R., Mannor, S., Tsitsiklis, J.: A contract-based model for directed network formation. Games and Economic Behavior 56, 201–224 (2006)
10. Johari, R., Tsitsiklis, J.: Efficiency loss in a network resource allocation game. Mathematics of Operations Research 29(3), 407–435 (2004)
11. Johari, R., Tsitsiklis, J.: Efficiency of scaler parameterized mechanisms. Operations Research 57(4), 823–839 (2009)
12. Kelly, F.: Charging and rate control for elastic traffic. Euro. Trans. on Telecommunications 8(1), 33–37 (1997)
13. Lazar, A., Semret, N.: Design and analysis of the progressive second price auction for network bandwidth sharing. Telecommunication Systems - Special issue on Network Economics (1999)
14. Luenberger, D.: Optimization by vector space methods. John Wiley and Sons (1969)
15. Maheswaran, R., Basar, T.: Social welfare of selfish agents: Motivating efficiency for divisible resources. In: Proc. CDC (2004)
16. Maille, P., Tuffin, B.: Multi-bid auctions for bandwidth allocation in communication networks. In: Proc. IEEE INFOCOM (2004)
17. Mas-Colell, A., Whinston, M., Green, J.: Microeconomic Theory, ch. 23. Oxford University Press (1995)
18. Shakkottai, S., Srikant, R.: Economics of network pricing with multiple ISPs. IEEE/ACM Transactions on Networking 14(6), 1233–1245 (2006)
19. Vickrey, W.: Counterspeculation, auctions, and sealed tenders. J. Finance 16, 8–37 (1961)
20. Yang, S., Hajek, B.: VCG-Kelly Mechanisms for allocation of divisible resources: Adapting VCG mechanisms to one-dimensional signals. IEEE J. Selected Areas of Communications 25, 1237–1243 (2007)

Local Public Good Provision in Networks: A Nash Implementation Mechanism

Shrutivandana Sharma[1] and Demosthenis Teneketzis[2]

[1] Electrical Engineering and Computer Science
University of Michigan, Ann Arbor, MI 48109, USA
svandana@umich.edu
[2] Electrical Engineering and Computer Science
University of Michigan, Ann Arbor, MI 48109, USA
teneket@eecs.umich.edu
http://www.eecs.umich.edu/~teneket

Abstract. In this paper we study resource allocation in decentralized information local public good networks. A network is a local public good network if each user's actions directly affect the utility of an arbitrary subset of network users. We consider networks where each user knows only that part of the network that either affects it or is affected by it. Furthermore, each user's utility and action space are its private information, and each user is a self utility maximizer. This network model is motivated by several applications including social networking, online advertising and wireless communications. For this network model we formulate a decentralized resource allocation problem and develop a decentralized resource allocation mechanism (game form) that possesses the following properties: (i) All Nash equilibria of the game induced by the mechanism result in allocations that are optimal solutions of the corresponding centralized resource allocation problem (Nash implementation). (ii) All users voluntarily participate in the allocation process specified by the mechanism (individual rationality). (iii) The mechanism results in budget balance at all Nash equilibria and off equilibrium.

Keywords: network, local public good, decentralized resource allocation, mechanism design, Nash implementation, budget balance, individual rationality.

1 Introduction

In networks individuals' actions often influence the performance of their directly connected neighbors. Such an influence of individuals' actions on their neighbors' performance can propagate through the network affecting the performance of the entire network. Examples include several real world networks. As an instance, when a jurisdiction institutes a pollution abatement program, the benefits also accrue to nearby communities. Or, when a university builds a new library, other colleges also benefit from the subscription to the new library. In online advertising, the utility (users' attention) that an advertisement gets may be increased

R. Jain and R. Kannan (Eds.): GameNets 2011, LNICST 75, pp. 27–42, 2012.

or decreased due to adjacent advertisements on the webpage. The influence of neighbors is also observed in the spread of information and innovation in social and research networks. Networks with above characteristics are called local public good networks.

A local public good network differs from a typical public good system in that a local public good (alternatively, the action of an individual) is accessible to and directly influences the utilities of individuals in a particular locality/neighborhood within a big network. On the other hand a public good is accessible to and directly influences the utilities of all individuals in the system ([9, Chapter 11]). Because of the localized interactions of individuals, in local public good networks (such as ones described above) the information about the network is often localized; i.e., the individuals are usually aware of only their neighborhoods and not the entire network. In many situations the individuals also have some private information about the network or their own characteristics that are not known to anybody else in the network. Furthermore, the individuals may also be selfish who care only about their own benefit in the network. Such a decentralized information local public good network with selfish users gives rise to several research issues. In the next section we provide a literature survey on prior research in local public good networks.

1.1 Literature Survey

There exists a large literature on local public goods within the context of local public good provision by various municipalities that follows the seminal work of [14]. These works mainly consider network formation problems in which individuals choose where to locate based on their knowledge of the revenue and expenditure patterns (on local public goods) of various municipalities. In this paper we consider the problem of determining the levels of local public goods (actions of network agents) for a given network; thus, the problem addressed in this paper is distinctly different from those in the above literature. Recently, Bramoullé and Kranton [1] and Yuan [3] analyzed the influence of selfish users' behavior on the provision of local public goods in networks with fixed links among the users. The authors of [1] study a network model in which each user's payoff equals its benefit from the sum of efforts (treated as local public goods) of its neighbors less a cost for exerting its own effort. For concave benefit and linear costs, the authors analyze Nash equilibria (NE) of the game where each user's strategy is to choose its effort level that maximizes its own payoff from the provision of local public goods. The authors show that at such NE, *specialization* can occur in local public goods provision. Specialization means that only a subset of individuals contribute to the local public goods and others free ride. The authors also show that specialization can benefit the society as a whole because among all NE, the ones that are "specialized" result in the highest social welfare (sum of all users' payoffs). However, it is shown in [1] that none of the NE of abovementioned game can result in a local public goods provision that achieves the maximum possible social welfare. In [3] the work of [1] is extended to directed networks where the externality effects of users' efforts on each others' payoffs can be unidirectional or bidirectional. The authors of [3] investigate the relation between the structure of directed networks and the existence and nature of Nash equilibria of users' effort levels in those networks. For that matter they introduce a notion of ergodic stability to study the influence of perturbation of users' equilibrium efforts on the stability of NE. However, none of the NE of the game analyzed in [3] result in a local public goods provision that achieves optimum social welfare.

In this paper we consider a generalization of the network models investigated in [1,3]. Specifically, we consider a fixed network where the actions of each user directly affect the utilities of an arbitrary subset of network users. In our model, each user's utility from its neighbors' actions is an arbitrary concave function of its neighbors' action profile. Each user in our model knows only that part of the network that either affects it or is affected by it. Furthermore, each user's utility and action space are its private information, and each user is a self utility maximizer. Even though the network model we consider has similarities with those investigated in [1,3], the problem of local public goods provision we formulate in this paper is different from those in [1,3]. Specifically, we formulate a problem of local public goods provision in the framework of implementation theory[1] and address questions such as – How should the network users communicate so as to preserve their private information, yet make it possible to determine actions that achieve optimum social welfare? How to provide incentives to the selfish users to take actions that optimize the social welfare? How to make the selfish users voluntarily participate in any action determination mechanism that aims to optimize the social welfare? In a nutshell, the prior work of [1,3] analyzed specific games for local public good provision, whereas our work focusses on *designing a mechanism that can induce (via taxation) "appropriate" games among the network users so as to implement the optimum social welfare in NE.*

Previous works on implementation approach (Nash implementation) for (pure) public goods can be found in [5,7,15,2]. For our work, we obtained inspiration from [7]. In [7] Hurwicz presents a Nash implementation mechanism that implements the Lindahl allocation (optimum social welfare) for a public good economy. Hurwicz' mechanism also possesses the properties of individual rationality (i.e. it induces the selfish users to voluntarily participate in the mechanism) and budget balance (i.e. it balances the flow of money in the system). A local public good network can be thought of as a limiting case of a public good network, in which the influence of each public good tends to vanish on a subset of network users. However, taking the corresponding limits in the Hurwicz' mechanism does not result in a local public good provision mechanism with all the original properties of the Hurwicz' mechanism. In particular, such a limiting mechanism does not retain the budget balance property which is very important to avoid any scarcity/wastage of money. In this paper we address the problem of designing a local public good

[1] Refer to [8,13,12] and [11, Chapter 3] and for an introduction to implementation theory.

provision mechanism that possesses the desirable properties of budget balance, individual rationality, and Nash implementation of optimum social welfare. The mechanism we develop is more general than Hurwicz' mechanism, and Hurwicz' mechanism can be obtained as a special case of our mechanism. To the best of our knowledge the resource allocation problem and its solution that we present in this paper is the first attempt to analyze a local public goods network model in the framework of implementation theory. Below we state our contributions.

1.2 Contribution of the Paper

The key contributions of this paper are: 1. The formulation of a problem of local public goods provision in the framework of implementation theory. 2. The specification of a game form[2] (decentralized mechanism) for the above problem that, (i) implements in NE the optimal solution of the corresponding centralized local public good provision problem; (ii) is individually rational;[3] and (iii) results in budget balance at all NE and off equilibrium.

The rest of the paper is organized as follows. In Section 2.1 we present the model of local public good network. In Section 2.2 we formulate the local public good provision problem. In Section 3.1 we present a game form for this problem and discuss its properties in Section 3.2. We conclude in Section 4 with a discussion on future directions.

Notation used in the paper

We use bold font to represent vectors and normal font for scalars. We use bold uppercase letters to represent matrices. We represent the element of a vector by a subscript on the vector symbol, and the element of a matrix by double subscript on the matrix symbol. To denote the vector whose elements are all x_i such that $i \in \mathcal{S}$ for some set $\mathcal{S}$, we use the notation $(x_i)_{i \in \mathcal{S}}$ and we abbreviate it as $\boldsymbol{x}_{\mathcal{S}}$. We treat bold $\mathbf{0}$ as a zero vector of appropriate size which is determined by the context. We use the notation $(x_i, \boldsymbol{x}^*/i)$ to represent a vector of dimension same as that of $\boldsymbol{x}^*$, whose ith element is x_i and all other elements are the same as the corresponding elements of $\boldsymbol{x}^*$. We represent a diagonal matrix of size $N \times N$ whose diagonal entries are elements of the vector $\boldsymbol{x} \in \mathbb{R}^N$ by $\text{diag}(\boldsymbol{x})$.

2 The Local Public Good Provision Problem

In this section we present a model of local public good network and formulate a resource allocation problem for it. We first describe the components of the model and the assumptions we make on the properties of the network. We then present the resource allocation problem and formulate it as an optimization problem.[4]

[2] See [11, Chapter 3] and [12,13,8] for the definition of "game form".

[3] Refer to [11, Chapter 3] and [12] for the definition of "individual rationality" and "implementation in NE."

[4] A discussion on applciations that motivate Model (M) can be found in [11].

2.1 The Network Model (M)

We consider a network consisting of N users and one network operator. We denote the set of users by $\mathcal{N} := \{1, 2, \ldots, N\}$. Each user $i \in \mathcal{N}$ has to take an action $a_i \in \mathcal{A}_i$ where $\mathcal{A}_i$ is the set that specifies user i's feasible actions. In a real network, a user's actions can be consumption/generation of resources or decisions regarding various tasks. We assume that,

Assumption 1. *For all $i \in \mathcal{N}$, $\mathcal{A}_i$ is a convex and compact set in $\mathbb{R}$ that includes 0.*[5] *Furthermore, for each user $i \in \mathcal{N}$, the set $\mathcal{A}_i$ is its private information, i.e. $\mathcal{A}_i$ is known only to user i and nobody else in the network.*

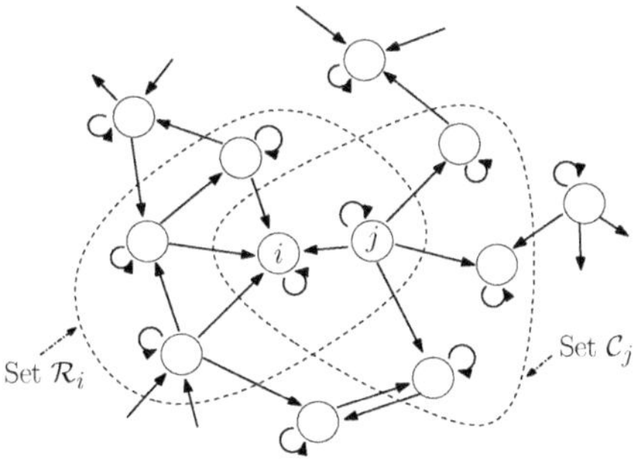

Fig. 1. A local public good network depicting the Neighbor sets $\mathcal{R}_i$ and $\mathcal{C}_j$ of users i and j respectively

Because of the users' interactions in the network, the actions taken by a user directly affect the performance of other users in the network. Thus, the performance of the network is determined by the collective actions of all users. We assume that the network is large-scale, therefore, every user's actions directly affect only a subset of network users in $\mathcal{N}$. Thus we can treat each user's action as a local public good. We depict the above feature by a directed graph as shown in Fig. 1. In the graph, an arrow from j to i indicates that user j affects user i; we represent the same in the text as $j \to i$. We assume that $i \to i$ for all $i \in \mathcal{N}$. Mathematically, we denote the set of users that affect user i by $\mathcal{R}_i := \{k \in \mathcal{N} \mid k \to i\}$. Similarly, we denote the set of users that are affected by user j by $\mathcal{C}_j := \{k \in \mathcal{N} \mid j \to k\}$. We represent the interactions of all network users together by a graph matrix $\boldsymbol{G} := [G_{ij}]_{N \times N}$. The matrix $\boldsymbol{G}$ consists of 0's and 1's, where $G_{ij} = 1$ represents that user i is affected by user j, i.e. $j \in \mathcal{R}_i$ and $G_{ij} = 0$ represents no influence of user j on user i, i.e. $j \notin \mathcal{R}_i$. Note that $\boldsymbol{G}$ need not be a symmetric matrix. Because $i \to i$, $G_{ii} = 1$ for all $i \in \mathcal{N}$. We assume that,

---

[5] In this paper we assume the sets $\mathcal{A}_i, i \in \mathcal{N}$, to be in $\mathbb{R}$ for simplicity. However, the decentralized mechanism and the results we present in this paper can be easily generalized to the scenario where for each $i \in \mathcal{N}$, $\mathcal{A}_i \subset \mathbb{R}^{n_i}$ is a convex and compact set in higher dimensional space $\mathbb{R}^{n_i}$. Furthermore, each space $\mathbb{R}^{n_i}$ can be of a different dimension n_i for different $i \in \mathcal{N}$.

Assumption 2. *The sets $\mathcal{R}_i, \mathcal{C}_i, i \in \mathcal{N}$, are independent of the users' action profile $\boldsymbol{a}_{\mathcal{N}} := (a_k)_{k\in\mathcal{N}} \in \prod_{k\in\mathcal{N}} \mathcal{A}_k$. Furthermore, for each $i \in \mathcal{N}$, $|\mathcal{C}_i| \geq 3$.*

For examples of applications where Assumption 2 holds, see [11, Chapter 5] and [3,1]. The performance of a user that results from actions taken by the users affecting it is quantified by a utility function. We denote the utility of user $i \in \mathcal{N}$ resulting from the action profile $\boldsymbol{a}_{\mathcal{R}_i} := (a_k)_{k\in\mathcal{R}_i}$ by $u_i(\boldsymbol{a}_{\mathcal{R}_i})$. We assume that,

Assumption 3. *For all $i \in \mathcal{N}$, the utility function $u_i : \mathbb{R}^{|\mathcal{R}_i|} \to \mathbb{R} \cup \{-\infty\}$ is concave in $\boldsymbol{a}_{\mathcal{R}_i}$ and $u_i(\boldsymbol{a}_{\mathcal{R}_i}) = -\infty$ for $a_i \notin \mathcal{A}_i$.[6] The function u_i is user i's private information.*

The assumptions that u_i is concave and is user i's private information are motivated by applications described in [11, Chapter 5] and [3,1]. The assumption that $u_i(\boldsymbol{a}_{\mathcal{R}_i}) = -\infty$ for $a_i \notin \mathcal{A}_i$ captures the fact that an action profile $(\boldsymbol{a}_{\mathcal{R}_i})$ is of no significance to user i if $a_i \notin \mathcal{A}_i$. We assume that,

Assumption 4. *Each network user $i \in \mathcal{N}$ is strategic and non-cooperative/selfish.*

Assumption 4 implies that the users have an incentive to misrepresent their private information, e.g. a user $i \in \mathcal{N}$ may not want to report to other users or to the network operator its true preference for the users' actions, if this results in an action profile in its own favor.

Each user $i \in \mathcal{N}$ pays a tax $t_i \in \mathbb{R}$ to the network operator. This tax can be imposed for the following reasons: (i) For the use of the network by the users. (ii) To provide incentives to the users to take actions that achieve a network-wide performance objective. The tax is set according to the rules specified by a mechanism and it can be either positive or negative for a user. With the flexibility of either charging a user (positive tax) or paying compensation/subsidy (negative tax) to a user, it is possible to induce the users to behave in a way such that a network-wide performance objective is achieved. For example, in a network with limited resources, we can set "positive tax" for the users that receive resources close to the amounts requested by them and we can pay "compensation" to the users that receive resources that are not close to their desirable ones. Thus, with the available resources, we can satisfy all the users and induce them to behave in a way that leads to a resource allocation that is optimal according to a network-wide performance criterion. We assume that,

Assumption 5. *The network operator does not have any utility associated with the users' actions or taxes. It does not derive any profit from the users' taxes and acts like an accountant that redistributes the tax among the users according to the specifications of the allocation mechanism.*

Assumption 5 implies that the tax is charged in a way such that

$$\sum_{i\in\mathcal{N}} t_i = 0. \tag{1}$$

[6] Note that a_i is always an element of $\boldsymbol{a}_{\mathcal{R}_i}$ because $i \to i$ and hence $i \in \mathcal{R}_i$.

To describe the "overall satisfaction" of a user from the performance it receives from all users' actions and the tax it pays for it, we define an "aggregate utility function" $u_i^A(\boldsymbol{a}_{\mathcal{R}_i}, t_i) : \mathbb{R}^{|\mathcal{R}_i|+1} \to \mathbb{R} \cup \{-\infty\}$ for each user $i \in \mathcal{N}$:

$$u_i^A(\boldsymbol{a}_{\mathcal{R}_i}, t_i) := \begin{cases} -t_i + u_i(\boldsymbol{a}_{\mathcal{R}_i}), \text{ if } a_i \in \mathcal{A}_i, a_j \in \mathbb{R}, j \in \mathcal{R}_i \backslash \{i\}, \\ -\infty, \text{ otherwise.} \end{cases} \quad (2)$$

Because u_i and $\mathcal{A}_i$ are user i's private information (Assumptions 1 and 3), the aggregate utility u_i^A is also user i's private information. As stated in Assumption 4, users are non-cooperative and selfish. Therefore, *the users are self aggregate utility maximizers.*

In this paper we restrict attention to static problems, i.e. we assume,

Assumption 6. *The set $\mathcal{N}$ of users, the graph $\boldsymbol{G}$, users' action spaces $\mathcal{A}_i, i \in \mathcal{N}$, and their utility functions $u_i, i \in \mathcal{N}$, are fixed in advance and they do not change during the time period of interest.*

We also assume that,

Assumption 7. *Every user $i \in \mathcal{N}$ knows the set $\mathcal{R}_i$ of users that affect it as well as the set $\mathcal{C}_i$ of users that are affected by it. The network operator knows $\mathcal{R}_i$ and $\mathcal{C}_i$ for all $i \in \mathcal{N}$.*

In networks where the sets $\mathcal{R}_i$ and $\mathcal{C}_i$ are not known to the users beforehand, Assumption 7 is still reasonable because of the following reason. As the graph $\boldsymbol{G}$ does not change during the time period of interest (Assumption 6), the information about the neighbor sets $\mathcal{R}_i$ and $\mathcal{C}_i, i \in \mathcal{N}$, can be passed to the respective users by the network operator before the users determine their actions. Alternatively, the users can themselves determine the set of their neighbors before determining their actions.[7] Thus, Assumption 7 can hold true for the rest of the action determination process. In the next section we present a local public good provision problem for Model (M).

2.2 The Decentralized Local Public Good Provision Problem (P_D)

For the network model (M) we wish to develop a mechanism to determine the users' action and tax profiles $(\boldsymbol{a}_{\mathcal{N}}, \boldsymbol{t}_{\mathcal{N}}) := ((a_1, a_2, \ldots, a_N), (t_1, t_2, \ldots, t_N))$. We want the mechanism to work under the decentralized information constraints of the model and to lead to a solution to the following centralized problem.

The centralized problem (P_C)

$$\begin{aligned} \max_{(\boldsymbol{a}_{\mathcal{N}}, \boldsymbol{t}_{\mathcal{N}})} \quad & \sum_{i \in \mathcal{N}} u_i^A(\boldsymbol{a}_{\mathcal{R}_i}, t_i) \\ \text{s.t.} \quad & \sum_{i \in \mathcal{N}} t_i = 0 \end{aligned} \quad (3)$$

[7] The exact method by which the users get information about their neighbor sets in a real network depends on the network characteristics.

$$\begin{aligned} &\equiv \max_{(\boldsymbol{a}_{\mathcal{N}},\boldsymbol{t}_{\mathcal{N}})\in\mathcal{D}} \sum_{i\in\mathcal{N}} u_i(\boldsymbol{a}_{\mathcal{R}_i}) \\ &\text{where, } \mathcal{D} := \{(\boldsymbol{a}_{\mathcal{N}},\boldsymbol{t}_{\mathcal{N}}) \in \mathbb{R}^{2N} \mid a_i \in \mathcal{A}_i \ \forall\, i \in \mathcal{N}; \sum_{i\in\mathcal{N}} t_i = 0\} \end{aligned} \tag{4}$$

The centralized optimization problem (3) is equivalent to (4) because for $(\boldsymbol{a}_{\mathcal{N}},\boldsymbol{t}_{\mathcal{N}}) \notin \mathcal{D}$, the objective function in (3) is negative infinity by (2). Thus $\mathcal{D}$ is the set of feasible solutions of Problem (P_C). Since by Assumption 3, the objective function in (4) is concave in $\boldsymbol{a}_{\mathcal{N}}$ and the sets $\mathcal{A}_i, i \in \mathcal{N}$, are convex and compact, there exists an optimal action profile $\boldsymbol{a}^*_{\mathcal{N}}$ for Problem (P_C). Furthermore, since the objective function in (4) does not explicitly depend on $\boldsymbol{t}_{\mathcal{N}}$, an optimal solution of Problem (P_C) must be of the form $(\boldsymbol{a}^*_{\mathcal{N}},\boldsymbol{t}_{\mathcal{N}})$, where $\boldsymbol{t}_{\mathcal{N}}$ is any feasible tax profile for Problem (P_C), i.e. a tax profile that satisfies (1).

The solutions of Problem (P_C) are ideal action and tax profiles that we would like to obtain. If there exists an entity that has centralized information about the network, i.e. it knows all the utility functions $u_i, i \in \mathcal{N}$, and all action spaces $\mathcal{A}_i, i \in \mathcal{N}$, then that entity can compute the above ideal profiles by solving Problem (P_C). Therefore, we call the solutions of Problem (P_C) optimal centralized allocations. In the network described by Model (M), there is no entity that knows perfectly all the parameters that describe Problem (P_C) (Assumptions 1 and 3). Therefore, we need to develop a mechanism that allows the network users to communicate with one another and that leads to optimal solutions of Problem (P_C). Since a key assumption in Model (M) is that the users are strategic and non-cooperative, the mechanism we develop must take into account the users' strategic behavior in their communication with one another. To address all of these issues we take the approach of implementation theory [8] for the solution of the decentralized local public good provision problem for Model (M). Henceforth we call this decentralized allocation problem as Problem (P_D). In the next section we present a decentralized mechanism (game form) for local public good provision that works under the constraints imposed by Model (M) and achieves optimal centralized allocations.

3 A Decentralized Local Public Good Provision Mechanism

For Problem (P_D), we want to develop a game form (message space and outcome function) that is *individually rational*, *budget balanced*, and that *implements in Nash equilibria* the goal correspondence defined by the solution of Problem (P_C).[8] Individual rationality guarantees voluntary participation of the users in the allocation process specified by the game form, budget balance guarantees that there is no money left unclaimed/unallocated at the end of the allocation process (i.e. it ensures (1)), and implementation in NE guarantees

[8] The definition of game form, goal correspondence, individual rationality, budget balance and implementation in Nash equilibria is given in [11, Chapter 3].

that the allocations corresponding to the set of NE of the game induced by the game form are a subset of the optimal centralized allocations (solutions of Problem (P_C)).

We would like to clarify at this point the definition of individual rationality (voluntary participation) in the context of our problem. Note that in the network model (M), the participation/non-participation of each user determines the network structure and the set of local public goods (users' actions) affecting the participating users. To define individual rationality in this setting we consider our mechanism to be consisting of two stages as discussed in [4, Chapter 7]. In the first stage, knowing the game form, each user makes a decision whether to participate in the game form or not. The users who decide not to participate are considered out of the system. Those who decide to participate follow the game form to determine the levels of local public goods in the network formed by them.[9] In such a two stage mechanism, individual rationality implies the following. If the network formed by the participating users satisfies all the properties of Model (M),[10] then, at all NE of the game induced by the game form among the participating users, the utility of each participating user will be at least as much as its utility without participation (i.e. if it is out of the system). This in turn imples that, if there are at least two other participating users that are affected by the actions of a user, then such a user voluntarily participates in the game form.

We would also like to clarify the rationale behind choosing NE as the solution concept for our problem. Note that because of assumptions 1 and 3 in Model (M), the environment of our problem is one of incomplete information. Therefore one may speculate the use of Bayesian Nash or dominant strategy as appropriate solution concepts for our problem. However, since the users in Model (M) do not possess any prior beliefs about the utility functions and action sets of other users, we cannot use Bayesian Nash as a solution concept for Model (M). Furthermore, because of impossibility results for the existence of non-parametric efficient dominant strategy mechanisms in classical public good environments [6], we do not know if it is possible to design such mechanisms for the local public good environment of Model (M). The well known Vickrey-Clarke-Groves (VCG) mechanisms that achieve incentive compatibility and efficiency with respect to non-numeraire goods, do not guarantee budget balance [6]. Hence they are inappropriate for our problem as budget balance is one of the desirable properties in our problem. VCG mechanisms are also unsuitable for our problem because they are direct mechanisms and any direct mechanism would require infinite message space to communicate the generic continuous (and concave) utility functions of

[9] This network is a subgraph obtained by removing the nodes corresponding to non-participating users from the original graph (directed network) constructed by all the users in the system.

[10] In particular, the network formed by the participating users must satisfy Assumption 2 that there are at least three users affected by each local public good in this network. Note that all other assumptions of Model (M) automatically carry over to the network formed by any subset of the users in Model (M).

users in Model (M). Because of all of above reasons, and the known existence results for non-parametric, individually rational, budget-balanced Nash implementation mechanisms for classical private and public goods environments [6], we choose Nash as the solution concept for our problem. Because NE in general describe strategic behavior of users in games of complete information, we interpret NE in the incomplete information environment of Model (M) in the way suggested by [6, Section 4] and [10]. Specifically, by quoting [6, Section 4], "we do not suggest that each user knows all of system environment when it computes its message. We do suggest, however, that the complete information Nash game-theoretic equilibrium messages may be the possible stationary messages of some unspecified dynamic message exchange process." Alternatively, by quoting [10], "we interpret our analysis as applying to an unspecified (message exchange) process in which users grope their way to a stationary message and in which the Nash property is a necessary condition for stationarity."

We next construct a game form for the resource allocation problem (P_D) that achieves the abovementioned desirable properties – Nash implementation, individual rationality, and budget balance.

3.1 The Game Form

In this section we present a game form for the local public good provision problem presented in Section 2.2. We provide explicit expressions of each of the components of the game form, the message space and the outcome function. We assume that the game form is common knowledge among the users and the network operator.

The message space
Each user $i \in \mathcal{N}$ sends to the network operator a message $\boldsymbol{m}_i \in \mathbb{R}^{|\mathcal{R}_i|} \times \mathbb{R}_+^{|\mathcal{R}_i|} =: \mathcal{M}_i$ of the following form:

$$\boldsymbol{m}_i := ({}^i\boldsymbol{a}_{\mathcal{R}_i}, {}^i\boldsymbol{\pi}_{\mathcal{R}_i}); \quad {}^i\boldsymbol{a}_{\mathcal{R}_i} \in \mathbb{R}^{|\mathcal{R}_i|},\ {}^i\boldsymbol{\pi}_{\mathcal{R}_i} \in \mathbb{R}_+^{|\mathcal{R}_i|}, \tag{5}$$

$$\text{where,} \quad {}^i\boldsymbol{a}_{\mathcal{R}_i} := ({}^ia_k)_{k\in\mathcal{R}_i} \text{ and } {}^i\boldsymbol{\pi}_{\mathcal{R}_i} := ({}^i\pi_k)_{k\in\mathcal{R}_i}, \quad i \in \mathcal{N}. \tag{6}$$

User i also sends the component $({}^ia_k, {}^i\pi_k), k \in \mathcal{R}_i$, of its message to its neighbor $k \in \mathcal{R}_i$. In this message, ${}^ia_k$ is the action proposal for user $k, k \in \mathcal{R}_i$, by user $i, i \in \mathcal{N}$. Similarly, ${}^i\pi_k$ is the price that user $i, i \in \mathcal{N}$, proposes to pay for the action of user $k, k \in \mathcal{R}_i$. A detailed interpretation of these message elements is given in Section 3.2.

The outcome function
After the users communicate their messages to the network operator, their actions and taxes are determined as follows. For each user $i \in \mathcal{N}$, the network operator determines the action $\hat{a}_i$ of user i from the messages communicated

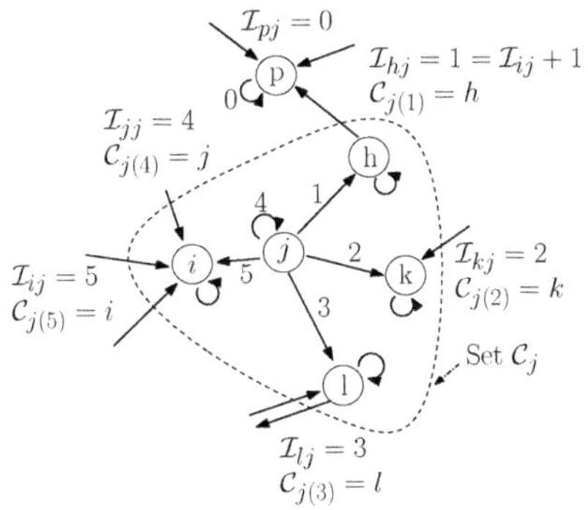

Fig. 2. Illustration of indexing rule for set $\mathcal{C}_j$ shown in Fig. 1. Index i_{rj} of user $r \in \mathcal{C}_j$ is indicated on the arrow directed from j to r. The notation to denote these indices and to denote the user with a particular index is shown outside the dashed boundary demarcating the set $\mathcal{C}_j$.

by its neighbors that are affected by it (set $\mathcal{C}_i$), i.e. from the message profile $\boldsymbol{m}_{\mathcal{C}_i} := (\boldsymbol{m}_k)_{k \in \mathcal{C}_i}$:

$$\hat{a}_i(\boldsymbol{m}_{\mathcal{C}_i}) = \frac{1}{|\mathcal{C}_i|} \sum_{k \in \mathcal{C}_i} {}^k a_i, \quad i \in \mathcal{N}. \tag{7}$$

To determine the users' taxes the network operator considers each set $\mathcal{C}_j$, $j \in \mathcal{N}$, and assigns indices $1, 2, \ldots, |\mathcal{C}_j|$ in a cyclic order to the users in $\mathcal{C}_j$. Each index $1, 2, \ldots, |\mathcal{C}_j|$ is assigned to an arbitrary but unique user $i \in \mathcal{C}_j$. Once the indices are assigned to the users in each set $\mathcal{C}_j$, they remain fixed throughout the time period of interest. We denote the index of user i associated with set $\mathcal{C}_j$ by i_{ij}. The index $\mathrm{i}_{ij} \in \{1, 2, \ldots, |\mathcal{C}_j|\}$ if $i \in \mathcal{C}_j$, and $\mathrm{i}_{ij} = 0$ if $i \notin \mathcal{C}_j$. Since for each set $\mathcal{C}_j$, each index $1, 2, \ldots, |\mathcal{C}_j|$ is assigned to a unique user $i \in \mathcal{C}_j$, therefore, $\forall\, i, k \in \mathcal{C}_j$ such that $i \neq k$, $\mathrm{i}_{ij} \neq \mathrm{i}_{kj}$. Note also that for any user $i \in \mathcal{N}$, and any $j, k \in \mathcal{R}_i$, the indices i_{ij} and i_{ik} are not necessarily the same and are independent of each other. We denote the user with index $k \in \{1, 2, \ldots, |\mathcal{C}_j|\}$ in set $\mathcal{C}_j$ by $\mathcal{C}_{j(k)}$. Thus, $\mathcal{C}_{j(\mathrm{i}_{ij})} = i$ for $i \in \mathcal{C}_j$. The cyclic order indexing means that, if $\mathrm{i}_{ij} = |\mathcal{C}_j|$, then $\mathcal{C}_{j(\mathrm{i}_{ij}+1)} = \mathcal{C}_{j(1)}$, $\mathcal{C}_{j(\mathrm{i}_{ij}+2)} = \mathcal{C}_{j(2)}$, and so on. In Fig. 2 we illustrate the above indexing rule for the set $\mathcal{C}_j$ shown in Fig. 1.

Based on the above indexing, the users' taxes $\hat{t}_i, i \in \mathcal{N}$, are determined as follows.

$$\begin{aligned} \hat{t}_i((\boldsymbol{m}_{\mathcal{C}_j})_{j \in \mathcal{R}_i}) = & \sum_{j \in \mathcal{R}_i} l_{ij}(\boldsymbol{m}_{\mathcal{C}_j})\, \hat{a}_j(\boldsymbol{m}_{\mathcal{C}_j}) + \sum_{j \in \mathcal{R}_i} {}^i\pi_j \left({}^i a_j - {}^{\mathcal{C}_{j(\mathrm{i}_{ij}+1)}} a_j \right)^2 \\ & - \sum_{j \in \mathcal{R}_i} {}^{\mathcal{C}_{j(\mathrm{i}_{ij}+1)}}\pi_j \left({}^{\mathcal{C}_{j(\mathrm{i}_{ij}+1)}} a_j - {}^{\mathcal{C}_{j(\mathrm{i}_{ij}+2)}} a_j \right)^2, \quad i \in \mathcal{N}, \end{aligned} \tag{8}$$

$$\text{where,} \quad l_{ij}(\boldsymbol{m}_{\mathcal{C}_j}) = {}^{\mathcal{C}_{j(\mathrm{i}_{ij}+1)}}\pi_j - {}^{\mathcal{C}_{j(\mathrm{i}_{ij}+2)}}\pi_j, \quad j \in \mathcal{R}_i, \quad i \in \mathcal{N}. \tag{9}$$

The game form given by (5)–(9) and the users' aggregate utility functions in (2) induce a game $(\times_{i\in\mathcal{N}}\mathcal{M}_i, (\hat{a}_i, \hat{t}_i)_{i\in\mathcal{N}}, \{u_i^A\}_{i\in\mathcal{N}})$. In this game, the set of network users $\mathcal{N}$ are the players, the set of strategies of a user is its message space $\mathcal{M}_i$, and a user's payoff is its utility $u_i^A\Big(\big(\hat{a}_j(\boldsymbol{m}_{\mathcal{C}_j})\big)_{j\in\mathcal{R}_i}, \hat{t}_i\big((\boldsymbol{m}_{\mathcal{C}_j})_{j\in\mathcal{R}_i}\big)\Big)$ that it obtains at the allocation determined by the communicated messages. We define a NE of this game as a message profile $\boldsymbol{m}_{\mathcal{N}}^*$ that has the following property:

$$\begin{aligned} &u_i^A\Big(\big(\hat{a}_j(\boldsymbol{m}_{\mathcal{C}_j}^*)\big)_{j\in\mathcal{R}_i}, \hat{t}_i\big((\boldsymbol{m}_{\mathcal{C}_j}^*)_{j\in\mathcal{R}_i}\big)\Big) \geq \\ &u_i^A\Big(\big(\hat{a}_j(\boldsymbol{m}_i, \boldsymbol{m}_{\mathcal{C}_j}^*/i)\big)_{j\in\mathcal{R}_i}, \hat{t}_i\big((\boldsymbol{m}_i, \boldsymbol{m}_{\mathcal{C}_j}^*/i)_{j\in\mathcal{R}_i}\big)\Big), \quad \forall\, \boldsymbol{m}_i \in \mathcal{M}_i, \ \forall\, i \in \mathcal{N}. \end{aligned} \tag{10}$$

As discussed earlier, NE in general describe strategic behavior of users in games of complete information. This can be seen from (10) where, to define a NE, it requires complete information of all users' aggregate utility functions. However, the users in Model (M) do not know each other's utilities; therefore, the game induced by the game form (5)–(9) and the users' aggregate utility functions (2) is not one of complete information. Therefore, for our problem we adopt the NE interpretation of [10] and [6, Section 4] as discussed at the beginning of Section 3. That is, we interpret NE as the "stationary" messages of an unspecified (message exchange) process that are characterized by the Nash property (10).

In the next section we show that the allocations obtained by the game form presented in (5)–(9) at all NE message profiles (satisfying (10)), are optimal centralized allocations.

3.2 Properties of the Game Form

We begin this section with an intuitive discussion on how the game form presented in Section 3.1 achieves optimal centralized allocations. We then formalize the results in Theorems 1 and 2.

To understand how the proposed game form achieves optimal centralized allocations, let us look at the properties of NE allocations corresponding to this game form. A NE of the game induced by the game form (5)–(9) and the users' utility functions (2) can be interpreted as follows: Given the users' messages $\boldsymbol{m}_k, k \in \mathcal{C}_i$, the outcome function computes user i's action as $1/|\mathcal{C}_i|\big(\sum_{k\in\mathcal{C}_i} {}^k a_i\big)$. Therefore, user i's action proposal ${}^i a_i$ can be interpreted as the increment that i desires over the sum of other users' action proposals for i, so as to bring its allocated action $\hat{a}_i$ to its own desired value. Thus, if the action computed for i based on its neighbors' proposals does not lie in $\mathcal{A}_i$, user i can propose an appropriate action ${}^i a_i$ and bring its allocated action within $\mathcal{A}_i$. The flexibility of proposing any action ${}^i a_i \in \mathbb{R}$ gives each user $i \in \mathcal{N}$ the capability to bring its allocation $\hat{a}_i$ within its feasible set $\mathcal{A}_i$ by unilateral deviation. Therefore, at any NE, $\hat{a}_i \in \mathcal{A}_i$, $\forall\, i \in \mathcal{N}$. By taking the sum of taxes in (8) it can further be seen, after some computations, that the allocated tax profile $(\hat{t}_i)_{i\in\mathcal{N}}$ satisfies (1) (even at off-NE messages). Thus, all NE allocations $\Big((\hat{a}_i(\boldsymbol{m}_{\mathcal{C}_i}^*))_{i\in\mathcal{N}}, \ (\hat{t}_i((\boldsymbol{m}_{\mathcal{C}_j}^*)_{j\in\mathcal{R}_i}))_{i\in\mathcal{N}}\Big)$ lie in $\mathcal{D}$ and hence are feasible solutions of Problem (P_C).

To see further properties of NE allocations, let us look at the tax function in (8). The tax of user i consists of three types of terms. The type-1 term is $\sum_{j\in\mathcal{R}_i} l_{ij}(\boldsymbol{m}_{\mathcal{C}_j})\ \hat{a}_j(\boldsymbol{m}_{\mathcal{C}_j})$; it depends on all action proposals for each of user i's neighbors $j \in \mathcal{R}_i$, and the price proposals for each of these neighbors by users other than user i. The type-2 term is $\sum_{j\in\mathcal{R}_i} {}^i\pi_j \left({}^ia_j - {}^{\mathcal{C}_{j(\mathrm{i}_{ij}+1)}}a_j\right)^2$; this term depends on ${}^i\boldsymbol{a}_{\mathcal{R}_i}$ as well as ${}^i\boldsymbol{\pi}_{\mathcal{R}_i}$. Finally, the type-3 term is the following: $-\sum_{j\in\mathcal{R}_i} {}^{\mathcal{C}_{j(\mathrm{i}_{ij}+1)}}\pi_j \left({}^{\mathcal{C}_{j(\mathrm{i}_{ij}+1)}}a_j - {}^{\mathcal{C}_{j(\mathrm{i}_{ij}+2)}}a_j\right)^2$; this term depends only on the messages of users other than i. Since ${}^i\boldsymbol{\pi}_{\mathcal{R}_i}$ does not affect the determination of user i's action, and affects only the type-2 term in $\hat{t}_i$, the NE strategy of user $i, i \in \mathcal{N}$, that minimizes its tax is to propose for each $j \in \mathcal{R}_i$, ${}^i\boldsymbol{\pi}_j = 0$ unless at the NE, ${}^ia_j = {}^{\mathcal{C}_{j(\mathrm{i}_{ij}+1)}}a_j$. Since the type-2 and type-3 terms in the users' tax are similar across users, for each $i \in \mathcal{N}$ and $j \in \mathcal{R}_i$, all the users $k \in \mathcal{C}_j$ choose the above strategy at NE. Therefore, the type-2 and type-3 terms vanish from every users' tax $\hat{t}_i, i \in \mathcal{N}$, at all NE. Thus, the tax that each user $i \in \mathcal{N}$ pays at a NE $\boldsymbol{m}^*_{\mathcal{N}}$ is of the form $\sum_{j\in\mathcal{R}_i} l_{ij}(\boldsymbol{m}^*_{\mathcal{C}_j})\ \hat{a}_j(\boldsymbol{m}^*_{\mathcal{C}_j})$. The NE term $l_{ij}(\boldsymbol{m}^*_{\mathcal{C}_j}), i \in \mathcal{N}, j \in \mathcal{R}_i$, can therefore be interpreted as the "personalized price" for user i for the NE action $\hat{a}_j(\boldsymbol{m}^*_{\mathcal{C}_j})$ of its neighbor j. Note that at a NE, the personalized price for user i is not controlled by i's own message. The reduction of the users' NE taxes into the form $\sum_{j\in\mathcal{R}_i} l_{ij}(\boldsymbol{m}^*_{\mathcal{C}_j})\ \hat{a}_j(\boldsymbol{m}^*_{\mathcal{C}_j})$ implies that at a NE, each user $i \in \mathcal{N}$ has a control over its aggregate utility only through its action proposal.[11] If all other users' messages are fixed, each user has the capability of shifting the allocated action profile $\hat{\boldsymbol{a}}_{\mathcal{R}_i}$ to its desired value by proposing an appropriate ${}^i\boldsymbol{a}_{\mathcal{R}_i} \in \mathbb{R}^{|\mathcal{R}_i|}$ (See the discussion in the previous paragraph). Therefore, the NE strategy of each user $i \in \mathcal{N}$ is to propose an action profile ${}^i\boldsymbol{a}_{\mathcal{R}_i}$ that results in an allocation $\hat{\boldsymbol{a}}_{\mathcal{R}_i}$ that maximizes its aggregate utility. Thus, at a NE, each user maximizes its aggregate utility for its given personalized prices. By the construction of the tax function, the sum of the users' tax is zero at all NE and off equilibria. Thus, the individual aggregate utility maximization of the users also result in the maximization of the sum of users' aggregate utilities subject to the budget balance constraint which is the objective of Problem (P_C).

It is worth mentioning at this point the significance of type-2 and type-3 terms in the users' tax. As explained above, these two terms vanish at NE. However, if for some user $i \in \mathcal{N}$ these terms are not present in its tax $\hat{t}_i$, then, the price proposal ${}^i\boldsymbol{\pi}_{\mathcal{R}_i}$ of user i will not affect its tax and hence, its aggregate utility. In such a case, user i can propose arbitrary prices ${}^i\boldsymbol{\pi}_{\mathcal{R}_i}$ because they would affect only other users' NE prices. The presence of type-2 and type-3 terms in user i's tax prevent such a behavior as they impose a penalty on user i if it proposes a high value of ${}^i\boldsymbol{\pi}_{\mathcal{R}_i}$ or if its action proposal for its neighbors deviates too much from other users' proposals. Even though the presence of type-2 and

[11] Note that user i's action proposal determines the actions of all the users $j \in \mathcal{R}_i$; thus, it affects user i's utility $u_i\Big(\big(\hat{a}_j(\boldsymbol{m}^*_{\mathcal{C}_j})\big)_{j\in\mathcal{R}_i}\Big)$ as well as its tax $\sum_{j\in\mathcal{R}_i} l_{ij}(\boldsymbol{m}^*_{\mathcal{C}_j})\ \hat{a}_j(\boldsymbol{m}^*_{\mathcal{C}_j})$.

type-3 terms in user i's tax is necessary as explained above, it is important that the NE price $l_{ij}(\boldsymbol{m}^*_{\mathcal{C}_j}), j \in \mathcal{R}_i$ of user $i \in \mathcal{N}$ is not affected by i's own proposal ${}^i\boldsymbol{\pi}_{\mathcal{R}_i}$. This is because, in such a case, user i may influence its own NE price in an unfair manner and may not behave as a price taker. To avoid such a situation, the type-2 and type-3 terms are designed in a way so that they vanish at NE. Thus, this construction induces price taking behavior in the users at NE and leads to optimal allocations.

The results that formally establish the above properties of the game form are summarized in Theorems 1 and 2 below.

Theorem 1. *Let $\boldsymbol{m}^*_{\mathcal{N}}$ be a NE of the game induced by the game form presented in Section 3.1 and the users' utility functions (2). Let $(\hat{\boldsymbol{a}}^*_{\mathcal{N}}, \hat{\boldsymbol{t}}^*_{\mathcal{N}}) := (\hat{\boldsymbol{a}}_{\mathcal{N}}(\boldsymbol{m}^*_{\mathcal{N}}), \hat{\boldsymbol{t}}_{\mathcal{N}}(\boldsymbol{m}^*_{\mathcal{N}})) := \Big((\hat{a}_i(\boldsymbol{m}^*_{\mathcal{C}_i}))_{i\in\mathcal{N}}, (\hat{t}_i((\boldsymbol{m}^*_{\mathcal{C}_j})_{j\in\mathcal{R}_i}))_{i\in\mathcal{N}}\Big)$ be the action and tax profiles at $\boldsymbol{m}^*_{\mathcal{N}}$ determined by the game form. Then,*

(a) Each user $i \in \mathcal{N}$ weakly prefers its allocation $(\hat{\boldsymbol{a}}^*_{\mathcal{R}_i}, \hat{t}^*_i)$ to the initial allocation $(\mathbf{0}, 0)$. Mathematically,

$$u_i^A\Big(\hat{\boldsymbol{a}}^*_{\mathcal{R}_i}, \hat{t}^*_i\Big) \geq u_i^A\Big(\mathbf{0},\, 0\Big), \qquad \forall\, i \in \mathcal{N}.$$

(b) $(\hat{\boldsymbol{a}}^*_{\mathcal{N}}, \hat{\boldsymbol{t}}^*_{\mathcal{N}})$ is an optimal solution of Problem (P_C). ⊠

Theorem 2. *Let $\hat{\boldsymbol{a}}^*_{\mathcal{N}}$ be an optimum action profile corresponding to Problem (P_C). Then,*

(a) There exist a set of personalized prices $l^*_{ij}, j \in \mathcal{R}_i, i \in \mathcal{N}$, such that

$$\hat{\boldsymbol{a}}^*_{\mathcal{R}_i} = \underset{\substack{\hat{a}_i \in \mathcal{A}_i \\ \hat{a}_j \in \mathbb{R},\, j \in \mathcal{R}_i \setminus \{i\}}}{\arg\max} \; - \sum_{j \in \mathcal{R}_i} l^*_{ij}\, \hat{a}_j + u_i(\hat{\boldsymbol{a}}_{\mathcal{R}_i}), \qquad \forall\, i \in \mathcal{N}.$$

(b) There exists at least one NE $\boldsymbol{m}^*_{\mathcal{N}}$ of the game induced by the game form presented in Section 3.1 and the users' utility functions (2) such that, $\hat{\boldsymbol{a}}_{\mathcal{N}}(\boldsymbol{m}^*_{\mathcal{N}}) = \hat{\boldsymbol{a}}^*_{\mathcal{N}}$. Furthermore, if $\hat{t}^*_i := \sum_{j\in\mathcal{R}_i} l^*_{ij}\hat{a}^*_j, i \in \mathcal{N}$, the set of all NE $\boldsymbol{m}^*_{\mathcal{N}} = (\boldsymbol{m}^*_i)_{i\in\mathcal{N}} = ({}^i\boldsymbol{a}^*_{\mathcal{R}_i}, {}^i\boldsymbol{\pi}^*_{\mathcal{R}_i})$ that result in $(\hat{\boldsymbol{a}}^*_{\mathcal{N}}, \hat{\boldsymbol{t}}^*_{\mathcal{N}})$ is characterized by the solution of the following set of conditions:

$$\frac{1}{|\mathcal{C}_i|} \sum_{k \in \mathcal{C}_i} {}^k a^*_i = \hat{a}^*_i, \quad i \in \mathcal{N},$$

$${}^{\mathcal{C}_{j(\mathrm{i}_{ij}+1)}}\pi^*_j - {}^{\mathcal{C}_{j(\mathrm{i}_{ij}+2)}}\pi^*_j = l^*_{ij}, \quad j \in \mathcal{R}_i,\; i \in \mathcal{N},$$

$${}^i\pi^*_j\left({}^ia^*_j - {}^{\mathcal{C}_{j(\mathrm{i}_{ij}+1)}}a^*_j\right)^2 = 0, \quad j \in \mathcal{R}_i, \quad i \in \mathcal{N},$$

$${}^i\pi^*_j \geq 0, \quad j \in \mathcal{R}_i, \quad i \in \mathcal{N}.$$

Because Theorem 1 is stated for an arbitrary NE $\boldsymbol{m}^*_{\mathcal{N}}$ of the game induced by the game form of Section 3.1 and the users' utility functions (2), the assertion of the theorem holds for all NE of this game. Thus, part (a) of Theorem 1 establishes that the game form presented in Section 3.1 is *individually rational*, i.e., at any NE allocation, the aggregate utility of each user is at least as much as its aggregate utility before participating in the game/allocation process. Because of this property of the game form, each user voluntarily participates in the allocation process. Part (b) of Theorem 1 asserts that all NE of the game induced by the game form of Section 3.1 and the users' utility functions (2) result in optimal centralized allocations (solutions of Problem (P_C)). Thus the set of NE allocations is a subset of the set of optimal centralized allocations. This establishes that the game form of Section 3.1 *implements in NE* the goal correspondence defined by the solutions of Problem (P_C). Because of this property, the above game form guarantees to provide an optimal centralized allocation irrespective of which NE is achieved in the game induced by it.

The assertion of Theorem 1 that establishes the above two properties of the game form presented in Section 3.1 is based on the assumption that there exists a NE of the game induced by this game form and the users' utility functions (2). However, Theorem 1 does not say anything about the existence of NE. Theorem 2 asserts that NE exist in the above game, and provides conditions that characterize the set of all NE that result in optimal centralized allocations of the form $(\hat{\boldsymbol{a}}^*_{\mathcal{N}}, \hat{\boldsymbol{t}}^*_{\mathcal{N}}) = (\hat{\boldsymbol{a}}^*_{\mathcal{N}}, (\sum_{j \in \mathcal{R}_i} l^*_{ij} \hat{a}^*_j)_{i \in \mathcal{N}})$, where $\hat{\boldsymbol{a}}^*_{\mathcal{N}}$ is any optimal centralized action profile. In addition to the above, Theorem 2 also establishes the following property of the game form. Since the optimal action profile $\hat{\boldsymbol{a}}^*_{\mathcal{N}}$ in the statement of Theorem 2 is arbitrary, the theorem implies that the game form of Section 3.1 can obtain each of the optimum action profiles of Problem (P_C) through at least one of the NE of the induced game. This establishes that the above game form is not biased towards any particular optimal centralized action profile.

Due to lack of space, we are not presenting the proofs of Theorem 1 and Theorem 2 here. The proofs are available in [11, Chapter 5].

4 Future Directions

The problem formulation and the solution of the local public goods provision problem presented in this paper open up several new directions for future research. First, the development of efficient mechanisms that can compute NE is an important open problem. To address this problem there can be two different directions. (i) The development of algorithms that guarantee convergence to Nash equilibria of the games constructed in this paper. (ii) The development of alternative mechanisms/game forms that lead to games with dynamically stable NE. Second, the network model we studied in this paper assumed a given set of users and a given network topology. In many local public good networks such as social or research networks, the set of network users and the network topology must be determined as part of network objective maximization. These situations give rise to interesting admission control and network formation problems many of which are open research problems. Finally, in this paper we focused on static resource allocation problem where the characteristics of the local public good network do not change with time. The development of implementation mechanisms under dynamic situations, where the network characteristics change during the determination of resource allocation, are open research problems.

Acknowledgements. The authors are grateful to Yan Chen and A. Anastasopoulos at the University of Michigan for stimulating discussions.

References

1. Bramoullé, Y., Kranton, R.: Public goods in networks. Journal of Economic Theory 135, 478–494 (2007)
2. Chen, Y.: A family of supermodular mechanisms implementing Lindahl allocations for quasilinear environments. Economic Theory 19(4), 773–790 (2002)
3. Chih Yuan, K.: Public goods in directed networks. Preprint
4. Fudenberg, D., Tirole, J.: Game theory. The MIT Press (1991)
5. Groves, T., Ledyard, J.: Optimal allocation of public goods: A solution to the 'free rider' problem. Econometrica 45, 783–809 (1977)
6. Groves, T., Ledyard, J.: Incentive compatibility since 1972. In: Groves, T., Radner, R., Reiter, S. (eds.) Information, Incentives, and Economic Mechanisms: Essays in Honor of Leonid Hurwicz, pp. 48–109. University of Minnesota Press, Minneapolis (1987)
7. Hurwicz, L.: Outcome functions yielding Walrasian and Lindahl allocations at Nash equilibrium points. Review of Economic Studies 46, 217–225 (1979)
8. Jackson, M.: A crash course in implementation theory. In: Social Choice and Welfare, pp. 655–708 (2001)
9. Mas-Colell, A., Whinston, M.D., Green, J.R.: Microeconomic theory. Oxford University Press (2002)
10. Reichelstein, S., Reiter, S.: Game forms with minimal message space. Econometrica 56(3), 661–692 (1988)
11. Sharma, S.: A mechanism design approach to decentralized resource allocation in wireless and large-scale networks: Realization and implementation. PhD thesis, University of Michigan, Ann Arbor (2009)
12. Sharma, S., Teneketzis, D.: A game-theoretic approach to decentralized optimal power allocation for cellular networks. Control Group Report, Dept. of EECS, Univ. of Michigan, CGR 08-05 (2008); To appear in GameComm Special Issue of Telecommunication Systems Journal
13. Stoenescu, T., Teneketzis, D.: Decentralized resource allocation mechanisms in networks: realization and implementation. In: Birkhřuser, E.H.A. (ed.) Advances in Control, Communiation Networks, and Transportation Systems, pp. 225–263 (2005), in honor of Pravin Varaiya
14. Tiebout, C.M.: A pure theory of local expenditures. The Journal of Political Economy 64(5), 416–424 (1956)
15. Walker, M.: A simple incentive compatible scheme for attaining Lindahl allocations. Econometrica 49, 65–71 (1981)

Channel Assignment on Wireless Mesh Network Backbone with Potential Game Approach

Pedro B.F. Duarte[1], Zubair Md. Fadlullah[1], Athanasios V. Vasilakos[2], and Nei Kato[1]

[1] Graduate School of Information Sciences (GSIS), Tohoku University, Japan
burjack@it.ecei.tohoku.ac.jp
[2] Department of Computer and Telecom. Engineering, University of Western Macedonia, Greece

Abstract. The Wireless Mesh Network (WMN) has already been recognized as a promising technology as broadband access network from both academic and industry points of view. In order to improve its performance, research has been carried on how to increase the number of simultaneous transmissions in the network while avoiding signal interference among radios. Considering WMNs based upon IEEE 802.11 b/g standards, lately most of researchers have been relying on the usage of orthogonal channels for solving the Channel Assignment (CA) problem. However, in this paper, we introduce a novel CA algorithm exploiting partially overlapped channels (POC) that overcome the common orthogonal channel approach. This algorithm is derived based on Game Theory framework using Potential Games and yields near optimum CA.

Keywords: Wireless Mesh Networks, channel assignment problem, partially overlapped channels, game theory, potential games.

1 Introduction

Wireless Mesh Networks (WMNs) have attracted interest from researchers, industry, and users [1]. Its multi-hop characteristics can greatly improve the network coverage area with lower transmission power and provide reliable broadband access services for campus and community networks. WMNs are considered to be a key technology in Next Generation Networks (NGNs) and aims to deploy ubiquitous Internet access. With such a promising future, several standards have been developed for different access ranges, namely IEEE 802.15.4, IEEE 802.11s and IEEE 802.16j, which target Wireless Personal Area Networks (WPANs), Wireless Local Area Networks (WLANs) and Wireless Metropolitan Area Networks (WMANs), respectively. This paper will focus on WMNs based on WLAN technology.

WMNs consist of a multi-hop environment nevertheless its concepts and targets differ from conventional Mobile Adhoc Networks (MANET). A WMN comprises two different types of nodes, namely Mesh Routers (MRs) and Mesh

R. Jain and R. Kannan (Eds.): GameNets 2011, LNICST 75, pp. 43–56, 2012.

Clients (MCs). The former is responsible for network routing and bridging while the latter, being a light-weight node, would perform just the routing function, if necessary. Moreover, MRs compose a backbone network and concerning mobility and battery life-time, they are usually static and have no constrains on energy consumption. Such differences between WMNs and MANETs led to novel protocols development to address specific challenges on WMNs.

As the users' requests for better services, e.g., higher transmission rates and lower delay networks, are an ever increasing demand, solutions to improve the network capacity are constantly addressed by researchers. On WMN, several solutions have been already proposed to improve its capacity, such as modified Medium Access Control (MAC) protocols, directional and Multiple Input Multiple Output (MIMO) antennas and Multi-Radio Multi-Channel (MRMC) topology.

Inside the MRMC field, one of the most promising techniques is Partially Overlapped Channel Assignment by using IEEE 802.11 b/g devices, which can increase the network throughput by exploiting more simultaneous transmissions. According to the afore-mentioned standard, there are 11 channels available to communication on the 2.4 GHz ISM band. Each of them has a bandwidth of 22 MHz and a center frequency distance of only 5 MHz. Hence, there are just three orthogonal[1] channels available, namely, channels 1, 6 and 11. Using these three channels configuration does not provide an efficient frequency-spatial reuse. However, by exploiting all eleven channels in a systematic approach to avoid the interference among adjacent channels, we are able to achieve a greater number of simultaneous transmissions rather than just three orthogonal channels. Nevertheless, this systematic approach is not trivial and if not well planned, it can actually severely degrade the network performance (throughput and delay) due to adjacent channel interference, that is considerably more harmful than co-channel interference.

In this work, we derive and investigate a novel near-optimal CA. In order to derive this near-optimal CA, we use a Game Theory approach. This mathematical tool is specially useful in the network engineering field to model high complex scenarios that may include complex traffic models, mobility, unpredictable link quality, in which pure mathematical analysis has met limited success [2]. Game Theory, a field of applied mathematics, suits this purpose because of its ability to model individual, independent decision makers, or players, that interacts and impact other decision makers, which closely resembles WMNs and MANETs dynamics[2].

The remainder of this paper is organized as follows. Section 2 surveys related works on solving MRMC CA problem, followed by Section 3 which reviews the interference model used for this article. The near-optimal CA is derived on Section 4. The performance of the proposed algorithm is evaluated in Section 5, and finally Section 6 concludes the paper.

---

[1] *Orthogonal* and *non-overlapping* channels are interchangeably used.

[2] *Players*, *nodes*, and *routers* are interchangeably used.

2 Related Works

A multi-channel MAC (MMAC) protocol for handling multi-channel assignment using a single radio was proposed in [3]. The protocol uses non-overlapping channels and it reserves one channel for control packets and two others for data packets.

Draves et al. [4] start employing multi-radio topology. In their contribution, a new routing metric Weighted Cumulative Expected Transmission Time (WCETT) is developed. In addition, they assume non-interfering channels and they employ fixed CA. In 2007, a survey on channel assignment was performed by Skalli et al. [5]. It reviews several CA strategies and also proposes a new one. Including the surveyed and proposed algorithms, all of them employ non-overlapping channels. According to this paper, "this leads to efficient spectrum utilization and increases the actual bandwidth available to the network".

More recently, Bukkapatanam et al. [6] using numerical analysis showed that the usage of overlapping channels achieves better performance than three non-overlapping channels in the backbone network, expanding the previous work of Mishra [7]. However, none of the three above cited works actually describes a novel CA algorithm exploiting POC.

Using the Game Theoretical perspective to address complex engineering related issues has attracted the attention of several researchers in the last decade and its applicability abound: power control in cellular radio systems [8], optimal routing control [9] and reputation mechanisms for ad hoc networks [10] are few examples.

Although works on Game Theory concerning CA are also numerous, usually POC are not considered by the proposed models. For example, in [11] the authors derive a perfectly fair CA using concepts as Nash equilibrium (NE) and fairness under a non-cooperative game, however their model is simply based on orthogonal channels and their simulations just evaluate a single collision domain. In [12], Gao and Wang, model the game as for the CA as a coalition rather than a non-cooperative game and than prove the existence of a Nash equilibrium under such conditions. That work also assumes non-overlapping channels and just single collision domain. In Zhang and Fang's research [13], a joint solution for channel and power allocation is studied from the Game Theoretical perspective. Differently from our work, they mainly focus on the access network issue rather than the backbone.

In short terms, in this paper, we address the CA problem on the WMN backbone and we develop a novel CA algorithm exploiting POC. We also employ Game Theory concepts to model MRs as decision makers of a cooperative game. The interaction among all MRs can be classified as an *identical interest game* as in [13]. Further, we introduce a negotiation-based CA algorithm for the frequency assignment that converges to a steady state (NE), and as a property of identical interest games, this condition implies achieving an optimum CA. Therefore, the contributions of this work are two fold: 1) develop a game theory mathematical model that comprises the interference models necessary to POC assignment. 2) Game theoretic CA algorithm that yields near-optimal throughput performance.

3 Interference Model

We may define the CA problem as an optimization one in terms of mapping available communication channels to network interfaces in order to maximize the communication capacity while minimizing signal interference. Interference range is defined as the distance within which interference occurs. Two different transmissions are considered to interfere to each other if they lie within interference range.

In a multi-channel environment, four different types of interference and their influence on the network capacity should be addressed. Here, we describe them in more details. For this description, we consider two pairs of nodes, each of them having a sender and a receiver. Let the sender and receiver of the first pair be denoted by S_1 and R_1, and those of the second pair be denoted by S_2 and R_2. All these nodes are positioned within the interference range.

- Co-channel Interference: consider that all four nodes are operating in the same channel. Because of CSMA/CA, this type of interference is less harmful for the network capacity than Adjacent Channel Interference. Consider the following scenario: node S_1 is starting to transmit a packet to R_1. It checks if the medium is busy or idle. If it is busy, the node will withdraw its transmission and postpone it. However, if the medium is idle, it will proceed with the transmission. Meanwhile S_1 is sending data to R_1, S_2 also attempts to send a packet to R_2. In this case, the medium will be busy. Hence, S_2 will withdraw the transmission attempt and wait over backoff period. Later on, it will attempt again and the transmission between S_1-R_1 will be already ceased. Then, S_2 will succeed with the signal transmission. In this scenario, we have a contention based access, in which a concurrent access to the medium occurs.
- Orthogonal Channels: In this scenario interference will not occur. Consider S_1-R_1 and S_2-R_2 using two orthogonal channels. Again, S_1 detects an idle medium and starts the packet transmission. Meanwhile, S_2 will also detect an idle medium since it is operating on a distinct channel. Both pairs are able to successfully transmit their packets simultaneously, because there is no overlapping frequency band between those channels.
- Adjacent Channel Interference (ACI): This kind of interference seriously degrade the network capacity. Here, we consider S_1-R_1 and S_2-R_2 assigned to channel 1 and 3, respectively. Following the same scheme in which S_1 begins transmitting first, S_2 will detect an idle medium in channel 3 and also starts to send its packet. However, since channels 1 and 3 share a common frequency band, the receivers will not be able to successfully decode the packets, causing a transmission error that severely degrades the network throughput. Although, it is important to note that the interference range for adjacent channels is inversely proportional to the actual channel separation.
- Self Interference: Self Interference is defined as a node itself causing interference to one of its own transmissions. This case will occur in multiple radio nodes using omni-directional antennas. To explain this case, consider S_1 with

two network interfaces, assigned to channel 1 and 3. Whenever S_1 tries to simultaneously send packets on both interfaces, the SIR (Signal to Interference Ratio) will be degraded no matter where the receiver node is. This type of interference can be avoided if no node has its interfaces assigned to overlapping channels. This means that in IEEE 802.11b/g, we can assemble at most 3 interfaces using orthogonal channels at any given node.

Considering the afore-mentioned types of interference, the authors in [14] developed a schematic procedure for CA. This model is named as I-Matrix and it determines whether it is possible or not to assign channels to a given link exploiting POC. We adopt the Interference Factor concept from their work, in order to devise our interference model.

3.1 Interference Factor

The *interference factor* $f_{i,j}$ takes as input parameters geographical distance and channel separation, and provides the effective spectral overlapping level between channels i and j. In order to calculate $f_{i,j}$, the experimental measurements showed in [15] are used and scaled by a factor of 10. We use the *interference range* (IR) table where δ is the channel separation $\delta = |i - j|$ and $IR(\delta)$ is the maximum geographical distance in which there will be interference between channels i and j.

Table 1. Interference Range (IR)

δ	0	1	2	3	4	5
$IR(\delta)$	132.6	90.8	75.9	46.9	32.1	0

Given the IR table, let d be the Euclidean distance between transceivers using channels i and j. Also by definition, considering the case that the transceivers are assembled in the same node, we define d being zero. To calculate $f_{i,j}$ we should consider the three following cases:

1. $f_{i,j} = 0$: when $\delta \geq 5$ or $d > IR(\delta)$
   In this case there will be no interference between the radios since either they are assigned orthogonal channels, or they are distant enough not to cause interference given IR for channel i and j.
2. $1 < f_{i,j} < \infty$: when $0 \leq \delta < 5$ and $d \leq IR(\delta)$
   Here we have two radios assigned to overlapping channels i and j, and also the distance between them is within the interference range. Thus, interference factor should be calculated as the following equation in which $f_{i,j}$ is inversely proportional to the distance between radios.

$$f_{i,j} = IR(\delta)/d : \tag{1}$$

3. $f_{i,j} = \infty$: when $0 \leq \delta < 5$ and $d = 0$
   As mentioned before, here we strictly exclude the self interference to happen. Two overlapping channels ($\delta < 5$) will not be assigned at a given node.

4 Modeling Channel Assignment Game

In this section, we model our MRs as players as in Game Theory. The main objective here is to derive an near optimal CA using the mathematical analyses provided by the Game Theory framework. Each MR is considered *player*, i.e. *decision maker* of the game, and we model the interactions among them as a cooperative channel assignment game (CoCAG). The game is composed of a finite set of players, $\mathbf{A} = \{a_1, a_2, \ldots, a_N\}$ and all the players have a common strategy space $\mathbf{S} = \mathbf{S_i}, \forall i$. In this context, we map the channel(s) assigned to any given MRs' radios as its chosen strategy. Formally, the strategy of ith player is $\mathbf{s_i} = \{k_{i,1}, \ldots, k_{i,c}, \ldots, k_{i,|C|}\}$, where $k_{i,c}$ is a binary value set to 1 if channel c is assigned to one of the player's radio, 0 otherwise, and $|C|$ is the number of channels for the channel set $\mathbf{C}$. The game profile is defined as the cartesian product of the players' strategy vector, $\mathbf{\Psi} = \times_{i \in \mathbf{A}} \mathbf{s_i} = \mathbf{s_1} \times \mathbf{s_2} \times \cdots \times \mathbf{s_N}$. Note that a game profile includes, one and only one strategy for each player. $\mathbf{s_{-i}}$ is specially defined as the strategy set chosen by all other players except player i.

The objective of the game is to maximize the network throughput. We define a joint metric M_i, for each player i, that correlates the links configuration and topology to a numerical value. This metric is directly proportional to the number of assigned links in each node. Each link capacity (transmission rate) is evaluated according to number of interfering links. Also two topology control factors, k and h, are included, since the network should not be evaluated only by its number of links but also how efficiently these links connects the MRs towards the GW, *i.e.* hop count.

$$M_i = k \frac{\sum_{j \in \mathbf{C}} \frac{R}{n_j}}{h} \tag{2}$$

where

- k is a connectivity factor set to 1 if the node is can indirectly reach the GW, 0 otherwise.
- R is the link data rate (Mbps).
- n is the number of interfering links.
- h is hop count from the node to the GW.

Finally, each player has its utility function dependent on its strategy and the other players' strategy $U_i(\mathbf{\Psi})$, and since we defined a cooperative game, the following holds and U_{NET} stands for utility of the network:

$$U_{NET}(\mathbf{\Psi}) = U_i(\mathbf{\Psi}) = \sum_{i \in \mathbf{A}} M_i, \forall i \tag{3}$$

Players will negotiate and change their interdependent strategies in $\mathbf{S}$ in order to achieve an optimal value for U_{NET}. Then two important issues arise: 1) whether they ever reach a consensus, or steady state 2) how efficient would be this steady state performance, if ever existent. The answers for these questions are the following:

In game theory, *Nash Equilibrium* is an important concept. The players will meet an agreement if NE exists. NE formal definition as in [16] is described bellow.

Definition 1. *strategy* $\mathbf{s}^* \in \mathbf{S}$ *is a Nash Equilibrium if*

$$U_i(\mathbf{s}^*) \geq U_i(\mathbf{s_i^{'}}, \mathbf{s_{-i}}) \ \forall \ \mathbf{s_i^{'}} \in \mathbf{S_i}, \forall i \in \mathbf{A} \tag{4}$$

According to this definition no player can benefit by deviating from its strategy if other player do not change hers[3]. In other words, this result guarantees an agreement for negotiations among players, although no optimal outcome or fairness is intrinsically guaranteed. Nevertheless, a specific type of game, denominated *potential games*, has very useful properties that address the outcome efficiency issue and the NE existence. For a potential game, the following holds:

- Every finite potential game possesses at least one pure strategy NE [17].
- All NE are either local or global maximizers of the utility function [17].
- There are well-known learning schemes to reach these function maximizers in the literature, namely *best response* and *better response* [2].

A potential game is defined as a game in which a *potential function* $P : \mathbf{S} \rightarrow \mathbf{R}$ exists

$$P(\mathbf{s}^{'}, \mathbf{s_{-i}}) - P(\mathbf{s}^{''}, \mathbf{s_{-i}}) = U_i(\mathbf{s}^{'}, \mathbf{s_{-i}}) - U_i(\mathbf{s}^{''}, \mathbf{s_{-i}}) \ \forall \ i, \mathbf{s}^{'}, \mathbf{s}^{''} \tag{5}$$

where $\mathbf{s}^{'}$ and $\mathbf{s}^{''}$ stands for two arbitrary strategies.

Lemma 1. *CoCAG is a potential game.*

Proof. It is straightforward that the network utility function (3) itself is a potential function for CoCAG. For the process of identifying if a given utility function is a potential function, we have the following definition from [16]. A coordination game is defined when all users have the same utility function. That is to say, $U_i(\mathbf{\Psi}) = P(\mathbf{\Psi})$. Since CoCAG is a coordination game, we thus conclude that (3) is a potential function for this game. □

Hence, we have:

$$P = U_i(\mathbf{\Psi}) = U_{NET}(\mathbf{\Psi}), \forall i \tag{6}$$

Making use of NE and potential games, we guarantee that our CoCAG will converge to an agreement among players and this point will be a utility function

[3] According to game-theoretic conventions *players* should be referred to with *female* pronouns.

maximizer. In the literature, there are two famous learning schemes to accomplish this purpose, namely best response (7) and better response (8). In the former scheme, during her turn to choose a strategy to play, the player searches her entire strategy space and selects the one that yields the best outcome considering the other players' strategy. This scheme provides a fast convergence. On the other hand, it requires intensive processing that grows exponentially according to the number of players. In the latter scheme, during its turn, each player selects a random strategy and keep it as long as it generates a better outcome than the previous one. Thus, better response provides less intensive computation at the cost of a slower convergence to the equilibrium.

$$\mathbf{s_i^{t+1}} = \arg\max_{\mathbf{s_j} \in \mathbf{S_i}} U_i(\mathbf{\Psi}) \tag{7}$$

$$\mathbf{s_i^{t+1}} = \begin{cases} \mathbf{s_i^{rand}} & \text{if } U_i(\mathbf{s_i^{rand}}, \mathbf{s_{-i}}) > U_i(\mathbf{s_i^t}, \mathbf{s_{-i}}) \\ \mathbf{s_i^t} & \text{otherwise} \end{cases} \tag{8}$$

Nevertheless, the equilibrium may happen at the local optimum of the utility function, instead of the global optimum. In this case, the system performance will be trapped in a sub-optimal state and, since this is one NE, no player will be able to increase her utility function by changing her strategy.

We propose the following negotiation based algorithm that converges to NE. We assume identical MRs, and each of them has an unique identification parameter a_iID for routing purpose. In addition, we generalize the finalization criteria (T). The finalization criteria can be met following different parameters, few possibilities are exemplified. In this article we will employ the maximum number of negotiations as finalization criteria.

- maximum number of negotiations;
- timely limited;
- utility function threshold.

Our algorithm has a distributed coordination and in order perform likewise the following mechanism should be met. The algorithm has two distinct steps, namely negotiation and operation phases. In the negotiation phase, all nodes operate using a common channel to exchange the messages, which guarantees the distributed coordination of the algorithm. This is necessary to avoid deafness problems, i.e., nodes trying to exchange control messages but, since they are operating in different channels, the message would not be detected by the destination nodes. During the negotiation phase, for each decision, in other words, selecting a strategy, the nodes have a conservative 200 ms window to broadcast their decisions. After the finalization criteria is met, the algorithm switches to operation phase. And just at this point, the nodes actually switch channels on the radios.

Consider the following scenario in order to illustrate the algorithm's execution. The scenario consists of a WMN backbone containing N MRs. As previously mentioned, during the negotiation phase all MRs should be assigned to a common

Algorithm 1. Near-optimal Partially Overlapping Channel Assignment (NPOCA)

$\mathtt{s_i} = \{0\}$ $\forall a_i \in \mathbf{A}$
while $T =$ false do
    Randomly select a_i with prob. $1/N$
    $\mathtt{s_i^{rand}} \leftarrow$ random strategy $\{k_{i,1}, \ldots, k_{i,c}, \ldots, k_{i,|C|}\}$
    while $\mathtt{s_i^{rand}} \neq$ valid strategy do ▷ Examine Interference constraints
        $\mathtt{s_i^{rand}} \leftarrow$ random strategy
    end while
    if $\mathtt{s_i^{t+1}} > \mathtt{s_i^{t}}$ then ▷ Eq. (8)
        $\mathtt{s_i^{t+1}} \leftarrow \mathtt{s_i^{rand}}$
    else
        $\mathtt{s_i^{t+1}} \leftarrow \mathtt{s_i^{t}}$
    end if
    Broadcast $a_i ID$ + $\mathtt{s_i^{t+1}}$
    Update T
end while

channel, for instance, channel 1. Initially, all MRs set their strategy vector $\mathbf{s_i}$ to zero. Then the gateway will randomly calculate the sequence in which each MR will play its strategy, and broadcast it. During its turn, each MR will play the game following the better response learning scheme. In other words, a MR randomly selects one CA for its radios, $\mathbf{s_i^{rand}}$, and checks if $\mathbf{s_i^{rand}}$ will not cause adjacent channel interference considering the strategy of all other players or self-interference. In case this strategy causes interference, another random strategy should be selected. Then, a_i verifies if $\mathbf{s_i^{rand}}$ yields a higher value for the utility function (3) than the previous strategy, $\mathbf{s_i^t}$. In the affirmative case, a_i will decide on the random strategy as his playing strategy ($\mathbf{s_i^{t+1}} \leftarrow \mathbf{s_i^{rand}}$), or it will remain with the previous strategy ($\mathbf{s_i^{t+1}} \leftarrow \mathbf{s_i^{t}}$), otherwise. Finally, a_i will broadcast its decision $\mathbf{s_i^{t+1}}$ to all other MRs. This loop of each player selecting an improving strategy or maintaining the previous one will occur until the finalization criteria T is met for the negotiations. Thereupon, all MRs will start the operation phase, in which they actually switch the channels on the radios according to the final selected strategy.

Other then having a distributed coordination function, note that NPOCA is also scalable and resilient. Negotiations to improve the network performance can restart whenever the algorithm is triggered. For instance, radio failure or addition of a MR. In case of failures, the strategy vector of each player should not be cleared to zero unless the failure caused a network partition. As for addition of MRs, there will be no necessity of clearing the strategy vector. By using the current CA when restarting the algorithm, a faster convergence can be achieved and there will be no deafness problem, since the MRs are already able to successfully send and receive broadcast messages.

5 Performance Evaluation

In this section, we evaluate the game theoretical CA that we will call from now on as NPOCA (Near-optimal Partially Overlapping Channel Assignment). We also add one more CA protocol called Hybrid Multi-Channel Protocol (HMCP) [18], which uses the non-overlapping CA, in order to compare our results. We evaluate both algorithms' performance using numerical analysis. The simulation scenarios were created using Java. A grid topology is constructed on the backbone.

5.1 NE and Better Response

In Fig. 1 we illustrate the negotiation process reaching the NE. This is a small topology containing just 5 nodes, we estimate the global optimum using a centralized brute force algorithm. The nodes were placed using a squared topology. Four nodes at the corners and one positioned at the center. The gateway is positioned at the corner to stimulate a multi-hop topology. MCS 6 Mbit/s is set as link data rate. After optimum estimation, we simulate NPOCA. We set T=50 iterations and repeat the simulation using 100 random seeds to calculate the average.

From the results depicted in Fig. 1 we can visualize the negotiation steps, in which the nodes are selecting random strategies. The utility function maintain its value when a random strategy yielded a worse channel strategy, since the nodes decide to maintain the previous strategy, as expected from eq. (8). The curve $NPOCA_{100}$ represents the average after 100 simulations. Our algorithm is classified as near-optimal because although it sometimes reaches global-optimum, it may also generates sub-optimal results when node find themselves trapped in a local optimum NE.

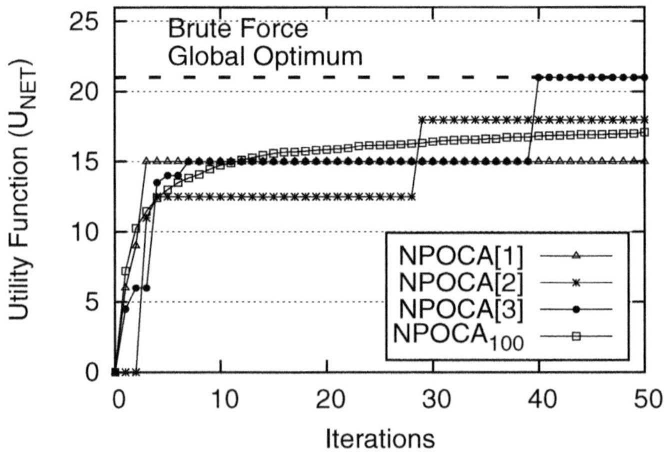

Fig. 1. NPOCA: 5 nodes topology

5.2 Random Topology

As mentioned previously in this work, the processing cost for the simulations exponentially increases with the number of players as a result of the exponential growth for existing game profiles ($\boldsymbol{\Psi}$). Consequently, estimating the global optimum using brute force algorithm for topologies containing various nodes becomes unfeasible. However, considering the results from the five nodes topology, we can assume that the following results generated by NPOCA are near optimal channel assignments.

In this section, we evaluate NPOCA's performance in random topologies. In this scenario, nodes are randomly placed in a field with squared dimensions of 100, 200, 300 and 400 m. The gateway is also randomly selected, hence it can be positioned any where in the field which yields that at a given simulation the gateway might be distant from most of the other nodes in the backbone. This occurrence is severely minimized as the number of nodes increases. We simulate topologies containing 10, 15, 20, 25 and 30 nodes. Since these scenarios contain more nodes, we increase the iterations (T=100), but we maintain the 100 random seeds.

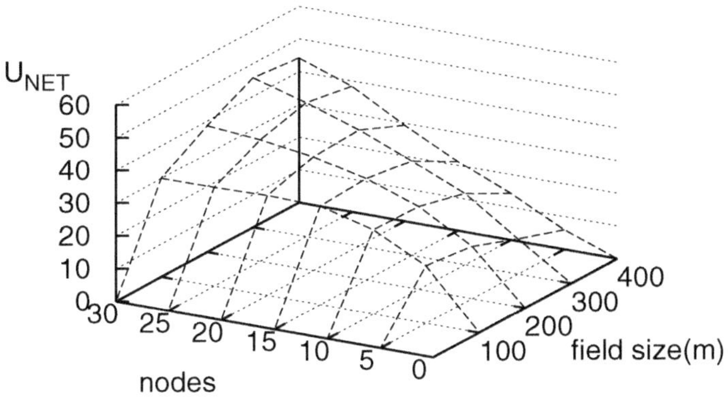

Fig. 2. NPOCA: random topology

First we analyze the results from the node axis' perspective. From Fig. 2 we notice that when there are 5 player, a field size of 100x100 m yields the best performance, hence this was the better node density for this experiment. Once the field size increases the nodes become to sparse degrading the network performance due to (dis)connectivity issues. On the other hand, regarding this simulation parameters, the bigger the field, the better performance we can observe for 30 nodes. Analyzing the results from the field size axis' perspective, we note that for 100 and 200 m, we reach a maximum desirable concentration of 20 nodes, since from this point on, increasing the number of nodes, do not increase the network overall performance. Although for 300 and 400 m field size, the network would improve its capacity if more nodes were added.

5.3 NPOCA and HMCP

In our last evaluation, we consider the improvements of our proposed CA against HMCP. In this scenario we use grid topologies to evaluate both protocols. The grid step is set to 120 m, which is the distance between adjacent nodes. The node positioned in the bottom right corner is assumed to be the gateway. MCS 6 Mbit/s is set as link data rate. In our experiments, we vary the grid size using 3x3, 3x4, 4x4, 4x5, and 5x5 arrangements that we will refer to as 9, 12, 16, 20, and 25 topologies, respectively. In this scenario we used T=100 and 100 random seeds.

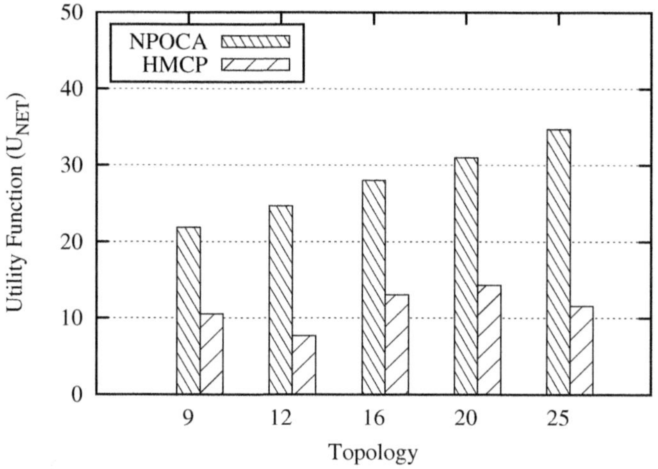

Fig. 3. NPOCA vs. HMCP: grid topology

In Fig. 3, we can notice a significant performance issue due to co-channel interference that occurs between nodes when HMCP is used. NPOCA, as expected, generates better results due to extensive negotiation iterations, in which at every iteration the nodes agree upon an improved CA topology. In addition, by using POCs, NPOCA is able to achieve a better spatial channel reuse factor compared to the traditional three orthogonal channels approach.

6 Conclusion

In this article, we developed a novel game-theoretical CA algorithm with near optimal performance for WMNs. In our algorithm, we exploited POC assignment following the latest research trends in the field. From the simulation results and analysis, we conclude that if well managed, overlapping channels can clearly overcome the overall performance of the common CA strategies using just the three orthogonal channels. Such improvements can be measured as network throughput, channel spatial re-use, non-interfering links.

Although our algorithm reaches near-optimal performance, it still can be improved. For example, in many simulations the algorithm reached low levels of performance, considering our utility function. One open issue in our research is how to devise a method to ensure that the nodes do reach these low level performance with very low probability during the strategy negotiation phase, hence our average performance would increase, approximating even more to the global optimum.

References

1. Akyildiz, I., Wang, X.: A survey on wireless mesh networks. IEEE Commun. Mag. 43(9), S23–S30 (2005)
2. Srivastava, V., Neel, J., Mackenzie, A., Menon, R., Dasilva, L., Hicks, J., Reed, J., Gilles, R.: Using game theory to analyze wireless ad hoc networks. IEEE Commun. Surveys Tutorials 7(4), 46–56 (2005)
3. So, J., Vaidya, N.H.: Multi-channel mac for ad hoc networks: handling multi-channel hidden terminals using a single transceiver. In: MobiHoc 2004: Proc. of the 5th ACM Int. Symp. on Mobile Ad Hoc Networking and Computing, Tokyo, Japan, pp. 222–233 (2004)
4. Draves, R., Padhye, J., Zill, B.: Routing in multi-radio, multi-hop wireless mesh networks. In: MobiCom 2004: Proc. of the 10th Annu. Int. Conf. on Mobile Computing and Networking, Philadelphia, PA, pp. 114–128 (2004)
5. Skalli, H., Ghosh, S., Das, S., Lenzini, L., Conti, M.: Channel assignment strategies for multiradio wireless mesh networks: Issues and solutions. IEEE Commun. Mag. 45(11), 86–95 (2007)
6. Bukkapatanam, V., Franklin, A., Murthy, C.: Using partially overlapped channels for end-to-end flow allocation and channel assignment in wireless mesh networks. In: ICC 2009: IEEE Int. Conf. on Commun., Dresden, Germany, pp. 1–6 (2009)
7. Mishra, A., Shrivastava, V., Banerjee, S., Arbaugh, W.: Partially overlapped channels not considered harmful. In: SIGMETRICS 2006: Proc. of the Joint Int. Conf. on Measurement and Modeling of Comput. Syst., Saint-Malo, France, pp. 63–74 (2006)
8. Xiao, M., Shroff, N., Chong, E.: Utility-based power control in cellular wireless systems. In: INFOCOM 2001: Proc. of 20th Annu. Joint Conf. of the IEEE Comput. and Commun. Soc., Anchorage, AK, vol. 1, pp. 412–421 (2001)
9. La, R., Anantharam, V.: Optimal routing control: repeated game approach. IEEE Trans. Automat. Control 47(3), 437–450 (2002)
10. Dewan, P., Dasgupta, P., Bhattacharya, A.: On using reputations in ad hoc networks to counter malicious nodes. In: Proc. 10th Int. Conf. on Parallel and Distributed Syst., ICPADS 2004, Newport Beach, CA, PP. 665–672 (2004)
11. Chen, T., Zhong, S.: Perfectly fair channel assignment in non-cooperative multi-radio multi-channel wireless networks. Comput. Commun. 32(6), 1058–1061 (2009)
12. Gao, L., Wang, X.: A game approach for multi-channel allocation in multi-hop wireless networks. In: MobiHoc 2008: Proc. of the 9th ACM Int. Symp. on Mobile Ad Hoc Networking and Computing, Hong Kong SAR, China, pp. 303–312 (2008)
13. Song, Y., Zhang, C., Fang, Y.: Joint channel and power allocation in wireless mesh networks: A game theoretical perspective. IEEE J. Sel. Areas Commun. 26(7), 1149–1159 (2008)

14. Hoque, M., Hong, X., Afroz, F.: Multiple radio channel assignement utilizing partially overlapped channels. In: GLOBECOM 2009: Proc. IEEE Global Telecommun. Conf., Honolulu, HI, pp. 1–7 (2009)
15. Feng, Z., Yang, Y.: How much improvement can we get from partially overlapped channels? In: WCNC 2008: Proc. of IEEE Wireless Commun. and Networking Conf., Las Vegas, NV, pp. 2957–2962 (2008)
16. Mackenzie, A.: Game Theory for Wireless Engineers, 1st edn. Morgan & Claypool Publishers (2006)
17. Monderer, S.: Potential games. Journal of Games and Economic Behavior 14, 124–143 (1996)
18. Kyasanur, P., Vaidya, N.H.: Routing and link-layer protocols for multi-channel multi-interface ad hoc wireless networks. SIGMOBILE Mob. Comput. Commun. Rev. 10(1), 31–43 (2006)

Game Based Self-Organizing Scheme for Femtocell Networks

Kwanghun Han[1], Seunghyun Choi[1], Du Ho Kang[2], and Sunghyun Choi[1]

[1] School of Electrical Engineering and INMC, Seoul National University, Korea
[2] Wireless@KTH, Royal Institute of Technology (KTH), Sweden
{khhan,shchoi}@mwnl.snu.ac.kr, schoi@snu.ac.kr

Abstract. A femto base station (BS) is an emerging candidate solution to guarantee wireless coverage and enhance capacity in indoor environments. Ideally, femto BSs should be designed to be installed by customers without their manual configuration. Hence, a femtocell network should be automatically organized by configuring the operating frequency channel and transmit power level of the femto BSs adaptively according to the interference environment. However, in order to enhance the capacity of femtocell users, the femto BSs in the network should be carefully configured since they can cause severe co-channel interference to the existing macrocell networks operating in the same frequency channel. In this work, we propose an automatic self-organizing scheme for a femtocell network by jointly considering transmit power control and dynamic frequency selection, which tries to maximize the mean sum downlink achievable rate of the femtocell users and to guarantee the performance of the macrocell users by limiting the co-channel interference from the femtocells. The proposed scheme is based on a potential game which guarantees a convergence property, and we enhance the scheme with a Tabu search, which attempts to achieve the optimality.

1 Introduction

As the demand of data traffic rapidly increases, securing wireless coverage and enhancing capacity in an indoor environment, where a big portion of wireless data service occurs, become more important. Conventionally, installing radio frequency (RF) repeaters is a widely adopted solution for it since it is simple and cheap. However, a repeater cannot fundamentally increase the wireless capacity since it simply amplifies and repeats the signal from the macro base station (BS). A small BS, which is commonly referred to as a femto BS, is a more attractive solution compared with the repeater since a femto BS can achieve a higher wireless capacity by achieving better spectral efficiency [1]. A femto BS is a small low-cost BS serving a small area referred to as a femtocell [2]. A femto BS is usually connected to a macrocell network via a broadband wired connection, e.g., an Internet protocol (IP) network over x digital subscriber line (xDSL) or a dedicated backhaul network. The emerging IMT-advanced candidate systems including 3GPP LTE-advanced and IEEE 802.16m also feature this femtocell technology [3, 4, 5].

R. Jain and R. Kannan (Eds.): GameNets 2011, LNICST 75, pp. 57–75, 2012.

Considering a large number of expected femto BS deployments, the desirable features of femto BSs include 1) minimizing the cost of installation and maintenance of a femto BS and 2) operating the femto BS without any in-depth knowledge of the network. Due to such features, self-organizing network (SON) is considered an indispensable technique for the success of femtocell networks, where SON is defined as a network which can be autonomously organized with a minimum intervention of the network operator/user. In a self-organizing femtocell network, each BS becomes self-organized by adaptively adjusting its transmit power and operating frequency channel.

For the femtocell network, in [6], the authors propose a heuristic algorithm for selecting the frequency band. The effect of network parameters, e.g., transmit power of femto BS, femtocell radius, and loading factor, are evaluated. Moreover, they propose an heuristic dynamic frequency selection (DFS) algorithm of the IEEE 802.16e femtocell systems in terms of the network coverage and the system capacity. In [7], the deployment of femto BSs is discussed for an intelligent cell planning. The authors propose a frequency assignment technique considering the distance to the macro BS. The main issue of this paper is to determine the distance threshold for the frequency channel assignment. In addition, transmit power control (TPC) for femtocell networks is investigated in [8, 9, 10, 11]. However, most of the papers assume the uplink transmission since it is easy to quantify and limit the sum interference to the macro BS. Moreover, the optimization goal of TPC is not to maximize the sum capacity of the network but to guarantee the target SINR, since such an objective is easy to solve, while the capacity maximization is more appropriate for the data network.

We in this work consider the joint problem of DFS and TPC as the goal of SON, and develop an operational procedure for a practical application of SON, which is referred to as a SON operation. For the algorithm, we consider the downlink system, which is different from the uplink system in that the victims of the co-channel interference are the macrocell users and femtocell users. The problem in consideration tries to maximize the mean sum downlink achievable rate of the femtocell users while guaranteeing the performance of the macrocell users by limiting the co-channel interference from femtocells. We develop an algorithm for a SON operation based on a potential game, which systematically has a strong convergence property. Even though the proposed algorithm cannot be fully distributed since it requires the network-wide information, the operation can be distributed with a proper initialization procedure. In addition, since the considered problem is known as NP-hard and the proposed algorithm does not guarantee optimality, the proposed algorithm might not result in an optimal solution. Consequently, the game algorithm is enhanced by applying a Tabu search, which attempts to obtain the optimal value, and we evaluate the proposed algorithm via simulations. To our best knowledge, the network-wide optimality is desirable but has not been handled so far since it is difficult to be modeled and achieved.

The rest of this paper is organized as follows. In Section 2, we describe a joint DFS/TPC problem in a self-organizing femtocell network, and then a system model is represented in Section 3. A potential game based SON algorithm is presented in

Section 4, and the detailed operation of the proposed scheme is explained in Section 5. In Section 6, the Tabu game algorithm is proposed, and we evaluate the proposed algorithm in Section 7. Finally, we conclude the paper in Section 8.

2 Self-Organizing Femtocell Network

Assume that femto BSs are arbitrarily distributed across the area in consideration, and each femto BS attempts to maximize the communication rate with its associated users. If multiple channels are available, the femto BSs prefer to utilize a frequency channel which is orthogonal to other femto BSs' since it is the simplest solution for the rate maximization. However, as the number of femto BSs increases more than the number of available frequency channels, femto BSs might inevitably interfere with each other. As a result, each femto BS should mitigate co-channel interference by controlling its transmit power (i.e., TPC) as well as changing its operating frequency channel (i.e., DFS) according to the interference environment. Moreover, considering the expected huge number of femto BSs, it is almost impossible to keep the network optimized via the manual setting by a human engineer as done in conventional cellular networks. Therefore, the femtocell network is desired to be a SON.

Developing a SON operation which tries to maximize the sum throughput of femtocells is challenging in that co-channel interference severely affects the sum throughput of femtocells. It is beacuse the signal-to-interference and noise ratio (SINR) of a femto BS is in proportion to the desired signal's power increment whereas it is inversely proportional to the transmit power increment of other BSs in the same frequency channel. Therefore, mitigating such co-channel interference is the key issue of the SON operation.

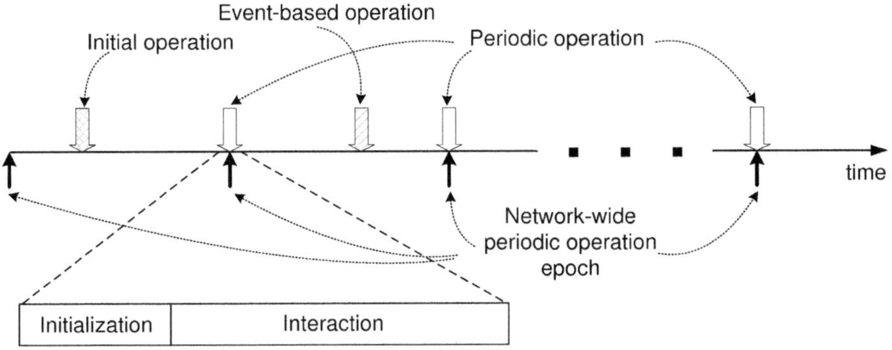

Fig. 1. The SON operation of a femto BS is depicted. For a periodic operation, the network determines the operation epoch, and the femto BSs conduct the SON operation. Such a periodic operation is the most essential and important SON operation for the network management. In addition, the initial and event-based SON operations are also considered. In this work, the SON operation is assumed to be composed of two procedures, i.e., the initial and interaction procedures.

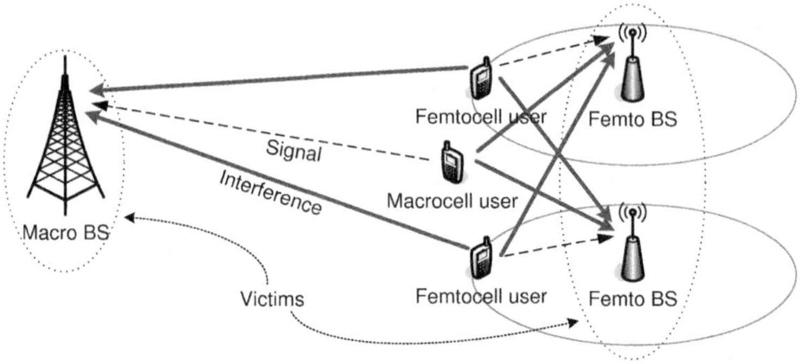

Fig. 2. During the uplink transmission, the macrocell and femtocell users transmit the data to the serving BSs. Since the receivers are the BSs, the victims by the interferences are also the BSs.

In this work, we mainly concentrate on the problem with a network-wide objective function, i.e., the mean sum downlink achievable rate of the femtocell network while maintaining the SINR of the macrocell users, represented as follows:

$$\max \sum_{i \in U_f} R_i \tag{1}$$

$$\text{s.t.} \quad \eta_j > \eta_{th} \qquad \forall j \in U_m \tag{2}$$

where R_i is the mean downlink achievable rate of femtocell user i and U_f is the index set of the femtocell users. η_j is the downlink SINR of macrocell user j and U_m is the index set of the macrocell users. η_{th} is the SINR threshold for guaranteeing the performance of the macrocell users. The problem modeling and formulation is presented in Section 3.

Lastly, we need to define the period of the SON operations. Practically, the SON operation does not need to be realtime, since the SON operation is not a part of resource scheduling but a part of network configuration. In Fig. 1, three types of operations are depicted. First, a femtocell network conducts the SON operation periodically in order to maintain the change of the network as time passes. During the periodic SON operation, each femto BS can reflect the SINR constraints of the macrocell users. The period of this operation might be enough to be done once a day or several days, for instance. Second, an initial SON operation is conducted for the initial setup of a femto BS which is newly added. When the initial operation is conducted, the femto BS might lack necessary information, e.g., channel information among other femto BSs and the SINR constraints of neighboring macrocell users. Then, it is hard to optimize the SON parameters. Third, the SON operation could be initiated based on the event of the network changes, e.g., neighboring femto BS addition and removal.

3 Modeling of Femtocell Network

In this work, we consider the downlink of a femtocell network and try to solve the problem which tries to maximize the sum rate of femto users while considering the constraints of the macrocell user's interference. In the femtocell network, the co-channel interference problems of downlink and uplink are different. As shown in Figs. 2 and 3, the main difference is that victims of the interference are different. For the uplink transmission, the victims of the interference are the BSs, i.e., macro BSs and femto BSs. Since the victims are static, the uplink problem is relatively easy to measure the interference level. Especially, the interference level of the macrocell can be measured only at the macro BS, and it is possible to limit the interference level under the interference threshold easily. Whereas, for the downlink transmission, the victims are the macrocell and femtocell users. Therefore, it is complicated to measure and limit the interference level since the users are ephemeral, distributed and even mobile, and we need to clearly analyze the victims to solve the problem and apply the solution to the practical system.

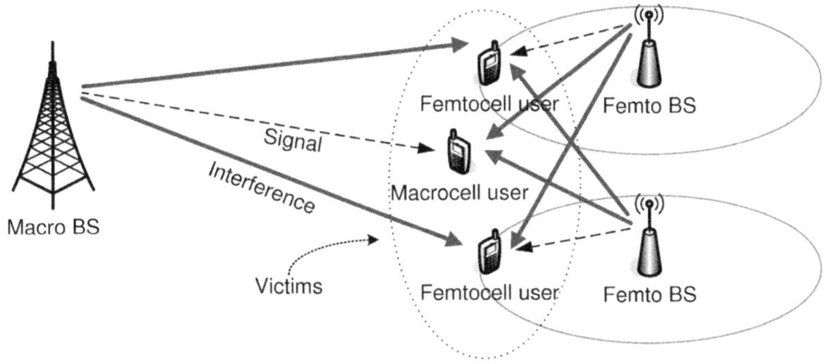

Fig. 3. During the downlink transmission, the macro and femto BSs transmit the data to the associated users. Since the receiver are the users, the victims by the interferences are also the users.

In Fig. 4, the downlink system model in consideration is depicted. In order to measure the performance of a femtocell, we consider the mean downlink achievable rate of an arbitrary user in the cell, $\tilde{R}$. As shown in the figure, a user is assumed to be located at the the arbitrary distance of r_u from the femto BS. Then, the achievable rate, $R_i(f_i, p_i)$, of femto BS i operating with transmit power of p_i in frequency channel f_i can be represented and calculated from the SINR of femtocell user j, $\gamma_j(f_i, p_i)$:

$$I(f_i) := \sum_{k \in B_m(f_i)} I_k(l_j) + \sum_{k \in B_f(f_i), k \neq i} I_k(l_j), \tag{3}$$

$$\gamma_j(f_i, p_i) = \frac{g_{ij} p_i}{N_0 + I(f_i)}, \tag{4}$$

$$R_i(f_i, p_i) = B \log\left(1 + \gamma_j(f_i, p_i)\right) \tag{5}$$

where the g_{ij} is the channel gain between the femto BS i and user j, and p_i is the transmit power of femto BS i. N_0 is the noise power, and $I_k(l_j)$ is the interference from macro or other femto BS k which is experienced by user j located at l_j.

$B_m(f_i)$ and $B_f(f_i)$ are the index sets of macro BSs and femto BSs, operating in frequency channel f_i, respectively. B is the bandwidth of the considered system.

Note that the interference needs to be averaged to get the mean downlink achievable rate, $\tilde{R}_i(f_i, p_i)$. Then, we average the interference for all the users at the arbitrary distance of r_u from the femto BS:

$$\tilde{I}(f_i) := \sum_{k \in B_m(f_i)} \frac{\int_C I_k(l_j) dl_j}{2\pi r_u} + \sum_{k \in B_f(f_i), k \neq i} \frac{\int_C I_k(l_j) dl_j}{2\pi r_u} \tag{6}$$

where $\int_C I_k(l_j) dl_j$ is the line integral of function $I_k(l_j)$ along a piecewise smooth curve C, and C in this work is a circle with radius r_u centered at femto BS i. We assume that arbitrary users are continuously located at point l_j on C.

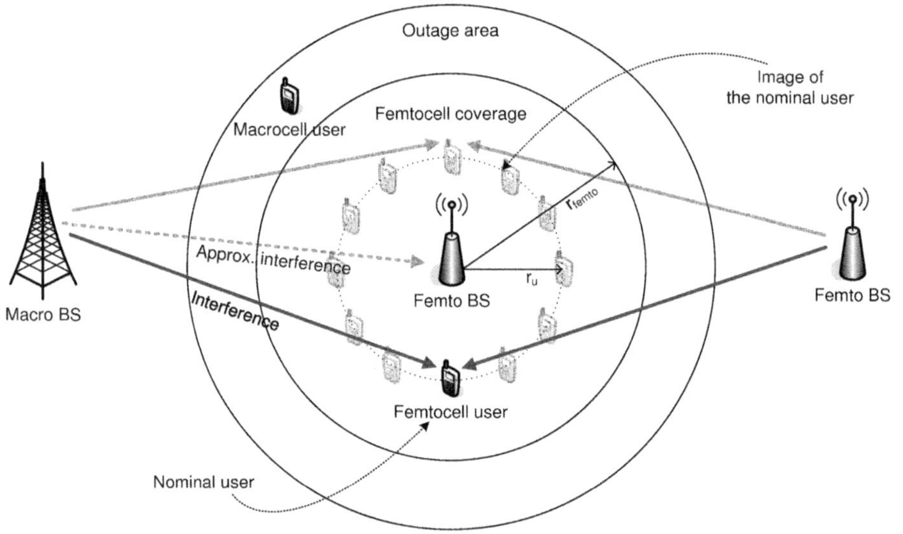

Fig. 4. The system model in consideration is depicted. To model the downlink performance of a femtocell, we use the mean downlink achievable rate of an arbitrary femtocell user located at a given distance. In addition, the macrocell users are considered to represent the interference to the macrocell.

For the simplicity, we approximate the mean interference as follows:

$$\tilde{I}(f_i) \approx \sum_{k \in B_m(f_i)} I_k(L_j) + \sum_{k \in B_f(f_i), k \neq i} I_k(L_j) \tag{7}$$

$$= \sum_{k \in B_m(f_i)} g_{ki} P + \sum_{k \in B_f(f_i), k \neq i} g_{ki} p_k \tag{8}$$

where L_j is the location of femtocell user j colocated with femto BS i, and P and p_k are the transmit power of a macro BS and femto BS k, respectively. g_{ki} is the channel gain between BSs k and i. We assume that the average amount of the interference of all the femtocell users at the arbitrary distance of r_u is the same as the amount of the interference to femtocell user j located at L_j. Note that the exact calculation of the average interference might be insignificant as long as the calculation method is identical to all the femto BSs. Consequently, the approximated mean SINR is represented as follows:

$$\widetilde{\gamma}_j(f_i, p_i) := \frac{g_{ij} p_i}{N_0 + \tilde{I}(f_i)} \tag{9}$$

where the g is identical since the distance from the serving femto BS to users is identical. Consequently, the mean downlink achievable rate of femto BS i is as follows:

$$\tilde{R}_i(f_i, p_i) := B \log_2 (1 + \widetilde{\gamma}_j(f_i, p_i)). \tag{10}$$

Even though the throughput might be more proper metric to represent the performance of a cell, the throughput is tightly related with scheduling algorithms and the user distribution. Then, it is not easy to calculate the throughput without considering a specific resource scheduler, and we will leave it as a future work. Instead, we use the achievable rate as a performance metric in this work.

As a main constraint, we need to limit the interference amount from the femto BSs to the macrocell users. In Fig. 4, the outage area is depicted, and the macrocell users in this area need to be considered as monitoring users. The interference level of the monitoring users needs to be limited under the threshold to guarantee the performance of the macrocell users. Then, we consider the feedback based reporting procedure of the monitoring users, and it will be detailed in Section 5.

4 Potential Game for SON Operation

The centralized optimization problem which maximizes the sum downlink rate of femtocell users requires high computational power and is not able if there is no central decision entity [12]. Thus, an efficient algorithm with distributed decision entities is more desirable even at the cost of optimality. In this section, we model the interaction among femto BSs as a constrained potential game which guarantees the convergence.

Game is a branch of applied mathematics, which describes and analyzes the problem of strategic interactions among multiple players who attempt to maximize their utility. Essentially, a game consists of 1) two or more players which interact one another, 2) utility functions which each player wants to maximize or minimize, and 3) action space from which players select their strategies. Then, a normal form game is represented as follows:

$$\Gamma = < N, A_i, u_i > \tag{11}$$

where $N = 1, 2, ..., |N|$, is a set of players of the game. A_i and u_i denote the action set and the utility function of player i, respectively. The action space of all available actions for all players is represented by $A = \times_{i \in N} A_i$, and $u_i : A \rightarrow \Re$,

Algorithm 1. Game Algorithm with Constraints

Initialization, $\forall i \in B_f$

$act_i \leftarrow (RND(f), RND(p))$, $act_i \in \mathbb{F} \times \mathbb{P}$
$titr_i \leftarrow 0$, $citr_i \leftarrow 0$, $cutil_i \leftarrow 0$

Iteration, $\forall i \in B_f$

while $titr_i < titr_{th}$ **and** $citr_i < citr_{th}$ **do**
  $titr_i \leftarrow titr_i + 1, butil_i \leftarrow 0, bact_i = (0,0), cret_i \leftarrow 0$
  Gets the current network state: $cst_i \leftarrow (\mathbf{f}_{-i}, \mathbf{p}_{-i})$
  Waiting for a decision epoch
  if $DEC(|B_f|) \neq$ success **then**
    continue
  else
    for $tact_i \leftarrow (f,p)$, $\forall (f,p) \in \mathbb{F} \times \mathbb{P}$ **do**
      if $CONST(tact_i, cst_i) = 1$ **then**
        continue
      if $CONST(tact_i, cst_i) = 0$ **and** $cret_i = 0$ **then**
        $cret_i \leftarrow 1$
      if $UTIL(tact_i, cst_i) > butil_i$ **then**
        $butil_i \leftarrow UTIL(tact_i, cst_i)$, $bact_i \leftarrow tact_i$
    if $cret_i = 0$ **then**
      $act_i \leftarrow (RND(f), RND(p))$, $act_i \in \mathbb{F} \times \mathbb{P}$
      continue
    if $cutil_i < butil_i$ **then**
      $citr_i \leftarrow 0, cutil_i \leftarrow butil_i$, $cact_i \leftarrow bact_i$
    else
      $citr_i \leftarrow citr_i + 1$
    $act_i \leftarrow bact_i$

where $\times$ means Cartesian product and $\Re$ is the set of real values. At each decision epoch of a game referred to as a stage, the players who are involved in a game attempt to maximize or minimize their utilities according to decision and order. In this work, we consider a best response policy as a decision rule. At each stage, player i determines an action $b_i \in A_i$ such that $u_i(b_i, a_{-i}) > u_i(a_i, a_{-i})$, while other players continue to implement the same actions. Here, $a_i \in A_i$ is an action of player i, and $a_{-i} \in \times_{j \neq i, j \in N} A_j$ is an action vector of all the players except player i. In addition, we assume that the order of decision is randomly determined before the game converges.

When a normal form game is conducted repeatedly, players determine actions that improve their utility functions at each stage. As the players adapt their actions, the game might reach a certain point where no players can increase their utilities without changing other players' actions. This point is called a Nash equilibrium, which is an action profile from which no player can improve its utility value by unilateral deviations, and an actual convergence point of a game. An action profile $a \in A$ is a NE if and only if $u_i(a) \geq u_i(b_i, a_{-i})\ \forall i \in N, a \in A, b_i \in A_i, a_{-i} \in \times_{j \neq i, j \in N} A_j$. However, it should be noted that a Nash equilibrium does not guarantee a point which maximizes the sum capacity of all players. Even worse, in general, a game does not always converge to a Nash equilibrium.

Fortunately, a special type of game, called potential game, is proven to converge to a Nash equilibrium [13]. A potential game is a normal form game such that any changes in the utility function of any player in the game due to a unilateral deviation by the player is reflected in a potential function. In other words, a player's individual choice increasing its own utility increases the whole network utility. Potential games with finite number of players and finite action space have been proved to converge to a Nash equilibrium when they follow a best (or better) response policy, and we also have an additional flexibility of the decision order [13]. In detail, at each decision epoch, the number of players can be more than one, and it is more desirable for the distributed operation.[1]

To resolve the convergence issue for the distributed decision structure, we can model the problem as a potential game. In our game, the femto BSs are regarded as players, and action space consists of discrete level of powers and frequency channels. Every femto BS i, which is a player, takes actions, i.e., changes its operating frequency channel f_i and its transmit power level p_i to maximize its utility function. Based on our problem definition, all players have an identical utility function of the mean sum downlink achievable rate:

$$U(\mathbf{f}, \mathbf{p}) = \sum_{i \in B_f} \tilde{R}_i(f_i, p_i) \tag{12}$$

where B_f is the index set of all the femto BSs in the game, and $\mathbf{f}$ and $\mathbf{p}$ are the vectors representing all the BSs' operating frequency channels and transmit power levels, respectively. In order to maximize the utility function, each player i chooses a best joint power-frequency action vector $a_i = (p_i, f_i)$ from action space $A_i = \mathbb{F} \times \mathbb{P}$ where $\mathbb{F}$ and $\mathbb{P}$ are sets of frequency channels and power levels, respectively.

For the constraints, the SINR of macrocell user j which is associated with macro BS i and located at l_j within the outage area of the considered femtocell, i.e., η_j, should be secured over η_{th} to ensure the link quality. It can be represented as follows:

$$I(f_i) := \sum_{k \in B_m(f_i)} I_k(l_j) + \sum_{k \in B_f(f_i)} I_k(l_j), \tag{13}$$

$$\eta_j(f_i) = \frac{g_{ij} P}{N_0 + I(f_i)} \tag{14}$$

where f_i is the operating frequency channel of macro BS i, and $I(f_i)$ is the interference from all the macro and femto BSs operating in f_i. We assume that the macro BS uses a fixed transmit power of P. In the viewpoint of the game,

[1] The order of action and the action policy can be selected more freely within some known policies. For instance, the potential game will converge where the order of players' actions is random (referred to as asynchronous order), and a player randomly chooses its action if the action increases its utility compared with the current utility (referred to as the random better response).

Algorithm 2. The Proposed Operation of SON Scheme

Initialization procedure

1: **for** $i \leftarrow 1,\ i \leftarrow i+1,\ i \leq |B_f|$ **do**
2: Femto BS i:
3: Transmits the preamble with its maximum power
4: Other femto BSs:
5: 1) Measure the channel gain from femto BS i
6: 2) Send the results to the management server

Interaction procedure

1: Do the interaction for the game, e.g., Algorithm 1

the constraints due to the macrocell users limit the action space of a femto BS to the feasible space. Then, the considered problem can be formulated as follows:

$$(\mathbf{f}^*, \mathbf{p}^*) = \arg\max_{\mathbf{f},\mathbf{p}} U(\mathbf{f}, \mathbf{p}) \tag{15}$$

$$\text{s.t.} \quad \eta_j(f_i) > \eta_{th} \qquad \forall j \in U_{out} \tag{16}$$

where U_{out} is the index set of macrocell users in the outage area of the considered femtocell.

In this work, the order of players' actions is assumed to be random. A player is randomly chosen in a round, and the player chooses the action which makes its utility maximized (referred to as best response policy). For the practical operation, we assume that there is a management server which is in charge of managing and exchanging the data, clustering SON groups, and supervising the SON operation for each group. Such a management server will be practically mandatory and can make the proposed scheme more applicable and practical. Summarily, the proposed game algorithm is explained in Algorithm 1. First, all the femto BSs initialize their action randomly, and prepare for the game by initializing their algorithm parameters. When the game starts, each femto BS increases its total iteration counter, $titr_i$, get the current network state, cst_i, by measurement, and waits for the decision epoch. At the decision epoch, each femto decides whether it should make a decision at this time by the decision order represented as $DEC(-)$. Each femto BS which decides to make a decision choose its action considering its macrocell user constraints, where $CONST(-)$ is the function to check whether the selected temporal action, $tact_i$, satisfies all the macrocell user constraints. In addition, $UTIL(-)$ is the function to calculate the utility. Next, if no action satisfies the constraints, i.e., $cret_i = 0$, the femto BS selects random action and waits for the next decision epoch. Otherwise, the femto BS checks whether its convergence utility, $cutil_i$, should be updated compared with the temporal best utility, $butil_i$. Then, it also waits for the next decision epoch.

5 Operation of the Proposed SON Scheme

For a distributed and practical operation of the proposed SON scheme, the operational procedure needs to be considered, and there mainly arise three issues;

1) the common utility function, $U(\mathbf{f}, \mathbf{p})$ in Eq. (12), should be shared among the femto BSs, 2) the decision results, i.e., $\mathbf{f}$ and $\mathbf{p}$, should be delivered to all the femto BSs, and 3) the macrocell users' feedback, η, should be reflected to the femto BSs' decision. For the first and second issues, we need to focus on U in Eq. (12) which is a function of $\widetilde{\gamma}$ in Eq. (9). Then, in order to build U, each femto BS should have all g and p values of other BSs in Eqs. (8) and (9). Basically, g in Eq. (9) is a given value for each femto BS. However, g's in Eq. (8) should be measured before being exchanged.

The proposed SON scheme can be divided into two procedures, i.e., initialization and interaction proceduresAlgorithm 2. During the initialization procedure, the femto BSs measure the channel gain values among them, i.e., g's in Eq. (8). In a round robin manner, each femto BS transmits its preamble with a maximum power and other femto BSs measure the channel gain from the femto BS. Such a measurement procedure repeats until all the channel gain information is measured. Based on the exchanged information, the required objective and constraints for the game can be generated. Next, during the interaction procedure, the femto BSs conduct the proposed game as described in Algorithm 1. The decision results of other femto BSs can be measured during the game, and then no more backhaul communication is required during the game.

The proposed SON operation could be completely applied for the periodic operation shown in Fig. 1. However, for the initial and event-based operation, it might not be proper since it is a burden to repeat the measurement procedure, for instance, whenever a new femto BS enters the network. Then, the corresponding femto BS, e.g., the newly entered femto BS, only selects its frequency channel and power lever without any interaction with other femto BSs. For instance, the femto BS can select the frequency channel which is most idle and set the transmit power level to a given value. Such femto BSs will join the next periodic SON operation, and interact with other femto BSs to optimize the network.

For the third issue, i.e., the macrocell users' feedback, the femto BSs need to cooperate with the macro BSs and macrocell users. For the operation, we assume that the femto BSs can have the scheduling information, e.g., a scheduling map of IEEE 802.16e, of the macro BS. Based on the scheduling information, the femto BSs can find the macrocell users in the outage area around their service coverage by detecting the uplink transmission of the macrocell users. Since we cannot directly find the macrocell users from the downlink transmission, such an uplink detection is one of the feasible and reasonable ways to find the macrocell users near a femto BS.

Based on such operations, the femto BS can obtain the channel gain between the femto BS itself and the macrocell users which are identified. The femto BSs use the channel gain information as the constraints to the macrocell users described in Eq. (16) when the femto BSs conducts the SON operation. Basically, the proposed SON scheme can work based on the described operation. However, as explained in Section 4, the performance of the proposed game is not optimal, and it can be improved.

Algorithm 3. Tabu Game Algorithm

Initialization, $\forall i \in B_f$

$\mathbf{T}_i \leftarrow \emptyset,\ act_i \leftarrow (RND(f), RND(p)),\ act_i \in \mathbb{F} \times \mathbb{P}$
$titr_i \leftarrow 0,\ citr_i \leftarrow 0,\ cutil_i \leftarrow 0,\ cact_i \leftarrow (0,0)$

Iteration, $\forall i \in B_f$

while $titr_i < titr_{th}$ **and** $citr_i < citr_{th}$ **do**
  $titr_i \leftarrow titr_i + 1, butil_i \leftarrow 0, bact_i = (0,0), cret_i \leftarrow 0$
  Gets the current network state: $cst_i \leftarrow (\mathbf{f}_{-i}, \mathbf{p}_{-i})$
  Waiting for a decision epoch
  if $DEC(|B_f|) \neq$ success **then**
    continue
  else
    for $tact_i \leftarrow (f,p),\ \forall (f,p) \in \mathbb{F} \times \mathbb{P}$ **do**
      if $CONST(tact_i, cst_i) = 1$ **then**
        continue
      if $CONST(tact_i, cst_i) = 0$ **and** $cret_i = 0$ **then**
        $cret_i \leftarrow 1$
      if $UTIL(tact_i, cst_i) > butil_i$ **then**
        $butil_i \leftarrow UTIL(tact_i, cst_i),\ bact_i \leftarrow tact_i$
    if $cret_i = 0$ **then**
      $act_i \leftarrow (RND(f), RND(p)),\ act_i \in \mathbb{F} \times \mathbb{P}$
      continue
    $tcnt_i \leftarrow 0$
    while $CTABU(\mathbf{f}, \mathbf{p}_{-i}) = 1$ **and** $tcnt_i < tcnt_{th}$ **do**
      $act_i \leftarrow (RND(f), RND(p)),\ act_i \in \mathbb{F} \times \mathbb{P}$
      $tcnt_i \leftarrow tcnt_i + 1$
    if $cutil_i < butil_i$ **then**
      $citr_i \leftarrow 0, cutil_i \leftarrow butil_i,\ cact_i \leftarrow bact_i$
    else
      $citr_i \leftarrow citr_i + 1$
    $act_i \leftarrow bact_i$
    $ATABU(\mathbf{T}_i, \mathbf{f}, \mathbf{p}_{-i})$

6 Tabu Search Extended Game Algorithm

Even though the proposed game algorithm in Section 4 converges to a Nash equilibrium, it does not guarantee a global optimality, since the problem in consideration is non-convex and NP-hard due to the non-linearity of the power control and the combinatorial property of the channel selection. Frankly speaking, even if we consider only the DFS problem, the problem is NP-hard where no algorithm can give the global optimal solution in polynomial time. Then, in computer science, metaheuristic is developed in order to improve the solutions of such combinatorial and/or non-linear problems [15]. In general, many metaheuristic algorithms mainly are based on the random processes in order to escape local optima even though such a randomness might deteriorates the objective values of the problem in consideration temporarily. They try to improve the

Table 1. System parameters for the simulation

P_{max}	20 dBm	P_{min}	−3 dBm
Power levels	10	Num. of freq. ch.	3
Size of open area	200 x 200 m^2	N_0	−95 dBm
Building size	50 x 50 m^2	Penetration loss	12 dB
$P_{macroBS}$	43 dBm	Interf. threshold	1 dB

solutions by iteratively solving the problem with manipulated solution spaces. However, such metaheuristic algorithms also cannot guarantee a global optimality, but they only give better solutions compared with local search algorithms. In this work, in order to attempt to achieve a global optimality, we enhance the proposed game algorithm by using a Tabu search, which is one of well-known metaheuristic algorithms [16].

In the viewpoint of an algorithm structure, a metaheuristic algorithm is an extended local search algorithm, which allows to choose a random solution under some given conditions. Especially, a Tabu search algorithm uses an additional memory referred to as a Tabu list which stores previously visited solutions and helps the algorithm find new solutions by preventing it from reselecting the stored solutions. The usage of a memory enables an effective search compared with the algorithms which heavily depends on a random process without a memory. In this work, the proposed game algorithm is considered a distributed local search algorithm, and its performance can be improved by using a Tabu search algorithm. Theoretically, the extension is simple, and it only changes the searching procedure.

The Tabu search integrated with a game has an effect that expands the searching space so as to leverage player's opportunity to reach the region where the optimal point lies over the Nash equilibrium points. The main point of the extension is considering the distributed property of the proposed game algorithm. In other words, a Tabu list also needs to be distributed, and then each femto BS has its own Tabu list and automatically updates its Tabu list. Note that, for a distributed Tabu list, the important issue is what should be elements of the Tabu list. In this work, we consider a vector which is composed of the decision maker's operating frequency channel and the interference environment. For instance, the Tabu list for femto BS i can be defined as $\mathbf{T}_i$, and an element for $\mathbf{T}_i$ is defined as follows:

$$(\mathbf{f}, \mathbf{p}_{-i}) \tag{17}$$

The reason why its transmit power level is not included in the vector is that the decision will be identical under the same decision rule when the femto BS meets the same interference environment. In addition, for such a case, it is more efficient to perturb the interference environment if the femto BS chooses a frequency channel among other frequency channels to find a new search space.

Algorithm 3 presents the proposed Tabu game. Basically, the operation procedure is almost same as Algorithm 1 since it is based on it. The difference from the original game is that the action set of a player is limited by its Tabu list, and the player can only choose its action which is not included in its Tabu list. For a

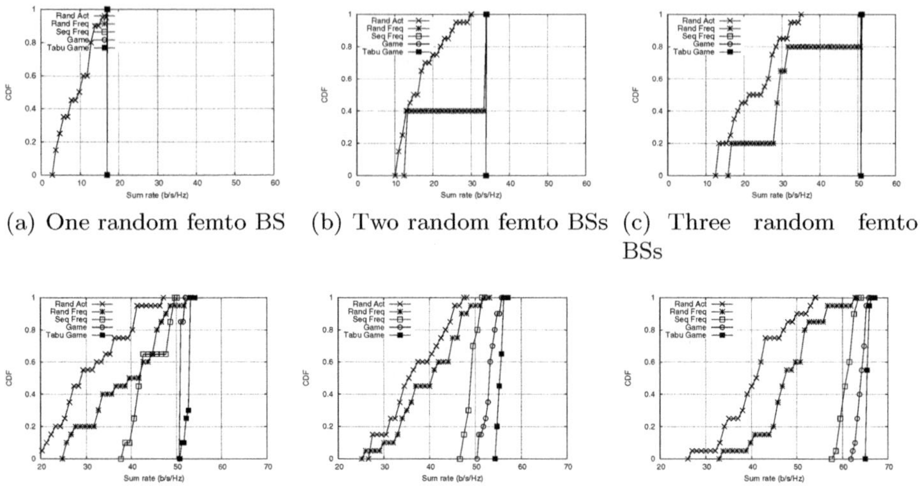

(a) One random femto BS (b) Two random femto BSs (c) Three random femto BSs

(d) Six random femto BSs (e) Nine random femto BSs (f) Twelve random femto BSs

Fig. 5. The proposed schemes are compared with other schemes with different numbers of the installed femto BSs.

Tabu game, all players additionally initialize their Tabu list in the initialization procedure, and starts an iteration procedure. During the iteration procedure, for instance, femto BS i additionally check its Tabu list by $CTABU(-)$ in order to decide its action. Then, if the selected action is in its Tabu list, $\mathbf{T}_i$, femto BS i randomly selects an action. At the end of the decision procedure, femto BS i adds the element in $\mathbf{T}_i$ by using $ATABU(-)$ to avoid the reselection of the action, and wait for the next decision epoch. Finally, a Tabu game algorithm remembers the best solution, and decides it as its final decision at the end of the iteration.

7 Performance Evaluation

In this section, we evaluate the proposed game and Tabu game algorithms. First, we compare the performance of the proposed algorithms with others. In order to show the pure performance gain of the proposed algorithms, the evaluation is conducted without considering the constraints of the macrocell users. Second, the proposed algorithm is evaluated in a more practical simulation environment to estimate the expected performance for the real installation. For this simulation, the simulation environment and the theoretical performance bound in [12] are considered for comparison.

For the first simulation, we assume that femto BSs are deployed in a 200 m square-shaped area. The transmit power of a femto BS is equally quantized into ten levels ranging from -3 dBm to 20 dBm in a dB scale. Three orthogonal frequency channels are considered, and all the channel models follow the

ITU-R M.1225 model [3]. Each femto BS adapts its transmit power and operating frequency channel by following the best response, and the order of decision is randomly determined. The size of each femto BS's Tabu list is assumed to be unlimited, since the price of the memory is very cheap compared with the past days. We assume that femto BSs are randomly located in the considered area, and there is no macro BS and macrocell user. Based on such assumptions, the performance of the proposed scheme could be purely compared with other schemes. Those parameters are summarized in Table 1.

For the comparison, we consider the following algorithms:

1. **Random frequency and power selection (Rand Act):** Each femto BS selects its transmit power and frequency channel randomly.
2. **Random frequency selection (Rand Freq):** Each femto BS selects its frequency channel randomly with its maximum transmit power.
3. **Sequential frequency selection (Seq Freq):** Femto BSs sequentially select their frequency channel with their maximum transmit power. When they select the power and frequency channel, they greedily try to maximize their own utility function.

In Fig. 5, The results of the performance comparison are depicted based on the CDF of the sum rate, i.e., the sum objective values of all the femto BSs when they converge. In the figure, there are six subfigures according to the number of femto BSs. When the number of femto BSs increases one to three, the number of femto BSs is less than or equal to the number of the orthogonal frequency channels. Then, the performance increases as the number of femto BSs increases. For the proposed algorithms, since all three femto BSs can orthogonally select one of the frequency channels, the obtained performance linearly increase according to the number of the frequency channels. For those cases, the performance of the proposed game and Tabu game algorithms is identical. However, other random algorithms cannot achieve the maximum performance since they do not guarantee the orhogonal channel selection.

When the number of femto BSs is greater than three, the performance dose not linearly increase since there is no more orthogonal frequency channel from the fourth femto BS. Nevertheless, the performance can increase as the number of femto BSs increases since the chance that more femto BSs are located in the preferred locations increases. However, the performance will be saturated eventually when the number of femto BSs is large enough. In these cases, the Tabu game algorithm outperforms all other algorithms. One interesting result is that the proposed game algorithm and the Seq Freq algorithm also give relatively good performance. Such a tendency becomes clearer as the number of femto BSs increases, especially for the Seq Freq algorithm. It means that as the number of femto BSs increases, the performance could be reasonably improved only by a proper frequency channel selection, even though high performance discrepancy still exists from the proposed algorithms. The proposed Tabu game algorithm always shows the best performance compared other algorithms, with a small variance of the objective values. On the other hand, the CDF graphs of two

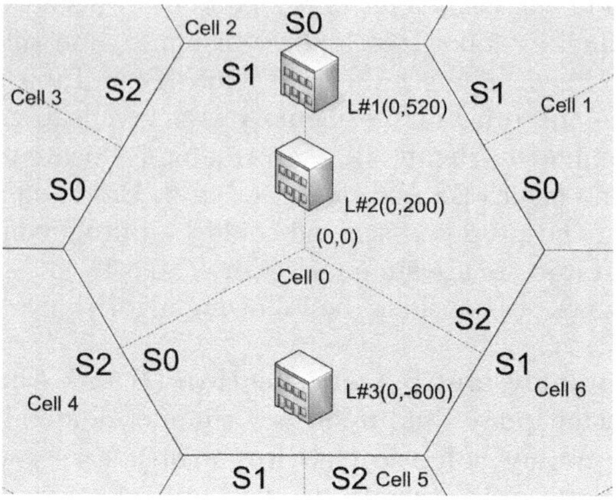

Fig. 6. Simulation environment for the second simulation

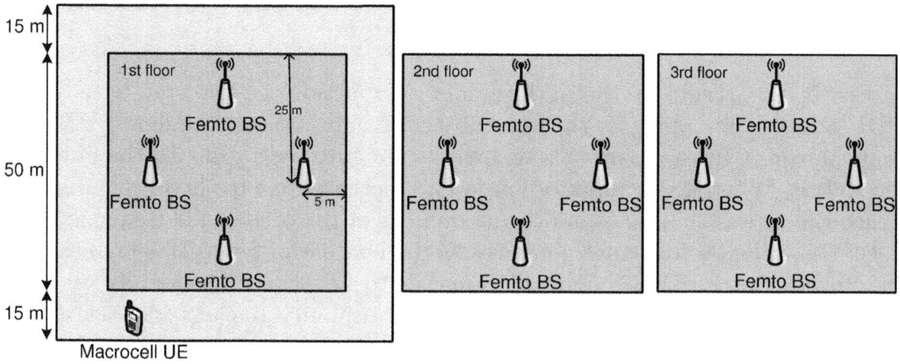

Fig. 7. Locations of femto BSs inside the building

random selection schemes are very similar and are widely spread, which are not proper for the practical environment.

In Fig. 8, the proposed Tabu game algorithm is evaluated in the same simulation environment of the referenced paper [12]. A conventional multi-cell honeycomb structure is considered, and each macrocell is divided into three sectors denoted by S0, S1, and S2 as shown in Fig. 6. The sector is labeled by Sx ($x = 0, 1, 2$) and it uses the $(x + 1)$th frequency channel out of three available channels. We assume that all macro BSs has 43 dBm transmit power. In this simulation, the building is assumed to be located in the L#2 location where the macrocell users are served via frequency channel 1 and 2. We assume that the size of building is 50 x 50 m^2, and the street width is 30 m where the street is the same as the outage area in this work. Twelve femto BSs are located in the three story building as shown in Fig. 7.

Assuming that both macrocell and femtocell networks are perfectly synchronized, the femto BS plays as a downlink interferer to users in a macrocell net-

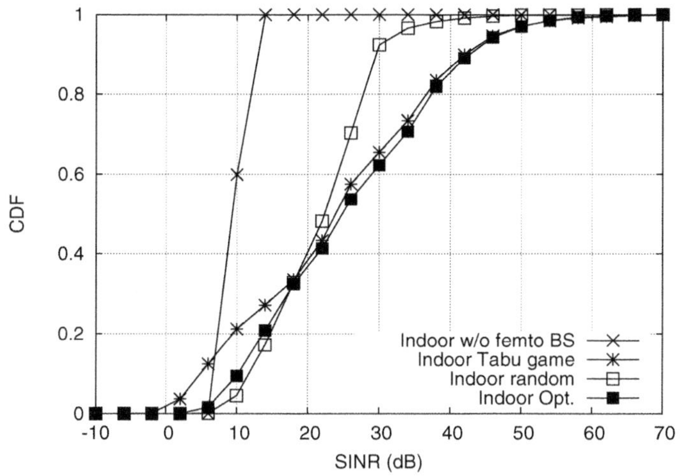

(a) The CDF of SINR, indoor case

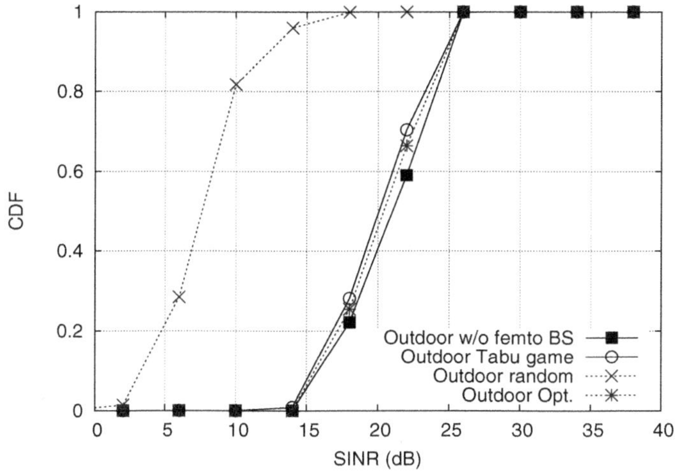

(b) The CDF of SINR, outdoor case

Fig. 8. The CDF of SINR, when the building is located at L#2

work, and this is the major difference of the simulation environment compared with the first simulation. Then, the femto BSs should satisfy the interference constraints to the macrocell users near the building. The interference threshold for the performance degradation of the macrocell users are assumed to be 1 dB. The penetration loss is considered 12 dB, and the simulation parameters are also summarized in Table 1.

In Fig. 8, four pairs of SINR CDF graphs are depicted for the cases without femto BSs, with femto BSs configured by a Tabu game algorithm, with femto BSs by a Rand Act algorithm, and with femto BSs configured optimally. As shown in Fig. 8(a), for the case when there is no femto BS in the building, the average SINR is low compared with other cases. However, the average SINR increases with the deployment of the femto BSs, and the Tabu game algorithm almost achieves the optimal performance. Interestingly, the Rand Act algorithm seems to achieve high performance, however, it is not desirable since it dose not consider the interference to the macrocell users. In Fig. 8(b), the CDF of outdoor SINR are depicted, except the Rand Act algorithm case, all other algorithms meet the interference threshold. As a conclusion, the Tabu game algorithm is expected to achieve the high performance for the practical environment.

8 Conclusion

In this work, we consider a self-organizing femtocell network performing TPC and DFS for the downlink of a femtocell network. First, the classification of the femtocell network is considered, and we model the downlink of the femtocell network in consideration. For the practical application of the self-organizing femtocell network, we describe the proposed operational procedures. Basically, we develop the self-organizing femtocell network based on a potential game which guarantees the convergence, and we develop a distributed strategy of femto BSs for adapting their air-parameters. Additionally, the global optimality issue is also addressed by proposing a Tabu search game algorithm. Compared with other expected self-organizing schemes, the evaluation shows that the proposed algorithms achieve the higher gain in average and small variance. In addition, based on the simulation results, the proposed game and Tabu game algorithms are highly expected to be proper for the practical deployment.

References

[1] Alouini, M., Goldsmith, A.: Area Spectral Efficiency of Cellular Mobile Radio Systems. IEEE Trans. Vehicular Technology 48(4), 1047–1066 (1999)

[2] Chandrasekhar, V., Andrews, J., Gatherer, A.: Femtocell Networks: A Survey. IEEE Communications Magazine 46(9), 59–67 (2008)

[3] Requirements Related to Technical Performance for IMT-Advanced Radio Interface(s), ITU-R Std. Report M.2134 (December 2008)

[4] The 3rd Generation Partnership Project; Technical Specification Group Radio Access Network; Evolved Universal Terrestrial Radio Access (E-UTRA); Medium Access Control (MAC) protocol specification (Release 8), 3GPP Std. TS 36.321 v8.1.0 (December 2007)

[5] IEEE 802.16m System Description Document (SDD), IEEE Std. IEEE 802.16m-09/0034r3 (June 2010)

[6] Zeng, H., Zhu, C., Chen, W.: System Performance of Self-Organizing Network Algorithm in WiMAX Femtocells. In: Proc. WICON 2008, Hawai, US (November 2008)

[7] Guvenc, I., Jeong, M., Watanabe, F., Inamura, H.: A Hybrid Frequency Assignment for Femtocells and Coverage Area Analysis for Co-Channel Operation. IEEE Communincations Letters 12(12), 880–882 (2008)

[8] Chandrasekhar, V., Andrews, J., Muharemovic, T., Shen, Z., Gatherer, A.: Power Control in Two-Tier Femtocell Networks. IEEE Trans. Wireless Communications 8(8), 4316–4328 (2009)
[9] Jo, H., Mun, C., Moon, J., Yook, J.: Interference Mitigation Using Uplink Power Control for Two-Tier Femtocell Networks. IEEE Trans. Wireless Communications 8(10), 4906–4910 (2009)
[10] Arulselvan, N., Ramachandran, V., Kalyanasundaram, S., Han, G.: Distributed Power Control Mechanisms for HSDPA Femtocells. In: Proc. IEEE VTC 2009-Spring, Barcelona, Spain (April 2009)
[11] Li, X., Qian, L., Kataria, D.: Downlink Power Control in Co-Channel Macrocell Femtocell Overlay. In: Proc. CISS 2009, Baltimore, MD (March 2009)
[12] Han, K., Choi, Y., Na, M., Kim, D., Choi, S., Han, K.: Optimization of Femtocell Network Configuration Under Interference Constraints. In: Proc. WiOpt, Seoul, Korea (June 2009)
[13] Neel, J.: Analysis and Design of Cognitive Radio Networks and Distributed Radio Resource Management Algorithms. Ph.D. dissertation, Virginia Polytechnic Institute and State University, Blacksburg, VA, US (September 2006)
[14] IEEE Standard for Local and Metropolitan Area Networks Part 16: Air Interface for Fixed and Mobile Broadband Wireless Access Systems Amendment 2: Physical and Medium Access Control Layers for Combined Fixed and Mobile Operation in Licensed Bands and Corrigendum 1, IEEE Std. 802.16e-2005 and 802.16-2004/Cor 1-2005 (February 2006)
[15] Bertsekas, D.: Nonlinear programming. Athena Scientific (1999)
[16] Glover, F., Laguna, M.: Tabu Search. Kluwer (1997)

Additively Coupled Sum Constrained Games

Yi Su and Mihaela van der Schaar

Department of Electrical Engineering, UCLA

Abstract. We propose and analyze a broad family of games played by resource-constrained players, which are characterized by the following central features: 1) each user has a multi-dimensional action space, subject to a single sum resource constraint; 2) each user's utility in a particular dimension depends on an additive coupling between the user's action in the same dimension and the actions of the other users; and 3) each user's total utility is the sum of the utilities obtained in each dimension. Familiar examples of such multi-user environments in communication systems include power control over frequency-selective Gaussian interference channels and flow control in Jackson networks. In settings where users cannot exchange messages in real-time, we study how users can adjust their actions based on their local observations. We derive sufficient conditions under which a unique Nash equilibrium exists and the best-response algorithm converges globally and linearly to the Nash equilibrium. In settings where users can exchange messages in real-time, we focus on user choices that optimize the overall utility. We provide the convergence conditions of two distributed action update mechanisms, gradient play and Jacobi update.

1 Introduction

Game theory provides a formal framework for describing and analyzing the interactions of multiple decision-makers. Recently, there has been a surge in research activities that adopt game theoretic tools to investigate a wide range of modern communications and networking problems. In resource-constrained communication networks, a user's utility is usually not only affected by its own action but also by the actions taken by all the other users sharing the same resources. Due to the mutual coupling among users, the performance optimization of multi-user communication systems is challenging. Depending on the characteristics of different applications, numerous game-theoretical models and solution concepts have been proposed to characterize the multi-user interactions and optimize the users' decisions in communication networks. A variety of game theoretic solutions have been developed to characterize the resulting performance of the multi-user interactions, including Nash equilibrium (NE) and Pareto optimality [1]. The purpose of this paper is to introduce and analyze a general framework that abstracts the common characteristics of this family of multi-user interaction scenarios, which includes, but is not limited to, the power control scenario. In particular, the main contributions of this paper are as follows.

R. Jain and R. Kannan (Eds.): GameNets 2011, LNICST 75, pp. 76–91, 2012.

First of all, we define the class of *Additively Coupled Sum Constrained Games* (ACSCG), which captures and characterizes the key features of several communication and networking applications. In particular, the central features of ACSCG are: 1) each user has a multi-dimensional strategy that is subject to a single sum resource constraint; 2) each user's payoff in each dimension is impacted by an additive combination of its own action in the same dimension and a function of the other users' actions; 3) users' utilities are separable across different dimensions and each user's total utility is the sum of the utilities obtained within each dimension.

Second, based on the feasibility of real-time information exchange, we provide the convergence conditions of various generic distributed algorithms in different scenarios. When no message exchanges between users are possible and every user maximizes its own utility, it is essential to determine whether a NE exist and if yes, how to achieve such an equilibrium. In ACSCG, a pure NE exists because ACSCG belongs to concave games [1] [2]. Our key contribution in this context is that we investigate the uniqueness of pure NE and consider the best response dynamics to compute the NE. We explore the properties of the additive coupling among users given the sum constraint and provide several sufficient conditions under which best response dynamics converges linearly[1] to the unique NE, for any set of feasible initialization with either sequential or parallel updates. When users can collaboratively exchange messages with each other in real-time, we present the sufficient convergence conditions of two alternative distributed pricing algorithms, including gradient play and Jacobi update, to coordinate users' action and improve the overall system efficiency. The proposed convergence conditions generalize the results that have been previously obtained in [8]- [13] for the multi-user power control problem and they are immediately applicable to other multi-user applications in communication networks that fulfill the requirements of ACSCG.

The rest of this paper is organized as follows. Section 2 defines the model of ACSCG. For ACSCG models, Sections 3 and 4 present several distributed algorithms without and with real-time information exchanges, respectively, and provide sufficient conditions that guarantee the convergence of the proposed algorithms. Conclusions are drawn in Section 5. Due to space limitations, the formal proofs and numerical examples are omitted; for these proofs and examples, the reader is referred to [19].

2 Game Model

2.1 Strategic Games, Nash Equilibrium, and Pareto Optimality

A strategic game is a suitable model for the analysis of a game where all users act independently and simultaneously according to their own self-interests and with no or limited a priori knowledge of the other users' strategies. This can be

[1] A sequence $x^{(k)}$ with limit x^* is linearly convergent if there exists a constant $c \in (0,1)$ such that $|x^{(k)} - x^*| \leq c|x^{(k-1)} - x^*|$ for k sufficiently large.

formally defined as a tuple $\Gamma = \langle \mathcal{N}, \mathcal{A}, u \rangle$. In particular, $\mathcal{N} = \{1, 2, \ldots, N\}$ is the set of decision makers. Define $\mathcal{A}$ to be the joint action set $\mathcal{A} = \times_{n \in \mathcal{N}} \mathcal{A}_n$, with $\mathcal{A}_n \subseteq \mathcal{R}^K$ being the action set available for user n. The vector utility function $u = \times_{n \in \mathcal{N}} u_n$ is a mapping from the individual users' joint action set to real numbers, i.e. $u : \mathcal{A} \to \mathcal{R}^N$. In particular, $u_n(\mathbf{a}) : \mathcal{A} \to \mathcal{R}$ is the utility of the nth user that generally depends on the strategies $\mathbf{a} = (\mathbf{a}_n, \mathbf{a}_{-n})$ of all users, where $\mathbf{a}_n \in \mathcal{A}_n$ denotes a feasible action of user n, and $\mathbf{a}_{-n} = \times_{m \neq n} \mathbf{a}_m$ is a vector of the actions of all users except n. We also denote by $\mathcal{A}_{-n} = \times_{m \neq n} \mathcal{A}_m$ the joint action set of all users except n. To capture the multi-user performance tradeoff, the utility region is defined as $\mathcal{U} = \{(u_1(\mathbf{a}), \ldots, u_N(\mathbf{a})) | \; \exists \; \mathbf{a} \in \mathcal{A}\}$. Various solutions, such as NE and Pareto optimality, were developed in the literature [1]. Significant research efforts have been devoted in the literature to constructing operational algorithms in order to achieve NE and Pareto optimality in various games with special structures of action set $\mathcal{A}_n$ and utility function u_n.

Nash Equilibrium: Definition, Existence, and Convergence. To avoid the overhead associated with exchanging information in real-time, network designers may prefer fully decentralized solutions in which the participating users simply compete against other users by choosing actions $\mathbf{a}_n \in \mathcal{A}_n$ to selfishly maximize their individual utility functions $u_n(\mathbf{a}_n, \mathbf{a}_{-n})$, given the actions $\mathbf{a}_{-n} \in \mathcal{A}_{-n}$. Most of these approaches focus on investigating the existence and properties of NE. NE is defined to be an action profile $(\mathbf{a}_1^*, \mathbf{a}_2^*, \ldots, \mathbf{a}_N^*)$ with the property that for every player, it satisfies $u_n(\mathbf{a}_n^*, \mathbf{a}_{-n}^*) \geq u_n(\mathbf{a}_n, \mathbf{a}_{-n}^*)$ for all $\mathbf{a}_n \in \mathcal{A}_n$, i.e. given the other users' actions, no user can increase its utility alone by changing its action. Many of the well-known results on NE rely on specific structural properties of action set $\mathcal{A}$ and utility function u in the investigated multi-user interactions. For an extensive discussion of the methodologies studying the existence, uniqueness, and convergence of various equilibria in communication networks, we refer the readers to [14].

Pareto Optimality and Network Utility Maximization. A profile of actions is Pareto optimal if there is no other profile of actions that makes every user at least as well off and at least one user strictly better off. It is important to note that operating at a NE will generally limit the performance of the user itself as well as that of the entire network, because the available network resources are not always effectively exploited due to the conflicts of interest occurring among users. As opposed to the NE-based approaches, there exists a large body of literature that focuses on studying how to compute Pareto optimal solutions in large-scale networks where centralized solutions are infeasible by optimizing a certain common objective function $f(u_1(\mathbf{a}), u_2(\mathbf{a}), \ldots, u_N(\mathbf{a}))$. This function represents the fairness rule based on which the system-wide resource allocation is performed. Different objective functions, e.g. sum utility maximization in which $f(u_1(\mathbf{a}), u_2(\mathbf{a}), \ldots, u_N(\mathbf{a})) = \sum_{n=1}^{N} u_n(\mathbf{a})$, can provide reasonable allocation outcomes by jointly considering fairness and efficiency. An important example is the NUM framework that develops distributed algorithms to solve network resource allocation problems [6]. The majority of the results in the existing NUM literature are based on convex optimization theory. It is well-known that, for convex optimization problems, users can collaboratively exchange price signals that reflect the "cost" for consuming the constrained resources and the Pareto optimal allocation that maximizes the network utility can be determined in a fully distributed manner [7].

Summarizing, these general structural results without and with real-time message exchange turn out to be very useful when analyzing various multi-user interactions in communication networks. In the remaining part of this paper, we will derive several structural results for a particular type of multi-user interaction scenario.

2.2 Additively Coupled Sum Constrained Games

Definition 1. *A multi-user interaction $\Gamma = \langle \mathcal{N}, \mathcal{A}, u \rangle$ is a ACSCG if it satisfies the following assumptions:*

A1: *$\forall n \in \mathcal{N}$, action set $\mathcal{A}_n \subseteq \mathcal{R}^K$ is defined as $\mathcal{A}_n =$* [2]

$$\left\{(a_n^1, a_n^2, \cdots, a_n^K) \mid a_n^k \in [a_{n,k}^{\min}, a_{n,k}^{\max}] \text{ and } \sum_{k=1}^{K} a_n^k \leq M_n\right\}. \tag{1}$$

A2: *There exist $h_n^k : \mathcal{R} \to \mathcal{R}$, $f_n^k : \mathcal{A}_{-n} \to \mathcal{R}$, and $g_n^k : \mathcal{A}_{-n} \to \mathcal{R}$, $k = 1, \ldots, K$, such that*

$$u_n(\mathbf{a}) = \sum_{k=1}^{K} \left[h_n^k(a_n^k + f_n^k(\mathbf{a}_{-n})) - g_n^k(\mathbf{a}_{-n})\right], \tag{2}$$

for all $\mathbf{a} \in \mathcal{A}$ and $n \in \mathcal{N}$. $h_n^k(\cdot)$ is an increasing, twice differentiable, and strictly concave function and $f_n^k(\cdot)$ and $g_n^k(\cdot)$ are both twice differentiable.

The ACSCG model defined by assumptions A1 and A2 covers a broad class of multi-user interactions. Assumption A1 indicates that each player's action set is a K-dimensional vector set and its action vector is sum-constrained. This represents the communication scenarios in which each user needs to determine its multi-dimensional action in various channels or networks while the total amount of resources it can consume is constrained. Assumption A2 implies that each user's utility is separable and can be represented by the summation of concave functions h_n^k minus "penalty" functions g_n^k across the K dimensions. In particular, within each dimension, the input of h_n^k is an additive combination of user n's action a_n^k and function $f_n^k(\mathbf{a}_{-n})$ that depends on the remaining users' joint action $\mathbf{a}_{-n}$. Since a_n^k only appears in the concave function h_n^k, it implies that each user's utility is concave in its own action, i.e. diminishing returns per unit

[2] We consider a sum constraint throughout the paper rather than a weighted-sum constraint, because a weighted-sum constraint can be easily converted to a sum constraint by rescaling $\mathcal{A}_n$. Besides, we nontrivially assume that $\sum_{k=1}^{K} a_{n,k}^{\max} \geq M_n$.

of user n's invested action $\mathbf{a}_n$, which is common for many application scenarios in communication networks.

Summarizing, the key features of the game model defined by A1 and A2 include: each user's action is subject to a *sum constraint*; users' utilities are impacted by *additive combinations* of a_n^k and $f_n^k(\mathbf{a}_{-n})$ through concave functions h_n^k. Therefore, we term the game Γ that satisfies assumptions A1 and A2 as ACSCG. In [19], we present several illustrative multi-user interaction examples that belong to ACSCG, including power control in frequency-selective Gaussian interference channel, delay minimization in Jackson networks, and asynchronous transmission in digital subscriber lines network.

2.3 Issues Related to ACSCG

Since ACSCG represents a good abstraction of numerous multi-user resource allocation problems, we aim to investigate the convergence properties of various distributed algorithms in ACSCG without and with real-time message passing.

ACSCG is a concave game [1] [2] and therefore, it admits at least one pure NE. In practice, we want to provide the sufficient conditions under which best response dynamics provably and globally converges to a pure NE. However, the existing literature, e.g. the diagonal strict concavity (DSC) conditions in [2] and the supermodular game theory [3]- [5], does not provide such convergence conditions for the general ACSCG model. On the other hand, if we want to maximize the sum utility by enabling real-time message passing among users, we also note that, the utility u_n is not necessarily jointly concave in $\mathbf{a}$ because of the existence of $g_n^k(\cdot)$. Therefore, the existing algorithms developed for the convex NUM are not immediately applicable either.

In the following sections, we will fully explore structures of ACSCG and address the convergence properties of various distributed algorithms in two different scenarios. Specifically, Section 3 investigates the scenarios in which each user n can only observe $\{f_n^k(\mathbf{a}_{-n})\}_{k=1}^K$ and cannot exchange any information with any other user. Section 4 focuses on the scenarios in which each user n is able to announce and receive information in real-time to and from the remaining users about $\frac{\partial u_n(\mathbf{a})}{\partial a_m^k}$ and $\frac{\partial u_m(\mathbf{a})}{\partial a_n^k}$, $\forall m \neq n, k = 1, \ldots, K$.

3 Scenario I: No Message Exchange among Users

In communication scenarios where users cannot exchange messages to achieve coordination, the participating users can simply choose actions to selfishly maximize their individual utility functions $u_n(\mathbf{a})$ by solveing the following optimization program:

$$\max_{\mathbf{a}_n \in \mathcal{A}_n} u_n(\mathbf{a}). \tag{3}$$

The steady state outcome of such a multi-user interaction is usually characterized as a NE.

3.1 Properties of Best Response Dynamics in ACSCG

In this subsection, we first focus on the scenarios in which $f_n^k(\mathbf{a}_{-n})$ is the linear combination of the remaining users' action in the same dimension k, i.e.

$$f_n^k(\mathbf{a}_{-n}) = \sum_{m \neq n} F_{mn}^k a_m^k \tag{4}$$

and $F_{mn}^k \in \mathcal{R}$, $\forall m, n, k$. In Section 3.2, we will extend the results derived for the functions $f_n^k(\mathbf{a}_{-n})$ defined in (4) to general $f_n^k(\mathbf{a}_{-n})$.

Since $h_n^k(\cdot)$ is concave, the objective in (3) is a concave function in a_n^k when the other users' actions $\mathbf{a}_{-n}$ are fixed. To find the globally optimal solution of the problem in (3), we can first form its Lagrangian

$$L_n(\mathbf{a}_n, \lambda) = u_n(\mathbf{a}) + \lambda(M_n - \sum_{k=1}^{K} a_n^k), \tag{5}$$

in which $a_n^k \in [a_{n,k}^{\min}, a_{n,k}^{\max}]$. By taking the first derivatives of (5), we have

$$\frac{\partial L_n(\mathbf{a}_n, \lambda)}{\partial a_n^k} = \frac{\partial h_n^k(a_n^k + \sum_{m \neq n} F_{mn}^k a_m^k)}{\partial a_n^k} - \lambda = 0. \tag{6}$$

Denote

$$l_n^k(\mathbf{a}_{-n}, \lambda) \triangleq \left[\left\{ \frac{\partial h_n^k}{\partial x} \right\}^{-1} (\lambda) - \sum_{m \neq n} F_{mn}^k a_m^k \right]_{a_{n,k}^{\min}}^{a_{n,k}^{\max}}, \tag{7}$$

in which $\left\{ \frac{\partial h_n^k}{\partial x} \right\}^{-1}$ is the inverse function[3] of $\frac{\partial h_n^k}{\partial x}$ and $[x]_b^a = \max\{\min\{x, a\}, b\}$. The optimal solution of (3) is given by $a_n^{*k} = l_n^k(\mathbf{a}_{-n}, \lambda^*)$, where the Lagrange multiplier λ^* is chosen to satisfy the sum constraint $\sum_{k=1}^{K} a_n^{*k} = M_n$.

We define the best response operator $B_n^k(\cdot)$ as

$$B_n^k(\mathbf{a}_{-n}) = l_n^k(\mathbf{a}_{-n}, \lambda^*). \tag{8}$$

We consider the best response algorithm in which each user updates its action using the best response strategy that maximizes its utility function in (2). We consider two types of update orders, including sequential update and parallel update. Specifically, in sequential update, individual players iteratively optimize in a circular fashion with respect to their own actions while keeping the actions of their opponents fixed. At stage t, user n chooses its action according to

$$a_n^{k,t} = B_n^k([\mathbf{a}_1^t, \ldots, \mathbf{a}_{n-1}^t, \mathbf{a}_{n+1}^{t-1}, \ldots, \mathbf{a}_N^{t-1}]). \tag{9}$$

On the other hand, players adopting the parallel update revise their actions at stage t according to

$$a_n^{k,t} = B_n^k(\mathbf{a}_{-n}^{t-1}). \tag{10}$$

[3] If $\nexists\, x = x^*$ such that $\frac{\partial h_n^k}{\partial x}|_{x=x^*} = \lambda$, we let $\left\{ \frac{\partial h_n^k}{\partial x} \right\}^{-1}(\lambda) = -\infty$.

We obtain several sufficient conditions under which best response dynamics converges. Similar convergence conditions are proved in [9]- [11] in which $h_n^k(x) = \log_2(\sigma_n^k + H_{nn}^k x)$. We consider more general functions $h_n^k(\cdot)$ and further extend the convergence conditions in [9]- [11]. The key differences among all the sufficient conditions which will be provided in this section are summarized in Table 1.

Table 1. Comparison among conditions (C1)-(C6)

Conditions	Assumptions about $f_n^k(\mathbf{a}_{-n})$	$h_n^k(x)$	Measure of residual error $\mathbf{a}_n^{t+1} - \mathbf{a}_n^t$	Contraction factor
(C1)	(4)	A2	1-norm	$2\rho(\mathbf{T}^{\max})$
(C2)	(4) and F_{mn}^k have the same sign for $\forall k, m \neq n$	A2	1-norm	$\rho(\mathbf{T}^{\max})$
(C3)	(4)	(13)	weighted Euclidean norm	$\rho(\mathbf{S}^{\max})$
(C4)	general	A2	1-norm	$2\rho(\bar{\mathbf{T}}^{\max})$
(C5)	$\frac{\partial f_n^k(\mathbf{a}_{-n})}{\partial a_m^{k'}}$ have the same sign for $\forall \mathbf{a} \in \mathcal{A}, k, k', m \neq n$	A2	1-norm	$\rho(\bar{\mathbf{T}}^{\max})$
(C6)	general	(13)	weighted Euclidean norm	$\rho(\bar{\mathbf{S}}^{\max})$

General $h_n^k(\cdot)$. The first sufficient condition is developed for the general cases in which the functions $h_n^k(\cdot)$ in the utilities $u_n(\cdot)$ are specified in assumption A2. Define

$$[\mathbf{T}^{\max}]_{mn} \triangleq \begin{cases} \max_k |F_{mn}^k|, & \text{if } m \neq n \\ 0, & \text{otherwise.} \end{cases} \tag{11}$$

and let $\rho(\mathbf{T}^{\max})$ denote the spectral radius of the matrix $\mathbf{T}^{\max}$.

Theorem 1. *If*

$$\rho(\mathbf{T}^{\max}) < \frac{1}{2}, \tag{C1}$$

then there exists a unique NE in game Γ and best response dynamics converges linearly to the NE, for any set of initial conditions belonging to $\mathcal{A}$ with either sequential or parallel updates.

Proof: This theorem is proved by showing that the best response dynamics defined in (9) and (10) is a contraction mapping under (C1). See Appendix A in [19] for details. ■

In multi-user communication applications, it is common to have games of *strategic complements* (or *strategic substitutes*), i.e. the marginal returns to any one component of the player's action rise with increases (or decreases) in the components of the competitors' actions [15]. For instance, in power control applications, increasing user n's transmitted power creates stronger interference to the other users and decreases their marginal achievable rates. Mathematically, if u_n

is twice differentiable, strategic complementarities (or strategic substitutes) can be described as

$$\frac{\partial^2 u_n(\mathbf{a}_n, \mathbf{a}_{-n})}{\partial a_n^j \partial a_m^k} \geq 0,\ \forall m \neq n, j, k, (\text{or } \frac{\partial^2 u_n(\mathbf{a}_n, \mathbf{a}_{-n})}{\partial a_n^j \partial a_m^k} \leq 0,\ \forall m \neq n, j, k). \quad (12)$$

For the ACSCG models that exhibit strategic complementarities (or strategic substitutes), the following theorem further relaxes condition (C1).

Theorem 2. *Let Γ be an ACSCG with strategic complementarities (or strategic substitutes), i.e. $F_{mn}^k \leq 0, \forall k, m \neq n$, (or $F_{mn}^k \geq 0, \forall k, m \neq n$). If*

$$\rho(\boldsymbol{T}^{\max}) < 1, \quad \text{(C2)}$$

then there exists a unique NE in game Γ and best response dynamics converges linearly to the NE, for any set of initial conditions belonging to $\mathcal{A}$ with either sequential or parallel updates.

Proof: This theorem is proved by adapting the proof of Theorem 1. See Appendix B in [19]. ■

Remark 1. (Implications of conditions (C1) and (C2)) Theorem 1 and Theorem 2 give sufficient conditions for best response dynamics to globally converge to a unique fixed point. Specifically, $\max_k |F_{mn}^k|$ can be regarded as a measure of the strength of the mutual coupling between user m and n. The intuition behind (C1) and (C2) is that, the weaker the coupling among different users is, the more likely that best response dynamics converges. Consider the extreme case in which $F_{mn}^k = 0, \forall k, m \neq n$. Since each user's best response is not impacted by the remaining users' action $\mathbf{a}_{-n}$, the convergence is immediately achieved after a single best-response iteration. If no restriction is imposed on F_{mn}^k, Theorem 1 specifies a mutual coupling threshold under which best response dynamics provably converge. The proof of Theorem 1 can be intuitively interpreted as follows. We regard every best response update as the users' joint attempt to approach the NE. Due to the linear coupling structure in (4), user n's best response in (7) contains a term $\sum_{m \neq n} F_{mn}^k a_m^k$ that is a linear combination of $\mathbf{a}_{-n}$. As a result, the residual error $|\mathbf{a}_n^{t+1} - \mathbf{a}_n^t|_1$, which is the 1-norm distance between the updated action profile $\mathbf{a}_n^{t+1}$ and the current action profile $\mathbf{a}_n^t$, can be upper-bounded using linear combinations of $|\mathbf{a}_m^t - \mathbf{a}_m^{t-1}|_1$ in which $m \neq n$. Recall that F_{mn}^k can be either positive or negative. We also note that, if $\mathbf{a}_m^t \neq \mathbf{a}_m^{t-1}$, $\mathbf{a}_m^t - \mathbf{a}_m^{t-1}$ contains both positive and negative terms due to the sum-constraint. In the worst case, the distance $|\mathbf{a}_n^{t+1} - \mathbf{a}_n^t|_1$ is maximized if $\{F_{mn}^k\}$ and $\{a_m^{k,t} - a_m^{k,t-1}\}$ are co-phase multiplied and additively summed, i.e. $F_{mn}^k(a_m^{k,t} - a_m^{k,t-1}) \geq 0$, for $\forall k = 1, \ldots, K, m \neq n$. After an iteration, all users except n contributes to user n's residual error at stage $t+1$ up to $\sum_{m \neq n} 2 \max_k |F_{mn}^k| |\mathbf{a}_m^t - \mathbf{a}_m^{t-1}|_1$. Under condition (C1), it is guaranteed that the residual error contracts. Theorem 2 focuses on the situations in which the signs of F_{mn}^k are the same, $\forall m \neq n, k$. In this case, $\{F_{mn}^k\}$ and $\{a_m^{k,t} - a_m^{k,t-1}\}$ cannot be co-phase multiplied. Therefore, the region of convergence enlarges and hence, condition (C2) stated in Theorem 2 is weaker than condition (C1) in Theorem 1.

Remark 2. (Relation to the results in references [9]- [11]) Similar to [9] [10], our proofs choose 1-norm as the distance measure for the residual errors $\mathbf{a}_n^{t+1} - \mathbf{a}_n^t$ after each best-response iteration. However, by manipulating the inequalities in a different way, condition (C2) is more general than the results in [9] [10], where they require $\max_k F_{mn}^k < \frac{1}{N-1}$. Interestingly, condition (C2) recovers the result obtained in [11] where it is proved by choosing the Euclidean norm as the distance measure for the residual errors $\mathbf{a}_n^{t+1} - \mathbf{a}_n^t$ after each best-response iteration. However, the approach in [11] using the Euclidean norm only applies to the scenarios in which $h_n^k(\cdot)$ is a logarithmic function. We prove that condition (C2) applies to any $h_n^k(\cdot)$ that is increasing and strictly concave.

A Special Class of $h_n^k(\cdot)$. In addition to conditions (C1) and (C2), we also develop a sufficient convergence condition for a family of utility functions parameterized by a negative number θ. In particular, $h_n^k(\cdot)$ satisfies[4]

$$h_n^k(x) = \begin{cases} \log(\alpha_n^k + F_{nn}^k x), \text{ if } \theta = -1, \\ \frac{(\alpha_n^k + F_{nn}^k x)^{\theta+1}}{\theta+1}, \quad \text{if } -1 < \theta < 0 \text{ or } \theta < -1. \end{cases} \tag{13}$$

and $\alpha_n^k \in \mathcal{R}$ and $F_{nn}^k > 0$. The interpretation of this type of utilities has been addressed in [16]. It is shown that varying the parameter θ leads to different types of fairness across $\alpha_n^k + F_{nn}^k(a_n^k + \sum_{m \neq n} F_{mn}^k a_m^k)$ for all k. In particular, $\theta = -1$ corresponds to the proportional fairness; if $\theta = -2$, then harmonic mean fairness; and if $\theta = -\infty$, then max-min fairness. In these cases, best response dynamics in equation (7) is reduced to

$$l_n^k(\mathbf{a}_{-n}, \lambda) = \left[\left(\frac{1}{F_{nn}^k}\right)^{1+\frac{1}{\theta}} \lambda^{\frac{1}{\theta}} - \frac{\alpha_n^k}{F_{nn}^k} - \sum_{m \neq n} F_{mn}^k a_m^k \right]_{a_{n,k}^{\min}}^{a_{n,k}^{\max}}, \tag{14}$$

Define $[\mathbf{S}^{\max}]_{mn} \triangleq$

$$\begin{cases} \frac{\sum_{k=1}^K (F_{mm}^k)^{1+\frac{1}{\theta}}}{\sum_{k=1}^K (F_{nn}^k)^{1+\frac{1}{\theta}}} \max_k \left\{ |F_{mn}^k| \left(\frac{F_{nn}^k}{F_{mm}^k}\right)^{1+\frac{1}{\theta}} \right\}, \text{ if } m \neq n \\ 0, \qquad\qquad \text{otherwise.} \end{cases} \tag{15}$$

For the class of utility functions in (13), Theorem 3 gives a sufficient condition that guarantees the convergence of the best response dynamics defined in (14).

Theorem 3. *For $h_n^k(\cdot)$ defined in (13), if*

$$\rho(\boldsymbol{S}^{\max}) < 1, \tag{C3}$$

then there exists a unique NE in game Γ and best response dynamics converges linearly to the NE, for any set of initial conditions belonging to $\mathcal{A}$ and with either sequential or parallel updates.

[4] If $\alpha_n^k + F_{nn}^k x \leq 0$, we let $h_n^k(x) = -\infty$. We assume for this class of $h_n^k(\cdot)$ that for $\forall \mathbf{a}_{-n} \in \mathcal{A}_{-n}$, there exists $\mathbf{a}_n \in \mathcal{A}_n$ such that $\alpha_n^k + F_{nn}^k x > 0$ for $\forall n, k$.

Proof: It can be proved by showing that the best response dynamics defined in (14) is a contraction mapping with respect to the weighted Euclidean norm. See Appendix C in [19] for details. ■

Remark 3. (Relation between conditions (C3) and the results in reference [11]) For power control in frequency-selective Gaussian interference channel, Scutari et al. established in [11] a sufficient condition under which the iterative water-filling algorithm converges. The iterative water-filling algorithm essentially belongs to best response dynamics. Specifically, in [11], Shannon's formula leads to $\theta = -1$ and cross channel coefficients satisfy $F_{mn}^k \geq 0, \forall k, m \neq n$. Equation (14) reduces to the water-filling formula

$$l_n^k(\mathbf{a}_{-n}, \lambda) = \left[\frac{1}{\lambda} - \frac{\alpha_n^k}{F_{nn}^k} - \sum_{m \neq n} F_{mn}^k a_m^k\right]_{a_{n,k}^{\min}}^{a_{n,k}^{\max}}, \tag{16}$$

and $[\mathbf{S}^{\max}]_{mn} = \max_k F_{mn}^k$. By choosing the weighted Euclidean norm as the distance measure for the residual errors $\mathbf{a}_n^{t+1} - \mathbf{a}_n^t$ after each best-response iteration, Theorem 3 generalizes the results in [11] for the family of utility functions defined in (13).

Remark 4. (Relation between conditions (C1), (C2) and (C3)) The connections and differences between conditions (C1), (C2) and (C3) are summarized in Table 1. We have addressed the implications of (C1) and (C2) in Remark 1. Now we discuss their relation with (C3). First of all, condition (C1) is proposed for general $h_n^k(\cdot)$ and condition (C3) is proposed for the class of utility functions defined in (13). However, Theorem 1 and Theorem 3 individually establish the fact that best response dynamics is a contraction map by selecting different vector and matrix norms. Therefore, in general, (C1) and (C3) do not immediately imply each other. Note that $[\mathbf{S}^{\max}]_{mn} \leq \zeta_{mn} \cdot \max_k |F_{mn}^k|$ in which ζ_{mn} satisfies

$$\zeta_{mn} = \frac{\sum_{k=1}^K (F_{mm}^k)^{1+\frac{1}{\theta}}}{\sum_{k=1}^K (F_{nn}^k)^{1+\frac{1}{\theta}}} \cdot \max_k \frac{(F_{nn}^k)^{1+\frac{1}{\theta}}}{(F_{mm}^k)^{1+\frac{1}{\theta}}} \in \left[1, \frac{\max_k (F_{nn}^k / F_{mm}^k)^{1+\frac{1}{\theta}}}{\min_k (F_{nn}^k / F_{mm}^k)^{1+\frac{1}{\theta}}}\right]. \tag{17}$$

The physical interpretation of ζ_{mn} is the similarity between the preferences of user m and n across the total K dimensions of their action spaces. Recall that both $\mathbf{S}^{\max}$ and $\mathbf{T}^{\max}$ are non-negative matrices and $\mathbf{S}^{\max}$ is element-wise less than or equal to $\max_{m \neq n} \zeta_{mn} \mathbf{T}^{\max}$. By the property of non-negative matrix and condition (C1), we can conclude $\rho(\mathbf{S}^{\max}) \leq \rho(\max_{m \neq n} \zeta_{mn} \mathbf{T}^{\max}) < \max_{m \neq n} \frac{\zeta_{mn}}{2}$. If users have similar preference in their available actions and the upper bound of ζ_{mn} that measures the difference of their preferences is below the following threshold:

$$\frac{\max_{k, m \neq n} (F_{nn}^k / F_{mm}^k)^{1+\frac{1}{\theta}}}{\min_{k, m \neq n} (F_{nn}^k / F_{mm}^k)^{1+\frac{1}{\theta}}} < 2, \tag{18}$$

we know that (C1) implies (C3) in this situation because $\rho(\mathbf{S}^{\max}) < \max_{m,n} \zeta_{mn} \cdot \rho(\mathbf{T}^{\max}) < 2 \cdot \frac{1}{2} = 1$. We also would like to point out that, the LHS of (18) is a function of θ and the LHS $\equiv 1$ if $\theta = -1$. When $\theta = -1$, $\mathbf{T}^{\max}$ coincides with $\mathbf{S}^{\max}$. Mathematically, in this case, (C3) is actually more general than (C2), because it still holds even if coefficients F_{mn}^{k} have different signs.

3.2 Extensions to General $f_n^k(\cdot)$

As a matter of fact, the results above can be extended to the more general situations in which $f_n^k(\cdot)$ is a nonlinear differentiable function, $\forall n, k$ and its input $\mathbf{a}_{-n}$ consists of the remaining users' action from all the dimensions. Accordingly, equation (7) becomes

$$l_n^k(\mathbf{a}_{-n}, \lambda) \triangleq \left[\left\{ \frac{\partial h_n^k}{\partial x} \right\}^{-1} (\lambda) - f_n^k(\mathbf{a}_{-n}) \right]_{a_{n,k}^{\min}}^{a_{n,k}^{\max}}. \tag{19}$$

The conclusions in Theorem 1, 2, and 3 can be further extended as Theorem 4, and 5, 6 that are listed below.

For general $f_n^k(\cdot)$, we denote

$$[\bar{\mathbf{T}}^{\max}]_{mn} \triangleq \begin{cases} \max_{\mathbf{a}\in\mathcal{A},k'} \sum_{k=1}^{K} \left| \frac{\partial f_n^k(\mathbf{a}_{-n})}{\partial a_m^{k'}} \right|, & \text{if } m \neq n \\ 0, & \text{otherwise.} \end{cases} \tag{20}$$

Besides, for $h_n^k(\cdot)$ defined in (13), we define $[\bar{\mathbf{S}}^{\max}]_{mn} \triangleq$

$$\begin{cases} \frac{\sum\limits_{k=1}^{K} (F_{mm}^{k})^{1+\frac{1}{\theta}}}{\sum\limits_{k=1}^{K} (F_{nn}^{k})^{1+\frac{1}{\theta}}} \max_{\mathbf{a}\in\mathcal{A},k'} \left\{ \sum_{k=1}^{K} \left| \frac{\partial f_n^k(\mathbf{a}_{-n})}{\partial a_m^{k'}} \right| \left(\frac{F_{nn}^{k'}}{F_{mm}^{k'}} \right)^{1+\frac{1}{\theta}} \right\}, & \text{if } m \neq n \\ 0, & \text{otherwise.} \end{cases} \tag{21}$$

Theorem 4. *If*

$$\rho(\bar{\boldsymbol{T}}^{\max}) < \frac{1}{2}, \tag{C4}$$

then there exists a unique NE in game Γ and best response dynamics converges linearly to the NE, for any set of initial conditions belonging to $\mathcal{A}$ with either sequential or parallel updates.

Proof: This theorem can be proved by combining the proof of Theorem 1 and the mean value theorem for vector-valued functions. See Appendix D in [19] for details. ■

Similarly as in Theorem 2, for the general ACSCG models that exhibit strategic complementarities (or strategic substitutes), we can relax condition (C4).

Theorem 5. *For Γ with strategic complementarities (or strategic substitutes), i.e. $\frac{\partial f_n^k(\mathbf{a}_{-n})}{\partial a_m^{k'}} \geq 0, \forall m \neq n, k, k', \boldsymbol{a} \in \mathcal{A}$, (or $\frac{\partial f_n^k(\mathbf{a}_{-n})}{\partial a_m^{k'}} \leq 0, \forall m \neq n, k, k', \boldsymbol{a} \in \mathcal{A}$), if*

$$\rho(\bar{\boldsymbol{T}}^{\max}) < 1, \tag{C5}$$

then there exists a unique NE in game Γ and best response dynamics converges linearly to the NE, for any set of initial conditions belonging to $\mathcal{A}$ with either sequential or parallel updates.

Theorem 6. *For $h_n^k(\cdot)$ defined in (13), if*

$$\rho(\bar{\boldsymbol{S}}^{\max}) < 1, \tag{C6}$$

then there exists a unique NE in game Γ and best response dynamics converges linearly to the NE, for any set of initial conditions belonging to $\mathcal{A}$ with either sequential or parallel updates.

Remark 5. (Implications of conditions (C4), (C5), and (C6)) Based on the mean value theorem, we know that the upper bound of the additive sum of first derivatives $\sum_{k=1}^{K} \left| \frac{\partial f_n^k(\mathbf{a}_{-n})}{\partial a_m^{k'}} \right|$ governs the maximum impact that user m's action can make over user n's utility. As a result, Theorem 4, Theorem 5, and Theorem 6 indicate that $\sum_{k=1}^{K} \left| \frac{\partial f_n^k(\mathbf{a}_{-n})}{\partial a_m^{k'}} \right|$ can be used to develop similar sufficient conditions for the global convergence of best response dynamics. Table 1 summarizes the connections and differences among all the aforementioned conditions from (C1) to (C6).

Remark 6. (Impact of sum constraints) An interesting phenomenon that can be observed from the analysis above is that, the convergence condition may depend on the maximum constraints $\{M_n\}_{n=1}^N$. This differs from the observation in [11] that the presence of the transmit power and spectral mask constraints does not affect the convergence capability of the iterative water-filling algorithm. This is because when functions $f_n^k(\mathbf{a}_{-n})$ are affine, the elements in $\bar{\mathbf{T}}^{\max}$ and $\bar{\mathbf{S}}^{\max}$ are independent of the values of $\{M_n\}_{n=1}^N$. Therefore, (C1)-(C6) are independent of M_n for affine $f_n^k(\mathbf{a}_{-n})$. However, for non-linear $f_n^k(\mathbf{a}_{-n})$, the values of $\{M_n\}_{n=1}^N$ specify the range of users' joint feasible action set $\mathcal{A}$, and this will affect $\bar{\mathbf{T}}^{\max}$ and $\bar{\mathbf{S}}^{\max}$ accordingly. In other words, in the presence of non-linearly coupled $f_n^k(\mathbf{a}_{-n})$, convergence may depend on the players' maximum sum constraints $\{M_n\}_{n=1}^N$.

4 Scenario II: Message Exchange among Users

In this section, our objective is to coordinate the users' actions in ACSCG to maximize the overall performance of the system, measured in terms of their total utilities, in a distributed fashion. Specifically, the optimization problem we want to solve is

$$\max_{\mathbf{a}\in\mathcal{A}} \sum_{n=1}^{N} u_n(\mathbf{a}). \tag{22}$$

We will study two distributed algorithms in which the participating users exchange price signals that indicate the "cost" or "benefit" that its action causes to the other users. Allocating network resources via pricing has been well-investigated for convex NUM problems [6], where the original NUM problem can be decomposed into distributedly solvable subproblems by setting price for each constraint resource, and each subproblem has to decide the amount of resources to be used depending on the charged price. However, unlike in the conventional convex NUM, pricing mechanisms may not be immediately applicable in ACSCG if the objective in (22) is not jointly concave in $\mathbf{a}$. Therefore, we are interested in characterizing the convergence condition of different pricing algorithms in ACSCG.

We know that for any local maximum $\mathbf{a}^*$ of problem (22), there exist Lagrange multipliers $\lambda_n, \nu_n^1, \cdots, \nu_n^N$ and $\nu_n'^1, \cdots, \nu_n'^N$ such that the following Karush-Kuhn-Tucker (KKT) conditions hold for all $n \in \mathcal{N}$:

$$\frac{\partial u_n(\mathbf{a}^*)}{\partial a_n^k} + \sum_{m \neq n} \frac{\partial u_m(\mathbf{a}^*)}{\partial a_n^k} = \lambda_n + \nu_n^k - \nu_n'^k, \ \forall n \tag{23}$$

$$\lambda_n \Big(\sum_{k=1}^{K} a_n^{k*} - M_n \Big) = 0, \ \lambda_n \geq 0 \tag{24}$$

$$\nu_n^k (a_n^{k*} - a_{n,k}^{\max}) = 0, \ \nu_n'^k (a_{n,k}^{\min} - a_n^{k*}) = 0, \ \nu_n^k, \nu_n'^k \geq 0. \tag{25}$$

Denote π_{mn}^k user m's marginal fluctuation in utility per unit decrease in user n's action a_n^k within the kth dimension

$$\pi_{mn}^k(a_m^k, \mathbf{a}_{-m}^k) = -\frac{\partial u_m(\mathbf{a})}{\partial a_n^k}, \tag{26}$$

which is announced by user m to user n and can be viewed as the cost charged (or compensation paid) to user n for changing user m's utility. Using (26), equation (23) can be rewritten as

$$\frac{\partial u_n(\mathbf{a}^*)}{\partial a_n^k} - \sum_{m \neq n} \pi_{mn}^k(a_m^{k*}, \mathbf{a}_{-m}^{k*}) = \lambda_n + \nu_n^k - \nu_n'^k. \tag{27}$$

If we assume fixed prices $\{\pi_{mn}^k\}$ and action profile $\mathbf{a}_{-n}^k$, condition (27) gives the necessary and sufficient KKT condition of the following problem:

$$\max_{\mathbf{a}_n \in \mathcal{A}_n} \ u_n(\mathbf{a}) - \sum_{k=1}^{K} a_n^k \cdot \Big(\sum_{m \neq n} \pi_{mn}^k \Big). \tag{28}$$

At an optimum, a user behaves as if it maximizes the differences between its utility minus its payment to the other users in the network due to its impact over the other users' utilities. Different distributed pricing mechanisms can be developed based on the individual objective function in (28) and the convergence conditions may also vary based on the specific action update equation.

We will investigate two distributed pricing mechanisms for non-convex ACSCG and provide two sufficient conditions that guarantee their convergence. Specifically, under these sufficient conditions, both algorithms guarantee that the total utility is monotonically increasing until it converges to a feasible operating point that satisfies the KKT conditions. Similarly as in Section 3.1, we first assume $f_n^k(\mathbf{a}_{-n})$ takes the form in (4) and users update their actions in parallel.

4.1 Gradient Play

The first distributed pricing algorithm that we consider is gradient play. The update iterations of gradient play need to be properly redefined in presence of real-time information exchange. Specifically, at stage t, users adopting this algorithm exchange price signals $\{\pi_{mn}^{k,t-1}\}$ using the gradient information at stage $t-1$. Within each iteration, each user first determines the gradient of the objective in (28) based on the price vectors $\{\pi_{mn}^{k,t-1}\}$ and its own utility function $u_n(\mathbf{a}_n, \mathbf{a}_{-n}^{t-1})$. Then each user updates its action a_n^t using gradient projection algorithm according to

$$a_n^{'k,t} = a_n^{k,t-1} + \kappa\Big(\frac{\partial u_n(\mathbf{a}_n, \mathbf{a}_{-n}^{t-1})}{\partial a_n^k} - \sum_{m\neq n} \pi_{mn}^{k,t-1}\Big). \tag{29}$$

and

$$\mathbf{a}_n^t = [a_n^{1,t} a_n^{2,t} \cdots a_n^{K,t}] = \Big[a_n^{'1,t} a_n^{'2,t} \cdots a_n^{'K,t}\Big]_{\mathcal{A}_n}^{\|\cdot\|_2}. \tag{30}$$

in which the stepsize $\kappa > 0$. The following theorem provides a sufficient condition under which gradient play will converge monotonically provided that we choose small enough constant stepsize κ.

Theorem 7. *If* $\forall n, k, \mathbf{x}, \mathbf{y} \in \mathcal{A}_{-n}$,

$$\inf_x \frac{\partial^2 h_n^k(x)}{\partial^2 x} > -\infty, \text{ and } \Big\| \nabla g_n^k(\mathbf{x}) - \nabla g_n^k(\mathbf{y})\Big\| \leq L'\big\|\mathbf{x}-\mathbf{y}\big\|, \tag{C7}$$

gradient play converges for a small enough stepsize κ.

Proof: This theorem can be proved by showing the gradient of the objective function in (22) is Lipschitz continuous and applying Proposition 3.4 in [17]. See Appendix E in [19] for details. ■

Remark 7. (Application of condition (C7)) A sufficient condition that guarantees the convergence of distributed gradient projection algorithm is the Lipschitz continuity of the gradient of the objective function in (22). For example, in the power control problem in multi-channel networks [12], we have $h_n^k(x) = \log_2(\alpha_n^k + H_{nn}^k x)$ and $g_n^k(\mathbf{P}_{-n}) = \log_2(\sigma_n^k + \sum_{m\neq n} H_{mn}^k P_m^k)$. For this configuration, we can immediately verify that condition (C7) is satisfied. Therefore, gradient play can be applied. Moreover, as in [12], if we can further ensure that the problem in (22) is convex for some particular utility functions, gradient play converges to the unique optimal solution of (22) at which achieving KKT conditions implies global optimality.

4.2 Jacobi Update

We consider another alternative strategy update mechanism called Jacobi update [18]. In Jacobi update, every user adjusts its action gradually towards the best response strategy. Specifically, the maximizer of problem (28) takes the following form

$$B_n^{'k}(\mathbf{a}_{-n}) = \left\{\frac{\partial h_n^k}{\partial x}\right\}^{-1}\left(\lambda_n + \nu_n^k - \nu_n^{'k} + \sum_{m\neq n}\pi_{mn}^k\right) - \sum_{m\neq n} F_{mn}^k a_m^k, \tag{31}$$

in which λ_n, ν_n^k, and $\nu_n^{'k}$ are the Lagrange multipliers that satisfy complementary slackness in (24) and (25), and π_{mn}^k is defined in (26). In Jacobi update, at stage t, user n chooses its action according to

$$a_n^{k,t} = a_n^{k,t-1} + \kappa\left[B_n^{'k}(\mathbf{a}_{-n}^{t-1}) - a_n^{k,t-1}\right], \tag{32}$$

in which the stepsize $\kappa \in (0,1]$. The following theorem establishes a sufficient convergence condition for Jacobi update.

Theorem 8. *If* $\forall n, k, \mathbf{x}, \mathbf{y} \in \mathcal{A}_{-n}$,

$$\begin{aligned} &\inf_x \frac{\partial^2 h_n^k(x)}{\partial^2 x} > -\infty,\ \sup_x \frac{\partial^2 h_n^k(x)}{\partial^2 x} < 0,\ \text{and} \\ &\left\| \nabla g_n^k(\mathbf{x}) - \nabla g_n^k(\mathbf{y}) \right\| \leq L' \left\| \mathbf{x} - \mathbf{y} \right\|, \end{aligned} \tag{C8}$$

Jacobi update converges if the stepsize κ is sufficiently small.

Proof: This can be proved using the descent lemma and the mean value theorem. The details of the proof are provided in Appendix F in [19]. ∎

Remark 8. (Relation between condition (C8) and the result in [13]) Shi et al. considered the power allocation for multi-carrier wireless networks with non-separable utilities. Specifically, $u_n(\cdot)$ takes the form

$$u_n(\mathbf{P}) = r_i\left(\sum_{k=1}^{K} \log_2\left(1 + \frac{H_{nn}^k P_n^k}{\sigma_n^k + \sum_{m\neq n} H_{mn}^k P_m^k}\right)\right), \tag{33}$$

in which $r_i(\cdot)$ is an increasing and strictly concave function. Since the utilities are non-separable, the distributed pricing algorithm proposed in [13], which in fact belongs to Jacobi update, requires only one user to update its action profile at each stage while keeping the remaining users' action fixed. The condition in (C8) gives the convergence condition of the same algorithm in ACSCG. We prove in Theorem 7 that, if the utilities are separable, convergence can still be achieved even if these users update their actions at the same time. Therefore, we do not need an arbitrator to select the single user that updates its action at each stage.

5 Conclusion

In this paper, we propose and investigate a new game model, which we refer to as additively coupled sum constrained games, in which each player is subject to a sum constraint and its utility is additively impacted by the remaining users' actions. The convergence properties of various generic distributed adjustment algorithms, including best response, gradient play, and Jacobi update, have been investigated. The sufficient conditions obtained in this paper generalize the existing results developed in the multi-channel power control problem and can be extended to other applications that belong to ACSCG.

References

1. Fudenberg, D., Tirole, J.: Game Theory. MIT Press, Cambridge (1991)
2. Rosen, J.: Existence and uniqueness of equilibrium points for concave n-person games. Econometrica 33, 520–534 (1965)
3. Topkis, D.M.: Supermodularity and Complementarity. Princeton Univ. Press, Princeton (1998)
4. Yao, D.: S-modular games with queueing applications. Queueing Syst. 21, 449–475 (1995)
5. Altman, E., Altman, Z.: S-modular games and power control in wireless networks. IEEE Trans. Automatic Control 48(5), 839–842 (2003)
6. Chiang, M., Low, S.H., Calderbank, A.R., Doyle, J.C.: Layering as optimization decomposition. Proceedings of the IEEE 95, 255–312 (2007)
7. Palomar, D.P., Chiang, M.: A tutorial on decomposition methods for network utility maximization. IEEE JSAC 24(8), 1439–1451 (2006)
8. Yu, W., Ginis, G., Cioffi, J.: Distributed multiuser power control for digital subscriber lines. IEEE JSAC 20(5), 1105–1115 (2002)
9. Chung, S.T., Seung, J.L., Kim, J., Cioffi, J.: A game-theoretic approach to power allocation in frequency-selective Gaussian interference channels. In: Proc. IEEE Int. Symp. on Inform. Theory, p. 316 (June 2003)
10. Cendrillon, R., Huang, J., Chiang, M., Moonen, M.: Autonomous spectrum balancing for digital subscriber lines. IEEE Trans. on Signal Process. 55(8), 4241–4257 (2007)
11. Scutari, G., Palomar, D.P., Barbarossa, S.: Optimal linear precoding strategies for wideband noncooperative systems based on game theory - Part II: Algorithms. IEEE Trans. Signal Process. 56(3), 1250–1267 (2008)
12. Huang, J., Berry, R., Honig, M.: Distributed interference compensation for wireless networks. IEEE JSAC 24(5), 1074–1084 (2006)
13. Shi, C., Berry, R., Honig, M.: Distributed interference pricing for OFDM wireless networks with non-seperable utilities. In: Proc. of Conference on Information Sciences and Systems (CISS), March 19-21, pp. 755–760 (2008)
14. Lasaulce, S., Debbah, M., Altman, E.: Methodologies for analyzing equilibria in wireless games. IEEE Signal Process. Magazine 26, 41–52 (2009)
15. Bulow, J., Geanakoplos, J., Klemperer, P.: Multimarket oligopoly: strategic substitutes and strategic complements. Journal of Political Economy 93, 488–511 (1985)
16. Mo, J., Walrand, J.: Fair end-to-end window-based congestion control. IEEE Trans. on Networking 8(5), 556–567 (2000)
17. Bertsekas, D.P., Tsitsiklis, J.N.: Parallel and Distributed Computation. Prentice Hall, Englewood Cliffs (1997)
18. La, R., Anantharam, V.: Utility based rate control in the internet for elastic traffic. IEEE/ACM Trans. Networking 10(2), 271–286 (2002)
19. Su, Y., van der Schaar, M.: Structural Solutions For Additively Coupled Sum Constrained Games, UCLA Tech. Rep. (2010), `http://arxiv.org/abs/1005.0880`

Technology Choices and Pricing Policies in Wireless Networks

Yuanzhang Xiao[1], William R. Zame[2], and Mihaela van der Schaar[1]

[1] Department of Electrical Engineering, UCLA, Los Angeles CA 90095, USA
{yxiao,mihaela}@ee.ucla.edu
[2] Department of Economics, UCLA, Los Angeles CA 90095, USA
zame@econ.ucla.edu

Abstract. This paper studies the provision of a wireless network by a monopolistic provider who may be either benevolent (seeking to maximize social welfare) or selfish (seeking to maximize provider profit). The paper addresses the following questions: Under what circumstances is it feasible for a provider, either benevolent or selfish, to operate a network in such a way as to cover costs? How is the optimal behavior of a benevolent provider different from the optimal behavior of a selfish provider, and how does this difference affect social welfare? And, most importantly, how does the medium access control (MAC) technology influence the answers to these questions? To address these questions, we build a general model, and provide analysis and simulations for simplified but typical scenarios; the focus in these scenarios is on the contrast between the outcomes obtained under carrier-sensing multiple access (CSMA) and outcomes obtained under time-division multiple access (TDMA). Simulation results demonstrate that differences in MAC technology can have a significant effect on social welfare, on provider profit, and even on the (financial) feasibility of a wireless network.

Keywords: network economics, pricing, wireless networks.

1 Introduction

There has been much recent debate about the deployment of wireless networks that would allow Internet access in public areas. Central to this debate is the tradeoff between costs and benefits. Surprisingly, this debate seems to have ignored that the costs and benefits of such wireless networks depend crucially on the technology that is or could be employed. The purpose of this paper is to provide a framework for exploring the influence of technology on the costs and benefits of wireless networks and to demonstrate in a simple scenario that the feasibility and desirability of such a network may depend on the technology chosen. We show that the analysis depends crucially on the technology layer, the application layer, and the economic layer, and most crucially of all, on the interactions between these layers.

To see why the analysis depends crucially on the interactions between the various layers, consider a simple scenario that seems typical. There are two classes

R. Jain and R. Kannan (Eds.): GameNets 2011, LNICST 75, pp. 92–107, 2012.

of (potential) users: *data users*, who are sensitive to throughput but relatively insensitive to delay, and *video users*, who are sensitive to both throughput and delay. In managing the network, the service provider can offer a pricing policy and a scheduling policy, but the service provider's range of choices depends on the technology – in particular, on the medium access control (MAC) protocol – employed. If time-division multiple access (TDMA) is employed, the service provider will be able to guarantee quality of service (QoS) and monitor the usage of each user in order to charge per bit. Hence, the service provider can use a tiered pricing policy to screen the users into a number of types and offer performance guarantees to those users willing to pay for such guarantees. If carrier-sensing multiple access (CSMA) is employed, the service provider will be unable to guarantee QoS. Absent such performance guarantees, video users who require higher throughput or less delay may be unwilling to pay more than data users who will accept lower throughput and more delay. In this case, it is more reasonable for the service provider to adopt a flat fee for both types of users. As we will show, there are large regions within the range of plausible parameters in which employing TDMA rather than CSMA makes possible large improvements in social welfare. Indeed, there are regions in which employing TDMA would be consistent with operating a self-financing network while employing CSMA would not be.

1.1 Related Work

Two substantial bodies of work in the engineering literature ask about optimal behavior of the provider of a wireless network. The first considers a benevolent provider whose objective is to maximize social welfare [1]- [6]; the second considers a selfish provider whose objective is to maximize profit [7]- [12]. What we do here is to ask different (although related) questions that do not seem to have been studied before: Under what circumstances is it possible for a provider to operate a network in such a way as to cover costs? How is optimal behavior of a benevolent provider different from optimal behavior of a selfish provider and how does the difference affect social welfare? And, perhaps most importantly, how does the MAC protocol influence the answers to these questions?

Among the papers that focus on optimal pricing in networks, Palomar and Chiang [1] and Kelly *et al.* [2] [3] consider a network with one service provider serving multiple users and propose charging in proportion to the flow rates of the users in order to maximize social utility. Johari and Tsitsiklis [5] [6] focus on the efficiency loss under this pricing scheme and its variant with price differentiation. Gibbens and Kelly [4] propose a packet-based pricing policy for more effective flow control. Under the same scenario, Basar *et al.* [7] [8] [9] propose linear and nonlinear differentiated pricing schemes to control the network usage and maximize the provider's revenue. For cellular networks, Mandayam *et al.* [10] and Alpcan *et al.* [11] propose pricing for power control to reduce interference. It should be noted, however, that the prices in the above papers are not actually paid by the users; rather, they are signals used for the purpose of controlling the network congestion. In Paschalidis and Tsitsiklis [12], which studies a dynamic network with users arriving and leaving the network and derive the optimal pricing strategy and its static approximation, prices are actually paid by users, but – as in [2]- [11] – the technology layer is highly abstracted (as a constraint on the resource allocation). Other papers use different models and have a different focus. Friedman and Parkes [13] study the existence of implementable mechanisms for the users to truthfully announce their arrivals in WiFi networks. Musacchio and Walrand [14] model WiFi pricing as a dynamic game involving one access point and one user, and study the Nash equilibrium (NE) of this game. van der Schaar [15] and Sarkar [16] focus on competition among multiple service providers with simplified user subscription models.

Our work differs from this literature in that we model prices as actually paid by users and collected by the service providers, and we provide a much more detailed and less abstracted description of technology. We make use of both of these differences to study the interaction between technology and pricing and their impacts on performance.[1] In particular, we consider various technologies and pricing policies (closely modeled as those used in the real world by wireless carriers) to study the interactions between technology and pricing.

The remainder of this paper is organized as follows. In Section 2 we introduce the system model for the three-layer network. In Section 3 we formulate the design problem for the benevolent and selfish providers and the decision process of the users as a two stage game (with the provider acting in the first stage and the users acting in the second stage). In Section 4, we focus our analysis on a typical scenario to gain insights into this problem, and provide simulation results in this typical scenario. Finally, in Section 5 we conclude the paper.

2 System Model

We consider a wireless network with a single access point (AP), created by a service provider to enable Internet connections to potential users. Keeping in mind that a single access point will typically serve a relatively small number of potential users who may come and go at any moment in time, we build a dynamic continuous-time framework in which a finite number of potential users arrive and depart randomly.

Before we begin with the description of the service provider, we first introduce the basic concept of the user type. The *users* are categorized into K types according to their utility functions and arrival and departure processes. There are N_k identical users of type k.

2.1 The Service Provider

The service provider must choose a MAC protocol and a pricing policy.

[1] The interplay of technology and pricing policy is discussed by Lehr *et al.* [17], but their paper provides no quantitative analysis. To our best knowledge, no previous work has ever mathematically modeled and explicitly studied this problem.

The Medium Access Control Protocol. The MAC protocol determines which users will have access to which resources in which way. In principle, the service provider might be able to choose among many MAC protocols. CSMA and TDMA are the canonical MAC protocols. CSMA is representative of the protocols without a central controller, where the packets contend to get access to the medium. TDMA is representative of the protocols with a central controller, where the packets access the medium in non-overlapping periods of time. The key difference between CSMA and TDMA is the ability to offer QoS guarantee, which will probably result in different selections of pricing policies. The lack of QoS guarantee in CSMA may prevent the provider from charging by bit. Imagine a video user who pays for some video frames but loses subsequent frames due to network congestion. Since those paid video frames may be useless because of the loss of subsequent frames, the video users may be unwilling to pay for those bits without QoS guarantee. *Therefore, the provider using TDMA is able to charge both a subscription fee and a per-bit fee, while the provider using CSMA is more likely to charge a subscription fee only.* We write θ for a particular protocol.

Pricing Plans, Pricing Policies, and Pricing States. A *pricing plan* is a schedule of charges to users. We assume that charges consist of a subscription fee (paid once per billing period) p_s and a per-bit surcharge q for usage in excess of some specified threshold number of bits β. Thus a pricing plan is a triple

$$\mathbf{p} = (p_s, q, \beta).$$

To allow for the possibility that some users choose not to belong to the network at all, let $\phi = (0, 0, 0)$ be a dummy plan that imposes no costs. A user choosing ϕ does not subscribe to the network.

A *pricing policy* is a vector of pricing plans; for simplicity, we assume here that each pricing policy is a vector of exactly $L+1$ pricing plans: $\mathbf{P} = (\mathbf{p}^0, \mathbf{p}^1, \ldots, \mathbf{p}^L)$; by convention we assume that $\mathbf{p}^0 = \phi$.

Given a pricing policy $\mathbf{P} = (\mathbf{p}^0, \mathbf{p}^1, \ldots, \mathbf{p}^L)$, each user type k chooses a pricing plan from $\mathbf{P}$ by randomizing over all the choices according to a probability distribution. We define the *pricing state* to be the vector $\mathbf{v} = (v^0, v^1, \ldots, v^L)$, where v^ℓ is the number of users who are currently online and choose the pricing plan $\mathbf{p}^\ell$. We write $\mathcal{V}$ for the set of pricing states.

2.2 Users

The users are characterized by their utility functions, arrival processes, and service times. Given user characteristics and the technology and the pricing policy adopted by the service provider, each user determines a probability distribution on the choices of pricing plans that maximizes its expected utility (which will depend on the choices of all the other users). At the beginning of time, each user chooses a pricing plan randomly according to the prescribed probability distribution, and every time a user arrives at the network, the user reports the chosen plan to the service provider. The service provider will make the scheduling according to the current pricing state and the choice of a particular user.

Choices of Pricing Plans. Users choose pricing plans to maximize their expected utility, given the menu of pricing plans, the MAC protocol of the provider and the choices of other users. We allow for the possibility that users *randomize*, so users of type k choose a probability distribution over pricing plans. We write $\pi_{k,\ell}$ for the probability that a user of type k chooses plan ℓ.

Allowing for randomization guarantees that equilibrium exists. We may interpret randomization literally: users who are indifferent over various plans break their indifference in a random way. Alternatively, we may interpret randomization simply as uncertainty in the minds of the provider and other users. If the number of users is large, we can also interpret the probability distribution over pricing plans as the distributions of plans among the population [20].

The randomization is realized at the beginning of time. Upon arrival, each user tells the service provider the pricing plan it chooses, and the provider uses this information for scheduling. Write $\pi_k = [\pi_{k,0}, \ldots, \pi_{k,L}]$ for the (random) action of users of type k, and $\pi = (\pi_1, \ldots, \pi_K)$ for the vector of actions of all users. Represent the result of the randomization by a set of vectors $\mathbf{n} = (\mathbf{n}_1, \ldots, \mathbf{n}_K) = ([n_{1,0}, \ldots, n_{1,L}], \ldots, [n_{K,0}, \ldots, n_{K,L}])$ with $n_{k,\ell}$ being the number of type-k users choosing plan ℓ.

System State. The *system state*, or the *true state*, is defined as the number of users of each type choosing each pricing plan. Specifically, the system state $\mathbf{X}(t)$ at time t is a $K \times (L+1)$ matrix, with $x_{k,\ell}$ as the element at the kth row and $(\ell+1)$th column, representing the number of type-k users who choose plan ℓ.

Arrival Process and Service Time. We use a continuous-time model for the arrival and departure processes[2] (reflecting the fact that users might arrive/depart at any moment); as in [21], we assume that the arrival process of type-k users choosing plan ℓ is Poisson with arrival rate

$$\lambda_{k,\ell}(t) = \lambda_k \cdot (n_{k,\ell} - x_{k,\ell}(t)),$$

where λ_k is the individual arrival rate of a type-k user. We also assume that the service time of one type-k user is exponentially distributed with mean $1/\mu_k$.

Billing Period. We fix a *billing period* of length ΔT, which is typically one month. Subscription fees are charged at the beginning of each billing period; other fees are charged at the end of each billing period. This is consistent with the usual billing methods: people pay a subscription fee prospectively and other charges retrospectively. For convenience, we assume that neither the provider nor the users discount utility and cost over the billing period.

Expected Utility. The service provider and the users evaluate the social welfare and their satisfaction, respectively, by the *expected utility*, defined as the

[2] Here, the arrival process characterizes the arrival of users, but not the arrival of users' packets. Similarly, the service time is the duration of users staying in the system.

expectation of the total utility over a billing period when the stochastic process of the system state $\mathbf{X}(t)$ reaches the steady state. Each user's total utility consists of two components: utility of use and disutility of cost. To keep the model simple, we assume that total utility is simply the sum of utility of use and disutility of cost and is linear in cost with marginal utility of cost equal to 1 [22]:

$$\text{total utility} \;=\; \text{utility of use} \;-\; \text{cost} \,. \tag{1}$$

We denote the expected utility of use of a type-k user by $U_k(\theta, \pi)$, if the MAC protocol is θ and the joint probability distribution over pricing plans is π. We can calculate the expected utility of use $U_k(\theta, \pi)$ as follows

$$U_k(\theta, \pi) = \sum_{\ell=1}^{L} \pi_{k,\ell} \cdot \sum_{\mathbf{n}: n_{k,\ell} \geq 1} \Pr(\mathbf{n}) \cdot V_k^{\ell}(\theta, \mathbf{n}), \tag{2}$$

where $\Pr(\mathbf{n})$ is the probability that the randomization results in $\mathbf{n}$, and $V_k^{\ell}(\theta, \mathbf{n})$ is the steady-state utility of use of a type-k user, if the MAC protocol is θ and the result of the randomization is $\mathbf{n}$.

We denote the expected cost of a type-k user by $C_k(\theta, \mathbf{P}, \pi)$, if the MAC protocol is θ, the pricing policy is $\mathbf{P}$, and the joint probability distribution over pricing plans is π. The details for the calculation of $\Pr(\mathbf{n})$, $V_k^{\ell}(\theta, \mathbf{n})$, and $C_k(\theta, \mathbf{P}, \pi)$ can be found in [23, Sec. II-B].

Users' Decision Process. Each user determines the randomizing probability that maximizes its own expected utility. The optimal action for a type-k user satisfies

$$\pi_k = \arg\max_{\pi'_k} \left\{ \widetilde{U}_k(\theta, (\pi; \pi'_k)) - \widetilde{C}_k(\theta, \mathbf{P}, (\pi; \pi'_k)) \right\}, \tag{3}$$

where $(\pi; \pi'_k)$ is the joint action profile π with one type-k user changing its action from π_k to π'_k, and $\widetilde{U}_k(\theta, (\pi; \pi'_k))$ and $\widetilde{C}_k(\theta, \mathbf{P}, (\pi; \pi'_k))$ are the utility of use and cost of that deviating user, respectively, calculated in [23, Sec. II-B].

Since each user maximizes their own expected utility, the outcome of the users' decision process is naturally the Nash equilibrium of the *plan selection game* defined as

$$\mathcal{G}_{\mathbf{P}} = \left\{ \mathcal{K} = \{1, \ldots, K\}, \{\pi_k\}_{k=1}^{K}, \{U_k - C_k\}_{k=1}^{K} \right\}.$$

Here we put $\mathbf{P}$ in the subscript of $\mathcal{G}$ to emphasize that the plan selection game depends on the pricing policy of the provider. We denote $\pi^{\mathrm{NE}}(\mathbf{P})$ as the Nash equilibrium of $\mathcal{G}_{\mathbf{P}}$.

Proposition 1. *There exists a symmetric Nash equilibrium in the plan selection game* $\mathcal{G}_{\mathbf{P}}$*.*

Proof. The plan selection game $\mathcal{G}_{\mathbf{P}}$ is a finite game; Nash [22], [24] shows that each such game has an Nash equilibrium in which players of the same type choose the same strategy.

3 Problem Formulation

In this section, we formulate the design problem of the service provider as a Stackelberg game. The service provider tries to find a MAC protocol θ and a pricing policy $\mathbf{P}$, so that at the equilibrium of the plan selection game $\mathcal{G}_{\mathbf{P}}$, the social welfare (for the benevolent provider) or the total revenue (for the selfish provider) is maximized, subject to the constraint that costs be covered.

Before doing this, however, we must note that our notion of solution assumes that the service provider knows the arrival rates, service times, and utility functions of all types of users (but does not know the type of a particular user), and foresees the behavior of the users. The users in turn must also know the behavior of other users. Implicitly, therefore, we view the outcome as involving some learning process that is not modeled here. We intend to address this issue in later work, while focusing on characterizing the system performance at the equilibria in this paper.

Under the above assumptions, we can formulate the design problem of the service provider as follows. For a benevolent service provider aiming at maximizing the social welfare, its design problem (PB) can be written as

$$\begin{aligned} \max_{\theta,\mathbf{P}} \quad & \sum_{k=1}^{K} \left(U_k(\theta, \pi^{\mathrm{NE}}(\mathbf{P})) - C_k(\theta, \mathbf{P}, \pi^{\mathrm{NE}}(\mathbf{P}))\right) \cdot N_k \\ s.t. \quad & \mathrm{IR} : \textstyle\sum_{k=1}^{K} C_k(\theta, \mathbf{P}, \pi^{\mathrm{NE}}(\mathbf{P})) \cdot N_k \geq C_0, \end{aligned}$$

where C_0 is the fixed cost for the service provider during a billing period due to the maintenance of the network. The objective function is the social welfare defined as the sum utility of all the users. The constraint is the individual rationality (IR) constraint (or participation constraint) for the service provider. The solution $\mathbf{P}^*$ to the above problem provides the users with a set of pricing plans to choose from. After each user chooses the pricing plan that maximizes its own expected utility, the system reaches the maximum social welfare.

Similarly, for a selfish service provider aiming at maximizing its own revenue, its design problem (PS) can be written as

$$\begin{aligned} \max_{\theta,\mathbf{P}} \; & \textstyle\sum_{k=1}^{K} C_k(\theta, \mathbf{P}, \pi^{\mathrm{NE}}(\mathbf{P})) \cdot N_k \\ s.t. \; & \mathrm{IR} : \textstyle\sum_{k=1}^{K} C_k(\theta, \mathbf{P}, \pi^{\mathrm{NE}}(\mathbf{P})) \cdot N_k \geq C_0. \end{aligned}$$

Here, the only difference between the problem (PB) and (PS) is the objective function.

Because our focus is the influence of technology on the economic layer and system performance, we will first find the optimal pricing policy of the problems (PB) and (PS) with fixed MAC protocol, and then compare the optimal pricing policies and the resulting system performance under different MAC protocols.

4 Two Simple Scenarios

In this section, we study two simple scenarios. In each scenario, there are two types of users: type-1 users are video users with stringent throughput and delay requirements, while type-2 users are data users, who require low throughput and can tolerate large delay. In the first scenario, the service provider uses CSMA and only charges the same subscription fee for all the active users. In the second scenario, the service provider uses TDMA and charges for a per-bit surcharge in addition to the subscription fee.

4.1 CSMA with Subscription Fee Only

The provider using CSMA offers the dummy pricing plan $\mathbf{p}^0 = \phi$ and a single non-dummy pricing plan $\mathbf{p}^1 = (p_s, 0, 0)$. The design problem of the provider can be analyzed using backward induction. In the plan selection game, there can be three types of Nash equilibria depending on the value of $\pi_{k,1}$: $\pi_{k,1} = 0$, $\pi_{k,1} = 1$, or $0 \leq \pi_{k,1} \leq 1$. We can calculate the optimal pricing policy that induces the desired equilibrium, and the corresponding social welfare and provider revenue. The benevolent (selfish) provider compares all the possible equilibria and adopts the subscription fee that induces the NE with the highest social welfare (revenue). In both cases, the constraint is that revenue must cover cost – else the network will not operate at all.

Theorem 1. *Suppose that the service provider uses CSMA and offers the following pricing policy*

$$\mathbf{P} = \left(\mathbf{p}^0 = \phi, \mathbf{p}^1 = (p_s, 0, 0)\right).$$

For the pure Nash equilibria, we show the optimal pricing policies of both providers and the resulting social welfare and provider revenue, as well as the existence conditions for the NE, as follows:

- *Type-1 NE:* $\pi_{1,1} = 1$, $\pi_{2,1} = 1$. *See Table 1.*
- *Type-2 NE:* $\pi_{k,1} = 1$, $\pi_{3-k,1} = 0$. *See Table 2 for the case with* $\pi_{1,1} = 1$ *and* $\pi_{2,1} = 0$. *The case with* $\pi_{1,1} = 0$ *and* $\pi_{2,1} = 1$ *is symmetric.*
- *Type-3 NE:* $\pi_{1,1} = 0$, $\pi_{2,1} = 0$. *This NE is a trivial one that can be achieved by setting the subscription fee high enough.*

Proof. See [23, Appendix A].

Remark 1: In the above theorem, we only characterize the system performance at the pure Nash equilibria, because pure Nash equilibria seem to be a more reasonable outcome in terms of information availability. As we can see from Table 1-2, information on the users' probability distribution over pricing plans π is *not* required for service providers and especially for the users. However, for the mixed Nash equilibrium, providers and users need to know the actions of all

Table 1. CSMA, Type-1 NE: $\pi_{1,1} = 1$, $\pi_{2,1} = 1$, $\mathbf{n} = ([0, N_1], [0, N_2])$, $k^\star = \arg\min_k V_k^1(\theta, \mathbf{n})$

Provider Type	Benevolent	Selfish
Pricing Policy	$p_s = \frac{C_0}{N_1+N_2}$	$p_s = V_{k^\star}^1(\theta, \mathbf{n})$
Social Welfare	$\sum_{i=1}^{2} V_i^1(\theta, \mathbf{n}) \cdot N_i - C_0$	$(V_{3-k^\star}^1(\theta, \mathbf{n}) - V_{k^\star}^1(\theta, \mathbf{n})) \cdot N_{3-k^\star}$
Provider Revenue	C_0	$V_{k^\star}^1(\theta, \mathbf{n}) \cdot (N_1 + N_2)$
Existence Conditions	$V_{k^\star}^1(\theta, \mathbf{n}) \cdot (N_1 + N_2) \geq C_0$	

Table 2. CSMA, Type-2 NE: $\pi_{1,1} = 1$, $\pi_{2,1} = 0$, $\mathbf{n} = ([0, N_1], [N_2, 0])$, $\mathbf{n}' = ([0, N_1], [N_2 - 1, 1])$

Provider Type	Benevolent	Selfish
Pricing Policy	$p_s = \max\left\{\frac{C_0}{N_1}, V_2^1(\theta, ([0, N_1], [N_2 - 1, 1]))\right\}$	$p_s = V_1^1(\theta, \mathbf{n})$
Social Welfare	$V_1^1(\theta, \mathbf{n}) \cdot N_1 - \max\left\{C_0, V_2^1(\theta, \mathbf{n}') \cdot N_1\right\}$	0
Provider Revenue	$\max\left\{C_0, V_2^1(\theta, \mathbf{n}') \cdot N_1\right\}$	$V_1^1(\theta, \mathbf{n}) \cdot N_1$
Existence Conditions	$V_1^1(\theta, \mathbf{n}) \cdot N_1 \geq C_0$, $V_1^1(\theta, \mathbf{n}) > V_2^1(\theta, \mathbf{n}')$	

the users. Take the equilibrium $\pi_{1,1} = 1$, $\pi_{2,1} \in (0, 1)$ for example. In this case, both benevolent and selfish providers should set the subscription fee as

$$p_s = \sum_{n_{2,1}=1}^{N_2} \binom{N_2 - 1}{n_{2,1} - 1} \pi_{2,1}^{n_{2,1}-1} (1 - \pi_{2,1})^{N_2 - n_{2,1}} V_2^1(\theta, \{\mathbf{n}_1, [N_2 - n_{2,1}, n_{2,1}]\}),$$

where $\pi_{2,1}$ is required to compute p_s. The same argument applies to Theorem 2, which only characterizes the pure NE.

Remark 2: As seems obvious, the benevolent provider charges as little as possible, subject to revenue being at least as great as cost; the selfish provider charges as much as possible, subject to the cost to each user being no greater than utility. As the simulations in Sec. IV-C make clear, there are ranges of the user number and demand parameters for which the outcome when the provider is benevolent and the outcome when the provider is selfish do not lead to the usage by the same types.

4.2 TDMA with Subscription Fee and Per-bit Surcharge

Similar to the case with CSMA, we can get the following theorem about the pure equilibria when the service provider uses TDMA and can charge a subscription fee plus a per-bit surcharge.

Theorem 2. *Suppose that the service provider uses TDMA and offers the following pricing policy*

$$\mathbf{P} = \left(\mathbf{p}^0 = \phi, \mathbf{p}^1 = (p_s^1, 0, 0), \mathbf{p}^2 = (p_s^2, q, \beta)\right).$$

For the pure Nash equilibria, we show the optimal social welfare and provider revenue, as well as the existence conditions for the NE, as follows:[3]

- *Type-1 NE:* $\pi_{1,2} = 1$, $\pi_{2,2} = 1$. *See Table 3.*
- *Type-2 NE:* $\pi_{k,2} = 1$, $\pi_{3-k,1} = 1$. *See Table 4 for the case with* $\pi_{1,2} = 1$ *and* $\pi_{2,1} = 1$. *The case with* $\pi_{1,1} = 1$ *and* $\pi_{2,2} = 1$ *is symmetric.*
- *Type-3 NE:* $\pi_{k,2} = 1$, $\pi_{3-k,0} = 1$. *See Table 5 for the case with* $\pi_{1,2} = 1$ *and* $\pi_{2,0} = 1$. *The case with* $\pi_{1,0} = 1$ *and* $\pi_{2,2} = 1$ *is symmetric.*
- *Type-4 NE:* $\pi_{1,0} = 1$, $\pi_{2,0} = 1$. *This NE is a trivial one that can be achieved by setting the subscription fees high enough.*

Table 3. TDMA, Type-1 NE: $\pi_{1,2} = 1$, $\pi_{2,2} = 1$, $\mathbf{n} = ([0,0,N_1],[0,0,N_2])$, $k^\star = \arg\min_k V_k^2(\theta,\mathbf{n})$, $j^\star = \arg\max_j \hat{B}_j^2(\theta,\mathbf{n})$

Provider Type	Benevolent	Selfish
Social Welfare	$\sum_{i=1}^{2} V_i^2(\theta,\mathbf{n}) \cdot N_i - C_0$	$(V_{3-j^\star}^2(\theta,\mathbf{n}) - V_{k^\star}^2(\theta,\mathbf{n})) \cdot N_{3-j^\star}$
Provider Revenue	C_0	$V_{j^\star}^2(\theta,\mathbf{n}) \cdot N_{j^\star} + V_{k^\star}^2(\theta,\mathbf{n}) \cdot N_{3-j^\star}$
Existence Conditions	$\exists\, i : \hat{B}_i^2(\theta,\mathbf{n}) \geq \hat{B}_{3-i}^2(\theta,\mathbf{n})$ and $V_i^2(\theta,\mathbf{n}) \cdot N_i + V_{k^\star}^2(\theta,\mathbf{n}) \cdot N_{3-i} \geq C_0$	

Table 4. TDMA, Type-2 NE: $\pi_{1,2} = 1$, $\pi_{1,1} = 1$, $\mathbf{n} = ([0,0,N_1],[0,N_2,0])$, $\mathbf{n}' = ([0,1,N_1-1],[0,N_2,1])$, $\gamma = \max\left\{0, V_1^1(\theta,\mathbf{n}') - V_2^1(\theta,\mathbf{n})\right\}$

Provider Type	Benevolent	Selfish
Social Welfare	$V_1^2(\theta,\mathbf{n}) \cdot N_1 +$ $V_2^1(\theta,\mathbf{n}) \cdot N_2 - C_0$	$\gamma \cdot N_1$
Provider Revenue	C_0	$\left(V_1^2(\theta,\mathbf{n}) - \gamma\right) \cdot N_1 +$ $V_2^1(\theta,\mathbf{n}) \cdot N_2$
Existence Conditions	$\left(V_1^2(\theta,\mathbf{n}) - \gamma\right) \cdot N_1 + V_2^1(\theta,\mathbf{n}) \cdot N_2 \geq C_0$	

Proof. See [23, Appendix B].

Remark 3: From the above theorem, we can predict the equilibrium point induced by both providers under TDMA. First, if the utility of one type of users alone in the system is higher than the sum utility of two types of users coexisting in the system, both providers will admit only the high-utility users (most likely the video users), resulting in the type-3 scenario. However, the type-3 scenario may not be common under TDMA, because the providers can charge video users for a high surcharge to control their data usage, such that they will

[3] In Table 3-5, $\hat{B}_k^\ell(\theta,\mathbf{n})$ is the expected amount of excessive data usage consumed by a type-k user choosing plan ℓ over a billing period at the steady state; see [23, Eqn. (5)] for the detailed definition and calculation.

Table 5. TDMA, Type-3 NE: $\pi_{1,2} = 1$, $\pi_{2,0} = 1$, $\mathbf{n} = ([0,0,N_1],[N_2,0,0])$, $\mathbf{n}' = ([0,0,N_1],[N_2-1,0,1])$

Provider Type	Benevolent	Selfish
Social Welfare	if $\hat{B}_1^2(\theta,\mathbf{n}) < \hat{B}_2^2(\theta,\mathbf{n}')$: $V_1^2(\theta,\mathbf{n})\cdot N_1 - C_0$; else : $V_1^2(\theta,\mathbf{n})\cdot N_1 - \max\left\{C_0, V_2^2(\theta,\mathbf{n}')\cdot N_1\right\}$.	0
Provider Revenue	0	$V_1^2(\theta,\mathbf{n})\cdot N_1$
Existence Conditions	$V_1^2(\theta,\mathbf{n})\cdot N_1 \geq C_0$, $\{\hat{B}_1^2(\theta,\mathbf{n}) < \hat{B}_2^2(\theta,\mathbf{n}')$ or $(\hat{B}_1^2(\theta,\mathbf{n}) \geq \hat{B}_2^2(\theta,\mathbf{n}')$ and$V_1^2(\theta,\mathbf{n}) \geq V_2^2(\theta,\mathbf{n}'))\}$	

not consume a large amount of data to congest the network. Both type-1 and type-2 scenarios characterize the cases when the providers admit both types of users. For type-1 scenario, both providers set a very high p_s^1 so that no users choose $\mathbf{p}^1$. Then the benevolent provider charges a small p_s^2 and q just to cover the cost, while the selfish one set appropriate p_s^2 and q so that both types of users receive zero total utility. The selfish provider can do that as long as the high-usage users have higher utility of use than the low-usage users. For type-2 scenario, both providers set appropriate plans so that low-usage users choose $\mathbf{p}^2$ and high-usage users choose $\mathbf{p}^1$.

Remark 4: By comparison between the scenarios under CSMA and TDMA, we can see that the feasible region under TDMA becomes larger because the service provider can measure the data usage and charge for the excessive bits used by the users. Intuitively, if the SP can only charge the same subscription fee for all the users, the high-usage users, such as the video users, will have the incentives to use unlimited amount of data, which will congest the network and result in a negative utility for the low-usage users that are online. By imposing the surcharge, the benevolent provider can charge less for the data users and more for the video users so that both types of users have positive utility. The selfish provider can use the surcharge to maximize its own revenue. In particular, if the high-usage users have higher utility of use than the low-usage users do, the selfish provider can gain so much revenue that both types of users get zero utility.

4.3 Numerical Simulation

Now we use numerical simulations to observe more details about the impact of the technology on the system performance. The key parameters in the simulation are described as follows:

- The service provider uses CSMA protocol with constant backoff window of 16ms or TDMA protocol.
- The pricing policy is $\mathbf{P} = (\phi, \mathbf{p}^1 = (p_s, 0, 0))$ for CSMA and $\mathbf{P} = (\phi, \mathbf{p}^1 = (p_s^1, 0, 0), \mathbf{p}^2 = (p_s^2, q, \beta))$ for TDMA.

- The total throughput of the AP is $B = 54\ Mbps$.
- The utility of type-1 users, the video users, is the Peak Signal-to-Noise Ratio (PSNR) of the video sequences. Here we use the *Foreman* video (CIF 15Hz), whose operational utility-rate-delay function is calculated by experiment. The details can be found in [25].
- The utility of type-2 users, the data users, is [8] [9]

$$u_2 = 10 \cdot \log(1 + \tau_2). \tag{4}$$

- The billing period is ΔT = 360 hours/month, namely 12 hours/day times 30 days/month.
- The cost of the service provider is $C_0 = 1000$.

In the simulation, we change the numbers and arrival rates of the users and solve the problem of the benevolent and selfish providers under different tuples of user numbers and arrival rates. The simulation results and the corresponding analysis is as follows.

Numbers of Users. Here we show the phase diagram of the types of users in the system at the equilibrium under different user numbers. The phase diagram here illustrates which type or types of users are admitted to the system, given different numbers of video and data users in the system with other parameters fixed.

Fig. 1 show the phase diagrams with low-demand video users and low-demand data users, low-demand video users and high-demand data users, high-demand video users and low-demand data users, and high-demand video users and high-demand data users, respectively. We can see from the figures that in general, the benevolent provider admits more types of users than the selfish one does, whenever it is possible. The phase diagram under TDMA with both users having low demand is also shown as a representative scenario under TDMA protocol. We omit the TDMA scenarios with other user demands here due to space limitation. More detailed analysis on each scenario is presented as below.

Fig. 1(a) shows that, when the video users and data users both have low demands, the benevolent provider tends to admit both types of users to maximize the social welfare if the numbers of both types of users are large. On the contrary, the selfish one tends to admit video users to give the entire bandwidth to the highly profitable video users and denies access for the data users with low utility.

Fig. 1(b) shows that, when the demand of video users remains low and the data users have higher demand, both providers begin to admit some data users, in addition to video users, to achieve larger social welfare or gain more revenue, since the data users occupy the channels more often and thus have higher total utility now. When the data users significantly outnumber the video users, the selfish provider will admit only the data users.

Fig. 1(c) and Fig. 1(d) show that, when the video users have high demand, the benevolent provider drops all the data users when their demand is low, and tries to admit some data users when their demand is high. This means that the benevolent provider chooses the high-utility video users, when both users have

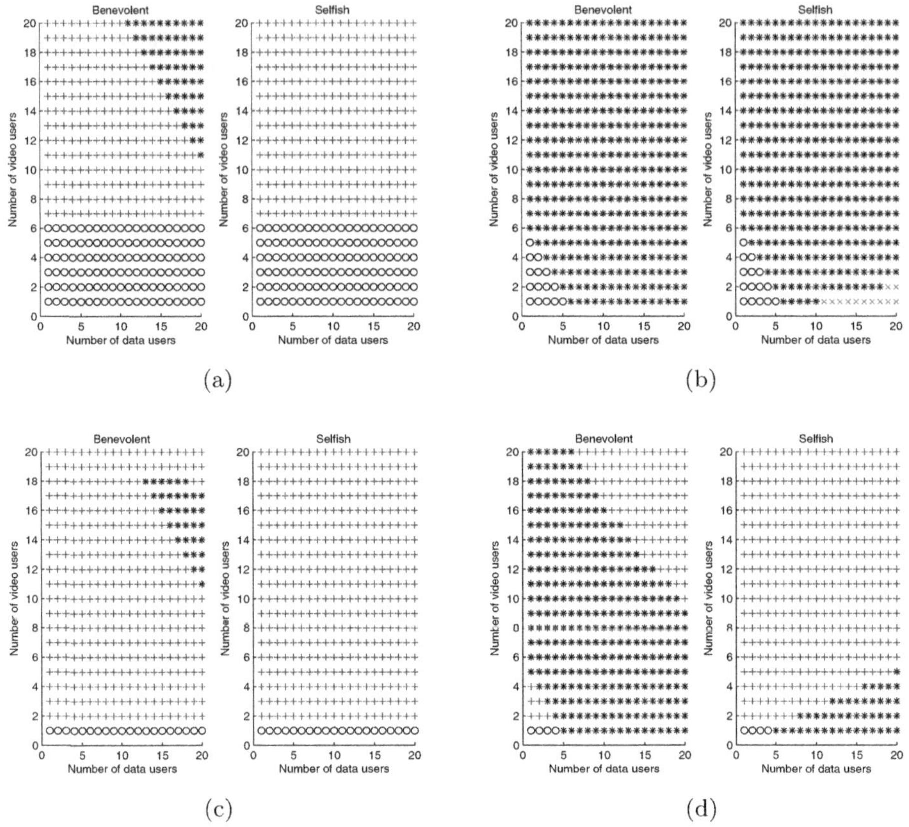

Fig. 1. Phase diagrams of the types of users in the system at the equilibrium under CSMA. 'blue *': both video and data users, 'red +': only video users, 'green ×': only data users, 'black ∘': none. (a): low-demand video users and low-demand data users; (b): low-demand video users and high-demand data users; (c): high-demand video users and low-demand data users; (d): high-demand video users and high-demand data users. Here low demand means $\lambda_1/\mu_1 = 0.1$ and high demand means $\lambda_1/\mu_1 = 1$.

high demands and it has to choose one from the two types to reduce congestion. For the selfish provider, it always tends to drop the data users to allocate the entire bandwidth to the video users to maximize the revenue.

We also show the phase diagram under TDMA protocol with low-demand video and data users in [23, Fig. 3], which we omit here due to space limit. We observe that both providers admit both types of users under most configurations of user numbers: the benevolent provider admits both users to maximize social welfare, while the selfish one admits both users to maximize revenue. Compared to CSMA, TDMA enables both providers to admit both users by setting different plans for different types of users, when the difference between the utility of different users is large. This trend of admitting more users remains the same with other user demands.

Arrival Rates of Users. In Fig. 2, we show the phase diagram on what types of users are in the system at the equilibrium under different arrival rates of users of both types. We fix the number of users of each type at 20. From the figure, we can see that the benevolent provider admits both types of users under a large range of arrival rates. In particular, when the data users have large arrival rates and the video users have medium arrival rates, the relatively low subscription fee set by the benevolent provider draws a large number of users, resulting in low throughput and thus low utility of use of video users. Hence, only data users choose to join the network in the charge of the benevolent provider. On the contrary, the selfish provider sets a high subscription fee to squeeze out the data users, leaving only video users in the system, in order to gain more revenue.

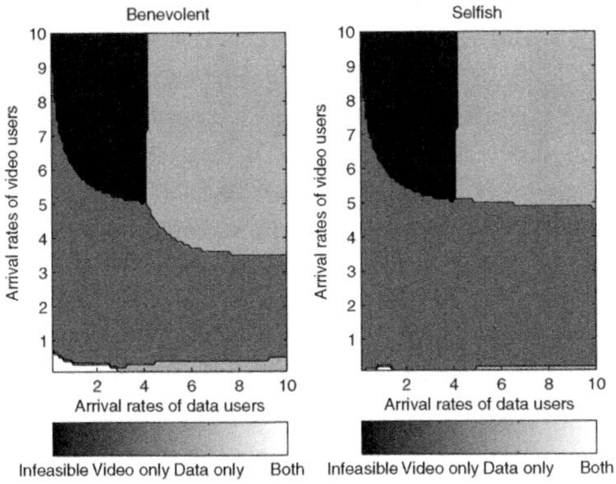

Fig. 2. Phase diagrams of the types of users in the system at the equilibrium with different arrival rates under CSMA. The number of users of each type is 20.

5 Conclusion

In this paper, we studied the provision of a public wireless network by a single (monopolistic) provider who may be either benevolent (seeking to maximize social welfare) or selfish (seeking to maximize provider profit). The paper presented a model for the public wireless network with three interdependent layers, namely the technology layer, the application layer, and the economic layer. Using the proposed model, we analyzed the influence of technology on the economic layer, and more importantly, the interaction of technology and economic layers that determines the feasibility and desirability of the network. We derived the feasibility conditions and the social welfare at the optimal operating points of the benevolent and selfish service providers for the public wireless network under different technologies. By simulation, we characterized different behaviors of a benevolent provider and a selfish provider at their optimal operating points, and the difference social welfare and revenue resulting from the different behaviors. Simulation results also demonstrated that differences in MAC technology can have a significant effect on the system performance. By using TDMA, which enables the providers to monitor the data usage of each user and charge per-bit rate, both the benevolent provider and the selfish provider can exploit the flexibility of differentiated pricing plans in order to maximize social welfare and revenue, respectively.

References

1. Palomar, D.P., Chiang, M.: A tutorial on decomposition methods for network utility maximization. IEEE J. Sel. Areas Commun. 24(8), 1439–1451 (2006)
2. Kelly, F.P.: Charging and rate control for elastic traffic. Eur. Trans. TeleCommn. 8, 33–37 (1997)
3. Kelly, F.P., Maulloo, A.K., Tan, D.K.H.: Rate control for communication networks: Shadow prices, proportional fairness and stability. J. Oper. Res. Soc. 49, 237–252 (1998)
4. Gibbens, R.J., Kelly, F.P.: Resource pricing and the evolution of congestion control. Automatica 35(12), 1969–1985 (1999)
5. Johari, R., Tsitsiklis, J.N.: Efficiency loss in a network resource allocation game. Math. Operations Research 29(3), 407–435 (2004)
6. Johari, R., Tsitsiklis, J.N.: Efficiency of scalar-parameterized mechanisms. Operations Research 57(4), 823–839 (2009)
7. Basar, T., Srikant, R.: Revenue-maximizing pricing and capacity expansion in a many-users regime. In: Proceedings IEEE INFOCOM 2002, pp. 1556–1563 (2002)
8. Shen, H., Basar, T.: Differentiated Internet pricing using a hierarchical network game model. In: Proc. 2004 American Control Conference, pp. 2322–2327 (2004)
9. Shen, H., Basar, T.: Optimal nonlinear pricing for a monopolistic network service provider with complete and incomplete information. IEEE J. Select. Areas Commun. 25, 1216–1223 (2007)
10. Saraydar, C.U., Mandayam, N.B., Goodman, D.J.: Efficient power control via pricing in wireless data networks. IEEE Trans. on Communications 50, 291–303 (2002)
11. Alpcan, T., Basar, T.: A hybrid noncooperative game model for wireless communications. In: Advances in Dynamic Games: Applications to Economics, Finance, Optimization, and Stochastic Control. Annals of Dynamic Games, vol. 9. Birkhauser (2006)
12. Paschalidis, I.C., Tsitsiklis, J.N.: Congestion-dependent pricing of network services. IEEE/ACM Trans. Networking 8(2), 171–184 (2000)
13. Friedman, E., Parkes, D.: Pricing WiFi at Starbucks - Issues in online mechanism design. Working Paper (2002), `http://www.eecs.harvard.edu/~parkes/pubs/online.pdf`
14. Musacchio, J., Walrand, J.: WiFi access point pricing as a dyanmic game. IEEE/ACM Trans. Networking 14(2), 289–301 (2006)
15. Ren, S., Park, J., van der Schaar, M.: User subscription dynamics and revenue maximization in communication markets. To appear in Infocom 2011 (2011)
16. Kasbekar, G., Sarkar, S.: Spectrum pricing games with bandwidth uncertainty and spatial reuse in cognitive radio networks. In: Proceedings of ACM MOBIHOC 2010, September 20-24 (2010)
17. Sirbua, M., Lehr, W., Gillett, S.: Evolving wireless access technologies for municipal broadband. Government Information Quarterly 23, 480–502 (2006)
18. IEEE 802.11b: Wireless LAN Medium Access Control (MAC) and Physical layer (PHY) Specifications, IEEE Standard (1999)

19. Draft Supplement to Part 11: WIreless Medium Access Control (MAC) and Physical Layer (PHY) Specifications: Medium Access Control (MAC) Enhancements for Quality of Service (QoS), IEEE 802.11e/D10.0 (November 2004)
20. Tembine, H., Altman, E., El-Azouzi, R., Hayel, Y.: Evolutionary games in wireless networks. IEEE Transactions on Systems, Man, and Cybernetics, Part B: Cybernetics 40(3), 634–646 (2009)
21. Ross, K.W., Tsang, D.: The stochastic knapsack problem. IEEE Trans. on Commun. 37(7), 740–747 (1989)
22. Mas-Colell, A., Whinston, M., Green, J.: Microeconomic Theory. Oxford Univ. Press, Oxford (1995)
23. Xiao, Y., Zame, W.R., van der Schaar, M.: Technology choices and pricing policies in public and private wireless networks, `http://arxiv.org/abs/1011.3580`
24. Nash, J.F.: Non-cooperative games. The Annals of Mathematics 54(2), 286–295 (1951)
25. van der Schaar, M., Andreopoulos, Y., Hu, Z.: Optimized scalable video streaming over IEEE 802.11 a/e HCCA wireless networks under delay constraints. IEEE Trans. Mobile Comput. 5(6), 755–768 (2006)

Network Regulations and Market Entry*

Galina Schwartz[1], John Musacchio[2], Mark Felegyhazi[3], and Jean C. Walrand[1]

[1] University of California, Berkeley, Cory Hall, Berkeley, CA 94720
{schwartz,wlr}@eecs.berkeley.edu
[2] University of California, Santa Cruz, 1156 High Street, Santa Cruz, CA 95064
johnm@soe.ucsc.edu
[3] Budapest Univ. of Tech. and Econ.
mfelegyhazi@crysys.hu

Abstract. This paper uses a two-sided market model to study if last-mile access providers (ISPs), should charge content providers (CPs), who derive revenue from advertisers, for the right to access ISP's end-users. We compare two-sided pricing (ISPs could charge CPs for content delivery) with one-sided pricing (neutrality regulations prohibit such charges). Our analysis indicates that number of CPs is lower, and the number of ISPs often higher, with two- rather than one-sided pricing. From our results the superiority of one regime over the other depends on parameters of advertising rates, end-user demand, CPs' and ISPs' costs, and relative importance of their investments. Thus, caution should be taken in designing neutrality regulations.

Keywords: network neutrality, two-sided markets, market entry.

1 Introduction

Today, an Internet Service Provider (ISP) charges only end-users, who subscribe to that ISP for Internet access, and content providers (CPs) connected to the Internet directly via that ISP. That is, each ISP generally charges only CPs who buy access from it. One of the focal questions in the network neutrality policy debate is whether current ISPs' charging practices should continue and be mandated by law, or ISPs ought to be allowed charging CPs for the delivery of content to the end-users. This question is part of the larger debate on network neutrality, which includes diverse issues such as whether service differentiation should be allowed, or whether charges for content constitute an impingement of freedom of speech [9,2].

In our past work [6] as well as here, we use a two-sided market model of interactions between ISPs, end-users, and CPs, with the ISPs playing the role of a "platform" intermediating the two sides: CPs and end-users. We model a "non-neutral" network as a market with two-sided pricing, meaning that each ISP charges his end-users and also charges CPs for delivering traffic to his end-users.

* This research is supported by NSF grants CNS-0910695, CNS-0953884, and CNS-0910711.

R. Jain and R. Kannan (Eds.): GameNets 2011, LNICST 75, pp. 108–123, 2012.

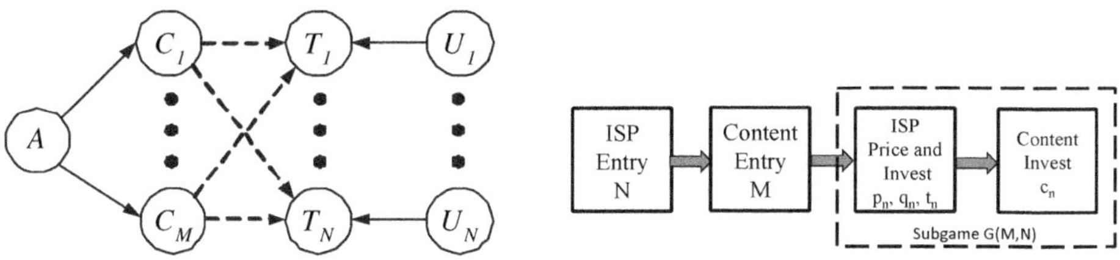

Fig. 1. Left panel: The direction of payments. Dotted lines reflect additional payments of the two-sided case. T_i and U_i are ISPs (transit providers) and each ISP's group of users, respectively. Right panel: The timing of the game G. The subgame $G(M, N)$ starts after M and N are chosen.

Conversely, a "neutral" network is modeled as a one-sided pricing regime in which an ISP is allowed to charge only CPs that buy their Internet access from it. We normalize this charge to zero, which allows us to model a neutral network as a one-sided regime in which each platform (ISP) charges his end-users only. After providers of both types choose to enter the market and sink irreversible entry costs, the ISPs simultaneously and independently choose their investments, end-user prices, and content-provider prices. Then, CPs choose their investments, simultaneously and independently. Our end-users demand for content variety has a flavor of the classical monopolistic competition model [3].

We explore how CP and ISP entry and investments differ with pricing regime. In [6], the numbers of ISPs and CPs are fixed, and we compare the social welfare in two-sided and one-sided pricing cases. In [6], we find that superior welfare regime depends on two key parameters: (i) the ratio of advertising rate to an end-user price sensitivity parameter, and (ii) the number of ISPs. Roughly, when (i) is extreme (large or small), a two-sided (non-neutral) pricing is welfare superior, but when (i) is mid-range, a one-sided (neutral) pricing is superior. An intuitive explanation for this is that when either the ISP or CPs have a much stronger ability to obtain revenue, a two-sided pricing which essentially allows more flexible revenue sharing between provider types, which allows to achieve a higher equilibrium welfare. Also, the parameter range for which a one-sided market is superior increases with the number of ISPs. An explanation for this is that with two-sided pricing, each ISP charges the CPs, and collectively the ISPs may "over-charge". This effect grows with the number of ISPs.

In this paper, we consider the entry decisions of potential ISPs and CPs, thus endogenizing the number of providers of each type. Since the results of [6] depend roughly on whether the ISP or CP market is more profitable in some sense, allowing more providers to enter a market that is highly profitable could alter the situation. Here we find that the two-sided (non-neutral) pricing is indeed welfare superior for parameters similar (and likely wider) than in [6] (industry structure is fixed). In contrast with [6], where Nash equilibrium symmetry is assumed, in this paper we prove the symmetry. We establish the existence of an equilibrium in the entry game in which potential ISPs and CPs make their irreversible entry decisions.

Our model is based on the ideas of two-sided markets, and there is a considerable literature on the subject. (See surveys [10] and [1].) Other researchers applied the ideas of two-sided markets to study network neutrality. Hermalin and Katz [4] model network neutrality as a restriction on the product space, and consider whether ISPs should be allowed to offer more than one grade of service. Hogendorn [5] studies two-sided markets where intermediaries sit between "conduits" and CPs. In his context, net-neutrality means that content has open access to conduits where an "open access" regime affords open access to the intermediaries. Njoroge at al. [7] and [8], consider two-sided market model with heterogeneous CPs and end-users, and the ISPs play the role of a platform. In [8], they find that social welfare is higher in the non-neutral regime. Work [12] discusses policy issues related to two-sided markets.

The paper is organized as follows. Section 2 presents a model that permits us to quantify the effects of network regime on player incentives to enter the network industry and invest. In Section 3, we analyze the two-sided pricing (non-neutral) one-sided pricing (neutral) regimes. In section 4 we discuss our findings and conclude. To ease the exposition, the proofs are relegated to the appendix, and to save space, not all the proofs are included; the complete technical details available in our technical report, see [11].

2 Model

The internet consists of local ISPs (i.e, residential ISPs who provide last-mile access for end users), and transit ISPs that mainly serve the internet backbone. We assume that the CPs connect to the internet via a transit ISP, whereas end-users are attached to a local ISP. Our assumption reflects that under current practices, the local ISPs do not charge CPs. Thus, in further analysis we abandon the distinction between local ISPs and transit ISPs, and focus on local ISPs only, which from now on we simply call ISPs.

First we consider the subgame $G(M,N)$ after M CPs and N ISPs after they entered the market. Later, we consider the game G which includes the stages in which providers decide to enter the market.

Fig. 1 illustrates our setting. Each ISP T_n charges his attached end-users U_n ($n = 1, \ldots, N$) an access price p_n per click; each ISP has a monopoly over its end-user base U_n. Thus, to reflect the market power of local ISPs, the end-users are divided between the ISPs, with each ISP having $1/N$ of the entire market. Each ISP T_n charges each CP C_m an amount equal to q_n per click. CP C_m invests c_m and ISP T_n invests t_n.

We normalize the CPs' access payment (for network attachment) to zero due to the ISPs' competition. To reflect the actual payment, we assume that the CP's access payment is subtracted from their charges to advertisers. Let B_n denote demand of end-users U_n; the B_n depends on p_n and provider investments:

$$B_n = \{\mu\eta(c_1^v + \cdots + c_M^v)t_n^w\}\, e^{-p_n/\theta}, \text{where } \mu = \frac{(1-e^{-kM})}{M^{1-v}}, \eta = \frac{1}{N^{1-w}}. \quad (1)$$

Here, $\theta > 0$, and $v, w \geq 0$ with $v + w < 1$, and $k \in (0, \infty)$. For a given network quality (the expression in the curly brackets), the end-user demand exponentially decreases with p_n; see [6] for the intuition about the normalization factor $\eta = 1/N^{1-w}$. The term μ is similar except that the factor $(1-e^{-kM})$ reflects that the end-users prefer a higher M, i.e. more variety in content. As M increases the "love for variety" diminishes. Let R_{mn} denote the end-user demand U_n to C_m, and D_m – the total demand for C_m:

$$R_{mn} = \frac{c_m^v}{c_1^v + \cdots + c_M^v} B_n, \text{ and } D_m = \sum_n R_{mn}. \tag{2}$$

Now we consider the game G, which includes the entry decisions of potential providers. The order of play is as follows. First, potential ISPs decide whether to enter; if they enter, the entry cost is t_e. Second, content providers observe the number of ISP entrants, and then, sink their entry costs c_e. Third, the ISPs announce end-user prices p_n (in the one-sided pricing case) and also content provider charges q_n in the two-sided case. Fourth, content providers choose their investments c_m, and ISPs – their investments t_n. Each CP's objective is to maximize its profit Π_{Cm} which is equal to the revenues net of its investment.

$$\Pi_{Cm} = \sum_{n=1}^{N} (a - q_n) R_{mn} - \beta c_m - c_e. \tag{3}$$

Here a is the amount that advertisers pay to CPs per unit of end-user demand; $\beta > 1$ is the outside option (alternative use of funds), and c_e – the CP's entry cost. Each ISP objective is to maximize profit Π_{Tn}:

$$\Pi_{Tn} = (p_n + q_n) B_n - \alpha t_n - t_e. \tag{4}$$

Where $\alpha > 1$ and t_e are respectively the ISP's outside option and entry cost.

3 Analysis

3.1 Analysis of the Subgame $G(M, N)$

Let $\Pi_C(M,N)$ and $\Pi_T(M,N)$ denote profits for each CP and ISP in the game $G(M,N)$. To compare one-sided and two-sided pricing (neutral and non-neutral networks), we make the following assumptions.

(a) One-sided pricing (neutral network): First, each T_n chooses (t_n, p_n). Here $q_n = 0$. Then, each C_m chooses c_m.
(b) Two-sided pricing (non-neutral network): First, each T_n chooses (t_n, p_n, q_n). Then, each C_m chooses c_m.

Two-Sided Pricing. In $G(M,N)$, in a network with two-sided pricing, each ISP chooses (t_n, p_n, q_n) and each CP chooses c_m. For a given (t_n, p_n, q_n), the optimal c_m maximizes (3). From the first order conditions,

$$\beta c_m^{1-v} = v\mu\eta \sum_n (a - q_n) t_n^w e^{-p_n/\theta} =: \beta c^{1-v}. \tag{5}$$

For that value of c_m, we find that:

$$\Pi_{Tn}(M,N) = M\mu\eta(q_n + p_n)t_n^w e^{-p_n/\theta}\left[\frac{v\mu\eta}{\beta}\sum_{k=1}^{N}(a - q_k)e^{-p_k/\theta}t_k^w\right]^{\frac{v}{(1-v)}} - \alpha t_n. \tag{6}$$

The n-th ISP chooses (t_n, p_n, q_n) that maximize (6). By analyzing IPSs best response functions we find:

Proposition 1. *With the two-sided pricing, in all $G(M,N)$ equilibria $t_n = t, p_n = p, q_n = q$ and $c_m = c$.*

The proof of Proposition 1 works in the following way. First, from the ISPs' FOCs wrt p_n reveal that $p_n = \theta - a$ for any n in equilibrium – thus equilibrium user prices are identical. Next, from the ISPs' FOCs wrt q_n one can infer that if $q_i \geq q_j$ it must be that $t_i \leq t_j$. But from the ISPs' FOCs wrt t_n we infer that if $q_i \geq q_j$ we must have $t_i \geq t_j$. This is possible only if $q_i = q_j$ and $t_i = t_j$. Thus, only a symmetric equilibrium could exist, and we demonstrate its existence by construction. We find the symmetric equilibrium by construction. It has the following form:

$$p_n = p^\ddagger = \theta - a, \text{ and } q_n = q^\ddagger = a - \theta\pi; \tag{7}$$

$$t_n = t^\ddagger = \left[(x^\ddagger)^{1-v} \cdot (y^\ddagger)^v \cdot e^{-(\theta-a)/\theta}\right]^{\frac{1}{(1-w-v)}}; \tag{8}$$

$$c_m = c^\ddagger = \left[(x^\ddagger)^w \cdot (y^\ddagger)^{1-w} \cdot e^{-(\theta-a)/\theta}\right]^{\frac{1}{(1-v-w)}} \times [\mu\eta N]^{\frac{1}{1-v}}; \tag{9}$$

$$\text{where } x^\ddagger = M\,(\mu\eta)^{\frac{1}{1-v}} \cdot \left(\frac{\theta w}{\alpha}\right) N^{\frac{v}{1-v}},\ y^\ddagger = \frac{\theta v}{\beta}\pi, \text{ and } \pi = \frac{v}{N(1-v)+v}. \tag{10}$$

From (7) - (10) and Proposition 1, the equilibrium uniqueness follows immediately. Proposition 1 is proved in the appendix.

One-Sided Pricing. The game $G(M,N)$ with one-sided pricing is similar to one with two-sided pricing, except that $q_n = 0$ for all n. Given a $\{q_n = 0, p_n, t_n\}$, the one finds that the CPs' best responses are identical and satisfy:

$$\beta c_m^{1-v} = v\mu\eta[\sum_k a t_k^w e^{-p_k/\theta}] =: \beta c^{1-v}.$$

Substituting into (4) we find that

$$\Pi_{Tn}(M,N) = M\mu\eta p_n t_n^w e^{-p_n/\theta}\left[\frac{v\mu\eta a}{\beta} \times \sum_{k=1}^{N} e^{-p_k/\theta}t_k^w\right]^{\frac{v}{(1-v)}} - \alpha t_n. \tag{11}$$

The ISP T_n chooses (t_n, p_n) that maximizes the above. Analysis of the above payoff function leads to the following result.

Proposition 2. *With one-sided pricing, in all $G(M, N)$ equilibria $t_n = t, p_n = p, q_n = 0$ and $c_m = c$.*

The intuition of the proof of Proposition 2 is similar to that of Proposition 1. We use Proposition 2 to construct a unique symmetric equilibrium from the FOCs of (11). We find:

$$p_n = p^\dagger = \theta(1 - \pi), \text{ and } q_n = q^\dagger = 0; \tag{12}$$

$$t_n = t^\dagger = \left[(x^\dagger)^{1-v}(y^\dagger)^v e^{-p^\dagger/\theta}\right]^{\frac{1}{1-v-w}}; \tag{13}$$

$$c_m = c^\dagger = \left[(x^\dagger)^w (y^\dagger)^{1-w} e^{-p^\dagger/\theta}\right]^{\frac{1}{1-v-w}} \cdot [\mu\eta N]^{\frac{1}{1-v}}; \tag{14}$$

$$\text{where } x^\dagger := x^\ddagger \text{ and } y^\dagger := \frac{av}{\beta}. \tag{15}$$

From (12) - (15) and Proposition 2, the equilibrium is unique. The proof of Proposition 2 is omitted because of space limitations, but it is similar to the proof of Proposition 1 provided in the appendix.

3.2 The Entry Game G

Since the equilibrium of $G(M, N)$ exists and is unique and symmetric, in any equilibrium in which (M, N) CPs and ISPs respectively enter, a necessary condition for equilibrium is that

$$\Pi_{Cm}(M, N) \geq c_e, \text{ and } \Pi_{Cm}(M + 1, N) < c_e.$$

if $M > 0$ otherwise $\Pi_{Cm}(1, N) < c_e$ if $M = 0$. Suppose that there is a unique $M(N)$ that satisfies the above for each N. (We show this is indeed true in our proof of Proposition 3 below.) Since the potential CPs get to observe the number N of ISPs that enter, another necessary condition[1] for equilibrium is that

$$\Pi_{Tn}(M(N), N) \geq t_e, \text{ and } \Pi_{Tn}(M(N + 1), N + 1) < t_e$$

if $N > 0$ otherwise $\Pi_{Tn}(M(1), 1) < t_e$ if $N = 0$. Together these conditions are necessary and sufficient for (M, N) to be an equilibrium of the game G. These conditions lead to the following propositions.

Proposition 3. *The equilibrium of the game G exists and is unique.*

Proposition 4. *Consider a game $\tilde{G}$, in which CPs and ISPs enter simultaneously rather than sequentially. Then, a pure strategy Nash equilibrium in $\tilde{G}$, provided it exists, coincides with the equilibrium of G.*

From Proposition 4, when a pure strategy Nash equilibrium in the game $\tilde{G}$ exists, it is unique. The proofs of both propositions are found in the appendix.

[1] If $M(N)$ were instead a set valued function there would have to exist elements of the sets $M(N)$ and $M(N+1)$ satisfying the inequality relation to support an equilibrium with N ISPs.

3.3 User Welfare and Social Welfare

We compute the consumer surplus (aka the end-user welfare) by taking the integral of the demand function from the equilibrium price to infinity. This yields

$$W_U(M,N) = NM^{=}\mu\eta\theta \cdot [(x)^w(y)^v]^{1/(1-w-v)}\, e^{-p/[\theta(1-w-v)]}.$$

3.4 Numerical Analysis

In this section, we numerically analyze some examples to illustrate the behavior of the model. We begin by studying the profits of CPs and ISPs in the post-entry game $G(M,N)$ as a function of M and N.. Fig. 2 shows the profits of CPs and ISPs for the two-sided (non-neutral) and one-sided (neutral) case. In this example we have chosen k to be large so that $(1-e^{-kM}) \approx 1$. Recall that this factor was included to model a preference among users for a larger number of CPs, so making k large is equivalent to removing this effect. Panel (a) of Fig. 2 shows that the CPs' profits decrease in both M and N in the two-sided case, while the lower left panel shows that the dependence of profits on N The "X" marks on the figure indicate the maximum number M that can enter for a given N and still generate positive profits for the CPs. Panels (b) and (e) of Fig. 2 shows that the transit provider profits decrease in N, but are independent of M, both for the one- and two-sided cases. Note this independence does not hold when k is not taken to be large. Panels (c) and (f) show the lines of the maximum M for CPs to be profitable given N ("C-line") and maximum N for ISPs to be profitable given M ("T-line"). The X on each of these graphs indicates the sequential equilibrium $(M^*(N^*), N^*)$. In this example the sequential equilibrium occurs at the intersection of the C-line and T-line. (In cases for which the "T-line" has M dependence – k not large – the sequential equilibrium need not occur at the intersection of the C-line and T-line.)

Fig. 3 studies how equilibrium consumer surplus and social welfare are effected by the transit entry cost t_e (upper plots) and advertising rate a (lower plots). The upper left plot shows that as t_e increases, the number of ISPs drops, though the drop is discontinuous because the number of providers is an integer. For the two-sided case, the social welfare decreases with the number of ISPs. Thus, when there are more ISPs trying to charge the CPs, they collectively "over-charge" the ISPs which in-turn discourages CP investment and reduces social welfare. In the one-sided case, we still observe that social welfare increases as t_e increases and the equilibrium number of ISPs decreases. An explanation for this is that the equilibrium consumer price p in the one-sided case increases as the ISP market becomes more fragmented – in other words there are more ISPs serving smaller and smaller subscriber bases. (See formula for π and $p^\dagger$). Thus as fewer ISPs enter, consumer prices go down, more usage (clicks) can occur, and thus more CPs enter.

In the lower plots of Fig. 3, we see that as a increases, the social welfare increases in the one-sided case. A higher a induces more CPs to enter the market. Generally, the social welfare in the two-sided case increases as well, except in

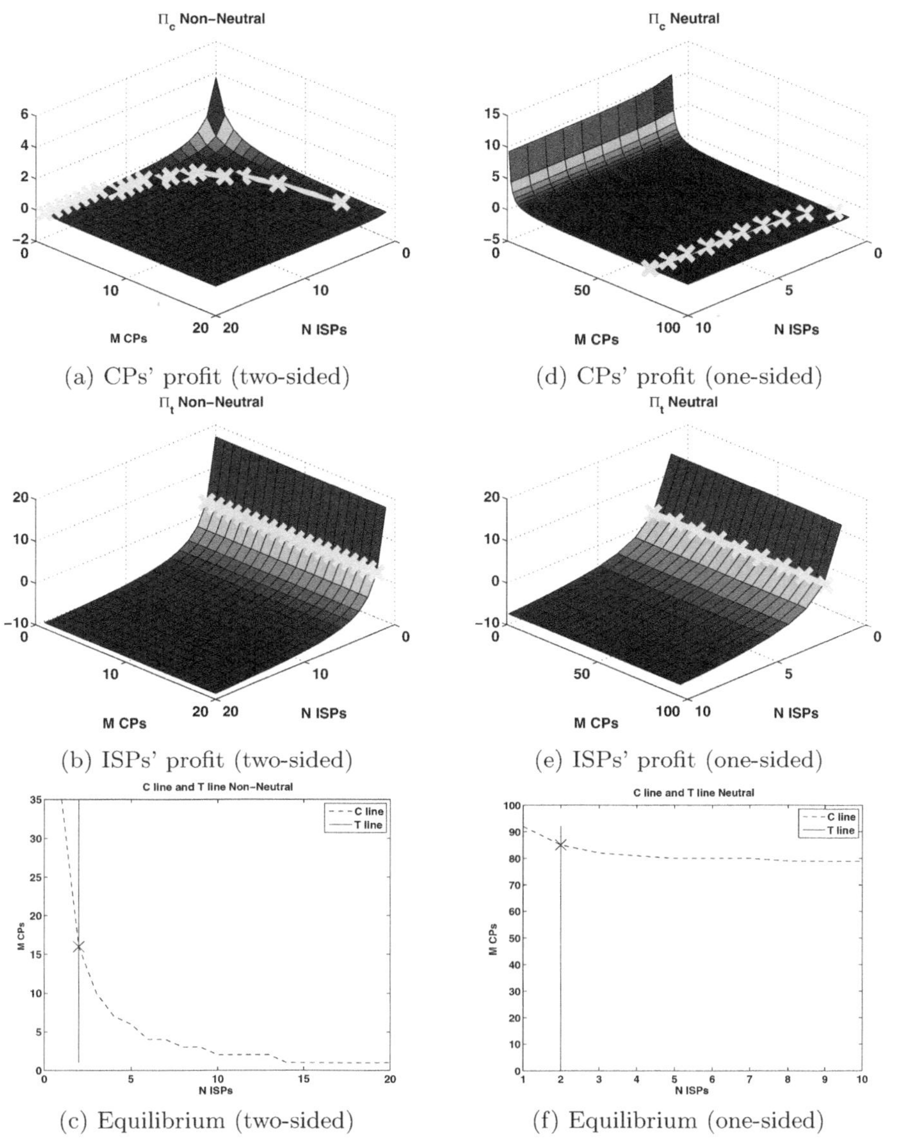

(a) CPs' profit (two-sided) (d) CPs' profit (one-sided)
(b) ISPs' profit (two-sided) (e) ISPs' profit (one-sided)
(c) Equilibrium (two-sided) (f) Equilibrium (one-sided)

Fig. 2. Profits in the two-sided market as a function of M and N. Parameters are as shown in Table 1.

"steps" where the number of equilibrium ISPs increases by 1. We also see that, roughly, the two-sided regime becomes social welfare superior to the one-sided regime when a is high. An explanation is that a high a creates more revenue potential and a two-sided market permits some of that revenue to be shared with ISPs so that they see an incentive to increase investment.

Table 1. Baseline parameters; k is large, thus $(1 - e^{-kM}) \approx 1$ for any $M \geq 1$

Parameter	v	w	a	θ	t_e	c_e	α	β	k
Value	0.1	0.3	15	50	8	0.1	1.2	1.2	"large"

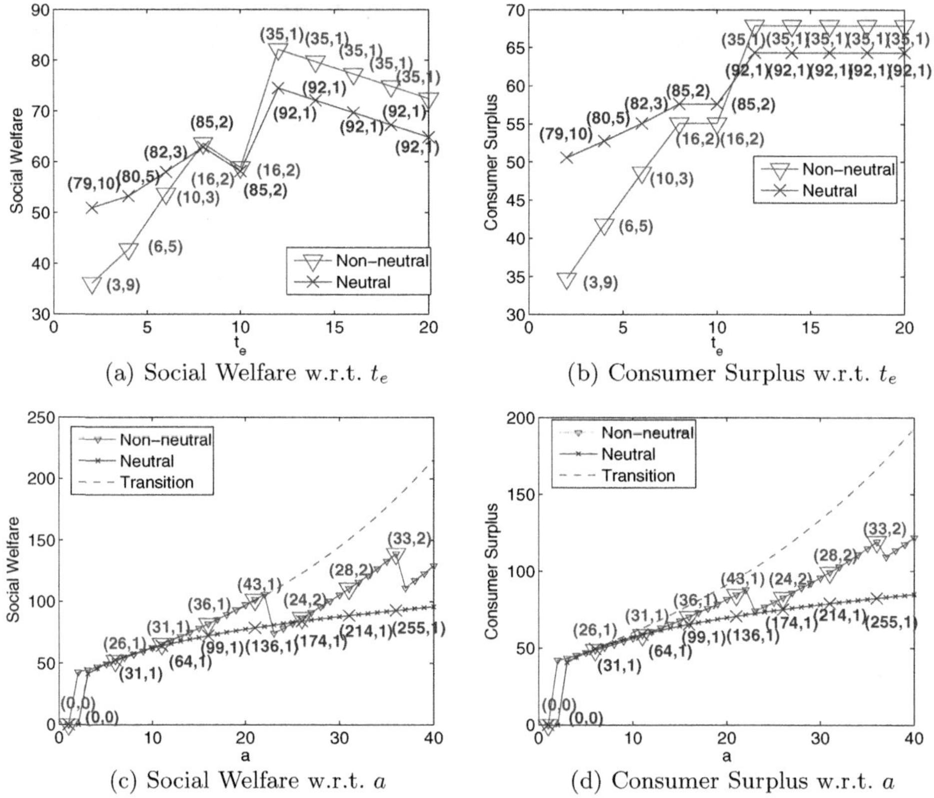

Fig. 3. Upper Row: The social welfare and consumer surplus as a function of t_e. Lower Row: The consumer surplus as a function of a. Non-neutral (two-sided) and neutral (one-sided) regimes are shown in both cases. A "transition" regime for which the equilibrium numbers of providers in a neutral (one-sided) regime are introduced to a (two-sided) regime before choosing prices and investments is also shown in the plots. Parameters other than the parameter being varied in each plot are as shown in Table 1. The number pairs in figure show the equilibrium number of CPs and ISPs.

Panels (a) and (b) of Fig. 4 depict the social welfare with respect to the CP entry cost c_e for two different values of $t_e = \{5, 15\}$. The figures show that social welfare decreases with c_e in both cases. The welfare superiority of neutral vs. non-neutral (one- vs. two-sided) is not changed significantly by changing c_e – neutral is better for $t_e = 5$ and non-neutral is better for $t_e = 15$. The last panel

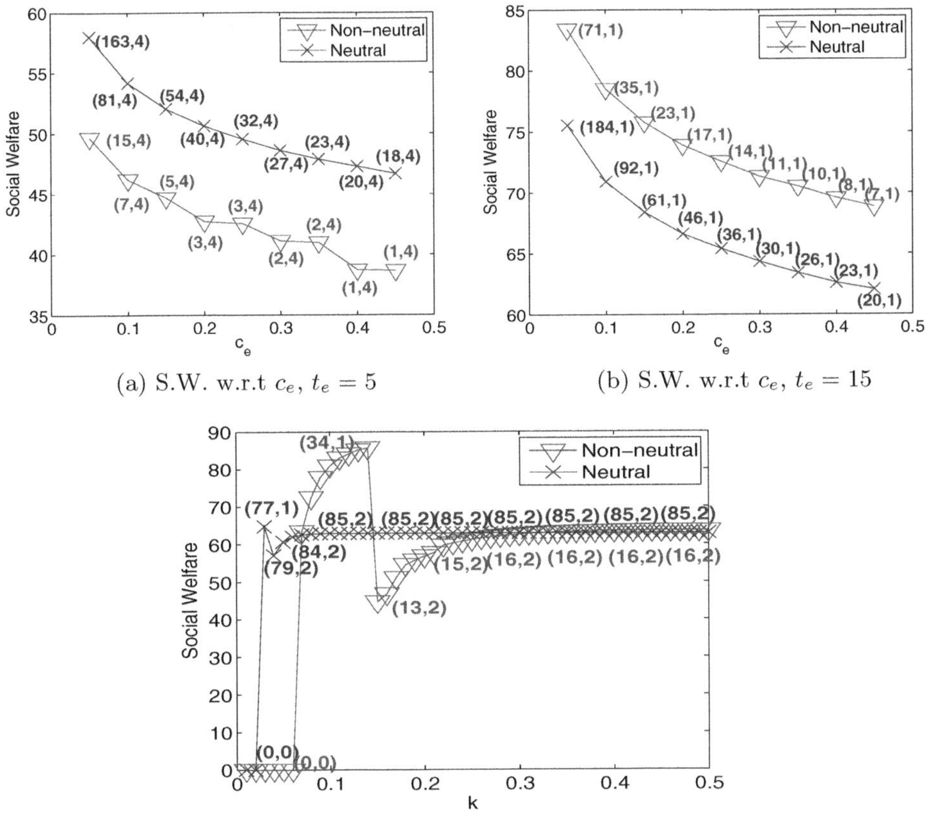

(a) S.W. w.r.t c_e, $t_e = 5$

(b) S.W. w.r.t c_e, $t_e = 15$

(c) S.W. w.r.t. k

Fig. 4. Panels (a) and (b): The social welfare and consumer surplus as a function of c_e for $t_e = 5$ and $t_e = 15$ respectively. Panel (c): The social welfare as k varies while $c_e = 0.1$ and $t_e = 8$.

of Fig. 4 illustrates how social welfare changes with respect to k. In the non-neutral case, for small k, no providers enter the market in equilibrium. When k passes a threshold, 1 ISP enters. As k rises further, a 2nd ISP enters and social welfare drops sharply, but then increases. Social welfare with respect to k in the neutral case seems to fluctuate less.

4 Conclusions

Our results suggest that welfare superior regime depends on the following key parameters. As in [6], we observe that a larger number of ISPs tend to reduce social welfare in the two-sided case. An explanation is that if a large number of ISPs try to extract revenue from each CP, the IPSs collective charges on the CPs exceed the socially optimal ones. This effect strengthens with a higher number of ISPs. Thus for low ISP entry costs t_e, we observe a significantly lower social welfare in the two-sided case.

A higher advertising rate a appears to favor the two-sided (non-neutral) case. An explanation for this is that if a large portion of the network's revenue is realized by advertising, allowing the ISPs to capture some of this revenue improves their incentive to invest, which leads to a higher social welfare. The effects of increased a and the effect of an increased number of ISPs in the two-sided case can interact in an interesting way. The lower panels of Fig. 3 show that the welfare of a two-sided market roughly increases as a increases, but it also exhibits step decreases every time the improved profitability of the ISP market permits an extra ISP entrant. Changes in the content provider entry cost c_e effect social welfare in both the one- and two-sided regimes, but it seems to have much less effect on the relative superiority of the two regimes than the parameters t_e or a.

Our results suggest that regulatory authorities should be cautious of restricting the pricing in the internet, and special attention ought to be paid to: the relative ability of CPs and ISPs to earn revenue (a in our model), the concentration of the *ISP* market (N), and the entry cost of ISPs.

References

1. Armstrong, M.: Competition in two sided markets. RAND Journal of Economics 37(3), 668–691 (2006)
2. Chong, R.: The 31 flavors of the net neutrality debate: Beware the Trojan horse. Advanced Communications Law and Policy Institute, Scholarship Series, New York Law School (December 2007)
3. Dixit, A., Stiglitz, J.: Monopolistic competition and optimum product diversity. The American Economic Review 67(3), 297–308 (1977)
4. Hermalin, B., Katz, M.: The economics of product-line restrictions with an application to the network neutrality controversy. Information Economics and Policy 19, 215–248 (2007)
5. Hogendorn, C.: Broadband internet: Net neutrality versus open access. International Economics and Economic Policy 4, 185–208 (2007)
6. Musacchio, J., Schwartz, G., Walrand, J.: A two-sided market analysis of provider investment incentives with an application to the net neutrality issue. Review of Network Economics 8(1), 22–39 (2009)
7. Njoroge, P., Ozdaglar, A., Stier, N., Weintraub, G.: Competition, market coverage, and quality choice in interconnected platforms. In: Proceedings of NetEcon Workshop, Stanford, CA (July 2009)
8. Njoroge, P., Ozdaglar, A.E., Stier-Moses, N.E., Weintraub, G.Y.: Investment in Two Sided Markets and the Net Neutrality Debate, SSRN eLibrary (2010)
9. Odlyzko, A.: Network neutrality, search neutrality, and the never-ending conflict between efficiency and fairness in markets. Review of Network Economics 8, 40–60 (2009)
10. Rochet, J.C., Tirole, J.: Two-sided markets: A progress report. RAND Journal of Economics 37(3), 655–667 (2006)

11. Schwartz, G., Musacchio, J., Felegyhazi, M., Walrand, J.: Network Regulations and Market Entry, EECS Department, University of California, Berkeley, UCB/EECS-2011-16 (March 2011), http://www.eecs.berkeley.edu/Pubs/TechRpts/2011/EECS-2011-16.html
12. Weiser, P.: Report from the center for the new west putting network neutrality in perspective. Center for the New West Discussion Paper (January 2007), http://www.centerfornewwest.org/pdf/TelecomSummary.pdf

5 Appendix

5.1 Analysis of the Two-Sided Case

Consider the game $G(M, N)$ (fixed N and M). We use Π_{Cm} (given by (3)), combined with R_{mn} defined by (2) and B_n is given by (1) to find:

$$\Pi_{Cm}(M,N) = \mu\eta c_m^v \sum_n (a - q_n) t_n^w e^{-p_n/\theta} - \beta c_m.$$

For a given (p_n, q_n, t_n), the c_m that maximizes the above satisfies

$$c_m = c = \left(\frac{v\mu\eta}{\beta} \sum_n (a - q_n) t_n^w e^{-p_n/\theta} \right)^{\frac{1}{(1-v)}}. \tag{16}$$

Substituting the above, $\Pi_{Tn}(M, N)$ assuming optimal content investment is

$$\Pi_{Tn}(M,N) = M\mu\eta(q_n + p_n) t_n^w e^{-p_n/\theta} \times \left[\frac{v\mu\eta}{\beta} \sum_{k=1}^{N} (a - q_k) e^{-p_k/\theta} t_k^w \right]^{\frac{v}{(1-v)}} - \alpha t_n. \tag{17}$$

If the derivative of (17) w.r.t. q_n is zero, then

$$A^{\frac{v}{(1-v)}} - (q_n + p_n) \frac{v}{1-v} A^{\frac{v}{(1-v)}-1} t_n^w e^{-p_n/\theta} = 0,$$

where $A := \sum_k (a - q_k) e^{-p_k/\theta} t_k^w$. Solving the above for A we have,

$$A = (q_n + p_n) \frac{v}{1-v} t_n^w e^{-p_n/\theta}. \tag{18}$$

The two aforementioned expressions for A imply

$$(q_n + p_n) \frac{v}{1-v} t_n^w e^{-p_n/\theta} = \sum_k (a - q_k) e^{-p_k/\theta} t_k^w, \tag{19}$$

and thus $(q_1 + p) t_1^w = (q_2 + p) t_2^w$. Thus $t_1 \leq t_2$ if $q_1 \geq q_2$.

Setting the derivative of (17) w.r.t. p_n is zero yields

$$e^{-p_n/\theta} A^{\frac{v}{(1-v)}} - \frac{1}{\theta}(q_n + p_n) e^{-p_n/\theta} A^{\frac{v}{(1-v)}} - \\ (q_n + p_n) e^{-p_n/\theta} \frac{v}{1-v} A^{\frac{v}{(1-v)}-1} \frac{1}{\theta} (a - q_n) t_n^w e^{-p_n/\theta} = 0.$$

Substituting relation (18) for A, multiplying the resulting expression by θ and dividing it by $e^{-p_n/\theta}A^{\frac{v}{(1-v)}}$, we find $p_n = \theta - a$, which is (7). Thus in equilibrium, all ISPs choose identical user prices.

Taking the derivative of (17) w.r.t t_n and setting it to zero, we find that

$$F \cdot \frac{1}{1-v} t_n^{\frac{wv}{(1-v)}} w t_n^{w-1} = \alpha, \tag{20}$$

$$\text{where } F = M\mu\eta \left[(q_n + p_n)e^{-p_n/\theta}\right]^{\frac{1}{(1-v)}} \left(\frac{v\mu\eta}{\beta}\frac{v}{1-v}\right)^{\frac{v}{(1-v)}}.$$

We are now ready to prove, Proposition 1, that the equilibrium of $G(M,N)$ is symmetric in the two-sided case.

Proof (Proposition 1). Suppose the proposition were not true and $q_1 > q_2$. Recall that from the ISPs' FOCs wrt q_n we have $(q_1 + p)t_1^w = (q_2 + p)t_2^w$. Therefore

$$t_1 < t_2 \text{ when } q_1 > q_2. \tag{21}$$

From the ISPs' FOCs wrt t_n (we raise (20) to the power $(1-v)$ and rearrange):

$$(q_1 + p)t_1^{wv} t_1^{(w-1)(1-v)} - (q_2 + p)t_2^{wv} t_2^{(w-1)(1-v)} = H,$$

$$\text{where } H = \left[\frac{\alpha}{Mw\mu\eta}\right]^{(1-v)} \left(\frac{\beta}{v\mu\eta}\frac{1-v}{v}\right)^{v} e^{p^{\ddagger}/\theta}. \tag{22}$$

Thus $(q_1+p)t_1^{w+v-1} = (q_2+p)t_2^{w+v-1}$, from which we must have $t_1 \geq t_2$ when $q_1 \geq q_2$. But from (21) we have $t_1 \leq t_2$ when $q_1 \geq q_2$, which is a contradiction, unless $t_1 = t_2$ and $q_1 = q_2$. Thus Proposition (1) is proven.

Having shown the equilibrium is unique, we turn to deriving expressions for the equilibrium. From (19) and the fact that $p_n = \theta - a$ we obtain

$$(q_n + \theta - a)\frac{v}{1-v} t_n^w - (a - q_n)t_n^w = \sum_{k \neq n} (a - q_k)t_k^w.$$

Thus, $t_n^w \left[\theta v + q_n - a\right] = (1-v)\sum_{k \neq n}(a - q_k)t_k^w$. Writing this for $n = 1, \dots N$ and summing we find $\sum_n t_n^w \left[\theta v + (q_n - a)(N(1-v) + v)\right] = 0$. Combining this with $q_n = q$ and $t_n = t$ we have $q^{\ddagger}$ as given in (7).

To find the optimal ISP investment $t = t^{\ddagger}$, we use $q_n = q$ and $t_n = t$ to express $\Pi_{Tn}(M,N)$ as $\Pi_{Tn}(M,N) = E \cdot t^w \cdot [Nt^w]^{\frac{v}{(1-v)}} - \alpha t_n$, where E is defined by

$$E^{1-v} = M^{1-v}(\mu\eta)(p+q)^{1-v} e^{-p/\theta} \left(\frac{v(a-q)}{\beta}\right)^{v}. \tag{23}$$

Writing the partial derivative of $\Pi_{Tn}(M,N)$ wrt t_n and equating it to zero, we find:

$$E \cdot \left[w t_n^{w-1} [t_1^w + \cdots + t_N^w]^{\frac{v}{(1-v)}} + t_n^w \frac{v}{1-v} w t_n^{w-1} [t_1^w + \cdots + t_N^w]^{\frac{v}{(1-v)}-1} \right] - \alpha = 0,$$

the symmetric solution $t_n = t$ is:

$$E \cdot \left[wt^{w-1}[Nt^w]^{\frac{v}{(1-v)}} + t^w \frac{v}{1-v} wt^{w-1}[Nt^w]^{\frac{v}{(1-v)}-1} \right] - \alpha = 0,$$

which simplifies to

$$E \cdot \left(N + \frac{v}{(1-v)} \right) wt^w t^{w-1} (Nt^w)^{\frac{v}{(1-v)}-1} = \alpha. \tag{24}$$

Lastly, we substitute our expressions for $p^{\ddagger}$ and $q^{\ddagger}$ into (24) to obtain

$$\begin{aligned} E^{1-v} &= (\mu\eta) \cdot \left[\frac{\theta N(1-v)}{N(1-v)+v} \right]^{1-v} e^{\frac{-(\theta-a)}{\theta}} \cdot \left(\frac{v}{\beta} \frac{\theta v}{N(1-v)+v} \right)^v \\ &= (\mu\eta)\, e^{\frac{-(\theta-a)}{\theta}} \cdot y^v \cdot \left[\frac{\theta N(1-v)}{N(1-v)+v} \right]^{1-v}. \end{aligned}$$

Then, we combine with (24) to find (8) with $x^{\ddagger}$and $y^{\ddagger}$ as defined in (10). Thus, we demonstrated that only a symmetric equilibrium with $p_n = p$, $q_n = q$, $t_n = t$ and $c_m = c$ exists, and it is unique (by construction). Finally, to calculate $c^{\ddagger}$, we substitute our expressions for equilibrium q and t into (16) to find (9).

5.2 Equilibrium Uniqueness in *G*

Before proceeding, we define some notation. Let π and δ be defined as:

$$\pi := \frac{v}{N(1-v)+v} \text{ and } \delta := \frac{a}{\theta}. \tag{25}$$

We assume that with today's Internet parameters, end user prices $p^{\dagger}$ are positive[2], which gives $p^{\dagger} = \theta(1-\pi) > 0$ and $p^{\ddagger} = \theta - a > 0$, and $q^{\ddagger} = a - \theta \frac{v}{N(1-v)+v} = a - \theta\pi$, from which $\pi < 1$, $\delta < 1$ and $\pi < \delta$. Also we infer $\pi < v$ and decreases with N In addition, from the expressions of the equilibrium parameters, one can show that the average (per each provider) returns in the equilibrium of the game $G(M, N)$ are:

$$\frac{\Pi_C(M,N)}{c} = \frac{\beta(1-v)}{v} \text{ and } \frac{\Pi_T(M,N)}{t} = \frac{\alpha}{w}[1 - w - \pi]. \tag{26}$$

We are ready to prove Proposition 3.

Proof (Proposition 3). Consider subgame $G(M, N)$. Since $\pi < \delta$ and $y^{\ddagger} = y^{\dagger}\frac{\pi}{\delta}$ we have $y^{\ddagger} < y^{\dagger}$. Using these equations, we obtain:

$$t^{\ddagger} = \frac{1}{N} \left[\pi^v \left(\frac{w}{\alpha} \right)^{1-v} \cdot \left(\frac{v}{\beta}\right)^v \cdot \theta e^{\frac{-(\theta-a)}{\theta}} \cdot \left[1 - e^{-kM}\right] \right]^{\frac{1}{(1-w-v)}},$$

[2] Despite widely available free internet access, average end-user access price is clearly strictly positive.

$$c^{\ddagger} = \frac{1}{M}\left[\pi^{1-w}\cdot\left(\frac{w}{\alpha}\right)^{w}\cdot(\frac{v}{\beta})^{1-w}\cdot\theta e^{\frac{-(\theta-a)}{\theta}}\cdot\left[1-e^{-kM}\right]\right]^{\frac{1}{(1-v-w)}}, \tag{27}$$

$$t^{\dagger} = \frac{1}{N}\left[\left(\frac{\theta w}{\alpha}\right)^{1-v}\cdot(\frac{av}{\beta})^{v}\cdot e^{\pi-1}\left[1-e^{-kM}\right]\right]^{\frac{1}{(1-w-v)}},$$

$$c^{\dagger} = \frac{1}{M}\left[\left(\frac{\theta w}{\alpha}\right)^{w}\cdot(\frac{av}{\beta})^{1-w}e^{\pi-1}\left[1-e^{-kM}\right]\right]^{\frac{1}{(1-v-w)}}, \tag{28}$$

and

$$t^{\ddagger} = \left(\frac{\beta}{\alpha}\frac{w}{v}\right)\cdot\pi^{-1}\frac{M}{N}c^{\ddagger};\quad t^{\dagger} = \frac{\theta}{a}\left(\frac{\beta}{\alpha}\frac{w}{v}\right)\cdot\frac{M}{N}c^{\dagger}.$$

Expression (28) and (27) have derivatives with respect to M that either transitions from being positive to being negative for one M, or is always negative. This property combined with (26) gives us that for any fixed N there exists unique $M(N)$ s.t.

$$\Pi_C(M(N),N) \geq c_e \text{ and } \Pi_C(M(N)+1,N) < c_e,$$

for $M(N) > 0$ or $\Pi_C(1,N) < c_e$ in the $M(N) = 0$ case. Moreover if $\tilde{M} > M(N)$, $\Pi_C(\tilde{M},N)$ decreases with $\tilde{M}$.

Next we claim $\frac{dM(N)}{dN} \leq 0$ by the following reasoning. Assume the reverse and let $N_1 < N_2$, and $M(N_1) < M(N_2)$. Then, since (28), (27) and (26) show that content provider profits decrease w.r.t. N for fixed M we have

$$\Pi_C(M(N_2),N_1) \geq \Pi_C(M(N_2),N_2) \geq c_e.$$

This contradicts the fact that $\Pi_C(\tilde{M},N)$ decreases with $\tilde{M}$ for any $\tilde{M} > M(N_1)$, and thus $\frac{dM(N)}{dN} \leq 0$ is proven. Lastly, we notice that in both regimes:

$$\frac{d\Pi_T(M(N),N))}{dN} = \frac{\alpha}{w}\frac{d\{[1-w-\pi(N)]t(M(N),N))\}}{dN} < 0,$$

which can be shown by differentiation. (We elaborate on this below.) The uniqueness and existence of the equilibrium follows immediately. Thus, we have proven Proposition 3.

To verify that $\frac{d\Pi_T(M(N),N))}{dN}$ is non-positive, we use $\frac{d\pi}{dN} = -\frac{(1-v)}{v}\pi^2$. In the two-sided case:

$$\frac{d\Pi_T(M(N),N))}{dN} = \left[1-e^{-kM}\right]^{\frac{1}{(1-w-v)}}\frac{\partial}{\partial N}\left\{\frac{1}{N}\left[1-w-\pi\right]\pi^{\frac{v}{1-w-v}}\right\}$$
$$+\left\{\frac{1}{N}\left[1-w-\pi\right]\pi^{\frac{v}{1-w-v}}\right\}\times\frac{1}{(1-w-v)}\left[1-e^{-kM}\right]^{\frac{1}{(1-w-v)}-1}k\underbrace{\frac{dM}{dN}}_{[-]},$$

where the [−] notation is a reminder that $\frac{dM}{dN}$ is nonpositive. The second curly brackets term is positive. Expanding the first curly brackets term we get

$$\frac{\partial}{\partial N}\left\{\frac{1}{N}\left[1-w-\pi\right]\pi^{\frac{v}{1-w-v}}\right\} = -\frac{1}{N^2}\left[1-w-\pi\right]\pi^{\frac{v}{1-w-v}} + \left\{\frac{1}{N}\left[1-w-\pi\right]\right\}\frac{v}{1-w-v}\pi^{\frac{v}{1-w-v}-1}\frac{d\pi}{dN} - \frac{1}{N}\pi^{\frac{v}{1-w-v}}\frac{d\pi}{dN}$$

Collecting terms, this becomes

$$-\frac{1}{N^2}\left[1-w-\pi\right]\pi^{\frac{v}{1-w-v}} + \left\{\frac{(1-v)}{v}\frac{1}{N}\pi^{\frac{1-w}{1-w-v}}\right\}\left\{-\frac{v\left[1-w-\pi\right]}{1-w-v}+\pi\right\} < 0,$$

because from $\pi < v < 1$ the last curly bracket is negative.

The derivations to show $\frac{d\Pi_T(M(N),N))}{dN}$ is negative for the one-sided case are similar so we omit them.

Proof (Proposition 4). Suppose there exists a pure strategy equilibrium of $\tilde{G}$, and let $(\tilde{M}^*, \tilde{N}^*)$ denote the respective equilibrium numbers of CPs and ISPs. Let (M^*, N^*) denote the respective number of CPs and ISPs in the unique equilibrium of the game G. Then it must be that $\tilde{M}^* = M(\tilde{N}^*)$, where $M(\cdot)$ is the function described in the proof of Prop. 3. The equilibrium actions and payoffs in the subgame $\tilde{G}(\tilde{M}^*, \tilde{N}^*)$ are the same as in subgame $G(\tilde{M}^*, \tilde{N}^*)$ since the games are identical after the entry stage and the subgame admits a unique equilibrium. From the proof of Prop. 3, per ISP profit $\Pi_T(M(N), N)$ decreases with N, and for any $\tilde{N}^* > N^*$, we have $\Pi_T(M(\tilde{N}^*), \tilde{N}^*) < t_e$, and thus, $\tilde{N}^* > N^*$ cannot occur as an equilibrium of the game $\tilde{G}$. If $\tilde{N}^* < N^* - 1$ another ISP will enter because the entering ISP sees profit of

$$\Pi_T(M(\tilde{N}^*), \tilde{N}^* + 1) \geq \Pi_T(M(\tilde{N}^* + 1), \tilde{N}^* + 1) > t_e$$

since $M(\cdot)$ and $\Pi_T(\cdot, N)$ are monotone. Finally, suppose $\tilde{N}^* = N^* - 1$. Then it must be that $\Pi_T(M(\tilde{N}^*), \tilde{N}^*+1) = \Pi_T(M(N^*-1), N^*) < t_e$ or else another ISP would have entered in the game $\tilde{G}$. But, $t_e \leq \Pi_T(M(N^*), N^*) \leq \Pi_T(M(N^* - 1), N^*)$ which is a contradiction.

Bandwidth Exchange for Fair Secondary Coexistence in TV White Space

Dan Zhang and Narayan Mandayam

WINLAB, Rutgers University
671 US-1 South, North Brunswick, NJ 08902
{bacholic,narayan}@winlab.rutgers.edu

Abstract. The recent ruling by the FCC has mandated a geo-location database approach to regulate the coexistence of primary and secondary users in TV white space. However, mechanisms for secondary coexistence have been left unspecified and prompts the design and study of the same. In this paper, we propose a mechanism to incentivize fair and efficient secondary user cooperation. Specifically, we assume the existence of cognitive radio equipped secondary users capable of OFDMA with the ability to dynamically exchange subcarriers among themselves. Based on a previously proposed incentive mechanism called Bandwidth Exchange (BE), we further leverage the capability of the mandatory geo-location database to enable fast negotiation between potential cooperation partners to realize a Nash Bargaining Solution (NBS) for secondary coexistence. We illustrate how this mechanism works when secondary users in an OFDMA access network form cooperation through BE for which the NBS is calculated based on information obtained from the database.

Keywords: Nash bargaining, White Space, bandwidth exchange, geo-location database, OFDMA.

1 Introduction

On Sept. 23, 2010, FCC released Second Memorandum Opinion and Order [1] which announced the official advent of TV White Space. Two classes of users are defined to operate in white space: the primary users such as TV stations and licensed wireless microphones; the secondary users that include many different white space devices (WSD) conforming to a number of rules to avoid interfering primary users. In the new ruling, plenty of protections and precautions are imposed to guarantee undisturbed operation of primary users. Though spectrum sensing is no longer required, a geo-location database that registers the locations of primary users has become a mandate. Every secondary user is required to query the database through internet to make sure it would not produce interference to nearby primary users before it starts transmission. The query is periodic in case some primary users want to initiate operation in the vicinity. These requirements provide a reliable shield between primary users and secondary users for the purpose of their coexistence. However, the new ruling

R. Jain and R. Kannan (Eds.): GameNets 2011, LNICST 75, pp. 124–136, 2012.

does not explicitly designate how coexistence is to be managed among the secondary users. This includes, but not limited to, how to control the interference between secondary users and how to incentivize cooperation among them. The latter issue is of critical importance since the usefulness of TV white space can only be realized if secondary users can coexist to form networks for carrying information.

Among other cooperation forms, cooperative forwarding [2] [3] is an essential technique to enhance connectivity and throughput. As forwarding usually incurs some sort of cost, incentive must be implemented in a certain way. In [4] the authors proposed an incentive mechanism called Bandwidth Exchange (BE) which essentially enables a user to delegate a portion of its bandwidth in exchange for forwarding. While some advantages of BE was outlined, such as its ability to find an efficient and proportionally fair joint cooperation strategy, the lack of central management forces each user to estimate the necessary bargaining [5] parameters through a lengthy process. With such slow estimation, bargaining can only be carried out pairwise while ignoring the fact that existence of other users may affect the bargaining power. This leads to approximation even for pairwise bargaining. In TV white space, with the presence of a geo-location database, it is possible to obtain the bargaining parameters using a reliable geometric channel model, so as to shorten or avoid the estimation process. This idea will form the basis for the study in this paper. We will discuss the database-assisted BE–NBS algorithm as an extension to the work presented in [4] for the specific scenario of white space.

While in [4] BE is introduced by exchanging generic bands of frequency, in this paper we assume an OFDMA system where users execute BE by exchanging (possibly noncontiguous) OFDM subcarriers for cooperation.

2 System Model and Bandwidth Exchange

2.1 System Model

Suppose we have M white space users (labeled $1, 2, \ldots, M$) transmitting to an AP (labeled as 0) using OFDMA. Each user is assigned N consecutive subcarriers. The OFDMA system is only generic thus we assume transmit power is evenly allocated to all subcarriers and only one modulation scheme is allowed for each of them. Each user has a minimum required rate $R_i^{\min}$ which with the generic OFDMA assumption translates into a minimum number of usable subcarriers. Let $L(d)$ denote the path loss (including fading) a subcarrier experiences in a slot, where d is the transmission distance. A subcarrier is usable if and only if $L(d) < L_{\text{th}}$ for some threshold L_{th}. The path loss is by assumption a random variable which are independent across subcarriers as well as across slots, yet it is identically distributed only across subcarriers of the same user. The identical distribution assumption is justified by noticing that for any frequency dependent channel model, the statistics vary little over a few Mega Hertz, which is the amount of bandwidth presumably allocated to a user in the target system.

Our objective is to maximize the average throughput from any node i to the AP, possibly by means of cooperative forwarding.

In what follows, subscript ij always stands for the link or request from i to j.

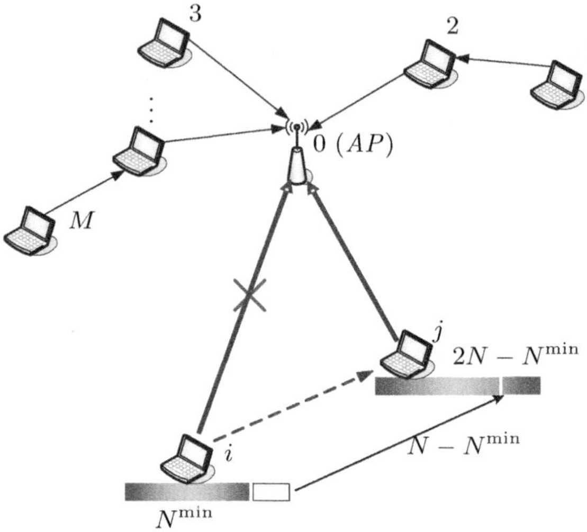

Fig. 1. When the direct link fails, user i tries to incentivise forwarding by delegating $N - N^{\mathrm{min}}$ subcarriers to user j

2.2 Bandwidth Exchange

At the beginning of every slot, user i first attempts to transmit directly to the AP, with X_{i0} ($\leq N$) usable subcarriers. If $X_{i0} < N^{\mathrm{min}}$, it broadcasts a cooperation request to its neighbors, expecting one of them to forward its data to AP, by means of BE. Specifically, it involves the following steps:

1. User i offers to delegate $N - N^{\mathrm{min}}$ subcarriers to user j as long as the remaining N^{min} subcarriers are usable.
2. With the offered subcarriers, j examines the number of usable subcarriers available to it and determine if a request is supportable. A request is considered supportable in two cases: (i) if initially j's direct link is dead, i.e., $X_{j0} < N^{\mathrm{min}}$, but with the added subcarrier, the direct link becomes alive, i.e., $X'_{j0} \geq N^{\mathrm{min}}$; (ii) if initially j's direct link is alive $X_{j0} \geq N^{\mathrm{min}}$, with the added subcarriers the direct link should be at least as good as to support both i and j, i.e., $X'_{j0} \geq 2N^{\mathrm{min}}$.
3. If the request is not supportable, the request is rejected; otherwise j chooses to cooperate with a probability P^{c}_{ij}. We assume there is no flow splitting and every forwarder serves at most one source.
4. If i receives multiple acknowledgements, it randomly picks one to follow in this slot.

This procedure is repeated for each slot.

3 Database Assisted Nash Bargaining for Bandwidth Exchange

We use NBS to determine P_{ij}^{c} in the course of BE. To simplify, we restrict ourselves to pairwise bargaining that regards the other users as (opportunistic) communication resource rather than bargaining participants. This is desired because literal N-user bargaining entails searching for the solution in a strategy space that scales exponentially with the number of users.

From the perspective of NBS, each slot corresponds to a stage game for user i and j is a supportable request is sent from i to j or vice versa. The potential forwarder j must decide whether the request can be granted, which depends on a number of factors:

- the probability P_{ij} of a supportable request from i to j and vice versa;
- the expected utilities available to both source and forwarder contingent on its decisions.

When request granted, the expected utility for source i is always $N^{\min}$, no matter how many positive acknowledgement i receives at last. When request rejected, the expected utility for forwarder j is always 0, which follows from an assumption that we will make shortly. We then denote the expected utility for j when request granted as u_{ij}^{f}, the expected utility for i when request rejected as u_{ij}^{s}. These notations are shown in Table 1.

Table 1. Expected utilities for source i and forwarder j if request supportable

	utility for i	utility for j
request granted	$N^{\min}$	u_{ij}^{f}
request rejected	u_{ij}^{s}	0

The pairwise NBS as presented in [4] follows the same methodology. However, it has several obvious drawbacks. First, the bargaining parameters such as P_{ij} and u_{ij}^{f} can only be estimated over time. This limits the applicability of BE to, at best, stationary or slowly moving users. If P_{ij} is very small, estimation can even fail in practice. Second, bargaining is restricted to pairwise, totally ignoring the effect the other users might have on the bargaining. For example, $u_{ij}^{\mathrm{s}} = 0$ in [4] even though i may still get cooperation from other users. While this significantly reduces the computation burden compared to the exact N-user NBS, it leads to incomplete consideration of bargaining power of different users as they interact. Now that a geo-location database is added and WSD's are required to consult it before and during operation, we have a better way to implement BE that partially alleviate the two issues associated with the original BE implementation. Specifically, we may resort to a geometric channel model possibly customized for the local transmission environment that enables a user to calculate the interesting bargaining parameters by itself. For this approach to be practical, it is necessary that WSD's engaging in cooperation based on BE register their (and the AP's) locations in the database. It would be more desirable that the registry contains additional information such as the frequency band a user is assigned to. The additional information makes applicable more sophisticated channel models such as those that are frequency dependent. At this point, FCC has not completely decided what information should/could be registered in the database. It is conceivable that a good deal of additional information other than the locations of primary users will be eventually allowed or incorporated in the database. Moreover, if the location information of secondary user is updated sufficiently frequently, BE as described here has a better chance to further support mobility.

As we focus on pairwise bargaining for the sake of its simplicity, the effect of existence of other users can be accounted for as bargain parameters instead of bargain participants. Suppose i is a requester and j is a potential forwarder. Intuitively, this implies even if j refuses to cooperate there is still a chance for i to get cooperation from other forwarders. Likewise, even if j agrees to forward for i, it is possible that i secures cooperation from another user, thus j's positive decision may bring it nothing in return.

However, the issue is complicated by the fact that for any potential forwarder, there could be multiple source users competing for its cooperation. To simplify, we notice that for a practical WSD, the outage probability of its direct link should be fairly low. Besides, the outage probabilities of different WSD's should be independent. This means the chance that user i needs to compete for cooperation is fairly low. Based on this approximate assumption, we also neglect the probability that a potential forwarder receives multiple request or two users send request to each other in a slot. This also explains when the request from source i is rejected, the expected utility for forwarder j is always 0.

3.1 Using Database to Obtain Bargaining Parameters

We demonstrate how to use the database and a channel model to calculate the pairwise request probability and the utility obtained by the forwarder if cooperation forms. These parameters do not depend on the existence and number of other users in the system.

First we note as the random path loss $L_i(d)$ is i.i.d. across the subcarriers of a user i, probability $q_d^i = P(L_i(d) < L_{\text{th}})$ is the probability that any subcarrier of i is usable. Given a number X of these subcarriers, the probability k of them are usable is of Binomial distribution, whose cumulative distribution function is

$$F(k, q_d^i, X) = \sum_{\ell=0}^{k} \binom{X}{k} (q_d^i)^k (1 - q_d^i)^{X-k}, \tag{1}$$

and the probability mass function is

$$\begin{aligned} P(k, q_d^i, X) &= \Delta F(k, q_d^i, X) \\ &= F(k, q_d^i, X) - F(k-1, q_d^i, X). \end{aligned} \tag{2}$$

This helps us write out various probabilities. For example, the probability that the direct link of i fails is given by $1 - F(N^{\min}, q_{i0}^{i}, N)$.

Define two disjoint events

$$A_{ij} = X'_{j0} \geq 2N^{\min} \wedge X_{j0} \geq N^{\min}, \tag{3}$$
$$B_{ij} = X'_{j0} \geq N^{\min} \wedge X_{j0} < N^{\min}. \tag{4}$$

We may calculate

$$\begin{aligned}
&P(A_{ij}) \qquad (5)\\
=&P(X_{j0} \geq N^{\min} \wedge \Delta X_{j0} + X_{j0} \geq 2N^{\min})\\
=&\sum_{k=N^{\min}}^{N} P(X_{j0} = k) \sum_{\ell=2N^{\min}-k}^{N-N^{\min}} P(\Delta X_{j0} = \ell)\\
=&\sum_{k=N^{\min}}^{N} \Delta F(N, q_{j0}^{j}, k)(1 - F(2N^{\min} - 1 - k, q_{j0}^{i}, N - N^{\min})),\\
&P(B_{ij}) \qquad (6)\\
=&P(X_{j0} < N^{\min} \wedge \Delta X_{j0} + X_{j0} \geq N^{\min})\\
=&\sum_{k=0}^{N^{\min}-1} P(X_{j0} = k) \sum_{\ell N^{\min}-k}^{N-N^{\min}} P(\Delta X_{j0} = \ell)\\
=&\sum_{k=0}^{N^{\min}-1} \Delta F(N, q_{j0}^{j}, k)(1 - F(N^{\min} - k - 1, q_{j0}^{i}, N - N^{\min})).
\end{aligned}$$

We then have

$$\begin{aligned}
P_{ij} &= P(X_{i0} < N_i^{\min} \wedge X_{ij} \geq N_i^{\min} \wedge (A_{ij} \vee B_{ij}))\\
&= P(X_{i0} < N_i^{\min})P(X_{ij} \geq N_i^{\min})(P(A_{ij}) + P(B_{ij}))\\
&= F(N_i^{\min} - 1, q_{i0}^{i}, N)\\
&\cdot (1 - F(N_i^{\min} - 1, q_{ij}^{i}, N))(P(A_{ij}) + P(B_{ij})).
\end{aligned} \tag{7}$$

Next we calculate v_{ij}^{f}, defined as the utility obtained by j if cooperation forms between i and j, First note

$$v_{ij}^{\mathrm{f}} = E[\Delta X_{ij} | A_{ij} \wedge B_{ij}] - N^{\min}, \tag{8}$$

then we calculate

$$E[\Delta X_{ij} | A_{ij} \wedge B_{ij}] = \frac{P(A_{ij})E[\Delta X_{ij}|A_{ij}] + P(B_{ij})E[\Delta X_{ij}|B_{ij}]}{P(A_{ij}) + P(B_{ij})}, \tag{9}$$

with

$$\begin{aligned}
&P(A_{ij})E[\Delta X_{ij}|A_{ij}] \\
&= \sum_{k=0}^{N-N^{\min}} kP(\Delta X_{ij} = k \wedge X_{j0} \geq N^{\min} \\
&\quad \wedge X_{j0} + \Delta X_{ij} \geq 2N^{\min}) \\
&= \sum_{k=0}^{N-N^{\min}} kP(\Delta X_{ij} = k \wedge X_{j0} \geq \max(N^{\min}, 2N^{\min} - k)) \\
&= \sum_{k=0}^{N-N^{\min}} k\Delta F(k, q^i_{j0}, N - N^{\min}) \\
&\quad \cdot (1 - F_{j0}(\max(N^{\min}, 2N^{\min} - k) - 1, q^j_{j0}, N - N^{\min}),
\end{aligned} \tag{10}$$

and similarly

$$\begin{aligned}
&P(B_{ij})E[\Delta X_{ij}|B_{ij}] \\
&= \sum_{k=0}^{N-N^{\min}} k\Delta F(k, q^i_{j0}, N{-}N^{\min})(F_{j0}(N^{\min}{-}1, q^j_{j0}, N{-}N^{\min}) \\
&\quad - F_{j0}(N^{\min} - k - 1, q^j_{j0}, N - N^{\min})).
\end{aligned} \tag{11}$$

3.2 Effect of Existence of Other Users

Let $P^{\mathrm{c}}_{i,-j}$ be the probability that i secures cooperation from some user ℓ, $\ell \neq i, j$ and $\alpha_{i,\ell}$ be the probability that i secures cooperation from user ℓ. Then

$$\begin{aligned}
\alpha_{i,\ell} &= P(X_{i\ell} \geq N^{\min} \wedge (A_{i\ell} \vee B_{i\ell}))P^{\mathrm{c}}_{i\ell} \\
&= P(X_{i\ell} \geq N^{\min})(P(A_{i\ell}) + P(B_{i\ell}))P^{\mathrm{c}}_{i\ell}, \qquad (12) \\
P^{\mathrm{c}}_{i,-j} &= 1 - \prod_{\ell \neq i,j} (1 - \alpha_{i\ell}). \qquad (13)
\end{aligned}$$

where

$$P(X_{i\ell} \geq N^{\min}) = 1 - F(N^{\min} - 1, q^i_{i\ell}, N) \tag{14}$$

and $A_{i\ell}$, $B_{i\ell}$ are defined in the same way as A, B. Let u^{r}_{ij} be the expected utility of i if j refuses to cooperation. Based on (13), we have

$$u^{\mathrm{s}}_{ij} = P^{\mathrm{c}}_{i,-j} N^{\min}. \tag{15}$$

The computation of expected utility for j, denoted by u^{f}_{ij}, when it agrees to forward for i is more complicated, because whether i takes this offer or not depends on how many acknowledgement it receives from all the potential forwarders.

Recall i would randomly select one according to the rule. Let $U = \{1, 2, \ldots, M\}$ denote the set of all users. We first calculate the probability that i takes j's offer, denoted by P^{o}_{ij},

$$P^{\mathrm{o}}_{ij} = \sum_{k=0}^{M-2} \left[\sum_{\substack{S \subset U \setminus \{i,j\} \\ |S|=k}} \prod_{\ell \in S} \alpha_{i\ell} \prod_{m \in S^c} (1 - \alpha_{im}) \right] \frac{1}{k+1}$$
$$= \int_0^1 \prod_{\ell \in U \setminus \{i,j\}} (1 - \alpha_{i\ell} + \alpha_{i\ell} x) \mathrm{d}x. \quad (16)$$

Then

$$u^{\mathrm{f}}_{ij} = P^{\mathrm{o}}_{ij} v^{\mathrm{f}}_{ij} \quad (17)$$

where v^{f}_{ij} is given by (8).

3.3 Pairwise Nash Bargaining Solution

Now we can draw the extensive form of the stage game as shown in Fig. 2, each leaf representing the expected utilities resulted from the respected decision – c for "cooperation" and n for "noncooperation".

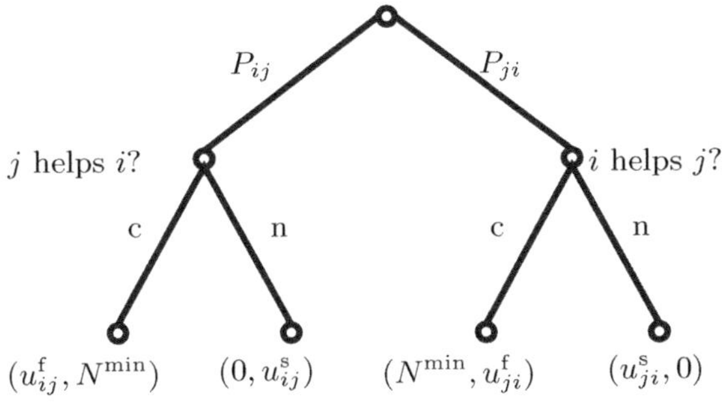

Fig. 2. Extensive form of the two-user stage game

The normal form of the game, as shown in Table 2, consists of four strategy profiles and their associated payoff profiles, denoted by $\langle \mathrm{n}, \mathrm{c} \rangle$, $\langle \mathrm{c}, \mathrm{c} \rangle$, $\langle \mathrm{c}, \mathrm{n} \rangle$ and $\langle \mathrm{n}, \mathrm{n} \rangle$, where $\langle \mathrm{n}, \mathrm{c} \rangle$ (abbreviation for $\langle$noncooperation, cooperation$\rangle$) means user j would choose not to forward for i if i requests its cooperation while i would choose to forward for j if j requests its cooperation. The two-user NBS is then a mixed strategy profile of these four that maximizes the proportional fairness metric, i.e.,

Table 2. Normal form of the stage game. The first component is j's average utility, the second component i's average utility, corresponding to the specific strategy profile.

j \ i	cooperation (c)	noncooperation (n)
c	$\begin{pmatrix} P_{ij}u_{ij}^{\mathrm{f}} + P_{ji}N^{\min} \\ P_{ij}N^{\min} + P_{ji}u_{ji}^{\mathrm{f}} \end{pmatrix}$	$\begin{pmatrix} P_{ij}u_{ij}^{\mathrm{f}} + P_{ji}u_{ji}^{\mathrm{s}} \\ P_{ij}N^{\min} \end{pmatrix}$
n	$\begin{pmatrix} P_{ji}N^{\min} \\ P_{ij}u_{ij}^{\mathrm{s}} + P_{ji}u_{ji}^{\mathrm{f}} \end{pmatrix}$	$\begin{pmatrix} P_{ji}u_{ji}^{\mathrm{s}} \\ P_{ij}u_{ij}^{\mathrm{s}} \end{pmatrix}$

$$\begin{aligned} \max_{\lambda_1,\lambda_2,\lambda_3,\lambda_4} \quad & u_i u_j, \\ \text{s.t.} \quad & u_j = \lambda_1(P_{ij}u_{ij}^{\mathrm{f}} + P_{ji}N^{\min}) + \lambda_2(P_{ij}u_{ij}^{\mathrm{f}} \\ & + P_{ji}u_{ji}^{\mathrm{s}}) + \lambda_3 P_{ji}N^{\min} + \lambda_4 P_{ji}u_{ji}^{\mathrm{s}}, \\ & u_i = \lambda_1(P_{ij}N^{\min} + P_{ji}u_{ji}^{\mathrm{f}}) + \lambda_2 P_{ij}N^{\min} \\ & + \lambda_3(P_{ij}u_{ij}^{\mathrm{s}} + P_{ji}u_{ji}^{\mathrm{f}}) + \lambda_4 P_{ij}u_{ij}^{\mathrm{s}}, \\ & \lambda_1 + \lambda_2 + \lambda_3 + \lambda_4 = 1, \quad \lambda_i \geq 0, \quad i = 1,2,3,4. \end{aligned} \tag{18}$$

The cooperation probabilities are then given by

$$P_{ij}^{\mathrm{c}} = \lambda_1 + \lambda_2, \quad P_{ji}^{\mathrm{c}} = \lambda_1 + \lambda_3. \tag{19}$$

Due to the recursive form as shown in (12), in practice the pairwise NBS may need to be evaluated repeatedly until it converges. This process is summarized in the following algorithm The simplicity of this algorithm is in contrast with

Algorithm 1. Algorithm for Computing NBS Based on BE

Require: initialize $\{P_{ij}^{\mathrm{c}}\}_{i\neq j}$
1: retrieve location information from database
2: compute $\{q_{ij}^{i}\}_{i\neq j}$ using a geometric channel model
3: compute $\{P(A_{ij})\}_{i\neq j}$ and $\{P(B_{ij})\}_{i\neq j}$ with (5) and (6)
4: compute $\{P_{ij}\}_{i\neq j}$ with (7)
5: compute $\{v_{ij}^{\mathrm{f}}\}_{i\neq j}$ with (8)
6: **repeat**
7: compute $\{\alpha_{ij}\}_{i\neq j}$ with (12)
8: compute $\{P_{i,-j}^{\mathrm{c}}\}_{i\neq j}$ with (13), then $\{u_{ij}^{\mathrm{s}}\}_{i\neq j}$ with (15)
9: compute $\{P_{ij}^{\mathrm{o}}\}_{i\neq j}$ with (16), then $\{u_{ij}^{\mathrm{f}}\}_{i\neq j}$ with (17)
10: compute P_{ij}^{c} by solving (18)
11: **until** $\{P_{ij}^{\mathrm{c}}\}_{i\neq j}$ converge

the complexity of the original BE algorithm presented in [4]. Initialization of P_{ij}^{c} can be arbitrary as experiments show that it has little effect on the outcome. In experiment, it has been observed that convergence of Alg. 1 is fast, usually taking less than 100 iterations for ≤ 50 users, though the proof or condition under which the convergence is unique is still ongoing work.

4 Performance Improvement

Performance improvement is measured in terms of outage probability and the average number of usable subcarriers (proportional to rate). Without BE, the outage probability for an arbitrary user i is given as

$$P_i^{\text{out}} = P(X_{i0} < N_{\min}) = F(N_{\min} - 1, q_{i0}^i, N). \tag{20}$$

With BE, the situation can be improved if, during a normal outage, any other user j could provide cooperation, which happens with probability α_{ij}. Therefore, with cooperation, the outage probability is given as

$$P_i^{\text{out, c}} = P_i^{\text{out}} \prod_{j \neq i} (1 - \alpha_{ij}), \tag{21}$$

i.e., the improvement is by a factor of $\prod_{i \neq j}(1 - \alpha_{ij})$. Without BE, the average number of usable subcarriers for an arbitrary user i is given as

$$m_i = \sum_{k=N_{\min}}^{N} kP(X_{i0} = k). \tag{22}$$

With BE, this number is

$$\begin{aligned} m_i &= P(X_{i0} \geq N_{\min}) E[X_{i0} | X_{i0} \geq N_{\min}] \\ &\quad + P(X_{i0} < N_{\min} \wedge \text{no cooperation}) N_{\min} \\ &= m_i + F(N_{\min} - 1, q_{i0}^i, N) \left(1 - \prod_{i \neq j} (1 - \alpha_{ij}) \right) N_{\min}, \end{aligned} \tag{23}$$

i.e., the improvement is given as

$$m_i^{\text{c}} - m_i = F(N_{\min} - 1, q_{i0}^i, N) \left(1 - \prod_{i \neq j} (1 - \alpha_{ij}) \right) N_{\min}. \tag{24}$$

5 Numerical Results

5.1 Simulation Model

For the purpose of illustration, we consider an infrastructure network in white space using OFDMA. We put 10 to 50 secondary users randomly in a 2000m $\times$ 2000m area where AP sits at the center. Each user is allocated with 40 subcarriers with 10KHz spacing. The total bandwidth $N \times 400$KHz will be taken from upper end of white space, i.e., from 698MHz down, i.e., user 1 is allocated 697.6-698MHz, user 2 allocated 697.2-697.6MHz, and so on. With the maximum number of 50 secondary users, TV white space channels 48 – 51 will provide enough bandwidth for our simulation.

We use the Hata urban model to simulate frequency dependent path loss. Given the antenna height at the AP ($h_A = 10$m) and antenna height at the secondary user ($h_U = 1$m), the path loss is modeled as

$$L(f) = 69.55 + 26.16 \log_{10} f - 13.82 \log_{10} h_A - C_A + (44.9 - 6.55 \log_{10} h_A) \log_{10} d + \Delta L, \quad (25)$$

$$C_A = 0.8 + (1.1 \log_{10} f - 0.7) h_U - 1.56 \log_{10} f \quad (26)$$

where d is the distance over which we make the measurement, ΔL is the log-normal shadowing term with zero mean and 8dB standard deviation, C_A is the correction term. Since each user has only (consecutive) 400KHz bandwidth, we can simplify the simulation by assuming the same path loss statistics for all subcarriers that belong to a single user.

We consider a subcarrier usable if the associated path loss is less than 130dB, a link usable if the number of usable subcarriers is no less than $N^{\min} = 10$.

5.2 Simulation Results

Fig. 3 shows the average outage probability with and without BE-based co-operation. As the number of users increase, the outage probability scales down exponentially demonstrating the power of user cooperation diversity incentivised by BE. Fig. 4 shows that BE also helps improve the rate. With BE, each user has on average additional 4 usable subcarriers. Considering that every user requires 10 subcarriers to have a sustainable connection, this improvement is substantial. Fig. 5 shows that the efficiency achieved by BE does not come at the cost of fairness. Specifically, with different number of total users in the system, we find the user with the highest/lowest outage probability and plot it against the left y-axis. At the same time we plot against the right y-axis the average number of additional subcarriers made available to it through BE and NBS. It is seen that the highest outage user receives nearly 10 additional subcarriers that help

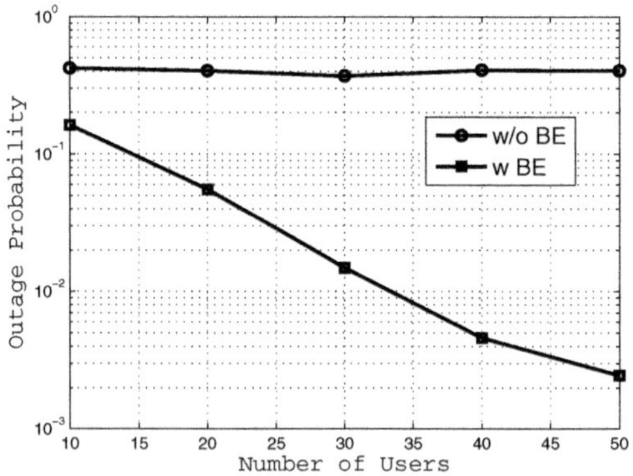

Fig. 3. Average outage probability with and without BE

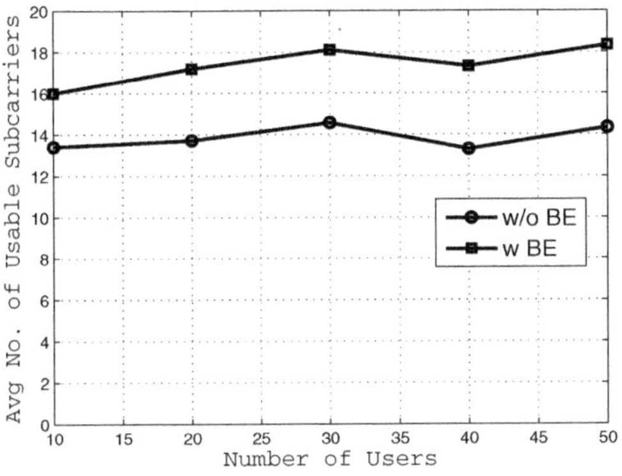

Fig. 4. Average available subcarriers with and without BE

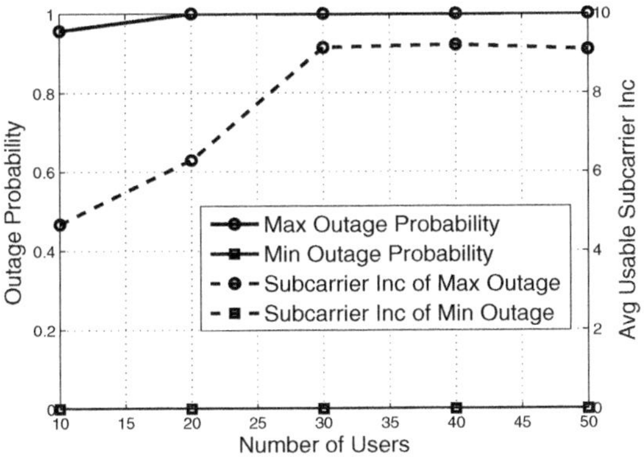

Fig. 5. User with the highest/lowest outage probability and the average extra number of subcarriers made available to it via BE

it almost always have a sustainable connection; the lowest outage user however does not benefit as much resulting in a higher social welfare as we expected.

6 Conclusion and Discussions

In the absence of a predefined coordination infrastructure, TV white space and WSD lack the mechanism that systematically brings self-conscious secondary users into synergy. In this paper we discussed a strategy to incentivize cooperative forwarding that constitutes an essential issue in secondary coexistence in white space. The strategy has been built with the particular notion of geo-location database in mind, which is mandatorily required to be consulted by all white space devices. Specifically for secondary users equipped with OFDM enabled radios, we showed that the database combined with a reliable channel model can be used to realize a form of cooperation called bandwidth exchange which promotes fair and efficient operation via a Nash bargaining framework. Though discussed in previous studies, the channel and network state information enabled by the database makes the bargaining process much faster and more accurate by taking into consideration the effect of existence of many other users. The numerical results show that bandwidth exchange dramatically improves the system performance in terms of outage and rate without compromising fairness.

References

1. FCC, Second Memorandum Opinion and Order (2010), http://www.fcc.gov/Daily_Releases/Daily_Business/2010/db0923/FCC-10-174A1.pdf
2. Biswas, S., Morris, R.: Opportunistic Routing in Multi-Hop Wireless Networks. In: ACM SIGCOMM (2004)
3. Dianati, M., Ling, X., Naik, K., Shen, X.: A Node-Cooperative ARQ Scheme for Wireless ad hoc Networks. IEEE Trans. Veh. Technol. 55(3), 1032–1044 (2006)
4. Zhang, D., Shinkuma, R., Mandayam, N.: Bandwidth Exchange: An Energy Conserving Incentive Mechanism for Cooperation. IEEE Trans. Wireless Commun. 9(6), 2055–2065 (2010)
5. Owen, G.: Game Theory, 3rd edn. Academic Press (1995)

Net Neutrality and Quality of Service

Eitan Altman[1], Julio Rojas[1,2], Sulan Wong[1,3],
Manjesh Kumar Hanawal[1,4], and Yuedong Xu[4]

[1] INRIA Sophia Antipolis, 2004 Route des Lucioles, France
[2] Dept. of Econ. and Bus. Sci. at Univ. of Barcelona, 08034 Barcelona, Spain
[3] Dept. of Law, Univ. of Coruña, 15071 A Coruña, Spain
[4] LIA, University of Avignon, 339, chemin des Meinajaries, Avignon, France
{eitan.altman,mhanawal}@inria.fr, yuedong.xu@gmail.com,
jrojasmo7@alumnes.ub.edu, swong@udc.es

Abstract. 2010 has witnessed many public consultations around the world concerning Net neutrality. A second legislative phase that may follow, could involve various structural changes in the Internet. The status that the Internet access has in Europe as a universal service evolves as the level of quality of service (QoS) to be offered improves. If guarantees on QoS are to be imposed, as requested by several economic actors, it would require introducing new indicators of quality of services, as well as regulation legislation and monitoring of the offered levels of QoS. This tendency in Europe may change the nature of the Internet from a best effort network to, perhaps, a more expensive one, that offers guaranteed performance. This paper presents an overview of the above issues as well as an overview of recent research on net-neutrality, with an emphasis on game theoretical approaches.

1 Introduction

Several public consultations on network neutrality policies have taken place in 2010. From them, regulation was introduced by the FCC in the USA [11], while the European authorities feel there is no need for it [12, p. 3] at this moment. France's Assembly, meanwhile, is discussing a bill on net neutrality [13]. The growing economic and social role of the Internet along with the fast evolution of its performance and of the services it offers, have triggered evolution of the legal status of the access to the Internet. Already on 2002, access to the Internet has become an universal service in the EU, which should imply guarantees on QoS. Even if data connections "should be capable of supporting data communications at rates sufficient for access to online services such as those provided via the public Internet", the European Parliament understands that due to the heterogeneous nature of the Internet "it is not appropriate to mandate a specific data or bit rate at Community level", leaving Member States with the responsibility of monitoring if the data rates provided by ISPs are "sufficient to permit functional Internet access" [15, Whereas 5]. Nonetheless, this guarantee only covers access to the narrowband Internet service, while the definition of the broadband Internet service as a universal service is currently in discussion [14].

R. Jain and R. Kannan (Eds.): GameNets 2011, LNICST 75, pp. 137–152, 2012.

From the debate on network neutrality we learn that new indicators of QoS are sought, and that we should expect an intensive work of regulating and standardization bodies on defining requested minimum values of performance measures. By actually requiring QoS to be guaranteed to the end user's actual experience (see e.g. [2] page 19), we may expect new legislation to create a new reliable and yet more expensive Internet, that would be different than the best effort type network that we have known.

This issue of a best effort versus a guaranteed performance network was not only present in the debate on network neutrality, but could be central in the legislation that would follow that debate. Another new comer issue is that of imposing taxes on content providers by the government. We provide more details on both issues in the next section.

The paper is organized as follows. The next section provides an overview on (i) the Internet as a universal service, (ii) the legislation concerning QoS, and on (iii) economic issues that arise in the debate on network neutrality. The following section provides a brief non exhaustive overview on mathematical models related to the neutral net question, with special focus on game theoretical models.

2 Overview

The Economic Aspects. The information and communication technologies have a central economic role. Indeed, "The information and communication technologies sector already generates revenue of 2,700 billion euros, or close to 7% of global GDP, and could account for 20% of GDP within the next 10 years" [2]. The latter reference specifies moreover: "Pyramid Research and Light Reading predict a rise in annual worldwide revenue for voice and data services of around 2.5% and 12.8%, respectively, between 2010 and 2013, while data traffic is forecast to increase by 131% during that same period. To give an example, in December 2009 ComScore reported that 5.4 billion videos had been watched in France that year (a 141% increase over the year before), of which 1.8 billion on YouTube.com between January and September".

In Europe, 5% of its GDP, i.e. 660 billion annually, comes from the IT sector, with 250 million daily Internet users and a penetration of the mobile market close to 100% [16].

Internet as a Tool for Exercising the Freedom of Speech. In 2009, France passed a law against non-authorized downloading of copyrighted material. Measures against file-sharing included disconnection from the Internet through an administrative order. The Constitutional Council went back to the **Declaration of the Rights of Man and of the Citizen** (from the time of the French revolution, two hundred years before the Internet was born) to conclude that freedom of speech could not be trusted to a new nonjudicial authority in order to protect holders of copyrights and neighboring rights. In their judgement, it recognized that the Internet is an instrument for exercising the freedom of speech. Similar relations between the Internet and the American constitution (and amendments) have also been made in the USA, see [3] for more details.

Recent events in the Arab world have shown us that people use Internet as a source not only to express their opinion on governments, but also to coordinate actions that allows overthrowing them. Governments have reacted, predictably, by blocking not only the Internet but also the mobile phone service. However, the effectiveness of these measures has been compromised with the release in media outlets throughout the world, of videos recorded in those countries.

Internet Access as a Universal Service. Directive 2002/22/EC of the European Union established the access to the Internet as a universal service, i.e. a global service guaranteed to all end users, regardless of their geographical location, at reasonable quality and reliability and at affordable price. Directive 2009/136/CD did not only request that access has a reasonable guaranteed service quality, it further complemented this request by giving ANRs the power to define the minimum QoS levels, in order to avoid service degradation, throttling and blocking.

To apply the law, one needs to constantly monitor the QoS in order to check if the targets values are indeed met. It is our conviction that this monitoring should involve not only those who offer the services and access to the Internet but also scientists that come from the public sector (research institutes and universities) who do not have direct economic interests.

2.1 New Indicators of Quality of Service Needed

Since 2002, there has been an exponential increase in the internet traffic, with new services and applications that appeared. In view of this huge growth, it has become clear that levels of QoS that have been appropriate for 2002 are no more sufficient, and moreover, there is a need in new indicators of QoS.

On the legislation side, there is a need to redefine the quality of services which the European universal service should offer. This can be done also in legislation at a state level. The first such initiative has been taken by Finland where already on October 2009 [18], followed by Spain in February 2010 [17, Art. 53]. Both of them require from providers a minimum guaranteed rate of 1 Mbps, a rate that makes difficult to see how the new multimedia services that are available through Internet could be provided.

Not surprisingly, we find the issue of QoS of today's and of future Internet in the center of the debate on network neutrality.

In [2], ARCEP (the French regulation body of Electronic Communications Markets) proposes six general policy directions for the network neutrality policy. The third one concerns QoS and it is summarized by: "3rd direction: A connection to the Internet must be provided with a sufficient and transparent QoS. To guarantee this, the Authority is launching sector-specific efforts to qualify the minimum QoS parameters for Internet access, and is working to implement specific indicators."

The document further specifies: "End users must be contractually informed of the technical properties of their Internet access, so that they can know the resources that have been assigned to them and the performance they can expect under "normal conditions" (i.e. "best effort" operations)... Work also needs to be done on the contribution of other players in the equation (ISPs[1], equipment manufacturers, software providers, etc.)".

How should one proceed to determine the required indicators of QoS as well as their minimum value?

In the paragraph on QoS (related to the third proposed policy direction in [2]), ARCEP first invites "operators and the associations that represent them to engage in sector-specific work devoted to setting minimum QoS parameters for "Internet access" (availability, bandwidth, latency, packet loss, jitter, etc.)."

They then propose that "this work could be the basis of exchanges with consumer associations and be enhanced by close collaboration with other relevant players, and particularly with ISPs since, as the designers of services and applications, they are particularly well suited to analyze user's qualitative experience."

Would it indeed be enough to put together the operators with the consumer associations in order to come up with indicators and minimum levels of QoS?

Do the consumer associations have the experts to understand the impact of choices of minimum values of quality of services, and of proposed indicators, on the quality they would perceive? Do the legislators have these?

We recommend to involve in this work a third actor, such as research institutes and universities etc., that has the required experts in answering these questions and has the commitment of contributing to progress of the society.

2.2 The Involvement of End-Users in Determining Policies in France

2010 has experienced dramatic events related to network neutrality question. Exceptional legislation initiatives have been taken, that may pave the way to shape a different future Internet. 2010 saw the first country, Chile, adopting a legislation that establishes network neutrality. At the same year, a USA Court shook the foundation of the Net neutrality in USA by denying the USA telecom regulation body, the FCC, the authority to take decisions and actions on that topic. In preparation to legislation on the topic, public consultations were launched in USA, France and the European Union (EU).

There is a huge difference in the number of participants between the American consultation, on one side, and the French and European ones, on the other. In total there are more than 89,000 filings in the American one, where as the French consultation was answered by only 121 stake holders [5] and the European one was answered 318 times [12]. As can be seen in the FCC web site, the vast majority of the answers to the consultation in USA came from individuals who used a web tool provided by the platform `savetheinternet.com` for the automation of this process. This tool had a very basic template with a short standard text in favor of net neutrality, in which the interested individual provided his name.

If the level of participation seems low for France, a country with close to 65 million people of which 68.9 % have access to Internet [6], we were shocked by

[1] Internet Service Providers.

the number of responses that the European one attracted, as the EU has an estimated two thirds more population than the USA.

Among the 121 responses made by stake holders [5] in the French consultation, eight came from ISPs, four from networking vendors, six from content production corporations, three from copyright collecting societies, eleven from software and content providers, six from user associations, three from public initiative networks, four from other kinds of professional associations, two from a group of experts gathered by Nathalie Kosciusko-Morizet[2], five from researchers, and 67 from individual citizens.

The European call also showed low level of interest among stake holders, attracting a total of 318 responses [12]. Of this relatively small participation, 34 answers came from ISPs, seven from infrastructure providers, two from mobile phone manufacturers, 16 from national and regional authorities, 38 from Internet-related organizations, 18 from media-related organizations, six from content providers, 42 from industrial organizations, social, consumer and non-governmental organizations, seven from other companies, ten from academic institutions, three from political parties, and 145 from individual citizens.

We can compare these figures with the consultation process in Canada that was initiated by the CRCT. It provided around three months for sending comments to the commission. In addition it had several days of hearing. The commission which received "437 initial comments, 35 reply comments, and 34 final replies from parties (companies and advocacy groups) and individuals. In addition, an online campaign resulted in over 13,000 email submissions to the Commission from individuals. At the oral hearing in July 2009, 26 presentations were made. Finally, an online consultation initiated by the Commission resulted in 1,400 additional individual comments" [7, §10]. Thus the response to both, the French and the European consultations, are also much lower than the one for the Canadian consultation.

2.3 Duration and Timing

Of the three public consultation processes, the French was the shortest, running for 39 days (April 9 to May 17, 2010), followed by the European, that extended for 93 days (June 30, 2010 to September 30, 2010), and then the longest is the American, which ran for 187 days (October 22, 2009 to April 26, 2010)[3]. The French consultation is not only much shorter than the American. The timing for the consultation was chosen to overlap the two weeks vacation period of the Eastern holidays in France, in which many French spending vacations with their families are disconnected from the politics. Since both the duration and the timing of the consultation are under the control of the government, it seems natural to speculate that the French government was not interested in having a large participation. The European consultation was not only half as long as the

[2] French State Secretary for the Digital Economy.

[3] Due to the "Comcast" decision[8], the FCC extended the deadline for filing reply comments in response to the NPRM from March 5 to April 26.

American one, it also had the same problem of the French one, as it ran over the summer holidays that usually spread from mid-July to mid-September, a period of almost 60 days of very little activity by the consultees.

2.4 The Impact of the Government Position

As both, the U.S. government led by President Obama and the FCC headed by the Commissioner Genachoswki, have been strong proponents of the consecration of the principle of net neutrality, either by an amendment of the Telecommunications Act, or by an administrative mandate issued by the FCC itself, the debate on net neutrality has been re-launched, achieving a media presence that is usually very difficult to reach for such a complex issue that weaves together three different areas of knowledge. In December 2010, the FCC issued a Report and Order [11] as the conclusive and regulatory document obtained from the NPRM. In it, the FCC keeps maintaining its authority to adopt rules on the open Internet [11, Part IV], but not unanimously, as two of its members believe that it does not so [11, p. 148-150, 188-193].

The European Union, both through the statement issued by the Commission under the Telecom Package[4] as well as through Commissioners Reding and Kroes, made it clear that the intention of the European government is to protect the neutrality of the network.

We recall that in France, the conditions chosen to launch the consultation seem to indicate that the government was not interested in receiving a large number of responses. Minimizing the dimensions of the public debate associated with the consultation may also be useful in order to avoid the French citizens questioning other aspects of the government policy on the Internet. In particular, the government was probably aware that public discussions on the HADOPI law[5] could be triggered by the fact that the questionnaire of the French consultation includes an important link between the HADOPI law and the proposed net neutrality. The relatively limited interest in France in getting feedback from end users should not be interpreted as a disinterest in the opinion of the various economic actors. Indeed, as we have already seen, there was a conference held by the ARCEP in which important economic actors participated[6].

2.5 Taxation Issues

Among the issues that Net neutrality is concerned with are relations between access and content providers along with related pricing issues, as well as the

[4] The Commission acknowledges in a declaration attached to the Telecom Package that net neutrality is "a policy objective and regulatory principle to be promoted by national regulatory authorities" [19].

[5] Adopted in France last year and which bans downloading unauthorized copyrighted content.

[6] We were surprised not to see among the participants speakers from French universities or research institutes. In fact, the only two talks from Professors from universities are from the USA. In addition, one can find video interviews of many stake holders in the conference's home page: `http://www.arcep.fr/index.php?id=10370`.

possibility of an access provider to have exclusive agreement with some content provider or some service provider.

In Europe, these issues take another dimension, due to the fact that many large content providers (such as google, facebook etc) are non-european: these are mainly American companies. These companies make large benefits from advertisement.

It is known that google pays very little taxes on this income, in contrast to what google pays in the USA. In a context where all other actors related to the Internet do pay taxes, applying network neutrality would mean favoring or subsidising these providers that do not pay. To be more precise, Google managed to cut 3.1 billion American dollars of taxes in the three last years by declaring its foreign profits (made in Europe) in Bermuda. This enabled Google to reduce its overseas tax rate to 2.4% . This is done using Irish law that allows to legally shuttle profits into and out of subsidiaries there, thus escaping the Irish 12.5 % tax. Facebook is preparing a similar strategy to shift declared benefits from Ireland to Caymans. For details, see [4].

The so-called "Zelnick Report" [9], which came out in France in January 2010, proposed to impose a tax on advertising revenue generated by the use of online services from France. According to estimates put forward by the authors, between 10 to 20 million euros would be collected mainly from U.S. content providers (Google, Microsoft, AOL, Yahoo and Facebook). It is pertinent to note that the report expresses concerns about the drop in advertising revenues of the French content providers, citing the poor state of competition in the French market for search engines, and certain behaviors (never clarified in the text) of Google. Later, French President Nicolas Sarkozy supported this proposal in a speech where he presented a set of policies to support the sector of cultural content creation.

The French position concerning google had perhaps some impact. In September 2010, Google CEO Eric Schmidt met with the French president Sarkozy to discuss the opening in Paris of a Google research center and the creation of a European cultural institute[10]. Schmidt said that the only reason for this initiative is economic, since it considers the French market for online searches as very dynamic. He added that in the meeting, the so-called "Google tax" was not discussed and that he has not met with the French competition authority.

2.6 Incentives for Investments

One of the issues in the debate on Network Neutrality has been incentives for investments in the infrastructure. Some argue that neutrality would create the incentives and some argue that only a non–neutral net would guarantee that. This problem has been partly resolved in the European Union when Internet access was declared as a universal service. Indeed, there are several possible ways to finance the cost of Providing communications services to all end-users comes. The Universal Service Directive allows providers to be compensated either from public funds or through a cost-sharing arrangement between providers if it is demonstrated that by complying with the universal service obligations they incur a loss or suffer net costs that exceed normal commercial standards [1].

The latter reference further says: "Member States are free to go beyond the minimum requirements laid down in the Directive, the only stipulation being that any additional obligation cannot be funded by a Levy on telecom providers."

We note however that in order to guarantee that the development of infrastructure would continue, it may be necessary to upgrade also the legislation on the universal service in order to apply it not just for the Internet access but also to broadband Internet services, which has not yet been declared universal service by the European community.

3 Mathematical Modeling of Network Neutrality Issues

3.1 Conclusions of the Models at a Glance

There is one particular economic issue that is at the heart of the conflict over network neutrality. Hahn and Wallsten [26] write that net neutrality "usually means that broadband service providers charge consumers only once for Internet access, do not favor one content provider over another, and do not charge content providers for sending information over broadband lines to end users." This motivates the recent studies on the implications of being non-neutral and of charging the content providers (CP).

Two central questions in the context of a possibly non-neutral Internet which many researchers have tried to to answer are [37]: i) who would gain or loser in the absence of neutrality; ii) will ISPs or CPs have more incentives to invest on the network infrastructure. The heated debate on these questions involve legal, economic and technological aspects, see e.g. [36]. There have not been much work which involves all these expertise. Moreover, some of the existing works draw contradictory conclusions due to their differences in the market modeling.

We first take a glance at the state of the art research before diving into the details. We concentrate on the profits of the economic actors as well as the incentive of investment when the network neutrality is abandoned. The main results of existing work are summarized in Table 1. The first column lists the recent work as well as the year of publication. The second column highlights the power of ISPs in an Internet market. The charging of CPs may have different impacts on network utility in the competing and the monopoly ISP markets. The symbol $\surd$ denotes that the performance metric is better off with network non-neutrality. On the contrary, the symbol $\times$ represents a disadvantage in the non-neutral networks. For the two questions raised above, only part of existing works present definite answers. We use the symbol $\star$ to denote the case where the authors provide a more complex result: they provide some conditions under which the network non-neutrality is beneficial and others for which it is harmful. The symbol $\varnothing$ means that the authors have not studied that specific metric. From this table, we can conclude that the available models do not seem to agree with each other. However, the ISPs have more incentives to invest if they are better off with no regulation, according to these studies (including [23,32]).

Table 1. Comparisons of Existing Work on Net non-neutrality

	Market	ISP Surplus	CP Surplus	User Surplus	Social Surplus	ISP Investment
ET[38]	One ISP	$\surd$	$\times$	$\surd$	$\times$	$\varnothing$
ET[38]	Two ISPs	$\times$	$\times$	$\surd$	$\times$	$\varnothing$
NOSW[34]	Two ISPs	$\surd$	$\surd$	$\surd$	$\surd$	$\surd$
CBG[37]	One ISP	$\surd$	$\times$	$\surd$	$\surd$	$\times$
JL[44]	One ISP	$\varnothing$	$\varnothing$	$\varnothing$	$\surd$	$\surd$
MSW[32]	More ISPs	$\star$	$\star$	$\star$	$\star$	$\star$
ALX[23]	One ISP	$\star$	$\star$	$\star$	$\star$	$\star$
HCCR[39]	One ISP	$\star$	$\star$	$\surd$	$\surd$	$\varnothing$
HCCR[39]	More ISPs	$\star$	$\star$	$\surd$	$\star$	$\varnothing$

Game theoretic modeling of neutral or non-neutral networks may consider as actors that are involved in strategic interactions not only service providers and content providers but also the users as well as the advertisement sector that often represents a major source of revenue to the content providers. Not all game theoretic models studied the net neutrality problem from a non-cooperative point of view. When considering the relation between service providers and content providers, some researchers have considered cooperative mechanisms to regulate the price that one provider pays to the other. The Nash bargaining solution was considered in [31,22,23] where as the Shapley value was considered in [27,28].

The Nash bargaining paradigm is also known as the proportional fair rate allocation in the traffic engineering [30]. It is the unique way of transferring utilities that satisfies a well known set of four axioms [29] related to fairness. In [27]-[28] the Shapley value (which is known to have some fairness properties [25]) has been used for deciding how revenues from end users should be split between the service and the content providers. Interestingly, it is the service provider that is seen to be the one that has to pay the content provider, which reflects the fact that the benefits of the service providers are obtained thanks to the presence of the content provider (assuming that users subscribe to the ISP in order to access the content of the CPs).

3.2 Literature Survey

We classify the existing work mainly based on types of game models that are used, i.e. noncooperative and the cooperative games. For the models concentrated on noncooperative price competition, we subdivide them into two classes. One of them assumes the same QoS for the packets of all content providers. The other, on the contrary, allows an ISP to provider premier QoS for the content providers that agree to pay to the ISP. Beside the game oriented studies, we further describe a work that considers the neutrality issue in network utility maximization.

Noncooperative Game without Considering QoS Differentiation. Economides and Tag [38] proposed a model of two-sided market in which the ISPs play the role of platform, collecting revenues from both the end users and the non-competing content providers. The quantities of content providers and end users are assumed to be normalized continuums. In the monopoly market, the demand generated by end users is increasing with regard to the quantity of the CPs, while decreasing along with the access price of the ISP. A CP has a positive externality of revenue from advertisers, at the cost of payment to the ISP. According to their analysis, the ISP and the users are better off, but the CPs' surplus as well as the social surplus are worse without network neutrality. In a duopoly market with two ISPs and multi-homing CPs, the quantity of end users subscribing to one ISP depends on not only the strategy of this ISP, but also that of its opponent. By using non-cooperative game tool, the authors find that the total surplus as well as the surplus of the CPs and the ISPs are better off at the equilibrium under the neutral regulation. This is opposite to the monopoly ISP market.

Musacchio, Schwartz and Walrand [32] investigate a two sided market where the CPs and the ISPs invest jointly on the network infrastructure. Each ISP is a monopoly over its end users and the CPs can be contacted by all the users. The total click rate (or flow rate equivalently) to the CPs is strictly increasing with regard to the investment of all the CPs and the ISPs. For a CP, a larger investment will attract more clicks, hence bringing more revenues from the advertisers. The major performance measure studied in this work is the social surplus. The authors indicate that the ratio between parameters characterizing advertising rates and end user price sensitivity plays a key role in choosing the one-sided or the two-sided pricing. If this ratio is either low or high, the two-sided pricing is more favorable, and vise versa.

In [40], Zhang et al. study the competition and innovation of the service-oriented Internet. This service-oriented Internet can be regarded as a two sided market composed of two CPs and two ISPs. The CPs charge end users based on their usage, while the ISPs charge them flat rate fees. The CPs engage in a Cournot competition where the price is determined by the total demand from the users. The ISPs engage in a Bertrand game so that they compete over the side payment from the CPs (while not the end users). The Cournot and the Bertrand competitions are tied together in a two-stage Stackelberg game. The authors indicate that the update of an ISP becomes profitable only when the increase of the marginal cost is upper bounded by an appropriate gain in its market share.

Motivated by [26], the authors of [21] investigate network non-neutrality with a monopoly ISP, one CP and a number of end users. The sources of income (other than side payments) are payments of end users (to both the ISP and the CPs), and some third party payments (e.g. publicity income) that the content providers receive. We formulate the price competition as a noncooperative game. The CP's strategy is its charging to the users, and the ISP's strategy is the charging from both the CP and the end user. In [21] we find that if the ISP has the power to decide the side payment, not only do the CP and the end users suffer, but also the ISP's utility degrades. More precisely, we show that the only possible equilibrium would be characterized by prices that will induce zero demand from the users. This phenomenon does not occur if the CP's payment is fixed by some regulators, or the ISP determines the payments from the CP and the users sequentially.

We extend [21] to incorporate the QoS of users provided by the ISP in the non-neutral model [23]. This QoS measure is different from those in [37,43] where more demands lead to a reduced QoS. We connect the QoS with the incentive of ISP's investment. A larger demand from the users means a larger revenue, resulting in a larger bandwidth provision of the ISP and a better QoS. The authors introduce a parameter called *relative price sensitivity* to model the difference of demand sensitivities to the price of the ISP and that of the CP. When the price paid by the CP to the ISP for per-unit of traffic is a constant, the qualitative impact of being non-neutral is decided by the *relative price sensitivity*. If this relative sensitivity is greater than 1, the users value the service of the ISP more than that of the CP. We show that a positive payment of the CP to the ISP leads to worse surplus of all parties involved and worse QoS of end users. Our analysis reveals an implication that the ISP may pay to the CP of high price sensitivity so that the CP is able to reduce its service price. This type of reverse payment is rarely discussed in the literature of network non-neutrality.

In [24], we explore the effects of content-specific (i.e. not *application* neutral) pricing, including multiple CPs providing different types of content. Also, we consider competition among multiple providers of the same type, including different models consumer stickiness (inertia or loyalty). In an on-going work, we are also considering providers' infrastructure and operating costs (as in, e.g. [32]), more complex models of end-user demand and their collective social welfare, and the effects of different options for flat-rate pricing (e.g. [33,35]).

Noncooperative Game with QoS Differentiation. Hermalin and Katz [41] consider the two sided market with a monopolistic ISP and continuum of CPs and end users. They compare the levels of profit, social welfare under neutral and non-neutral regime. In the non-neutral regime they assume that the ISP can offer a range of differentiated quality of connection qualities (e.g., an ISP can offer different combinations of bandwidth, latency, and packet loss rate) to the CPs and charges them depending on the type of quality of connections they opt for. Their analysis suggests that any restriction on the ISP's choice to offer differentiated services will often result in poor social welfare and it improves social welfare only under few conditions. They further observe that the small scale content providers-the ones who are intended to benefit from regulations-are almost always harmed by the regulations. They extend the analysis of non-neutral regime to a Hotelling duopoly ISP model and observe that welfare results of monopoly ISP carry over.

Economides and Hermalin [42] study a two sided market similar to the one in [41] by considering the affect of network congestion. They allow the amount of information purchased by the users to vary. Their work shows restriction on granting or selling of priority services, i.e., neutral regime leads to superior social welfare. Further they show that the incentive to invest is ambiguous under non neutral regime. The investment by the ISP helps to improve the overall quality of the network and thus reducing to some degree the difference among the services offered at discriminated prices, thus reducing the ISP's income.

Cheng et al. [37] study a market with one ISP, two competing content providers and a finite number of end users. The monopoly ISP provides two type of services, the preferential and the non-preferential delivery. The content providers can pay the ISP a fixed fee for preferential service, which implies a non-neutral network. The authors model the QoS by the M/M/1 queueing delay, and the competition of the CPs by a hotelling framework. In this paper, the principle with no regulation is beneficial to the surplus of the ISP, while harmful to those of the CPs. The social benefit is improved when one of the CP pays to the ISP, but remains unchanged when both CPs join the preferential service. The non-neutrality might lead to better QoS for a majority of users and worse QoS of the others if the social surplus is better. The authors also observe that the ISP has less incentive for capacity expansion in a non-neutral network.This is because if the bandwidth is upgraded, more users experience less congestion and switch to the non-preferential service. Thus, the difference of aggregate surplus of the ISP before and after capacity expansion becomes smaller in a non-neutral regime than in a neutral regime, resulting in less incentives to invest.

Choi and Kim [43] study the investment incentives with and without network regulation. They consider a monopolist ISP and two CPs. Prioritization of delivery of packets by assigning "fast lane" to one of the CPs who agree to pay to the ISP is considered as the main mode of the non neutrality. The congestion is taken into account by modeling the network as M/M/1 queue. They study the neutral and non neutral regime by comparing the market equilibrium in the short run (fixed capacity) and investment incentives in long run. They observe that in the short run the CPs will face the prisoner's dilemma to get access to the fast lane and will be worse off. The social welfare improves in the non-neutral network when there is significant asymmetry across the content providers. In the long run they argue that contrary to ISP's claim that net neutrality regime will have adverse affect on their expansion, they may not have investment incentive in the non-neutral regime. The CPs also may not have investment incentives as they may fear that the ISP can expropriate some of the benefits made by them. Their analysis yields ambiguous conclusions on investment incentives.

Njoroge et al. [34] consider a network where two interconnected ISPs compete for the users and the CPs over the quality and the price. The definition of "neutrality" is different from that of [38]. In a neutral model, a CP pays to one ISP for Internet access, but does not pay to the other ISP without direct connection. In a non-neutral model, a CP has to pay to the ISP without direct connection in order to be reachable by its end users. The inter-ISP link is bandwidth limited so that the quality of a user-CP connection is decided by this bottleneck. The authors model the price and quality competition as a six-stage sequential game and solve it using backward induction. They show that the non-neutrality is able to improve the surplus of the ISPs, the CPs and the users. The social surplus is also better, and the ISPs have higher incentives of investment.

Regulation Mechanisms Based on Cooperative Games. Giving the full control of a market, an ISP can charge an arbitrary price from CPs for the delivered contents. However, there does not exist such a "dictatorship" status in reality. Recent work introduces cooperative game tools, such as Nash bargaining game and Shapley value, to study the revenue splitting issues among the players.

Shapley value is a well known concept in cooperative game theory that provides a way of splitting revenues obtained by the cooperating players[7]. It satisfies important properties like fairness, efficiency, symmetry, additivity, etc. In [27] Ma et al. explore Shapley pricing mechanism [45] to share the revenues from Internet subscribers among the service providers that peer each others' traffic. The authors show that if Shapley value based revenue sharing is enforced at the global level the selfish ISPs, at the Nash equilibrium point, will opt for strategies that will maximize the aggregate network profits. Further extending their work, in [28] the authors consider three type of ISPs: content ISPs, eye ball ISPs and transit ISPs. They obtain closed form expressions of the ISPs' revenues for the bipartite topologies (each type of ISP nodes can be separated) and give dynamic programming procedures to calculate the Shapley revenues for the general internet topologies. With the Shapley value solution they suggest the appropriate pricing structure for the differentiated services (non-neutral regime) that improves social welfare.

Saavedra in [31] initially uses Nash bargaining game to study joint investment in a non-neutral regime with one CP and two ISPs. The CP is able to negotiate with one or the both ISPs via contracts. The author highlights the impact of the CP's bargaining power on QoS agreements. If the ISPs allow lower quality of services in the non-neutral regime, the CPs with low bargaining power can enter into the exclusive contract with ISP and improve their bargaining position. Building upon [31], Altman et al. focus In [22] on mechanisms of setting the price that one provider would pay to another based on the Nash bargaining paradigm. The authors address a problem and a model that deffer however, from [31]. Two bargaining concepts, the *pre-bargaining* and the *post-bargaining* are presented to characterize the exact sequence of decisions that determine the side payment as well as the other prices. The first concept is used when bargaining over the side payment takes place before the user prices are determined. The second one models the occurrence of bargaining after the user price competition. The authors point out under what situations the *pre-bargaining* or the *post-bargaining* is preferable. They further study another aspect of non-neutral behavior which is the possibility for an ISP and a CP to collude together which could result in better performances for them but a worse performance for other competing CPs.

Network Neutrality in Network Utility Maximization Issues. In this survey, we do not restrict to the papers that use game theoretic models. We

[7] On the usefulness of this concept to split profits or costs we can learn from [46] where K Binmore writes [46]: "I was once summoned urgently to London to explain what the French government was talking about when it suggested that the costs of a proposed tunnel under the English Channel be allocated to countries in the European Union using the Shapley value".

mention one more reference whose model includes some aspects of competition. Hande et al. [39] consider the two-sided market in a quite different context. Instead of using the predefined price-demand curve, they look into the network utility maximization with the participation of the CPs. This model mitigates the key drawback of economic studies that does not deal with the engineering aspect of rate allocation. They define the utility of users as a function of flow rates. The concept "non-neutrality" refers to the restriction on the maximum price that the ISPs can extract from the CPs. This price restriction exhibits different impacts on the profits concerning the power of the ISPs. In a market of competitive ISPs, the users' benefit and the social benefit increase as the price restriction is relaxed. If we understand neutrality as zero payment for per-unit of delivered content, the non-neutral regime is always favorable in terms of the social profit and the users' profits.

4 Conclusion

We summarized some recent aspects in the development of the Internet as well as in the debate on its neutrality. We cover both legal aspects, economic ones as well as technological issues. We then present an overview of research papers on the net non-neutrality problem. We focus in particular on issues related QoS. As a universal service, the Internet is required to guarantee certain QoS levels. We related this to the legislation over network neutrality and to the question of the incentives for investing in the infrastructure. We then provided an overview of the research studies on network neutrality issues with an emphasis on using game theoretical tools.

Acknowledgement. This work was performed within the Web Science Institute of Avignon University (SFR. ST des CSN). It was partly supported by the INRIA ARC Meneur on Network Neutrality.

References

1. http://ec.europa.eu/information_society/policy/ecomm/current/consumer_rights/universal_service/index_en.htm
2. ARCEP, Neutralité des réseaux: Les actes du colloque (2010), http://www.arcep.fr/index.php?id=8652
3. Wong, S., Altman, E., Rojas-Mora, J.: Internet access: Where law, economy, culture and technology meet. Computer Networks 55(2) (2011)
4. Drucker, J.: Google 2.4% Rate Shows How $ 60 Billion Lost to Tax Loopholes, October 21(2010) published in Bloomberg, http://www.bloomberg.com/news/2010-10-21/google-2-4-rate-shows-how-60-billion-u-s-revenue-lost-to-tax-loopholes.html
5. Ministère de l'Économie, de l'industrie et de l'emploi. Consultation publique sur la neutralité du Net. telecom.gouv.fr, June 21 (2010)
6. Internet World Stats. Internet users in the european union, June 30 (2010), http://www.internetworldstats.com/stats9.htm

7. Canadian Radio-television and Telecommunications Commission. Telecom Regulatory Policy CRTC 2009-657 (2009)
8. United States Court of Appeals, District of Columbia Circuit. COMCAST Corporation (Petitioner) v. Federal Communications Commission and USA (Respondents) and NBC Universal, et al. (Intervernors), 600 F.3d 642 (2010)
9. Zelnik, P., Toubon, J., Cerutti, G.: Creation et Internet. Technical report, Ministre de la Culture et de la Communication, République Française (2010)
10. Ducourtieux, C.: Le patron de Google reçu par Nicolas Sarkozy. Le Monde.fr, September 9 (2010)
11. Federal Communications Commission. Report and Order, FCC 10-201 (2010)
12. European Commission: Information Society and Media Directorate-General. Report on the public consultation on the open Internet and net neutrality in Europe (2010)
13. Ayrault, J.-M., et al.: Proposition de loi N 3061 relative à la neutralité de l'internet (2010)
14. Kroes, N.: Who pays what? Broadband for all and the future of Universal Service Obligations (2010)
15. European Parliament and Council of the European Union. Directives 2009/136/EC, 2002/22/EC, 2002/58/EC of the European Parliament and of the Council
16. European Commission A Digital Agenda for Europe (COM(2010) 245) (2010)
17. Ministerio de Economía y Hacienda del Reino de España Proyecto de Ley de Economía Sostenible (2011)
18. Ministry of Transport and Communications of Finland Decree on the minimun rate of a funcional Internet access as a universal service (732/2009) (2009)
19. European Commission Telecom Reform 2009: Commission Declaration on Net Neutrality (2009)
20. Wong, S., Rojas-Mora, J., Altman, E.: Public Consultations on Net Neutrality 2010. In: Proc. of NetCoop 2010, Ghent, Belguim (November 2010)
21. Altman, E., Bernhard, P., Caron, S., Kesidis, G., Rojas-Mora, J., Wong, S.L.: A Study of Non-Neutral Networks with Usage-based Prices. In: The 3rd ETM Workshop, Amsterdam (2010), Longer version: INRIA research report 00481702
22. Altman, E., Hanawal, M.K., Sundaresan, R.: Non-neutral network and the role of bargaining power in side payments. In: NetCoop, Ghent, Belgium (November 2010)
23. Altman, E., Legout, A., Xu, Y.D.: Network Non-neutrality Debate: An Economic Analysis. In: IFIP Networking 2011, Valencia, Spain (May 2011)
24. Caron, S., Kesidis, G., Altman, E.: Application neutrality and a paradox of side payments. In: The 3rd Int. Workshop on Re-Architecting the Internet (ReArch 2010), Philadelphia, USA, November 30 (2010); collocated with ACM CoNEXT
25. Winter, E.: The Shapley value. In: Aumann, R.J., Hart, S. (eds.) The Handbook of Game Theory, ch. 53, vol. 3. North-Holland (2002)
26. Hahn, R., Wallsten, S.: The Economics of Net Neutrality. The Berkeley Economic Press Economists' Voice 3(6), 1–7 (2006)
27. Ma, T.B., Chiu, D.M., Lui, J.C.S., Misra, V., Rubenstein, D.: Interconnecting eyeballs to content: A shapley value perspective on ISP peering and settlement. In: Proc. of ACM NetEcon 2008, pp. 61–66 (2008)
28. Ma, T.B., Chiu, D.M., Lui, J.C.S., Misra, V., Rubenstein, D.: On cooperative settlement between content, transit and eyeball internet service providers. In: Proc. of ACM CoNext 2008, New York, USA (2008)

29. Nash, J.F.: The bargaining problem. Econometrica 18, 155–162 (1950)
30. Kelly, F.P., Maulloo, A., Tan, D.: Rate control in communication networks: shadow prices, proportional fairness and stability. J. Oper. Res. Society 49, 237–252 (1998)
31. Saavedra, C.: Bargaining, power and the net neutrality problem. Presented at NEREC Research Conference on Electronic Communications, Edcole Polytechnique, September 11-12 (2009) (manuscript)
32. Musacchio, J., Schwartz, G., Walrand, J.: A two-sided market analysis of provider investment incentives with an application to the net-neutrality issue. Review of Network Economics 8(1) (2009)
33. Odlyzko, A.: Internet pricing and history of communications. Computer Networks 36(5-6), 493–518 (2001)
34. Njoroge, P., Ozdagler, A., Stier-Moses, N., Weintraub, G.: Investment in two-sided markets and the net-neutrality debate. Decision, Risk, and Operations Working Papers Series, DRO-2010-05. Columbia Business School (July 2010)
35. Kesidis, G., Das, A., de Veciana, G.: On Flat-Rate and Usage-based Pricing for Tiered Commodity Internet Services. In: Proc. CISS. Princeton (March 2008)
36. Economides, N.: Net Neutrality, Non-Discrimination and Digital Distribution of Content Through the Internet. Journal of Law and Policy for the Information Society 4(2), 209–233 (2008)
37. Cheng, H.K., Bandyopadyay, S., Guo, H.: The debate on Net Neutrality: A policy Perspective. Information Systems Research, March 1 (2010)
38. Economides, N., Tag, J.: Net Neutrality on the Internet: A Two-Sided Market Analysis. NET Institute Working Paper No. 07-45, SSRN: `http://ssrn.com/abstract=1019121`
39. Hande, P., Chiang, M., Calderbank, R., Rangan, S.: Network Pricing and Rate Allocation with Content Provider Participation. In: Proc. of IEEE Infocom 2009, pp. 990–998 (2009)
40. Zhang, Z.L., Nabipay, P., Odlyzko, A.: Interaction, Competition and Innovation in a Service-Oriented Internet: An Economic Model. In: Proc. of IEEE Infocom 2010 (2010)
41. Hermalin, B.E., Katz, M.L.: The Economics of product-line restrcitions with an applications to the neutrality debate. AEI-brookings joint center for regulatory studies, `http://ssrn.com/abstract=1003391`
42. Economides, N., Hermalin, B.E.: The Economics of Netwotk Neutrality. NET Institute Working Papers, No.10-25, `http://works.bepress.com/economides/38/`
43. Choi, J.P., Kim, B.C.: Net neutrality and Investment incentives. To appear in Rand Journal of Economics
44. Jan, K., Lukas, W.: Network Neutrality and Congestion-Sensitive Content Providers: Implications for Service Innovation, Broadband Investment and Regulation, MPRA Paper No. 22095 (2010)
45. Shapley, L.S.: A Value for n-person Games. In: Kuhn, H.W., Tucker, A.W. (eds.) Contributions to the Theory of Games, volume II. Annals of Mathematical Studies, vol. 28, pp. 307–317. Princeton University Press (1953)
46. Binmorme, K.: Game theory, a very short introduction. Oxford Univ. Press (2007)

The Sharing-Mart System: Digital Content Sharing, Online Auctions, and Incentives

Christopher Leberknight[1], Ranjan Pal[2], Mung Chiang[1], and Harold Vincent Poor[1]

[1] Electrical Engineering Department
Princeton University
[2] Computer Science Department
University of Southern California

Abstract. This paper introduces *Sharing-Mart* (S-Mart), an online digital trading platform developed at Princeton University to perform social file sharing experiments on top of technological networks as overlays. It describes the S-Mart system, the experiments conducted, and incentivization aspects which can be investigated using S-Mart. In the first part of the paper, the S-Mart system and the experiments conducted are explained, and the economic behaviors and dynamics of package auctions run on S-Mart are described. The major experimental observation that stands out here is that Internet users are less incentivized to share content on competitive applications, whose success depends on the co-operation of other users in the system. To alleviate incentivization issues in these applications, in the second part of the paper a mathematical framework is proposed that derives user population threshold values, which hint at the necessity of a certain base population strength in S-Mart for co-operation to take place amongst all the users. An outline of two experiments to validate the theory is presented.

Keywords: Sharing-Mart, Auctions, Incentives.

1 Introduction

The combination of existing business models based on digital content and the proliferation of Web 2.0 technologies such as blogs and social networking sites (i.e., YouTube, Vimeo, and Flickr) suggests the potential and feasibility of a market for user generated content. The value and demand for user generated content is much more complex to quantify and raises several challenges that are fundamental to classical economics. First, determining the value of user generated content is akin to monetizing information which is highly subjective and therefore cannot be accurately defined using a fixed or marked price. Second, classical economics emphasizes methods to efficiently allocate scarce resources primarily comprised of *private tangible goods*. However, user generated content within an electronic market place is more aptly characterized as a *public or intangible good* subject to *multiplicity* and *abundance*. To effectively address these two

R. Jain and R. Kannan (Eds.): GameNets 2011, LNICST 75, pp. 153–168, 2012.

challenges, an open market pricing mechanism is required to enable consumers to provide the appropriate signaling function, specify preferences, and ration resources. We have developed a fully operational online virtual money file sharing system, known as **Sharing-Mart** (`http://sharingmart.princeton.edu`), which enables the transaction of digital goods using fixed price and multi-winner auctions. Sharing-Mart is the content equivalent of physical exchange markets such as eBay and Amazon and provides the opportunity to examine many different theoretical dimensions of auction theory, which can be substantiated or falsified through human subject experiments. In this paper, we touch upon the following four important facets of the Sharing-Mart (S-Mart) system.

- We briefly describe the Sharing-Mart system, its architecture, and the principles behind its operation. (See Section 2.)
- We give an overview of our experimental designs that were conducted on the S-Mart system, i.e., we describe the auction games played by users in each experiment and how subjects for the experiments were recruited via *Facebook*. (See Section 3.)
- Through our experiments we study the economic behaviors and dynamics of package auctions for public goods in a virtual economy. We also observe through our preliminary experiments that cooperation amongst users (a desired property for the successful[1] working of the system) is not perceived as a dominant strategy, thereby indicating that our provided incentive to make users sell/contribute, is not powerful enough. (See Section 4.)
- In order to overcome the problem of strongly incentivizing users to contribute content, we propose voluntary as well as involuntary settings on the S-Mart framework, which are both realistic and practically implementable, and promote users to contribute content unselfishly. We give an explanation of our settings, and describe the computation of population threshold parameters based on the settings. The population threshold parameters provide the S-Mart administrator with a valuable estimate of the mandatory presence of a certain number of users for an application in the S-Mart system so that the application is deemed effective in regard to unselfish content contribution. Our incentive settings and related analysis will also prove useful to knowledge management system (KMS) networks [4,5,6], in which a major challenge is to incentivize knowledge sharing amongst users by accounting for the dynamics of competition and cooperation at an organizational level. (See Section 5.)

2 Sharing-Mart System

Sharing-Mart is a virtual money based file sharing system developed at Princeton University, which allows different digital rights (e.g., view only, download, and

[1] We emphasize here that for an application like *sharing course notes,* cooperation amongst the users is essential for increasing the overall knowledge of students, although the application is inherently competitive. Thus, in such an application, students need to contribute/sell their knowledge content for the better working of the system but may not, if strong incentives are not provided.

resell rights) of various file types (e.g., video, audio, graphics and documents) to be traded by means of different transaction styles (e.g., marked price transactions and multi-winner auctions). S-Mart has recently been integrated with Facebook to enable rapid and increased interaction and analysis of users within the S-Mart social network. Anyone with a Facebook account can use their Facebook identity to access S-Mart. The Sharing-Mart system can be accessed, either by visiting the main website `http://sharingmart.princeton.edu` and clicking on the Facebook icon, or by adding the application to a user's application list on his/her main page within Facebook, by selecting the Sharing-Mart application from the Application Directory. The first time the system is accessed, a user will need to complete a brief registration form which will also initialize the user's personal homepage or directory within the system. Subsequently, each time a user accesses the application she will be presented with her personal login screen which will display information regarding past purchases and sales of digital content as well as other statistics regarding usage and popular content. Sharing-Mart has several features such as the *ability to view the most popular content, search for content, user statistics, and the ability to sell and purchase content using online auctions.* The full list of features as well as a user manual can be found at `http://sharingmart.princeton.edu/HTML-User-Manual-v1.htm`. The Sharing-Mart auction mechanism is implemented as a Vickrey-Clarke-Grove (VCG) auction, and at the same time enables the sale of multiple objects similar to package auctions [7].

3 Experimental Designs

In this section, we briefly describe two experimental designs that we conducted using the Sharing-Mart system. These experiments extend our previous research (open only to Princeton students) [1], in which agent-based competition was examined among seven graduate students at Princeton as a part of a homework problem set. Further details regarding the experimental analysis in this research are available in our technical report [20].

While auctions are generally competitive by nature, we further amplify the inherent competition through experiments which investigate the desire of students to cooperate with other students by contributing their knowledge in the form of digitized course notes. This allows us to not only examine different auction configurations such as package auctions [7] but also different patterns of human behavior.

3.1 Subject Recruitment

Subjects were recruited for two experiments from two undergraduate courses in social networking and systems analysis and design. There were 25 student subjects in each class. Additional subjects were also recruited using Facebook. A user account on Facebook, Share Mart, was used to recruit one hop neighbors from 5 students. A total of 64 members joined the Share Mart group and advertisements were broadcast to all Share Mart friends.

Both experiments were based on a word game, and each subject was provided with details of the experiment and rules for winning the game. Winners in the first and second experiment would receive a monetary reward of up to $50.00 and $25.00, respectively.

3.2 Game Description

The objective of the word game used in each experiment is to purchase and correctly organize all the letters for a hidden word. Each subject is instructed to sell and buy letters using an online auction in Sharing-Mart to collect all the letters. The winner of the game received a monetary prize if she/he is the first player to determine the word, and/or the player who maximizes his/her token balance by selling letters. Therefore, players can win the game by selling and/or buying letters and there is no limit on the number of times each letter can be sold.

4 Experimental Results

This section presents the results from two experiments in which subjects were required to purchase and sell four different files using the Sharing-Mart online auction module. Each experiment consisted of two rounds with multiple auctions in which subjects were required to purchase *all* the files. This type of auction configuration is most similar to a package auction [7]. The reason for designing the game as a package auction or auction of multiple objects with two rounds was to automate and simplify the process of generating items of interest for all subjects, and to discourage subjects from bidding their entire token balance in one auction. If the game consisted only of one auction, and the winner of the auction received the reward, then every player would be incentivized to bid his/her entire token balance to win the game and the demand function would simply be defined by the maximum token balance (all players have the same initial token balance). Since the tokens are virtual currency and do not represent any real value to the players, requiring players to purchase all files discourages players from bidding their entire token balance for one file and encourages them to sell the files they have won in previous auctions if their token balance drops below a certain threshold. This is the basis for designing the game as a package auction and motivating or incentivizing players to participate in the game. The two rounds are used to "seed" the players with items of interest to all players and then to let the dynamics of the game drive the activity in the system. In the first round the files were individually auctioned by the S-Mart Operator and subjects were required to bid on each file. This process "seeded" the group with digital content which would be in high demand by all players in the system. Since it is highly unlikely that one player would win all auctions in the first round, a player who won and paid too much for a file would have to sell her file to replenish her token balance in order to purchase the remaining files in the second round. This motivated players to contribute to the system by selling their files and let the dynamics of supply and demand drive the activity in the system. Also even more noteworthy is the implicit motivation for subjects to contribute content in the second round. This corresponds to encouraging cooperation among players in a competitive environment and is analogous to students contributing or selling their course notes to other students in the same class. Incentivizing cooperation is one of the main challenges in this research and also a main challenge for many other file-sharing systems and is typicallly known as the free rider problem [14,15,16,17,18,19].

The hidden word for Experiment I was "sink" and the hidden word for Experiment II was "calf." In both experiments all subjects had the same initial token balance. In Experiment I, all subjects had an *initial balance of 1200 tokens* and in Experiment II, all subjects had an *initial balance of 500 tokens.* The letters were sold in four separate auctions in Round 1 with different initial prices (reserve prices). In addition, the reward for winning the game in Experiment I was equivalent to $50.00 and the reward for winning the game in Experiment II was $25.00. Details of initial price (I-Price) and final price (F-Price) for Round I of each experiment is provided in Table 1.

We have developed several research questions to examine the issue of incentivizing content contribution in competitive environments. Three research questions form the basis for understanding economic behavior subject to the experimental configuration using the Sharing-Mart system. Our first research question, **Q1**, investigates the final price subjects paid for content as a function of the initial price charged for the content and the subjects' token balances. Specifically, *are subjects with greater purchasing power ultimately more active in the system, and are they more incentivized to engage in future contributions compared to subjects who have less purchasing power?*

Table 1. Initial and Final Prices for Experiments I and II

Auction# (Filename)	**Letter**	**I-Price**	**F-Price**	**Letter**	**I-Price**	**F-Price**
1 (1.pdf)	N	16	450	F	2	325
2 (2.pdf)	S	14	550	C	3	500
3 (3.pdf)	K	11	600	A	8	500
4 (4.pdf)	I	19	850	L	4	490
Experiment	I Token Balance = 1200 Reward = $50.00			II Token Balance = 500 Reward = $25.00		
Round I	Keyword = "SINK"			Keywork = "CALF"		

Our corresponding first hypothesis, **H1**, states: *Subjects who pay more for files perceive greater benefit and therefore will contribute/participate more in the system.* We therefore expect that since players in Experiment I had larger initial token balances compared to players in Experiment II, the winners in Experiment I would be more active participants in the game compared to the winners in Experiment II. For Experiment I, subject 1 and subject 7 won the auctions in

Round I and for Experiment II subject 7 and subject 12 won the auctions in Round I. Based on the results from Figure 1a and Figure 1b it can be observed that subjects who won auctions in the first round did contribute/participate more in the system compared to other subjects. Specifically, the results indicate winners in Experiment I Round I were more active throughout the entire game compared to the winners in Round I of Experiment II. This supports our first hypothesis and suggests individuals with a larger token balance may participate more in the system compared to individuals with smaller token balances. We may therefore expect that one parameter that may influence contributions to the system is an individual's token balance. Individuals with larger token balances may have a greater perceived benefit of using the system compared to individuals who have smaller token balances. To confirm this conjecture we might investigate varying token balances across all users in the system to test whether subjects are incentivized to participate based on their initial token balances.

A secondary goal of this research is to determine whether the economic behavior of players in the experiments follow real world patterns that are observed in real economies and auctions. In a real economy it is expected that consumers who have larger budgets will bid higher than those with a smaller budget if the "true value" of the item auctioned is public information. Therefore, in these experiments involving the transaction of digital goods we expect that Experiment I, with a reward of $50.00, will have higher final prices than the Experiment II with a reward of $25.00. That is, files will have a higher demand and consequently higher sale price in Experiment I compared to Experiment II. Based on the results in Table 1, the average final price over all four files is 612.5 tokens in Experiment I compared to 453.75 tokens in Experiment II. Therefore, the higher token balance in Experiment I of 1200 tokens resulted in a higher average final price (612.5) for all items compared to Experiment II which had a lower token balance of 500 tokens and lower average final price (453.75) for all items. However, it is unclear thus far from the analysis whether the higher final prices were due to the higher reward, higher initial token balance or arrangement of initial prices for each file. In addition, since the tokens merely serve as a proxy for the reward, the analysis of the reward amount may prove more useful. As a result, in future experiments the reward amounts will be reversed keeping all other parameters (token balance, initial price sequence) constant. This will help to clarify whether the reward amount or token balance is the stronger predictor of final prices.

Thus far, the analysis has examined only the activity corresponding to Round I in which players were only allowed to purchase files. Round II accounts for situations in which players are free to choose whether they would like to purchase or sell content. Even though the reward, token balance, and auction duration are greater in Experiment I, there is more activity or demand in Experiment II, Round II compared toExperiment I, Round II. This contradicts the previous results observed between Experiments I and II in Round I. In Round I, longer auction durations and higher initial token balances and reward amounts, are associated with higher final prices. Therefore, within group (Experiment I and

II) demand was greater in Round I compared to Round II, but between group demand (i.e. between experiments) demand was greater in Round I for Experiment I compared to Experiment II Round I, and less in Experiment I Round II compared to Experiment II Round II. The differences in demand within and between experiments for both rounds is presented in Figure 1. The total number of bids per round and experiment is depicted in Figure 1. Here it can be observed that a total of 66 bids were placed in Round II in Experiment II and only 46 bids were placed in Round II in Experiment I. In addition, the higher number of bids within each experiment between Round I and Round II, undescore the challenge of incentivizing users to contribute content. More people bid in Round I when there were only buy options. In Round II, players could buy and sell and there were fewer bids which suggest fewer players chose not to bid.

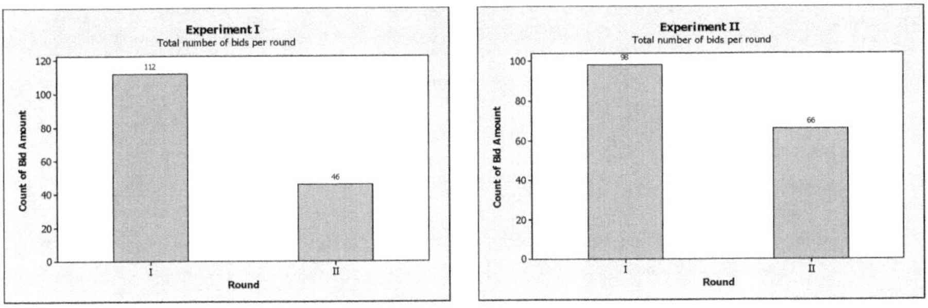

(a) (b)

Fig. 1. Total number of bids per round and experiment

To gain deeper insight and a possible explanation for the difference in demands, the next step in this analysis examines the activity in Round II to explicitly understand why lower values (token balance, reward amounts and auction durations) generated more activity in Round II, Experiment II compared to Round II, Experiment I. An analysis of the number of bids per file per round and the number of times files were sold per round is provided in Figure 2. The results highlight which files were in greatest demand and the saturation of files in the system. While the results indicate there are fewer bids in Round II of Experiment I compared to Round II in Experiment II, Figures 2(a) and 2(c), also illustrate that there were more winners in Round II of Experiment I compared to Round II in Experiment II, Figures 2(b) and 2(d). The larger number of transactions or bids observed in Round II of Experiment II suggests that demand for the files was greater compared to the demand for files in Round II of Experiment I. This is because even though the incentive was lower in Experiment II (i.e. the rewards was $25.00 compared to $50.00 in Experiment I) it is likely that more bids were placed based on the observation that there were fewer winners. That is the supply of files in the system was lower for Experiment II, Round II compared to Experiment I, Round II. Fewer winners sggests the market was not as saturated with files since fewer subjects obtained the files which. According to the fundamental principles of supply and demand a lower supply is typically associated with a higher demand and higher price. An analysis of the ratio of the average initial prices to the token balance confirms that the behavior observed in these experiments corresponds to the expected behavior in a real economy. Specifically, since the ratio in Experiment II Round II is 16.8% compared to a ratio of 12.58% in Experiment I Round I, average initial prices for files relative to the respective initial token balances demonstrate that greater demand does indeed correspond to higher prices for intangible goods such as digital content.

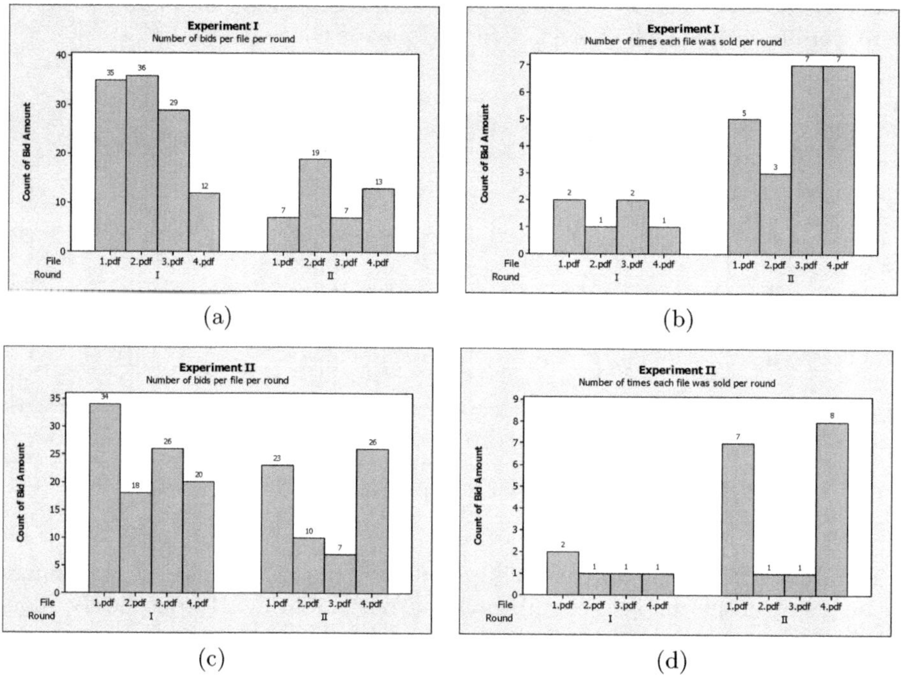

Fig. 2. Number of bids vs. files sold

Our analysis now shifts focus from investigating the issues surrounding content contribution to examining the activity associated with winners of the game. Our ssecond research question, **Q2** is: *Do players that bid the most number of times win the game?*. This question is examined with the corresponding hypothesis, **H2**, which claims *the higher the number of bids the higher the probability of winning the game.* The intuition is that more aggressive or motivated players will bid more during the auctions compared to players who are less motivated, and are therefore more likely to win the game. However, we see that the three winners of the game in Experiment I, Figure 3(b) are subjects 8, 9 and 12 and only subject 8 is among the top three players who bid the most number of times (Figure 3(a) subjects who bid the most are subjects 7, 8 and 11). In Experiment

II also showed similar results. In Experiment II there was only one winner, subject 7, who was not among the subjects who bid the most number of times (Figure 3(c) and 3(d)). This suggests that other factors or behavior may have contributed to winning strategies.

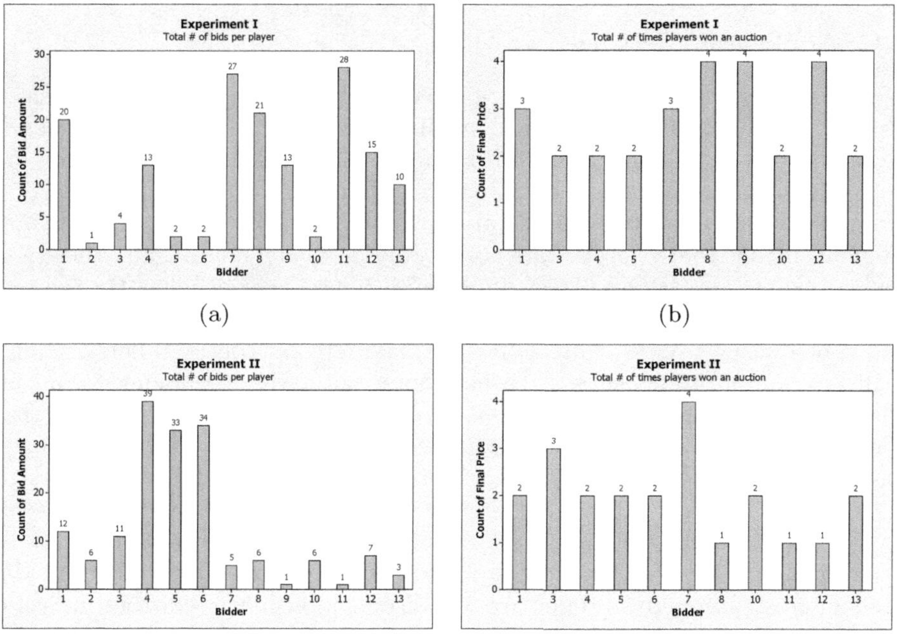

(a) (b) (c) (d)

Fig. 3. Number of bids vs. number of auctions won

5 Incentivizing User Cooperation

The major experimental observation that stands out from our experimental results in the previous section is the challnege of calibrating a proper incentive for motivating Internet users to share content in competitive applications, whose success depends on the co-operation of other users in the system. In this section, we propose a mathematical framework that derives user population threshold values, which hint at the necessity of a certain base population strength in S-Mart for co-operation to take place amongst all the users. We compute (1) the *contributor threshold value* (CTV), which is defined as the minimum number of S-Mart users required to contribute valuable content (without any social influence) for all the S-Mart users to willfully contribute valuable content on a given topic, and (2) the *socially influenced population threshold* (SIPT), which we define to be the population of S-Mart users on a given topic required, in order to maximize the incentive of each S-Mart user to contribute valuable content on the given topic. Complete details regarding the notations and derivation of CTV and SIPT are available in our technical report [20]. We describe our application setting as applied to the S-Mart system. Next, we outline two experiments that we wish to conduct in the near future, to validate the theory as proposed in the mathematical framework.

5.1 Application Settings

We assume the following two types of settings: 1) S-Mart users share content on a particular topic without experiencing any regulation, social influence, quality demands, or central monitoring. This implies that S-Mart users may or may not contribute content, depending on their free will, and social friends of S-Mart users logged on to S-Mart *cannot* influence their friends to contribute. Content, if contributed by S-Mart users could be of any quality, and there is no central monitoring taking place to test S-Mart user misbehavior. By the term "misbehavior", we mean either "withholding" behavior, or "cheating" behavior, i.e., contributing useless content, despite having good content. This setting is realistic of applications such as casual course notes sharing amongst students, where there is no pressure on anyone to contribute and 2) S-Mart users share content on a particular topic without any regulation, but there is a central monitoring system in place to detect[2] with a certain probability of success on whether S-Mart users withhold information or share low quality content despite having good quality content. Once user misbehavior above a certain level is detected, the S-Mart system can impose certain punishments on all the S-Mart users interested in a given topic. We will discuss punishments further in Section 5. In this setting, S-Mart users may be influenced by social friends on not to withhold information so as to avoid global punishment laid down by S-Mart on all users relevant to a given topic. We assume here that social influence *always* motivates users to act altruistically. This influence can be exerted via *Facebook* like social sites, given that the S-Mart application is embedded in a social networking site. An example of an application fitting this setting is *faculty administered collaborative learning*. In this application, students of a class (ex., MATH 101) share course documents (related to homework sets) with other fellow students, and are awarded positive points for sharing valuable content, but the whole class gets negative points for exceeeding a certain degree of misbehavior (if detected)[3]. The points contribute to the final grades of the students in the class.

---

[2] The system can detect misbehavior based on information from other users, or by its own monitoring.

[3] Punishing everyone is a strategy to induce each user to contribute valuable content. Given that users are generally social and that Facebook like social websites can host an S-Mart application, it may not be that difficult to identify users who have cheated. Eventually the cheaters would end up losing their social value amongst friends due to the whole class suffering because of them.

5.2 Computing CTV

Suppose there are n Sharing-Mart users comprising of content producers and consumers, on a certain topic of common interest. We assume that each producer in S-Mart has a certain initial amount of content with itself regarding the topic. Producers could be consumers and vice-versa. By the term 'topic', we refer to a subject, information about which is useful to the members of S-Mart interested in the subject, ex., the topic could be Lagrange Multipliers in a MATH 101 calculus course. Let $U_i(NC|\gamma)$ be the utility of a non-cooperative user, i, in S-Mart, when γ members in S-Mart decide to *contribute* content on a topic. Here, a non-cooperative user is an S-Mart user who either withholds information or provides low-quality content inspite of having better quality content. The *contributors* are assumed to be *altruistic* and share the best content they have with the S-Mart users. Similarly, we denote by $U_i(C|\gamma)$ the utility of the same non co-operative user, i, in S-Mart, when it turns co-operative (contributes), and γ members in S-Mart decide to contribute something on a topic. Throughout the rest of the paper, we use the terms 'co-operation' and 'contribution' interchangeably. We state the following relationship on an *individual* level:

$$U_i(NC|\gamma) > U_i(C|\gamma),\ 0 \leq \gamma < n. \tag{1}$$

The above inequality states that on an individual level, a non-cooperative S-Mart user is better off withholding content rather than sharing it with others, as it diminishes the user's strategic advantage. By withholding content, a non cooperative user enjoys all the benefits of other's contributions without giving anything away itself.

However, on the *group* level we derive the following relationship:

$$U_{grp}(0) < U_{grp}(k),\ k_t < k \leq n. \tag{2}$$

This equation implies that a group of size k greater than a threshold k_t, benefits in co-operation more that when no one in the group co-operates, because if everyone were to withhold content, there would be no benefit to the group, and in turn to any individual. We consider a group utility function to be the utility of S-Mart system. Thus, from equations (1) and (2), we observe that a user will not want to contribute individually, but might not benefit anything if all members in the group behave in the same manner. In this section, we propose a way to reverse the sense of inequality (1) such that S-Mart members are individually incentivized to contribute valuable content for the benefit of the system.

An individual user i's utility function when it contributes

$$U_i(C) = d_i \cdot NUC_i - f_i \sum_{j=1}^{z_i} c_{ij} = B_i - f_i \sum_{j=1}^{z_i} c_{ij}. \tag{3}$$

It is evident that when user i decides to be non-cooperative, its utility function, $U_i(NC)$ equals $d_i \cdot NUC_i = B_i$. Thus $U_i(NC) > U_i(C)$. En-route to computing Contributor Threshold Value (CTV), we execute the following two steps: 1)

We derive k_t, the *minimum* number of S-Mart users amongst the n users, whose positive contribution results in the group utility being more than the utility when none of the users co-operate, i.e., $U_{grp}(0)$. k_t from equation 2 arises due to the fact that contributing content places a cost on users, and as a result the benefit due to co-operation amongst a certain number of users should exceed the cost of contribution before any group activity to take place and 2) Having executed step 1, we ensure that contribution is efficient on the group level beyond a certain size. However, it does not help reverse the sense of inequality (1). In this step, we propose a system that provides bonuses to users who contribute, such that they can be compensated for their contribution costs. The system reverses the sense of inequality (1) and individually incentivizes S-Mart users to contribute content.

We omit the detailed derivation steps to computing CTV due to lack of space. The reader is referred to [20] for more details. In a system with n users, in which each user has z units of content about a particular TOI, each with identical value v; each user incurs a cost c for contributing a unit of content and contributes a fraction, f of its total accumulation; the degree of content overlap between any two users on a particular TOI is cov, and the degree to which any user i gains from NUC_i is d; and the events of content overlap between users i and j, and between users i and k are statistically independent, the closed form expression for CTV is given as

$$CTV = \frac{\ln(1 - cov(1 + \frac{2c}{vd}))}{\ln(1 - cov)} \tag{4}$$

Sensitivity Analysis. Based on the CTV expression, we observe that CTV values increase with increasing $\frac{c}{v}$ values. This is intuitive as the cost incurred by a user for sharing topic information increases w.r.t. the benefits obtained, and as a result users are less incentivized to contribute and the critical number increases. We also observe that CTV values increases with increasing d values. This result is intuitive as well because higher values of d imply that a user benefits more from the shared information pool and this happens only when the critical number increases.

5.3 Computing SIPT

In this section, we study the role of *social influence* and S-Mart punishments in ensuing co-operative behavior amongst S-Mart users in socially selfish applications (e.g., sharing course notes/lectures). It is evident that if *every* S-Mart user w.r.t to a TOI contributes, we are guaranteed a successful operating S-Mart system with every user doing its best to help the other users gain knowledge. However, in reality this is hardly the case. Users are non-cooperative by nature and do not want to share valuable content with others. In such situations, social influence from friends, or imposing punishments upon detecting selfish behavior could change user mindset in favor of contributing valuable content. Given the tremendous popularity of social networking websites, its not difficult to embed and administer educational S-Mart applications on a site like Facebook (refer to application setting 2 in Section I). In such cases, it is important that *each* user is incentivized to contribute for the benefit of the whole system. The entire system could represent a course in an university, in which one of the main goals of the instructor is to facilitate collaborative learning amongst students for altruistic knowledge dissemination. In this section, we compute the socially influenced population threshold (SIPT), which we define to be the number of members(users) needed in a system functioning on the S-Mart framework such that each user in the system is maximally incentivized to co-operate.

Let W_i be the probability that user i withholds or cheats on valuable information. Let D_i denote the probability that the S-Mart system detects this misbehavior. We define P_i to be the probability that user i withholds information and the system (we use the term 'system' and 'S-Mart system' interchangeably) detects it. We assume independence of the events that users cheat and the system detects, and denote P_i to be the product of W_i and D_i. We also assume that the S-Mart system imposes a punishment if it detects any user misbehaving. Thus,the probability P that at least one user withholds content $1 - \prod_{i=1}^{n}(1 - W_i \cdot D_i)$. Given that the system punishes all the users once it detects any misbehavior, a user could either be 1) insensitive to any punishment, or 2) concerned about the punishment. Let U_i^{nc} denote the utility of a user not concerned with punishments imparted by S-Mart. We formulate U_i^{nc} as $(U_i + PM_i)P + U_i(1 - P)$, where U_i denotes the individual utility of an S-Mart user when it decides to co-operate, i.e., $U_i(C)$, or the utility when it chooses not to co-operate, i.e., $U_i(NC)$, and $PM_i < 0$ is the punishment imparted to user i by S-Mart. A user concerned with punishments would try its best to avoid it. Individually, he would not want to cheat, and would also want others not to cheat. One way a user could prevent others from misbehaving is by influencing its friends, who in turn influence their friends, and so on. Given that the S-Mart application is embedded in a social networking site friend influence should be possible. Let U_i^c denote the utility of a user concerned with punishments imparted by S-Mart. We formulate U_i^c as $(U_i+PM_i-CSI_i)P'+(U_i-CSI_i)(1-P')$, where C_i is the social influence index of user i, where $C_i \in [0, 1]$. This quantity indicates the degree to which a user is influenced by his friends to not withhold valuable content for the benefit of S-Mart. CSI_i denotes the cost to user i for not withholding valuable information/content due to social influence, when in fact he would have preferred selfish behavior without the social influence. P' is the probability that at least one user, after being socially influenced, is caught misbehaving in the system. We denote P' by $1 - \prod_{i=1}^{n} W_i \cdot (1 - C_i) \cdot D_i$. The difference in utility, U_i^{diff}, between user i's mindset of being concerned and unconcerned about S-Mart punishments is given as

$$U_i^{diff} = U_i^c - U_i^{nc}. \tag{5}$$

A user prefers being concerned about punishments to being unconcerned if $U_i^{diff} \geq 0$. We perform an utility analysis and derive the value of SIPT as

$$SIPT = \frac{\ln\{\frac{\ln(1-W\cdot D)}{\ln[1-W\cdot(1-C)\cdot D]}\}}{\ln\{\frac{[1-W\cdot(1-C)\cdot D]}{1-W\cdot D}\}} \tag{6}$$

The details of the derivation is omitted due to lack of space. The readers are referred to [20] for further details.

Sensitivity Analysis. Based on the SIPT expression, we observe that SIPT values decrease with increase in the values of D. The intuition behind this result is the fact that with increasing values of D - the detection probability, the users would willingly contribute for the fear of punishments, even in the case of a low number of users present in the system. We also observe that the SIPT values decrease with increase in the C values. This result is also intuitive as with increasing social influence, it requires fewer of users to be present so as to maximize user willingness to contribute content. However, we see that the SIPT values increase with increase in W because an increase in the withholding probability of users implies the requirement of greater number of users in the system to maximize user willingness to contribute content.

5.4 Future Experiments

In this section we give an experimental outline of how to go about determining CTV and SIPT values empirically. Our goal of conducting the experiments is to validate the theory proposed in the mathematical framework.

Experiment outline to measure CTV: Assume a class assignment, for which students are required to write a class report regarding a given topic. (E.g., a survey paper on routing protocols in wireless networks.) The students are evaluated based on the quality of the report, which is determined by the number of salient points in the report. Apart from some very common information, students would vary w.r.t. one another in terms of topic points. We assume that there is a class organizer such as the Professor or the teaching assistant (TA). We plan to conduct the experiment in two rounds. In the first round, the Professor/TA gathers topic points separately from each student. In the second round, the Professor/TA creates an online discussion board, in which students could share their topic points. The sharing is not made compulsory; however, if students share their knowledge, they are awarded a certain number of points for their contributions. The flip side to this benefit is that students might lose a competitive advantage to other competitors. We measure this loss of advantage in terms of a cost. The CTV value could be estimated by the Professor/TA in the second round by observing the rate at which students upload content points. We expect a sudden surge of content uploads over time. We need to keep track of the time when the surge occurs, and identify the number of users just before the surge occurs. This number will give us an estimate of the CTV.

Experiment outline to measure SIPT: We design a similar experiment to measure SIPT. The only difference is that we incorporate social influence, misbehavior detection, and punishments. Social influence is a natural property and is not within the control of the Professor/TA. We assume here that a student may be positively influenced through chat or Facebook like mechanisms to willfully contribute content. We capture misbehavior detection via student complaints to the Professor/TA about someone having information and not sharing it, or someone willfully sharing wrong information. Punishments are computed in the form of points deducted from every student in the class. Similar to the experiment to measure CTV, the experiment here will have two rounds and the Professor/TA can estimate the SIPT from the second round by observing the contributing population count at the time when a surge of content contribution occurs.

6 Conclusion

In this paper, we have described Sharing-Mart, a virtual file sharing platform, and have investigated whether similar economic behavior is observed in the virtual economy as in a real economy. Two research questions and hypotheses (H1, and H2) have been presented to understand the economic behaviors and dynamics of the package auction for public goods using the Sharing-Mart system. H1, which states subjects who initally pay more for content are more likley to particiapte more in the system compared to subjects who pay less, appears to be true. However, analysis of the results for H2 indicate the hypothesis is not supported. Therefore, while higher bid amounts may correpsond to high activity or participation the most active bidders and users who paid the most for content do not necessarily win the game. The experimental observations also highlight the challenge of calibrating proper incentives to motivate participation and content contribtuion in competitive applications, whose success depends on mutual cooperation amongst the users. To alleviate this problem we have proposed a mathematical framework that derives user population threshold values, which hint at the necessity of a certain base population strength in S-Mart for co-operation to take place amongst all the users.

Acknowledgement. We would like to thank Felix Ming-Fai Wong for many suggestions and edits while writing this paper. This work was in part supported NSF-CNS-0905086.

References

1. Inaltekin, H., Ogzencil, H., Xu, H., Chiang, M., Poor, H.V.: Online content-trading: Sharing Mart system and auction experiments. In: Proc. IEEE Workshop on Social Networking, Cape Town, South Africa (May 2010)
2. Alavi, M., Leidner, D.E.: Review: Knowledge Management and Knowledge Management Systems: Conceptual Foundations and Research Issues. MIS Quarterly 25(1), 107–136 (2001)
3. Alavi, M., Leidner, D.E.: Knowledge management systems: Issues, challenges, and benefits. Communications of the AIS 1, 7 (1999)

4. King, W.R., Marks Jr., P.V.: Motivating knowledge sharing through a knowledge management system. Omega 36, 131–146 (2008)
5. Lin, T.-C., Huang, C.-C.: Understanding knowledge management system usage antecedents: an integration of social cognitive theory and task technology fit. Information & Management 45(6), 410–417 (2008)
6. Kankanhalli, A., Tan, B.C.Y., Wei, K.-K.: Contributing Knowledge to Electronic Knowledge Repositories: An Empirical Investigation. MIS Quarterly 29(1), 113–143 (2005)
7. Milgrom, P.: Putting Auction Theory to Work: The Simultaneous Ascending Auction. Journal of Political Economy 108, 245–272 (2000)
8. Krishna, V.: Auction Theory. Academic Press, San Diego (2002)
9. Milgrom, P.: Putting Auction to Work. Cambridge University Press (2004)
10. Bichler, M.: Trading Financial Derivatives on the Web. An Approach Towards Automating Negotiations on the OTC Markets. Information Systems Frontiers 1(4) (2000)
11. Klemperer, P.D. (ed.): The Economic Theory of Auctions. Edward Elgar, Cheltenham (2000a)
12. Klemperer, P.D.: Auction Theory: A Guide to the Literature. Journal of Economic Surveys 13(3), 227–286 (1999a)
13. Edelman, B., Ostrovsky, M., Schwarz, M.: Internet Advertising and the Generalized Second-Price Auction: Selling Billions of Dollars Worth of Keywords. American Economic Review 97(1), 242–259 (2007)
14. Adar, E., Huberman, B.A.: Free riding on Gnutella, Technical report, Xerox PARC (August 2000)
15. Hughes, D., Coulson, G., Walkerdine, J.: Free riding on Gnutella revisited: The bell tolls? IEEE Dist. Systems Online 6(6) (2005)
16. Jun, S., Ahamad, M.: Incentives in BitTorrent induce free riding. In: Proceedings of the 2005 ACM SIGCOMM Workshop on Economics of Peer-to-Peer Systems, Philadelphia, PA, pp. 116–121 (2005)
17. Liogkas, N., Nelson, R., Kohler, E., Zhang, L.: Exploiting BitTorrent for fun (but not profit). In: Proceedings of the International Workshop on Peer-to-Peer Systems (IPTPS), Santa Barbara, CA (2006)
18. Piatek, M., Isdal, T., Anderson, T., Krishnamurthy, A., Venkatararamani, A.: Do incentives build robustness in BitTorrent? In: Proc. of 4th USENIX Symposium on Networked Systems Design & Implementation, Cambridge, MA (April 2007)
19. Sirivianos, M., Park, J.H., Chen, R., Yang, X.: Free-riding in BitTorrent networks with the large view exploit. In: Proceedings of the 6th International Workshop on Peer-to-Peer Systems (IPTPS), Bellevue, WA (2007)
20. Leberknight, C.S., Pal, R., Chiang, M., Poor, H.V.: Sharing-Mart: Online Auctions for Digital Content Trading. Princeton University, Technical Report (February 2011), http://scenic.princeton.edu/research/researchSharingMart.html/leberknight-content-contribution-2011.pdf

Selfish Random Access: Equilibrium Conditions and Best-Response Learning*

Hazer Inaltekin[1], Mung Chiang[2], and Harold Vincent Poor[2]

[1] The University of Melbourne, Parkville, VIC 3010, Australia
hazeri@unimelb.edu.au
[2] Princeton University, Princeton, NJ 08544, USA
{chiangm,poor}@princeton.edu

Abstract. This paper studies a class of random access games for wireless channels with multipacket reception. First, necessary and sufficient equilibrium conditions for a contention strategy profile to be a Nash equilibrium for general wireless channels are established. Then, applications of these equilibrium conditions for well-known channel models are illustrated. Various engineering insights and design ideas are provided. Finally, the results are extended to an incomplete information game setting, and best-response learning dynamics leading to Nash equilibria are investigated.

Keywords: Slotted ALOHA, multipacket reception, game theory, contention control, medium access control.

1 Introduction

1.1 Background and Contributions

Game theory and the related field of mechanism design have the potential to guide engineering efforts to overcome potential design challenges in fourth generation (4G) wireless networks by providing a bottom-up analytical and principled approach to design local operation rules and to verify resulting collective network behavior through equilibrium analyses. In particular, it is illustrated in recent works [1]-[3] that game theory provides new insights to reverse/forward engineer existing medium access control (MAC) protocols, better fairness and service differentiation, higher throughput and a mechanism to decouple contention control from handling failed packets for a class of multiple access networks. This paper also focuses on layer-2 MAC protocols for wireless networks, and provides new equilibrium results and design insights based on noncooperative game theory.

Wireless channels are broadcast channels by their nature. Therefore, transmissions through them must be coordinated to control multiple access interference (MAI). Contention based random access approaches, which we consider in this

* This research was supported in part by the U.S. National Science Foundation under Grant CNS-09-05086, and by the Australian Research Council under Grant DP-11-0102729.

R. Jain and R. Kannan (Eds.): GameNets 2011, LNICST 75, pp. 169–181, 2012.

paper, include slotted ALOHA, CSMA/CA and IEEE 802.11 DCF, and do not require centralized scheduling. A key design degree of freedom in random access is channel access probability determination, or contention resolution, to mitigate MAI. Different protocols differ in how they implement contention resolution such as window based and persistence probability based approaches. This paper analyzes equilibrium channel access probabilities (equivalently called contention resolution strategies, or transmission probabilities) and the resulting network performance for a wide range of channel models and communication scenarios.

Our contributions can be summarized as follows. We focus on the contention resolution problem over wireless channels with multipacket reception capability, and characterize the set of Nash equilibria by providing necessary and sufficient conditions for a Nash equilibrium contention strategy profile. Multipacket reception capability is an important feature of our model to capture probabilistic receptions in wireless multiple access. We demonstrate applications of these results in practical communication scenarios, and provide engineering insights to achieve optimal throughput. Finally, random access games with incomplete information structure are analyzed by obtaining the form of equilibrium contention resolution strategies, establishing existence and uniqueness results and illustrating learning dynamics on the best-response path.

1.2 Related Work

There is a large and growing body of work applying game-theoretic techniques for contention resolution in wireless networks. Here, we mention the ones that are most relevant to this work.

We focus on a class of random access games, first introduced in [4] and then further improved in [5]. When compared to these works, we provide more detailed equilibrium conditions for a contention strategy profile to be a Nash equilibirum for more general wireless channels.

In [1]-[3] and [6], the main focus is on the dynamics of iterative strategy update mechanisms such as best-response, gradient, or Jacobi play, achieving a desired equilibrium point over collision channels. Unlike these works, the physical layer model considered in this paper is more general, including the collision channel model as a special case. Except for the incomplete game formulation, our main focus here is on necessary and sufficient conditions to be satisfied by steady-state equilibrium contention resolution strategies, rather than on the transient network behavior. These equilibrium conditions can be solved either analytically or numerically to obtain equilibrium transmission probabilities, and then the network can be readily stabilized to a desired equilibrium by broadcasting these probabilities to users. For the incomplete game formulation, even though the existence and uniqueness of the equilibrium can be established, such closed-form or numerical solutions are not readily available, and therefore the transient network behavior is also investigated by studying learning dynamics on the best-response path.

Similar to our problem set-up, multipacket reception capability in the random access setting is also considered in [7] and [8]. As compared to the existence results, which provide limited information about the structure of equilibrium strategies, appearing in these works, we obtain more detailed necessary and sufficient equilibrium conditions, and solve them, either analytically or numerically, to derive the shape of equilibrium contention resolution strategies.

2 System Model

Consider a wireless multiple access communication network in which N selfish mobile users, indexed by $\mathcal{I} = \{1, 2, \cdots, N\}$, are contending for channel access to communicate with a common base station (BS). The wireless channel is characterized by the stochastic reception matrix

$$\boldsymbol{R} = (r_{n,k})_{\substack{1 \leq n \leq N, \\ 0 \leq k \leq N}}, \tag{1}$$

where $r_{n,k}$ represents the probability that k packets are received correctly given n of them are transmitted. By default, $r_{n,k}$ is set to zero if $k > n$. If $r_{1,0} > 0$, we say that the channel is *imperfect* (or, noisy). All users have identical packet success probabilities given by

$$\gamma_n = \frac{1}{n} \sum_{k=1}^{n} k r_{n,k}. \tag{2}$$

It is assumed that $\gamma_{n+1} \leq \gamma_n$ to model destructive effects of MAI on packet receptions. It is also assumed that γ_{n+1} is strictly smaller than γ_n for at least one n in $\{1, 2, \cdots, N-1\}$ in order to avoid trivialities.

If mobile user i transmits a packet successfully, it receives a normalized utility of 1 unit. If the transmission fails, it receives a utility of $-c_i$ units, where $c_i > 0$ is interpreted as the cost of packet failure for user i. If it waits, it receives a utility of 0 units. These utilities are necessary for setting up a selfish random access utility maximization problem as well as allowing us to model different battery levels, delay and quality-of-service requirements of different users. We let $\boldsymbol{u}_i = (1, 0, -c_i)$. The *random access game* $\mathcal{G}$ is defined to be the triple $\mathcal{G} = \langle \mathcal{I}, \{\mathcal{S}_i\}_{i \in \mathcal{I}}, \{\boldsymbol{u}_i\}_{i \in \mathcal{I}} \rangle$, where $\mathcal{S}_i = [\alpha_i, \beta_i] \subseteq [0, 1]$ is the set from which user i chooses a transmission probability (i.e., a contention resolution strategy) to access the wireless channel.

The average utility that a user receives is a function of her transmission probability and the transmission probabilities of other users. Let $\mathcal{S} = \prod_{i \in \mathcal{I}} \mathcal{S}_i$ be the product set of user strategies, $\boldsymbol{p}$ be a vector of transmission probabilities in $\mathcal{S}$, and $U_i(\boldsymbol{p})$ be the expected utility that the ith user receives as a function of $\boldsymbol{p}$. The selfish optimization problem to be solved by user i is to find $p_i^\star$ such that $U_i(p_i^\star, \boldsymbol{p}_{-i}) \geq U_i(p_i, \boldsymbol{p}_{-i})$ for all $p_i \in \mathcal{S}_i$, where $\boldsymbol{p}_{-i}$ represents the vector of transmission probabilities of other users. We say that $\boldsymbol{p}^\star = (p_i^\star)_{i \in \mathcal{I}}$ is a *Nash equilibrium* if and only if $U_i(p_i^\star, \boldsymbol{p}_{-i}^\star) \geq U_i(p_i, \boldsymbol{p}_{-i}^\star)$ for all $i \in \mathcal{I}$ and $p_i \in \mathcal{S}_i$.

3 Equilibrium Contention Resolution Strategies: General Wireless Channels

In this section, we analyze equilibrium contention resolution strategies for a wireless channel with a general multipacket reception model, and obtain necessary and sufficient equilibrium conditions to be satisfied by a Nash equilibrium transmission probability vector. In the next section, we will illustrate the applications of our results in more specific communication scenarios by solving these equilibrium conditions.

The first critical issue to resolve is the existence of a Nash equilibrium. To this end, a positive existence result directly follows from the Glicksberg fixed point theorem [9]. (See also Theorem 1.2 in [10].) However, such existence results provide limited information about the structure of the Nash equilibria, and selfish transmission probabilities at these equilibria. Therefore, we establish necessary and sufficient conditions to be satisfied by a Nash equilibrium transmission probability vector in the following theorem.

Theorem 1. *For a given contention strategy profile* $\boldsymbol{p}$, *let*

$$\Gamma_i\left(\boldsymbol{R}, \boldsymbol{p}_{-i}\right) = \sum_{n=1}^{N} \sum_{\substack{\mathcal{I}_n \subseteq \mathcal{I}: \\ i \in \mathcal{I}_n, |\mathcal{I}_n| = n}} \gamma_n \prod_{j \in \mathcal{I}_n - \{i\}} p_j \prod_{j \in \mathcal{I} - \mathcal{I}_n} (1 - p_j). \tag{3}$$

Then, $\boldsymbol{p}^\star$ *is a Nash equilibrium if and only if the following equilibrium conditions hold for all* $i \in \mathcal{I}$.

(i) *If* $\Gamma_i\left(\boldsymbol{R}, \boldsymbol{p}^\star_{-i}\right) > \frac{c_i}{1+c_i}$, *then* $p_i^\star = \beta_i$.
(ii) *If* $\Gamma_i\left(\boldsymbol{R}, \boldsymbol{p}^\star_{-i}\right) = \frac{c_i}{1+c_i}$, *then* $\alpha_i \leq p_i^\star \leq \beta_i$.
(iii) *If* $\Gamma_i\left(\boldsymbol{R}, \boldsymbol{p}^\star_{-i}\right) < \frac{c_i}{1+c_i}$, *then* $p_i^\star = \alpha_i$.

Proof. For a given $\boldsymbol{p}$, it is easy to see that $\Gamma_i\left(\boldsymbol{R}, \boldsymbol{p}_{-i}\right)$ is the probability that a packet transmission from user i becomes successful given the channel reception matrix $\boldsymbol{R}$ and other users' transmission probabilities $\boldsymbol{p}_{-i}$. Then, the expected utility that user i receives when the random access game is played according to $\boldsymbol{p}$ is equal to

$$U_i\left(p_i, \boldsymbol{p}_{-i}\right) = p_i\left(\left(1 + c_i\right)\Gamma_i\left(\boldsymbol{R}, \boldsymbol{p}_{-i}\right) - c_i\right). \tag{4}$$

Now, suppose that $\boldsymbol{p}^\star$ is a Nash equilibrium. Firstly, $p_i^\star$ must be β_i if $\Gamma_i\left(\boldsymbol{R}, \boldsymbol{p}^\star_{-i}\right) > \frac{c_i}{1+c_i}$ since $U_i\left(p_i, \boldsymbol{p}^\star_{-i}\right)$ is a linear function of p_i when $\boldsymbol{p}^\star_{-i}$ is fixed. Secondly, $p_i^\star$ must be α_i if $\Gamma_i\left(\boldsymbol{R}, \boldsymbol{p}^\star_{-i}\right) < \frac{c_i}{1+c_i}$. Finally, $p_i^\star$ can be set to any value in $[\alpha_i, \beta_i]$ if $\Gamma_i\left(\boldsymbol{R}, \boldsymbol{p}^\star_{-i}\right) = \frac{c_i}{1+c_i}$. This completes the proof for the *only if* part of the theorem. The other direction also follows from similar arguments.

For a given $\boldsymbol{R}$, $\Gamma_i\left(\boldsymbol{R}, \boldsymbol{p}_{-i}\right)$ can be interpreted as the contention signal that user i receives when the contention resolution strategies of other users are given by $\boldsymbol{p}_{-i}$. A higher contention signal received by user i means that less MAI is generated by other users, and therefore the higher the channel access and the resulting packet success probabilities of user i are.

In the next theorem, we establish a symmetry property for equilibrium contention resolution strategies in symmetric random access games.

Theorem 2. *Assume all users have the same strategy set* $[\alpha, \beta] \subseteq [0,1]$, *and the same cost of packet failure* $c > 0$. *If* $\mathbf{p}^\star$ *is a Nash equilibrium, then* $p_i^\star = p_j^\star$ *for all* $p_i^\star$ *and* $p_j^\star$ *in* (α, β). *In particular, if* $p_i^\star \in (\alpha, \beta)$ *for all* $i \in \mathcal{I}$, *then all users access the channel with the same transmission probability* $p^\star$ *solving*

$$J(p^\star) = \frac{c}{1+c}, \tag{5}$$

where $J(p) = \sum_{n=0}^{N-1} \gamma_{n+1} \binom{N-1}{n} p^n (1-p)^{N-1-n}$.

Proof. Let $\mathcal{I}_{-\{i,j\}} = \mathcal{I} - \{i,j\}$ and $\boldsymbol{p}_{-\{i,j\}} = (p_k)_{k \in \mathcal{I}_{-\{i,j\}}}$. Let $\Gamma_i\left(\boldsymbol{R}, \boldsymbol{p}_{-i}\right)$ be defined as in the proof of Theorem 1. We can expand $\Gamma_i\left(\boldsymbol{R}, \boldsymbol{p}_{-i}\right)$ as a function of p_j as

$$\Gamma_i\left(\boldsymbol{R}, \boldsymbol{p}_{-i}\right) = \Gamma_i\left(\boldsymbol{R}, p_j, \boldsymbol{p}_{-\{i,j\}}\right) =$$

$$p_j \sum_{n=0}^{N-2} \sum_{\substack{\mathcal{I}_n \subseteq \mathcal{I}_{-\{i,j\}}: \\ |\mathcal{I}_n| = n}} \gamma_{n+2} \prod_{k \in \mathcal{I}_n} p_k \prod_{k \in \mathcal{I}_{-\{i,j\}} - \mathcal{I}_n} (1-p_k)$$
$$+ (1-p_j) \sum_{n=0}^{N-2} \sum_{\substack{\mathcal{I}_n \subseteq \mathcal{I}_{-\{i,j\}}: \\ |\mathcal{I}_n| = n}} \gamma_{n+1} \prod_{k \in \mathcal{I}_n} p_k \prod_{k \in \mathcal{I}_{-\{i,j\}} - \mathcal{I}_n} (1-p_k).$$

The last equation implies the relation $\Gamma_i\left(\boldsymbol{R}, p_i, \boldsymbol{p}_{-\{i,j\}}\right) = \Gamma_j\left(\boldsymbol{R}, \boldsymbol{p}_{-j}\right)$. We also have

$$\frac{\partial \Gamma_i\left(\boldsymbol{R}, \boldsymbol{p}_{-i}\right)}{\partial p_j} = -\sum_{n=0}^{N-2} \sum_{\substack{\mathcal{I}_n \subseteq \mathcal{I}_{-\{i,j\}}: \\ |\mathcal{I}_n| = n}} (\gamma_{n+1} - \gamma_{n+2}) \prod_{k \in \mathcal{I}_n} p_k \prod_{k \in \mathcal{I}_{-\{i,j\}} - \mathcal{I}_n} (1-p_k),$$

which is strictly smaller than zero. Thus, $\Gamma_i\left(\boldsymbol{R}, p_j, \boldsymbol{p}_{-\{i,j\}}\right)$ is strictly decreasing in p_j for any given fixed $\boldsymbol{p}_{-\{i,j\}}$. Let $\boldsymbol{p}^\star$ be a Nash equilibrium such that there exist $p_i^\star$ and $p_j^\star$ in (α, β) and $p_i^\star \neq p_j^\star$. By Theorem 1, this can happen only if $\Gamma_i\left(\boldsymbol{R}, p_j, \boldsymbol{p}^\star_{-\{i,j\}}\right)$ crosses $\frac{c}{1+c}$ at two points $p_j^\star$ and $p_i^\star$; but this contradicts the strictly decreasing nature of $\Gamma_i\left(\boldsymbol{R}, p_j, \boldsymbol{p}^\star_{-\{i,j\}}\right)$ as a function of p_j. Equation (5) follows after some simplifications.

4 Applications and Discussion

We will now demonstrate some applications of the above general equilibrium results in two specific communication scenarios. Further applications are also possible. We start our discussion with equilibrium contention resolution strategies for imperfect collision channels.

4.1 Selfish Random Access over Imperfect Collision Channels

In the collision channel model, a packet transmission is assumed to be successful only if there is no other user attempting to transmit simultaneously. Hence, $r_{n,k} = \delta_{0,k}$ if $n \geq 2$, where $\delta_{i,j} = 1$ if $i = j$, and zero otherwise. We let $r_{1,0} = \theta$ and $r_{1,1} = 1 - \theta$ for some $\theta \in [0, 1]$. Here, the parameter θ can be interpreted as a measure of the noise level summarizing all random factors such as background noise, fading and path-loss affecting packet receptions. The smaller the θ is, the less noise is present in the system, and a packet transmission is more likely to be successful if there is no other transmission attempt. On the other hand, if θ is large, it is more likely that a packet fails even if there is no other user transmitting simultaneously.

By setting the strategy sets to $[0, 1]$, we can simplify the equilibrium conditions in Theorem 1 as follows: $\boldsymbol{p}^\star$ is a Nash equilibrium if and only if, for all $i \in \mathcal{I}$, it satisfies (i) $p_i^\star = 1$ if $(1 - \theta) \prod_{j \neq i} (1 - p_j^\star) > \frac{c_i}{1+c_i}$, (ii) $p_i^\star \in [0, 1]$ if $(1 - \theta) \prod_{j \neq i} (1 - p_j^\star) = \frac{c_i}{1+c_i}$, and (iii) $p_i^\star = 0$ if $(1 - \theta) \prod_{j \neq i} (1 - p_j^\star) < \frac{c_i}{1+c_i}$.

To simplify these conditions further, we will focus on the most interesting case in which $c_i < \frac{1-\theta}{\theta}$ for all $i \in \mathcal{I}$. Other cases can be analyzed similarly. Let $\boldsymbol{p}^\star$ be a Nash equilibrium such that a subset $\mathcal{I}_0$ of users in $\mathcal{I}$ transmit with positive probability, while others exercise zero transmission probability as their contention resolution strategies. Then, contention resolution strategies of users in $\mathcal{I}_0$ must satisfy the second equilibrium condition, which leads to closed form expressions

$$p_i^\star = 1 - \frac{1 + c_i}{c_i} \left(\frac{1}{1 - \theta} \varphi(\mathcal{I}_0) \right)^{\frac{1}{|\mathcal{I}_0| - 1}}, \forall i \in \mathcal{I}_0, \tag{6}$$

where the set function $\varphi : 2^{\mathcal{I}} - \{\emptyset\} \mapsto \mathbb{R}_+$ is defined as $\varphi(\mathcal{I}_0) = \prod_{i \in \mathcal{I}_0} \frac{c_i}{1+c_i}$ for all non-empty subsets $\mathcal{I}_0$ of $\mathcal{I}$. Note that such a solution is feasible only if $\left(\frac{1}{1-\theta} \varphi(\mathcal{I}_0) \right)^{\frac{1}{|\mathcal{I}_0| - 1}} \leq \frac{c_i}{1+c_i} < 1$, which further implies $\varphi(\mathcal{I}_0) < 1 - \theta$. Therefore, when $|\mathcal{I}_0| = 1$, we set $p_i^\star$ to 1 for $i \in \mathcal{I}_0$ without causing any ambiguity. Since transmission probabilities for users with different cost values are different, these transmission probabilities also indicate how different services are provisioned to different users.

In Fig. 1, we plot the equilibrium transmission probabilities and the equilibrium throughput for the homogenous case versus c by setting N to 5. We focus only on the equilibrium where all users transmit with the same positive probability, which corresponds to the fair allocation of communication resources. The equilibrium transmission probability is, then, given by $p^\star = 1 - \sqrt[N-1]{\frac{1}{1-\theta} \frac{c}{1+c}}$. As expected, when the noise level θ increases, transmission probabilities and the system throughput decrease. In all cases, small values of c lead to high transmission probabilities, which in turn results in excessive packet collisions and low

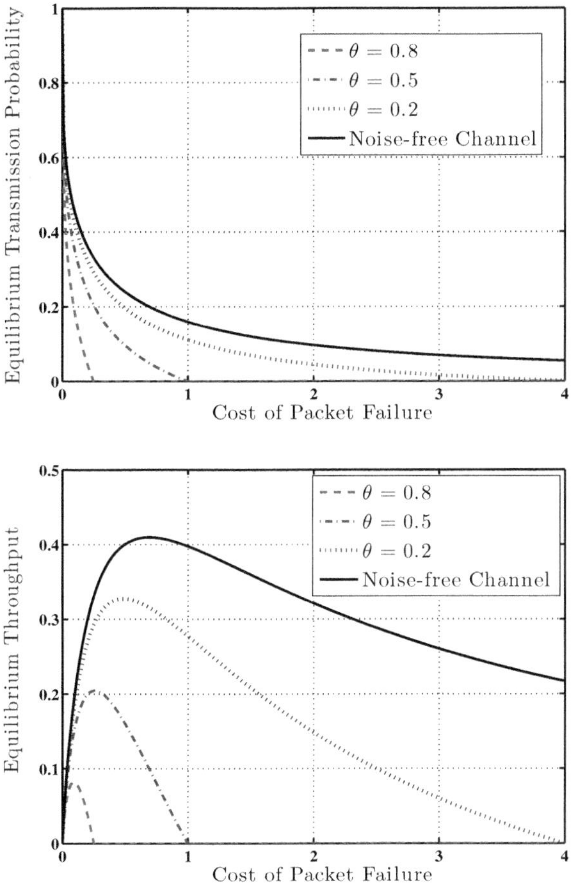

Fig. 1. Equilibrium contention resolution strategies (top figure) and the corresponding system throughput (bottom figure). Imperfect collision channels with $N = 5$.

throughput. Similarly, large values of c result in channel under-utilization, and therefore low throughput. In the middle-ground, there exists an optimal level of c maximizing the system throughput. It is easy to see that this maximum throughput is also the best that we can achieve via a central controller since transmission probabilities are continuous functions of costs. Therefore, there is no loss from selfish operation if selfish transmission probabilities can be manipulated to drive the system to the optimal operating point.

For example, Fig. 1 suggests that when c is small, a central controller can use the parameter θ as a signaling device to manipulate transmission probabilities, and drive the system to the optimal operating point, either by declaring a fictitious noise level to be greater than the true noise level, or by introducing artificial noise during the channel estimation phase. This approach will decrease

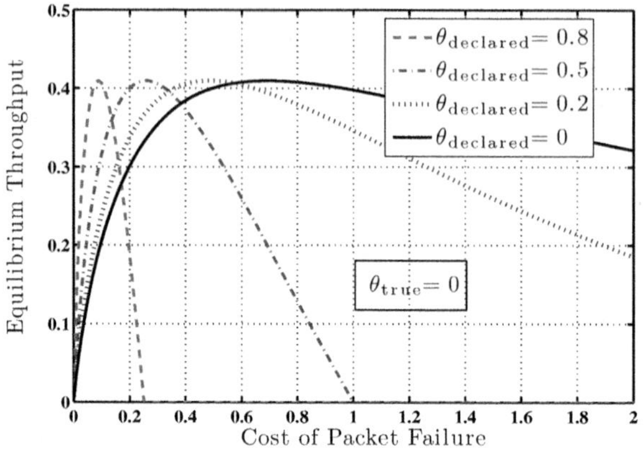

Fig. 2. System throughput when selfish users are manipulated by declaring fictitious noise levels. Perfect collision channel with $N = 5$.

users' greediness, eliminate excessive collisions and increase the system throughput. This operation can also be considered as a design process for user utility functions based on changing the effective value of c to achieve optimal performance. Figure 2 illustrates that the throughput increases significantly, and the same maximum throughput can be achieved by declaring fictitious noise levels $0.8, 0.5$ and 0.2 when c is around $0.1, 0.27$ and 0.5, respectively, for a noise-free channel.

4.2 Selfish Random Access for T-Out-of-N Channels

The second application of our results will be to a special type of multipacket reception channel in which all packets can be reconstructed successfully with probability $1-\theta_n$ if the collision size n is smaller than or equal to $T \in \{1, 2, \cdots, N\}$. On the other hand, if $n > T$, all packets are destroyed together. If the noise parameter θ_n is 0 for all n, then this is the channel model studied in [11] and [12]. Such channels can be implemented by using T-out-of-N codes [13].

We will focus only on the homogenous case and the Nash equilibrium at which all users access the channel with positive probability for illustrative purposes, but a similar analysis can be conducted for the heterogenous case and other equilibria as in the collision channel model above. We set the strategy sets to $[0, 1]$. In this case, $J(p)$ in (5) is given by $J(p) = \sum_{n=0}^{T-1} (1-\theta_{n+1}) \binom{N-1}{n} p^n (1-p)^{N-1-n}$. The common equilibrium transmission probability $p^\star$ is obtained by solving $J(p^\star) = \frac{c}{1+c}$. If $T < N$, then $J(1) = 0$, and it is enough to have $J(0) = 1-\theta_1 \geq \frac{c}{1+c}$ for the existence of $p^\star$ solving $J(p^\star) = \frac{c}{1+c}$. Otherwise, $J(1) = 1-\theta_N$, and we require $1-\theta_1 \geq \frac{c}{1+c} \geq 1-\theta_N$. Note also that if $1-\theta_N > \frac{c}{1+c}$ (for $T = N$), then users transmit with probability one, and if $1-\theta_1 < \frac{c}{1+c}$ (for $T \leq N$), they never transmit.

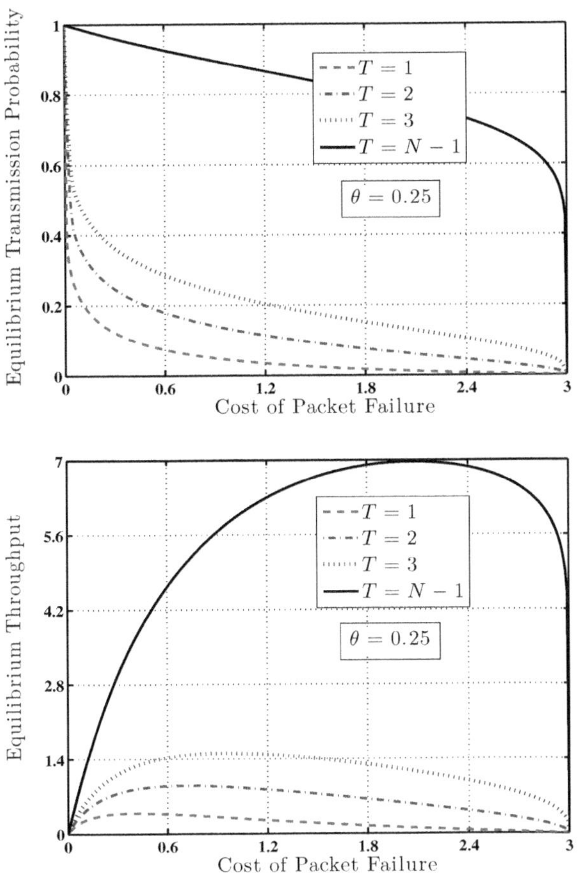

Fig. 3. Equilibrium contention resolution strategies (top figure) and the corresponding system throughput (bottom figure) for T-out-of-N channels. $\theta = 0.25$ and $N = 10$.

In Fig. 3, we plot the equilibrium strategies and the corresponding system throughput for the channels with the common noise parameter $\theta = 0.25$ when $N = 10$. Similar conclusions continue to hold for other values of N and θ_n varying with n. As expected, equilibrium transmission probabilities and the corresponding throughput increase with T. More importantly, maximum achievable throughput increases more than linearly with T. We have this maximum throughput to be around 0.39, 0.9, 1.51 and 6.97 for $T = 1, 2, 3$ and 9, respectively. For T large, we also observe a severe cut-off in transmission probabilities and a corresponding sharp decrease in the equilibrium throughput when the cost of packet failure comes close to the critical level $\frac{1-\theta}{\theta}$. On the other hand, it can be shown that throughput does not exhibit such an abrupt decrease with increasing cost for the noise-free channel. This indicates the importance of the calibration of costs and noise levels in order to avoid high penalty in equilibrium throughput for noisy T-out-of-N channels with large multipacket reception capability.

5 Imperfect Information Random Access Games

Now, we turn our attention to imperfect information random access games in which c_i is randomly distributed according to a cost distribution F_i but is perfectly known by user i before the start of a transmission. On the other hand, user i does not know the cost values of other users exactly but only has a set of belief distributions $\{F_j\}_{j\in\mathcal{I}-\{i\}}$ to predict them.

In this Bayesian game setting, the strategy of user i is a function s_i that maps $c_i \in (0,\infty)$ to a transmission probability $p_i \in [\alpha_i, \beta_i]$. With a slight abuse of notation, we will still represent the strategy set of user i by $\mathcal{S}_i$. As is standard, a strategy profile $\boldsymbol{s}^\star$ is said to be a Nash equilibrium if $s_i^\star$ is a solution of the selfish utility maximization problem $\max_{s_i\in\mathcal{S}_i} U_i\left(s_i, \boldsymbol{s}_{-i}^\star\right)$ for all $i \in \mathcal{I}$. In contrast to our analysis in Section 3, this optimization problem is now over the infinite dimensional function spaces. However, as established in the next theorem, equilibrium strategy profiles can be identified by using a threshold vector $\boldsymbol{\tau}^\star$ in $\mathbb{R}_+^N$. We will skip the proof of Theorem 3 since it is similar to the proofs given above for Theorems 1 and 2.

Theorem 3. *Let $\tilde{\Gamma}_i(\boldsymbol{R}, \boldsymbol{s}_{-i})$ be given as*

$$\tilde{\Gamma}_i(\boldsymbol{R}, \boldsymbol{s}_{-i}) = \sum_{n=1}^{N} \sum_{\substack{\mathcal{I}_n\subseteq\mathcal{I}: \\ i\in\mathcal{I}_n, |\mathcal{I}_n|=n}} \gamma_n \prod_{j\in\mathcal{I}_n-\{i\}} \int_0^\infty s_j(c_j)\, dF_j(c_j) \prod_{j\in\mathcal{I}-\mathcal{I}_n} \left(1 - \int_0^\infty s_j(c_j)\, dF_j(c_j)\right)$$

for all $i \in \mathcal{I}$. Then, a strategy profile $\boldsymbol{s}^\star$ is a Nash equilibrium if and only if $s_i^\star$ is a threshold strategy in the form

$$s_i^\star(c_i) = \beta_i 1_{\{c_i<\tau_i^\star\}} + p_i 1_{\{c_i=\tau_i^\star\}} + \alpha_i 1_{\{c_i>\tau_i^\star\}} \tag{7}$$

almost surely (with respect to F_i) for all i, where $\tau_i^\star = \frac{\tilde{\Gamma}_i(\boldsymbol{R},\boldsymbol{s}_{-i}^\star)}{1-\tilde{\Gamma}_i(\boldsymbol{R},\boldsymbol{s}_{-i}^\star)}$ and $p_i \in [\alpha_i, \beta_i]$.

Theorem 3 allows us to restrict the search for equilibrium strategies to only threshold strategies. Therefore, by viewing the random access game as a game in which users choose a threshold, we can write the best-response function $\boldsymbol{B}: \mathbb{R}_+^N \mapsto \mathbb{R}_+^N$, with a slight abuse of notation, as

$$\boldsymbol{B}(\boldsymbol{\tau}) = \left(\frac{\tilde{\Gamma}_1(\boldsymbol{R}, \boldsymbol{\tau}_{-1})}{1-\tilde{\Gamma}_1(\boldsymbol{R}, \boldsymbol{\tau}_{-1})}, \frac{\tilde{\Gamma}_2(\boldsymbol{R}, \boldsymbol{\tau}_{-2})}{1-\tilde{\Gamma}_2(\boldsymbol{R}, \boldsymbol{\tau}_{-2})}, \ldots, \frac{\tilde{\Gamma}_N(\boldsymbol{R}, \boldsymbol{\tau}_{-N})}{1-\tilde{\Gamma}_N(\boldsymbol{R}, \boldsymbol{\tau}_{-N})}\right).$$

By restricting thresholds to take values only from $[0, \Delta]$ for some large but finite positive constant Δ and assuming all belief distributions are continuous, we can use the Brouwer fixed point theorem to conclude that $\boldsymbol{B}$ has at least one fixed point, which is the Nash equilibrium of the random access game. Moreover, by appealing to [14], we can also ensure the uniqueness of the Nash equilibrium if the Jacobian of $\boldsymbol{B}$ does not have an eigenvalue equal to 1 for all $\boldsymbol{\tau} \in [0, \Delta]^N$.

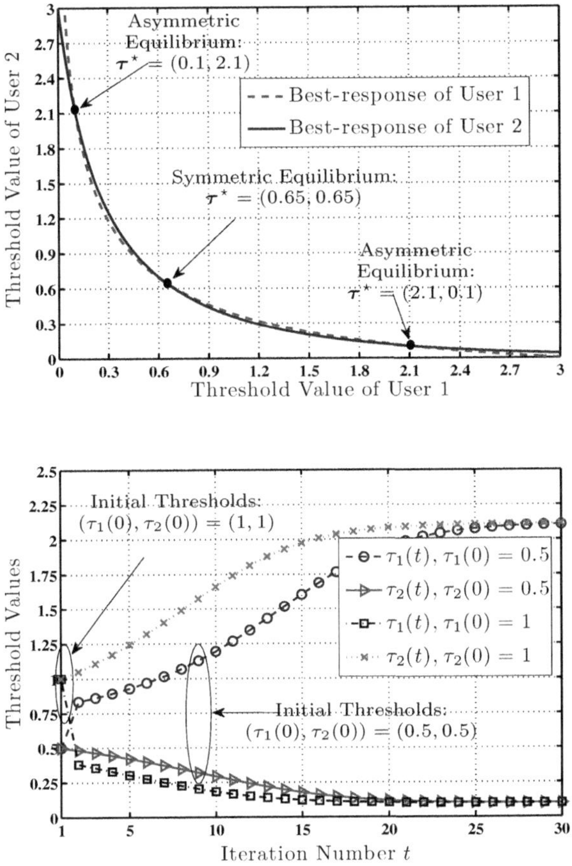

Fig. 4. For the imperfect collision channel with $r_{1,1} = 0.75$, the top figure illustrates the best-response functions of users 1 and 2, while the bottom figure illustrates the learning process leading to Nash equilibria. Belief distributions are exponential with the same parameter $\lambda = 1$.

Now, we focus on a numerical application of Theorem 3 to imperfect collision channels with two users having the same exponential belief distribution with parameter $\lambda > 0$, i.e., $F_1(c) = F_2(c) = 1 - e^{-\lambda c}$. We set $r_{1,1}$ to 0.75.

The resulting network behavior can be quite complicated. To start with, depending on the value of λ, the equilibrium does not need to be unique or symmetric. For example, in Fig. 4, we show the communication scenario in which $\lambda = 1$ and the best-response functions of users intersect at three different points, two of which correspond to asymmetric equilibria at which one user sets its threshold to 0.1, while the other one sets it to 2.1. Moreover, the symmetric equilibrium is unstable, i.e., small perturbations to this equilibrium will lead the system to converge to an asymmetric equilibrium when there are multiple equilibria. That is, starting from any initial condition except for the symmetric equilibrium, the learning process on the best-response path converges to only one of these asymmetric equilibria. For example, at the bottom in Fig. 4, we show two learning curves with different initial conditions leading to these asymmetric equilibria. At iteration $t+1$, user 1 moves first and updates its threshold to $\tau_1(t+1) = \frac{\tilde{\Gamma}_1(\boldsymbol{R}, \tau_2(t))}{1-\tilde{\Gamma}_1(\boldsymbol{R}, \tau_2(t))}$ by estimating $\tilde{\Gamma}_1(\boldsymbol{R}, \tau_2(t))$. Then, user 2 moves and updates its threshold to $\tau_2(t+1) = \frac{\tilde{\Gamma}_2(\boldsymbol{R}, \tau_1(t+1))}{1-\tilde{\Gamma}_2(\boldsymbol{R}, \tau_1(t+1))}$ by estimating $\tilde{\Gamma}_2(\boldsymbol{R}, \tau_1(t+1))$. Iteration $t+1$ terminates after user 2 updates its threshold. As a result, the time-scale of each iteration corresponds to several tens of time-slots in a physical system allowing users to accurately estimate the congestion signals. Based on these observations, we conclude that the symmetric equilibrium, desirable for fairness purposes, exists but may never appear even in symmetric communication scenarios when there are multiple equilibria.

On the other hand, the Nash equilibrium is unique for small values of λ, e.g., $\lambda = 0.5$. Such an equilibrium is automatically symmetric due to the symmetry in the problem, and our analysis indicates that it is also the stable equilibrium for noisy collision channels. However, it should be noted that a Nash equilibrium may not be stable even if it is unique for other channel models, e.g., noise-free collision channels [15].

6 Conclusions

In this paper, we have focused on layer-2 contention resolution strategies for wireless networks with multipacket reception. We have obtained necessary and sufficient conditions for a Nash equilibrium strategy. Applications of these equilibrium conditions have been illustrated for specific channel models along with the resulting network performance analysis and various engineering insights. Finally, we have examined the contention resolution problem with imperfect information, derived the form of equilibrium strategies, their existence and uniqueness, and analyzed a strategy update mechanism based on best-response dynamics converging to an equilibrium.

References

1. Lee, J.-W., Tang, A., Huang, J., Chiang, M., Calderbank, A.R.: Reverse-engineering MAC: A Non-cooperative Game Model. IEEE J. Sel. Area Commun. 25, 1135–1147 (2007)
2. Cui, T., Chen, L., Low, S.H.: A Game-theoretic Framework for Medium Access Control. IEEE J. Sel. Areas Commun. 26, 1116–1127 (2008)
3. Chen, L., Low, S.H., Doyle, J.C.: Random Access Game and Medium Access Control Design. IEEE/ACM Trans. Netw. 18, 1303–1316 (2010)
4. MacKenzie, A.B., Wicker, S.B.: Selfish Users in Aloha: A Game-theoretic Approach. In: IEEE Vehicular Technology Conf., Atlantic City, NJ, pp. 1354–1357 (2001)

5. Inaltekin, H., Wicker, S.B.: The Analysis of Nash Equilibria of the One-shot Random Access Game for Wireless Networks and the Behavior of Selfish Nodes. IEEE/ACM Trans. on Netw. 16, 1094–1107 (2008)
6. Jin, Y., Kesidis, G.: Equilibria of a Noncooperative Game for Heterogenous Users of an ALOHA Network. IEEE Commun. Lett. 6, 282–284 (2002)
7. MacKenzie, A.B., Wicker, S.B.: Stability of Multipacket Slotted Aloha with Selfish Users and Perfect Information. In: 22nd Annual Joint Conference of the IEEE Computer and Communications Societies, San Francisco, CA, pp. 1583-1590 (2003)
8. Ngo, M.H., Krishnamurthy, V.: Game Theoretic Cross-layer Transmission Policies in Multipacket Reception Wireless Networks. IEEE Trans. Signal Process. 55, 1911–1926 (2007)
9. Glicksberg, I.L.: A Further Generalization of the Kakutani Fixed Point Theorem with Application to Nash Equilibrium Points. Proc. American Mathematical Society 3, 170–174 (1952)
10. Fudenberg, D., Tirole, J.: Game Theory. MIT Press, Cambridge (1991)
11. Mahravari, N.: Random-access Communication with Multiple Reception. IEEE Trans. Inf. Theory 36, 614–622 (1990)
12. Tsybakov, B.S., Mikhailov, V.A., Likhanov, N.B.: Bounds for Packet Transmission Rate in a Random Access System. Problemy Peredachi Informatsii 19, 61–81 (1983)
13. Mathys, P.: A Class of Codes for a T Active Users out of N Multiple-access Communication System. IEEE Trans. Inf. Theory 36, 1206–1219 (1990)
14. Kellogg, R.B.: Uniqueness in the Schauder Fixed Point Theorem. Proc. American Mathematical Society 60, 207–210 (1976)
15. Inaltekin, H., Wicker, S.B.: Random Access Games: Selfish Nodes with Incomplete Information. In: IEEE Military Communications Conf., Orlando, FL, pp. 1–6 (2007)

Multi-portfolio Optimization: A Potential Game Approach

Yang Yang[1], Francisco Rubio[1], Gesualdo Scutari[2], and Daniel Palomar[1]

[1] Department of Electronic and Computer Engineering
The Hong Kong University of Science and Technology
{yangyang,eerubio,palomar}@ust.hk
[2] Department of Electrical Engineering
State University of New York (SUNY) at Buffalo
gesualdo@buffalo.edu

Abstract. Trades from separately managed accounts are usually pooled together for execution and the transaction cost for a given account may depend on the overall level of trading. Multi-portfolio optimization is a technique for combing multiple accounts at the same time, considering their joint effects while adhering to account-specific constraints. In this paper, we model multi portfolio optimization as a game problem and adopt as a desirable objective the concept of Nash Equilibrium (NE). By formulating the game problem as a potential game, we are able to provide a complete characterization of NE and derive iterative algorithms with a distributed nature and satisfactory convergence property.

1 Introduction

In a couple of ground-breaking articles [1,2] laying down the foundations of modern portfolio theory, Markowitz introduced half a century ago a fundamental framework for solving the canonical problem of how an individual account allocates wealth across a portfolio of risky assets by optimizing the associated risk-return tradeoff. Since then, numerous generalizations, such as limitations on transaction costs and other portfolio characteristics, have been proposed in order to effectively model realistic operating conditions underlying the practice of mean-variance optimization of single portfolio.

In a practical framework, trades of diverse clients are usually pooled and executed simultaneously for the sake of efficiency. But trading one account raises the marginal transaction costs for other accounts, so a particularly relevant problem is that of realistically modeling the trading costs incurred when rebalancing multiple accounts, and more specifically their market impact cost. Indeed, in the multi-portfolio rebalancing problem, the market impact cost associated with a given account depends on the overall level of trading of all accounts and not just on its specific trading requirements. As a consequence, the actual market impact cost of trading multiple accounts is typically larger than the sum of the estimated market impact costs of trading each account separately.

R. Jain and R. Kannan (Eds.): GameNets 2011, LNICST 75, pp. 182–189, 2012.

To the best of our knowledge, O'Cinneide and his collaborators [5] are the first to introduce the simultaneous rebalancing of multiple accounts into a multi-portfolio optimization problem. Their approach is based on the maximization of the total welfare over all accounts, namely the Pareto optimal solution to the well-known "social planner" problem. Another methodology frequently used is known as the Nash Equilibrium (NE) approach [4,7]. In NE approach, rather than colluding to maximize total welfare, each account optimizes its own welfare, under the assumption that the trade decisions of other accounts have been made and are fixed. A NE is achieved when no account has an incentive to unilaterally deviate from it. Although this has been done in [7], the authors do not provide any characterizations of the NE. Besides, each account is only subject to short-selling constraint, which is hardly enough to model various complicated regulations in practice.

One common shortcoming of [5,7] is that an all-at-once approach is used to generate optimal trades for all accounts simultaneously. Compared with centralized approach, distributed algorithms are more suitable for implementation, especially when the number of accounts is large. During each iteration, accounts are rebalanced independently but taking into account the market impact of the desired trades of other accounts.

In this paper, the multi-portfolio optimization problem is modeled as a game problem and NE is adopted as the desirable objective. We first consider a Nash Equilibrium Problem where one player's feasible strategy set is independent of other players. We then generalize the problem by incorporating global constraints imposed on all accounts, which may arise due to practical considerations such as transaction size constraint over multiple accounts. In both cases, we give a complete characterizations of the NE and derive iterative algorithms that can be implemented in a distributed manner.

2 Problem Formulation

Treating the market-impact in a single account optimization as if it is the only account being traded underestimates the true trading cost of rebalancing each account. Instead, the market-impact caused by all accounts being optimized simultaneously should be considered. Under this consideration, the utility function for account n is defined as [7]

$$u_n(\mathbf{w}_n, \mathbf{w}_{-n}) = \boldsymbol{\alpha}_n^T \mathbf{w}_n - \frac{1}{2}\rho \cdot \mathbf{w}_n^T \mathbf{Q} \mathbf{w}_n - \frac{1}{2}\delta \cdot \mathbf{w}_n^T \boldsymbol{\Omega} \left(\sum_{m=1}^{N} \mathbf{w}_m \right), \tag{1}$$

where for account n, $\boldsymbol{\alpha}_n^T \mathbf{w}_n$ is the expected return, $\rho \mathbf{w}_n^T \mathbf{Q} \mathbf{w}_n$ represents the penalization for risk, $\delta \cdot \mathbf{w}_n^T \boldsymbol{\Omega} \left(\sum_{m=1}^{N} \mathbf{w}_m \right)$ is a nonlinear market-impact function while aggregate effects generated by other accounts are taken into account.

The maximization of account n's utility function (1), however, is subject to one or various kinds of constraints due to practical considerations. Some of them include short-selling constraint, holding constraint, budget constraint and cardinality constraint. Since our discussion does not depend on the type of constraints, we use a simple notation $\mathcal{K}_n$ to denote the set of account n's feasible strategies. We further assume that $\mathcal{K}_n$ is a non-empty, closed and convex set, and it is independent of the other accounts' strategies.

Given the utility fuction (1) and strategy set $\mathcal{K}_n$, we formulate the system design as a Nash Equilibrium Problem (NEP) using as desirable criterion the concept of Nash Equilibrium (NE). Specifically, we consider a strategic noncooperative game in which the players are the accounts. Each player n competes against the others by choosing a strategy $\mathbf{w}_n$ that maximizes his own utility function. In other words, given the strategy of other players, player n solves the following optimization problem:

$$\left.\begin{array}{l}\underset{\mathbf{w}_n}{\text{maximize}}\ u_n(\mathbf{w}_n, \mathbf{w}_{-n}) \\ \text{subject to } \mathbf{w}_n \in \mathcal{K}_n\end{array}\right\} \forall n. \tag{2}$$

A NE is achieved when no player has an incentive to deviate unilaterally, which is formally defined as follows:

Definition 1. *A (pure) strategy profile* $\mathbf{w}^\star = (\mathbf{w}_n^\star)_{n=1}^N$ *is a NE of NEP (2) if*

$$u_n(\mathbf{w}_n^\star, \mathbf{w}_{-n}^\star) \geq u_n(\mathbf{w}_n, \mathbf{w}_{-n}^\star), \forall \mathbf{w}_n \in \mathcal{K}_n, \forall n,$$

with $\mathbf{w}_{-n} = (\mathbf{w}_m)_{m=1, m\neq n}^N$.

3 Potential Game and Its Characterizations

To deal with NEP (2), we use in this paper a framework given by potential game theory [3,8], which allows us to infer the properties of NEP by solving a single optimization problem.

To begin with, a potential game is formally defined below.

Definition 2. *A strategic game (2) is called an exact potential game if there exists a function* $\mathcal{P} : \mathcal{K} \to \mathbb{R}$ *such that for all* n *and* $(\mathbf{w}_n^1, \mathbf{w}_{-n}), (\mathbf{w}_n^2, \mathbf{w}_{-n}) \in \mathcal{K}$*:*

$$u_n(\mathbf{w}_n^1, \mathbf{w}_{-n}) - u_n(\mathbf{w}_n^2, \mathbf{w}_{-n}) = \mathcal{P}(\mathbf{w}_n^1, \mathbf{w}_{-n}) - \mathcal{P}(\mathbf{w}_n^2, \mathbf{w}_{-n}). \tag{3}$$

With the definition of potential game in (3), it is easy to see that the set of NE for NEP (2) remains the same when all utility functions in (2) are replaced with $\mathcal{P}(\mathbf{w})$. This implies that we can study the properties of NEs using a single function that does not depend on the particular player. Furthermore, it is natural to ask what is the relationship between NEs of NEP (2) and the maxima of $\mathcal{P}$, which can be obtained by solving:

$$\begin{array}{l}\underset{\mathbf{w}}{\text{maximize}}\ \mathcal{P}(\mathbf{w}) \\ \text{subject to } \mathbf{w} \in \mathcal{K}.\end{array} \tag{4}$$

Lemma 1. *[8] Let NEP (2) be a potential game with potential function $\mathcal{P}$. If $\mathbf{w}^\star$ is a maximum of $\mathcal{P}$, then it is a NE of NEP (2). If $\mathcal{K}$ is a convex set with a Cartesian structure, i.e., $\mathcal{K} = \mathcal{K}_1 \times \mathcal{K}_2 \times \ldots \times \mathcal{K}_N$, and $\mathbf{w}^\star$ is a NE of NEP (2), then $\mathbf{w}^\star$ is a maximum of $\mathcal{P}$.*

Recall that in Section 2, we have assumed that $\mathcal{K}_n$ is independent of other players. Then Lemma 1 indicates that $\mathbf{w}^\star$ is a NE of NEP (2) if and only if it is a maximum of NLP (4). This provides us with new intuition to study the potential game, which is the framework of standard optimization theory applied to the potential function.

To make use of Lemma 1, we first define a function $\theta_1(\mathbf{w})$ as

$$\theta_1(\mathbf{w}) \triangleq \frac{1}{2}\mathbf{w}^T\mathbf{M}_1\mathbf{w} - \boldsymbol{\alpha}, \tag{5}$$

where $\mathbf{M}_1 \triangleq \mathbf{I} \otimes (\rho\mathbf{Q} + \delta\boldsymbol{\Omega}) + \frac{\delta}{2} \cdot \mathbf{S} \otimes \boldsymbol{\Omega}$, $\boldsymbol{\alpha} \triangleq (\alpha_n)_{n=1}^N$, $\mathbf{w} \triangleq (\mathbf{w}_n)_{n=1}^N$, and

$$\mathbf{S} \triangleq \begin{bmatrix} 0\,1 \ldots 1 \\ 1\,0 \ldots 1 \\ \vdots\;\vdots\;\ddots\;\vdots \\ 1\,1 \ldots 0 \end{bmatrix}. \tag{6}$$

Using the definition of potential function, we can readily show that NEP (2) is a potential game, as stated in the following theorem.

Theorem 1. *[10] Suppose each player's strategy set $\mathcal{K}_n$ is independent of other players' strategies and $\mathcal{K}$ has a Cartesian structure, NEP (2) is equivalent to the following optimization problem:*

$$\begin{aligned} &\underset{\mathbf{w}}{\text{maximize}}\ \theta_1(\mathbf{w}) \\ &\text{subject to}\ \mathbf{w} \in \mathcal{K}. \end{aligned} \tag{7}$$

The equivalence of NEP (2) and NLP (7) enables us to explore the existence and uniqueness of NE of NEP (2) by considering NLP (7). The result is stated in the following theorem.

Theorem 2. *[10] Suppose each player's strategy set $\mathcal{K}_n$ is independent of other players' strategies and $\mathcal{K}$ has a Cartesian structure, the Nash Equilibrium Problem (2) always has a unique NE.*

We mention that the pareto-optimal solution of NEP (2), i.e., the optimal solution to the sum-utility maximization problem, can be interpreted as the NE of a NEP with a modified objective function. Interested readers are referred to [10] for details.

Given the existence and uniqueness of NE, a natural question is that is there any algorithm that can be implmented in a distributed manner and has satisfactory convergence behavior? We consider best-response based iterative algorithms with both sequential (Gauss-Seidel) and simultaneous (Jacobi) update, as described in Algorithm 1.

Algorithm 1. Iterative Best Response Algorithm

Data : Choose any $\mathbf{w}_n^0 \in \mathcal{K}_n$ for all $n = 1, 2, \ldots, N$, and set $q = 0$.
Step 1: If $\mathbf{w}^q$ satisfies a suitable termination criterion: STOP.
Step 2: Sequentially or Simultaneously for $n = 1, 2, \ldots, N$, update $\mathbf{w}_n^{q+1}$ as follows:

Sequential (Gauss-Seidel) Update:
$\mathbf{w}_n^{q+1} \triangleq \arg\min_{\mathbf{w}_n \in \mathcal{K}_n} u_n(\mathbf{w}_1^{q+1}, \ldots, \mathbf{w}_{n-1}^{q+1}, \mathbf{w}_n, \mathbf{w}_{n+1}^q, \ldots, \mathbf{w}_N^q)$.
Simultaneous (Jacobi) Update:
$\mathbf{w}_n^{q+1} \triangleq (1 - \frac{1}{N})\mathbf{w}_n^q + \frac{1}{N} \cdot \arg\min_{\mathbf{w}_n \in \mathcal{K}_n} u_i(\mathbf{w}_1^q, \ldots, \mathbf{w}_{n-1}^q, \mathbf{w}_n, \mathbf{w}_{n+1}^q, \ldots, \mathbf{w}_N^q)$.

Step 3: Set $q \leftarrow q + 1$; and go to **Step 1**.

Theorem 3. *[10] Any sequence $\{\mathbf{w}^q\}_{q=0}^{\infty}$ generated by the sequential and simultaneous update of iterative best-response algorithm in Algorithm 1 converges to the unique NE of NEP (2), for any given updating order of the users.*

4 Generalized Nash Equilibrium Problem

In all previous developments we have assumed that one player's strategy set is independent of the rival players' actions, but this is not always the case. There are many applications of interest where the feasible sets naturally depend on the variables of the player's rivals. In this section, we consider the NEP (2) with global constraints as in (8). This results in a generalized Nash Equilibrium Problem (GNEP), which are formally described as follows:

$$\left.\begin{array}{l} \underset{\mathbf{w}_n}{\text{maximize}}\ \boldsymbol{\alpha}_n^T \mathbf{w}_n - \frac{1}{2}\rho \mathbf{w}_n^T \mathbf{Q} \mathbf{w}_n \\ \text{subject to}\ \sum_{m=1}^{N} \left| w_{n,i} \right| \le C_i, \forall i = 1, \ldots, I \\ \qquad \sum_{m=1}^{N} \sum_{j \in \mathcal{J}_l} \left| w_{m,j} \right| \le U_l, \forall l = 1, \ldots, L \\ \qquad \left(\sum_{m=1}^{N} \mathbf{w}_m \right)^T \delta \boldsymbol{\Omega} \left(\sum_{m=1}^{N} \mathbf{w}_m \right) \le T \\ \qquad \mathbf{w}_n \in \mathcal{K}_n \end{array}\right\} \forall n, \tag{8}$$

Note that we have preserved $\mathcal{K}_n$ to exclusively denote one player's individual constraints such as budget constraint. The first and second global constraint in (8) represents the transaction size constraint over multiple accounts and limitations on the amount invested over groups of assets with related characteristics, respectively. In this formulation, we remove the market-impact function from the objective and incorporate it as the third global constraint in (8).

We call a game problem with coupled constraint sets as defined in (8) a Generalized Nash-Equilibrium Problem (GNEP). To analyze the GNEP (8), we can follow a similar approach as that in Section 3. After some elementary algebra, it can be shown that GNEP (8) is a potential game with the following constrained optimization problem:

$$\begin{aligned} \underset{\mathbf{w}}{\text{maximize}} \ & \boldsymbol{\alpha}^T \mathbf{w} - \tfrac{1}{2} \mathbf{w}^T \left(\mathbf{I} \otimes \rho \mathbf{Q} \right) \mathbf{w} \\ \text{subject to} \ & \mathbf{w} \in \mathcal{K}_1 \times \ldots \times \mathcal{K}_N \\ & \mathbf{g}(\mathbf{w}) \leq \mathbf{0}, \end{aligned} \tag{9}$$

where

$$\mathbf{g}(\mathbf{w}) \triangleq \begin{bmatrix} \left(\sum_{n=1}^{N} |w_{n,i}| - C_i \right)_{i=1}^{I} \\ \left(\sum_{n=1}^{N} \sum_{j \in \mathcal{J}_l} |w_{n,j}| - U_l \right)_{l=1}^{L} \\ \left(\sum_{n=1}^{N} \mathbf{w}_n \right)^T \delta \boldsymbol{\Omega} \left(\sum_{n=1}^{N} \mathbf{w}_n \right) - T, \end{bmatrix}. \tag{10}$$

We denote the feasible set of (9) as $\mathcal{K} \triangleq \{\mathbf{w} : \mathbf{w} \in \mathcal{K}_1 \times \ldots \times \mathcal{K}_N, \mathbf{g}(\mathbf{w}) \leq \mathbf{0}\}$. It is easy to see that NLP (9) is a strongly convex optimization problem, and it has a unique maximum.

Note that in NLP (9), $\mathcal{K}$ does not have a Cartesian structure. The equivalence between GNEP (8) and NLP (9) as indicated by Lemma 1 may not hold any more. As shown in [9,6,8], a NE of GNEP (8) is not necessarily a maximum of NLP (9). Nash Equilibriums of the GNEP (8) that are also maxima of NLP (9) are termed as Variational Equilibriums (VE). In another word, GNEP (8) and NLP (9) are equivalent in the sense of VE. From now on, we will focus on the VE of the GNEP (8) and give detailed analysis on its existence, uniqueness and algorithms.

Theorem 4. *[10] There always exists a unique variational equilibrium of GNEP (8).*

The potential game formulation of GNEP (8), i.e., NLP (9), not only serves as a direct way to characterize the VE, but also provides us with some intuition to devise distributed algorithms achieving the VE. First we derive a result that is valid for all potential games.

Theorem 5. *[10] Consider a Nash Equilibrium game where each players solves the following convex optimization problem*

$$\left.\begin{aligned} \underset{\mathbf{w}_n}{\text{maximize}} \ & f_n(\mathbf{w}_n, \mathbf{w}_{-n}) \\ \text{subject to} \ & \mathbf{w}_n \in \mathcal{K}_n, \end{aligned}\right\} \forall n \tag{11}$$

with a concave potential function $\mathcal{P}(\mathbf{w})$. Now suppose a global convex constraint $\mathbf{g}(\mathbf{w}) \leq \mathbf{0}$ is imposed on all players, i.e., each player solves the following optimization problem

$$\left.\begin{aligned} \underset{\mathbf{w}_n}{\text{maximize}} \ & f_n(\mathbf{w}_n, \mathbf{w}_{-n}) \\ \text{subject to} \ & \mathbf{w}_n \in \mathcal{K}_n \\ & \mathbf{g}(\mathbf{w}) \leq \mathbf{0}. \end{aligned}\right\} \forall n. \tag{12}$$

Then $\mathbf{w}^\star$ is a Variational Equilibrium of GNEP (12) if and only if it is a Nash Equilibrium of the following NEP

$$\left.\begin{array}{l}\underset{\mathbf{w}_n}{\text{maximize}}\ f_n(\mathbf{w}_n, \mathbf{w}_{-n}) - \boldsymbol{\lambda}^T \cdot \mathbf{g}(\mathbf{w}) \\ \text{subject to}\ \mathbf{w}_n \in \mathcal{K}_n\end{array}\right\} \forall n \tag{13}$$

with $\boldsymbol{\lambda}$ *such that* $\mathbf{0} \leq \boldsymbol{\lambda} \perp \mathbf{g}(\mathbf{w}^\star) \leq \mathbf{0}$.

Thanks to Theorem 5, we have transformed GNEP (12) with coupled strategy set into NEP (13) with no coupling in strategy set. The transformation is beneficial because we can solve GNEP (12) in a distributed manner. Specifically, we can design a double-loop algorithm, where in the outer loop, the price tuple $\boldsymbol{\lambda}$ is updated by sub-gradient method, and in the inner loop, NEP (13) is solved using Algorithm 1. We summarize this double-loop algorithm in Algorithm 2.

Algorithm 2. Sub-gradient Algorithm

Data: Choose any $\boldsymbol{\lambda}^{(0)} \geq \mathbf{0}$, and set $q = 0$.
Step 1: If $\boldsymbol{\lambda}^{(q)}$ satisfies a suitable termination criterion: STOP.
Step 2: Compute the unique NE $\mathbf{w}^\star\left(\boldsymbol{\lambda}^{(q)}\right)$ of NEP (13) using Algorithm 1.
Step 3: $\boldsymbol{\lambda}^{(q+1)} = \left[\boldsymbol{\lambda}^{(q)} - \gamma^{(q)} \cdot \mathbf{g}\left(\mathbf{w}^\star\left(\boldsymbol{\lambda}^{(q)}\right)\right)\right]$, where $\gamma^{(q)}$ is the q-th stepsize.
Step 4: $q \leftarrow q + 1$; go to **Step 1**.

A common criterion for choosing the stepsize $\gamma^{(q)}$ in Algorithm is that $\gamma^{(q)}$ must be square summable, but not absolute summable. The convergence property of Algorithm 2 is given by the following theorem.

Theorem 6. *[10] Algorithm 2 solving GNEP (12) converges as long as Algorithm 1 solving NEP (11) converges.*

Theorem 6 indicates that the introduction of global convex constraints does not require stricter convergence conditions. For GNEP (8), as we have already proved in Theorem 3 that Algorithm 1 always converges, we can therefore conclude that Algorithm 2 can surely converge to the unique VE of GNEP (8).

5 Discussions and Conclusions

In Figure 1, we show the convergence of Algorithm 1 and Algorithm 2. From Figure 1(a), the sequential update of best-response iterative algorithm converges to the unique NE very fast. On the other hand, the convergence speed of the simultaneous update of best-response iterative algorithm depends on N, the number of accounts. When there are a large number of accounts, its convergence speed is typically small.

Figure 1(b) shows that the outer-loop price tuple $\boldsymbol{\lambda}$ converges with a satisfactory convergence speed. As we have pointed out before, the convergence of inner-loop best-response iterative algorithm guarantees the convergence of Algorithm 2.

In conclusion, we have modeled the multi-portfolio optimization problem as a Nash Equilibrium problem and analyze it under the framework of potential game.

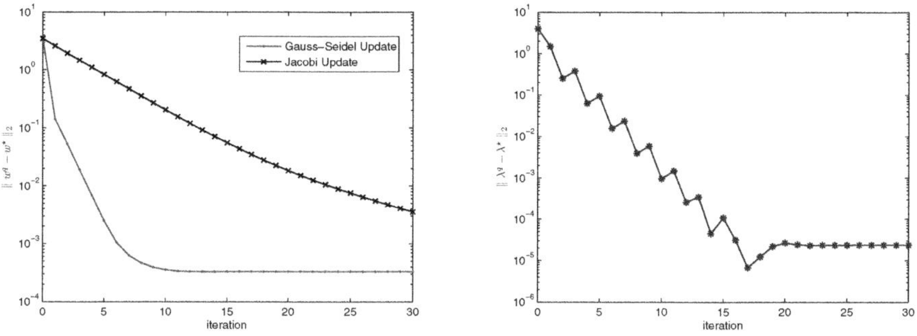

(a) Algorithm 1 (b) Algorithm 2

Fig. 1. Convergence Behavior of Algorithm 1 and Algorithm 2

Specifically, we consider both NEP with uncoupled strategy set and generalized NEP with global constraints imposed on all players. We then give a complete characterizations of NE of NEP and VE of GNEP. We further derive iterative algorithms that can be implemented in a distributed manner and has satisfactory convergence behavior.

References

1. Markovitz, H.: Portfolio selection. Journal of Finance 7(1), 77–91 (1952)
2. Markowitz, H.M.: Portfolio selection: Efficient diversification of investments. Wiley (1959)
3. Monderer, D., Shapley, L.S.: Potential games. Games and Economic Behavior 14, 124–143 (1996)
4. Nash, J.: Non–cooperative games. Annals of Mathematics 54(2), 286–295 (1951)
5. O'Cinneide, C., Scherer, B., Xu, X.: Pooling trades in a quantitative investment process. The Journal of Portfolio Management 32(4), 33–43 (2006)
6. Pang, J.S., Scutari, G., Palomar, D., Facchinei, F.: Design of cognitive radio systems under temperature-interference constraints: A variational inequality approach. IEEE Transactions on Signal Processing 58(6), 3251–3271 (2010)
7. Savelsbergh, M.W.P., Stubbs, R.A., Vandenbussche, D.: Multiportfolio optimization: A natural next step. In: Guerard, J.B. (ed.) Handbook of Portfolio Construction, pp. 565–581. Springer, US (2010)
8. Scutari, G., Barbarossa, S., Palomar, D.: Potential games: A framework for vector power control problems with coupled constraints. In: ICASSP 2006 Proceedings, vol. 4 (2006)
9. Scutari, G., Facchinei, F., Pang, J.S., Palomar, D.P.: Monotone communication games: Theory, algorithms, and models. Submitted to IEEE Transactions on nformation Theory (2010)
10. Yang, Y., Rubio, F., Scutari, G., Palomar, D.: Multi-portfolio optimization: A potential game approach (2011) (in preparation)

Analyzing the Dynamics of Evolutionary Prisoner's Dilemma on Structured Networks

Ahmet Yasin Yazicioglu, Xiaoli Ma, and Yucel Altunbasak

School of Electrical and Computer Engineering,
Georgia Institute of Technology, Atlanta, GA 30332, USA
{yasin,xiaoli,yucel}@ece.gatech.edu

Abstract. The spread of cooperation in the evolutionary dynamics of social dilemma games such as Prisoner's Dilemma can be facilitated by various means such as topological heterogeneities, a high benefit-to-cost ratio, or asymmetric interactions. In evolutionary dynamics, the agents adopt the strategies of neighbors with higher payoffs with a probability proportional to the payoff difference. In this study, we analyze evolutionary dynamics of mixed strategies in the Prisoner's Dilemma game through the expected value of the payoff difference for arbitrary neighbors and the evolutionary advantage/disadvantage of nodes due to the degree distributions within their neighborhoods. Simulation results for various networks and game parameters are also presented.

1 Introduction

Social systems consist of various individuals who try to benefit from interactions among themselves. In such systems, the emergence of situations in which collective interests contradict private interests are inevitable. These situations are known as social dilemmas. When a social dilemma occurs, related individuals usually have two options: they either cooperate or defect, and their decisions affect the overall outcome. In this manner, while the cooperators represent the people who contribute to the collective behavior at personal expense, whereas the defectors represent the ones who do not. In game theory, social dilemmas are analyzed through widely-used metaphors such as the Prisoner's Dilemma, Stag-Hunt, and Hawk-Dove (also known as Chicken or Snowdrift) games [1,2,3]. While the payoffs taken from possible cooperate-defect combinations differ among the models, in all of these games, agents get higher payoffs (rewarded) when they both cooperate rather than both defect.

When the game models are considered for very large populations, the rationality assumption is relatively controversial [4]. For such populations, evolutionary game theory is used in the analysis of population dynamics. In evolutionary games, instead of being rational maximizers, players enter the game with particular strategies attached to them. Players then confront other players who are programmed to play their own strategies and accumulate payoffs from these interactions. At the end of each time period (generation), strategies that provide higher payoffs may be adopted by neighboring agents. However, the number of

R. Jain and R. Kannan (Eds.): GameNets 2011, LNICST 75, pp. 190–204, 2012.

players who utilize strategies that provide lower payoffs decline. This process is similar to evolution in biology. While biological evolution may occur through natural selection, for socio-economic phenomena evolution mostly arises as a consequence of imitation or learning. Eventually, a population attains an equilibrium state, namely an *evolutionary stable configuration.* A strategy is called an "evolutionary stable strategy" (ESS) if a whole population using that strategy cannot be invaded by a small group with a mutant genotype that uses a different strategy. In this context, ESS is a refined form of the Nash equilibrium, which provides the main solution basis for classical analysis. While all evolutionary stable strategies are also Nash equilibria, the converse is not necessarily true [5].

Conventional evolutionary games treat a population as an infinite and homogeneous structure, in which every agent has an equivalent position on a corresponding network. However, this is not true for real populations [8,9] in which interactions and connections among the agents are determined by social and spatial constraints. Hence, most real populations have heterogeneous network topologies that may significantly affect the overall behavior of the corresponding population. For instance, in the Prisoner's Dilemma game, since cooperation is a strictly dominated strategy, it can never invade a population in the evolutionary sense for infinite homogeneous societies [1]. However, recent studies in the literature show that the topological properties of structured networks and game parameters [10] - [15], and heterogeneities and asymmetries in the interactions of agents [16,17] can facilitate the spread of cooperation through evolutionary dynamics. In analyzing the cooperative behavior of a network, we utilize mixed strategies as they provide a higher resolution in the quantitive comparison of different cases and reveal certain information that can not be extracted via pure strategy analysis [18].

In this work, we analyze the evolutionary dynamics on structured networks for the Prisoner's Dilemma game. Such an analysis may be used for various purposes. For instance, biological, economic, political studies and many others may benefit from estimating the level and the strength of cooperative behavior in a particular population. Moreover, when it is possible to manipulate the topology or game parameters, this analysis can be used for design purposes. In problems such as designing an organizational structure of a company, transportation planning or the design of many other systems in which a large number of autonomous agents will participate, it is desirable to attain a topology that implies a more cooperative behavior since it increases the overall utility of the population. The expected probability of cooperation at steady state provides a measure for quantifying the evolutionary favorability of cooperation. Due to the complex nature of dynamics on heterogeneous, large-scale networks, this value can be estimated through simulation results rather than analytical or numerical methods. However, particularly for design purposes, it is crucial to explicitly represent the influence of topological parameters and game parameters on the evolutionary favorability of cooperation. To this end, we use analytical methods for analysis on the micro level. In [18] we present the initial steps of our approach by examining the expected value of payoff difference for arbitrary neighbors. In

Table 1. Game Table for Prisoner's Dilemma

		Player 2	
		Cooperate	*Defect*
Player 1	*Cooperate*	(*b*-*c*, *b*-*c*)	(-*c*, *b*)
	Defect	(*b*, -*c*)	(0, 0)

this work, we extend our analysis and discussion related to the expected value of payoff difference and also investigate the evolutionary advantage/disadvantage of nodes due to the degree distributions within their neighborhoods. Based on the analysis, we discuss how topology and game parameters can influence the cooperative behavior of the network.

The organization of this paper is as follows: Section 2 presents the game model and the evolutionary dynamics. Section 3 analyzes the dynamics at the micro level along with the conditions for favorability of cooperation, and Section 4 presents the simulation results and discussions. Finally, Section 5 concludes the paper with some remarks and indicates possible future directions.

2 Game Model and Evolutionary Dynamics

Prisoner's Dilemma is perhaps the most-widely used metaphor representing the social dilemmas. In Prisoner's Dilemma, a cooperator pays a cost, c, for the other player to receive a benefit, b, where $b > c$. By contrast, a defector does not pay any cost and does not distribute any benefit. This scheme is depicted by the game table shown in Table 1.

Some social dilemmas in a population can be represented as rounds of Prisoner's Dilemma played among the people who interact with each other. The social group can be represented via a graph in which the individuals in the network occupy the vertices and links exist among the nodes that play the game against each other. At each round, the nodes play their current strategies against all of their neighbors and get accumulated payoffs. If mixed strategies are played, the payoffs are random rather than deterministic variables and the strategy of a node can be defined by the probability that it chooses to cooperate.

Analysis of this game on large populations can be obtained through evolutionary game theory [6,7]. Classical evolutionary game dynamics are defined for infinite well-mixed populations and known as "replicator dynamics". Replicator dynamics are defined by the following differential equation:

$$\dot{P}_i = P_i(U_i - \bar{U}) \tag{1}$$

where P_i is the fraction of phenotype i in a population, U_i is the fitness of this phenotype, defined as the average accumulated payoff for the members of this phenotype, and $\bar{U}$ is the average accumulated payoff in the entire population. For finite populations, replicator dynamics do not apply directly and analogous dynamics, which converge to the replicator dynamics in the limit of infinite complete graphs, should be used [3]. These dynamics involve the following events: At the beginning of each time step (generation), nodes play a single round of Prisoner's Dilemma against each of their neighbors and they accumulate the resulting payoffs. After that, each node, x, randomly picks one of its neighbors, y, and compares their accumulated payoffs U_x and U_y. Node x adopts the strategy of node y only if $U_y > U_x$ with the transition probability that increases monotonically to 1 as $U_y - U_x$ increases. Different functions can be used to define transition probabilities. One possible option is $1/(1 + exp[-(U_y - U_x)/K])$ where K characterizes possible noise effects [20]. Alternatively $(U_y - U_x)/(max(k_x, k_y)(b+c))$, where k_x and k_y are node degrees, can be used [12].

Let us assume that at the end of a time step an arbitrary node, x picks a random neighbor y to possibly adopt its strategy. As the essential property of any transition probability function is to equal zero when $U_y < U_x$ and monotonously increase as $U_y - U_x$ gets larger, in analyzing evolutionary dynamics we need to consider the difference of accumulated payoffs for these nodes. These accumulated payoffs can be represented as:

$$U_x = \sum_{i=1}^{k_x-1} b_{xi} - \sum_{i=1}^{k_x-1} c_{xi} + \frac{b}{c}c_{yx} - c_{xy},$$

$$U_y = \sum_{i=1}^{k_y-1} b_{yi} - \sum_{i=1}^{k_y-1} c_{yi} + \frac{b}{c}c_{xy} - c_{yx}, \tag{2}$$

where b_{xi} represents the benefit node x receives from its ith neighbor, and c_{xi} represents the cost of cooperation that node x pays in its interaction with its ith neighbor. Note that b_{xi}'s are independent variables which equals b with probability q_{xi} (cooperation probability of corresponding neighbor) and equals 0 with probability $1 - q_{xi}$. Random variables c_{xi} are also independent and each realization is either c (with probability p_x) or 0 (with probability $1 - p_x$). Random variables b_{yi} and c_{yi} similarly represent the interaction of node y with its neighbors. Furthermore, outcome of the round where x and y play against each other are depicted by the variables c_{xy}, the cost of cooperation node x pays, and c_{yx}, the cost of cooperation node x pays. Benefits received by the other player are deterministic functions of these variables since if one pays the cost the other receives the corresponding benefit. Note that $c_{xy} = c$ with probability p_x and $c_{xy} = 0$ with probability $1 - p_x$. Similarly we have $c_{yx} = c$ with probability p_y and $c_{yx} = 0$ with probability $1 - p_y$.

3 Micro Level Analysis

Our micro level analysis has two parts. First we consider the expected value of payoff difference, which determines, for arbitrary neighbors, whose strategy is more likely to fare better. This part of the analysis involves the game parameters and the topological parameters. Next we consider the evolutionary advantage of nodes. In that context we focus on the probability that a particular node is randomly picked by its neighbors to possibly adapt its strategy. This probability affects the chances of a node to spread its strategy and solely depends on the network topology.

3.1 Expected Value of Payoff Difference

For arbitrary neighbors x and y, transition probabilities are determined by the accumulated payoff difference of nodes, $U_y - U_x$. The strategy that provides its player a higher accumulated payoff will be evolutionary favored through the connection between x and y. Based on the expected value of accumulated payoff difference, $E[U_y - U_x]$, it is possible to say which strategy, on the average, will be favored [18]. If this value is negative, strategy of x provides a higher expected payoff and will be favored. On the other hand, strategy of y will be favored if $E[U_y - U_x]$ is positive. These two cases are separated by the condition where $E[U_y - U_x] = 0$. In light of Eq. 2, $E[U_y - U_x] = 0$ is attained when the following is satisfied:

$$c(k_x p_x - k_y p_y) + b(\bar{q}_y - \bar{q}_x + p_x - p_y) = 0, \tag{3}$$

where $\bar{q}_x = \sum_{i=1}^{k_x-1} q_{xi}$, $\bar{q}_y = \sum_{i=1}^{k_y-1} q_{yi}$. As b, c and k_y are positive, we can divide the inequality by $-(b + ck_y)$ and rearrange to obtain the condition as

$$p_y - \frac{b + ck_x}{b + ck_y} p_x + \frac{b}{b + ck_y}(\bar{q}_x - \bar{q}_y) = 0. \tag{4}$$

For the real variables p_x and p_y, Eq. (4) defines a line in two-dimensional space, $\mathbb{R}^2$. Note that for static network topology and constant game parameters this line has a constant slope but the intercept may change in time as the strategy of their neighbors, and consequently $\bar{q}_x$ and $\bar{q}_y$, may change during the evolutionary dynamics. For the points (p_x,p_y) which are located below this line we have p_y being favored as $E[U_y - U_x] > 0$. On the other hand, for the points (p_x,p_y) which are located above this line we have $E[U_y - U_x] < 0$ and p_x is favored. Note that as p_x and p_y are probabilities, they are bounded within interval $[0, 1]$. Depending on the parameters of the $E[U_y - U_x] = 0$ line, the feasible region where $p_x, p_y \in [0, 1]$ can at most be divided into four separate regions. Let us define these regions as follows: Region I is the region where $E[U_y - U_x] > 0$ and $p_y < p_x$. Region II is the region where $E[U_y - U_x] > 0$ and $p_y > p_x$. Region III is the region where $E[U_y - U_x] < 0$ and $p_y > p_x$. Finally, Region IV is the region where $E[U_y - U_x] < 0$ and $p_y < p_x$. In this context, Regions I and III are the regions where defective strategy has evolutionary advantage. Regions

II and IV, on the other hand, are the regions where cooperative strategy has evolutionary advantage. Existences and sizes of these regions are determined by the $E[U_y - U_x] = 0$ line. Possible cases can be classified under 9 major groups. Examples of cases in each group are depicted in Fig. 1.

Let us consider how the 9 groups shown in Fig. 1 are characterized. In light of Eq. (4) it can be seen that, as $(b + ck_x)/(b + ck_y)$ is always positive, the slope of the $E[U_y - U_x] = 0$ line is always positive. This constraint implies that either both x and y intercepts of the line are 0 or one of them is positive whereas the other one is negative. So initially we can separate the possible cases into 3 groups depending on the y intercept (or x intercept). As the $\frac{b}{b+ck_y}$ term is always positive, y intercept's being negative, positive or equal to zero is solely determined by $(\bar{q}_x - \bar{q}_y)$.

If $(\bar{q}_x - \bar{q}_y) < 0$, then we have positive y intercept and we can further obtain 3 distinct groups based on p_x^*, the value of p_x on the $E[U_y - U_x] = 0$ line for $p_y = 1$: $p_x^* < 0$ as shown in Fig. 1 (a), $0 < p_x^* < 1$ as shown in Fig. 1 (b), and $1 < p_x^*$ as shown in Fig. 1 (c).

If $(\bar{q}_x - \bar{q}_y) = 0$, then we have both intercepts equal to zero and we can further obtain 3 distinct groups based on the value of the slope: slope is greater than 1 $(k_x > k_y)$ as shown in Fig. 1 (d), slope is equal to 1 $(k_x = k_y)$ as shown in Fig. 1 (e), and slope is smaller than 1 $(k_x < k_y)$ as shown in Fig. 1 (f).

If $(\bar{q}_x - \bar{q}_y) > 0$, then we have positive x intercept and we can further obtain 3 distinct groups based on the value of p_y^*, the value of p_y on the $E[U_y - U_x] = 0$ line for $p_x = 1$: $p_y^* < 0$ as shown in Fig. 1 (g), $0 < p_y^* < 1$ as shown in Fig. 1 (h), and $1 < p_y^*$ as shown in Fig. 1 (i).

For arbitrary neighbors x and y, valuable information about their evolutionary interaction is encoded in the group to which the $E[U_y - U_x] = 0$ line belongs. For instance if this line belongs to the groups shown in Fig. 1 (a) or Fig. 1 (i), we have a significantly influential node, x or y, as it accumulates higher expected payoff than the other node for every possible value of p_x and p_y in $[0, 1]$. In this case the possible evolutionary interaction may favor the cooperation if the influential node is utilizing a cooperative strategy. Alternatively, if this line belongs to the group shown in Fig. 1 (e), which consists of a single line that is coincident with $p_x = p_y$ line, neither Region II nor Region IV exists hence it is not possible to expect an evolutionary interaction that will favor the cooperation. Once the particular $E[U_y - U_x] = 0$ line is classified, one can also quantify the favorability of cooperation through the interaction of x and y by using the areas of the regions. This quantification can easily be obtained through geometry and it can be used to compare the favorability of cooperation in different cases. To this end, one may compare the area of the regions where cooperative strategy is favored (areas of Regions II and IV) to the area of regions where defective strategy is favored (areas of Regions I and III). Since these areas can explicitly be represented in terms of game parameters and node degrees, one can relate these parameters to the expected evolutionary outcome of the interaction between nodes x and y.

Evolutionary interaction at any time instant between two neighbors, x and y, is determined by the current $E[U_y - U_x] = 0$ line. As shown in Eq. (4) parameters

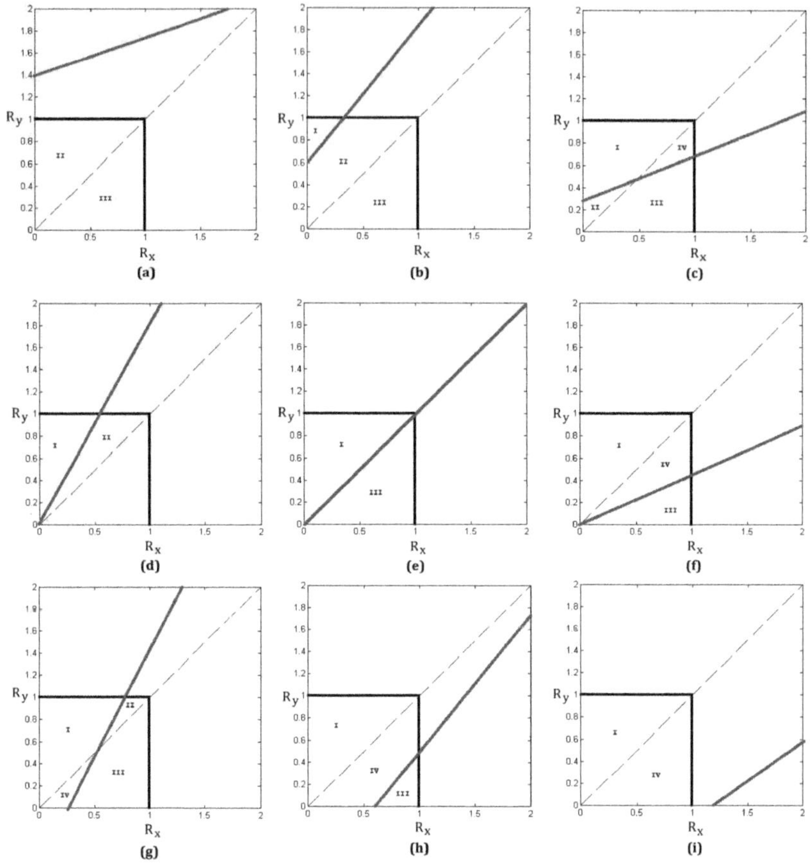

Fig. 1. Examples for each of the 9 major groups of possible cases determined by the $E[U_y - U_x] = 0$ line for two meeting agents, x and y. Upper bound of 1 for p_x and p_y on the axes of $\mathbb{R}^2$ (R_x and R_y) are marked with solid lines. Each group is characterized by the existence and structure of Region I ($E[U_y - U_x] > 0$, $p_y < p_x$), Region II ($E[U_y - U_x] > 0$, $p_y > p_x$), Region III ($E[U_y - U_x] < 0$, $p_y > p_x$) and Region IV ($E[U_y - U_x] < 0$, $p_y < p_x$). The $p_x = p_y$ line is also shown as dashed, whereas the line corresponding to $E[U_y] = E[U_x]$ is solid.

of this line depends on the topology and game parameters. To interpret Eq. (4) and the effect of different parameters on the cooperation let us first consider the example of a well-mixed homogeneous network where the cooperative behavior is known to be eliminated through evolution. For such a population, every node has the same degree (k_x=k_y) and every node is connected to all other nodes ($\bar{q}_x$=$\bar{q}_y$), which makes the $E[U_y - U_x] = 0$ line coincident with $p_y = p_x$ line, as depicted in Fig. 1 (e), for every neighbor x and y throughout the whole evolutionary dynamics. Since neither Region II nor Region IV exists for this case, it is impossible to expect an evolutionary interaction that is likely to favor the cooperation. Since $E[U_y] > E[U_x]$ can only occur when $p_y < p_x$, cooperation is strongly opposed by evolution. Note that the situation where the $E[U_y] = E[U_x]$ line coincident with $p_y = p_x$ line is the only case where the area of the portion where cooperative strategy is favored is zero and there is no combination of p_x and p_y where the cooperative strategy provides a higher expected payoff. In order to promote the cooperative behavior, the $E[U_y - U_x] = 0$ line needs to deviate from the $p_x = p_y$ line. This deviation can be achieved through the game parameters and heterogeneity in neighbor degrees. This way, for arbitrary neighbors it is possible to have some cases where more cooperative strategy provides a higher expected payoff and possibly be adopted by the other node.

3.2 Evolutionary Advantage of Nodes

In the previous section we presented the analysis and discussions on the evolutionary dynamics of cooperation through the expected value of accumulated payoff differences for arbitrary neighbors. However, the spread of a particular strategy from a node depends on the possibility of having a neighbor node considering to adopt its strategy. For a node y to have its strategy adopted by node x, first of all node x should pick node y to possibly adopt its strategy. Hence, in addition to the ares of different regions as proposed in previous part, the probabilities that one node picks the other node should also be considered. If the interaction of two nodes at any time instant is considered, this can be incorporated into the analysis by weighing the regions I and III with probability of node y picking node x and weighing the regions II and IV with probability of node x picking node y. Based on the assumption that nodes pick any of their neighbors with equal probability these probabilities are $1/k_y$ and $1/k_x$ respectively.

When we consider the overall network, although the final step in adoption of a strategy depends on the accumulated payoffs, clearly some nodes have higher chance than the others as being considered by more nodes at each generation. Intuitively, one may expect that the nodes with higher degrees are favored in this context and it is partly true. However, only having large degree is not enough for a node to have this topological advantage. Consider a scenario, where a particular node has a very large degree but its neighbors also have very large degrees. For this particular node the probability of being chosen by any of its neighbors is quite small and it may not acquire a significant advantage from the topology.

Let us consider the topological significance of a node for evolutionary dynamics as the expected number of nodes per generation which pick it and may adopt its strategy, and denote it as n. Based on the assumption that nodes pick any of their neighbors with equal probability, for a particular node x we have:

$$n_x = \sum_{i=1}^{k_x} \frac{1}{k_{yi}} \tag{5}$$

where k_{yi} is the degree of the ith neighbor of x. For a complete graph with N nodes, nodes have equivalent positions and they are equally significant as all having $n = 1$. Any given graph can be considered as the outcome of a procedure where certain edges are removed from a complete graph. As these edges are removed, some nodes increase their chances to spread their strategies whereas others have their chances decreased. While the total significance (N) is distributed uniformly among the nodes for a complete graph, an arbitrary topology is likely to result in a different distribution.

Having a significance value greater than 1 provides an evolutionary advantage for the particular node. Note that regardless of the topology every node definitely picks one neighbor to possibly adopt its strategy, which can be considered as the measure of the local influence on the evolution of a node. Hence for a network with N nodes we have:

$$\sum_{i=1}^{N} n_i = N \tag{6}$$

Every node is equally influenced from its neighborhood in the evolution process, as each of them considers to adopt exactly one of its neighbors at each generation. However, nodes do not influence their neighborhoods equally. Nodes with significance values greater than 1 can have more influence on their neighborhoods than their neighborhoods have on them. On the other hand nodes with significance values smaller than 1 can not influence the evolution in their neighborhoods as much as they are influenced by that.

To see how this topological effect influences the evolutionary spread of a node's strategy, let us consider a simple scenario as shown in Figure 2. In this figure a small graph with 15 nodes is given. The nodes initially have pure cooperation or pure defection strategies assigned to them. Given that game parameters satisfy $b > 6c$ the evolution on this graph definitely ends up at a state where all nodes cooperate as cooperators accumulate higher payoffs than their defecting neighbors. Throughout the evolutionary process, nodes 1 and 2 spread their strategies. Both nodes have degree of 6, and their clustering coefficients are also equal. However, when we consider their chances to spread their strategies it can be seen that node 1 spreads its strategy much faster. If we consider the first generation in evolutionary dynamics and pick the transition probability as in [12], neighbors of node 1 and node 2 adopt their strategies with equal probability of transition, $p_t = (b-3c)/(3b+3c)$. However, expected number of neighbors that pick node 1 to adopt its strategy is larger than the expected number of neighbors that pick node 2. It can easily be shown that at the first time step, expected number of defectors that adopt cooperation from node 1 is $4p_t/3$ whereas expected number of defectors that adopt cooperation from node 2 is $3p_t/4$.

Nodes with large n values are much efficient in spreading their strategies and have an important role in spreading cooperative behavior as cooperators exploit having large n value better than defectors. This is due to the fact that as a cooperator converts a defecting neighbor it increases its fitness, its chance to survive and spread. As long as all of the defector neighbors of a cooperator are doing better, it will eventually adopt the defection strategy. However the interesting condition occurs when a portion of defecting neighbors are doing worse. In this case if the cooperator can convert any of those defectors, it increases its fitness

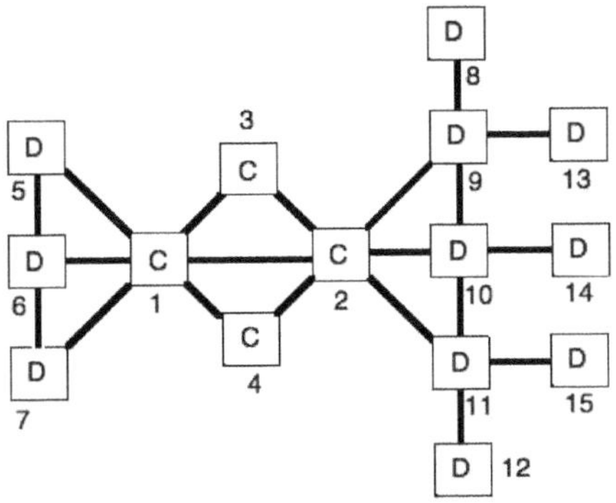

Fig. 2. A simple example to see how topology provides an evolutionary advantage to certain nodes. While for the game parameters $b > 6c$ this system converges to all cooperators, node 1 spreads cooperation faster than node 2.

and the portion of its defecting neighbors that are doing worse increases, as long as all other defecting neighbors are not also neighbors of the converted node. As n increases, the chance of being picked by such a defecting neighbor significantly increases and so does the survival potential of a cooperator. For defectors, on the other hand, this process works the opposite way. When the defector converts any of its cooperating neighbors, it decreases its fitness and the portion of its cooperating neighbors that are doing worse decreases, as long as all other cooperating neighbors are not also neighbors of this new defecting node. Eventually, the defector may end up at a state where it has a certain amount of cooperating neighbors doing better than him and adopt their strategy. Moreover when a defector has low n value, its chance of converting cooperating neighbor decreases hence it does not cause a significant decrease in its own fitness. However, this time it is more likely to adopt the strategy of one of its cooperating neighbors, especially if some of them have large n values and are able to spread cooperation in their neighborhood. Low n values have even worse effect on the cooperators as their survival chances are greatly attenuated when their spreading probability decreases. Based on this discussion we can say that when there is a heterogeneous distribution of n values in the network, this works in favor of the cooperators. Although, they have their reduced survival chances for small n values they can benefit greatly from large n values. Defectors, on the other hand, do not obtain a great survival advantage from neither small nor large n values.

4 Simulation Results

Simulations were carried out for various networks and different game parameters. First, we consider mixed strategy evolutions for various cases on Erdös-Rényi random, Watts-Strogatz small world and Barabási-Albert scale-free topologies with 1000 nodes. We check for the expected probability of cooperation at steady state for varying average degree and game parameters. Networks with average degrees (k) 4,6,8,10 and 12 are generated for each topology. Game parameters are normalized by setting $b = 1$ and c is varied in $0.02 - 0.2$ interval with 0.02 increments. For each combination of k and c, 10 simulations are run and averaged result is reported. Steady state values are obtained through averaging of 1000 time steps after a warm up period of 10,000 time steps, starting from a uniform distribution of strategies among the nodes. Transition probabilities are computed as presented in [12].

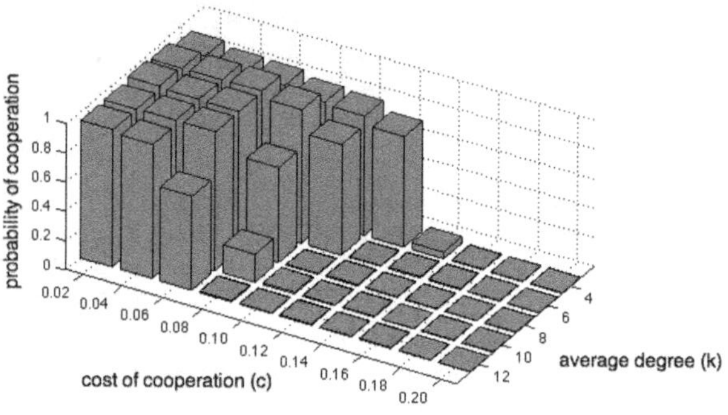

Fig. 3. Expected probability of cooperation at steady state on Erdös-Rényi random networks for various average degrees and costs of cooperation

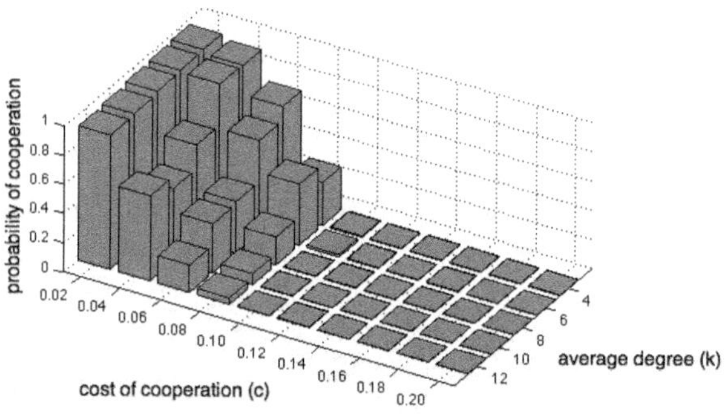

Fig. 4. Expected probability of cooperation on Watts-Strogatz small world networks for various average degrees and costs of cooperation

As it can be seen from the results, expected probability of cooperation is highly dependent on the network topology. For the same average degree and cost of cooperation the steady state behaviors are quite different among the three topologies. As Barabási-Albert scale-free being the most heterogeneous topology, it shows a more cooperative behavior in a wider range of k and c. In this topology, direct links among the hubs also help to promote a certain level cooperation. Note that, in this topology hubs mostly have high significance values (n) as a

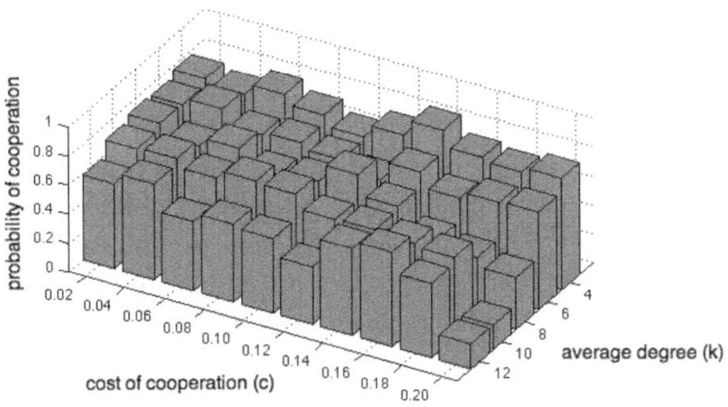

Fig. 5. Expected probability of cooperation on Barabási-Albert scale-free networks for various average degrees and costs of cooperation

significant amount of their neighbors have very low degrees compared to them. When two hubs are connected, the one with higher probability of cooperation gets an advantage. This is due to the discussion we presented about the effect of large n values for cooperators and defectors. Both hubs are likely to be imitated by most of their low degree neighbors through evolution, however as this happens the more cooperative hub creates itself a more cooperative neighborhood resulting in higher payoffs and whereas the less cooperative hub reduces its own payoff as it spreads it strategy. Ultimately the defective hub adopts the strategy of cooperative hub and its neighbors are also likely to adopt this strategy in the following generations. Note that in this fashion hubs can also convert their neighbors with higher probability of cooperation and this is the reason why at very low values of c and k, the other two topologies can display more cooperative behaviors. In a sense, the strategy with the highest probability of cooperation that is played by a hub and also is able to keep that hub resistive (depending on the parameters b and c) to imitate its low degree neighbors has high chance of spreading throughout the population for Barabási-Albert scale-free topology. While connections among the hubs favor this more cooperative strategy, high significance value differences between hubs and low degree neighbors can also cause the elimination of even more cooperative strategies which were initially utilized by low degree nodes.

Results also depict that the Watts-Strogatz small world networks result in the least cooperative behavior among the three topologies and cooperation easily dies out with increasing average degree or cost of cooperation. This is expected since the small world networks are obtained from rewiring (with a certain probability of rewiring) of regular networks, hence their heterogeneity lies somewhere between the two.

Next we consider the effect of significance values on the spread of cooperation. To this end we simulate the small network in Fig. 2 starting from initial condition as shown there. This network is specially generated to highlight the influence of significance values (n) on the spreading chance and speed of strategies. Network is structured so that for $b > 6c$ it converges to all cooperators starting from the particular initial condition. Nodes 1 and 2 are the main sources for the spread of cooperation among the other nodes. They have the same degree, clustering coefficients and almost identical except the degree of their neighbors, hence n values. We simulate this system for $b = 1$ and $c = 0.15$ for 100 repetitions and compute the average probability of cooperation for each node versus time. Note that there are only pure strategies in this simulation hence p values are either 1 or 0 for each node and computing the average probability of cooperation at a particular time step is equivalent to the percentage of simulations where that node acts as a cooperator at that time step. Each simulation is run for 500 time steps as it has been observed to be enough to have all nodes settled as cooperators throughout the 100 simulations. Results are shown in Fig. 6.

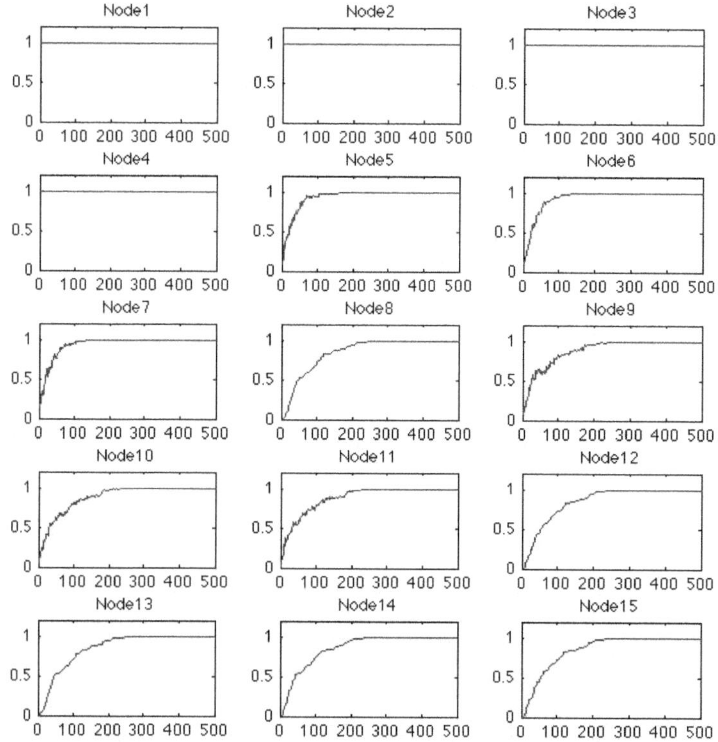

Fig. 6. Average probability of cooperation against time for 100 realization of the simple example with 15 nodes, starting from initial condition as shown in Fig. 2

Spreading speed and efficiency of nodes 1 and 2 can be compared through the comparison of average probability of cooperation of nodes 5, 6, 7 to average probability of cooperation of nodes 9, 10, 11 as these are the one hop neighbors or node 1 and node 2. Results depict that nodes 9, 10, 11 are quite slower than the nodes 5, 6, 7 in adopting and settling at the cooperator state.

5 Conclusion and Future Directions

In this paper, mixed strategy evolution in PD game on structured networks was considered. Evolutionary dynamics are analyzed on the micro level for arbitrary neighbors. The expected value of accumulated payoff differences and the expected number of neighbors that pick a particular node to possibly adopt its strategy were considered as significant factors in the evolutionary process and were analyzed. Evolution dynamics and influences of the network topology and game parameters were presented along with simulation results for various cases. Random, small world and scale-free topologies were simulated for different average degrees and costs of cooperation. Scale-free topology presented a higher robustness against changes in game and network parameters and more cooperative behaviors for a wide range of these parameters. By contrast, increases in the average degree or the cost of cooperation highly attenuate tendency to cooperate for small world topology due to the higher regularity of this topology. However, it was also shown that, for very small values of the cost of cooperation, it is possible to observe more cooperative behaviors with small world or random topologies as the robustness of scale-free topology may be limiting for such cases due to the the imitation of hubs by lower degree neighbors who initially utilize strategies with higher probability of cooperation. Furthermore, it was shown that the expected number of neighbors that pick a particular node to possibly adopt its strategy has an important role in that node's chance to spread its strategy. When a node has higher fitness, this value affects the speed of spread from that node. The spreading speed is important for the survival of cooperators since their survival greatly depend on the number of their cooperating neighbors.

As a future work, evolutionary dynamics for other widely used social dilemma games can be explored. Moreover, the dependency of evolutionary dynamics on the network topology and other variables can be analyzed in further details. Also similar analysis can be studied for dynamic topologies where edges and nodes are added/removed.

References

1. Dixit, A., Skeath, S.: Games of Strategy, 2nd edn. W. W. Norton, New York (2004)
2. Skyrms, B.: The Stag Hunt and the Evolution of Social Structure. Cambridge University Press, Cambridge (2003)
3. Santos, F.C., Pacheco, J.M., Lenaerts, T.: Evolutionary dynamics of social dilemmas in structured heterogeneous populations. PNAS 103, 3490–3494 (2006)
4. Kahneman, D.: New challenges to the rationality assumption. Journal of Institutional and Theoretical Economics 150, 18–36 (1994)
5. Kandori, M.: Evolutionary game theory in economics. In: Kreps, Wallis (eds.) Advances in Economics and Econometrics: Theory and Applications. Cambridge University Press (1997)
6. Axelrod, R., Hamilton, W.D.: The evolution of cooperation. Science 211, 1390–1396 (1981)
7. Nowak, M.A., May, R.M.: Evolutionary games and spatial chaos. Nature 359, 826–829 (1992)

8. Barabási, A.L., Albert, R.: Emergence of scaling in random networks. Science 286, 509–512 (1999)
9. Watts, D.J., Strogatz, S.H.: Collective dynamics of small-world networks. Nature 393, 440–442 (1998)
10. Killingback, T., Doebeli, M.: Spatial evolutionary game theory: Hawks and Doves revisited. Proc. R. Soc. Lond. B 263, 1135–1144 (1996)
11. Hauert, C., Doebeli, M.: Spatial structure often inhibits the evolution of cooperation in the snowdrift game. Nature 428, 643–646 (2004)
12. Pacheco, J.M., Santos, F.C.: Network dependence of the dilemmas of cooperation. In: AIP Conf. Proc., (776), pp. 90–100 (2005)
13. Ohtsuki, H., Hauert, C., Lieberman, E., Nowak, M.A.: A simple rule for the evolution of cooperation on graphs and social networks. Nature 441, 502–505 (2006)
14. Santos, F.C., Santos, M.D., Pacheco, J.M.: Social diversity promotes the emergence of cooperation in public goods games. Nature 454(7201), 213–216 (2008)
15. Garcin, F., Manshaei, M.H., Hubaux, J.-P.: Cooperation in underwater sensor networks. In: Proc. ICST/IEEE Intl. Conf. on Game Theory for Networks (GameNets), pp. 540–548 (2009)
16. Gaoa, J., Lia, Z., Wub, T., Wan, L.: Diversity of contribution promotes cooperation in public goods games. Physica A: Stat. Mech. Appl. 389(16), 3166–3171 (2010)
17. Peng, D., Yang, H.-X., Wang, W.-X., Chen, G.R., Wang, B.-H.: Promotion of cooperation induced by nonuniform payoff allocation in spatial public goods game. Eur. Phys. J. B(73), 455–459 (2010)
18. Yazicioglu, A.Y., Ma, X., Altunbasak, Y.: Evolution of mixed strategies for social dilemmas on structured networks. In: Proc. IEEE Intl. Conf. on Networking, Sensing and Control (ICNSC), pp. 1–6 (2011)
19. Perc, M., Szolnoki, A.: Social diversity and promotion of cooperation in spatial prisoner's dilemma game. Phys. Rev. E 77, 011904 (2008)
20. Szabó, G., Töke, C.: Evolutionary prisoner's dilemma game on a square lattice. Phys. Rev. E 58, 69–73 (1998)

Spatio-temporal Control for Dynamic Routing Games

Manjesh Kumar Hanawal[1,2], Eitan Altman[1],
Rachid El-Azouzi[2], and Balakrishna J. Prabhu[3,4]

[1] Maestro group, INRIA, 2004 Route des Lucioles, Sophia Antipolis, France
[2] LIA, University of Avignon, 339, chemin des Meinajaries, Avignon, France
[3] CNRS; LAAS; 7 avenue du Colonel Roche, F-31077 Toulouse, France
[4] Université de Toulouse; UPS, INSA, INP, ISAE; LAAS; F-31077 Toulouse, France
{mhanawal,eitan.altman}@inria.fr, rachid.elazouzi@univ-avignon.fr,
balakrishna.prabhu@laas.fr

Abstract. In this paper, we study dynamic routing games where the decision of an user is spatio-temporal control. Each user ships its demand over time on a shared resource. We investigate the equilibrium of such systems and show the existence and uniqueness of equilibrium. In the second part, we study a stochastic congestion games where there is only one shared resource and the traffic is indivisible. The information structure that we consider is such that each user knows the state of its own buffer but not aware of states and the actions taken by other users. The game can be described as a game with random environment. We characterize the structure of equilibria policies using linear programming. We also study the properties of equilibrium considering another model for stochastic congestion game in which a fixed amount of divisible demand arrives each day. This demand can shipped to destination by sending some part today and remaining the next day.

1 Introduction

Routing games are concerned with one or more classes of individuals, where each class is characterized by a source-destination pair and a demand function. A given network is shared by the users. Routes are chosen by the players so as to minimize the delays. In road traffic engineering, routing games appeared in 1952 [5]. Non-cooperative routing has long been studied both in the framework of road-traffic as well as in the framework of telecommunication networks. Such frameworks allow us to model the flow configuration that results in networks in which routing decisions are made in a non-cooperative and distributed manner between the users. There can be finite or infinite number of users.

In the case of a infinite number of players each player is assumed to be atomless. By atomless we mean that the impact of routing choices of a single player on the utilities of other players is negligible. The resulting flow configuration corresponds to the Wardrop equilibrium [5]. This concept, has long been studied in the context of road traffic where there is an infinite number of players (drivers)

R. Jain and R. Kannan (Eds.): GameNets 2011, LNICST 75, pp. 205–220, 2012.

[3]. In the telecommunication community, the mostly used routing game model introduced by Orda, Rom and Shimkin [2] is that in which the number of players are finite, where a player (typically corresponding to a service provider) takes the routing decisions for the whole class of users that it controls. It then decides on how to split the demand it controls between various possible routes. They establish existence and uniqueness of Nash equilibrium over large class of general cost functions. This approach also appeared in the road traffic literature (e.g. [1]) but was not much used there. Such a routing game may be handled by models similar to [6] in the special case of a topology of parallel links.

An alternative class of routing games is the one in which a player has to route all the demand it controls through the same path. A special case of such framework is the "congestion games" introduced by Rosenthal in [4]. All the above works have been well studied in time-invariant networks.

In this paper, we study a dynamic routing game where the decision of a user is spatio-temporal. The demand has to be split not only over space but also over time. As an example, assume that N players have each its own demand which should be shipped within a week from a given source to a destination. Thus a player has to split its demand into that corresponding to each of the days of the week. At each day, the route corresponding to the daily demand of each player should be determined. Examples of such games in road traffic appear in [13].

A dynamic routing game over T days can often be transformed into an equivalent static one. Indeed, in the equivalent static model, we make T replicas of the original network, one corresponding to each day. The source node for a player in each of the replicated networks is connected to a node corresponding to the source for that player (and we do the same with the destination). The fact that the game is repeated over time allows often to observe what other players did in the past.

The paper is organized as following. In section 2 we briefly overview some results in dynamic games that are relevant to the dynamic routing game: the existence of equilibria within open-loop as well as closed-loop strategies and procedures for computing them. We obtain explicit expression for equilibrium considering polynomial cost functions and study its structure. In section 3 a simple stochastic congestion game is modeled in which only one unit of demand can arrive to each player with some probability. The state of a player evolves according to whether they decide to transmit or not. We study the existence and structure of stationary policies. In section 4 we consider another model of stochastic congestion game in which the fixed demand arrives each day that has to be shipped within two days. We end with conclusion in section 5.

2 Dynamic Game with Fixed Demand

Assume that there is a fixed amount of demand that has to be shipped over a link within a fixed time, say T days. We consider two non-cooperative scenarios. In the first case the demand is considered as infinitely many users. In other case we consider finite number of players, where each tries to ship their demand over the shared link in a selfish way so that the total cost incurred by them is minimized. We study structure of spatio temporal equilibrium flows in both cases using notion of Wardrop equilibrium in the former case and Nash equilibrium in the latter.

2.1 Non-atomic Routing Games

In the context of road traffic, Wardrop [5] proposed the following definition of equilibrium

"The journey times on all the routes actually used are equal, and less than those which would be experienced by a single vehicle on any unused route"

Consider an amount d of demand that has to traverse a common shared link. Each player has to determine at what day within a common finite set S to leave. Let x_i be the amount of players that leave at day i. A player leaving in day i pays a congestion cost $f(x_i)$ and a delay cost of i units. The vector $x = (x_1, ..., x_T)$ is said to be feasible if $\sum_{i=1}^{T} x_i = d$. It is a spatio temporal Wardrop equilibrium if there is some constant $\alpha > 0$ such that for each $i = 1, ..., T$, we have

$$f(x_i) + i \geq \alpha \quad \text{and} \quad x_i(f(x_i) + i - \alpha) = 0$$

Equivalently,

$$\begin{array}{ll} \text{all days for which } x_i > 0, & f(x_i) + i = \alpha \\ \text{and for all other } i\text{'s}, & f(0) + i \geq \alpha. \end{array} \tag{1}$$

and for all $i \in I$ we have $x_i = f^{(-1)}(\alpha - i)$.

Thus if f is increasing then so is $f^{(-1)}$ and then x_i is decreasing in i at equilibrium. Thus there is a threshold policy such that till day i^*, some traffic is shipped everyday, and after that threshold nothing is shipped.

2.2 Atomic Routing Games

In this section we consider finite number of players N each competing for the link. The decision of each player influences the cost of the others. Let the demand of each player $i = 1, 2, \cdots, N$ be $d_i > 0$ that has to be shipped to destination over a period of T days. The amount of flow sent by player i in the jth day is denoted by x_j^i and the vector $\mathbf{x}^i = (x_1^i, x_2^i, \cdots, x_T^i)$ denotes the flow of user i. The vector $\mathbf{x}^i$ is said to be feasible if $\sum_{j=1}^{T} x_j^i = d_i$. Let the vector $\mathbf{x}_j = (x_j^1, x_j^2, \cdots, x_j^N)$ denote the amount of flow sent by each user on day j. The total flow on day j is denoted as $x_j = \sum_{i=1}^{N} x_j^i$. For a given flow configuration of users $(\mathbf{x}^1, \mathbf{x}^2, \cdots, \mathbf{x}^N)$, user i pays a congestion cost of $f(x_j)$ and delay cost of j per unit of its flow on day j. The objective of each user is to minimizes his cost given by

$$J^i(\mathbf{x}^1, \mathbf{x}^2, \cdots, \mathbf{x}^N) = \sum_{j=1}^{T} x_j^i (f(x_j) + j)$$

subjected to its demand constraints. A feasible vector $\mathbf{x}^* = (\mathbf{x}^{*1}, \mathbf{x}^{*2}, \cdots, \mathbf{x}^{*N})$ is said to be Nash equilibrium if for each user $i = 1, 2, \cdots, T$.

$$J^i(\mathbf{x}^{*1}, \mathbf{x}^{*2}, \cdots, \mathbf{x}^{*i}, \cdots, \mathbf{x}^{*N}) \leq J^i(\mathbf{x}^{*1}, \mathbf{x}^{*2}, \cdots, \mathbf{x}^{i}, \cdots, \mathbf{x}^{*N}) \quad \text{for all feasible } \mathbf{x}^i.$$

The above spactio temporal dynamic game is completely equivalent to a static parallel link problem studied by [2] with link cost $J_i(x_i)$ taking value $J_i(x_i) = f(x_i) + i$. When f is a positive, strictly increasing, convex and continuously differentiable all the assumptions in type-B functions in [2] are satisfied. Hence the existence and uniqueness of Nash equilibrium hold. Further the following property hold for the equilibrium flow.

Proposition 1. *The total flow x_j is non-increasing in each day.*

Proof. Let $\mathcal{L}_j$ denote the set of all users who put a positive flow on day j, i.e., $\mathcal{L}_j = \{i : x_j^i > 0\}$. From the KKT conditions for all $i \in \mathcal{L}_j$ there exists a λ^i such that (see, [2][eq. 2,3])

$$f(x_j) + x_j^i f'(x_j) + j = \lambda^i \quad \text{and} \quad f(x_k) + x_k^i f'(x_k) + k \geq \lambda^i \quad \text{for all} \quad k \neq j.$$

Summing over the set $\mathcal{L}_j$ $(j > 1)$ we get

$$\begin{aligned} f(x_j)|\mathcal{L}_j| + x_j f'(x_j) &\leq f(x_{j-1})|\mathcal{L}_j| + \sum_{i \in \mathcal{L}_j} x_{j-1}^i f'(x_{j-1}) \\ &\leq f(x_{j-1})|\mathcal{L}_j| + x_{j-1} f'(x_{j-1}) \end{aligned}$$

By the assumption that f and f' are monotonically increasing we obtain $x_{j-1} > x_j$ for all j.

Ayesta *et al.* [7] showed while studying the load balancing non-cooperative game that the worst case Nash equilibrium occurs when each user have the same amount of demand, i.e, when users are symmetric. Next we will calculate the Nash equilibrium for this case.

Symmetric Users: Consider that every user has a fixed demand $d_i = d$ to route. Cominetti, Correa, and Stier-Moses [8][Sec. 4.2] showed that the game with symmetric users is a potential game [9] and the Nash equlilibrium can be obtained as the solution of a single optimization problem as following. If $\mathbf{x}^* = (x_1^*, x_2^*, \cdots, x_T^*)$ is the solution of the following optimization problem

$$\begin{aligned} &\underset{\mathbf{x}=(x_1,x_2,\cdots,x_T)}{\text{minimize}} && \sum_{i=1}^{T} x_i(f(x_i) + i) + (N-1)\sum_{i=1}^{T} \int_0^{x_i} (f(x) + i)dx \\ &\text{subject to} && -x_i \leq 0, \ i = 1, 2, \cdots, T \ \text{and} \ \sum_{i=1}^{T} x_i = d. \end{aligned}$$

then flow profile at Nash equilibrium is given by $\mathbf{x}^i = \{\frac{x_1^*}{N}, \frac{x_2^*}{N}, \cdots \frac{x_T^*}{N}\}$ for each user $i = 1, 2, \cdots, N$.

We consider the special case when the congestion cost is polynomial, i.e., of the form $f(x) = ax^p + b$, for some $a, b > 0$ and $p > 1$. Polynomial functions were introduced for congestion cost originally in the context of road traffic. In [11] existence of equilibrium and its uniqueness are studied for the general network topology with polynomial congestion cost. The following proposition gives values of the flow at equilibrium.

Proposition 2. *For polynomial congestion cost the equilibrium flow is*

$$x_j^* = \begin{cases} \left(\frac{\beta - j - b}{a + \frac{ap}{N}}\right)^{\frac{1}{p}} & \text{if } j < \beta - b \\ 0 & \text{otherwise} \end{cases}$$

where β is such that it satisfies

$$\sum_{j:j<\beta-b} \left(\frac{\beta - j - b}{a + \frac{ap}{N}}\right)^{\frac{1}{p}} = d$$

Proof. See Appendix A.

Next we consider the case when players co-operate among themselves and try to minimize the total cost.

2.3 Global Optimum

Consider the scenario as in the previous sub section 2.2. The objective of each user when the total demand D is

$$\underset{(x_1, x_2, \cdots, x_T)}{\text{minimize}} \quad \sum_{j=1}^{T} x_j (f(x_j) + j)$$

$$\text{subject to} \quad -x_j \leq 0, \; j = 1, 2, \cdots, T. \text{ and } \quad \sum_{j=1}^{T} x_j = D.$$

Let $(x_1^*, x_2^*, \cdots, x_T^*)$ be the optimal value of flows that achieve minima. Then by the necessary KKT conditions there exist a $\alpha > 0$ such that the following hold

$$\begin{aligned} f(x_j^*) + x_j^* f'(x_j^*) + j &= \alpha \text{ if } \; x_j^* > 0 \text{ and} \\ f(x_j^*) + x_j^* f'(x_j^*) + j &\geq \alpha \text{ if } \; x_j^* = 0 \end{aligned} \tag{2}$$

Since we assumed that f is monotonically increasing and convex, the function $F(x) = f(x) + xf'(x)$ is monotonically increasing. Also, by writing $x_j^* = F^{-1}(\alpha - j)$ we observe that flow is monotonically decreasing till day j^* and after that nothing is shipped.

From equations (2) and (1) the following lemma is easy to verify.

Lemma 1. *Let $\mathbf{x}^* = (x_1^*, x_2^*, \cdots, x_T^*)$ denote the minimum of the global optimization problem $\sum_{i=1}^{T}(g(x_i) + i \cdot x_i)$ subjected to constraints, where g is a differentiable function. Define $f(\cdot) = g'(\cdot)$ then $\mathbf{x}^*$ is Wardrop equilibrium of the game with cost function f if and only if it is a global optimum.*

More general versions of the above results appear in [12][Chapter 18].

Again, considering the polynomial cost structure, optimal flows can be characterized as in the following proposition.

Proposition 3. *For the polynomial congestion cost optimal flow is*

$$x_j^* = \begin{cases} \left(\frac{\beta-j-b}{a+ap}\right)^{\frac{1}{p}} & \text{if } j < \beta - b \\ 0 & \text{otherwise,} \end{cases}$$

where β is such that it satisfies

$$\sum_{j:j<\beta-b} \left(\frac{\beta - j - b}{a + ap}\right)^{\frac{1}{p}} = D.$$

With the Nash equilibrium flow and optimal flow in proposition 2 and proposition 3, we can compute social cost at equilibrium and optimal cost and hence price of anarchy(PoA). For the polynomial cost function we obtain an upper bound of $\left(\frac{1+p}{1+p/N}\right)^{1+1/p}$. For detailed proof see Appendix B. The proof method is similar to that in [7].

3 Stochastic Congestion Game

3.1 Motivating Examples

The game is motivated by the following scenario: N transport companies share a network of roads over which they want to schedule their traffic on a day-to-day basis. The traffic to be routed by each company varies on a day-to-day basis according to a given stochastic process. The objective of each company is to route its traffic in such a way so as to minimize the average delay of its traffic. The cost incurred by a player, which represents the delay due to congestion on the roads over which they schedule the traffic, depends not only on the amount it routes but also on the amount the other players route thereby leading to a game.

Another practical scenario where this game is played is that of a set of universities who share a cluster of processors over which they want to schedule tasks which arrive according to a given stochastic process. At the beginning of each day, each player has to decide the tasks it schedules over each of the processors with the knowledge that the delay incurred by its tasks depends also on the scheduling policy of the other players.

In this section we assume that there is only one shared resource (road or processor), and that the traffic is *indivisible*, that is, a player cannot divide its traffic over two or more slots. It should either route the entire traffic or wait. Unlike in the previous section we consider the case where game is repeated infinitely many times. Description of our game model is as following. We largely follow the notation used in [10].

3.2 Model

1. **State space:** At the beginning of a slot, player i can be either *active* or *inactive.* In the active state, the player has some amount of traffic to ship from some source to some destination. It stays in this state until it routes this traffic and then moves to the inactive state. The player then stays in the inactive state until a new unit of traffic arrives.
We shall assume that, when player i is in the inactive state, a new unit of traffic arrives with probability p_i as a result of which the player becomes active. Let x_t^i denote the state of player i in slot t, and let $\mathbf{x}_t$ denote the state vector at the beginning of slot t. We denote state space of ith player as $X^i = \{0,1\}$. Active state corresponds to 1 and inactive state to 0.
2. **Action set:** At the beginning of a slot, an active player can decide either to route its traffic or to postpone. The actions available to player i in slot t, a_t^i, is thus a subset of $\{0,1\}$. A player sends all of its traffic when it decides to route in the active mode. We shall denote the action set of player i when it is in state x as $A^i(x)$. Specifically, $A^i(x=0)=0$ and $A^i(x>0)=\{0,1\}$. Let $\mathbf{a}_t = (a_t^1, a_t^2, \cdots, a_t^N) \in \prod_{i=1}^N A^i(x_t^i)$ denote the action vector at the beginning of slot t when the state vector is x_t. Define the local set of sate-action pair for player i as $\mathcal{K}^i = \{(x^i, s^i) : x^i \in X^i, s^i \in A^i(x^i)\}$, and $\mathcal{K}^{-i}$ denote the state-action set of all other users, i.e., $\mathcal{K}^{-i} = \prod_{j\neq i}^N \mathcal{K}^j$
3. **Dynamics of the state:** The state of player i, x_t^i evolves according to a Markov chain whose transition matrix in slot t depends on the action a_t^i and the current state. For each user we denote the transition probability from state $x \in X^i$ to $y \in X^i$ when user takes action $a \in S^i(x)$ as P^i_{xay}. When a player i is in inactive mode it can get a unit of traffic with probability p_i, hence takes values $x_{t+1}^i = 1$ with probability p_i or remains in the inactive mode with probability $1-p_i$. Similarly, in the active mode if the action is to transmit then a new state becomes active with probability p_i and inactive with probability $1-p_i$. If the state is active and action is not to transmit than state remains active. Transition probability matrix is given by

$$P^i_{(\cdot)a(\cdot)} = \begin{bmatrix} 1-p_i & p_i \\ (1-p_i)a & 1-(1-p_i)a \end{bmatrix} \tag{3}$$

4. **Cost functions:** In a given slot t, an active player who routes its traffic incurs a cost which is a function of the number of players who routed their traffic in that slot. On the other hand, if an active player decides to postpone, then it incurs a fixed cost d which can be inferred as a penalty for the increased delay incurred by the traffic. Let $c_t^i : \prod_{i=1}^N \mathcal{K}_i \to \mathcal{R}$ be the cost incurred by player i in slot t, then

$$c_t^i(\mathbf{x}_t, \mathbf{a}_t) = f\left(\sum_i x_t^i a_t^i\right) + d\left((1-a_t^i)x_t^i\right). \tag{4}$$

5. **Policies and Information:** Define the history of player i at time t to be a sequence of its past state and actions including the current state

$\mathbf{h}_t^i = (x_1^i, a_1^i, x_2^i, a_2^i, \cdots x_{t-1}^i, a_{t-1}^i)$. Let H_t^i denote the collection of all such histories. The policy of a players i is a sequence of maps $(u_1^i, u_2^i, \cdots)$, where the maps are given by $u_t^i : H_t^i \to M(A^i)$ with $M(A^i)$ denoting the probability vector on the set A^i. Set of all policies of player i is denoted as U^i and the collection $U = \prod_{i=1}^N U^i$ is called multi-polices. A **stationary policy** for a player i is a function $u^i : \mathbf{X}^i \to M(A^i)$ so that $u^i(\cdot|x^i) \in M(A^i(x^i))$. Set of all stationary policies for player i is denoted as U_S^i and the collection of all users stationary policies as $U_S = \prod_{i=1}^N U_S^i$. Note that when the user is in inactive mode then only possible action is 0, hence $u^i(0|x^i = 0) = 1$. In our model a stationary policy is completely characterized by the probability of transmission in active state, i.e., $q_i := u^i(1|1)$. We use both u^i and its associated q_i to denote a stationary policy interchangeably. The transition probability matrix when the ith users stationary policy is q_i is given by

$$P_i(q_i) = \begin{bmatrix} 1 - p_i & p_i \\ q_i(1 - p_i) & 1 - q_i(1 - p_i) \end{bmatrix} \tag{5}$$

For any multi policy $u \in U$, let u^{-i} denote the multi policy without the policy u^i and $[u^{-i}|v^i]$ denote the multi policy with the policy u^i of ith user replaced by v^i. We assume that each user i has information of its own state and makes decision only based on this information.

6. **Utility and the objective:** The objective of player i is to select its strategies so as to minimize the total cost it incurs over the horizon. Let β denote the distribution of the initial state. Let $\{X_t, A_t\}_{t>0}$ denote the sequence of random variable on state and action sets, where $X_t = \{X_t^1, X_t^2, \cdots, X_t^N\}$ and $A_t = \{A_t^1, A_t^2, \cdots, A_t^N\}$. The state action pair $\{X_t, A_t\}$ evolve according to the distribution which is determined by multi policy $u \in U$, the transition probabilities and β. We denote this distribution by P_β^U and expectation with respect to this distribution as $\mathbb{E}_\beta^U$. Each user $i = 1, 2, \cdots, N$ seeks to minimize his average expected cost.

$$C^i(U, \beta) = \limsup_{T \to \infty} \frac{1}{T} \mathbb{E}_\beta^U \sum_{t=1}^{T} c_t^i(X_t, A_t) \tag{6}$$

Below we restate the definition in [10][Def 2.1] for Nash equilibrium and optimal response for a stochastic game.

Definition 1. *1. A multi policies $u \in U$ is said to be Nash equilibrium if $i = 1, 2, \cdots, N$ and for any $v^i \in U^i$*

$$C^i(\beta, u) \leq C^i(\beta, [u^{-i}|v^i]) \text{ for any } v^i \in U^i i = 1, 2, \cdots, N. \tag{7}$$

2. For any multi policy u, policy u^i of user i is said to be optimal response against u^{-i} if (7) holds for any $v^i \in U^i$.

3. For any multi polices $u, v \in U$, v is said to be optimal response against u if for each user v^i is the optimal response against u^{-i}.

Next we will study the existence of stationary Nash equilibrium and its properties.

3.3 Equilibrium and Properties

Let $\pi(q_i) = (\pi(1|q_i), \pi(0|q_i))$ denote the stationary distribution on the states of user i when he/she uses the stationary policy $q_i \in U_S^i$, where the component $\pi(x|q_i)$ denotes the stationary probability of state $x \in A^i$. Markov chain of each user is irreducible with the stationary distribution as a function of policy given by

$$\pi(1|q_i) = \frac{q_i(1-p_i)}{q_i + p_i(1-q_i)} \quad \text{and} \quad \pi(0|q_i) = \frac{p_i}{q_i + p_i(1-q_i)}. \tag{8}$$

The stochastic congestion game defined above satisfies all the assumptions in Theorem 2.1 in [10]. Hence, we have the following existence result.

Proposition 4. *The stochastic game has a stationary multi policy $u \in U_S$ which is a Nash equilibrium.*

The optimal response of the ith user when the other user use stationary policies can be computed from the linear programming.

3.4 Linear Programming

For any state action pair $(x, a) \in \mathcal{K}_i$ the expected cost incurred by the the i player when other user use stationary policy $u^{-i} \in U_S^{-i}$ is

$$c^{i,u}(x,a) = \sum_{(x^{-i},a^{-i}) \in \mathcal{K}^{-i}} \left(\prod_{j \neq i} u^j(a^j|x^j)\pi(x^j|q_j) \right) c^i(\mathbf{x}, \mathbf{a}), \tag{9}$$

where q_j denotes the probability associated with stationary strategy u^j and $\mathbf{x} = [x^{-i}|x^i]$ $\mathbf{a} = [a^{-i}|a^i]$.
Let $z_{i,u}^* = \{z_{i,u}^*(x,a); (x,a) \in \mathcal{K}_i\}$ be such that it minimizes

$$\begin{aligned} &\sum_{(x,a)\in\mathcal{K}_i} c^{i,u}(x,a) z_{i,u}(x,a) \quad \text{subjected to} \\ &\sum_{(x,a)\in\mathcal{K}_i} z_{i,u}(x,a)[\delta_r(x) - P^i_{xar}] \quad \forall r \in X^i \\ &z_{i,u}(x,a) \geq 0 \quad \forall\, (x,a) \in \mathcal{K}_i \quad \text{and} \quad \sum_{(x,a)\in\mathcal{K}_i} z_{i,u}(x,a) = 1 \end{aligned} \tag{10}$$

Then the optimal policy is given by

$$u^{i*}(a|x) = \frac{z_{i,u}^*(x,a)}{\sum_{a\in A^i} z_{i,u}^*(x,a)}. \tag{11}$$

Next, we will consider symmetric users and characterize the Nash equilibrium.

3.5 Symmetric Users

Let p denote the probability that a packet arrives to each players. From equation (9) observe that $c^{i,u}(1,0) = d(1)$ and $c^{i,u}(0,0) = 0$. By substituting the the expected cost of ith player when the others use stationary strategy u^{-i} in (10) the best response of ith player is given by the following linear programming below. Let $x := z_{i,u}(1,1)$,

$$\begin{aligned} &\text{minimize} \qquad && c^{i,u}(1,1)x + c^{i,u}(1,0)\big(1 - \frac{x}{p}\big) \\ &\text{subjected to} && 0 \leq x \leq p \end{aligned} \tag{12}$$

If x^* is the optimal value of the above optimization problem than from equation (11) the best stationary response of ith player is given by $q_i = x^*/\big(x^* + (1 - x^*/p)\big)$.

Now, considering two player case we characterize the Nash equilibrium as following.

Proposition 5. *For any given functions f, d and packet arrival probability p, if there exists $q \in [0\ 1]$ that satisfies*

$$q\pi(1|q)f(2) + (1 - q\pi(1|q))f(1) - \frac{d(1)}{p} = 0,$$

then (q,q) is the stationary symmetric Nash equilibrium (SSNE). Further, if $f(1) > \frac{d(1)}{p}$ then 'no-transmit' for both player is SSNE. If $(1-p)f(2) + pf(1) < \frac{d(1)}{p}$ than 'transmit' for both players is SSNE.

Proof. See Appendix C.

4 Stochastic Congestion Game with Fixed Demand

In this section we consider another version of stochastic congestion games with one player every day. Each day the arriving player has to decide how much of the traffic to send that day and how much to send the next day. The model is as following.

– One player arrives each day with a fixed demand of $\phi > 0$ units. This demand has to be shipped to the destination using a shared link. $\phi - \lambda$ units of demand is urgently required at the destination and hence needs to be shipped the same day by every player, where $0 \leq \lambda \leq \phi$. The remaining λ units of demand can be shipped either entirely in that same day or entirely at the next day, but no later. The ith player ships the remaining λ units of demand with probability p_i the same day. We denote the amount of demand shipped by the ith player at its arrival date as X_i and it takes values as following

$$X_i = \begin{cases} \phi, & \text{w.p } p_i \\ \phi - \lambda, & \text{w.p } 1 - p_i \end{cases} \tag{13}$$

- **Cost function:** Each player has to pay for the usage of the shared link. We assume that the amount paid depends on the total units of demand shipped on that day (*congestion cost*). Let f denote the congestion function. If the player decides to send λ units of demand the next day he/she has to pay a *storage cost* of d units per demand and the congestion cost the next day. Note that the cost incurred by the ith player depends only on shipping policy of the players of the previous day and that of the next day. We denote the total cost incurred by the ith player as J_i and it is given by

$$J_i(X_{i-1}, X_i, X_{i+1}) = X_i f(X_i + \phi - X_{i-1}) + (\phi - X_i)(d + f(\phi - X_i + X_{i+1}))$$

- **Strategy and utility:** Each player has to decide whether to ship the entire demand in one shot or ship only $(\phi - \lambda)$ at its arrival day and the remaining λ units of demand the next day. The strategy of the ith player is the probability with which it transmits the entire demand in one shot. We denote it by $p_i \in [0\ 1]$. We are interested in symmetric equilibria, so we assume that there is some constant p_{-i} such that any player other than i ships its entire demand with the same probability p_{-i}. The objective of each player is to choose a strategy $p_i \in [0\ 1]$, for day i, such that it minimizes his/her expected cost given that other players use p_{-i}. Player i faces the following optimization given that all other days the strategy used is p_{-i}.

$$\min_{0 \le p_i \le 1} \mathbf{E}_{p_i, p_{-i}}[J_i(X_{i-1}, X_i, X_{i+1})] \quad i = 1, 2, \cdots$$

$\mathbf{E}_{p_i, p_{-i}}$ denotes that expectation is taken by assigning probability p_i to X_i.

4.1 Equilibrium Strategies

We will be interested in studying the stationary equilibrium of the above stochastic congestion game. In particular, we will study the stationary symmetric Nash equilibrium (SSNE) assuming linear congestion cost, i.e., $f(x) = \beta x$, for some $\beta > 0$. For any given strategy $\{p_i, p_{-i}\}$ profile of the players, the utility function for the ith player is given by

$$\begin{aligned}
&\mathbf{E}_{p_i, p_{-i}}[J_i(X_{i-1}, X_i, X_{i+1})] \\
&= p_i \phi \mathbf{E}_{p_{i-1}}[f(2\phi - X_{i-1})] + (1 - p_i)\mathbf{E}p_{i-1}[(\phi - \lambda) f(2\phi - \lambda - X_{i-1})] \\
&\quad + (1 - p_i) a \mathbf{E}p_{i+1}[d + f(\lambda + X_{i+1})] \\
&= p_i \Big(2\beta\phi^2 - \beta\phi \mathbf{E}_{p_{i-1}}[X_{i-1}] - \beta(\phi - \lambda)(2\phi - \lambda - \mathbf{E}_{p_{i-1}}[X_{i-1}]) - \lambda d - \beta\lambda^2 \\
&- \beta\lambda \mathbf{E}_{p_{i+1}}[X_{i+1}] \Big) + \beta(\phi - \lambda)(2\phi - \lambda - \mathbf{E}_{p_{i-1}}[X_{i-1}]) + \lambda d + \beta\lambda^2 + \beta\lambda \mathbf{E}_{p_{i+1}}[X_{i+1}]
\end{aligned}$$

Now assume that all player $j \neq i$ use the strategy $p_j = \overline{p}$. Then it is clear that $\mathbf{E}[X_{i+1}] = \mathbf{E}[X_{i-1}] = (\phi - \lambda) + \lambda\overline{p}$. Substituting in the above equations and continuing the chain of equalities we have

$$\mathbf{E}_{p_i,p_{-i}}[J_i(X_{i-1},X_i,X_{i+1})]$$
$$= p_i(-2\beta\lambda^2\overline{p} + \beta\lambda\phi - \lambda d) + \beta\phi^2 + \lambda d + \beta\lambda(2\lambda - \phi)\overline{p} \quad (14)$$

SSNE are characterized in the following proposition.

Proposition 6

- *When* $d \leq \beta\phi$
  - $p_i^* = 0$ *for all* i *is SSNE.*
  - $p_i^* = (\beta\phi - d)/2\lambda\beta$ *for all* i *is also SSNE.*
- *When* $d > \beta\phi$
  $p_i^* = 1$ *for all* i *is SSNE.*

Proof. First assume that $d \leq \beta\phi$. Suppose $p_i > 0$ for player$i$ and $p_j = 0$ for all other players $j \neq i$. Then from equation (14) it can be seen that the utility of the ith player is given by

$$\mathbf{E}_{p_i,0}[J_i] = p_i(\beta\lambda\phi - \lambda d) + \beta\phi^2 + \lambda d,$$

which is strictly larger than the case when $p_i = 0$ is chosen. Hence $p_i^* = 0$ for all i is a SSNE. To see that $p_i^* = (\beta\phi - d)/2\lambda\beta$ is also an equilibrium, substitute this value in equation (14) and notice that it becomes independent of p_i. For the case $d > \beta\phi$ the claim can be verified by by substituting $\overline{p} = 1$ in equation (14) and noticing that the coefficient of p_i is negative.

5 Conclusion

We studied the dynamic routing game in which there is both congestion and delay cost. We established the existence of equilibrium and gave its properties considering both cases of atomic and non-atomic players. When the demand of players is random, we considered a simple stochastic congestion game and investigated the existence of stationary equilibrium policies and its properties. Its interesting to consider more general model in stochastic game scenario and study its equilibrium properties. For example, when buffer length is more than one. We also considered a simple congestion game in which the demand that arrives is not random but fixed and characterized its equilibrium.

Appendix A: Proof of Proposition 2

Proof

$$\underset{\mathbf{x}=(x_1,x_2,\cdots,x_T)}{\text{minimize}} \quad \sum_{j=1}^{T} \frac{x_j}{N}(ax_j^p + b + j) + \frac{N-1}{N}\sum_{j=1}^{T}\int_0^{x_j}(ax^p + b + j)dx$$

$$\text{subject to} \quad -x_j \leq 0,\ j = 1,2,\cdots,T \text{and} \sum_{j=1}^{T} x_j = d.$$

If $\mathbf{x}^* = (x_1^*, x_2^*, \cdots, x_T^*,)$ is the optimal solution than by the necessary conditions of the KKT theorem there exits $\lambda_j \geq 0$ for $j = 1, 2, 3 \dots T$ and $\beta \in R$ such that the following holds

$$ax_j^{*p} + b + j + \frac{ap}{N} x_j^{*p} = \beta + \lambda_j \quad \text{for} \ \ j = 1, 2, \cdots, T \tag{15}$$

$$x_j^* \lambda_i = 0 \quad \text{for} \ \ i = 1, 2, \cdots, T \tag{16}$$

$$\sum_{i=1}^{T} x_j^* = d \tag{17}$$

substituting the value of λ_j from equation (15) into equation (16) we get

$$x_j^* \cdot (ax_j^{*p} + b + j + \frac{ap}{N} x_j^{*p} - \beta) = 0.$$

if $b + j \geq \beta$ than it is clear that $x_j^* = 0$. Now consider the case $b + j < \beta$, we have either $x_j^* 0$ or $x_j^* > 0$. Now suppose that $x_j^* = 0$, then from equation (15) and the fact that $\lambda_j \geq 0$ we get $b + j \geq \beta$, which is a contradiction. Hence, for the case $b + j < \beta$, x_j^* is strictly positive and can be obtained as the solution of

$$ax_j^{*p} + b + j + \frac{ap}{N} x_j^{*p} - \beta = 0 \quad \text{, i.e.,} \ \ ax_j^{*p} = \left(\frac{\beta - j - b}{a + \frac{ap}{N}} \right)^{\frac{1}{p}}.$$

Substituting the above in equation (17) we get

$$\sum_{j: j < \beta - b} \left(\frac{\beta - \alpha i - b}{a + \frac{ap}{N}} \right)^{\frac{1}{p}} - d = 0.$$

This concludes the proof.

Appendix B: Price of Anarchy

We began by studying the properties of the lagrangian multiplier associated with equality constraints (β) as a function of number of user. In this regard define a function W_i for each $i = 1, 2, \cdots, T$ as following

$$W_i(N, \gamma) = \mathbb{I}_{\{i + b < \gamma \leq (i+1) + b\}} \left(\frac{\gamma - i - b}{a + \frac{ap}{N}} \right)^{\frac{1}{p}}.$$

Also define $W(N, \gamma) = \sum_{i=1}^{T} W_i(N, \gamma)$. Some of the properties of the function W are summarized in the following lemma.

Lemma 2. *The function W is such that*

1. *For a fixed N, W is continuous and monotonically increasing in γ*
2. *For a fixed γ, W is monotonically increasing N.*

3. *For a fixed* N, $W(N,\gamma)=0$ *has a unique solution denoted as* $\gamma(N)$ *in the interval* $(1+b,\infty)$

Proof. First, notice that for each i, W_i has a right limit at the point $i+b$ given by $W_i(N,i+b)$. It is clear that W is continuous in each interval $(i+b<\gamma<(i+1)+b)$, we need to verify its continuity only at the boundaries. Fix j and consider the difference

$$\begin{aligned}
&\lim_{\gamma\to((j+1)+b)^+} W(N,\gamma) - \lim_{\gamma\to((j+1)+b)^-} W(N,\gamma)\\
&= W_{j+1}(N,(j+1)+b) - W_j(N,(j+1)+b)\\
&= \sum_{i=1}^{j+1}\left(\frac{(j+1)+b-i-b}{a+\frac{ap}{N}}\right)^{\frac{1}{p}} - D - \sum_{i=1}^{j}\left(\frac{(j+1)+b-i-b}{a+\frac{ap}{N}}\right)^{\frac{1}{p}} + D = 0
\end{aligned}$$

Hence W is continuous. Also, W_i is strictly increasing in the interval $(i+b,(i+1)+b)$ for each i which implies that W is strictly increasing. The second claim is straightforward. To prove the third part of the lemma notice that W starts from $-D<0$ at $\gamma=1+b$, then the claim follows from the first part of the lemma.

We next show that the unique zero of the function W is increasing in N.

Lemma 3. $\gamma(N)$ *is monotonically decreasing in* N

Proof. Let $\gamma(N1)$ and $\gamma(N2)$ denote the unique roots of $W(N1,\gamma)$ and $W(N2,\gamma)$ respectively, and assume that $N_1<N_2$. We have

$$W(N1,\gamma(N1)) = 0 = W(N2,\gamma(N2)) \geq W(N1,\gamma(N2)),$$

where the last inequality follows from second part of lemma 2. Now the claim directly follows from first part of the lemma 2.

From the previous lemma 3 it is easy to observe that as the number of symmetric users increases (with total demand held constant), positive flow is sent on lesser number of days at equilibrium. Let L_N denote set of days on which total flow sent by N users is positive, i.e., $L_N=\{i : i+b<\gamma(N)\}$.Then the observation made form the previous lemmas can be written as $L_{N+1}\subset L_N$. In particular, when the there is just one user, i.e., centrally controlled system, the positive flow is send on maximum number of days.

Equilibrium social cost with N users is given by

$$\begin{aligned}
D_N = \sum_{i\in L_N} x_i^*(a(x_i^*)^p+b+i) &= \sum_{i\in L_N}\left(\frac{\gamma(N)-i-b}{a+\frac{ap}{N}}\right)^{\frac{1}{p}}\left(a\left(\frac{\gamma(N)-i-b}{a+\frac{ap}{N}}\right)+b+i\right)\\
&= \sum_{i\in L_N}\left(\frac{\gamma(N)-i-b}{a+\frac{ap}{N}}\right)^{\frac{1}{p}}\left(\frac{\gamma(N)+\frac{p}{N}(i+b)}{1+\frac{p}{N}}\right),
\end{aligned}$$

and optimal social cost is given by

$$D_1 = \sum_{i \in L_1} x_i^*(a(x_i^*)^p + b + i) = \sum_{i \in L_1} \left(\frac{\gamma(1) - i - b}{a + ap}\right)^{\frac{1}{p}} \left(a\left(\frac{\gamma(1) - i - b}{a + ap}\right) + b + i\right)$$
$$= \sum_{i \in L_1} \left(\frac{\gamma(1) - i - b}{a + ap}\right)^{\frac{1}{p}} \left(\frac{\gamma(1) + p(i + b)}{1 + p}\right)$$

$$\text{PoA} = \frac{\sum_{i \in L_N} \left(\frac{\gamma(N) - i - b}{a + \frac{ap}{N}}\right)^{\frac{1}{p}} \left(\frac{\gamma(N) + \frac{p}{N}(i+b)}{1 + \frac{p}{N}}\right)}{\sum_{i \in L_1} \left(\frac{\gamma(1) - i - b}{a + ap}\right)^{\frac{1}{p}} \left(\frac{\gamma(1) + p(i+b)}{1 + p}\right)}$$
$$\leq \frac{a^{\frac{1}{p}}(1+p)^{1+\frac{1}{p}} \sum_{i \in L_1} \left(\gamma(1) - i - b\right)^{\frac{1}{p}} \left(\gamma(1) + \frac{p}{N}(i+b)\right)}{a^{\frac{1}{p}}(1+\frac{p}{N})^{1+\frac{1}{p}} \sum_{i \in L_1} \left(\gamma(1) - i - b\right)^{\frac{1}{p}} \left(\gamma(1) + p(i+b)\right)} \leq \left(\frac{1+p}{1+\frac{p}{N}}\right)^{1+\frac{1}{p}}$$

Where the first inequality follows as $\gamma(N) \leq \gamma(1)$ and $L_N \subset L_1$.

Appendix C: Proof of Proposition 5

Let q be denote the stationary strategy of player-1, from (9) the value of the expected cost incurred by player-2 when his/her action is to transmit in active state is given by

$$c^{2,u}(1,1) = q\pi(1|q)f(2) + (1-q)\pi(0|q)f(1) + \pi(0|q)f(1).$$

Substituting the above in the objective function in (12), and differentiating with respect to x we get the following first order optimality condition

$$q\pi(1|q)f(2) + (1 - q\pi(1|q))f(1) - \frac{d(1)}{p} = 0. \tag{18}$$

Suppose player-1 chooses q such that the above equation is satisfied than any values of $x \in [0\ p]$ is optimal and any values of $q_2 \in [0\ 1]$ is optimal response for player-2. Hence (q, q) is the stationary symmetric Nash equilibrium. To see the other part of the proposition, substitute the expression for stationary distributions in (8), when the stationary policy is q, in (12) to get

$$\left(\frac{q^2(1-p)}{q + p(1-q)} f(2) + \left(1 - \frac{q^2(1-p)}{q + p(1-q)}\right) f(1) - \frac{d(1)}{p}\right) x - d(1).$$

If player-1 chooses $q_1 = 0$, $x^* = 0$ is the minimizer of the above equation when $f(1) > \frac{d(1)}{p}$, hence $(0,0)$ is the SSNE. By similar arguments we can show that $(1,1)$ is the SSNE when $(1-p)f(2) + pf(1) < \frac{d(1)}{p}$.

Acknowledgement. The first author like to thank Hamidou Tembine for many useful discussions. This work is supported by ANR EcoSCells project.

References

1. Haurie, A., Marcotte, P.: On the relationship between Nash-Cournot and Wardrop equilibria. Networks 15, 295–308 (1985)
2. Orda, A., Rom, R., Shimkin, N.: Competitive routing in multi-user environments. IEEE/ACM Trans. on Networking, 510–521 (1993)
3. Patriksson, M.: The traffic assignment problem: Models and methods. VSP, Utrecht (1994)
4. Rosenthal, R.W.: A class of games possessing pure strategy Nash equilibria. Int. J. Game Theory 2, 65–67 (1973)
5. Wardrop, J.G.: Some theoretical aspects of road traffic research communication networks. Proc. Inst. Civ. Eng., part 2, 1, 325–378 (1952)
6. Park, K., Sitharam, M., Chen, S.: Quality of service provision in noncooperative networks with diverse user requirements. Decis. Support Syst. 28(1-2), 101–122 (2000)
7. Ayesta, U., Brun, O., Prabhu, B.J.: Price of Anarchy in Non-Cooperative Load Balancing. In: Proceedings of IEEE INFOCOM, San Diego (March 2010)
8. Cominetti, R., Correa, J.R., Stier-Moses, N.E.: The Impact of Oligopolistic Competition in Networks. Operation Reserach, INFORMS 57(6), 1421–1437 (2009)
9. Monderer, D., Shapley, L.S.: "Potential Games". Games and Economic Behavior 14, 124–143 (1996)
10. Altman, E., Avrachenkov, K., Bonneau, N., Debbah, M., El-Azouzi, R., Sadoc Menasche, D.: Constrained Cost-Coupled Stochastic Games with Independent State Processes. Operations Research Letters 36, 160–164 (2008)
11. Altman, E., Basar, T., Jimenez, T., Shimkin, N.: Competitive routing in networks with polynomial cost. IEEE Trans. on Automatic Control 47, 92–96 (2002)
12. Nissan, N., Roughgarden, T., Tardos, E., Vazirani, V.V.: Algorithmic Game Theory. Cambrige University Press (2009)
13. Wie, B.W., Friesz, T.L., Tobin, R.L.: Dynamic User Optimal Traffic Assignment on Congested Multidestination Networks. Transportation Research 24B(6), 431–442 (1990)

Designing Incentive Schemes Based on Intervention: The Case of Imperfect Monitoring

Jaeok Park and Mihaela van der Schaar

University of California, Los Angeles (UCLA), Los Angeles CA 90095, USA

Abstract. In this paper, we propose a class of incentive schemes based on intervention. We develop a general game-theoretic framework for the design of intervention schemes under imperfect monitoring. We examine a model of slotted multiaccess communication to illustrate our framework. In this model, an intervention device monitors the behavior of agents for a period called the test phase and takes an intervention action which affects agents for the remaining period called the intervention phase. We analyze the problems of designing an optimal intervention rule given a length of the test phase and choosing an optimal length of the test phase. Intervention schemes can induce cooperative behavior by applying intervention following signals with a high likelihood of deviation. Increasing the length of the test phase has two counteracting effects: It improves the quality of signals, but at the same time it weakens the impact of intervention due to increased delay.

Keywords: intervention, incentive schemes, slotted multiaccess communication, game theory.

1 Introduction

In a system where selfish agents compete for available resources, it is common that the resources are not utilized optimally from a system-wide point of view. This calls for an incentive scheme to drive selfish agents towards the system objective. In this paper, we propose a class of incentive schemes, called intervention schemes. An intervention scheme can be implemented by augmenting the system with an intervention device that can monitor the behavior of agents and intervene in their resource usage. To analyze interaction under an intervention scheme, we formulate a class of games called intervention games and propose a solution concept called intervention equilibrium. In an intervention game, a system manager specifies an intervention rule used by the intervention device and the actions of agents. The intervention rule prescribes an action taken by the intervention device following each possible signal it can observe regarding the actions of agents. By increasing the intensity of intervention following signals that suggest a deviation from the specified actions, an intervention scheme can provide incentives for agents to follow the actions recommended by the manager. At an intervention equilibrium, the manager chooses a pair of an intervention

R. Jain and R. Kannan (Eds.): GameNets 2011, LNICST 75, pp. 221–232, 2012.

rule and an action profile that optimizes the system objective while making the action profile self-enforcing given the intervention rule.

The idea of using intervention to provide incentives can be found in [1], [2]. Garg et al. [1] consider a congestion control problem and analyze performance under different scheduling mechanisms, which can be considered as intervention schemes. With a scheduling mechanism that assigns higher priority to flows with a smaller rate, the input rates of users can be restrained voluntarily. In other words, users do not send their traffic excessively in their self-interest because doing so will increase the probability that their packets are dropped. In our previous work [2], we consider a slotted multiaccess communication network with an intervention device that can jam the packets of users. We showed that intervention schemes can successfully regulate the transmission probabilities of selfish users in the case of perfect monitoring, where the intervention device observes the transmission probabilities of users. In the current paper, we provide a general framework for the design of intervention schemes, unlike the two previous papers that focus on specific communication scenarios. Then we illustrate our framework with a model of slotted multiaccess communication, as considered in [2], while relaxing the monitoring requirement. That is, we consider the case of imperfect monitoring, where the intervention device obtains only imperfect information about the transmission probabilities of users.

In an intervention game, the manager chooses an intervention rule before agents take their actions, and thus the manager can be considered as a leader and agents as followers. Such a hierarchical structure in the interaction between the leader and the followers has been analyzed using a Stackelberg game in the literature (see, for example, [3]–[5]). In [3]–[5], the leader uses a Stackelberg strategy, simply taking an action before the followers take theirs. In contrast, in an intervention game, the manager chooses an intervention rule, which is a complete contingent plan for intervention actions to be taken given each possible signal about the actions of agents. Thus, an intervention rule requires more overhead in terms of monitoring capability for implementation than a Stackelberg strategy. However, when the manager does not value his resource usage, intervention schemes have an advantage over Stackelberg strategies. In intervention schemes, intervention actions can be adjusted to the observed behavior of agents, and thus intervention can be applied only when it is necessary. On the contrary, Stackelberg strategies lack such adaptivity.

Pricing offers an alternative means that the manager can use to provide incentives. In pricing schemes, the manager collects payments from agents based on their resource usage. Due to its foundations in market economies, pricing has received a significant amount of attention in the literature (see, for example, [6], [7]). The main difference between intervention and pricing is that intervention affects the resource usage of agents inside of the system whereas pricing affects the payoffs of agents through an outside instrument. Intervention schemes are robust in the sense that agents cannot avoid intervention as long as they use resources in the system, while pricing is not effective if agents can evade payments while still using resources. Moreover, even in the case where payments can be enforced technologically, there are situations where pricing is undesirable due to policy considerations. For example, it has been argued in a public policy debate that access to the Internet should be provided as a public good by a public authority rather than as a private good in a market [8]. In addition, intervention schemes have informational advantages over pricing schemes. Intervention affects the payoffs of agents through physical quantities (e.g., data rates and delay) that agents care about, and the impacts of intervention actions on such physical quantities are relatively easy to measure. On the contrary, finding out the impact of pricing on agents requires the manager to know agents' subjective valuation of payments relative to physical quantities, which is difficult to measure. Below we summarize related work discussed so far in Table 1.

Table 1. Summary of related work

Intervention	Stackelberg strategy	Pricing
[1]: congestion control	[3]: congestion control	[6]: congestion control
[2]: medium access control	[4]: medium access control	[7]: cognitive radio
	[5]: power control	

In game theory, repeated games have been studied as a method to provide incentives in a long-run relationship [9]. In a repeated game, agents monitor the actions of other agents and choose their actions depending on their past observations. Hence, the burden of monitoring and executing reward and punishment is distributed to agents in a repeated game, while it is imposed solely on the intervention device in an intervention game. In other words, compared to repeated game strategies, intervention schemes require a central infrastructure in the system while reducing a burden on agents. We need repeated interaction among agents to apply a repeated game strategy, while intervention schemes are applicable to systems with a dynamically changing agent population (e.g., mobile networks). In the repeated game literature, reputation schemes are used to overcome limited observation due to infrequent interaction (see, for example, [10]). A reputation scheme usually has a central infrastructure, but its role is limited to collecting observations from agents, processing the collected observations, and disseminating the processed information.

An active area of research that focuses on designing incentive schemes is mechanism design. In a standard mechanism design problem [11, Chap. 23], the manager can decide an outcome that agents care about but has incomplete information about the types of agents. Relying on the revelation principle, standard mechanism design is concerned about incentives for agents to reveal their types truthfully. In contrast, our focus in designing intervention schemes is on providing incentives for agents to choose desirable actions. In this paper, we assume that the manager has complete information, knowing the action spaces and the payoff functions of agents. We can extend our framework to study the design of intervention schemes when the manager has incomplete information. We leave this extension for future research.

The remainder of the paper is organized as follows. In Sect. 2, we develop a game-theoretic framework for the design of intervention schemes. In Sect. 3, we apply the framework to a model of slotted multiaccess communication and analyze optimal intervention schemes. In Sect. 4, we conclude.

2 Framework for the Design of Intervention Schemes

We consider a system (e.g., a communication network) where N agents and an intervention device interact. The set of the agents is denoted by $\mathcal{N} = \{1, \dots, N\}$. The action space of agent i is denoted by A_i, and an action for agent i is denoted by $a_i \in A_i$, for all $i \in \mathcal{N}$. An action profile is represented by a vector $a = (a_1, \dots, a_N) \in A := \prod_{i \in \mathcal{N}} A_i$. An action profile of the agents other than agent i is written as $a_{-i} := (a_1, \dots, a_{i-1}, a_{i+1}, \dots, a_N)$ so that a can be expressed as $a = (a_i, a_{-i})$. Once an action profile of the agents is chosen, a signal is realized from a finite set of signals, denoted by Y, and is observed by the intervention device. The probability that a signal $y \in Y$ is realized given an action profile $a \in A$ is denoted by $\rho(y|a)$. After observing the realized signal, the intervention device takes an action, called an intervention action. We use a_0 and A_0 to denote an intervention action and the action space of the intervention device, respectively.

Since the intervention device chooses its action after observing a signal, a strategy for it can be represented by a mapping $f : Y \rightarrow A_0$, which is called an *intervention rule*. The set of all possible intervention rules is denoted by $\mathcal{F}$. There is a system manager who determines the intervention rule used by the intervention device. We assume that the manager can commit to an intervention rule, for example, by using a protocol embedded in the intervention device. The payoffs of the agents and the manager are determined by the intervention action, the action profile of the agents, and the realized signal. Thus, we denote the payoff function of agent $i \in \mathcal{N}_0 := \mathcal{N} \cup \{0\}$ by $u_i : A_0 \times A \times Y \rightarrow \mathbb{R}$. The payoff of the manager, u_0, can be interpreted as a measure of system performance. The game played by the manager and the agents is formulated as an *intervention game*, which is summarized by the data

$$\Gamma = \langle \mathcal{N}_0, (A_i)_{i \in \mathcal{N}_0}, (u_i)_{i \in \mathcal{N}_0}, (Y, \rho) \rangle \ . \tag{1}$$

Given an intervention rule and an action profile, expected payoffs can be computed by taking expectations of payoffs over signals. The expected payoff function of agent $i \in \mathcal{N}_0$ is denoted by a function $v_i : \mathcal{F} \times A \rightarrow \mathbb{R}$, which can be computed as

$$v_i(f, a) = \sum_{y \in Y} \rho(y|a) u_i(f(y), a, y) \ . \tag{2}$$

Once the manager chooses an intervention rule f, the agents play a simultaneous game, whose normal form representation is given by

$$\Gamma_f = \langle \mathcal{N}, (A_i)_{i \in \mathcal{N}}, (v_i(f, \cdot))_{i \in \mathcal{N}} \rangle \ . \tag{3}$$

We predict actions chosen by the agents given an intervention rule f by applying the solution concept of Nash equilibrium [12] to the induced game Γ_f.

Definition 1. *An intervention rule $f \in \mathcal{F}$* sustains *an action profile $a^* \in A$ if a^* is a Nash equilibrium of the game Γ_f, i.e.,*

$$v_i(f, a_i^*, a_{-i}^*) \geq v_i(f, a_i, a_{-i}^*) \quad \forall a_i \in A_i \ , \ \forall i \in \mathcal{N} \ . \tag{4}$$

Let $\mathcal{E}(f) \subseteq A$ be the set of action profiles sustained by f. The manager aims to maximize his expected payoff by specifying an intervention rule for the intervention device and an action profile for the agents sustained by the intervention rule. The manager's problem leads to the following solution concept for intervention games.

Definition 2. *$(f^*, a^*) \in \mathcal{F} \times A$ is an* intervention equilibrium *if $a^* \in \mathcal{E}(f^*)$ and*

$$v_0(f^*, a^*) \geq v_0(f, a) \quad \textit{for all } (f, a) \textit{ such that } a \in \mathcal{E}(f) \ . \tag{5}$$

$f^* \in \mathcal{F}$ is an optimal intervention rule *if there exists an action profile $a^* \in A$ such that (f^*, a^*) is an intervention equilibrium.*

An intervention equilibrium solves the following optimization problem:

$$\max_{f \in \mathcal{F}} \max_{a \in \mathcal{E}(f)} v_0(f, a) \ . \tag{6}$$

An intervention equilibrium can be considered as a subgame perfect equilibrium (or Stackelberg equilibrium) of an intervention game, with an implicit assumption that the manager can induce the agents to choose the best Nash equilibrium for him in case of multiple Nash equilibria. In our interpretation, the manager specifies an intervention equilibrium (f^*, a^*) to the agents so that a^* becomes a self-enforcing operating point for them given the intervention rule f^*. In other words, the manager can make a^* a focal point [12] of the game Γ_{f^*} by recommending it to the agents.

3 Intervention Schemes in a Slotted Multiaccess Communication Network

3.1 Model

There is a communication channel shared by N users. Time is divided into slots of equal length, and packets have a length that can be transmitted within a time slot. In each time slot, a user can attempt to transmit its packet or wait. If there is only one transmission attempt in a slot, the packet is successfully transmitted. If there is more than one transmission attempt in a slot, packets collide and no transmission is successful. For simplicity, we assume that each user can choose one of two transmission probabilities p_l and p_h, where $p_\text{l} = 1/N < p_\text{h} < 1$. Note that each user choosing p_l maximizes the total throughput, defined as the average number of successfully transmitted packets per time slot, assuming that all the users choose the same transmission probability [13].

We consider a period consisting of T consecutive time slots, and analyze interaction in the period without any consideration of past or future periods. We assume that the number of users and their transmission probabilities are fixed throughout a period. Then the action space of user i is given by $A_i = \{p_l, p_h\}$, for all $i \in \mathcal{N}$. The payoff of user i is given by the number of its successfully transmitted packets per time slot. Then the expected payoff of user i is given by the probability of its successful transmission, $a_i \prod_{j \in \mathcal{N} \setminus \{i\}} (1 - a_j)$. It is easy to see that the action p_h is a dominant strategy for every user. Hence, $(p_h, \ldots, p_h)$ is the unique Nash equilibrium, which yields a lower total throughput than the symmetric social optimum $(p_l, \ldots, p_l)$.

In order to improve the inefficiency of Nash equilibrium, we introduce an intervention device in the system. The intervention device is capable of monitoring the actions of the users and interfering in the transmission of the users. In each slot, the intervention device can sense the channel to learn whether the channel is idle (i.e., no user attempts to transmit its packet) or busy (i.e., at least one user attempts to transmit its packet). After observing channel states for the first t slots, where $1 \leq t \leq T$, the intervention device chooses its transmission probability, which remains fixed until the end of the period. We assume that, unlike the users, the intervention device can choose any transmission probability in $[0, 1]$. Thus, we have $A_0 = [0, 1]$. The first t slots and the remaining $(T - t)$ slots are called the test phase and the intervention phase, respectively.

Let $S = \{\text{idle}, \text{busy}\}$ be the set of channel states that can be observed in a slot. Then the set of all possible signals that the intervention device can obtain in the test phase is S^t. Since the transmission probabilities of the users do not change in a period, there is no gain for the manager to treat channel states from different slots differently. Hence, we focus on the class of intervention rules that use only the number of idle slots, which allows us to work with a smaller signal space $Y = \{0, 1, \ldots, t\}$ instead of S^t. Channel states are independent across slots, and the probability of an idle state in a slot given an action profile a is given by $q(a) := \prod_{i \in \mathcal{N}} (1 - a_i)$. Thus, the probability that k idle slots arise in the test phase is $\rho(k|a) = \binom{t}{k} q(a)^k (1 - q(a))^{t-k}$, for $k = 0, 1, \ldots, t$. Note that monitoring is imperfect in that the intervention device cannot observe the action profile of the users but obtains only imperfect information about the action profile.

The sequence of events in a period can be listed as follows.

1. At the beginning of the period, the users choose their transmission probabilities $a \in A$, which are used from slot 1 to slot T, knowing the intervention rule f adopted by the intervention device.
2. The intervention device collects observations of channel states from slot 1 to slot t (test phase).
3. The intervention device intervenes using the transmission probability prescribed by the intervention rule f from slot $t + 1$ to slot T (intervention phase).

The payoff of user i given intervention action a_0 and action profile a conditional on signal k (i.e., k idle slots in the test phase) is given by

$$
\begin{aligned}
& u_i(a_0, a, k) && (7)\\
&= \frac{t-k}{T} \frac{a_i \prod_{j \in \mathcal{N} \setminus \{i\}} (1-a_j)}{1-q(a)} + \frac{T-t}{T} a_i (1-a_0) \prod_{j \in \mathcal{N} \setminus \{i\}} (1-a_j) && (8)\\
&= \left[\frac{t-k}{T} \frac{1}{1-q(a)} + \frac{T-t}{T}(1-a_0) \right] a_i \prod_{j \in \mathcal{N} \setminus \{i\}} (1-a_j) \,. && (9)
\end{aligned}
$$

The expected payoff of user i given intervention rule f and action profile a can be computed using (2):

$$
v_i(f, a) = \left[1 - \frac{T-t}{T} \sum_{k=0}^{t} \binom{t}{k} q(a)^k (1-q(a))^{t-k} f(k) \right] a_i \prod_{j \in \mathcal{N} \setminus \{i\}} (1-a_j) \,. \quad (10)
$$

Note that $\sum_{k=0}^{t} \binom{t}{k} q(a)^k (1-q(a))^{t-k} f(k)$ is the expected transmission probability of the intervention device while $(T-t)/T$ is the weight on the intervention phase.[1]

3.2 Formulation of the Design Problem

For notation, let us define

$$
\lambda(k;t) = \binom{t}{k} \left[(1-p_\mathrm{l})^N \right]^k \left[1 - (1-p_\mathrm{l})^N \right]^{t-k} , \quad (11)
$$

$$
\mu(k;t) = \binom{t}{k} \left[(1-p_\mathrm{l})^{N-1}(1-p_\mathrm{h}) \right]^k \left[1 - (1-p_\mathrm{l})^{N-1}(1-p_\mathrm{h}) \right]^{t-k} , \quad (12)
$$

for $k = 0, 1, \ldots, t$, and let $\tau_\mathrm{c} = p_\mathrm{l}(1-p_\mathrm{l})^{N-1}$ and $\tau_\mathrm{d} = p_\mathrm{h}(1-p_\mathrm{l})^{N-1}$. $\lambda(k;t)$ is the probability of k idle slots arising in the t slots of the test phase when every user cooperates (i.e., chooses p_l), while $\mu(k;t)$ is that when exactly one user defects (i.e., chooses p_h). τ_c is the cooperation throughput that each user obtains when all the users choose p_l, while τ_d is the defection throughput that a user obtains when it deviates to p_h unilaterally. Note that an idle slot is more likely to occur when every user cooperates than when some user defects. Also, note that $\tau_\mathrm{d} > \tau_\mathrm{c}$, which reflects the positive gain from defection when there is no intervention.

Suppose that the objective of the manager is to maximize the sum of the payoffs (i.e., total throughput) while sustaining cooperation among the users. Formally, the payoff function of the manager can be written as

$$
u_0(a_0, a, k) = \begin{cases} \sum_{i \in \mathcal{N}} u_i(a_0, a, k) \,, & \text{if } a_i = p_\mathrm{l} \ \forall i \in \mathcal{N} \,, \\ -\infty \,, & \text{otherwise} \,. \end{cases} \quad (13)
$$

[1] For simplicity, we assume that the users value a successful transmission equally across slots. Introducing time discounting in the model will make the weight on the intervention phase smaller, since a successful transmission in later slots yields less value than that in earlier slots.

Provided that other users cooperate, the payoff to a cooperating user is given by

$$\left[1 - \frac{T-t}{T} \sum_{k=0}^{t} \lambda(k;t) f(k)\right] \tau_{\mathrm{c}} \, , \tag{14}$$

while that to a defecting user is

$$\left[1 - \frac{T-t}{T} \sum_{k=0}^{t} \mu(k;t) f(k)\right] \tau_{\mathrm{d}} \, . \tag{15}$$

Hence, the incentive constraint to sustain cooperation can be written as

$$\left[1 - \frac{T-t}{T} \sum_{k=0}^{t} \lambda(k;t) f(k)\right] \tau_{\mathrm{c}} \geq \left[1 - \frac{T-t}{T} \sum_{k=0}^{t} \mu(k;t) f(k)\right] \tau_{\mathrm{d}} \, , \tag{16}$$

and the problem of designing an optimal intervention rule can be expressed as

$$\max_{f} N \left[1 - \frac{T-t}{T} \sum_{k=0}^{t} \lambda(k;t) f(k)\right] \tau_{\mathrm{c}} \tag{17}$$

subject to

$$\left[1 - \frac{T-t}{T} \sum_{k=0}^{t} \lambda(k;t) f(k)\right] \tau_{\mathrm{c}} \geq \left[1 - \frac{T-t}{T} \sum_{k=0}^{t} \mu(k;t) f(k)\right] \tau_{\mathrm{d}} \, , \tag{18}$$

$$0 \leq f(k) \leq 1 \quad \forall k = 0, \ldots, t \, . \tag{19}$$

3.3 Analysis of the Design Problem

The design problem (17)–(19) can be rewritten as a linear programming (LP) problem as follows:

$$\min_{f} \sum_{k=0}^{t} \lambda(k;t) f(k) \tag{20}$$

$$\text{subject to} \; \frac{T-t}{T} \sum_{k=0}^{t} \left[\tau_{\mathrm{d}} \mu(k;t) - \tau_{\mathrm{c}} \lambda(k;t)\right] f(k) \geq \tau_{\mathrm{d}} - \tau_{\mathrm{c}} \, , \tag{21}$$

$$0 \leq f(k) \leq 1 \quad \forall k = 0, \ldots, t \, . \tag{22}$$

The LP problem (20)–(22) is to minimize the expected transmission probability of the intervention device while satisfying the incentive constraint and the probability constraints. Exerting intervention following some signals is necessary to punish a deviation, but at the same time intervention incurs efficiency loss under imperfect monitoring. Thus, the manager wants to use the minimum possible intervention level while providing the incentive for cooperation. The left-hand side of the incentive constraint (21) is the expected loss from deviation due to the change in the probability distribution of signals induced by deviation, while the right-hand side is the gain from deviation.

Lemma 1. *Suppose that an optimal solution to the LP problem (20)–(22) exists. Then the incentive constraint (21) is satisfied with equality at the optimal solution.*

Proof. Let f^* be an optimal solution. Suppose that $[(T-t)/T]\sum_{k=0}^{t}[\tau_{\mathrm{d}}\mu(k;t) - \tau_{\mathrm{c}}\lambda(k;t)]f^*(k) > \tau_{\mathrm{d}} - \tau_{\mathrm{c}}$. Since $\tau_{\mathrm{d}} > \tau_{\mathrm{c}}$, there exists k' such that $\tau_{\mathrm{d}}\mu(k';t) - \tau_{\mathrm{c}}\lambda(k';t) > 0$ and $f^*(k') > 0$. Then we can reduce $f^*(k')$ while satisfying the incentive constraint and the probability constraint for k', which decreases the objective value since $\lambda(k;t) > 0$ for all k. This contradicts the optimality of f^*. □

Lemma 1 validates the intuition that the manager wants to use a punishment just enough to prevent deviation. The following proposition provides a necessary and sufficient condition for the LP problem to have a feasible solution, and the structure of an optimal solution.

Proposition 1. *Let* $k_0 = \max\{k : \tau_{\mathrm{d}}\mu(k;t) - \tau_{\mathrm{c}}\lambda(k;t) > 0\}$. *Then the LP problem has a feasible solution if and only if*

$$\frac{T-t}{T}\sum_{k \le k_0}[\tau_{\mathrm{d}}\mu(k;t) - \tau_{\mathrm{c}}\lambda(k;t)] \ge \tau_{\mathrm{d}} - \tau_{\mathrm{c}} \; . \tag{23}$$

Moreover, if the LP problem has a feasible solution, then there exists a unique optimal solution f^* described by

$$f^*(k) = \begin{cases} 1 \; , & \text{if } k < \bar{k} \; , \\ \frac{1}{\tau_{\mathrm{d}}\mu(\bar{k};t) - \tau_{\mathrm{c}}\lambda(\bar{k};t)}\left[\frac{T}{T-t}(\tau_{\mathrm{d}} - \tau_{\mathrm{c}}) - \sum_{k=0}^{\bar{k}-1}[\tau_{\mathrm{d}}\mu(k;t) - \tau_{\mathrm{c}}\lambda(k;t)]\right] \; , & \\ \qquad \text{if } k = \bar{k} \; , \\ 0 \; , & \text{if } k > \bar{k} \; , \end{cases} \tag{24}$$

where

$$\bar{k} = \min\left\{k' : \frac{T-t}{T}\sum_{k \le k'}[\tau_{\mathrm{d}}\mu(k;t) - \tau_{\mathrm{c}}\lambda(k;t)] \ge \tau_{\mathrm{d}} - \tau_{\mathrm{c}}\right\} \; . \tag{25}$$

Proof. Define the likelihood ratio of signal k by

$$L(k;t) = \frac{\mu(k;t)}{\lambda(k;t)} = \left(\frac{1-p_{\mathrm{h}}}{1-p_{\mathrm{l}}}\right)^k \left(\frac{1-(1-p_{\mathrm{l}})^{N-1}(1-p_{\mathrm{h}})}{1-(1-p_{\mathrm{l}})^N}\right)^{t-k} \; . \tag{26}$$

It is easy to see that $L(0;t) > 1$, $L(t;t) < 1$, and $L(k;t)$ is monotonically decreasing in k. Note that $\tau_{\mathrm{d}}\mu(k;t) - \tau_{\mathrm{c}}\lambda(k;t) > 0$ if and only if $L(k;t) > p_{\mathrm{l}}/p_{\mathrm{h}}$. Hence, k_0 is well-defined, and $\tau_{\mathrm{d}}\mu(k;t) - \tau_{\mathrm{c}}\lambda(k;t) > 0$ if and only if $k \le k_0$. If (23) is satisfied, then $\tilde{f}$ defined by $\tilde{f}(k) = 1$ for all $k \le k_0$ and $\tilde{f}(k) = 0$ for all $k > k_0$ is a feasible solution. To prove the converse, suppose that a feasible solution, say f, exists. Then we have

$$\frac{T-t}{T}\sum_{k \le k_0}[\tau_{\mathrm{d}}\mu(k;t) - \tau_{\mathrm{c}}\lambda(k;t)] \ge \frac{T-t}{T}\sum_{k=0}^{t}[\tau_{\mathrm{d}}\mu(k;t) - \tau_{\mathrm{c}}\lambda(k;t)]f(k) \tag{27}$$

and

$$\frac{T-t}{T}\sum_{k=0}^{t}[\tau_{\rm d}\mu(k;t)-\tau_{\rm c}\lambda(k;t)]f(k)\geq\tau_{\rm d}-\tau_{\rm c}\ , \tag{28}$$

and combining the two yields (23).

To prove the remaining result, suppose that the LP problem has a feasible solution. Then there exists a feasible solution, say f, that satisfies the incentive constraint with equality. Define the likelihood ratio of f by

$$l(f)=\frac{\sum_k \mu(k;t)f(k)}{\sum_k \lambda(k;t)f(k)}\ . \tag{29}$$

Then the objective value in (20) at f can be expressed as

$$\frac{T}{T-t}\frac{\tau_{\rm d}-\tau_{\rm c}}{\tau_{\rm d}l(f)-\tau_{\rm c}}\ . \tag{30}$$

Hence, the objective value decreases as f has a larger likelihood ratio. To optimize the objective function, f should put the probabilities on the signals starting from signal 0 to signal 1, and so on, until the incentive constraint is satisfied with equality. Thus, we obtain $\bar{k}$ in (25), where $0\leq\bar{k}\leq k_0$, associated with the unique optimal solution. □

Since a smaller number of idle slots gives a higher likelihood ratio, an intervention rule yields a smaller efficiency loss when intervention is exerted following a smaller number of idle slots. Put differently, signal k provides a stronger indication of defection as k is smaller. However, using only signal 0 may not be sufficient to provide the incentive for cooperation, in which case other signals need to be used as well. Using signal k with $k\leq k_0$ contributes to the incentive for cooperation, although the "quality" of the signal decreases as k increases. Hence, it is optimal for the manager to use signals with smaller k primarily, which yields a threshold $\bar{k}$.

Timing of Intervention. So far we have analyzed the problem of designing an optimal intervention rule given the length of the test phase, t. Now we consider a scenario where the manager can choose a length of the test phase as well as an intervention rule. In this scenario, there are two counteracting effects of increasing t. First, note that the objective function in (17) can be expressed as

$$N\left[1-\frac{\tau_{\rm d}-\tau_{\rm c}}{\tau_{\rm d}l(f)-\tau_{\rm c}}\right]\tau_{\rm c}\ , \tag{31}$$

which shows that increasing t affects the objective value only through f. Since $L(k;t_1)>L(k;t_2)$ for all t_1,t_2 such that $t_1>t_2$ and for all $k\leq t_2$, increasing t increases the likelihood ratios of existing signals. At the same time, it adds new signals available for the manager. Thus, a larger likelihood ratio $l(f)$ can be achieved with larger t. In other words, as the intervention device collects more observations, the information about deviation becomes more accurate (quality effect). On the other hand, increasing t decreases the weight given on the intervention phase, which makes the impact of intervention weaker and the incentive constraint harder to satisfy (delay effect).

Let $\tau^*(t)$ be the optimal value of the design problem (17)–(19) with the length of the test phase t, where we set $\tau^*(t) = Np_{\mathrm{h}}(1-p_{\mathrm{h}})^{N-1}$ if there is no feasible solution with t. The problem of finding an optimal length of the test phase can be written as $\max_{t\in\{1,\ldots,T\}} \tau^*(t)$. In general, $\tau^*(t)$ is a non-monotonic function of t, and we provide a numerical example to illustrate non-monotonicity. We consider system parameters $N = 5$, $p_{\mathrm{l}} = 1/N = 0.2$, $p_{\mathrm{h}} = 0.8$, and $T = 100$. Then we have $\tau_{\mathrm{c}} = 0.08$ and $\tau_{\mathrm{d}} = 0.33$. The numerical results show that the LP problem is infeasible for $t = 1$ and $t \geq 21$. With $t = 1$, there is not sufficient information based on which intervention can provide the incentive for cooperation. With $t \geq 21$, the delay effect is too strong to have the incentive constraint satisfied. Figure 1 plots $\tau^*(t)$ for $t = 2, \ldots, 20$. We can see that $\tau^*(t)$ is non-monotonic while reaching the maximum at $t = 18$ with $\tau^*(18) = 0.37$. In the plot, the dotted line represents the total throughput at $(p_{\mathrm{l}}, \ldots, p_{\mathrm{l}})$, $N\tau_{\mathrm{c}}$. The difference between $\tau^*(t)$ and $N\tau_{\mathrm{c}}$ can be interpreted as the efficiency loss due to imperfect monitoring.[2] Lastly, we note that $\bar{k}$ in Proposition 1 is non-decreasing in t, with $\bar{k} = 1$ for $t = 2, \ldots, 7$, $\bar{k} = 2$ for $t = 8, \ldots, 13$, $\bar{k} = 3$ for $t = 14, \ldots, 18$, and $\bar{k} = 4$ for $t = 19, 20$.

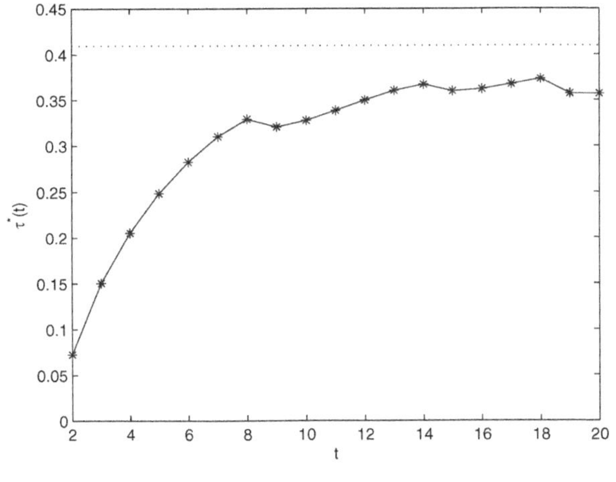

Fig. 1. Plot of $\tau^*(t)$ for $t = 2, \ldots, 20$

4 Conclusion

In this paper, we have proposed a class of incentive schemes, called intervention schemes. We have presented a general game-theoretic framework for the

[2] If the intervention device can observe the actions of the users immediately (i.e., perfect monitoring), it can sustain cooperation without incurring an efficiency loss by using the threat of transmitting with probability 1 when a deviation is detected.

design of intervention schemes under imperfect monitoring. In order to illustrate our framework and obtain concrete results, we have analyzed a simple model of slotted multiaccess communication. Our results suggest that we can design an intervention scheme that sustains an action profile from which a deviation yields a sufficiently distinct distribution of signals. When the manager cares about efficiency, it is optimal to use a punishment just enough to prevent deviation, in order to minimize the efficiency loss due to imperfect monitoring. Also, in a scenario where the manager can decide the timing of intervention, we have identified the two counteracting effects of having a longer test phase. Our framework of intervention schemes can be potentially applied to any application scenario in which individual objectives are in conflict with the system objective and some monitoring is possible. Investigating intervention schemes in various settings will provide us with insights into the properties of optimal intervention schemes as well as the capabilities and limitations of intervention schemes.

References

1. Garg, R., Kamra, A., Khurana, V.: A game-theoretic approach towards congestion control in communication networks. Comput. Commun. Rev. 32, 47–61 (2002)
2. Park, J., van der Schaar, M.: Stackelberg contention games in multiuser networks. EURASIP J. Adv. Sig. Process., article ID 305978, 15 pages (2009)
3. Korilis, Y.A., Lazar, A.A., Orda, A.: Achieving network optima using Stackelberg routing strategies. IEEE/ACM Trans. Netw. 5, 161–173 (1997)
4. Ma, R.T., Misra, V., Rubenstein, D.: An analysis of generalized slotted-Aloha protocols. IEEE/ACM Trans. Netw. 17, 936–949 (2009)
5. Su, Y., van der Schaar, M.: A new perspective on multi-user power control games in interference channels. IEEE Trans. Wireless Commun. 8, 2910–2919 (2009)
6. Başar, T., Srikant, R.: A Stackelberg network game with a large number of followers. J. Opt. Theory Applicat. 115, 479–490 (2002)
7. Bloem, M., Alpcan, T., Başar, T.: A Stackelberg game for power control and channel allocation in cognitive radio networks. In: Proc. ValueTools (2007)
8. Hallgren, M.M., McAdams, A.K.: The economic efficiency of Internet public goods. In: McKnight, L.W., Bailey, J.P. (eds.) Internet Economics, pp. 455–478. MIT Press, Cambridge (1997)
9. Mailath, G., Samuelson, L.: Repeated Games and Reputations: Long-run Relationships. Oxford Univ. Press, Oxford (2006)
10. Kandori, M.: Social norms and community enforcement. Rev. Econ. Stud. 59, 63–80 (1992)
11. Mas-Colell, A., Whinston, M.D., Green, J.R.: Microeconomic Theory. Oxford Univ. Press, Oxford (1995)
12. Fudenberg, D., Tirole, J.: Game Theory. MIT Press, Cambridge (1991)
13. Massey, J.L., Mathys, P.: The collision channel without feedback. IEEE Trans. Inf. Theory 31, 192–204 (1985)

How to Choose Communication Links in an Adversarial Environment?

Assane Gueye, Jean C. Walrand, and Venkat Anantharam

University of California at Berkeley, EECS Department, Berkeley CA 94720, USA
{agueye,wlr,ananth}@eecs.berkeley.edu

Abstract. Given the topology of a network, characterized by an undirected graph, we consider the following game situation: a network manager is choosing (as communication infrastructure) a spanning tree of the graph, and an attacker is trying to disrupt the communication tree by attacking one link of the network. Attacking a link has a certain cost for the attacker who also has the option of not attacking. We model the interaction between the network manager and the attacker as a bimatrix game and study the set of mixed strategy Nash equilibria. We define the notion of critical subset of links and determine the structure of a particular set of Nash equilibria when the attack cost is nonzero. In each NE of this set, the attacker targets edges in critical subsets and all edges in the same critical subset are attacked with the same probability. For the game of zero cost of attack considered in [8], we characterize the set of all Nash equilibria. Some implications of the results are discussed and a detailed proof of the NE theorem is provided.

Keywords: Network Topology, Connectivity, Graph Vulnerability, Spanning Trees, Minimum Cut Set, Game Theory, Nash Equilibrium, Linear Programming, Blocking pairs of polyhedra.

1 Introduction

In [8], we have studied the *strategic* interaction between a network manager whose goal is to choose a spanning tree of a network as communication infrastructure, and an attacker who tries to disrupt the communication tree by attacking one link in the network. Therein, we assumed that the cost of attack is equal to zero for the attacker and we discussed the notions of vulnerability and critical subset of links. We have also shown that there always exists a Nash equilibrium under which the attacker targets uniformly, at random, links in a critical set.

In the present paper, we generalize our results to the case where the attacker incurs a positive cost by attacking a given link of the graph. We revisit the notions of *vulnerability* and *criticality* of a subset of links and show that the critical subset attack theorem in [8] generalizes to the present case. We determine a particular set of Nash equilibria for the game of positive attack cost, and for the game with zero cost of attack, we characterize the set of all NE. We also provide a unifying proof of the Nash equilibrium theorem that applies both to the game presented in [8], and the one studied here.

This paper is organized as follows. We present the model and the environment of the game in Section 2. The notion of critical subset is discussed in section 3 followed by the

R. Jain and R. Kannan (Eds.): GameNets 2011, LNICST 75, pp. 233–248, 2012.

main result of this paper in section 4. We provide a proof of the main theorem in section 6. The proof requires the notion of blocking pair of matrices. Appendix B gives a brief introduction to this notion and presents a lemma that unifies the proofs of the theorems in [8] and in this paper. The implications of the result as well as illustrative examples are presented in section 5. This paper ends with concluding remarks discussed in section 7.

2 Model

The network topology is given by a connected undirected graph $G = (\mathcal{V}, \mathcal{E})$ with $|\mathcal{E}| = m$ links and $|\mathcal{V}| = n$ nodes. Let $\mathcal{T}$ be the set of spanning trees, and let $N = |\mathcal{T}|$. Each edge $e \in \mathcal{E}$ is associated with some cost $\boldsymbol{\mu}(e)$ that an attacker needs to spend to successfully attack that link. Each tree T has the same cost for the network manager that we assume to be equal to 1.

To get all nodes connected in a cycle-free way, the network manager chooses a spanning tree $T \in \mathcal{T}$ of the graph. The attacker simultaneously chooses a link $e \in \mathcal{E}$ to attack. The attacker wins if the attacked link belongs to the chosen spanning tree, otherwise the network wins. More precisely, for a choice pair (T, e) of tree and edge, the attack loss is $L(T, e) = \mathbf{1}_{e \in T}$ for the network, while the net attack reward is equal to $R(T, e) = \mathbf{1}_{e \in T} - \boldsymbol{\mu}(e)$ for the attacker.

The manager picks a spanning tree according to a chosen distribution $\boldsymbol{\alpha}$ on $\mathcal{T}$ to minimize the expected attack loss. Similarly, the attacker chooses a link according to some distribution $\boldsymbol{\beta}$ on $\mathcal{E}$ to maximize the expected attack reward. We assume that the attacker has the option of not attacking, which results in a zero net reward for the attacker and a zero loss for the manager.

We formulate this interaction as a one-shot 2-player game between the network manager and the attacker. Their respective pure strategy sets are the set $\mathcal{T}$ of spanning trees and the set $\mathcal{E}$ of edges of the graph. We are interested in analyzing mixed strategy Nash equilibria of this game.

Let $\mathcal{A} := \{\boldsymbol{\alpha} \in \Re_+^N \mid \sum_{T \in \mathcal{T}} \boldsymbol{\alpha}_T = 1\}$ be the set of mixed strategies for the network manager, and $\mathcal{B} := \{\boldsymbol{\beta} \in \Re_+^m \mid \sum_{e \in \mathcal{E}} \boldsymbol{\beta}_e = 1\}$ the set of mixed strategies for the attacker. Define A as the loss matrix for the manager, with $A_{T,e} = 1_{e \in T}$ and B is the reward matrix of the attacker with $B_{T,e} = 1_{e \in T} - \boldsymbol{\mu}(e)$. The average expected loss $L(\alpha, \beta)$ for the manager, and reward $R(\alpha, \beta)$ for the attacker are given by

$$L(\boldsymbol{\alpha}, \boldsymbol{\beta}) = \boldsymbol{\alpha}' A \boldsymbol{\beta} = \sum_{T \in \mathcal{T}} \boldsymbol{\alpha}_T \left(\sum_{e \in T} \boldsymbol{\beta}_e \right), \tag{1}$$

$$R(\boldsymbol{\alpha}, \boldsymbol{\beta}) = \boldsymbol{\alpha}' B \boldsymbol{\beta} = \sum_{e \in \mathcal{E}} \boldsymbol{\beta}_e \left(\sum_{T \ni e} \boldsymbol{\alpha}_T - \boldsymbol{\mu}(e) \right). \tag{2}$$

3 Critical Subsets

In this section we define the notions of vulnerability and critical subset of a graph.

Definition 1. *For any nonempty subset of links $E \subseteq \mathcal{E}$, define*

$$\mathcal{M}(E) := \min_{T \in \mathcal{T}} | T \cap E |, \quad \textit{and} \quad \theta(E) := \frac{\mathcal{M}(E) - \boldsymbol{\mu}(E)}{|E|}. \tag{3}$$

$\theta(E)$ is called the vulnerability *of E. It is the minimum fraction of edges that E has in common with any tree minus the average cost of attacking E. Here and throughout the paper, we use the notation $\boldsymbol{\mu}(E) = \sum_{e \in E} \boldsymbol{\mu}(e)$.*
A nonempty subset of edges E is said to be critical *if it has maximum* vulnerability*: $\theta(E) = max_{E' \subseteq \mathcal{E}} \{\theta(E')\}$. We let $\mathcal{C}$ denote the set of all critical subsets.*
The vulnerability of the graph θ is defined to be equal to the vulnerability of its critical subset(s).

The examples shown in Figure 1 illustrate the definitions presented above. The network in Figure 1(a) has a vector of attack cost $\boldsymbol{\mu} = [0.5, 0.5, 0.5, 2, 0.5, 0.5, 0.5]$. It contains a bridge that has a relatively high cost of attack ($\boldsymbol{\mu}(4) = 2$). As a consequence it is not critical. There are two critical subsets $E_1 = \{1, 2, 3\}$ and $E_2 = \{5, 6, 7\}$ shown respectively by the dashed and dash-dotted lines. This example illustrates the impact of the attack cost. When a link is too costly to attack, it becomes less critical.

Figures 1(b) and 1(c) show the same network topology with different costs of attack. In the first one, the attack costs are $\boldsymbol{\mu} = [5, 3, 3, 5, 5, 4, 3, 3, 5, 5, 4, 5, 5, 3]/14$. For these values of the costs of attack, the minimum cutset of the graph (links 6 and 8) is critical. If the attack costs are equal to $\boldsymbol{\mu} = [2, 5, 1, 2, 1, 1, 6, 5, 3, 7, 1, 4, 3, 6]/21$ (second case), the minimum cutset is no longer critical. It has vulnerability $\theta(6, 8) = \frac{1-(4+3)/14}{2} = 1/4$. One critical subset of the graph is given by the set $E = \{1, 2, 3, 4, 5, 6, 7, 8, \}$. Its vulnerability is $\theta(E) = 0.3631$.

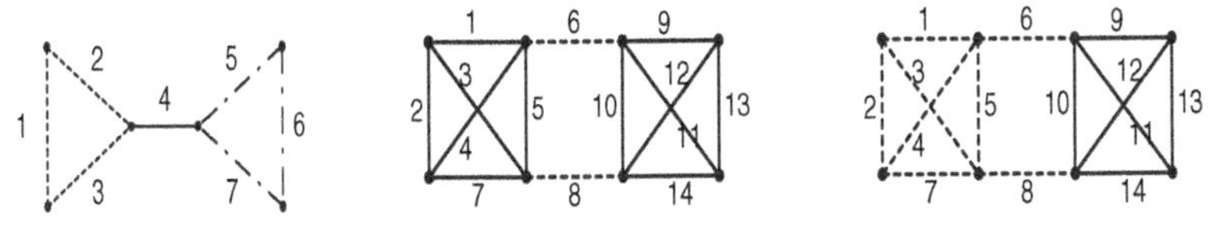

(a) Graph with bridge (b) Network where minimum cutset is critical (c) Minimum cutset is not critical

Fig. 1. Examples illustrating definition 1. The vector of attack costs are $\boldsymbol{\mu} = [0.5, 0.5, 0.5, 2, 0.5, 0.5, 0.5]$ for Figure 1(a), $\boldsymbol{\mu} = [5, 3, 3, 5, 5, 4, 3, 3, 5, 5, 4, 5, 5, 3]/14$ for Figure 1(b), and $\boldsymbol{\mu} = [2, 5, 1, 2, 1, 1, 6, 5, 3, 7, 1, 4, 3, 6]/21$ for Figure 1(c). Each set of dashed (or dash-dotted) lines is a critical subset.

4 Critical Subset Attack Theorem

Now, we give the critical subset attack theorem for the game defined above, a proof of which is provided in section 6.

Theorem 1 (Critical Subset Attack Theorem). *For the game defined in section 2 with attack costs $\boldsymbol{\mu}$, the following always holds.*

1. *If* $\theta = \max_{E \subseteq \mathcal{E}} \left(\frac{\mathcal{M}(E) - \boldsymbol{\mu}(E)}{|E|} \right) \leq 0$, *then there is a NE under which the attacker will opt to not launch an attack. The equilibrium strategy* $(\boldsymbol{\alpha}_T,\ T \in \mathcal{T})$ *for the defender is such that*

$$\boldsymbol{\alpha}(e) := \sum_{T \ni e} \boldsymbol{\alpha}_T \leq \boldsymbol{\mu}(e), \ \ \forall e \in \mathcal{E}. \tag{4}$$

The corresponding payoff is 0 for both players.

2. *If* $\theta \geq 0$, *then for every probability distribution* $(\gamma_E, E \in \mathcal{C})$ *on the set* $\mathcal{C}$ *of critical subsets, the attacker's strategy* $(\boldsymbol{\beta}(e), e \in \mathcal{E})$ *defined by*

$$\boldsymbol{\beta}(e) = \sum_{E \in \mathcal{C}} \gamma_E \frac{1_{e \in E}}{|E|}, \tag{5}$$

is in Nash equilibrium with any strategy $(\boldsymbol{\alpha}_T, T \in \mathcal{T})$ *of the defender that satisfies the following properties:*

$$\begin{cases} \boldsymbol{\alpha}(e) - \boldsymbol{\mu}(e) = \theta \ \text{for all} \ e \in \mathcal{E} \ \text{such that} \ \boldsymbol{\beta}(e) > 0. \\ \boldsymbol{\alpha}(e) - \boldsymbol{\mu}(e) \leq \theta \ \text{for all} \ e \in \mathcal{E}. \end{cases} \tag{6}$$

Furthermore, there exists at least one such strategy $\boldsymbol{\alpha}$.
The corresponding payoffs are θ *for the attacker, and* $r(\gamma)$ *for the defender, where*

$$r(\gamma) := \sum_{E \in \mathcal{C}} \gamma_E \frac{\mathcal{M}(E)}{|E|}. \tag{7}$$

3. *If* $\boldsymbol{\mu} = 0$, *then every Nash equilibrium pair of strategies for the game is of the type in (2.) above.*

5 Analyzing the NE Theorem

We discuss the NE theorem by considering a game on the graph shown in Figure 2. Table 1 shows the parameters and results of the game. The first column shows different values of the attack costs $\boldsymbol{\mu}$ and the second column shows the corresponding critical subset(s). The third column displays the vulnerability of the graph. For each vector of attack costs, we compute the Nash equilibria of the game. The next two columns of the table show the Nash equilibrium strategies, respectively $\boldsymbol{\alpha}$ for the network manager, and $\boldsymbol{\beta}$ for the attacker. The equilibrium payoffs are displayed in the last column. In all equilibria, we have chosen the distribution γ_E to only focus on a particular critical subset (the ones shown on the table). Note that we have not shown all Nash equilibria.

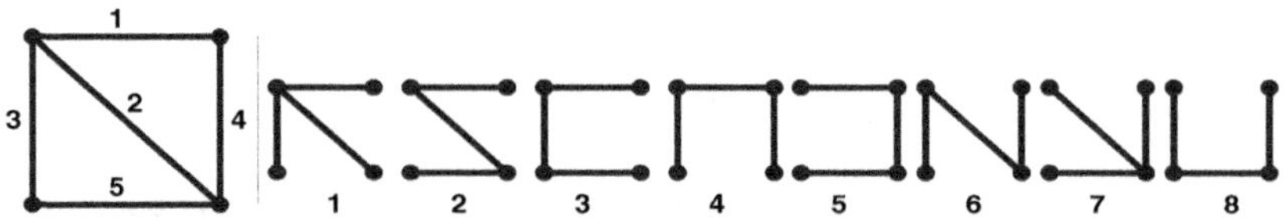

Fig. 2. Example of graph and its spanning trees. The left figure is the original graph with the 5 edges labeled with their number. The right figures are the 8 spanning trees of the graph also labeled with their numbers.

Table 1. Game with positive attack cost played for different values of the cost of attack μ

Attack Cost μ	Critical Set E_c	Vulnerability θ	Nash Equilibria α	β	Payoffs $(\lambda_{min}, \mu_{max})$
$[3\ 3\ 5\ 3\ 4]/5$	(1,4)	-0.1	$[0\ 0\ 2\ 0\ 0\ 1\ 0\ 2]/5$	No Attack	$(0, 0)$
			$[0\ 0\ 2\ 1\ 0\ 0\ 0\ 2]/5$	No Attack	$(0, 0)$
$[5\ 2\ 3\ 2\ 3]/5$	(1-5)&(2-4)	0	$[0\ 2\ 1\ 2\ 0\ 0\ 0\ 0]/5$	$[1\ 1\ 1\ 1\ 1]/5$	$(0.6, 0)$
			$[0\ 2\ 1\ 2\ 0\ 0\ 0\ 0]/5$	$[0\ 1\ 1\ 1\ 1]/4$	$(0.5, 0)$
			$[0\ 2\ 1\ 2\ 0\ 0\ 0\ 0]/5$	No Attack	$(0, 0)$
$[5\ 4\ 2\ 4\ 2]/8$	(3,5)	0.25	$[0\ 1\ 0\ 0\ 1\ 2\ 0\ 0]/4$	$[0\ 0\ 1\ 0\ 1]/2$	$(0.5, 0.25)$
			$[0\ 1\ 0\ 1\ 0\ 1\ 1\ 0]/4$	$[0\ 0\ 1\ 0\ 1]/2$	$(0.5, 0.25)$
$[4\ 3\ 2\ 4\ 3]/8$	(1-5)	0.2	$[0\ 11\ 1\ 5\ 11\ 12\ 0\ 0]/40$	$[1\ 1\ 1\ 1\ 1]/5$	$(0.6, 0.2)$
			$[0\ 11\ 1\ 16\ 0\ 1\ 11\ 0]/40$	$[1\ 1\ 1\ 1\ 1]/5$	$(0.6, 0.2)$
			$[1\ 10\ 1\ 16\ 0\ 0\ 12\ 0]/40$	$[1\ 1\ 1\ 1\ 1]/5$	$(0.6, 0.2)$

- The first game considers a case where $\mu = [3\ 3\ 5\ 3\ 4]/5$. Here, edge 3 has a particularly high cost (equal to the cost of a tree). In this case, the vulnerability of the graph ($\theta = -0.1$) is negative and the attacker does not make any gain by attacking. Her best strategy is to "*not attack*" and the network manager chooses a tree according to a distribution $\boldsymbol{\alpha}$ that satisfies (4). There exist many such distributions $\boldsymbol{\alpha}$; two of which are shown in the table. Since there is no attack, each player gets a payoff of zero.

  This game models scenarios where attacking requires an investment from the attacker that is larger than the maximum possible expected reward. As a rational decision maker, the attack will opt to not attack. The network manager needs to randomize his choice of trees to deter the attacker from attacking. In fact, if the network were to pick a fixed tree, then the attacker could get a positive reward by attacking the cheapest link (of cost 3/5) of that tree. In other word, the randomization is necessary for the NE to hold.
- In the next game (second row of the table), the cost of attack is $\mu = [5\ 2\ 3\ 2\ 3]/5$. In this case, the maximum attack reward is exactly equal to zero, and it can be achieved by several attack strategies as can be seen in the table (column 5). Although the attacker cannot gain by launching an attack, the damage she can cause to the network varies depending on the attack she launches.

  This game illustrates the importance of knowing the *type/nature* of an opponent. For example, if the attacker is a competitor who also wants to maximize the loss to the network, then, she will likely attack a link at random with the same probability (which gives a loss of 0.6). However, if the attacker is just interested in her own payoff, then she will probably not launch an attack.
- From these two examples and the first part of the theorem, one can infer that if the network manager is able to influence the attack costs μ, for example making the links harder to attack by investing on security (physical protection, Firewalls, Intrusion Prevention Systems (IPS) to avoid Denial of Service (DoS), etc...), then he can deter the attacker from attacking. This can be done by investing on the links to the point that $\mathcal{M}(E) \leq \mu(E)$ for all subsets of edges $E \subseteq \mathcal{E}$. One can compute the optimal investment by solving an *optimal reinforcement* like problem. The network

reinforcement problem of [4] is related to minimizing the price of increasing the cost of attack of individual edges in order to achieve a target vulnerability (here 0) for the graph. For details see [4]. If the cost of attack can be estimated by some means, this can be a very good candidate for preventive security.

- The last two games are examples where the maximum attack reward is strictly positive. In the first one, the attacker only targets the links that are less costly which turn out to be the minimum cutset of the graph (seen by the attacker). In the second example, the minimum cut seen by the attacker corresponds to links 3 and 5. However, the attack's reward is maximized by targeting all the links with the same probability as it is shown in the table.
- If all links of the graph have the same cost $\boldsymbol{\mu}(e) = \mu$, then the vulnerability of a subset E (defined in equation 3) becomes $\theta(E) = \frac{\mathcal{M}(E)-\mu(E)}{|E|} = \frac{\mathcal{M}(E)}{|E|} - \mu$, and a critical subset is one that maximizes the ratio $\frac{\mathcal{M}(E)}{|E|}$. This definition of criticality corresponds to the one given in [8] where the cost of attack was assumed to be zero. Theorem 1 implies that if the cost of attack μ is larger than $\frac{\mathcal{M}(E)}{|E|}$ for all E, the attacker will not attack. In fact, the net gain of attacking will be negative. If, in the other hand $\mu > \max_{E\subseteq\mathcal{E}}\left(\frac{\mathcal{M}(E)}{|E|}\right)$, then the second part of Theorem 1 corresponds to the critical subset attack theorem in [8] with $\gamma_E = 1$ for some critical subset E. The attacker can take any convex combination of uniform attack on the links in a critical subset, and the manager will choose trees according to (6).
- If $\gamma_{E_c} = 1$ for a some critical subset E_c, we have that the corresponding attack is to target uniformly links in E_c. The defense strategy should verify $\sum_{T\ni e}\boldsymbol{\alpha}_T - \boldsymbol{\mu}(e) \leq \frac{\mathcal{M}(E_c)}{|E_c|}$ for all $e \in \mathcal{E}$, and equality holds for each $e \in E_c$. Also, by the Nash equilibria conditions it must be that for any spanning tree T

$$\sum_{e\in T}\boldsymbol{\beta}(e) = \sum_{e\in T}\frac{1_{e\in E_c}}{|E_c|} = \frac{|E_c \cap T|}{|E_c|} \geq \frac{\mathcal{M}(E_c)}{|E_c|}. \tag{8}$$

The minimum value in the equation above is achieved at each T for which $\boldsymbol{\alpha}_T > 0$.

Since the defender's payoff is equal to $\frac{\mathcal{M}(E_c)}{|E_c|}$, we have that $\mathcal{M}(E_c) = \min_T(|E \cap T|) = |E_c \cap T|$ for each T for which $\boldsymbol{\alpha}_T > 0$. In other words, the defender will select only spanning trees that cross the critical subset in the minimum number of links. Furthermore, the net reward $(\sum_{T\ni e}\boldsymbol{\alpha}_T - \boldsymbol{\mu}(e))$ is the same at each link e of the critical subset E_c. This quantity is equal to θ, the vulnerability of the subset E_c. For any other link, this quantity should be less than θ.
- We have seen in [8] that if the cost of attack is zero, the attacker targets edges on a given critical subset with the same probability. The theorem of this paper tells that this still holds even with positive cost of attack. The attack operates by taking convex combination of uniform strategies on critical subsets.

This uniformity of attack on critical subsets comes from the geometry of the blocker $\mathcal{P}_A^b$ of the spanning tree polyhedron $\mathcal{P}_A$ induced by the defender's payoff matrix A (which is the spanning tree incidence matrix – see appendix B). The attack is no longer uniform if the payoff matrix changes. To see this, assume for example that the defender incurs a certain operation cost η_T by choosing spanning tree T. In this case his payoff matrix is given as $A_{T,e} = 1_{e \in T} + \eta_T$. We consider the simplest none-trivial topology of two nodes connected by two parallel edges e_1 and e_2. With this topology, edge e_1 corresponds to tree T_1. Similarly for link e_2 and tree T_2. We further simplify by assuming that the attack cost $\mu = 0$.

Letting $(\alpha, 1-\alpha)$ and $(\beta, 1-\beta)$ respectively be the defender's and the attacker's strategy, the expected attack loss for the defender can be written as $L(\alpha, \beta) = \alpha(2\beta + \eta_1 - \eta_2 - 1) + 1 - \beta + \eta_2$. The attacker's expected reward is $R(\alpha, \beta) = \beta(2\alpha - 1) + 1 - \alpha$. By analyzing these payoff functions, we see that the NE is given as follow. If $\eta_1 \geq 1 + \eta_2$, then $\alpha = 0$ and $\beta = 0$. If $\eta_2 \geq 1 + \eta_1$, then $\alpha = 1$ and $\beta = 1$. If $0 < |\eta_1 - \eta_2| < 1$, then $\alpha = 1/2$ and $\beta = \frac{\eta_2 - \eta_1 + 1}{2}$. Hence, the attacker's mixed strategy equilibrium is not in general uniform. We get the uniform distribution only if $\eta_1 = \eta_2$.

This shows the importance of the geometry of the problem (namely the polyhedron induced by the payoff matrix and its blocker) for the determination of the NE structure. The authors have found that [7] for the quasi-zero-sum game defined in section 2 with arbitrary nonnegative payoff matrix A, and attack cost $\mu \geq 0$, the attacker's NE strategies are obtained by normalizing *critical vertices* of the blocker polyhedron $\mathcal{P}_A^b$. In the case of the spanning tree game, the blocker is such that the normalized vertices correspond to uniform distributions. For a general payoff matrix, normalized vertices can give arbitrary distribution.

- The Nash equilibria characterization provided in this paper (and which the authors have studied in a more general setting [7], [6]) can be considered as an application of the result in [1] to the particular case of *quasi zero-sum game*. Although Avis *et al.* were not interested in characterizing Nash equilibria (which would be very laborious for an arbitrary two-player matrix game) and did not explicitly consider the notion of blockers, all the ingredients we have used in our NE characterization can be derived from their results. Our use of the combinatorial notion of blocker was the key to our success in characterizing the mixed strategy Nash equilibria of the game. To our knowledge, such notion was not used before in the context of computing Nash equilibria.

6 Proof of the Critical Subset Attack Theorem

In this section we provide a proof of the Nash equilibrium theorem presented in section 4. In the first part of the proof, we argue that the strategies given in the theorem for $\theta \leq 0$ and $\theta \geq 0$ are best responses to each other. The second part shows the existence of a distribution $\boldsymbol{\alpha}$ that satisfies (4) if $\theta \leq 0$ and (6) if $\theta \geq 0$. The last part of the proof shows that when $\mu = 0$, all Nash equilibria have the form given in part (2) of the theorem. The proof requires the notion blocking pair of polyhedra that we define in the appendix section B.

6.1 Best Responses

First, notice that if the attacker chooses to not attack, then any $\boldsymbol{\alpha}$ will result to the minimum loss of zero for the defender (in particular the one given in the theorem).

Also, if $\boldsymbol{\alpha}$ is such that $\boldsymbol{\alpha}(e) - \boldsymbol{\mu}(e) \leq 0,\ \forall\, e \in \mathcal{E}$, one can easily see from (2) that not attacking is a *dominant* strategy for the attacker. Thus, if $\theta \leq 0$, the strategies given in the theorem are best responses to each other.

Next, we show that if $\theta \geq 0$ the strategies given in (5) and (6) are best response to each other. We start by showing that:

Lemma 1. *If* $\theta \geq 0$, *then not attacking is a* dominated *strategy for the attacker. The domination is strict if* $\theta > 0$.

The Lemma implies that if $\theta \geq 0$ the attacker can always at least do as well as than not attacking (and better strictly better if $\theta > 0$).

Proof Sketch: The proof follows from the fact that if $\theta \geq 0$, then, the attacker can always get a nonnegative attack reward by uniformly targeting the edges of a critical subset E. Indeed, there always exists at least one critical subset. The reward of such attack is lower bounded by $\frac{\mathcal{M}(E) - \boldsymbol{\mu}(E)}{|E|}$, which is greater than zero under the assumption that $\theta \geq 0$. The bound is strict if $\theta > 0$.

Now, suppose that the defender plays a strategy $\boldsymbol{\alpha}$ that satisfies (6). Then, any distribution $\boldsymbol{\beta}$ of the form $\boldsymbol{\beta}(e) = \sum_{E \in \mathcal{C}} \gamma_E \frac{1_{e \in E}}{|E|}$ for some distribution $\gamma = (\gamma_E,\ E \in \mathcal{C})$, achieves a reward of θ. This is the maximum possible reward that the attacker can get. To see this, observe that for any $\boldsymbol{\beta}$,

$$R(\boldsymbol{\alpha}, \boldsymbol{\beta}) = \sum_{e \in \mathcal{E}} \boldsymbol{\beta}(e) \left(\boldsymbol{\alpha}(e) - \boldsymbol{\mu}(e)\right) \leq \sum_{e \in \mathcal{E}} \boldsymbol{\beta}(e)\theta = \theta. \tag{9}$$

The upper bound of θ is achieved by any $\tilde{\boldsymbol{\beta}} = (\frac{1_{e \in E}}{|E|}, e \in \mathcal{E})$ uniform on a critical subset $E \in \mathcal{C}$. In fact, replacing such $\tilde{\boldsymbol{\beta}}$ in (2) and reordering the terms, we get

$$R(\boldsymbol{\alpha}, \tilde{\boldsymbol{\beta}}) = \sum_{T \in \mathcal{T}} \boldsymbol{\alpha}_T \left(\sum_{e \in \mathcal{E}} \frac{1_{e \in E}}{|E|} 1_{e \in T} - \sum_{e \in \mathcal{E}} \frac{1_{e \in E}}{|E|} \boldsymbol{\mu}(e) \right) \tag{10}$$

$$= \sum_{T \in \mathcal{T}} \boldsymbol{\alpha}_T \left(\frac{|E \cap T|}{|E|} - \sum_{e \in \mathcal{E}} \frac{1_{e \in E}}{|E|} \boldsymbol{\mu}(e) \right) \tag{11}$$

$$\geq \sum_{T \in \mathcal{T}} \boldsymbol{\alpha}_T \frac{\mathcal{M}(E)}{|E|} - \frac{\boldsymbol{\mu}(E)}{|E|} = \frac{\mathcal{M}(E)}{|E|} - \frac{\boldsymbol{\mu}(E)}{|E|} = \theta, \tag{12}$$

where in the last step we use the fact that E is critical.
As a consequence, any distribution of the form $(\frac{1_{e \in E}}{|E|}, e \in \mathcal{E})$ for $E \in \mathcal{C}$ critical is a best response and any convex combination of those distributions is also a best response.

Now assume that $\boldsymbol{\beta}$ is given as in (5) for some distribution $(\gamma_E,\ E \in \mathcal{C})$. Then, the distribution $(\boldsymbol{\alpha}_T,\ T \in \mathcal{T})$ in (6) achieves a loss of $r(\gamma) = \sum_{E \in \mathcal{C}} \gamma_E \frac{\mathcal{M}(E)}{|E|}$. This is the minimum possible loss. To see this, use this expression for $\boldsymbol{\beta}$ to rewrite the expected loss (1) (for any $\boldsymbol{\alpha}$) as

$$L(\boldsymbol{\alpha},\boldsymbol{\beta}) = \sum_{T\in\mathcal{T}} \alpha_T \left(\sum_{E\in\mathcal{C}} \gamma_E \left(\sum_{e\in\mathcal{E}} \frac{1_{e\in E}}{|E|} 1_{e\in T}\right)\right) \tag{13}$$

$$\geq \sum_{T\in\mathcal{T}} \alpha_T \left(\sum_{E\in\mathcal{C}} \gamma_E \frac{\mathcal{M}(E)}{|E|}\right) \tag{14}$$

$$= \sum_{T\in\mathcal{T}} \alpha_T r(\gamma) = r(\gamma). \tag{15}$$

To get equation (13) from (1), we have reversed the order of the summations over $\mathcal{E}$ and over $\mathcal{C}$.

The lower bound $r(\gamma)$ can be achieved by choosing $\boldsymbol{\alpha}$ such that $\sum_{T\in\mathcal{T}} \alpha_T 1_{e\in T} = \theta + \boldsymbol{\mu}(e)$ for each $e \in \mathcal{E}$ such that $\boldsymbol{\beta}(e) > 0$ (the existence of such $\boldsymbol{\alpha}$ is shown in the second part of the theorem). This can be seen by using $\sum_{T\in\mathcal{T}} \alpha_T 1_{e\in T} = \theta + \boldsymbol{\mu}(e)$ and $\boldsymbol{\beta}(e) = \sum_{E\in\mathcal{C}} \gamma_E \frac{1_{e\in E}}{|E|}$ in (1) to get

$$L(\boldsymbol{\alpha},\boldsymbol{\beta}) = \sum_{e\in\mathcal{E}} \boldsymbol{\beta}(e) \left(\sum_{T\in\mathcal{T}} \alpha_T 1_{e\in T}\right) = \sum_{e\in\mathcal{E}} \boldsymbol{\beta}(e)\,(\theta + \boldsymbol{\mu}(e)) \tag{16}$$

$$= \theta + \sum_{e\in\mathcal{E}} \left(\sum_{E\in\mathcal{C}} \gamma_E \frac{1_{e\in E}}{|E|} \boldsymbol{\mu}(e)\right) \tag{17}$$

$$= \theta + \sum_{E\in\mathcal{C}} \gamma_E \left(\sum_{e\in\mathcal{E}} \frac{1_{e\in E}}{|E|} \boldsymbol{\mu}(e)\right) \tag{18}$$

$$= \theta + \sum_{E\in\mathcal{C}} \gamma_E \left(\frac{\boldsymbol{\mu}(E)}{|E|}\right) \tag{19}$$

$$= \theta + \sum_{E\in\mathcal{C}} \gamma_E \left(\frac{\mathcal{M}(E)}{|E|} - \theta\right) \tag{20}$$

$$= \sum_{E\in\mathcal{C}} \gamma_E \frac{\mathcal{M}(E)}{|E|} = r(\gamma) \tag{21}$$

This implies that the distribution $(\alpha_T,\ T \in \mathcal{T})$ in (6) is a best response to $\boldsymbol{\beta}$ given in (5).

6.2 Existence of the Equilibrium Distribution $\boldsymbol{\alpha}$

In the previous section we have shown that the strategies given in the theorem are best responses to each other. The distribution in (5) exists by definition. However, a priori, one does not know if there exists a probability distribution that satisfies (4) if $\theta \leq 0$. Similarly, if $\theta \geq 0$, one needs to show the existence of a distribution that verifies the conditions in (6). Using the results discussed in appendix B, we show the existence of such distributions. More concretely, we will show that:

- if $\theta \leq 0$, there exists $\boldsymbol{\alpha}$ verifying, $\boldsymbol{\alpha} \geq 0$, $\mathbf{1}_{\mathcal{T}}'\boldsymbol{\alpha} = 1$, and $A'\boldsymbol{\alpha} \leq \boldsymbol{\mu}$,
- if $\theta \geq 0$, there exists $\boldsymbol{\alpha}$ verifying, $\boldsymbol{\alpha} \geq 0$, $\mathbf{1}_{\mathcal{T}}'\boldsymbol{\alpha} = 1$, and $A'\boldsymbol{\alpha} \leq \theta \mathbf{1}_{\mathcal{E}} + \boldsymbol{\mu}$, with equality in the constraints for each e such that $\boldsymbol{\beta}(e) > 0$.

Recall that 'A' is the tree-link incidence matrix $A_{T,e} = 1_{e \in T}$. Also, the spanning tree polyhedron $\mathcal{P}_A$ is characterized by (see appendix B, [8], and [3])

$$\mathcal{P}_A = \{\mathbf{x} \in R_+^m \mid \mathbf{x}(\mathcal{E}(P)) \geq |P| - 1, \text{ for all feasible partitions } P = \{V_1, V_2, \dots, V_{|P|}\}\}. \tag{22}$$

P is said to be a *feasible* partition of the nodes $\mathcal{V}$ of G if each V_i induces a connected subgraph $G(V_i)$ of G. We let $\mathcal{E}(P)$ denote the set of edges going from one member of the partition to another, and $G_{\bar{\mathcal{E}}(P)}$ be the graph obtained by removing from G the edges going across P. The number of connected components of $G_{\bar{\mathcal{E}}(P)}$ is denoted $Q(G_{\bar{\mathcal{E}}(P)})$ and is equal the size of the partition P. We have also shown in [8] that $\mathcal{M}(E) = Q(G_{\bar{E}}) - 1$ for all $E \subseteq \mathcal{E}$.

Now, we claim that,

Lemma 2. – *If* $\theta \leq 0$, *then* $\boldsymbol{\mu} \in \mathcal{P}_A$.
– *If* $\theta \geq 0$, *then* $(\theta \mathbf{1}_{\mathcal{E}} + \boldsymbol{\mu}) \in \mathcal{P}_A$.

Using the first part of this lemma, and Lemma 3 of Appendix B, we conclude that if $\theta \leq 0$, the value of the following LP is greater than 1.

$$\text{Maximize } \mathbf{1}_{\mathcal{T}}'\mathbf{x}, \quad \text{subject to } A'\mathbf{x} \leq \boldsymbol{\mu}, \text{ and } \mathbf{x} \geq \mathbf{0}. \tag{23}$$

Using this, we can construct a distribution $\boldsymbol{\alpha}$ satisfying (4) by normalizing any solution of this LP.

Similarly, if $\theta \geq 0$, we can construct a distribution $\boldsymbol{\alpha}$ that satisfies $A'\boldsymbol{\alpha} \leq \theta \mathbf{1}_{\mathcal{E}} + \boldsymbol{\mu}$. This gives an $\boldsymbol{\alpha}$ for which we still need to show that equality holds whenever $\boldsymbol{\beta}(e) > 0$, where $\boldsymbol{\beta}$ is a distribution of the form (5). For that, we make the following additional claims.

Theorem 2. *Let* $\mathbf{x}^*$ *be the solution of the following LP:*

$$\begin{aligned} &\textit{Maximize } \mathbf{1}_{\mathcal{T}}'\mathbf{x} \\ &\textit{subject to } A'\mathbf{x} \leq \mathbf{b}, \ \mathbf{x} \geq \mathbf{0}. \end{aligned} \tag{24}$$

where $\mathbf{b} = \theta \mathbf{1}_{\mathcal{E}} + \boldsymbol{\mu}$. *Then,*
a) $\mathbf{1}_{\mathcal{T}}'\mathbf{x}^* \leq 1$;
b) $A'\mathbf{x}^*(e) = \mathbf{b}(e), \forall\, e \in \mathcal{E}$ *for which* $\boldsymbol{\beta}(e) > 0$, *where* $\boldsymbol{\beta}$ *is given in (5).*

Notice that, from Lemma 2 we have that the value of the linear program is greater than 1. This, combined with part a) of the theorem, imply that the value of the LP is exactly 1. Part b) of the theorem gives the equality conditions that we needed. As a consequence, $\mathbf{x}^*$ satisfies (6) and implies the existence of the NE distribution $\boldsymbol{\alpha}$ when $\theta \geq 0$.

6.2.1 Proof of Lemma 2

– By definition of θ, we have $\theta \leq 0 \Leftrightarrow \boldsymbol{\mu}(E) \geq \mathcal{M}(E)$ for all $E \subseteq \mathcal{E}$. [8, Lemma 1] gives $\mathcal{M}(E) = Q(G_{\bar{E}}) - 1$, where $Q(G_{\bar{E}})$ is the number of connected components of the graph G when all edges in E are removed. Thus, $\theta \leq 0 \Leftrightarrow \boldsymbol{\mu}(E) \geq Q(G_{\bar{E}}) - 1$ for all $E \subseteq \mathcal{E}$.

Now, let P be a *feasible* partition of the nodes $\mathcal{V}$ of G. Using the above observations, we can conclude that

$$\theta \leq 0 \Leftrightarrow \boldsymbol{\mu}\left(\mathcal{E}(P)\right) \geq Q(G_{\bar{\mathcal{E}}(P)}) - 1 = |P| - 1 \tag{25}$$

Since the partition P is feasible, $\boldsymbol{\mu}\left(\mathcal{E}(P)\right) \geq |P| - 1$ implies that $\boldsymbol{\mu} \in \mathcal{P}_A$, which ends the proof of the first part of the lemma.

- To prove that the vector $\mathbf{b} = \theta \mathbf{1}_{\mathcal{E}} + \boldsymbol{\mu} \geq \mathbf{0}$ belongs to the polyhedron $\mathcal{P}_A$ whenever $\theta \geq 0$, we argue that

$$\mathbf{b}\left(\mathcal{E}(P)\right) \geq |P| - 1, \quad \text{for all feasible partitions } P. \tag{26}$$

Recall, from the above that for all feasible partitions P

$$\mathcal{M}(\mathcal{E}(P)) = |P| - 1. \tag{27}$$

Now, assume that $\mathbf{b}$ does not verify (26)– i.e $\mathbf{b}\left(\mathcal{E}(P)\right) < |P| - 1$, for some feasible partition P. Then one must have,

$$\begin{aligned} |P| - 1 > \sum_{e \in \mathcal{E}(P)} \mathbf{b}_e &= \theta\left(\mathcal{E}(P)\right) \sum_{e \in \mathcal{E}(P)} 1 + \sum_{e \in \mathcal{E}(P)} \boldsymbol{\mu}(e) & (28)\\ &= \theta\left(\mathcal{E}(P)\right) |\mathcal{E}(P)| + \boldsymbol{\mu}\left(\mathcal{E}(P)\right) & (29)\\ &= \mathcal{M}\left(\mathcal{E}(P)\right) - \boldsymbol{\mu}\left(\mathcal{E}(P)\right) + \boldsymbol{\mu}\left(\mathcal{E}(P)\right) = \mathcal{M}\left(\mathcal{E}(P)\right) & (30) \end{aligned}$$

which contradicts (27). Thus, $\mathbf{b}\left(\mathcal{E}(P)\right) \geq |P| - 1$ for all feasible P, or equivalently $\mathbf{b} \in \mathcal{P}_A$.

6.2.2 Proof of Theorem 2

a) To prove that $\mathbf{1}_{\mathcal{T}}' \mathbf{x}^* \leq 1$, we first observe that

$$\begin{aligned} \boldsymbol{\beta}' A' \mathbf{x} &= \sum_{T \in \mathcal{T}} \mathbf{x}_T \left(\sum_{e \in \mathcal{E}} \boldsymbol{\beta}(e) 1_{e \in T} \right) & (31)\\ &= \sum_{T \in \mathcal{T}} \mathbf{x}_T \left(\sum_{e \in \mathcal{E}} \left(\sum_{E \in \mathcal{C}} \gamma_E \frac{1_{e \in E}}{|E|} \right) 1_{e \in T} \right) & (32)\\ &= \sum_{T \in \mathcal{T}} \mathbf{x}_T \left(\sum_{E \in \mathcal{C}} \gamma_E \left(\sum_{e \in \mathcal{E}} \frac{1_{e \in E}}{|E|} 1_{e \in T} \right) \right) & (33)\\ &= \sum_{T \in \mathcal{T}} \mathbf{x}_T \left(\sum_{E \in \mathcal{C}} \gamma_E \left(\frac{|E \cap T|}{|E|} \right) \right) & (34)\\ &\geq \sum_{T \in \mathcal{T}} \mathbf{x}_T \left(\sum_{E \in \mathcal{C}} \gamma_E \frac{\mathcal{M}(E)}{|E|} \right) & (35)\\ &= \sum_{T \in \mathcal{T}} \mathbf{x}_T r(\gamma) & (36)\\ &= r(\gamma) \mathbf{1}_{\mathcal{T}}' \mathbf{x} & (37) \end{aligned}$$

On the other hand, from the constraints $A'\mathbf{x} \leq \mathbf{b} = \theta \mathbf{1}_{\mathcal{E}} + \boldsymbol{\mu}$ and using the same arguments as in (16)-(21), we have that

$$\boldsymbol{\beta}' \Lambda' \mathbf{x} \leq \boldsymbol{\beta}' \left(\theta \mathbf{1}_{\mathcal{E}} + \boldsymbol{\mu}\right) = \theta + \boldsymbol{\beta}' \boldsymbol{\mu} = r(\gamma). \tag{38}$$

Combining (37) and (38), it follows that,

$$r(\gamma)\mathbf{1}_{\mathcal{T}}' \mathbf{x} \leq \boldsymbol{\beta}' \Lambda' \mathbf{x} \leq r(\gamma). \tag{39}$$

Thus $\mathbf{1}_{\mathcal{T}}' \mathbf{x} \leq 1$ for all feasible $\mathbf{x}$, i.e. the value of the program is at most 1.

b) Notice from the above and from the conclusion of Lemma 2 that for $\theta \geq 0$ the value of the LP defined in Theorem 2 is exactly equal to 1. Thus,

$$\boldsymbol{\beta}' A' \mathbf{x}^* = r(\gamma)\mathbf{1}_{\mathcal{T}} \mathbf{x}^* = r(\gamma). \tag{40}$$

Also, $A'\mathbf{x}^* \leq \theta \mathbf{1}_{\mathcal{E}} + \boldsymbol{\mu}$ by the constraints of the primal LP above.
Now, assume that $A'\mathbf{x}^*(e) < \theta + \boldsymbol{\mu}(e)$ for some $e \in \mathcal{E}$ with $\boldsymbol{\beta}(e) > 0$. Then,

$$\begin{aligned}
\boldsymbol{\beta}' A' \mathbf{x}^* &= \sum_{e \in \mathcal{E}} \boldsymbol{\beta}(e) A' \mathbf{x}^*(e) && (41)\\
&< \sum_{e \in \mathcal{E}} \boldsymbol{\beta}(e)(\theta + \boldsymbol{\mu}(e)) && (42)\\
&= \theta + \sum_{e \in \mathcal{E}} \boldsymbol{\beta}(e) \boldsymbol{\mu}(e) && (43)\\
&= r(\gamma), && (44)
\end{aligned}$$

where the last equality is obtained by using the same arguments as in (16)-(21). This contradicts observation (40). As a consequence, $A'\mathbf{x}^*(e) = \theta + \boldsymbol{\mu}(e)$ for all $e \in \mathcal{E}$ with $\boldsymbol{\beta}(e) > 0$.

This ends the proof of the theorem and establishes the existence of an $\boldsymbol{\alpha}$ satisfying (6) for any $\boldsymbol{\beta}$ defined as in (5).

6.3 Enumerating All Nash Equilibria

In this section, we consider the zero-sum game where $\boldsymbol{\mu} = 0$ and show that all Nash equilibria of the game have the form given in Theorem 1 equations (5) and (6).

In this case, since there is no cost of attack, $\theta > 0$. We claim that for any strategy pair $(\boldsymbol{\alpha}_T, T \in \mathcal{T})$ and $(\boldsymbol{\beta}(e), e \in \mathcal{E})$ that are in Nash equilibrium, it must be the case that $(\boldsymbol{\beta}(e), e \in \mathcal{E})$ is given by

$$\boldsymbol{\beta}(e) = \sum_{E \in \mathcal{C}} \gamma_E \frac{1_{e \in E}}{|E|}, \tag{45}$$

for some probability distribution $(\gamma_E, E \in \mathcal{C})$ on the set of critical subsets.

As a consequence of this, we will conclude that $\boldsymbol{\alpha}$ must be in the form given in the Nash equilibrium theorem.

Because of space limitations, we describe the main points of the proof in appendix A and for the full proof, we refer the interested reader to [6] and [7].

7 Conclusion and Future Work

This paper studies a generalization of the topology design game defined in [8], where a network manager is choosing a spanning tree of a graph as communication infrastructure, and an attacker is trying to disrupt the communication tree by attacking one link of the graph. Assuming that the attacker incurs a positive cost by attacking any given link of the network, we revisit the notions of vulnerability and criticality of a subset of links.

We have determined the values of the attack costs for which a rational attacker will opt to not launch an attack. When the attacker decides to attack, we have shown that there always exists a NE under which she attacks randomly, with the same probability, links in a given critical subset. The randomization can also be done across critical subsets. The network manager chooses only spanning trees that cross the critical set in the minimum number of edges, and such that the sum of the probabilities of all trees going through any link in the critical set minus the cost of attacking that link, is the same.

For the game of zero costs of attack studied in [8], we have characterized the set of all Nash equilibria. The NE strategies are such that the attacker will always target links in critical subsets and attacks all links in the same critical subset with the same probability.

We have shown, by a simple example, that the uniformity of the attack on each critical subset is a consequence of the geometry of the problem. Mainly, the vertices of the blocker of the spanning tree polyhedron are such that if normalized, they results to uniform distribution. This is not always the case. For instance, if the defender incurs different cost of choosing different spanning trees, the attack strategies are no longer uniform on critical subsets.

The proof concepts presented in this paper have been generalized to identify Nash equilibria for a class of *quasi zero-sum* games. For details of the general study, we refer the interested readers to [7] and [6].

Acknowledgments. The authors would like to thank members of the Berkeley MURI and Netecon groups for their valuable input. Our special thanks go to Prof. Dorit Hochbaum for suggesting a set of very related papers. The work of the authors was supported by the ARO MURI grant W911NF-08-1-0233 and by the NSF grants CNS-0627161 and its continuation, CNS-0910702.

References

1. Avis, D., Rosenberg, G., Savani, R., von Stengel, B.: Enumeration of Nash Equilibria for Two-Player Games. Economic Theory 42, 9–37 (2010)
2. Boyd, S., Vandenberghe, L.: Convex Optimization. Cambridge University Press (March 2004)
3. Chopra, S.: On the Spanning Tree Polyhedron. Operations Research Letters 8(1), 25–29 (1989)
4. Cunningham, W.H.: Optimal Attack and Reinforcement of a Network. J. ACM 32(3), 549–561 (1985)
5. Fulkerson, D.R.: Blocking and Anti-Blocking Pairs of Polyhedra. Math. Programming (1), 168–194 (1971)

6. Gueye, A.: A Game Theoretical Approach to Communication Security. PhD dissertation, University of California, Berkeley, Electrical Engineering and Computer Sciences (March 2011)
7. Gueye, A., Walrand, J.C., Anantharam, V.: Bloking Games. Technical report, University of California, Berkeley (December 2010), `http://www.eecs.berkeley.edu/~agueye/index.html`
8. Gueye, A., Walrand, J.C., Anantharam, V.: Design of Network Topology in an Adversarial Environment. In: Alpcan, T., Buttyán, L., Baras, J.S. (eds.) GameSec 2010. LNCS, vol. 6442, pp. 1–20. Springer, Heidelberg (2010)
9. Wolsey, L.A., Nemhauser, G.L.: Integer and Combinatorial Optimization, 1st edn. Wiley-Interscience (November 1999)

A Proof Sketch of the NE Enumeration Claim

Theorem 1, tells that if the attack costs $\mu = 0$, then all Nash equilibrium pairs $(\boldsymbol{\alpha}, \boldsymbol{\beta})$ of the game have the form given in (6) for $\boldsymbol{\alpha}$ and in (5) for $\boldsymbol{\beta}$. To show this, we claim that for any strategy pair $(\boldsymbol{\alpha}_T, T \in \mathcal{T})$ and $(\boldsymbol{\beta}(e), e \in \mathcal{E})$ that are in Nash equilibrium, it must be the case that $\boldsymbol{\beta}$ is given by

$$\boldsymbol{\beta}(e) = \sum_{E \in \mathcal{C}} \gamma_E \frac{1_{e \in E}}{|E|}\,, \tag{46}$$

for some probability distribution $(\gamma_E, E \in \mathcal{C})$ on the set of critical subsets.

We prove this claim by scaling any mixed strategy $\boldsymbol{\beta}$ (seen as a vector in $R_+^{|\mathcal{E}|}$) with a proper constant such that it belongs to the blocker of the spanning tree polyhedron. The proof is based on the following ideas.

- Since the spanning tree polyhedron $\mathcal{P}_A$ and its blocker $\mathcal{P}_A^b$ are given in terms of feasible partitions, we establish a correspondence between feasible partitions and critical subsets of the graphs. Basically, we show that every critical subset is the set of edges going across the elements of some feasible partition. We define the notion of critical partitions Π_C (corresponding to critical subsets) and show the following equivalent claim:
$$\boldsymbol{\beta}(e) = \sum_{P \in \Pi_C} \gamma_P \frac{1_{e \in \mathcal{E}(P)}}{|\mathcal{E}(P)|}\,, \tag{47}$$
where γ is now viewed as a distribution on the critical partitions.
- Because the game is zero-sum, we know that all NE $(\boldsymbol{\alpha}, \boldsymbol{\beta})$ have payoff $\theta > 0$ which is given as:
$$\theta = \sum_{T \in \mathcal{T}} \boldsymbol{\alpha}_T \left(\sum_{e \in \mathcal{E}} \boldsymbol{\beta}(e) 1_{e \in T} \right) > 0. \tag{48}$$
We argue that $\sum_{e \in \mathcal{E}} \boldsymbol{\beta}(e) 1_{e \in T} > 0$ for all T, and can be scaled by a constant κ so that $\sum_{e \in \mathcal{E}} \kappa \boldsymbol{\beta}(e) 1_{e \in T} \geq 1$. This means that the vector $\kappa\boldsymbol{\beta}$ belongs to the blocker of the spanning tree polyhedron (see Theorem 3 of the appendix). Recall that [8] the vertices of this blocker are vectors of the form $(\frac{1_{e \in \mathcal{E}(P)}}{|P|-1}, e \in \mathcal{E})$, for some feasible partitions P.

We argue that by making the proper choice of κ the vector $\kappa\boldsymbol{\beta}$ can be written as

$$\kappa\boldsymbol{\beta}(e) = \sum_{P\in\Pi_C} \gamma_P \frac{1_{e\in\mathcal{E}(P)}}{|P|-1}, \tag{49}$$

and show that the proper κ must be in the form $\kappa = \frac{|P|-1}{|\mathcal{E}(P)|}$ for some critical partition $P \in \Pi_C$. But all critical partitions (subsets) have the same ratio $\frac{|P|-1}{|\mathcal{E}(P)|}$. By dividing the equation by κ we get

$$\boldsymbol{\beta}(e) = \sum_{P\in\Pi_C} \gamma_P \frac{1_{e\in\mathcal{E}(P)}}{|\mathcal{E}(P)|}. \tag{50}$$

Using the correspondence between critical partitions and critical subsets, we get the claim in (46).

B Blocking Pair of Matrices

The discussion in this appendix section is mostly based on [9, pp. 99-101].
Let A be a $r \times m$ nonnegative matrix. The polyhedron $\mathcal{P}_A$ associated with A is defined as the vector sum of the convex hull of its rows $(\mathbf{a}_1, \ldots, \mathbf{a}_r)$ and the nonnegative orthant:

$$\mathcal{P}_A = \text{conv.hull}\,(\mathbf{a}_1, \ldots, \mathbf{a}_r) + \mathbb{R}^m_+. \tag{51}$$

A row $\mathbf{a}_i$ of A is said to be *inessential* if it dominates a convex combination of other rows of A. If all the rows of A are essential, we say that A is *proper*. In this discussion we will assume that A is proper. For example, if A is the tree-link incidence matrix of the spanning trees of a graph, then A is a proper matrix and $\mathcal{P}_A$ defines the spanning polyhedron of the graph.
Next we define the blocker of the polyhedron $\mathcal{P}_A$.

Definition 2. *The* blocker $\mathcal{P}_A^b$ *of* $\mathcal{P}_A$ *is defined as:*

$$\mathcal{P}_A^b = \left\{\mathbf{x} \in \mathbb{R}^m_+ :\ \mathbf{x}\cdot\mathbf{y} \geq 1,\ \ \forall\, \mathbf{y} \in \mathcal{P}_A\right\} \tag{52}$$

We are interested in characterizing the polyhedron $\mathcal{P}_A$ and its blocker $\mathcal{P}_A^b$. This is given by the following theorem by Fulkerson [5]. It is based on the fact that there is a one-to-one correspondence between the rows of A and the extreme points of $\mathcal{P}_A$.

Theorem 3. *Let the* r-by-m *matrix* A *be proper with rows* $\mathbf{a}_1, \ldots, \mathbf{a}_r$*, and let the polyhedron* $\mathcal{P}_A$ *be defined as in (51). Let* $\mathbf{b}_1, \ldots, \mathbf{b}_s$ *be the extreme points of* $\mathcal{P}_A^b$*, and let* B *be the matrix having those points as rows. Then,*

i. The blocker $\mathcal{P}_A^b$ *of* $\mathcal{P}_A$ *is given by* $\mathcal{P}_A^b = \left\{\mathbf{x} \in \mathbb{R}^m_+ :\ A\mathbf{x} \geq 1\right\}$.
ii. B *is proper, and the polyhedron* $\mathcal{P}_A$ *can be described as* $\mathcal{P}_A = \left\{\mathbf{x} \in \mathbb{R}^m_+ : B\mathbf{x} \geq 1\right\}$.
iii. The blocker of the blocker $\mathcal{P}_A^b$ *verifies* $\left(\mathcal{P}_A^b\right)^b = \mathcal{P}_A$.

A and B are said to form a blocking pair of matrices.

Blocking pairs of matrices play an important role in the combinatorial problem of *maximum packing* (see Fulkerson[5]). In this paper, we use the theory of blocking pair to provide an easy argument for the existence of a probability distribution that satisfies a certain number of constraints.
Consider the following linear program:

$$\begin{aligned}&\text{Maximize } \mathbf{1}'\mathbf{x}\\&\text{subject to } A'\mathbf{x} \leq \mathbf{w}, \text{ and } \mathbf{x} \geq \mathbf{0},\end{aligned} \tag{53}$$

where the constraint A' is a nonnegative matrix.

We are interested to knowing whether the value of the program is greater than 1 or not. The following lemma gives an answer to that question.

Lemma 3. *The value of the LP in (53) is greater than 1 if and only if* $\mathbf{w}$ *belongs to the polyhedron* $\mathcal{P}_A$ *defined by* A.

Proof. The proof of the lemma is as follow.

First notice that strong duality holds for this LP. In fact, Slater's condition [2] is satisfied for any nonnegative $\mathbf{w}$. The dual of the LP is given as:

$$\begin{aligned}&\text{Minimize } \mathbf{w}'\mathbf{y}\\&\text{subject to } A\mathbf{y} \geq \mathbf{1}, \text{ and } \mathbf{y} \geq \mathbf{0}.\end{aligned} \tag{54}$$

The constraints of the dual program (54) define the blocker $\mathcal{P}_A^b = \{\mathbf{y} \in \mathbb{R}_+^m : A\mathbf{y} \geq 1\}$ of the polyhedron $\mathcal{P}_A$.

Now, if $\mathbf{w}$ belongs to $\mathcal{P}_A$, then for all $\mathbf{y} \in \mathcal{P}_A^b$, we have that $\mathbf{w}'\mathbf{y} \geq 1$.

Conversely, if $\mathbf{w}'\mathbf{y} \geq 1$ for all $\mathbf{y} \in \mathcal{P}_A^b$, then $\mathbf{w}$ must be in the blocker of $\mathcal{P}_A^b$ which is $\mathcal{P}_A$.

Adversary Games in Secure/Reliable Network Routing

Gruia Calinescu, Sanjiv Kapoor, Michael Quinn, and Junghwan Shin

Illinois Institute of Technology, Illinois, Chicago 60616, USA
{calinescu,kapoor,mquinn4,jshin7}@iit.edu

Abstract. In this paper, we consider security aspects of network routing in a game-theoretic framework where an attacker is empowered with an ability for intrusion into edges of the network; on the other hand, the goal of designer is to choose routing paths.

We interpret the secure routing problem as a two player zero sum game. The attacker can choose one or more edges for intrusion, while the designer has to choose paths between source-destination pairs for a given set of pairs. We give polynomial-time algorithms for finding mixed Nash equilibria if 1) the attacker is limited to a one-edge attack (for arbitrary number of source-destination pairs), 2) the designer has two source-destination pairs while the attacker is either limited to c edges, for given c, or the attacker incurs a cost for each edge attacked. Previous work gave an algorithm for one source-destination pair and multiple edge attacks.

Keywords: zero-sum games, Nash equilibrum, network flows, concurrent flows.

1 Introduction

1.1 Motivation

Current routing protocols utilize single paths and are typically prone to failures or malicious attacks. The fast growth of the Internet underscores the need for Network Security. With the current emphasis on distributed computing the network is further exposed to numerous dangerous situations, e.g., security attacks, eavesdropping and so on. An early illustration of this is [24] which considers various *attacks* and *defenses*.

In this paper we consider the security issues modeled as a game between the network designer and the malicious attacker. Attackers may be any malicious users attempting to intercept or eavesdrop packets on the network or impact physical constraints e.g., disconnection, congestion , etc. The network designer has to route data flow between multiple source-sink pairs and is allowed to utilize multiple paths for routing. The attacker selects edges for attack. The attacks can be on either single or multiple edges simultaneously. Similarly, the designer

R. Jain and R. Kannan (Eds.): GameNets 2011, LNICST 75, pp. 249–264, 2012.

can choose single or multiple paths simultaneously. In either case, the choice of path(s) is considered to be stochastic, i.e. one or more paths are chosen from a given set with a specific probability distribution.

An earlier illustration of this approach in the context of network routing is illustrated by the result on Game Theoretic Stochastic Routing (GTSR), which describes a particular approach to multi-path routing that provides a rigorous way to determine which routes should be used and with what frequency[6][4]. GTSR finds all paths between a source-destination pair and computes next hop probabilities, i.e., the probabilities that a packet takes a particular next-hop. This contrasts with single-path algorithms that simply determine the next-hop. Based on stochastic routing, this paper consider two alternatives : offline games where the attacker starts by selecting one link or a particular node and online games where the attacker scans on physical interface at all nodes [4]. They present results only for single source-sink pairs.

A different consideration of security for network systems via a game theoretic framework can be found in [26]. In this paper they model a security attack as a 2-player stochastic game. They utilize non-linear programming to determine the Nash equilibrium. However, they consider a set of attacks on a node of the computer network instead of an attack on the entire network.

In an earlier paper, Washburn and Wood [25] considers a similar model but in the context of providing security for defense/policing application. They consider game-theoretic models where an *evader* selects and attempts to traverse a path through a network from node s to node t without being detected by an *interdictor*. They have shown that optimal strategies can be obtained by solving a maximum flow problem on the network. Another result that measures the reliability of a general transport network but via only single edge attacks can be found in [3], where a two player game is formulated. The goal of the designer is to find a least cost path and the attacker utilizes a single edge attack. Linear programming formulations are provided when the link costs are fixed.

Multi-path routing is useful for enhanced reliability in the Internet. Traditionally, a single link failure, by a malicious user or a physical reason, can take a small but significant amount time to detect and correct. Further, another disadvantage is that malicious behaviors can be realized easily due to a prediction of paths. On the other hand, multi-path routing provides a set of possible paths that causes hard-prediction of forwarding packets. Moreover, it ensures that data throughput increases[7], traffic congestion decreases[9], network utilization is improved[20][12] and network security increases.[6][21][14][4]. In wireless, adhoc and sensor networks, routing with vulnerable links(interpreted as malicious attackers) is a challenging problem. Multi-path routing can ease unreliability and vulnerability[8][22][17]. Related game theoretic research on computer security includes the work of [10][18][19][8].

As in [25], the results in [5] consider the case of routing a single source-destination path under single edge attacks. Both papers formulate an LP solution and show that the value of the game is $1/f$, where f is a value of max flow.

The case where multiple source-destination paths are to be found is not considered in these papers.

In this paper we consider the case where the designer has to select k paths, each from given source s_i to given destination t_i. The attacker can select a set of c edges, with c also given. We give a polynomial-time algorithm for finding mixed Nash equilibria in directed graphs when $c = 1$ by using concurrent flows linear programs. We also give a polynomial-time algorithm for finding mixed Nash equilibria when the input graph is undirected and $k = 2$, for arbitrary c not exceeding the minimum edge-cut of the graph, using Hu's [11] two commodity flows. We leave open the case $k \geq 3$ and $c \geq 2$. Our approach does not extend to this case due to the fact that maxflow/mincut type conditions do not hold.

We also consider the case where the attacker can select an arbitrary set of edges incurring a cost of attack, which is dependant on the edge, and show how to compute the equilibrum in polynomial time for two commodities in undirected graphs, again using [11].

We remark that all the games considered, here and in previous work, are linear programs, but with exponential-size matrices. The challenge is finding polynomial-time algorithms for solving such linear programs; thus the matrices must be considered implicitly.

The paper outline is as follows. In Section 2, we describe the previously known game [25][5] where a designer uses one path; on the other hand, an attacker attempts to destroy one edge. While the pay-offs for these case are known, we illustrate the methodology linking these games to network flows. In Section 3, we consider the game where a designer is required to find k paths for given source-destination pairs, while the attacker still attempts to destroy one edge. In Section 4 we describe results for two source-destination pairs, while the attacker can choose c edges. In Section 5 we present polynomial-time algorithms for the case where the attacker incurs a cost for each edge attacked (dependant on the edge) while the designer chooses two paths, from given s_1 to t_1 and s_2 to t_2.

1.2 Notations

Given the directed or undirected graph, $G = (V, E)$, where V is a set of vertices and E is a set of edges we consider the network $N(G, S-T)$ where $S-T$ is a set of k source-destination pairs $(s_i, t_i), i = 1 \dots k, s_i, t_i \in V$. Let us define $\mathcal{P}$ and $\mathcal{P}^i$ as the set of all possible source-destination paths and the set of possible $s_i - t_i$ paths, respectively. We sometimes call an $s_i - t_i$ paths a path for commodity i. We let S_A be the strategy set of the attacker and S_D be the strategy set of the designer. Each strategy $s \in S_A$ of the attacker is a subset of edges of size c, i.e. $s = \{e_1 \dots e_c\}, e_i \in E$, which he chooses to attack. And each strategy of the designer $p \in S_D$ is a set of paths, one for each source-sink pair, i.e $p = \{P_1, \dots, P_k\}, P_i \in \mathcal{P}^i$. We use the following payoff matrix, for $s \in S_A$ and $p \in S_D$: the attacker gains $|\{i \in \{1, 2, \dots, k\} \mid P_i \cap s \neq \emptyset\}|$. That is, the attacker's pay-off is the number of intercepted paths (and intercepting a path more than once does not help him). We make this a zero-sum gain so the designer's payoff is the negation of the attacker's payoff; a designer therefore would minimize the number of intercepted paths. A mixed strategy of the attacker is denoted by the vector x, where $x = (x_1, x_2, ..., x_r)$ where $r = |S_A|$ and y denotes the mixed strategy of the designer, where $y = (y_1, y_2, ..., y_t)$ $t = |S_D|$.

In another version of the game, we use the following payoff matrix, for $s \in S_A$ and $p \in S_D$: the attacker gains $|\{i \in \{1, 2, \ldots, k\} \mid P_i \cap s \neq \emptyset\}| - \sum_{e \in s} c(e)$. That is, the attacker's payoff is the number of intercepted paths, less the total cost of the edges attacked. We consider this to be a zero-sum game so the designer's payoff is the negation of the attacker's payoff; the designer, therefore, would minimize the payoff of the attacker. A mixed strategy of the attacker is denoted by the vector x, where $x = (x_1, x_2, ..., x_r)$ where $r = |S_A|$ and y denotes the mixed strategy of the designer, where $y = (y_1, y_2, ..., y_t)$ and $t = |S_D|$.

Note that the Nash equilibrium is the solution of a linear program, which in our case has exponentially many variable and constraints. We are able to find an equilibrium point in polynomial time only for the case of one or two commodities. We directly describe the two-commodity case; the simpler one-commodity case can be solved by maximum flow and the methods of the two-commodity case, and can also be reduced to the two-commodity case by setting $s_2 = t_2$.

The attacker will pick, without loss of generality, all the edges of cost 0. Therefore, for the purpose of computing the equilibrium, we are only concerned with those commodity i for which there exist an $s_i - t_i$ path all whose edges have strictly positive costs. We discard from now on the commodities for which no such path exist as whatever path a designer chooses for such a pair $s_i - t_i$, it will be intercepted. For the remaining commodities, without loss of generality, the designer will never choose a path with an edge with zero cost, and therefore from now on we assume $c(e) > 0$ for all $e \in E$.

2 The (One Commodity, One Edge)-Problem

We consider the simplest version of the problem where $S - T$ comprises one source-sink pair (s, t). The attacker is allowed to attack one edge and the designer has to choose one path. The strategy sets are $S_A = E$ and $S_D = \mathcal{P}$ where $\mathcal{P}$ is the set of paths from s to t. We consider a zero-sum game where the attacker gains 1 unit if his chosen edge successfully intersects the designer path and the designer loses 1 unit. In the case when his chosen edge does not intersect the designer path, the designer and attacker loose or gain nothing.

We consider the pay-off matrix A of size $(m \times q), q = |\mathcal{P}|$, where player 1 (attacker) has m strategies (number of edges) and player 2 (designer) has q strategies (number of $s - t$ paths). Then, we can assign a payoff to each element of $A = a_{ij}$ as follows: when the attacker chooses a strategy i and the designer chooses a strategy j : if $e_i \in P_j$ then $a_{ji} = 1$. Otherwise, $a_{ji} = 0$.

2.1 Reduction to the Max-flow Min-cut Problem

Let us define the LPs for player 1 and player 2 as *LP1* and *LP2*, respectively, as follows :

LP1: Maximize w subject to

$$\sum_{i=1}^{m} x_i a_{ji} \geq w \; \forall 1 \leq j \leq q$$
$$\sum_{i=1}^{m} x_i = 1$$
$$x_i \geq 0 \qquad \forall i \in E$$

LP2: Minimize λ subject to

$$\sum_{j=1}^{q} a_{ji} y_j \leq \lambda \; \forall 1 \leq i \leq m$$
$$\sum_{j=1}^{q} y_j = 1$$
$$y_j \geq 0 \qquad \forall 1 \leq j \leq q$$

where x_i and y_j represent probabilities to choose an edge and a path, respectively. a_{ij} represents the payoff obtained when the designer uses strategy i and the attacker uses strategy j. In this game, a Nash equilibrium set of strategies is a probability distribution on the set of edges corresponding to the attacker choice of edges and a probability distribution on the possible flow paths chosen by the designer.

Theorem 1. *[23] (*Menger's theorem *(directed vertex-disjoint version)). Let $D = (V, A)$ be a digraph and let $S, T \subseteq V$. Then the maximum number of vertex-disjoint $S - T$ paths is equal to the minimum size of an $S - T$ disconnecting vertex set.*

Theorem 2. *[23] A maximum collection of arc-disjoint $s-t$ paths and a minimum-size $s - t$ cut can be found in time $O(m^2)$.*

2.2 Algorithm

1. Attacker : find min-cut, C, which separates V into two disjoint subset of vertices. The attacker chooses a strategy that selects each edge in C with probability $1/|C|$.
2. Designer : construct a undirected graph from a given network instance. choose r-disjoint paths which are equally assigned $1/r$, where $|C| = r$. By Theorem 1, we know that *the maximum number of arc-disjoint s-t paths is equal to the minimum size of an s-t edge-cut.* The theorem holds in the case of undirected graphs as follows. The undirected vertex-disjoint version follows immediately from Theorem 1 by replacing each undirected edge by two oppositely oriented arcs. Next, the undirected edge-disjoint version follows from the undirected vertex-disjoint version. We can find arc-disjoint $s - t$ paths in time polynomial by Theorem 2.

We can show the correctness by showing the solutions are feasible for the LP1 and LP2 and have the same payoff. Recall that a profile S of mixed strategies is a mixed strategy Nash equilibrium if and only if every player's mixed strategy is a best response to the other player's mixed strategies.

Theorem 3. *[5][25] A strategy (α^*, β^*) given by the algorithm above is a mixed Nash equilibrium for the attacker and designer.*

3 The (k Commodities, One Edge)-Problem

In this section we consider the problem where the designer chooses paths for multiple source-sink pairs, while the attacker is allowed one edge for intrusion. We present the result for directed graphs, although the results hold for undirected graphs as well.

Let A be the payoff matrix. With q_i denoting the number of $s_i - t_i$ paths, A has q rows, where $q = \Pi_{i=1}^{k} q_i$, and m columns, where $m = |E|$ (recall that in this section, an attacker's strategy is a single edge). Let P_j^i be the jth path from s_i to t_i, for $1 \le i \le k$ and $1 \le j \le q_i$. Let r be some correspondence from $\{1, 2, \ldots, q\}$ to $\{1, 2, \ldots, q_1\} \times \{1, 2, \ldots, q_2\} \times \cdots \times \{1, 2, \ldots, q_k\}$, with $r_i(j)$ giving the ith component of r $(1 \le i \le k)$. We index the columns of A by e, for $e \in E$. Then, based on previous discussion, we have for $e \in E$ and $j \in \{1, 2, \ldots, q\} : A_{je} = |\{i \in \{1, 2, \ldots, k\} \mid e \in P^i_{r_i(j)}\}|$.

Based on standard game theory, to find a Nash equilibrium the attacker must solve the following linear program with variables w and x_e, for $e \in E$:

LP3: Maximize w subject to

$$\sum_{e \in E} A_{je} x_e \ge w \ \forall 1 \le j \le q \quad (1)$$
$$\sum_{i=1}^{m} x_e = 1 \quad (2)$$
$$x_e \ge 0 \qquad \forall e \in E. \quad (3)$$

We denote the program above *LP3*. The designer must solve the linear program below, with variable λ and y_j:

LP4: Minimize λ subject to

$$\sum_{j=1}^{q} A_{je} y_j \le \lambda \ \forall e \in E \quad (4)$$
$$\sum_{j=1}^{q} y_j = 1 \quad (5)$$
$$y_j \ge 0 \qquad \forall j \in \{1, 2, \ldots, q\} \quad (6)$$

We denote the program above *LP4*. LP3 and LP4 are duals. With the goal of solving the exponentially large programs LP3 and LP4, we introduce the following linear program, with variables α and f_P, for $P \in \mathcal{P}$.

LP5: Minimize α subject to

$$\sum_{P \in \mathcal{P} \mid e \in P} f_P \le \alpha \ \forall e \in E \quad (7)$$
$$\sum_{j=1}^{q_i} f_{P_j^i} = 1 \qquad \forall i \in \{1, 2, \ldots, k\} \quad (8)$$
$$f_P \ge 0 \qquad \forall P \in \mathcal{P} \quad (9)$$

Recall that $\mathcal{P}$ is the set of all source-destination paths, while for $i \in \{1, 2, \ldots, k\}$, the set of $s_i - t_i$ paths is $\{P_1^i, P_2^i, \ldots, P_{q_i}^i\}$. We denote the program above *LP5*. LP5 is also exponentially large, but it is known that it can be solved in polynomial time as it is a concurrent flow problem [23]. Precisely, LP5 is solved by the standard decomposition into path-flows of the solution of the following linear program, with variables f_e^i for $e \in E$ and $i \in \{1, 2, \ldots, k\}$:

$$\textbf{LP7:}\ \text{Minimize } \gamma \text{ subject to} \quad \sum_{e\in\delta^-(u)} f_e^i = \sum_{e\in\delta^+(u)} f_e^i \quad \forall i \wedge \forall u \in V \setminus \{s_i, t_i\}$$

$$\sum_{e\in\delta^+(s_i)} f_e^i - \sum_{e\in\delta^-(s_i)} f_e^i = 1\ \forall i$$

$$\textstyle\sum_{i=1}^k f_e^i \le \gamma \qquad \forall e \in E$$

$$f_e^i \ge 0 \qquad \forall e \in E \wedge i$$

Here $\delta^-(u)$ denotes the set of edges entering u, and $\delta^+(u)$ denotes the set of edges leaving u. We denote the program above *LP7*.

The dual of LP5, denoted as *LP6*, given below, can also be solved in polynomial-time by solving the dual of the concurrent flow program above. LP6 has variables d_i, for $i \in \{1, 2, \ldots, k\}$, and l_e for $e \in E$; it has exponentially many constraints:

$$\textbf{LP6:}\ \text{Maximize } \sum_{i=1}^k d_i \text{ subject to} \quad d_i \le \sum_{e\in P_j^i} l_e \ \forall i \wedge \forall j\} \tag{10}$$

$$\textstyle\sum_{e\in E} l_e = 1 \tag{11}$$

$$l_e \ge 0 \qquad \forall e \in E \tag{12}$$

As an aside, LP6 is a fractional version of Sparsest Cut [15,13,16,2,1]. After optimally solving LP5 and LP6, we assign to the variables of LP3 and LP4 values as follows. For LP3, we set $w = \sum_{i=1}^k d_i^*$ and $x_e = l_e^*$, where d_i^*, for $i \in \{1, 2, \ldots, k\}$, and l_e^* for $e \in E$ are an optimal solution to LP6. For LP4, we set $\lambda = \alpha^*$ and $y_j = \Pi_{i=1}^k f^*_{P^i_{r_i(j)}}$, where α^* and f_P^* are an optimal solution to LP5. As the number of nonzero y_j can be exponentially large, it has to be remembered implicitly, and indeed the designer can generate one random strategy by drawing independently at random for each $i \in \{1, 2, \ldots, k\}$ one P_j^i with probability $f^*_{P_j^i}$. There is an alternate way for the designer, used in [6] (but not for this game): at each node of the network, the packets going from s_i to t_i are not forwarded deterministically but exit node u on edge e with probability $\frac{f_e^i}{\sum_{e\in\delta^+(u)} f_e^i}$, where f_e^i are from an acyclic optimum to LP7; in this case the game is played separately for each packet.

We have not proved yet that our w and x_e are feasible for LP3, and λ and y_j are feasible for LP4; however once we do so they are both optimal, since the objective functions of the two dual linear programs LP3 and LP4 match: $w = \lambda$ (as $\alpha^* = \sum_{i=1}^k d_i^*$).

Claim. As defined above, w and x_e are feasible for LP3.

Proof. We have to verify the constraints of LP3. Indeed, constraints 3 follow immediately from constraints 12, and 2 follows from 11. As for constraints 1, let $e \in E$ and $j \in \{1, 2, \ldots, q\}$. We have

$$\sum_{e\in E} A_{je}x_e = \sum_{e\in E} x_e |\{i \in \{1,2,\ldots,k\} \mid e \in P^i_{r_i(j)}\}| = \sum_{i=1}^{k} \sum_{e\in P^i_{r_i(j)}} x_e$$

$$= \sum_{i=1}^{k} \sum_{e\in P^i_{r_i(j)}} l^*_e \geq \sum_{i=1}^{k} d^*_i = w,$$

where the inequality follows from constraint 10 of LP6. ■

Claim. As defined above, λ and y_j are feasible for LP4.

Due to the page limitation, we omit the long technical proof. A similar proof appears later for Lemma 1.

4 The (2 Commodities, c Edges)-Problem

In this section we assume the input graph is undirected and c does not exceed the minimum edge-cut of the graph. Let A be the payoff matrix. With q_i denoting the number of $s_i - t_i$ paths, A has q rows, where $q = q_1 \cdot q_2$, and $\hat{m}$ columns, where $\hat{m} = \binom{m}{c}$. Let P^i_j be the jth path from s_i to t_i, for $1 \leq i \leq 2$ and $1 \leq j \leq q_i$. Let r be some correspondence from $\{1,2,\ldots,q\}$ to $\{1,2,\ldots,q_1\} \times \{1,2,\ldots,q_2\}$, with $r_i(j)$ giving the ith component of r $(1 \leq i \leq 2)$.

Let us use $\mathcal{T}$ to denote the set of all sets $S \subset E$ with $|S| = c$. We index the columns of A by $S \in \mathcal{T}$. Then, based on previous discussion, we have for $S \in \mathcal{T}$ and $j \in \{1,2,\ldots,q\} : A_{jS} = |\{i \in \{1,2\} \mid P^i_{r_i(j)} \cap S \neq \emptyset\}|$.

Based on standard game theory, to find a Nash equilibrium the attacker must solve the following linear program with variables w and x_S, for $S \in \mathcal{T}$:

LP8: Maximize w subject to

$$\textstyle\sum_{S\in\mathcal{T}} A_{jS}x_S \geq w \;\forall 1 \leq j \leq q \quad (13)$$

$$\textstyle\sum_{i=1}^{\hat{m}} x_S = 1 \quad (14)$$

$$x_S \geq 0 \qquad \forall S \in \mathcal{T}. \quad (15)$$

We denote the program above *LP8.* The designer must solve the linear program below, with variable λ and y_j, for $1 \leq j \leq q$:

LP9: Minimize λ subject to

$$\textstyle\sum_{j=1}^{q} A_{jS}y_j \leq \lambda \;\forall S \in \mathcal{T} \quad (16)$$

$$\textstyle\sum_{j=1}^{q} y_j = 1 \quad (17)$$

$$y_j \geq 0 \qquad \forall j \in \{1,2,\ldots,q\} \quad (18)$$

We denote the program above *LP9.* LP8 and LP9 are duals. With the goal of solving the exponentially large programs LP8 and LP9, we introduce the following linear program, with variables α and f_P, for $P \in \mathcal{P}$.

LP10: Minimize α subject to

$$\textstyle\sum_{P\in\mathcal{P} \mid e\in P} f_P \leq \alpha \;\forall e \in E \quad (19)$$

$$\textstyle\sum_{j=1}^{q_i} f_{P^i_j} = 1 \qquad \forall i \in \{1,2\} \quad (20)$$

$$f_P \geq 0 \qquad \forall P \in \mathcal{P} \quad (21)$$

Recall that $\mathcal{P}$ is the set of all source-destination paths, while for $i \in \{1, 2\}$, the set of $s_i - t_i$ paths is $\{P_1^i, P_2^i, \ldots, P_{q_i}^i\}$. We denote the program above *LP10* and note that the only difference between LP5 and LP10 is that LP10 has $k = 2$ and an undirected graph instead of arbitrary k and directed graph. LP10 is also exponentially large, but it is known that it can be solved in polynomial time [11], [23].

We show later how to extract a feasible solution for LP8 from an LP10 solution. However, the dual of LP10 does not immediately give a solution to LP9, and we do not use it. Instead, we use a sort of integral dual of LP10, which is what [11] provides. Precisely, let $C_1 \subset E$ be the minimum cut separating s_1 from t_1, let $C_2 \subset E$ be the minimum cut separating s_2 from t_2, and let $C_3 \subset E$ be the minimum cut separating both $s_i - t_i$ pairs. In other words, in the graph $(V, E \setminus C_3)$ there exist neither $s_1 - t_1$ nor $s_2 - t_2$ paths; C_3 can be computed by computing two minimum cuts: one separating s_1, s_2 from t_1, t_2, and the other separating s_1, t_2 from t_1, s_2. Then Hu's theorem states:

Theorem 4. *If* $\min(|C_1|, |C_2|) \leq |C_3|/2$, *then there exist a concurrent two-commodity flow in* G *shipping* $\min(|C_1|, |C_2|)$ *from* s_1 *to* t_1 *and* $\min(|C_1|, |C_2|)$ *from* s_2 *to* t_2. *Otherwise (*$|C_3|/2 < |C_1|$ *and* $|C_3|/2 < |C_2|$*), there exist a concurrent two-commodity flow in* G *shipping* $|C_3|/2$ *units of flow from* s_1 *to* t_1 *and* $|C_3|/2$ *units of flow from* s_2 *to* t_2. *Moreover, these two-commodity flows can be found in time polynomial in the size of* G.

It is clear (and known) that in LP10 we have for any feasible solution $\alpha \geq 1/|C_1|$, $\alpha \geq 1/|C_2|$, and $\alpha \geq 2/|C_3|$. Let us consider three cases, with the second being symmetric to the first.

Case 1. In the first case, $|C_1| \leq |C_2|$ and $|C_1| \leq |C_3|/2$. Then Hu's theorem and algorithm finds a feasible solution to LP10 with $\alpha = 1/|C_1|$. In LP9, we assign $\lambda = c \cdot \alpha$ and $y_j = f_{P^1_{r_1(j)}} \cdot f_{P^2_{r_2(j)}}$.

Lemma 1. *λ and y_j as described above make a feasible solution to LP9.*

Proof. The constraints 18 follow immediately from the definition of y_j and constraints 21. Regarding constraint 17, we have:

$$\sum_{j=1}^{q} y_j = \sum_{l_1=1}^{q_1} \sum_{j \in \{1,2,\ldots,q\} \mid r_1(j)=l_1} y_j = \sum_{l_1=1}^{q_1} \sum_{l_2=1}^{q_2} \sum_{j \in \{1,2,\ldots,q\} \mid r_1(j)=l_1 \wedge r_2(j)=l_2} y_j$$

$$= \sum_{l_1=1}^{q_1} \sum_{l_2=1}^{q_2} f_{P^1_{l_1}} \cdot f_{P^2_{l_2}} = \sum_{l_1=1}^{q_1} f_{P^1_{l_1}} \cdot \sum_{l_2=1}^{q_2} f_{P^2_{l_2}}$$

$$= \sum_{l_1=1}^{q_1} f_{P^1_{l_1}} = 1,$$

where we used the constraints 20 of LP10. Finally, to verify Constraints 16, we let $S \in \mathcal{T}$ be arbitrary. For path P and subset $S \subseteq E$, we define the matrix $B_{P,S}$ as follows:

$$B_{P,S} = \begin{cases} 1 \text{ if } P \cap S \neq \emptyset \\ 0 \text{ otherwise} \end{cases}$$

Then:

$$\begin{aligned}
\sum_{j=1}^{q} A_{jS} y_j &= \sum_{j=1}^{q} A_{jS} \cdot f_{P^1_{r_1(j)}} \cdot f_{P^2_{r_2(j)}} \\
&= \sum_{j=1}^{q} \left(B_{P^1_{r_1(j)},S} + B_{P^2_{r_2(j)},S} \right) \cdot f_{P^1_{r_1(j)}} \cdot f_{P^2_{r_2(j)}} \\
&= \sum_{l_1=1}^{q_1} \sum_{l_2=1}^{q_2} \left(B_{P^1_{l_1},S} + B_{P^2_{l_2},S} \right) \cdot f_{P^1_{l_1}} \cdot f_{P^2_{l_2}} \\
&\le \sum_{l_1=1}^{q_1} \sum_{l_2=1}^{q_2} \sum_{e \in S} \left(B_{P^1_{l_1},\{e\}} + B_{P^2_{l_2},\{e\}} \right) \cdot f_{P^1_{l_1}} \cdot f_{P^2_{l_2}} \\
&= \sum_{e \in S} \sum_{l_1=1}^{q_1} \sum_{l_2=1}^{q_2} \left(B_{P^1_{l_1},\{e\}} + B_{P^2_{l_2},\{e\}} \right) \cdot f_{P^1_{l_1}} \cdot f_{P^2_{l_2}} \\
&= \sum_{e \in S} \sum_{l_1=1}^{q_1} \sum_{l_2=1}^{q_2} B_{P^1_{l_1},\{e\}} \cdot f_{P^1_{l_1}} \cdot f_{P^2_{l_2}} \\
&\quad + \sum_{e \in S} \sum_{l_1=1}^{q_1} \sum_{l_2=1}^{q_2} B_{P^2_{l_2},\{e\}} \cdot f_{P^1_{l_1}} \cdot f_{P^2_{l_2}} \\
&= \sum_{e \in S} \sum_{l_1=1}^{q_1} \left(B_{P^1_{l_1},\{e\}} \cdot f_{P^1_{l_1}} \cdot \sum_{l_2=1}^{q_2} f_{P^2_{l_2}} \right) \\
&\quad + \sum_{e \in S} \sum_{l_2=1}^{q_2} \left(B_{P^2_{l_2},\{e\}} \cdot f_{P^2_{l_2}} \cdot \sum_{l_1=1}^{q_1} f_{P^1_{l_1}} \right) \\
&= \sum_{e \in S} \sum_{l_1=1}^{q_1} B_{P^1_{l_1},\{e\}} \cdot f_{P^1_{l_1}} + \sum_{e \in S} \sum_{l_2=1}^{q_2} B_{P^2_{l_2},\{e\}} \cdot f_{P^2_{l_2}} \\
&= \sum_{e \in S} \left(\sum_{l_1=1 \mid e \in P^1_{l_1}}^{q_1} f_{P^1_{l_1}} + \sum_{l_2=1 \mid e \in P^2_{l_2}}^{q_1} f_{P^2_{l_2}} \right) \\
&= \sum_{e \in S} \sum_{P \in \mathcal{P}} f_P \le \sum_{e \in S} \alpha \le c \cdot \alpha = \lambda
\end{aligned}$$

where we used Constraint 20 to replace $\sum_{l_2=1}^{q_2} f_{P^2_{l_2}}$ and $\sum_{l_2=1}^{q_2} f_{P^2_{l_2}}$ by 1, and Constraint 19 in the last line. ■

Also, in LP8, assign $x_S = 1/\binom{|C_1|}{c}$ to each $S \subseteq C_1$ with $|S| = c$, and $x_S = 0$ to all other $S \in \mathcal{T}$. Note that we made the assumption that c does not exceed any cut in G and therefore $\sum_{S \in \mathcal{T}} x_S = 1$, showing constraint 14 is satisfied. We also assign $w = c/|C_1|$. We proceed to show that this assignment is feasible for LP8; the fact that we found both equilibria follows from the fact that $w = \lambda$. Constraints 14 and 15 are immediate. We must verify constraints 13, so let

$j \in \{1, 2, \ldots, q\}$ be arbitrary, and let e_j be an edge of $C_1 \cap P^1_{r_1(j)}$ (such an edge must exist since C_1 is an $s_1 - t_1$ cut). Then indeed:

$$\sum_{S \in \mathcal{T}} A_{jS} x_S = \sum_{S \in \mathcal{T}} x_S |\{i \in \{1,2\} | P^i_{r_i(j)} \bigcap S \neq \emptyset\}| \geq \sum_{S \in \mathcal{T}} x_S |\{e_j\} \cap S|$$

$$= \sum_{S \mid e_j \in S} \frac{1}{\binom{|C_1|}{c}} = \frac{|\{S \subseteq C_1 \mid e_j \in S\}|}{\binom{|C_1|}{c}} = \frac{\binom{|C_1|-1}{c-1}}{\binom{|C_1|}{c}} = c/|C_1| = w.$$

Case 2. In the second case, $|C_2| \leq |C_1|$ and $|C_2| \leq |C_3|/2$. This is symmetric to the first case.
Case 3. In the third case, $|C_3| < 2|C_1|$ and $|C_3| < 2|C_1|$. Then Hu's theorem and algorithm finds a feasible solution to LP10 with $\alpha = 2/|C_3|$. In LP9, we assign $\lambda = c \cdot \alpha$ and $y_j = f_{P^1_{r_1(j)}} \cdot f_{P^2_{r_2(j)}}$, just as in the first case.

Claim. λ and y_j as described above make a feasible solution to LP9.

Proof. Let $f^1(e_i) = \sum_{P^1_{r_1(j)} \ni e_i} f_{P^1_{r_1(j)}}$ and $f^2(e_i) = \sum_{P^2_{r_2(j)} \ni e_i} f_{P^2_{r_2(j)}}$. Let $S_1 = \{e \in S : f^1(e) > 0 \text{ and } f^2(e) > 0\}$, and let $k = |S_1|$.

$$\begin{aligned}
\sum_{j=1}^{q} A_{jS} y_j &= \sum_{j=1}^{q} A_{jS} f_{P^1_{r_1(j)}} f_{P^2_{r_2(j)}} \\
&= \sum_{e \in S_1} f^1(e)(1 - f^2(e)) + \sum_{e \in S_1} f^2(e)(1 - f^1(e)) \\
&\quad +2 \sum_{e \in S_1} f^1(e) f^2(e) + \sum_{e \in S \setminus S_1} f^1(e) \sum_{j=1}^{q_2} f_{P^2_{r_2(j)}} \\
&= \sum_{e \in S_1} (f^1(e) + f^2(e)) + \sum_{e \in S \setminus S_1} f^1(e) = 2k/|C_3| + 2(c-k)/|C_3| \\
&= 2c/|C_3| = \lambda
\end{aligned}$$

■

Also, in LP8, assign $x_S = 1/\binom{|C_3|}{c}$ to each $S \subseteq C_3$ with $|S| = c$, and $x_S = 0$ to all other $S \in \mathcal{T}$. Note that we made the assumption that c does not exceed any cut in G and therefore $\sum_{S \in \mathcal{T}} x_S = 1$, showing constraint 14 is satisfied. We also assign $w = 2c/|C_3|$. We proceed to show that this assignment is feasible for LP8; the fact that we found both equilibrium follows from the fact that $w = \lambda$. Constraints 14 and 15 are immediate. We must verify constraints 13, so let $j \in \{1, 2, \ldots, q\}$ be arbitrary. Let e^1_j be an edge of $C_3 \cap P^1_{r_1(j)}$ and e^2_j be an edge of $C_3 \cap P^1_{r_2(j)}$ (such edges must exist since C_3 is separating both pairs $s_1 - t_1$ and $s_2 - t_2$). It may be that $e^1_j = e^2_j$. Then indeed:

$$\begin{aligned}
\sum_{S \in \mathcal{T}} A_{jS} x_S &= \sum_{S \in \mathcal{T}} x_S |\{i \in \{1,2\} | P^i_{r_i(j)} \bigcap S \neq \emptyset\}| \\
&\geq \sum_{S \in \mathcal{T}} x_S (|\{e^1_j\} \cap S| + |\{e^2_j\} \cap S|)
\end{aligned}$$

$$= \sum_{S \in \mathcal{T}} x_S |\{e_j^1\} \cap S| + \sum_{S \in \mathcal{T}} x_S |\{e_j^2\} \cap S|)$$
$$= \sum_{S \mid e_j^1 \in S} \frac{1}{\binom{|C_3|}{c}} + \sum_{S \mid e_j^2 \in S} \frac{1}{\binom{|C_3|}{c}}$$
$$= \frac{|\{S \subseteq C_3 \mid e_j^1 \in S\}|}{\binom{|C_3|}{c}} + \frac{|\{S \subseteq C_3 \mid e_j^2 \in S\}|}{\binom{|C_3|}{c}}$$
$$= 2\frac{\binom{|C_3|-1}{c-1}}{\binom{|C_3|}{c}} = 2c/|C_3| = w.$$

In conclusion, we provided a polynomial-time algorithm for computing the Nash equilibrium in the game where there are two commodities and the attacker can attack any subset of c edges, provided c does not exceed the minimum cut of undirected graph G.

5 Budgeted Attacks: Costs on Edges

In this case, we consider the problem where $S - T$ consists of two commodities with sources s_1 and s_2, and sinks t_1 and t_2. The attacker may choose to attack any number of edges, and the designer chooses two paths, one for each source-sink pair. Let A be the payoff matrix. Let q_i denote the number of $s_i - t_i$ paths. A will have q rows, where $q = q_1 \cdot q_2$, and h columns, where $h = 2^{|E|}$ is the cardinality of the power set of E. Let P_j^i be the jth path from s_i to t_i, for $i \in \{1, 2\}$, where $1 \le j \le q_i$. Let r be some correspondence from $\{1, 2, \ldots, q\}$ to $\{1, 2, \ldots, q_1\} \times \{1, 2, \ldots, q_2\}$, with $r_i(j)$ giving the ith component of r $(1 \le i \le 2)$. Let $\mathcal{T}$ denote the power set of E. The columns of A will be indexed by $S \in \mathcal{T}$. Thus: $A_{jS} = |\{i \in \{1, 2\} \mid P^i_{r_i(j)} \cap S \neq \emptyset\}| - \sum_{e \in S} c(e)$. We seek a Nash equilibrium. To get one, the attacker should solve the following linear program:

LP11: Maximize w subject to
$$\sum_{S \in \mathcal{T}} A_{jS} x_S \ge w \ \forall 1 \le j \le q \quad (22)$$
$$\sum_{S \in \mathcal{T}} x_S = 1 \quad (23)$$
$$x_S \ge 0 \qquad \forall S \in \mathcal{T}. \quad (24)$$

The above linear program shall be called LP11. The path designer must solve the dual of this linear program, denoted LP12, described below:

LP12: Minimize λ subject to
$$\sum_{j=1}^{q} A_{jS} y_j \le \lambda \ \forall S \in \mathcal{T} \quad (25)$$
$$\sum_{j=1}^{q} y_j = 1 \quad (26)$$
$$y_j \ge 0 \qquad \forall j \in \{1, 2, \ldots, q\} \quad (27)$$

Our general plan for solving these programs is as follows. We compute four minimum costs cuts in the network $N = (V, E, c)$. The attacker will choose one of these cuts, or the empty set as his strategy; note that this is a pure strategy. This choice will be function of the costs of these cuts; also this choice gives us a feasible w in LP11. For the designer, we use Hu's [11,23] result to obtain certain flows matching the cuts and use the value of these flows to assign values (probabilities) to the variables y_j from LP12. Finally, we show that LP12 is feasible with $\lambda = w$ and these value for y_j; matching primal and dual objective functions implies we have optimum solutions for both LP11 and LP12.

We continue with the detailed description of this algorithm. Let $C_1 \subset E$ be the minimum cost cut separating s_1 from t_1 in N, let $C_2 \subset E$ be the minimum cost cut separating s_2 from t_2 in N, and let $C_3 \subset E$ be the minimum cost cut separating both $s_i - t_i$ pairs. In other words, in the graph $(V, E \setminus C_3)$ there exist neither $s_1 - t_1$ nor $s_2 - t_2$ paths; C_3 can be computed by computing two minimum cost cuts: one separating s_1, s_2 from t_1, t_2, and the other separating s_1, t_2 from t_1, s_2.

As usual, for $S \subseteq E$, we use $c(S) := \sum_{e \in S} c(e)$ with the usual convention that $c(\emptyset) = 0$. We assign $\lambda = w = \max(0 - c(\emptyset), 1 - c(C_1), 1 - c(C_2), 2 - c(C_3))$, and the attacker picks as his strategy the set of edges above achieving the maximum. We now verify that this is indeed a feasible solution for LP11. Constraints 23 and 24 are clearly satisfied. For Constraint 22, we must consider the four cases, depending on how w is selected. If $w = 0$, then, as for any j $A_{j\emptyset} = 0 - 0 = 0$, we see that 22 is satisfied. If $w = 1 - c(C_1)$, then we note that the strategy of the attacker, C_1, being a set of edges separating s_1 from t_1 in N, satisfies , for any designer strategy indexed by j, $A_{jC_1} \geq 1 - c(C_1)$ and this is the same as 22. The case $w = 1 - c(C_2)$ is symmetric to the previous case, while if $w = 2 - c(C_3)$, then we note that the strategy of the attacker, C_3, being a set of edges separating both s_1 from t_1 and s_2 from t_2 in N, satisfies, for any designer strategy indexed by j, $A_{jC_1} \geq 2 - c(C_3)$ and this is the same as 22. For simplicity of notation, let $c_1 = c(C_1)$, $c_2 = c(C_2)$, and $c_3 = c(C_3)$. For the designer, we use the following result of Hu [11](corollary 71.1b in [23]):

Theorem 5. *Let $G = (V, E)$ be a graph, let s_1, t_1 and s_2, t_2 be pairs of vertices of G, let $c : E \to \mathbb{R}_+$, and let $d_1, d_2 \in \mathbb{R}_+$. Then there exists a 2-commodity flow subject to c and with value d_1, d_2 if and only if the following "cut condition" is satisfied: for any $U \subset V$, we have $\sum_{i \mid |U \cap \{s_i, t_i\}| = 1} d_i \leq \sum_{e \in E \mid |U \cap e| = 1} c(e)$ (here, to make notation compact, we use an undirected edge e as a set of two vertices). Moreover, these two-commodity flows can be found in time polynomial in the size of G.*

When applying this theorem 5, we will choose $0 < d_1 \leq 1$ and $0 < d_2 \leq 1$ depending on how λ is selected. We will verify that the cut condition is satisfied, and get a two-commodity flow f^1 and f^2 as in the theorem. Then in all four cases, we proceed as follows.

We algorithmically decompose the $s_1 - t_1$ flow f^1 into at most $|E|$ path flows f_j^1 for some paths $P_j^1 \in \mathcal{P}^1$, and the $s_2 - t_2$ flow f^2 into at most $|E|$ path flows f_j^2 for some paths $P_j^2 \in \mathcal{P}^2$; we implicitly (for the purpose of the proof, and not as part of the algorithm) keep $f_j^1 = 0$ if P_j^1 is not one of the above paths, and $f_j^2 = 0$ if P_j^2 is not one of the above paths. Thus we have the flow value constraint for the first and second commodity and the capacity constraint as follows :

$$\forall i \in \{1,2\} : \sum_{j=1}^{q_i} f_j^i = d_i \tag{28}$$

$$\forall e \in E : \sum_{j \mid e \in P_j^1} f_j^1 + \sum_{j \mid e \in P_j^2} f_j^2 \leq c(e) \tag{29}$$

Recall that we have $q = q_1 \cdot q_2$ variables y_j in LP2, and r is a correspondence from $\{1,2,\ldots,q\}$ to $\{1,2,\ldots,q_1\} \times \{1,2,\ldots,q_2\}$, with $r_i(j)$ giving the ith component of r $(1 \leq i \leq 2)$. We set $y_j = \frac{1}{d_1 \cdot d_2} f^1_{r_1(j)} f^2_{r_2(j)}$; algorithmically this is done only for those at most $|E| \cdot |E|$ indices j for which both $f^1_{r_1(j)}$ and $f^2_{r_2(j)}$ are strictly positive.

It is immediate that Constraint 27 holds, and we prove that 26 holds below:

$$\sum_{j=1}^{q} y_j = \frac{1}{d_1 \cdot d_2} \sum_{l_1=1}^{q_1} \sum_{l_2=1}^{q_2} f^1_{l_1} f^2_{l_2} = \frac{1}{d_1 \cdot d_2} \left(\sum_{l_1=1}^{q_1} f^1_{l_1} \right) \left(\sum_{l_2=1}^{q_2} f^2_{l_2} \right)$$
$$= \frac{1}{d_1 \cdot d_2} (d_1 \cdot d_2) = 1$$

where the third equality follows from Equations 28.

Also common to all four cases is the following inequality (whose long proof similar to Lemma 1 we omit), an upper bound on the left-hand side of Constraint 25:

$$\sum_{j=1}^{q} A_{jS} y_j \leq \left(\frac{1}{d_1} - 1 \right) d_1 + \left(\frac{1}{d_2} - 1 \right) d_2 = (1 - d_1) + (1 - d_2) \tag{30}$$

Only now we consider the four cases, function of how λ is selected. In the first case, $\lambda = 0$ and we choose $d_1 = d_2 = 1$; note that $0 < d_i \leq 1$. It is easy to verify then that the cut condition is satisfied, as $c_1 \geq 1$, $c_2 \geq 1$, and $c_3 \geq 2$ (this is how λ was selected). As for Constraint 25, from Equation 30 we deduce that indeed $\sum_{j=1}^{q} A_{jS} y_j \leq 0 = \lambda$.

In the second case, $\lambda = 1 - c_1$ and we choose $d_1 = c_1$ and $d_2 = 1$; note that $0 < d_i \leq 1$ since we made the assumption that $c(e) > 0$ for all e and there exists an $s_1 - t_1$ path in G. For the cut condition, notice that $d_1 \leq c_1$, and $d_2 \leq c_2$ follows from $1 - c_1 \geq 2 - c_3$ (this is how λ was selected) and $c_3 \leq c_1 + c_2$ (as $C_1 \cup C_2$ is a candidate for C_3). Also, $d_1 + d_2 = 1 + c_1 \leq c_3$ (this is how λ was selected). Therefore the cut condition is satisfied. As for Constraint 25, from Equation 30 we deduce that indeed $\sum_{j=1}^{q} A_{jS} y_j \leq 1 - c_1 = \lambda$.

The third case is symmetric with the second one. $\lambda = 1 - c_2$ and we choose $d_1 = 1$ and $d_2 = c_2$, and all the arguments from the second case hold.

In the fourth case, $\lambda = 2 - c_3$. We have two subcases. If $c_3/2 \leq c_1$ and $c_3/2 \leq c_2$, we use $d_1 = d_2 = c_3/2$. Note that $0 < c_3 \leq 2$ (from the selection of λ) so $0 < d_i \leq 1$. It is immediate to verify that the cut condition is satisfied.

As for Constraint 25, from Equation 30 we deduce that indeed $\sum_{j=1}^{q} A_{jS} y_j \leq (1 - c_3/2) + (1 - c_3/2) = 2 - c_3 = \lambda$.

The second case of the fourth case has $\lambda = 2 - c_3$. And either $c_3/2 > c_1$ or $c_3/2 > c_2$, and we assume by symmetry that $c_3/2 > c_1$. We use $d_1 = c_1$ and $d_2 = c_3 - c_1$. It is immediate that $d_1 > 0$, and $d_1 = c_1 < c_3/2 \leq 1$, since the choice of λ implies $2 - c_3 \geq 0$. Also, $c_3 - c_1 > c_3/2$ and therefore $d_2 > 0$. Also $d_2 = c_3 - c_1 \leq 1$ since the choice of λ implies $2 - c_3 \geq 1 - c_1$. The cut condition is satisfied since in the only non-trivial case, $d_2 = c_3 - c_1 \leq c_2$ since as argued above $c_3 \leq c_1 + c_2$ (as $C_1 \cup C_2$ is a candidate for C_3). As for Constraint 25, from Equation 30 we deduce that indeed $\sum_{j=1}^{q} A_{jS} y_j \leq (1 - c_1) + (1 - (c_3 - c_1)) = \lambda$.

We have checked that in all cases, we get in polynomial time a feasible solution to LP12. Moreover, we found in polynomial time in all four cases a feasible solution to LP11 of the same objective vlaue. Thus we have proven:

Theorem 6. *For two commodities, there exists a polynomial-time algorithm to find a mixed Nash equilibrium for the budgeted attacker/designer game.*

6 Conclusions

We note that our approach for two commodities works whenever an equivalent of Theorem 5 holds for the input graph. It is known that Theorem 5 does not hold for three commodities in undirected graphs, or two commodities in directed graphs.

Acknowledgements. This work was supported in part by NSF grant NeTS-0916743 (ARRA - NSF CNS-0916743).

References

1. Arora, S., Rao, S., Vazirani, U.: Expander flows, geometric embeddings and graph partitioning. J. ACM 56, 5:1–5:37 (2009)
2. Aumann, Y., Rabani, Y.: An $O(\log k)$ Approximate Min-Cut Max-Flow Theorem Approximation Algorithm. SIAM J. Comput. 27(1), 291–301 (1998)
3. Bell, M.G.H.: The measurement of reliability in stochastic transport networks. In: Proceedings of 2001 IEEE Intelligent Transportation Systems (2001)
4. Bohacek, S., Hespanha, J., Lee, J., Lim, C., Obraczka, K.: Game theoretic stochastic routing for fault tolerance and security in communication networks. IEEE/ACM Trans. on Parallel and Distributed Systems (2007)
5. Bohacek, S., Hespanha, J.P., Obraczka, K.: Saddle policies for secure routing in communication networks. In: Proc. of the 41st Conf. on Decision and Contr., pp. 1416–1421 (2002)
6. Bohacek, S., Hespanha, J.P., Obraczka, K., Lee, J., Lim, C.: Enhancing security via stochastic routing. In: Proceedings of the 11th IEEE International Conference on Computer Communications and Networks (2002)
7. Chen, J., Chan, S.-H.G., Li, V.O.K.: Multipath routing for video delivery over bandwidth-limited networks. IEEE Journal on Selected Areas in Communications 22(10), 1920–1932 (2004)

8. Chen, L., Leneutre, J.: On multipath routing in multihop wireless networks: security, performance, and their tradeoff. EURASIP J. Wirel. Commun. Netw. 2009, 6:1–6:13 (2009)
9. Elwalid, A., Jin, C., Low, S., Widjaja, I.: Mate: Mpls adaptive traffic engineering (2001)
10. Gueye, A., Walrand, J.C.: Security in networks: A game-theoretic approach. In: 47th IEEE Conference on Decision and Control, CDC 2008, pp. 829–834 (2008)
11. Hu, T.C.: Multi-commodity network flows. Operations Research 11, 344–360 (1963)
12. Lee, S.J., Gerla, M.: Split multipath routing with maximally disjoint paths in ad hoc networks (2001)
13. Klein, P., Rao, S., Agrawal, A., Ravi, R.: An approximate max-flow min-cut relation for undirected multicommodity flow, with applications. Combinatorica 15(2), 187–202 (1995)
14. Lee, P.P.C., Misra, V., Rubenstein, D.: Distributed algorithms for secure multipath routing. In: Proceedings of IEEE INFOCOM (2005)
15. Leighton, F., Rao, S.: Multicommodity max-flow min-cut theorems and their use in designing approximation algorithms. Journal of the ACM, 787–832 (1999)
16. Linial, N., London, E., Rabinovich, Y.: The geometry of graphs and some of its algorithmic applications. Combinatorica 15(2), 215–245 (1995)
17. Lou, W., Liu, W., Zhang, Y., Fang, Y.: Spread: Improving network security by multipath routing in mobile ad hoc networks. Wireless Networks 15, 279–294 (2009)
18. Mavronicolas, M., Papadopoulou, V., Philippou, A., Spirakis, P.: A Graph-Theoretic Network Security Game. In: Deng, X., Ye, Y. (eds.) WINE 2005. LNCS, vol. 3828, pp. 969–978. Springer, Heidelberg (2005)
19. Mavronicolas, M., Papadopoulou, V., Philippou, A., Spirakis, P.: A network game with attackers and a defender. Algorithmica 51, 315–341 (2008)
20. Mirrokni, V.S., Thottan, M., Uzunalioglu, H., Paul, S.: A simple polynomial time framework for reduced-path decomposition in multi-path routing. In: Proceedings of IEEE INFOCOM, pp. 183–194 (2004)
21. Papadimitratos, P., Haas, Z.J.: Secure message transmission in mobile ad hoc networks. Ad Hoc Networks Journal 1(1), 193–209 (2003)
22. Parissidis, G., Lenders, V., May, M., Plattner, B.: Multi-path routing protocols in wireless mobile ad hoc networks: A quantitative comparison
23. Schrijver, A.: Combinatorial Optimization. Springer, Heidelberg (2003)
24. Spafford, E.H.: Crisis and aftermath. Commun. ACM 32, 678–687 (1989)
25. Washburn, A., Wood, K.: Two-Person Zero-Sum Games for Network Interdiction. Operations Research 43(2), 243–251 (1995)
26. Lye, K.W., Wing, J.M.: Game strategies in network security (2002)

A Network Security Classification Game*

Ning Bao[1], O. Patrick Kreidl[2], and John Musacchio[1]

[1] University of California Santa Cruz, Santa Cruz, CA 95064, USA
{nbao,johnm}@soe.ucsc.edu
[2] BAE Systems–Technology Solutions, Burlington, MA 01803, USA
pat.kreidl@baesystems.com

Abstract. We consider a network security classification game in which a strategic defender decides whether an attacker is a strategic spy or a naive spammer based on an observed sequence of attacks on file- or mail-servers. The spammer's goal is attacking the mail-server, while the spy's goal is attacking the file-server as much as possible before detection. The defender observes for a length of time that trades-off the potential damage inflicted during the observation period with the ability to reliably classify the attacker. Through empirical analyses, we find that when the defender commits to a fixed observation window, often the spy's best response is either full-exploitation mode or full-confusion mode. This discontinuity prevents the existence of a pure Nash equilibrium in many cases. However, when the defender can condition the observation time based on the observed sequence, a Nash equilibrium often exists.

Keywords: network security, classification game, sequential detection.

1 Introduction

In many network security situations, an operator of a network (the defender) may need to discern between different types of attackers. An attempt at espionage needs to be treated differently than an attack by a spammer for instance. Because of this, defenders will want to employ intrusion detection systems and related software to look for attack signatures and/or apply statistical tests. Knowing that a defender is trying to classify attacks, an attacker is likely to change the way they attack in order to make it more difficult to be classified correctly. These games can be quite complicated because they have both asymmetric information and they happen over time. In this work, we consider a very simple model of such a classification game, and with it we extract some key insights.

1.1 Basic Model

The model is illustrated in Fig. 1. The defender faces an attacker of two possible types: a spy or spammer with probabilities p and $1-p$ respectively. The defender has two servers that can be attacked, a File-Server (FS) and a Mail-Server (MS).

* Research supported by AFOSR grant FA9550-09-1-0049.

R. Jain and R. Kannan (Eds.): GameNets 2011, LNICST 75, pp. 265–280, 2012.

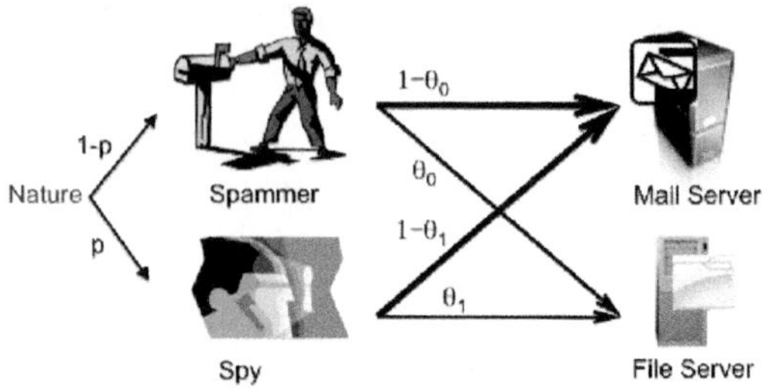

Fig. 1. An illustration of the classification game

We suppose that spammers attack the MS most often because they want to send spam and to get the addresses of potential victims. However, a spammer occasionally hits the FS as he explores the defender's network looking for other potential targets. We suppose time is discrete, and in each period k, a spammer hits the FS with probability θ_0 and otherwise he hits the MS. The attacks are restricted to be i.i.d. Bernoulli in each period. Moreover, we suppose the defender observes the sequence of attacks $z_k \in \{\text{MS}, \text{FS}\}$. Spammers are assumed to be non-strategic, so θ_0 is taken as a fixed parameter.

A spy chooses the frequency with which to hit the FS, which is the target with the information he wants. However, he can strategically choose to hit the MS during some time periods to make it more difficult for the defender to distinguish him from a spammer. We assume that the spy's single choice variable is θ_1, the probability of hitting the FS in any period. We restrict θ_1 to be larger than θ_0. By picking θ_1, the spy commits to attacking the FS according to a Bernoulli process with parameter θ_1. The spy's tradeoff is that if he picks θ_1 too high, then it is easy for the defender to distinguish him from a spammer, while if he lowers θ_1, he reduces the frequency with which he attacks his desired target.

The defender decides in each period whether to classify the attacker as a spammer or a spy, or to continue observing. While observing, the defender incurs a cost c_0 and c_1 for each MS hit by a spammer and FS hit by a spy, respectively, the latter a reward c_1 to the spy. The defender incurs a cost F if he mis-classifies a spammer as a spy. If he mis-classifies a spy as a spammer, we suppose that the spy can then continue to attack with impunity and thus earns a reward equal to the discounted net present value of an endless stream of FS attacks that happen with probability θ_1 in each period. This mis-classification reward to the spy, like the spy's rewards for all preceding FS attacks, appears as a cost to the defender.

1.2 Summary of Analysis and Results

We consider two versions of our classification game, differing in the class of strategies available to the defender. In the first version, the "commit to N" game (Sect. 3), the defender must commit to positive integer N, the number of observations he will take before making a classification. We show that at the end of N periods the defender should employ the well-known Likelihood Ratio Test

(LRT) [1], in which an optimal classification reduces to comparing the number of observed FS attacks to a certain threshold. The second version we call the "dynamic N" game (Sect. 4), as the defender can now decide in each period whether to continue to observe depending on what has been observed so far. Equivalently, the defender chooses a policy that maps observation sequences to control decisions (`continue`, `classify-spy`, `classify-spammer`). We show that the defender's best response to a spy's choice of θ_1 takes the form of the well-known Sequential Probability Ratio Test (SPRT) [2].

In both versions of our game, we focus on finding pure strategy Nash equilibria [3], all players playing pure strategies that are best responses to one another. Existence of such an equilibrium implies that it might be possible for the game to settle to a stable situation in which players behave predictably. In the specific context of our game, if we fix a defender strategy optimized for a particular hypothesis on θ_1 (i.e., if in the commit to N game we fix a particular observation window & LRT threshold or in the dynamic N game we fix a particular pair of SPRT thresholds), then it is possible that the spy's best response is to play with a θ_1 that does *not* match what the defender is expecting: a Nash equilibrium of the game would be a point where the spy's θ_1 and the defender's hypothesis $\hat{\theta}_1$ do match. Through a set of numerical experiments, in each case computing firstly the defender's best response to a hypothesis $\hat{\theta}_1$ and secondly the spy's best response θ_1 to that defender's strategy, we find that the commit to N game often has no pure Nash equilibrium whereas the dynamic N game often does.

1.3 Related Work

There is a growing body of work on attacker-defender security games, and much of it surveyed in [4]. Lye and Wing [5] propose a stochastic game model to study the behaviors of an administrator and an attacker in a Local Area Network (LAN). In a series of papers [6,7,8], Alpcan and Başar introduce a game-theoretic framework to model the interaction between the intrusion detection system (IDS) and attackers. Our game-theoretic framework focuses on attacker classification rather than intrusion detection. Our game also connects to the conclusions in [9], which showed that immediate expulsion is not the best response for all types of attackers. Statistical tests have been widely utilized in intrusion detection problems e.g., Jung and others [10] offer an on-line detection algorithm that identifies malicious port-scans using the seminal sequential hypothesis testing approach by Wald [2]. In the second version of our classification game, a similar but more general version of the Wald problem is studied and Wald's solution serves as the defender's best response function. Nelson and others [11] also study challenges to statistical classification when the defender faces a strategic attacker, but their focus is on vulnerabilities during the training of a classifier.

2 Detailed Model

This section formally describes the model introduced in Sect. 1, which involves prior probability $p \in (0, 0.5]$, the spammer's per-period probability $\theta_0 \in (0, 0.5]$

of a hit on the FS, and the spy's choice variable $\theta_1 \in (\theta_0, 1]$. Fig. 1 illustrates the situation. Also recall the positive-valued costs incurred by the defender: c_0 for each MS attack by a spammer, c_1 for each FS attack by a spy and F for mis-classifying a spammer as a spy. The positive-valued cost of mis-classifying a spy as a spammer will be expressed below in terms of other parameters.

2.1 Cost Functions of Defender and Spy

The cost functions of both players take the form of an expected total discounted cost with discount factor $\delta \in (0,1)$. In the commit to N game, integer N is determined before play begins, and thus is not a function of the defender's observations. Because we assume the consequence of mis-classifying a spy is that the spy continues to attack the FS with a Bernoulli process of parameter θ_1, this mis-classification cost is $\sum_{k=N}^{\infty} \delta^k c_1 \theta_1 = \delta^N c_1 \theta_1/(1-\delta)$. Now define the following two (conditional) probabilities of making an error:

$$\alpha = \mathbf{P}\left[U = 1 \mid X = 0\right] \quad \text{and} \quad \beta = \mathbf{P}\left[U = 0 \mid X = 1\right], \tag{1}$$

where U denotes the classification decision of the defender after N observations and $X \in \{0,1\}$ denotes the true type of the attacker being spammer or spy, respectively. It is standard [1] to call α the *false-alarm* rate (Type-I error probability) and β the *mis-detection* rate (Type-II error probability). Altogether, the defender's expected total discounted cost is given by

$$J^D = p\left\{\beta\delta^N \frac{c_1\theta_1}{1-\delta} + \sum_{k=0}^{N-1} \delta^k c_1\theta_1\right\} + (1-p)\left\{\alpha\delta^N F + \sum_{k=0}^{N-1} \delta^k c_0(1-\theta_0)\right\}, \tag{2}$$

which includes costs incurred due to (i) mis-detection of a spy, (ii) FS attacks by a spy, (iii) false-alarm of a spammer and (iv) MS attacks by a spammer. Similarly, the spy's expected total discounted cost is given by

$$J^A = -\beta\delta^N \frac{c_1\theta_1}{1-\delta} - \sum_{k=0}^{N-1} \delta^k c_1\theta_1, \tag{3}$$

which includes the reward of FS attacks (i) after the defender's mis-detection and (ii) before the defender's classification action.

When the defender's strategy is generalized to allow the observation sequence $(z_0, z_1, \ldots, z_k) \in \{\text{MS}, \text{FS}\}^{k+1}$ to influence when (and not just how) the classification is made, integer N becomes a random variable with distribution depending on both players' strategies. The two players' objectives are essentially the same as expressed by (2) and (3), but with the underlying expectations suitably generalized. Specifically, consider any positive integer n such that $\mathbf{P}\left[N = n\right]$ is nonzero and condition on the event that $(X, N) = (x, n)$. We first generalize the error probabilities in (1) to be $\bar{\alpha}(n) = \mathbf{P}\left[U = 1 \mid X = 0, N = n\right]$ and $\bar{\beta}(n) = \mathbf{P}\left[U = 0 \mid X = 1, N = n\right]$, respectively. We next generalize, for each

$k = 0, 1, \ldots, n-1$, the period-k probability of a FS attack by $\bar{\theta}_x(k|n) = \mathbf{P}\left[Z_k = \text{FS} \mid X = x, N = n\right]$, which we note is not a constant by virtue of conditioning on the event $N = n$. We define

$$G_1^D(n) = \bar{\beta}(n)\delta^n \frac{c_1\theta_1}{1-\delta} + \sum_{k=0}^{n-1} \delta^k c_1 \bar{\theta}_1(k|n),$$

$$G_0^D(n) = \bar{\alpha}(n)\delta^n F + \sum_{k=0}^{n-1} \delta^k c_0 \left[1 - \bar{\theta}_0(k|n)\right],$$

$$G^A(n) = -\bar{\beta}(n)\delta^n \frac{c_1\theta_1}{1-\delta} - \sum_{k=0}^{n-1} \delta^k c_1 \bar{\theta}_1(k|n)$$

to be the defender's cost conditioned on the event $(X, N) = (1, n)$, the defender's cost conditioned on the event $(X, N) = (0, n)$ and the spy's cost conditioned on $N = n$, respectively. Altogether, the two players' cost functions are then defined by a final expectation over the stopping time N i.e.,

$$J^D = p \sum_{n=1}^{\infty} G_1^D(n) \mathbf{P}\left[N = n \mid X = 1\right] + (1-p) \sum_{n=1}^{\infty} G_0^D(n) \mathbf{P}\left[N = n \mid X = 0\right] \quad (4)$$

and

$$J^A = \sum_{n=1}^{\infty} G^A(n) \mathbf{P}\left[N = n \mid X = 1\right]. \quad (5)$$

2.2 Assumed Behavior of Spammer and Spy

Our model makes the assumption that either type of attacker is restricted to hitting his desired target according to a Bernoulli process. This is reasonable for a spammer, who is taken to be non-strategic and happens upon a FS rather than a MS simply by mistake. The restriction of the spy's strategy space to picking the rate of a Bernoulli process is indeed a simplifying assumption, but it has some justifications (and implications) that we now discuss.

First consider an alternative formulation in which the spy's strategy space were the set of all binary sequences (with 1 and 0 corresponding to FS and MS hits respectively). In equilibrium, if a spy were playing a mixed strategy, every sequence he assigns positive probability would have to have the same expected payoff. In our formulation with the strategy restricted to be Bernoulli, any finite sequence will have positive probability, but all sequences will not lead to the same payoffs. Therefore the restriction of the spy's choice to a single variable θ_1 – the parameter of the Bernoulli process – is structurally different than allowing the attacker to choose his sequence directly.

One interpretation is that the spy has a commitment device, like a computer program, that will pick the actual attack sequence once the spy has chosen a θ_1. The restriction of the attacker's strategy space to a single dimension greatly simplifies the game. Moreover the spy can make his attack much less predictable, and thus possibly more effective on average, with a commitment device like this. Another issue is how this commitment can be credible. One possible argument is that if the spy overrode his device when it picked an unfavorable sequence, he might increase his payoff on single attack but then become more predictable in future attacks. Making this argument formal is outside the scope of this paper.

As for using a Bernoulli process, suppose instead the commitment device were made to produce a different random process. In particular, suppose the spy considers any distribution for the commitment device so long as it achieves an expected number of FS hits of $N\theta_1$ over some period N. In any equilibrium (if it exists), the defender would eventually come to know the choice of distribution and play a best response to it. The defender could use this distribution in a likelihood ratio test of the spy vs. spammer hypotheses. The expected log-likelihood ratio, under the hypothesis that the attacker is a spy, is just the K-L divergence between the spy and spammer distributions. It turns out that this quantity is minimized, subject to his FS hit rate constraint, by choosing a Bernoulli process. To make this precise, let $B_N(\theta)$ denote the binomial distribution on length-N binary-valued sequences with success probability $\theta \in (0, 1)$.

Proposition 1. *Distribution $B_N(\theta_1)$ minimizes K-L divergence $D(P||B_N(\theta_0))$ over P subject to the constraint that the expected number of successes is $N\theta_1$.*

The proof is in Appendix A.1. Proposition 1 addresses why a spy might want to use a Bernoulli process during the classification period, but it does not address the issue of why the spy would not change the process after the defender has made a classification decision. This is a more difficult question. For instance, if a spy could know he has just been misclassified, then he would want to change θ_1 to 1 and exploit the defender's mistake. But in game play, the spy would not know this had happened. If he were selecting a best response to a particular choice of N in the commit to N game, he could set his commitment device to increase θ_1 to 1 after time N. While we only allow a defender to classify once in our model, in a real situation a defender would likely reconsider the classification upon detecting such an abrupt change. A much more complex model would be needed to analyze this game. Therefore, we have elected to first understand a simpler model, limiting the spy's best response to a set of stationary policies.

3 Commit to N Game

In the commit to N game, the defender makes his classification decision after a fixed number N of observations, while the spy picks the probability θ_1 of hitting the FS in each attack. The problem (from the defender's point-of-view) turns out to be a standard binary hypothesis test for which we show that, for any N, the applicable Likelihood-Ratio Test (LRT) reduces to a comparison between the number of observed FS attacks and a certain threshold m. In turn, the defender's best response to a hypothesized strategy $\hat{\theta}_1$ is a pair of integers (N, m). Recall that under simultaneous-play assumptions the spy has no obligation (and, in fact, generally has incentives not) to behave as hypothesized by the defender.

A pure Nash equilibrium is a point for which the defender hypothesizes a value for $\hat{\theta}_1$, and designs his observation window N and LRT threshold m accordingly, such that the spy's best response to that integer pair (N, m) yields $\theta_1 = \hat{\theta}_1$.

3.1 Defender's Best Response

For any choice of N, the defender must decide between spammer or spy based on the observed sequence $z^N = (z_0, z_1, \ldots, z_{N-1})$ of server attacks. Given the spy's choice of probability θ_1 (and also values for model parameters p, θ_0, c_0, c_1, F and δ in (2)), this decision becomes equivalent to a binary hypothesis test between two binomial distributions $B_N(\theta)$: a spammer $H_0 : \theta = \theta_0$ versus a spy $H_1 : \theta = \theta_1$. The likelihood ratio is given by

$$L\left(z^N\right) = \frac{\mathbf{P}\left[z^N \mid X = 1\right]}{\mathbf{P}\left[z^N \mid X = 0\right]} = \left(\frac{\theta_1}{\theta_0}\right)^{\bar{z}} \left(\frac{1-\theta_1}{1-\theta_0}\right)^{N-\bar{z}} \tag{6}$$

where $\bar{z}$ denotes the number of FS attacks in the given sequence z^N. By the Neyman-Pearson lemma, a decision rule of the form "reject H_0 if $L(z^N) > M$" subject to a Type-I error probability α is a level-α Uniformly-Most-Powerful (UMP) test [1], achieving for some M the minimum Type-II error probability β (and, in turn, a minimum in (2)) associated with the level α. It is easy to verify that, in the case of (6) with $\theta_1 > \theta_0$, the condition that $L(z^N) > M$ for any M is equivalent to the test $\bar{z} \geq m$ for some integer m. The following proposition provides, for any given N, the minimizing integer threshold m^* in closed form.

Proposition 2. *Fix strategy $\theta_1 \in (\theta_0, 1]$ for the spy. For any observation window N, the defender's optimal decision (with respect to minimizing (2)) is to classify the attacker as a spy if the observed sequence z^N contains a number of FS attacks*

$$\bar{z} \geq m^*(N) = \left\lceil \frac{\log\left(\frac{(1-p)F(1-\delta)}{pc_1\theta_1}\right) - N\log\left(\frac{1-\theta_1}{1-\theta_0}\right)}{\log\left(\frac{\theta_1}{\theta_0}\right) - \log\left(\frac{1-\theta_1}{1-\theta_0}\right)} \right\rceil ;$$

otherwise, he classifies the attacker as a spammer.

The proof is in Appendix A.2. Recall that the defender's best response involves not just the choice of threshold m but also the choice of observation window N. Proposition 2 only tells us how to do the former given the latter, so the best response to a particular spy strategy θ_1 still involves a direct search on N. Moreover, in actual game play, the defender will not know the true value of θ_1. One could view this (for fixed N) as the defender carrying out a one-sided test $H_0 : \theta \leq \theta_0$ vs $H_1 : \theta > \theta_0$. However, by properly choosing an alternative hypothesis $H_1' : \theta = \hat{\theta}_1$, the defender may effectively transform the one-sided test into a simple-vs-simple one. By the Karlin-Rubin theorem [1], the decision rule of Proposition 2 (for any given N) still achieves the smallest mis-detection rate β among all tests with the desired false-alarm rate α. Thus, for the purposes of looking for Nash equilibria, we can view the defender's best response strategy as being characterized by the hypothesis $\hat{\theta}_1$ for which the integer pair $(N(\hat{\theta}_1), m(\hat{\theta}_1))$ is an optimal choice.

3.2 Spy's Best Response

Consider the spy's best response $\theta_1 \in (\theta_0, 1]$ to a given defender's strategy, or integer pair (N, m). Substitution of the achieved mis-detection probability $\beta(N, m) = \sum_{i=0}^{m-1} \binom{N}{i} \theta_1^i (1 - \theta_1)^{N-i}$ into (3) yields a spy's cost function that is an Nth-order polynomial in parameter θ_1. The minimizing argument will thus be either one of $N-1$ roots of the derivative polynomial or the boundary value of $\theta_1 = 1$, a total of up to N possibilities for J^A that can be compared numerically.

3.3 Numerical Experiments

The numerical procedure for each scenario is as follows. We search through a grid of possible $\hat{\theta}_1$ on the interval $(\theta_0, 1]$. For each $\hat{\theta}_1$, we search for the defender's best observation window N, always supposing that the LRT threshold m is chosen according to Proposition 2. This leads to a best response pair (N, m) against which we evaluate the spy's best response θ_1 via minimization of the Nth-order polynomial as described above.[1] Finally, we plot the best response θ_1 against $\hat{\theta}_1$, any point(s) where the two coincide being a Nash equilibrium.

Results for three representative scenarios are shown in Fig. 2, all assuming that (i) probability $p \leq 0.1$, or that spammers are much more common than spies, (ii) probability $\theta_0 = 0.1$, or that spammers mistakenly hit a FS only 10% of the time, (iii) costs $c_0 = 0.01$ and $c_1 = 1$, viewing a spied-FS much more costly than a spammed-MS, and (iv) discount factor $\delta = 0.99$ while cost $F \leq 10$, viewing a mis-detection at least as costly as a false-alarm. Each upper plot shows how the defender's strategy changes with his hypothesis $\hat{\theta}_1$, while each bottom plot shows the spy's best response θ_1 to that defender's strategy. In the first two scenarios, the spy's curve crosses the $\theta_1 = \hat{\theta}_1$ line at a discontinuity, and thus no Nash equilibrium exists; the third is a (not easily found) counter-example to any claim that a Nash equilibrium never exists in the commit to N game.

Examining Fig. 2 in more detail also exposes a confusion/exploitation tradeoff and interesting dynamics between defender and spy with respect to this tradeoff. Consider all three scenarios when $\hat{\theta}_1$ is near θ_0, or when the defender hypothesizes a spy favoring confusion over exploitation: the defender plays $(N, m) = (1, 2)$, corresponding to immediately choosing `classify-spammer`. That is, when the defender finds classification difficult and not worth the time expenditure required, he resorts to a trivial strategy that will ignore observations and rather base classification simply upon the relative costs between the two types of errors. The spy's best response, given that the defender is not even bothering to classify,

[1] If there are cases where there are multiple optimal pairs (N, m) for a given $\hat{\theta}_1$, it could be that some of these pairs support a Nash equilibrium and some do not. In our numerical experiments, however, we encountered no such cases.

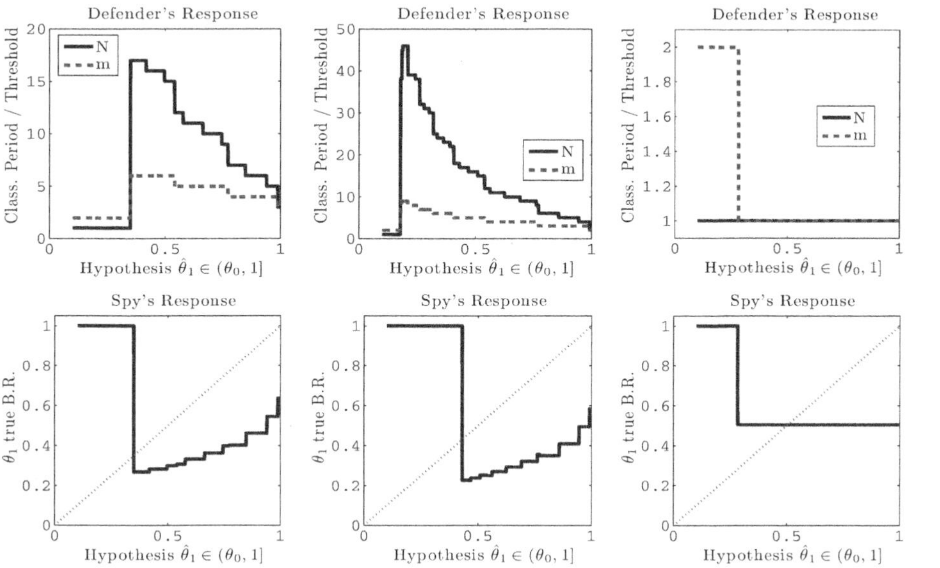

(a) $p = 0.01$ and $F = 10$ (b) $p = 0.1$ and $F = 10$ (c) $p = 0.001$ and $F = 0.08$

Fig. 2. Numerical results for three scenarios of the commit to N game with parameters $\theta_0 = 0.1$, $c_0 = .01$, $c_1 = 1$ and $\delta = 0.99$, while parameters p and F vary as indicated. No Nash equilibrium exists in (a) or (b), but in (c) is a (rare) case with an equilibrium.

is to attack the FS at full rate. Next consider the scenarios as $\hat{\theta}_1$ increases up to where the defender first chooses to employ a non-trivial classification strategy. In the first scenario (with $p = 0.01$), the defender first chooses $N = 17$ at $\hat{\theta}_1 \approx 0.375$ and the spy's best response switches discontinuously to $\theta_1 \approx 0.269$ in order to allow some chance of being misclassified. In the second scenario (with $p = 0.1$) when the defender expects more spies, he first employs nontrivial classification at a lower $\hat{\theta}_1$ and, in turn, chooses a larger N to allow for adequate classification given that presumed stealthier spy; the spy chooses to exploit this long observation window and still attack at full rate, viewing the long stretch of FS attacks worth the sure detection–it is not until $\hat{\theta}_1 \approx 0.431$ that N is chosen small enough for the spy to make that first abrupt switch. In both scenarios, as $\hat{\theta}_1$ increases beyond the occurrence of the spy's first abrupt switch, the defender continues to choose a smaller N because under his hypothesis it is decreasingly difficult to discriminate between the two attackers; with this decreasing power of the defender's test, the spy is indeed choosing to increase his attack rate, but he continues to confuse by staying well below the increasing rates hypothesized by the defender. In the third scenario, when false alarms are very cheap, the first and only nontrivial classification strategy remains very simple as $\hat{\theta}_1$ increases: immediately classify spammer or spy in accordance with whether one attack is against the MS or FS, respectively. The spy, in turn, holds steady at $\theta_1 \approx 0.5$, giving rise to the equilibrium at $\theta_1^* \approx 0.5$ with $N(\theta_1^*) = m(\theta_1^*) = 1$.

A final point concerns the restriction that the spy picks θ_1 greater than θ_0. In the scenarios we discussed, this constraint was never active—the spy would not have picked $\theta_1 \leq \theta_0$ even if it were allowed. There are scenarios, particularly with larger θ_0, where the constraint could become active. Allowing $\theta_1 \in [0, 1]$ only modestly complicates the game: the defender's best response to $\theta_1 < \theta_0$ still uses Proposition 2 but with the `classify-spy` decision made if $\bar{z} < m^*(N)$; in turn, the spy's best response to $\hat{\theta}_1 < \theta_0$ involves substitution of $\beta(N, m) = \sum_{i=m}^{N} \binom{N}{i} \theta_1^i (1 - \theta_1)^{N-i}$ into (3), which remains an Nth-order polynomial in θ_1. Here (and in the next section), we considered only $\theta_1 > \theta_0$ for ease of exposition.

4 Dynamic *N* Game

In this section, we remove the restriction that the defender commits to a fixed observation window. As in the famous Wald problem [2], the number of observations N before classification depends not just on the two players' strategies but also on the particular observation sequence $z^N = (z_0, z_1, \ldots, z_{N-1})$. While our problem (from the defender's point-of-view) turns out to be a minor variation of the famous Wald problem, we show that the defender's best response, given θ_1 chosen by the spy, is in the form of a Sequential Probability Ratio Test (SPRT). The spy's choice variable is θ_1 just as it is for the commit to N game.

4.1 Defender's Best Response

Let us first show that, given the spy's choice of probability θ_1 (and also values for model parameters p, θ_0, c_0, c_1, F and δ in (4)), the defender can access a best response strategy in the family of Wald's SPRT solutions [2]. An SPRT strategy in each period k can be parametrized by two probability thresholds we will denote by $\eta_k \in [0, 1]$ and $\xi_k \in [\eta_k, 1]$ i.e., initialize probability $b_{-1} = p$ and, in each period $k = 0, 1, 2 \ldots$, first apply the probabilistic state recursion

$$\mathbf{P}\left[X = 1 \mid z^{k+1}\right] \equiv b_k = \begin{cases} \dfrac{(1-\theta_1)b_{k-1}}{(1-\theta_0)(1-b_{k-1}) + (1-\theta_1)b_{k-1}} & \text{, if } z_k = \text{MS} \\[2ex] \dfrac{\theta_1 b_{k-1}}{\theta_0(1-b_{k-1}) + \theta_1 b_{k-1}} & \text{, if } z_k = \text{FS} \end{cases} \tag{7}$$

and then choose to `classify-spammer` if $b_k \leq \eta_k$, to `classify-spy` if $b_k \geq \xi_k$, and to `continue` otherwise. Much is known for Wald's problem: for instance, under the criterion to minimize the expected infinite-horizon total cost, the optimal SPRT thresholds are stationary (i.e., neither η nor ξ varies its value with period k) [12].

The defender's problem in our model is similar to Wald's problem in most ways that general results in the field of stochastic dynamic programming have been organized i.e., both are infinite-horizon optimal stopping problems involving a partially-observable two-state system with bounded cost per stage [12]. One difference is that Wald's problem uses an expected total cost criterion without discounting. The other difference is that our single-stage cost of taking another observation also depends on the (unobservable) type of attacker. Even so, the defender's best response strategy remains in the set of stationary SPRTs.

Proposition 3. *Fix strategy $\theta_1 \in (\theta_0, 1]$ for the spy. A best response strategy for the defender (with respect to minimizing (4)) exists in the set of all stationary SPRT policies i.e., the set of all lower and upper thresholds $\eta \in [0,1]$ and $\xi \in [\eta, 1]$ applied in every period k to the probabilistic state b_k as described in (7).*

The proof is in Appendix A.3. While Proposition 3 gives us the form of the defender's response function, computing the actual SPRT thresholds via dynamic programming can only be done approximately because of the need to discretize the probabilistic state space $[0, 1]$. A uniform discretization is always an alternative, but in certain problem instances a non-uniform discretization, favoring finer intervals in some sub-regions (e.g., around the concentration of the probabilistic state distribution, around the boundaries of the optimal thresholds), can significantly improve solution accuracy. Recognizing the defender's best response model as a special case of the well-studied Partially Observable Markov Decision Process (POMDP), we leverage a publicly available POMDP solver's implementation of such a non-uniform discretization.[2] Moreover, in actual game play, the defender will not know the true spy strategy but rather optimize SPRT thresholds and evolve the probabilistic state b_k via (7) based on a hypothesis $\hat{\theta}_1$.

4.2 Spy's Best Response

Consider the spy's best response to a given defender's strategy, or a given hypothesis $\hat{\theta}_1$ and the associated lower and upper thresholds $\eta(\hat{\theta}_1)$ and $\xi(\hat{\theta}_1)$ in the SPRT. A direct optimization of J^A in (5) over θ_1 is more complicated than was the case in the commit to N game. This is essentially because of the expectation with respect to N, whose distribution cannot be derived in closed form. Another complication is when the spy chooses $\theta_1 \neq \hat{\theta}_1$, the defender is not only employing sub-optimal SPRT thresholds but is also erroneously evolving his probabilistic state. The latter cannot be captured in the tractable dynamic programming reduction of Wald's problem.

Our approximation of the spy's best response is found by numerical searching over $\theta_1 \in (\theta_0, 1]$. For each value of θ_1, the spy's cost is found by constructing a finite-state Markov chain representation that exploits two key properties of the defender's SPRT strategy. Firstly, the probabilistic state recursion in (7) can (until classification) be equated with a random walk along the real line involving the defender's log-likelihood ratio (LLR)

$$R_k = \log\left(L(z^{k+1})\right) = \begin{cases} R_{k-1} + \log\left(\frac{1-\hat{\theta}_1}{1-\theta_0}\right) & \text{, if } z_k = \text{MS} \\ R_{k-1} + \log\left(\frac{\hat{\theta}_1}{\theta_0}\right) & \text{, if } z_k = \text{FS} \end{cases},$$

[2] Cassandra's implementation (see http://www.pomdp.org) of the "witness" algorithm [13] suited our setup particularly well.

starting from the origin $R_{-1} = 0$. In turn, the SPRT thresholds (and prior probability p) determine the segments of the real-line corresponding to the three control actions available to the defender i.e., choose to `classify-spammer` if $R_k \leq \log\left(\frac{(1-p)\eta(\hat{\theta}_1)}{p[1-\eta(\hat{\theta}_1)]}\right)$, to `classify-spy` if $R_k \leq \log\left(\frac{(1-p)\xi(\hat{\theta}_1)}{p[1-\xi(\hat{\theta}_1)]}\right)$, and to `continue` otherwise. Secondly, the spy's strategy θ_1 alters the statistics of this random walk (but *not* the increments $R_k - R_{k-1}$ themselves, which derive from $\hat{\theta}_1$), where lower (higher) values increase the chances that the LLR first exits the continue region at the lower (upper) end of the real-line. The Markov chain defines $Q + 3$ states, Q of them indexing the levels of a uniform quantization of the LLR continue region, one indexing an initial state and two indexing terminal states (one per classify decision). The transition probabilities reflect not only the true θ_1 and the increments $R_k - R_{k-1}$ of the defender's LLR walk based on hypothesis $\hat{\theta}_1$, but also the noise introduced by the quantization. The transition costs reflect the spy's rewards from file-server attacks and evading detection. A cost function is then defined on the state space by writing one-step calculations for each state. By solving this system of linear equations, we find the cost associated with the origin.

4.3 Numerical Experiments

The numerical procedure is analogous to that described for the commit to N game. For each hypothesis $\hat{\theta}_1$, we firstly employ the POMDP solver to obtain a particular SPRT threshold pair (η, ξ) for the defender and, secondly, employ the method described above to obtain a particular attack rate θ_1 for the spy. Any point(s) on the spy's response curve satisfying $\theta_1 = \hat{\theta}_1$ is (within the approximations discussed above) a Nash equilibrium. The results to be discussed used $Q = 100$ in the spy's response approximation and a precision of 0.001 to identify equilibria.

Fig. 3 shows results for three scenarios, the first two also considered for the commit to N game in Fig. 2. The key difference here is the smoother confusion versus exploitation tradeoff exhibited in the spy's best response curves. For example, let us compare the first scenario more closely. For hypotheses $\hat{\theta}_1$ closest to θ_0, the defender's response in both games are such that a spammer classification is made immediately regardless of the first observation; in turn, the spy's response is to hit the FS at every opportunity. However, in the dynamic N game, the defender's response first moves away from this trivial classification at a smaller $\hat{\theta}_1$ than in the commit to N game; at this point the spy's response also shifts away from his full rate but, different from the commit to N game, to rates $\theta_1 > \hat{\theta}_1$ that still exploit the confusion-oriented defense but not with full strength. For hypotheses $\hat{\theta}_1$ well away from θ_0, where an exploitative spy is anticipated, in both games the defender allows for time to reliably classify; in turn, the spy's response is to confuse (i.e., choose $\theta_1 < \hat{\theta}_1$) and better evade detection. Only the dynamic N game features a smooth transition between these two ends of play, at equilibrium $\theta_1^* \approx 0.244$ neutralizing the spy's incentive to either confuse an exploitation-oriented defense or to exploit a confusion-oriented defense.

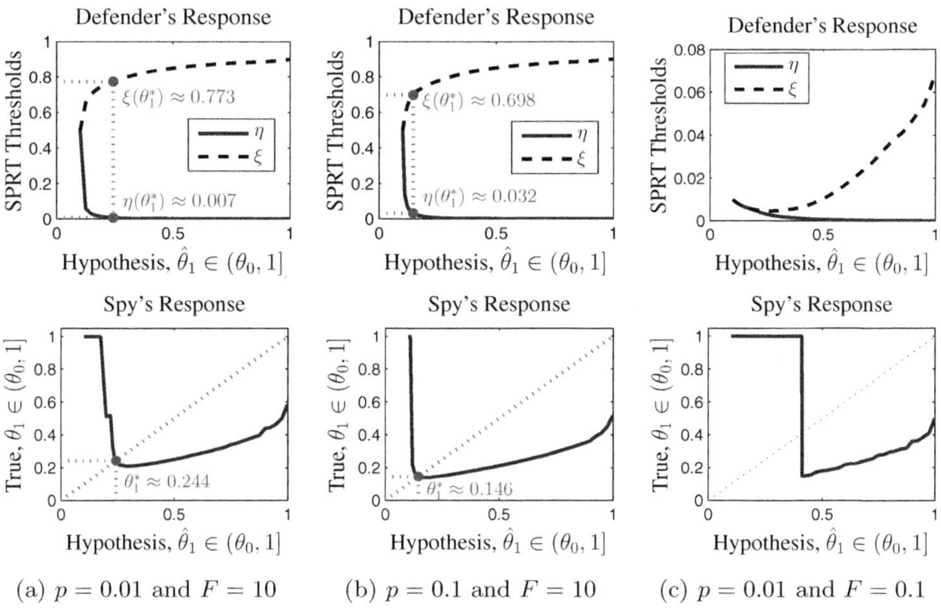

Fig. 3. Numerical results for the dynamic N game in three scenarios, using the same parameters as in Fig. 2 (for the commit to N game) except for p and F in (c). A Nash equilibrium exists in (a) and (b), but in (c) is a (rare) case with no such equilibrium.

Comparison between the two games is similar in the second scenario (with larger p), finding in the dynamic N game a lower equilibrium point than that found in the first scenario. The third scenario is a (not easily found) counter-example to any claim that a Nash equilibrium always exists in the dynamic N game.

5 Conclusion

This work developed a classification game in the network security context. The defender tries to reliably classify the attackers (spammer or spy) while controlling the damage during evidence gathering. A strategic spy faces the trade-off between (i) exploiting the defender's observation time by attacking aggressively and (ii) confusing the defender by mixing attacks and enjoying the benefits of mis-detection. The frequent non-existence of pure Nash equilibrium in our commit to N game suggests that an over-simplified strategy adopted by the defender will prohibit the emergence of a stable situation where both players behave predictably. This problem is mitigated by allowing the defender to make decisions in every period as in our dynamic N game, which dis-incentivizes the spy's response from shifting drastically between exploitation mode and confusion mode.

Our game provides a new perspective to study the classification problem in network security, capturing many subtle yet important interplays between defender and attackers. The results to date are largely empirical, and we plan to explore a number of theoretical questions in future work. For the commit to N game, we desire a proof for the frequent non-existence of a pure Nash equilibrium, providing more insight to the limitations of this restriction on the defender's strategy; for the dynamic N game, it would be useful to isolate precise conditions on the game parameters for a pure Nash equilibrium to exist. Many model extensions are also possible, such as richer action spaces (e.g., sandboxing by the defender and disconnecting by the spy) or continuous-time variants.

Acknowledgment. The authors thank Gregory Frazier, Patrick Loiseau and Jean Walrand for conversations about this work, and Lemonia Dritsoula for help with the experiments.

References

1. Casella, G., Berger, R.: Statistical Inference. Duxbury Press (2002)
2. Wald, A.: Sequential Analysis. Wiley (1947)
3. Fudenberg, D., Tirole, J.: Game Theory. MIT Press (1991)
4. Alpcan, T., Başar, T.: Network Security: A Decision and Game-Theoretic Approach. Cambridge University Press (2011)
5. Lye, K., Wing, J.: Game Strategies in Network Security. Int. J. Information Security 4, 71–86 (2005)
6. Alpcan, T., Başar, T.: A Game Theoretic Approach to Decision and Analysis in Network Intrusion Detection. In: 42nd IEEE Conf. Decision and Control, pp. 2595–2600 (2003)
7. Alpcan, T., Başar, T.: A Game Theoretic Analysis of Intrusion Detection in Access Control Systems. In: 43nd IEEE Conf. Decision and Control, pp. 1568–1573 (2004)
8. Alpcan, T., Başar, T.: An Intrusion Detection Game with Limited Observations. In: 12th Int. Symp. Dynamic Games and Applications (2006)
9. Bao, N., Musacchio, J.: Optimizing the Decision to Expel Attackers from an Information System. In: Allerton Conf. on Comms., Control and Computing (2009)
10. Jung, J., et al.: Fast Portscan Detection Using Sequential Hypothesis Testing. In: IEEE Symp. Security and Privacy (2004)
11. Nelson, B., et al.: Misleading Learners: Co-opting Your Spam Filter. In: Tsai, J., Yu, P. (eds.) Machine Learning and Cyber Trust, pp. 17–51. Springer, Heidelberg (2009)
12. Bertsekas, D.P.: Dynamic Programming and Optimal Control, vols. 1 & 2. Athena Scientific (1995)
13. Kaebling, L., Littman, M., Cassandra, A.: Planning and Acting in Partially Observable Stochastic Domains. Artificial Intelligence 101, 99–134 (1998)
14. Boyd, S., Vandenberghe, L.: Convex Optimization. Cambridge University Press (2004)

Appendices

A.1 Proof of Proposition 1: The problem for sequences of length N is

$$\min_{P(\cdot)} \sum_{s\in\mathcal{S}} P(s) \log\left(\frac{P(s)}{\binom{N}{\bar{s}}\theta_0^{\bar{s}}(1-\theta_0)^{N-\bar{s}}}\right)$$

$$\text{s.t.} \sum_{s\in\mathcal{S}} P(s) = 1, \qquad \sum_{s\in\mathcal{S}} P(s)\bar{s} = N\theta_1, \qquad \text{and} \quad P(s) \geq 0 \ \ \forall s \in \mathcal{S}. \tag{8}$$

Here, $\bar{s}$ denotes the number of successes in any sequence s and S the set of all length-N sequences. This is a convex optimization problem (by convexity of K-L divergence) and the equality constraints are affine, in which case the KKT conditions are both necessary and sufficient for a global optimum [14]. Thus, distribution P is optimal if and only if

$$\log P(s) = \bar{s}\log(\theta_0) + (N-\bar{s})\log(1-\theta_0) + \mu + \lambda\bar{s} - \gamma_s - 1, \ \gamma_s \geq 0, \ \gamma_s P(s) = 0$$

and constraints (8) are met. The possible solution $P = B_N(\theta_1)$ with appropriately chosen KKT multipliers satisfies these conditions.

A.2 Proof of Proposition 2: With probability θ_1 and N fixed, each choice of integer m in a rule of the form $\bar{z} \geq m$ leads to a particular pair of error probabilities α_m and β_m. (For instance, $m = 0$ corresponds to "always classify as spy" and thus $\alpha = 1$ and $\beta = 0$.) To achieve values of α that do not correspond to integer m, one can introduce randomized decision rules, effectively mixing between the thresholds of m and $m+1$ for which $\alpha_{m+1} < \alpha < \alpha_m$. Thus, for a given N, the curve of achievable pairs of (α, β) (i.e., the "error curve") form a piecewise-linear curve. By writing the slope of each line segment explicitly, one can show that this error curve is also convex. Because the defender's objective J^D in (2) is linear in both α and β, the fundamental theorem of linear programming [14] implies that an optimal point always occurs at one of the vertices of the error curve. (In the degenerate case, when the slope of an error curve segment exactly matches the equal cost contours of J^D, a randomized decision rule can also be optimal, but gains nothing over the deterministic rule at either vertex).

The setting is illustrated in Fig. 4(a). Now consider the slope of a line perpendicular to the gradient of J^D (i.e., solve for $\frac{d\beta}{d\alpha}$ in the equation $\nabla_\alpha J^D = 0$). This slope can be no steeper (and no shallower) in magnitude than the slope of the error curve's segment to the left (and to the right, respectively) of m^*, so

$$\left|\frac{\beta_{m^*} - \beta_{m^*-1}}{\alpha_{m^*} - \alpha_{m^*-1}}\right| \leq \frac{(1-p)F(1-\delta)}{p\theta_1 c_1} \leq \left|\frac{\beta_{m^*+1} - \beta_{m^*}}{\alpha_{m^*+1} - \alpha_{m^*}}\right|. \tag{9}$$

The log of the term on the right, after replacing m^* by m, is $m\log(\theta_1/\theta_0) + (N-m)\log((1-\theta_1)/(1-\theta_0))$. Equating this expression to the log of the middle term in (9), one can solve for an m that may not be integer. But with $\theta_1 > \theta_0$ this expression is monotone increasing in m, so taking $m^* = \lceil m \rceil$ insures that the right-side inequality of (9) is satisfied. Similar reasoning shows that this choice $m^* = \lceil m \rceil$ also satisfies the left-side inequality of (9).

A.3 Proof of Proposition 3: From Sect. 5.4 in Vol. 1 of [12], the imperfect state information problem involving a two-state system can be reduced to a perfect state information problem involving the probabilistic state recursion in (7). Like the partially-observable problem, the reformulated problem is an infinite-horizon discounted problem with bounded cost per stage, so from Sect. 1.2 in Vol. 2 of [12] the Bellman equation for all $b \in [0, 1]$ specializes to

$$J^*(b) = \min \left\{ b\frac{\theta_1 c_1}{1-\delta}, (1-b)F, g_C(b) + \delta \mathbf{E}\left[J^*(f_C(b, Z))\right] \right\}$$

in which $g_C(b) = (1-b)(1-\theta_0)c_0 + b\theta_1 c_1$, the function f_C denotes the recursion of (7) and the expectation is with respect to the (mixed Bernoulli) distribution $\mathbf{P}[Z = z] = (1-b)\mathbf{P}[Z_k = z \mid X = 0] + b\mathbf{P}[Z_k = z \mid X = 1]$. This is the same Bellman equation obtained for the standard infinite-horizon Wald problem (see Sect. 3.4 in Vol. 2) *except* that $g_C(b)$ is affine in b (rather than just a constant) and δ is not unity. It is easy to see that $J^*(0) = J^*(1) = 0$, and that $J^*(b)$ is bounded above by $\min\left\{b\frac{c_1\theta_1}{1-\delta}, (1-b)F\right\}$ on $[0, 1]$. All arguments that lead to concavity of $J^*(b)$ on $[0, 1]$ also still hold i.e., starting with $J^0(b) = 0$, function J^* is viewed as the point-wise limit of a sequence of functions $\{J^k\}$ resulting from repeated iterations of the Bellman equation in which (i) the monotonicity property of dynamic programming ensures that $J^k(b) \geq J^{k-1}(b)$ on $[0, 1]$ and (ii) Sect. 5.5 in Vol. 1 ensures that if $J^{k-1}(b)$ is concave on $[0, 1]$ then so is $\mathbf{E}\left[J^{k-1}(f_C(b, Z)\right]$ given f_C and $\mathbf{P}[Z = z]$ above, and thus so is $J^k(b)$. In turn, the reasoning to optimality of a stationary SPRT also still holds; see Fig. 4(b).

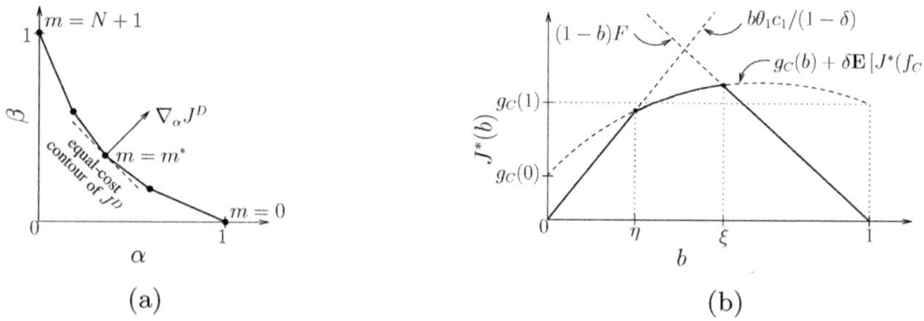

Fig. 4. Illustrations of (a) the error curve discussed in Appendix A.2 and (b) the optimal cost-to-go function discussed in Appendix A.3

Convergence Dynamics of Resource-Homogeneous Congestion Games*

Richard Southwell and Jianwei Huang

Information Engineering Department, The Chinese University of Hong Kong
Shatin, New Territories, Hong Kong
{richardsouthwell254,jianweihuang}@gmail.com

Abstract. Many resource sharing scenarios can be modeled as congestion games. A nice property of congestion games is that simple dynamics are guaranteed to converge to Nash equilibria. Loose bounds on the convergence time are known, but exact results are difficult to obtain in general. We investigate congestion games where the resources are homogeneous but can be player-specific. In these games, players always prefer less used resources. We derive exact conditions for the longest and shortest convergence times. We also extend the results to games on graphs, where individuals only cause congestions to their neighbors. As an example, we apply our results to study cognitive radio networks, where selfish users share wireless spectrum opportunities that are constantly changing. We demonstrate how fast the users need to be able to switch channels in order to track the time-variant channel availabilities.

Keywords: congestion game, resource allocation, cognitive radio, games on graphs, convergence time.

1 Introduction

Congestion games can be used to model a myriad of systems in biology, engineering, and the social sciences. In congestion games, players select resources to maximize their payoffs, while considering the congestion due to resource sharing. A key feature of congestion games is the *finite improvement property*: asynchronous player updating (where players switch resources to increase payoffs) always converges to a Nash equilibrium [1]. The finite improvement property is important due to the wide applicability of congestion games [2]. In many situations, such as drivers choosing routes or wireless users picking channels, it is useful to know that distributed and selfish behaviors always lead to a stable system state.

In this paper, we shall focus on the following question: *how long does convergence take in a congestion game?* Understanding this issue is critical for real-time resource allocation. Unfortunately, determining the convergence time

* This work is supported by the General Research Funds (Project Number 412509) established under the University Grant Committee of the Hong Kong Special Administrative Region, China.

R. Jain and R. Kannan (Eds.): GameNets 2011, LNICST 75, pp. 281–293, 2012.

of congestion games can be a computationally hard problem [3], [4]. The worst case convergence time can be exponential in the number of players n [5], although a polynomial convergence is possible for certain games [6], [7].

In this paper, we will analyze the convergence speed of a simple kind of congestion game, where *every resource has the same intrinsic value.* The players may have complicated and *different* payoff functions, but this does not affect the convergence dynamics. All that matters is that players prefer less congested resources. Despite the simplicity, this type of game can model a wide range of scenarios where selfish individuals share homogenous resources. In Section 5, we will illustrate why our games are particularly useful for analyzing cognitive radio networks.

To study the convergence of congestion games, the usual approach is to gain *loose bounds* by constructing a potential function. The special structure of our games allows us to take a more geometric approach, and derive many new (and sometimes *exact*) results about convergence (see Section 3). In Section 4, we will analyze more general congestion games, where the players are spatially distributed and can access different sets of resources.

Our main results include the following:

- An *exact* characterization of the *fastest* convergence time from *any* given initial state (Section 3.1);
- A *simple* updating mechanism, which *guarantees* the fastest convergence time (Section 3.1);
- *Exact* results and a *linear* bound on the slowest possible convergence time (Section 3.3);
- Characterizations of the convergence speed to Nash equilibrium on spatially extended congestion games (Section 4).

As a concrete application, in Section 5 we discuss how the analytical framework can be applied to spectrum sharing in cognitive radio networks. As the available resources (channels) come and go rapidly in cognitive radio networks (e.g., [8,9]), it is critical to understand how selfish users behave and whether they can adapt fast enough compared with the environment. **Proofs of our results are included in the online technical report [10].**

2 The Model

In an n-player congestion game, each player's strategy involves a set of resources. The payoff that a player p gets from using a resource i is described by a strictly decreasing function $f_i^p(x_i)$, where x_i is the number of players using i. In this paper, we are only concerned with *singleton congestion games*, where each player uses exactly one resource at any given time [6]. We further assume that resources are symmetric, i.e., $f_i^p = f_j^p = f^p$ for any resources i, j and any player p, so all resources are equivalent from a player's perspective. Note, however, that the payoffs are player-specific. This means players could have different tastes of the same resource, due to technological, psychological, or economic reasons.

A central idea in congestion game dynamics is the **better response switch**, i.e., a player increases its payoff by switching to a better (less congested) resource. A common way to study congestion games is to imagine that players' strategies evolve through time via asynchronous better response switches (i.e., *one player switches to a better decision at each discrete time step*).[1] When no player can increase their payoff by switching, the system reaches a *Nash equilibrium.*

To summarize, our systems are defined by a set of n players, a set of r resources, and a strictly decreasing payoff function f^p for each player p. Every player uses one resource from the set $\{1, 2, ..., r\}$ at any time step. The system *state vector* is $\boldsymbol{x} = (x_1, x_2, ..., x_r)$, where x_i is the number of players using resource i. If a player using resource i switches to resource j, then x_i decreases by 1 and x_j increases by 1. We refer to this action as an $i \to j$ *switch.* We suppose that one and only one player switches to a better response every time step. A key fact about our games is Theorem 1.

Theorem 1 (Better Response Switch)
A switch $i \to j$ *is a better response switch if and only if* $x_j + 1 < x_i$.

Proof. Consider a player p using resource i with a payoff $f^p(x_i)$. Switching to j is a better response if and only if $f^p(x_j + 1) > f^p(x_i)$. Now, since f^p is strictly decreasing, we have $f^p(x_j + 1) > f^p(x_i)$ if and only if $x_j + 1 < x_i$.

Theorem 1 has powerful implications: the better response switches (through which our system evolves) are independent of the payoff functions and the identity of the players. All that matters (with respect to the dynamics) is that one player decreases its congestion level by switching. A state $\boldsymbol{x}$ is a Nash equilibrium if and only if no better response switches can be performed.

Theorem 2 (Nash Equilibrium). *A state vector* $\boldsymbol{x}$ *is a Nash equilibrium if and only if* $(n \bmod r)$ *of* $\boldsymbol{x}$*'s entries are equal to* $\lceil \frac{n}{r} \rceil$, *whilst the remaining* $r - (n \bmod r)$ *entries of* $\boldsymbol{x}$ *are equal to* $\lfloor \frac{n}{r} \rfloor$.

3 Convergence Time to a Nash Equilibrium

Starting from an arbitrary state $\boldsymbol{x}$, the players can reach a Nash equilibrium through several routes (sees figures 1, 2 and 3). This means the convergence time depends upon the ways players choose to switch. The convergence time is an important measure of how quickly the players organize themselves. Since players' identities and payoff functions are irrelevant to the convergence dynamics, we will only work with the number of players n, the number of resources r, and system state $\boldsymbol{x}$ from now on.

[1] We are subscribing to the *elementary step hypothesis* [5], that *one and only one* player improves their strategy at each time step. This is commonly used to model situations where simultaneous updating is unlikely.

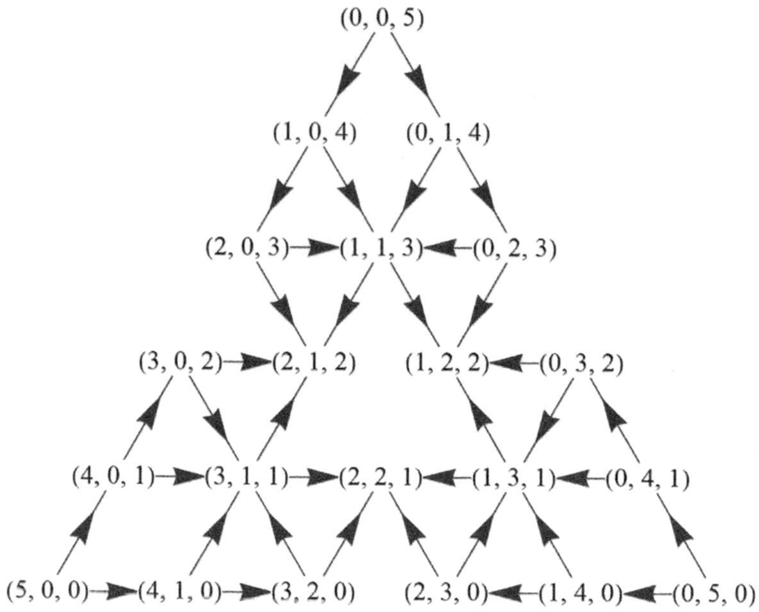

Fig. 1. The state space of a game with $n = 5$ players and $r = 3$ resources. The points represent state vectors and the arrows represent the *state transitions* which can be achieved through better response switches. For example, there is an arrow from $(5, 0, 0)$ to $(4, 1, 0)$ because the better response switch $1 \to 2$ converts $(5, 0, 0)$ into $(4, 1, 0)$. The Nash equilibria in this game are $(1, 2, 2)$, $(2, 1, 2)$, and $(2, 2, 1)$.

3.1 The Fastest Convergence

We want to determine the best way a group of players can switch their choices of resources to reach a Nash equilibrium. We will first study how many switches (*of any kind*) are required to convert one state into another.

Theorem 3 (The Switching Distance). *For any two states $\boldsymbol{x}$ and $\boldsymbol{y}$, the minimal number of switches (of any kind), $d(\boldsymbol{x}, \boldsymbol{y})$, required to convert $\boldsymbol{x}$ into $\boldsymbol{y}$ is*

$$d(\boldsymbol{x}, \boldsymbol{y}) = \sum_{i=1}^{r} \max\{0, y_i - x_i\} = \sum_{i=1}^{r} \max\{0, x_i - y_i\} = \frac{\sum_{i=1}^{r} |x_i - y_i|}{2}.$$

The **fastest convergence time**, $b(\boldsymbol{x})$, is the least number of better response switches that are required to convert state $\boldsymbol{x}$ into a Nash equilibrium.

Theorem 4 (Fastest Convergence Time). *For any state $\boldsymbol{x}$, the best (fastest) convergence time to a Nash equilibrium is*

$$b(\boldsymbol{x}) = \left(\sum_{i=1}^{r} \max\left\{0, x_i - \left\lfloor \frac{n}{r} \right\rfloor \right\} \right) - \min\left\{ \left| \left\{ i \in \{1, 2, ..., r\} : x_i \geq \left\lceil \frac{n}{r} \right\rceil \right\} \right|, n \bmod r \right\}.$$

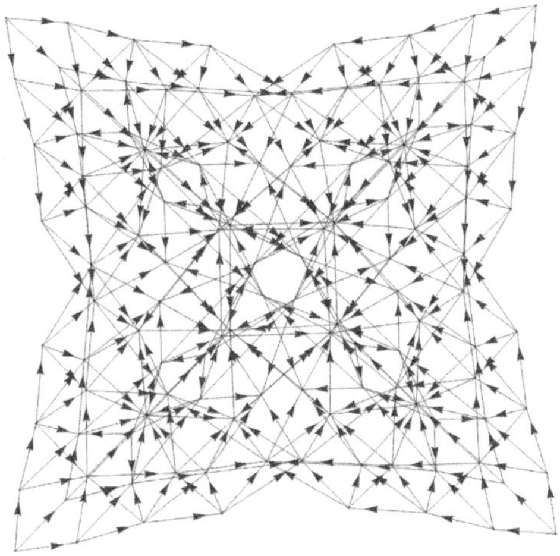

Fig. 2. The state space of a game with $n = 7$ players and $r = 4$ resources. The points represent state vectors and the arrows represent the *state transitions* which can be achieved through better response switches.

Fig. 3. The state space of a game with $n = 5$ players and $r = 5$ resources. The points represent state vectors and the arrows represent the *state transitions* which can be achieved through better response switches. Every path of sufficient length in this graph terminates at the central vertex -which represents the Nash equilibrium $(1, 1, 1, 1, 1)$.

Theorem 4 essentially states that $b(\boldsymbol{x})$ *is the minimal number of switches (of any kind) required to convert* $\boldsymbol{x}$ *into a Nash equilibrium.* To explain how a Nash equilibrium can be reached in this minimal number of switches, we define the strong switch. Suppose $\boldsymbol{x}$ is not a Nash equilibrium. We say an $i \to j$ switch is a **strong switch** of $\boldsymbol{x}$ if x_i is one of $\boldsymbol{x}$'s maximal entries and x_j is one of $\boldsymbol{x}$'s minimal entries. For example, the $2 \to 4$ switch is a strong switch of $\boldsymbol{x} = (5,5,3,1)$, while the $2 \to 3$ switch is not. Theorem 4 is proved by showing that strong switching is *optimal*, in the sense that it leads to a Nash equilibrium within the minimum possible number of switches.

Corollary 1. *Starting from an arbitrary state* $\boldsymbol{x}$*, any sequence of* $b(\boldsymbol{x})$ *strong switches leads to a Nash equilibrium.*

Theorem 4 implies that the maximum value of $b(\boldsymbol{x})$ occurs when all the players use the same resource (e.g., $\boldsymbol{x} = (n,0,0,...,0)$). Hence $b((n,0,0,...,0)) = n - \lceil \frac{n}{r} \rceil$ is the largest number of better response switches ever *required* to reach a Nash equilibrium.

3.2 Average Fastest Convergence from Random Initial Conditions

Often we do not have a choice of the initial system state. Thus it is useful to understand how fast we can reach a Nash equilibrium from random initial conditions. This gives a *global average* measure of the best case performance of our systems. Suppose the players select their initial resources from $\{1,2,...,r\}$ uniformly at random. In this case, we let $\beta_r(n)$ denote the expected number of strong switches required to reach Nash equilibrium (for a game with r resources and n players). In other words, $\beta_r(n)$ is the expected value of $b(\boldsymbol{x})$, when $\boldsymbol{x}$ is generated by allocating resources randomly and uniformly.

Theorem 5 (Average Fastest Convergence Time)
Suppose n is divisibly by r, in this case[2] *we have,*

$$\beta_r(n) = \left(1 - \frac{1}{r}\right)^{n-\frac{n}{r}} \left(\frac{1}{r}\right)^{\frac{n}{r}} \left(n - \frac{n}{r}\right) \frac{n!}{(n-\frac{n}{r})!\frac{n}{r}!},$$

$$\lim_{n\to\infty} \beta_r(n) = \sqrt{\frac{n(r-1)}{2\pi}}.$$

3.3 The Slowest Convergence

In reality, players may not switch in the optimal way. Often it is more likely that random players perform random better response switches. In this case, there are many ways the system can evolve. Here we discuss the **slowest convergence time** $w_r(n)$ for a system with r resources and n players. We define $w_r(n)$ to be largest number of better response switches that drives some initial state into a Nash equilibrium.

[2] There is a strong evidence that this asymptotic form is also accurate when n is not divisible by r (see technical report [10]).

Theorem 6 (Slowest Convergence Time). *The worst (slowest) convergence time, $w_r(n)$, has the following properties;*

1. $\frac{n(r-1)}{2} - \left(\frac{2.r^3-3.r^2+r}{6}\right) \leq w_r(n) \leq \frac{n(r-1)}{2}$.
2. $\lim_{n\to\infty} w_r(n) = \frac{n(r-1)}{2}$.
3. $w_2(n) = \lfloor \frac{n}{2} \rfloor$.
4. $w_3(n) = n - 1$.
5. $w_4(n) = \delta_{n,1} + \delta_{n,2} + \frac{3.n}{2} - (n \mod 2)\frac{5}{2} - (n+1 \mod 2)\left(2 + \left(\lfloor \frac{n}{2} \rfloor \mod 2\right)\right)$ *where $\delta_{i,j}$ is the Kronecker delta.*
6. *If $n \leq r$ then* $w_r(n) = \sum_{k=0}^{n-1} \left\lfloor \frac{\sqrt{1+8k}-1}{2} \right\rfloor$.

Our state space can be regarded as a directed graph $\mathcal{D}$, with directed edges representing the state transitions that can be achieved through better response

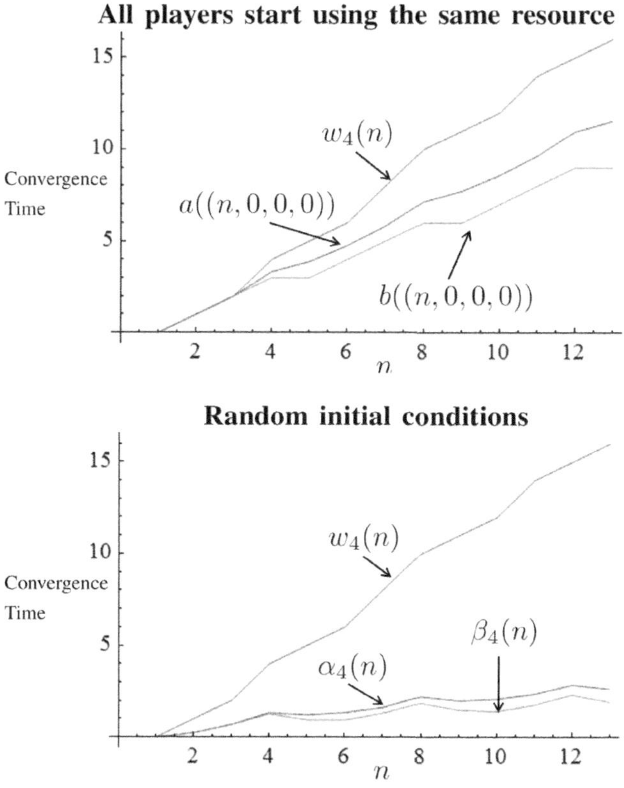

Fig. 4. Convergence times with $r = 4$ resources. The top line in both figures depicts the slowest run time $w_4(n)$. In the upper figure, the bottom line depicts the fastest run time $b((n,0,0,0))$, whilst the middle line depicts the expected run time, $a((n,0,0,0))$, of the random better response system. In the lower figure, the bottom line depicts the average fastest run time, $\beta_4(n)$, whilst the middle line depicts expected run time, $\alpha_4(n)$, of the random better response system (starting from a random resource allocation).

switches (see figures 1, 2 and 3). Here $w_r(n)$ is equal to the length of the longest path in $\mathcal{D}$. We proved Theorem 6 by considering the *reduced state space*, $\mathcal{D}^*$, which is subgraph of $\mathcal{D}$ induced upon the states with their entries in descending order. Here $w_r(n)$ is equal to the longest path from $(n, 0, 0, ...0)$ to $(\lceil \frac{n}{r} \rceil, ..., \lceil \frac{n}{r} \rceil, \lfloor \frac{n}{r} \rfloor, ..., \lfloor \frac{n}{r} \rfloor)$ in $\mathcal{D}^*$. Also $\mathcal{D}^*$ is equivalent to the r-part partition lattice [11]. The full proof of Theorem 6 can be found in the technical report [10].

3.4 Average Convergence, with Random Better Response Switches

In many scenarios, the players will update in some sort of random order. Understanding of random cases gives an insight into more realistic systems, and tests the relevance of our performance bounds. We consider a simple random updating model, where at each time step a single random "unsatisfied" player (i.e, one that *can* perform a better switch) performs a random better switch. Let $a(\boldsymbol{x})$ denote the **expected value of the convergence time of $\boldsymbol{x}$ under this "random better response system"**. We simulate and then compare with our other convergence time measures in Figure 4. Simulations suggest that the random convergence time $a(\boldsymbol{x})$ is often close to the fastest convergence time $b(\boldsymbol{x})$, and is much smaller than the worst case convergence time $w_r(n)$ (when the number of players n is reasonable large). This implies that $b(\boldsymbol{x})$ is a good estimation of the real convergence time and $w_r(n)$ is too pessimistic.

4 Spatial Variations on Our Models

So far we have assumed that each player interacts with all other players in the system. In many situations, the players are distributed over space, and can only cause congestion to their neighbors. This is the case in wireless networks, where only users close-by may cause significant mutual interferences. In this section, we look at the convergence of the *spatial congestion games* proposed in our previous work [9], where a graph structure $\mathcal{G}$ is used to represent the spatial relationship between players. This spatial game model is a generalization of the one defined in Section 2 with the added dimension of space.

Figure 5 shows an example of graph $\mathcal{G}$. Each vertex in $\mathcal{G}$ represents a player. A pair of players are connected by an edge in $\mathcal{G}$ when they are close enough to *potentially* cause congestion for one another. Each player p has a strictly decreasing payoff function, $f^p(N_i^p + 1)$, where N_i^p is the number of player p's neighbors that use the same resource i as player p. Just like before, the details of the payoff functions are irrelevant with respect to the convergence dynamics. All that matters is that players always prefer resources that are used by less of their neighbors.

In this general model, a *state* is an assignment of resources on the graph, one resource to each vertex (player). A *conflict* happens when a vertex uses the same resource as one of its neighbors. We still assume the resources are homogenous, so every vertex simply wishes to minimize the number of conflicts they incur. As before, the system evolves via asynchronous better response switches. Every

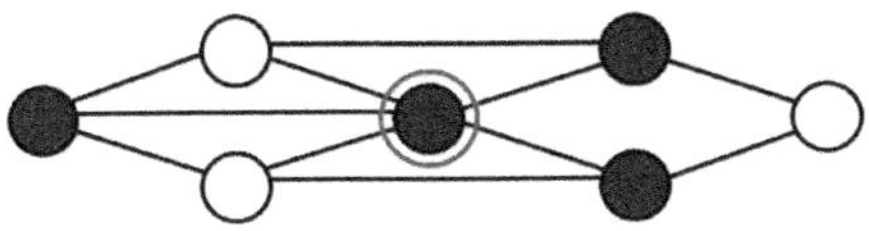

Fig. 5. An example of a state (resource allocation) of a spatial congestion game. Every vertex can access two resources: "white" and "black". The only unsatisfied player (vertex) is the circled one in the middle. This player currently suffers 3 conflicts. After this player does a better response switch (changing from black to white), the system will be in a Nash equilibrium.

time step, one vertex switches to a resource shared by less of its neighbors. A Nash equilibrium is a system state where no vertex can make a better response switch. We further generalize the model by assuming that different players may have *different sets of resources* available to them due to spatial variations. Formally, each player p has a specific resource set $R(p)$. Despite the generality of the model, convergence to a Nash equilibrium through better response switches is still guaranteed. Next we provide characterize the convergence speed and equilibrium characteristics. Theorem 7 is an extension of a result from [9].

Theorem 7. *Consider a general spatial congestion game with heterogeneous resource availabilities defined on a graph $\mathcal{G}$. Any sequence of $||\mathcal{G}||$ asynchronous better response switches leads to a Nash equilibrium, where $||\mathcal{G}|| \leq \frac{n(n-1)}{2}$ is the number of edges in the graph $\mathcal{G}$.*

Proof. The worst case initial state is where every vertex uses the same resource. In this case there are $||\mathcal{G}||$ conflicts. The number of conflicts in a given state forms a potential function, which decreases with better response switches. It follows that the system will converge to Nash equilibrium in at most $||\mathcal{G}||$ time steps.

Our next result bounds the worst case performance of a player at Nash equilibrium.

Theorem 8. *At a Nash equilibrium of a general spatial congestion game with heterogeneous resource availabilities, a player p will not suffer more than $\left\lfloor \frac{d(u)}{|R(p)|} \right\rfloor$ conflicts, where $d(p)$ is p's degree (number of neighbors) and $|R(p)|$ is the number of resources available to p.*

Proof. Let us prove by contradiction. Suppose player p suffers $K > \left\lfloor \frac{d(p)}{|R(p)|} \right\rfloor$ conflicts. Since the system is in a Nash equilibrium, player p cannot benefit by switching to a different resource. This means for any resource $i \in R(p)$, player p has at least K neighbors using resource i. It follows that p has degree greater than or equal to $|R(p)|K > d(p)$, which contradicts our assumption that player p has a degree $d(p)$.

Theorem 8 implies that when every player can access more resources than its degree, the system will always converge to a Nash equilibrium which involves no conflicts whatsoever.

5 Application: Cognitive Radio Networking

A good application for the above analytical framework is spectrum sharing in cognitive radio networks. Most of the usable wireless spectrum is owned by *license holders* who possess exclusive transmission rights. However, measurements show that many wireless channels are heavily under-utilized most of the time. Cognitive radio technology allows unlicensed *users* to opportunistically access these channels when the license holders are absent. One of the central questions of such spectrum sharing is: *how should users access the channels in a distributed fashion?*

Congestion games are a natural way to model how users switch channels to minimize their interference. The models discussed in this paper are very useful here due to several reasons. (1) In most wireless communication standards, spectrum is divided into equal bandwidth channels. Interleaving techniques can further homogenize the qualities of the channels. This means that channels are homogeneous to the same user. Different users, however, can achieve different data rates due to different choices of coding and modulation schemes. (2) A wireless user only generates significant interferences to close-by users, so spatial information is important. (3) License holders are spatially located and may have different activities; unlicensed users at different locations may be able to access different channels. (4) License holders often have stochastic traffic, meaning the channel availability is time-varying.

5.1 The Significance of Previous Results

Fast convergence is essential in cognitive radio networks due to the time variability of channels. Corollary 1 shows that the fastest convergence is achieved with strong switches, in which case no user needs to switch channel more than once. This is desirable as switching channel causes a costly disruption to quality of service. Theorem 6 states the upper-bound of the worst case convergence time $w_r(n)$ is linear in r and n. This suggests that congestion games remains a viable method for spectrum sharing in large networks. Theorem 6 also suggests that a user does not have to scan every channel; it is enough to locate one better channel before switching. This is useful as channel scanning is often time consuming in wireless communications.

The spatial congestion game makes a more realistic model for cognitive radio networks. This fits well into the commonly used protocol interference model, where the interference relationships among users is modeled by a graph. Theorems 7 and 8 guarantee that users in the network quickly converge to an efficient Nash equilibria. Theorem 8 is especially encouraging, because it implies that the Nash equilibrium will be interference free if users have enough channels to choose from.

5.2 Modeling the Dynamic Radio Environment

As licence holders enter and leave the system, the set of available channels changes. As this happens the Nash equilibria of the system changes and the users must adapt to this. Figure 6 illustrates how the ability of users to adapt quickly influences their performance. Here we study how fast the (unlicensed) users should adapt in order to catch up with channel dynamics.

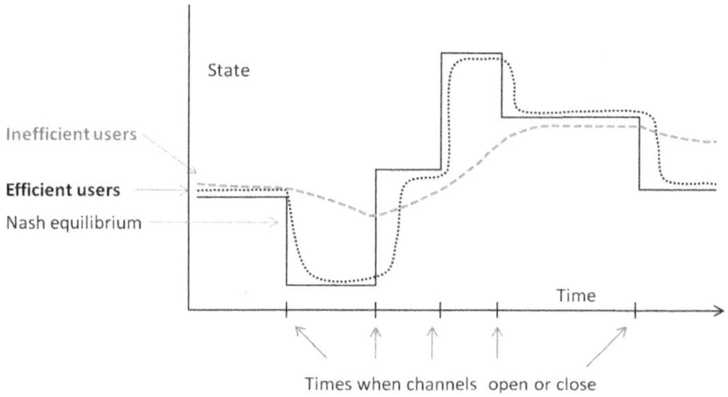

Fig. 6. An illustration of the dynamic spectrum allocation problem. As time goes by channels open and close. This changes the position of the Nash equilibrium, so the population of users must adapt their state accordingly. If the users converge quickly (dotted line) they will be able to stay close to the Nash equilibrium. If the users cannot adapt fast enough (dashed line) they will spend most of their time away from Nash equilibrium (note that in reality the Nash equilibrium will be moving around in high dimensional space).

We assume that the license holders enter and leave channels randomly, so the availability of each channel can be described by a two-state Markov chain. On average there are c channels open to the users in the network. In any given time step, a new channel opens with a probability p_1, an old channel closes with a probability p_0, and nothing happens with a probability $1 - p_0 - p_1$. We assume the time scale is so small that multiple channels will not open or close simultaneously.[3]

Necessary Switching Rate. The users must be able to perform at least $A = p_1 \left\lfloor \frac{n}{c+1} \right\rfloor$ switches per time step in order to always stay at a Nash equilibrium. The reasoning is as follows. When a new channel opens, it requires at least $\left\lfloor \frac{n}{c+1} \right\rfloor$ switches to fill the new channel with the proper number of users to reach a Nash equilibrium. When an old channel closes, the disruption depends upon

[3] In this section we are considering a "time step" to be defined with respect to the channel dynamics, and we are considering how fast the users must be able to update with respect to *this* time scale.

how many users occupied it and how they evacuate. If the users do not have time to perform A switches per time step, then they will drift away from the new Nash equilibria.

Sufficient Switching Rate. If users can do $B = \max\{p_1 w_{c+1}(n), p_0 w_{c-1}(n)\} \leq \frac{n}{2}\max\{p_1 c, p_0(c-2)\}$ switches per time step, they will almost certainly be able to stay at the Nash equilibrium. To see this, note that every sequence of better response switches $w_c(n)$ will converge to Nash equilibrium. Then users can "track" the equilibria if they can do B switches every time step.

6 Conclusion

The resource-homogeneous nature of our models has allowed us to investigate the convergence dynamics at a deeper level. This is satisfying because, although simple, our systems can be used to model a wide range of phenomena. Exact results about convergence rates are rare, and we hope that the results presented here will aid the study of more general systems. Perhaps our most critical modeling assumption is the elementary step hypothesis [5], that one player updates their strategy every time step. This could be unrealistic in many scenarios with large numbers of players. In the future we will study the dynamics of other simple congestion games with different updating mechanisms. There are many other directions we wish to take in future research. One interesting direction, for cognitive radio, will be to incorporate *switching costs* into our models. **Proofs of our results are included in the online technical report [10].**

Acknowledgments. The authors wish to thank Jing Ji for her helpful input into this research.

References

1. Rosenthal, R.: A class of games possessing pure-strategy Nash equilibria. International Journal of Game Theory 2, 65–67 (1973)
2. Vöcking, B., Aachen, R.: Congestion games: Optimization in competition. In: Proceedings of the Second ACiD Workshop (2006)
3. Fabrikant, A., Papadimitriou, C., Talwar, K.: The Complexity of Pure Nash Equilibria. In: Proceedings of STOC 2004 (2005)
4. Fotakis, D., Kontogiannis, S., Spirakis, P.: Atomic Congestion Games Among Coalitions. In: Bugliesi, M., Preneel, B., Sassone, V., Wegener, I. (eds.) ICALP 2006, Part I. LNCS, vol. 4051, pp. 572–583. Springer, Heidelberg (2006)
5. Fotakis, D., Kaporis, A.C., Spirakis, P.G.: Atomic Congestion Games: Fast, Myopic and Concurrent. In: Monien, B., Schroeder, U.-P. (eds.) SAGT 2008. LNCS, vol. 4997, pp. 121–132. Springer, Heidelberg (2008)
6. Ieong, S., Mcgrew, R., Nudelman, E., Shoham, Y., Sun, Q.: Fast and Compact: A Simple Class of Congestion Games. In: Proceedings of AAAI 2005 (2005)

7. Ackermann, H., Röglin, H., Vöcking, B.: On the Impact of Combinatorial Structure on Congestion Games. In: Proceedings of FOCS 2006 (2006)
8. Law, L., Huang, J., Liu, M., Li, S.: Price of Anarchy for Cognitive MAC Games. In: Proceedings of GLOBECOM 2009 (2009)
9. Ahmad, S., Tekin, C., Liu, M., Southwell, R., Huang, J.: Spectrum Sharing as Spatial Congestion Games. arXiv:1011.5384v1 [cs.GT] (2010)
10. Southwell, R., Huang, J.: Online technical report, `http://richardsouthwell.wordpress.com/technicalreport/`
11. Wan, H., Wootton, J.: A global compositional complexity measure for biological sequences: AT-rich and GC-rich genomes encode less complex proteins. Computers and Chemistry 24, 71–94 (2000)

Bottleneck Routing Games on Grids

Costas Busch, Rajgopal Kannan, and Alfred Samman

Department of Computer Science, Louisiana State University, Baton Rouge, LA 70803, USA
{busch,rkannan,samman}@csc.lsu.edu

Abstract. We consider routing games on grid network topologies. The social cost is the worst congestion in any of the network edges (bottleneck congestion). Each player's objective is to find a path that minimizes the bottleneck congestion in its path. We show that the price of anarchy in bottleneck games in grids is proportional to the number of bends β that the paths are allowed to take in the grids' space. We present games where the price of anarchy is $\widetilde{O}(\beta)$. We also give a respective lower bound of $\Omega(\beta)$ which shows that our upper bound is within only a poly-log factor from the best achievable price of anarchy. A significant impact of our analysis is that there exist bottleneck routing games with small number of bends which give a poly-log approximation to the optimal coordinated solution that may use an arbitrary number of bends. To our knowledge, this is the first tight analysis of bottleneck games on grids.

Keywords: algorithmic game theory, bottleneck games, routing games, price of anarchy, grid networks.

1 Introduction

We consider non-cooperative routing games with n players, where each player's *pure strategy set* consists of a set of paths in the network. A player selfishly selects a strategy (a single path) that maximizes the player's utility cost function. Such games are also known as *atomic* or *unsplittable-flow* games. We focus on *bottleneck routing games* where the objective for the social outcome is to minimize the bottleneck congestion C, the maximum congestion on any edge. Each player's objective is also to select a path with the smallest bottleneck congestion along the selected path's edges. Typically, the congestion on a edge is a non-decreasing function on the number of paths that use the edge; here, we consider the congestion to be simply the number of players that use the edge.

Bottleneck routing games have been studied in the literature [1,3,2]. In [1] the authors observe that bottleneck games are important in networks for various practical reasons. In wireless networks, the maximum congested link is related to the lifetime of the network since the nodes adjacent to high congestion links transmit large number of packets which results to higher energy depletion. High congestion links also result to congestion hot-spots which may slow-down the network throughput. Hot spots also increase the vulnerability of the network to malicious attacks which aim to increase the congestion of links in the hope to

R. Jain and R. Kannan (Eds.): GameNets 2011, LNICST 75, pp. 294–307, 2012.

bring down the network. Bottleneck games are also important from a theoretical point of view since the bottleneck congestion is immediately related to optimal packet scheduling. In a seminal result, Leighton *et al.* [9] showed that there exist packet scheduling algorithms that deliver the packets along their chosen paths in time very close to $C + D$, where D is the maximum chosen path length. When $C \gg D$, the congestion becomes the dominant factor in the packet scheduling performance. Thus, smaller bottleneck congestion C immediately implies faster packet delivery time.

A natural problem that arises in games concerns the effect of the players' selfishness on the welfare of the whole system measured with the *social cost* C. We examine the consequence of the selfish behavior in pure *Nash equilibria* which are stable states of the game in which no player can unilaterally improve her situation. We quantify the effect of selfishness with the *price of anarchy* (*PoA*) [8,13], which expresses how much larger is the worst social cost in a Nash equilibrium compared to the social cost in the optimal coordinated solution in the strategy space. The price of anarchy provides a measure for estimating how closely do Nash equilibria of bottleneck congestion games approximate the optimal C^* of the respective coordinated optimization problem in the player's strategy set.

Ideally, the price of anarchy should be small. However, the current literature results have only provided weak bounds for bottleneck games. In [1] it is shown that if the resource congestion delay function is bounded by some polynomial with degree k then $PoA = O(|E|^k)$, where E is the set of edges in the graph. In [3] it is shown that if $k = 1$ there are game instances with $PoA = \Omega(|E|)$. A natural question that we explore here is the circumstances under which there are bottleneck games with alternative and better price of anarchy bounds.

1.1 Contributions

We consider grid network topologies in which the nodes are placed in a d-dimensional array and each node connects with edges to at most $2d$ neighbors. The number of nodes is $n^d = N$. Grid networks have been used as interconnection networks in parallel multiprocessor computer architectures [10]. In wireless networks 2-dimensional grids provide a framework for formulating and analyzing wireless communication problems. In other communication networks routing and scheduling algorithms are typically first tested and analyzed on grids and then extended to arbitrary network topologies [4].

We explore games where the price of anarchy is expressed in terms of the numbers of bends that the paths use in the grid. A *bend* is a node in a path which connects two path segments in different dimensions. We explore games where the strategies of the players consists of paths whose bends are bounded by β, where β can be any number of nodes up to N. We first examine basic bottleneck games on grids with at most β bends for the paths. We show that there are instances in the 2-dimensional grid with $\beta = O(1)$ and price of anarchy $\Omega(\sqrt{N})$. However, this is not satisfactory. In order to obtain price of anarchy bounded by β, we explore two alternative games.

In the first game we utilize *channels*, where path segments on straight lines are routed in different channels according to their lengths. An edge accommodates $\alpha = \log n$ channels (logarithms are base 2), such that channel j is used by path segments of length in range $[2^j, 2^{j+1} - 1]$. Channels do not interfere with each other so that congestion can be created only by path segments in the same channel. Channels can be implemented with different frequencies in the physical communication medium, or with time division multiplexing, or with other means of signal multiplexing. The use of channels enables us to control the price of anarchy. We show that in *channel bottleneck games* if paths are allowed to use at most β bends, the price of anarchy is $PoA = O((\beta/d)\log N)$. We also provide a lower bound $PoA = \Omega(\beta)$. Thus, for constant d, the upper bound is tight within a $\log n$ factor.

We then explore games which use only one channel. Now, in order to control the price of anarchy we split the path segments into different grid lines according to the lengths of the segments. Odd lines with index $2i+1$ are used to route path segments of length in range $[2^{i \bmod \alpha}, 2^{(i \bmod \alpha)+1} - 1]$, where $\alpha = \log n$ (logarithms are base 2). Even index lines are used to route paths segments with length at most $2\alpha - 1$. Even index lines are uses to route paths close to the source and destination and when path segments switch to different lengths. This gives $\alpha + 1$ different types of lines. Thus, path segments are separated in space, and a single channel suffices. Note that we can still perform routing from every node to any other node without significantly increasing the number of bends, compared to a routing mode without space separated path segments. We show that in the respective *split bottleneck games* if paths are allowed to use at most β bends, the price of anarchy is $PoA = O((\beta/d^2)\log^2 N)$. We also provide a lower bound $PoA = \Omega(\beta)$. Thus, for constant d, the upper bound is tight within a $\log^2 n$ factor.

1.2 Impact of Games with Small Number of Bends

We demonstrate that Nash equilibria of bottleneck games with small number of bends can approximate efficiently the best coordinated solution that uses an arbitrary number of bends. Assuming that every path in the network can be used, there exist oblivious routing algorithms in grids which find paths with $O(d\log N)$ bends and achieve $O(d\log N)$ approximation to the optimal solution that uses an arbitrary number of bends [4]. Let $\widehat{C}$ denote the solution returned by the oblivious algorithm and $\widehat{C}^*$ denote the global optimal solution with an arbitrary number of bends. Clearly, $\widehat{C}/\widehat{C}^* = O(d\log N)$.

Consider now channel bottleneck games where the strategy of each player contains all possible paths in the grid with $\beta = O(d\log N)$ bends. Let C^* denote the smallest social cost. Clearly, $C^* \le \widehat{C}$. Let C be any Nash equilibrium of the game. Since $C/C^* \le PoA$, and $PoA = O((\beta/d)\log N) = O(\log^2 N)$, we obtain $C/\widehat{C}^* = O(d\log^3 N)$. Therefore, Nash equilibria of channel bottleneck games with small number of bends provide good approximations to the optimal coordinated solution with arbitrary number of bends.

We can obtain a similar result for split bottleneck games. Note that any solution of an oblivious routing algorithm with congestion C' and x bends is translated to a solution with congestion $C' \cdot \log n$ and $O(x)$ bends in the split grid, since some of the path segments have to be rerouted to nearby lines that accommodate their length. Since $PoA = O((\beta/d^2)\log^2 N) = O((1/d) \cdot \log^3 N)$, we obtain: $C/\widehat{C}^* = O((1/d)\log^5 N)$.

1.3 Related Work

Congestion games were introduced and studied in [12,14]. In [14], Rosenthal proves that congestion games have always pure Nash equilibria. Koutsoupias and Papadimitriou [8] introduced the notion of price of anarchy in the specific *parallel link networks* model in which they provide the bound $PoA = 3/2$. Roughgarden and Tardos [17] provided the first result for splittable flows in general networks in which they showed that $PoA \leq 4/3$ for a player cost which reflects to the sum of congestions of the resources of a path. Pure equilibria with atomic flow have been studied in [3,5,11,19] (our work fits into this category), and with splittable flow in [15,16,17,18]. Most of the work in the literature uses a player cost metric related to the aggregate sum of congestions on all the edges of the player's path; and the social cost metric is also an aggregate expression of all the edge congestions [5,16,17,18,19].

Bottleneck routing games have been studied in [1], where the authors consider the maximum congestion metric in general networks with splittable and atomic flow. They prove the existence and non-uniqueness of equilibria in both the splittable and atomic flow models. They show that finding the best Nash equilibrium that minimizes the social cost is a NP-hard problem. Further, they show that the price of anarchy may be unbounded for specific resource congestion functions. In [3], the authors consider bottleneck routing games in general networks where they prove that $\ell \leq PoA \leq c(\ell^2 + log^2|V|)$, where ℓ is the size of the largest edge-simple cycle in the graph and c is a constant. In [2] the authors consider bottleneck games with the $C + D$ metric. In [6], the authors prove the existence of strong Nash equilibria (which concern coalitions of players) for games with the lexicographic improvement property; such games include the bottleneck routing games that we consider here. In [7], the authors provide games with the bottleneck social cost which achieve low price of anarchy when the players use a cost function which is an aggregate exponential expression of the congestions of the edges in their selected paths.

Outline of Paper: In Section 2 we give basic definitions. In Section 3 we present a basic bottleneck routing game with high price of anarchy. In Sections 4 and 5 we present the channel and split bottleneck games, respectively, which achieve price of anarchy bounded by the number of bends β. We finish with providing lower bounds in Section 6.

2 Definitions

Grids: The *d-dimensional grid* $G = (V, E)$ consists $N = |V| = n^d$ nodes arranged in a grid of d dimensions with side length n in each dimension. There is an edge connecting a node with each of its $2d$ neighbors (except for the nodes at the boundaries of the grid). Each node has a coordinate $(a_1, a_2, \dots, a_d)$, where $a_i \in [0, n-1]$ denotes the position in the ith dimension. An example of a 2-dimensional grid is shown in Figure 1. A line segment with x edges in the kth dimension is a sequence of nodes $(a_1, \dots, a_k, \dots, a_d), \dots, (a_1, \dots, a_k + x, \dots, a_d)$.

Routings: Let $\Pi = \{\pi_1, \dots, \pi_\kappa\}$ be a set of players such that each π_i corresponds to a path request from a source u_i and destination v_i. A *routing* $\mathbf{p} = [p_1, p_2, \cdots, p_\kappa]$ is a collection of paths, where p_i is a path for player π_i from u_i to v_i. For any routing $\mathbf{p}$ and any edge $e \in E$, the *edge-congestion* $C_e(\mathbf{p})$ is the number of paths in $\mathbf{p}$ that use edge e. For any path q, the *path-congestion* $C_q(\mathbf{p})$ is the maximum edge congestion over all edges in q, namely, $C_q(\mathbf{p}) = \max_{e \in q} C_e(\mathbf{p})$. Player's π_i congestion is denoted as $C_{\pi_i}(\mathbf{p}) = C_{p_i}(\mathbf{p})$. The *network (bottleneck) congestion* $C(\mathbf{p})$ is the maximum edge-congestion over all edges in E, that is, $C(\mathbf{p}) = \max_{e \in E} C_e(\mathbf{p})$.

We denote the length (number of edges) of any path p as $|p|$. For a grid G, the path p consists of a sequence path segments which change dimensions. A *bend* of a path is a node that connects two consecutive path segments in different dimensions. By default, we take the source and destination nodes to be bends.

Routing Games: A routing game in graph G is a tuple $\mathcal{R} = (G, \Pi, \mathcal{P})$, where $\Pi = \{\pi_1, \pi_2, \dots, \pi_\kappa\}$ is the set of players such that each player π_i has a source node u_i and destination v_i. The set $\mathcal{P}$ is the strategy state space of the game, $\mathcal{P} = \mathcal{P}_1 \times \mathcal{P}_2 \times \cdots \times \mathcal{P}_\kappa$, where $\mathcal{P}_i$ is the *strategy set* of player π_i which is a collection of available paths in G for player i from u_i to v_i. Any path $p \in \mathcal{P}_i$ is a *pure strategy* available to player π_i. A *pure strategy profile (or game state)* is any routing $\mathbf{p} = [p_1, p_2, \cdots, p_\kappa] \in \mathcal{P}$.

For game $\mathcal{R}$ and routing $\mathbf{p}$, the *social cost* (or *global cost*) is a function of routing $\mathbf{p}$, and it is denoted $SC(\mathbf{p})$. The *player or local cost* is also a function on $\mathbf{p}$ denoted $pc_i(\mathbf{p})$. We use the standard notation $\mathbf{p}_{-i}$ to refer to the collection of paths $\{p_1, \cdots, p_{i-1}, p_{i+1}, \cdots, p_\kappa\}$, and $(p_i; \mathbf{p}_{-i})$ as an alternative notation for $\mathbf{p}$ which emphasizes the dependence on p_i. A *greedy move* is available to player π_i if the player can obtain lower cost by changing the current path from p_i to p'_i. Specifically, the greedy move takes the original routing $\mathbf{p} = (p_i; p_{-i})$ to $\mathbf{p}' = (p'_i; p_{-i})$ (in which path p_i is replaced by p'_i) such that $pc_i(\mathbf{p}') < pc_i(\mathbf{p})$.

Player i is *locally optimal* (or *stable*) in routing $\mathbf{p}$ if $pc_i(\mathbf{p}) \leq pc_i(p'_i; \mathbf{p}_{-i})$ for all paths $p'_i \in \mathcal{P}_i$. In other words, no greedy move is available for a locally optimal player. A routing $\mathbf{p}$ is in a Nash Equilibrium if every player is locally optimal. Nash Equilibria quantify the notion of a stable selfish outcome. A routing $\mathbf{p}^* \in \mathcal{P}$ is an *optimal pure strategy profile* if it has minimum attainable social cost: for any other pure strategy profile $\mathbf{p} \in \mathcal{P}$, $SC(\mathbf{p}^*) \leq SC(\mathbf{p})$.

We quantify the quality of the Nash Equilibria with the *price of anarchy* (PoA) (sometimes referred to as the coordination ratio) and the *price of stability* (PoS). Let $\mathbf{Q}$ denote the set of distinct Nash Equilibria, and let SC^* denote the social cost of the optimal routing $\mathbf{p}^* \in \mathcal{P}$. Then,

$$PoA = \sup_{\mathbf{p} \in \mathbf{Q}} \frac{SC(\mathbf{p})}{SC^*}, \qquad PoS = \inf_{\mathbf{p} \in \mathbf{Q}} \frac{SC(\mathbf{p})}{SC^*}.$$

3 Basic Game

Consider a routing game $\mathcal{R} = (G, \Pi, \mathcal{P})$ in a d-dimensional grid $G = (V, E)$, where each path in $\mathcal{P}_i$ is allowed to have at most β bends. For any routing $\mathbf{p} = [p_1, p_2, \cdots, p_\kappa] \in \mathcal{P}$, the social cost function is the bottleneck congestion, $SC(\mathbf{p}) = C(\mathbf{p})$, and the player cost function is the bottleneck congestion of its path, $pc_i(\mathbf{p}) = C_{\pi_i}(\mathbf{p}) = C_{p_i}(\mathbf{p})$.

We first show that such (basic) games have always Nash equilibria and the price of stability is 1. However, there are game instances where the price of anarchy is very large compared to the number of bends β. For this reason we explore alternative games with low price of anarchy in Sections 4 and 5.

The stability of the above basic game follows from techniques in [3,6] related to potential functions based on lexicographic ordering. We give the details here for completeness. For routing $\mathbf{p}$, the *congestion vector* $M(\mathbf{p}) = [m_0(\mathbf{p}), m_1(\mathbf{p}), \ldots, m_\kappa(\mathbf{p})]$ is defined such that each component $m_j(\mathbf{p})$ is the number of edges with congestion j. Note that $\sum_j m_j(\mathbf{p}) = |E|$. The network congestion $C(\mathbf{p})$ is the maximum index j for which $m_j > 0$. We define a lexicographic total order on routings according to their congestion vectors. Let $\mathbf{p}$ and $\mathbf{p}'$ be two routings, with $M(\mathbf{p}) = [m_0, m_1, \ldots, m_\kappa(\mathbf{p})]$, and $M(\mathbf{p}') = [m'_0, m'_1, \ldots, m'_\kappa(\mathbf{p})]$. Two routings are equal, written $\mathbf{p} = \mathbf{p}'$, if and only if $m_j = m'_j$ for all $j \geq 0$. Routing $\mathbf{p}$ is smaller than $\mathbf{p}'$, written $p < p'$, if and only if there is some $j \in [0, \kappa]$ such that $m_j < m'_j$ and $\forall j' > j, m_{j'} = m'_{j'}$.

It is easy to verify that for any greedy move of a player from a routing $\mathbf{p}$ to routing $\mathbf{p}'$ it holds that $\mathbf{p}' < \mathbf{p}$, since a lower index vector position increases in $M(\mathbf{p}')$ and a higher index vector position decreases in $M(\mathbf{p}')$ with respect to $M(\mathbf{p})$. Let $\mathbf{p}^* \in \mathcal{P}$ be the minimum routing (according to the total lexicographic order) in the available game state set. Routing $\mathbf{p}^*$ is a Nash equilibrium since no player can perform a greedy move to improve its cost. Further, $\mathbf{p}^*$ has optimal social cost, since if there was another state with smaller social cost then $\mathbf{p}^*$ wouldn't be minimum. Therefore, we obtain:

Theorem 1. *Any basic bottleneck game instance $\mathcal{R}$ has at least one Nash Equilibrium and $PoS(\mathcal{R}) = 1$.*

Next, we show that there are instances of the basic bottleneck game with large price of anarchy even when β is small.

Theorem 2. *There is a basic bottleneck game instance $\mathcal{R}$ in the 2-dimensional grid, with $\beta = 0(1)$ bends, such that $PoA(\mathcal{R}) = \Omega(\sqrt{N})$.*

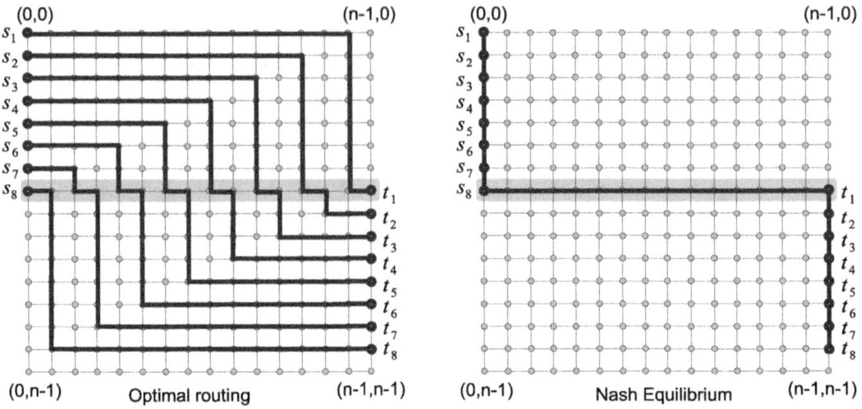

Fig. 1. A game with large price of anarchy and small number of bends

Proof. Consider an $n \times n$ grid. In the game there are $\kappa = n/2$ players, where each player π_i has source s_i in node $(0, i-1)$ of the column 0, and destination t_i in node $(n-1, i+n/2-2)$ of column $n-1$ (see Figure 1). The strategy set of player π_i consists of two paths $\mathcal{P}_i = \{p_i^1, p_i^2\}$. Both of the paths cross row $r = n/2 - 1$ (the row is highlighted in Figure 1). Path p_i^1 uses one "dedicated" edge in row r, so that the dedicated edges of different players do have any common nodes (see left of Figure 1). The remaining path segments of p_i^1 are used to connect the source and destination so that the first strategy paths of the players are disjoint. Note that path p_i^1 consists of at most five path segments (6 bends). Path p_i^2 uses all the edges of row r, and it consists of at most three path segments (4 bends), one in column 0, one in row r, and one in column $n-1$ (see right of Figure 1).

The routing with the first path choices $\mathbf{p}^1 = [p_1^1, p_2^1, \ldots, p_\kappa^1]$ is optimal, since the congestion is $C(\mathbf{p}^1) = 1$. The routing with the second path choices $\mathbf{p}^2 = [p_1^2, p_2^2, \ldots, p_\kappa^2]$ has congestion $C(\mathbf{p}^2) = \kappa$ and every player has cost $pc_i(\mathbf{p}^2) = \kappa$, due to the path segments in row r. Routing $\mathbf{p}^2$ is a Nash equilibrium, since if any player π_i switches to path p_i^1, then its cost remains κ because it still uses the dedicated edge in row r. Therefore: $PoA(\mathcal{R}) \geq C(\mathbf{p}^2)/C(\mathbf{p}^1) = \kappa = n/2 = \Omega(\sqrt{N})$.

4 Channel Game

Let $G = (V, E)$ be a d-dimensional grid, with $n^d = N$ nodes. We consider bottleneck routing games where each path is allowed to have at most β bends, and achieve price of anarchy bounded by β. In order to get this price of anarchy we use $\log n$ *channels*, as we describe below.

We can write any path p as a sequence of path segments $p = (q_1, q_2, \ldots, q_k)$, where each q_l is in a line which is in a different dimension than q_{l+1}, where $1 \leq l < k$. The number of nodes (bends) in the path p is bounded by $k + 1 \leq \beta$; thus, the number of path segments is $k \leq \beta - 1$.

Let $\alpha = \log n$. Each edge e accommodates $\alpha = \log n$ distinct *channels* $A_0, A_1, \ldots, A_{\alpha-1}$. The purpose of the channels is to route path segments of different lengths separately. A path segment q whose length is in range $|q| \in [2^j, 2^{j+1} - 1]$ uses channel A_j; we also say that the channel of q is $A(q) = A_j$. Note that a path may use multiple channels according to the lengths of its constituent segments.

Consider a routing $\mathbf{p}$. For any edge e denote by $C_e^{A_j}(\mathbf{p})$ the congestion caused by the path segments of channel A_j, which is equal to the number of path segments in channel A_j that use edge e. The congestion of a path segment q is $C_q(\mathbf{p}) = \max_{e \in q} C_e^{A(q)}(\mathbf{p})$, which is the maximum edge congestion along the path segment and its respective channel. Given a path $p = (q_1, q_2, \ldots, q_k)$, we denote the congestion of the path as the maximum congestion along any of its path segments, namely, $C_p(\mathbf{p}) = \max_{q_i \in p} C_{q_i}(\mathbf{p})$. Using this notion of path congestion all the congestion definitions in Section 2 can be extended in a grid with channels.

We are now ready to define the *channel bottleneck game* $\mathcal{R} = (G, \Pi, \mathcal{P})$. As in the basic bottleneck game, there is limit β on the allowed number of bends in a selected path. The social and player cost functions are also similar, $SC(\mathbf{p}) = C(\mathbf{p})$, and $pc_i(\mathbf{p}) = C_{\pi_i}(\mathbf{p}) = C_{p_i}(\mathbf{p})$, where all congestions are calculated using the channel model of the grid. Similar to the basic congestion game we obtain:

Theorem 3. *Any channel bottleneck game instance $\mathcal{R}$ has at least one Nash Equilibrium and $PoS(\mathcal{R}) = 1$.*

4.1 Price of Anarchy Analysis for Channel Game

Consider a Nash equilibrium $\mathbf{p} \in \mathcal{P}$. Let $\mathbf{p}^* = [p_1^*, p_2^*, \ldots, p_\kappa^*] \in \mathcal{P}$ be an optimal routing with lowest congestion $C^* = C(\mathbf{p}^*)$. Consider a set of players $\Pi' \subseteq \Pi$ such that the smallest congestion of any player of Π' in routing $\mathbf{p}$ is at least C'. Since $\mathbf{p}$ is an equilibrium, each player $\pi_i \in \Pi'$ has congestion at least $C' - 1$ in its optimal path p_i^*, namely, $C_{p_i^*}(\mathbf{p}) \geq C' - 1$. The $C' - 1$ congestion in p_i^* is due to some path segment $q_i^* \in p_i^*$ with congestion at least $C' - 1$, namely, $C_{p_i^*}(\mathbf{p}) \geq C_{q_i^*}(\mathbf{p}) \geq C' - 1$. Thus, there is an edge $e \in q_i^*$ such that $C_e^{A(q_i^*)} \geq C' - 1$. We call e the *special edge* of player π_i and the respective channel $A(q_i^*)$ the *special channel* of player π_i. Note that a player could have multiple special edges and respective special channels, in which case we choose one of them arbitrarily.

We say that two edges e_1 and e_2 are *far-apart with respect to channel A_j* if the edges are in different dimensions, or if the edges are in the same dimension and in different lines, or if the edges are in the same line and the shortest path length that connects any of their adjacent nodes is at least $2^{j-1} - 1$. If two edges are not far-apart with respect to channel A_j, then we say that they are *close* with respect to channel A_j.

Let $\widehat{A}(\Pi')$ denote the channel which is special for the majority of the players in Π'. Let $B(\Pi')$ be the subset of players in Π' with special channel $\widehat{A}(\Pi')$. Clearly, since there are α channels, $|B(\Pi')| \geq |\Pi'|/\alpha$. Let $\Gamma(\Pi')$ denote the set of special edges for the players in $B(\Pi')$. Let $\Delta(\Pi')$ denote a maximum set of edges such that $\Delta(\Pi') \subseteq \Gamma(\Pi')$, and each pair of edges in $\Delta(\Pi')$ is far-apart with respect to channel $\widehat{A}(\Pi')$. Let $\Phi(\Pi')$ denote the set of players which in routing $\mathbf{p}$ use an edge in $\Delta(\Pi')$ such that the path segment that crosses the edge belongs to channel $\widehat{A}(\Pi')$. Each player $\pi_i \in B(\Pi')$ has either (i) its special edge $e \in \Delta(\Pi')$, or (ii) there is an edge $e' \in \Delta(\Pi')$ such that e' is close to e with respect to channel $\widehat{A}$. In either case, we say that player π_i is *assigned* to respective edge e or e' of $\Delta(\Pi')$.

Lemma 1. *For any set of players $\Pi' \subseteq \Pi$, each edge in $\Delta(\Pi')$ has assigned to it at most $5C^*$ players of Π' in routing* $\mathbf{p}$.

Proof. Suppose that the channel $\Delta(\Pi')$ is in dimension x. Assume that there is an edge $e \in \Delta(\Pi')$ such that there are at least $z \geq 5C^* + 1$ players assigned to it. Let X be the set of players in $B(\Pi')$ which are assigned to e because e is their special edge (case (i) above). Let Y be the number of players in $B(\Pi')$ which are assigned to e because e is near their special edge (case (ii) above). We have that $z = |X| + |Y|$. If $|X| > C^*$, then the edge e is used in the optimal path of at least $C^* + 1$ players, which is impossible since the optimal congestion is C^*. Therefore, $|X| \leq C^*$, and hence $|Y| \geq 4C^* + 1$.

For ease of presentation, assume without loss of generality that x is the horizontal dimension. For any player $\pi_i \in Y$ we say that its special edge is in the *first (second)* part of its optimal path segment if it is positioned in the left (right) half of its optimal path segment (if the special edge is positioned exactly in the middle of the path segment then it is simultaneously in the first and second parts). Let Y_l and Y_r denote the players whose special edges appear on the left and right of e, respectively. Without loss of generality, assume that $|Y_l| \geq |Y|/2$. Without loss of generality, assume also that at least half of the special edges in Y_l are in the first half of their respective optimal segments. Denote by Y_l' these players. We have that $|Y_l'| \geq |Y|/4$. By the positions of the special edges of Y_l' all their optimal path segments intersect, which implies that there is an edge on same line with e which in the optimal routing $\mathbf{p}^*$ has congestion at least $|Y_l'| \geq |Y|/4 \geq (4C^* + 1)/4 > C^*$. This is a contradiction.

From Lemma 1, each edge in $\Delta(\Pi')$ is assigned at most $5C^*$ players of $B(\Pi')$. Since $|B(\Pi')| \geq |\Pi'|/\alpha$, we have:

Corollary 1. *For any set of players $\Pi' \subseteq \Pi$, $|\Delta(\Pi')| \geq |\Pi'|/(5\alpha C^*)$.*

Lemma 2. *For any set of players $\Pi' \subseteq \Pi$ with congestion at least C', $|\Phi(\Pi')| \geq (C' - 1)|\Pi'|/(20\alpha\beta C^*)$.*

Proof. Each edge in $\Delta(\Pi')$ is special for some player in $B(\Pi')$. Without loss of generality, let $\widehat{A}(\Pi') = A_j$. Then, $2^{j+1} - 1$ is the maximum path segment of any path that uses channel $\widehat{A}(\Pi')$. By the definition of the special edges $\Delta(\Pi')$, each path segment of channel $\widehat{A}(\Pi')$ can have at most four special edges. Since each player in $\Phi(\Pi')$ has at most β path segments each using at most four special edges in $\Delta(\Pi')$, and each special edge in $\Delta(\Pi')$ is used by at least $C' - 1$ players in $\Phi(\Pi')$ (since the edge e has congestion $C' - 1$ in channel $\widehat{A}(\Pi')$), from Corollary 1 we obtain: $|\Phi(\Pi')| \geq (C' - 1)|\Delta(\Pi')|/(4\beta) \geq (C' - 1)|\Pi'|/(20\alpha\beta C^*)$

Theorem 4. $C(\mathbf{p}) \leq 40\alpha\beta C^* + \log(5\alpha d n^d C^*)$.

Proof. Suppose that $C(\mathbf{p}) > 40\alpha\beta C^* + \log(5\alpha d n^d C^*)$. There is a player $\pi_i \in \Pi$ with congestion $C_{\pi_i}(\mathbf{p}) = C(\mathbf{p})$. We define recursively a sequence of player sets $\Pi_0, \Pi_1, \ldots, \Pi_k$, where $k = \log(5\alpha d n^d C^*)$ as follows. We define $\Pi_0 = \{\pi_i\}$. Suppose we have defined the set Π_t, where $t \geq 1$; we define $\Pi_{t+1} = \Phi(\Pi_t)$. From the above definition of Π_t, we have that for each $\pi_j \in \Pi_t$, $C_{\pi_j}(\mathbf{p}) \geq C(\mathbf{p}) - t \geq C(\mathbf{p}) - k \geq 40\alpha\beta C^* + 1$. From Lemma 2, $|\Pi_{t+1}| \geq 2|\Pi_t|$. Therefore, $|\Pi_k| \geq 2^k \geq 5\alpha d n^d C^*$. Consequently, from Corollary 1, $|\Delta(\Pi_k)| \geq |\Pi_k|/(5\alpha C^*) \geq dn^d$. However, we have a contradiction, since $|\Delta(\Pi_k)| \leq |E| < dn^d$.

From Theorem 4, since $\alpha = O(\log n)$ and $N = n^d$, we obtain the following corollary:

Corollary 2. *For any channel bottleneck game $\mathcal{R}$ in the d-dimensional grid which allows paths with at most β bends, $PoA(\mathcal{R}) = O((\beta/d)\log N)$.*

5 Split Game

We describe a way to split the path segments of a path in different lines according to their lengths. In this way we only need to use a single channel that all players can share. For ease of presentation, we first describe the respective game in the 2-dimensional grid, and then explain below how it can be extended to higher dimensions.

Let $G = (V, E)$ be a 2-dimensional $n \times n$ grid. Let $\alpha = \log n$. For convenience take n to be a multiple of $2\log n$. The odd index rows (columns) $1, 3, \ldots, n-1$ are used to route horizontal (vertical) path segments of lengths ranging from 2 to $n - 1$. In particular, row $(2i + 1) \bmod \alpha$ (column $2i \bmod \alpha$), where $i \in [0, n/2 - 1]$, is used for horizontal (vertical) path segments whose length is in range $[2^{i \bmod \alpha}, 2^{(i \bmod \alpha)+1} - 1]$. The even rows (columns) $0, 2, \ldots, n - 2$ are reserved to route horizontal (vertical) path segments whose length is in range $[1, 2\alpha - 1]$. Note that path segments in range $[2, 2\alpha - 1]$ have a chance to be routed either in even or odd rows and columns. We say that an odd row (column) $2i + 1$ $(2i)$ is of *type-($i \bmod \log n$)*, while any even row (column) is of the *local*-type. Note that there are $\alpha + 1$ types in total. Any edge $e \in E$ has the same type of the row or column that it belongs to. Note that with splitting the path segments into different rows we achieved to have a single channel that all players can share.

We are now ready to define the *split bottleneck game* $\mathcal{R} = (G, \Pi, \mathcal{P})$. As in the basic bottleneck game, there is limit β on the number of bends of a path. Each path has to follow the rules for using the appropriate rows and columns for its segments as described above. The social and player cost functions are similar, $SC(\mathbf{p}) = C(\mathbf{p})$, and $pc_i(\mathbf{p}) = C_{\pi_i}(\mathbf{p}) = C_{p_i}(\mathbf{p})$. Similar to the basic congestion game we obtain:

Theorem 5. *Any split bottleneck game instance $\mathcal{R}$ has at least one Nash Equilibrium and $PoS(\mathcal{R}) = 1$.*

5.1 Price of Anarchy Analysis for Split Game

Consider a Nash equilibrium $\mathbf{p} \in \mathcal{P}$. Consider a set of players $\Pi' \subseteq \Pi$. We can define the special edge and special type for a player in the same way as we did for channel bottleneck games. The only difference is that instead of the notion of the channel we use the notion of the type. Let $\tau(\Pi')$ be the type which is special for the majority of the players in Π'. Using $\tau(\Pi')$ we can define the sets: $B(\Pi')$, $\Gamma(\Pi')$, $\Delta(\Pi')$, and $\Phi(\Pi')$, as we did in Section 4, where $\tau(\Pi')$ plays the role of $\widehat{A}(\Pi')$. We have that $|B(\Pi')| \geq |\Pi'|/(\alpha+1)$, since there are $\alpha+1$ types.

Lemma 3. *For any set of players $\Pi' \subseteq \Pi$, each edge $e \in \Delta(\Pi')$ has assigned to it at most $c_1\alpha C^*$ players of Π' in routing $\mathbf{p}$, for some constant c_1.*

From Lemma 3, each edge in $\Delta(\Pi')$ is assigned at most $c_1\alpha C^*$ players of $B(\Pi)$. Since $|B(\Pi')| \geq |\Pi'|/(\alpha+1)$, we have $|\Delta(\Pi')| \geq |\Pi'|/((\alpha+1)\cdot(c_1\alpha C^*))$. Therefore,

Corollary 3. *For any set of players $\Pi' \subseteq \Pi$, $|\Delta(\Pi')| \geq |\Pi'|/(c_2\alpha^2 C^*)$, for some constant c_2.*

Lemma 4. *For any set of players $\Pi' \subseteq \Pi$ with congestion at least C', $|\Phi(\Pi')| \geq (C'-1)|\Pi'|/(c_3\alpha^2\beta C^*)$, for some constant c_3.*

Theorem 6. $C(\mathbf{p}) \leq 2c_3\alpha^2\beta C^* + \log(2c_2\alpha^2 n^2 C^*)$.

From Theorem 6, since $\alpha = \log n$ and $N = n^2$, we obtain the following corollary:

Corollary 4. *For any split bottleneck game $\mathcal{R}$ in the 2-dimensional grid which allows paths with at most β bends, $PoA(\mathcal{R}) = O(\beta \log^2 N)$.*

5.2 Split Game in the *d*-Dimensional Grid

We can extend the split games to a grid with d dimensions. The first dimension takes the role of the horizontal dimension, and the second dimension takes the role of the vertical dimension. Any other dimension (third and above) uses the first dimension to split the path segments. For example, in the 3-dimensional grid, a path segment q in the third dimension is a sequence of nodes with coordinates $q = (x, y, z), \ldots, (x, y, z+k)$. This path segment is placed in an appropriate odd first coordinate $x = 2^i + 1$ if $k \in [2^{i \bmod \alpha}, 2^{(i \bmod \alpha)+1} - 1]$, and if $k \leq 2\alpha - 1$ then it could use an even first coordinate $x = 2^i$. In this way we can characterize q as *type-i*, or local type, respectively. The total number of types for the d-dimensional grid remains $\alpha + 1$.

The main difference in the price of anarchy analysis is that Theorem 6 now returns $C(\mathbf{p}) \leq 2c_3\alpha^2\beta C^* + \log(c_2\alpha^2 d n^d C^*)$. Since $\alpha = O(\log n)$ and $N = n^d$, Corollary 4 now becomes:

Corollary 5. *For any split bottleneck game $\mathcal{R}$ in the d-dimensional grid which allows paths with at most β bends, $PoA(\mathcal{R}) = O((\beta/d^2)\log^2 N)$.*

6 Lower Bounds

Here, we give lower bounds in terms of bends for the price of anarchy for the channel and split games.

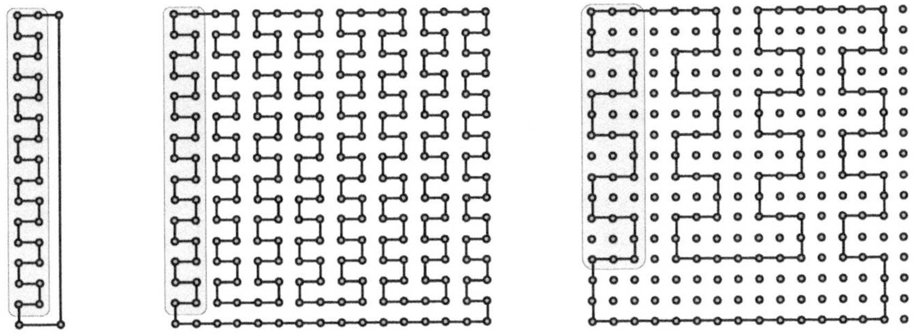

Fig. 2. Zig-zag path and cycles

Theorem 7. *In the d-dimensional grid with N nodes, given any $\beta \leq c'N$, for a specific constant c', there is a channel bottleneck game instance R with at most β bends, such that $PoA(R) = \Omega(\beta)$.*

Proof. We present the result for the 2-dimensional $n \times n$ grid G, and it can be extended to the d-dimensional grid. We define a game along a cycle c of the grid. The main building block of the cycle is the *zig-zag* path which is formed in two consecutive columns, by alternating edges between the columns and rows, as shown highlighted in the left of Figure 2. A x-zig-zag path contains x horizontal edges and $x - 1$ vertical edges, giving $2x - 2$ bends (without counting the end nodes). Given an x-zig-zag path we can build a cycle by closing the end points with 4 additional bends, giving a cycle with total $2x + 2$ bends. Since $x < n - 1$ (last row is reserved to close the cycle), the maximum number of bends that a single zig-zag path can provide is bounded by $2(n - 1) + 2 = 2n$.

In order to obtain a cycle with larger number of bends, we combine multiple zig-zag paths, as shown in the middle of Figure 2. The largest cycle is formed by using $n/2$ instances of $(n - 1)$-zig-zag paths by combining their original version and their horizontal mirrors, and connecting them with bridge edges in rows 0 and $n - 2$ and closing the loop with a path in row $n - 1$ and bridge edges in the bottoms of columns 0 and $n - 1$. This construction gives a cycle with total $\ell = (2(n - 1) + 2) \cdot n/2 + 4 = n^2 + n + 4$ bends. Using the above construction and adjusting appropriately the sizes of the zig-zag paths it is possible to obtain a cycle with any number of bends β up to ℓ. Clearly, the total number of edges in the cycle is $|c| = \Theta(\beta)$.

We define now a channel bottleneck game $\mathcal{R} = (G, \Pi, \mathcal{P})$. Let Z denote the set of edges in the zig-zag paths, excluding the edges adjacent to the end nodes of each zig-zag path. The game has $\kappa = |Z|$ players $\Pi = \{\pi_1, \ldots, \pi_\kappa\}$. Player π_i has two strategy sets: $\mathcal{P}_i = \{p_i^1, p_i^2\}$, where p_i^1 consists only of edge $e_i = (u_i, v_i) \in Z$ in a zig-zag path, and path p_i^2 consists of the alternate path in the cycle from v_i to u_i that traverses all the edges of c except e_i. The edges $e_i \in Z$ are chosen so that different players use different edges. Note that the first path has 2 bends, while the second path has β bends.

The optimal routing $\mathbf{p}^* \in \mathcal{P}$ is the one where each player π_i uses strategy p_i^1, namely, $\mathbf{p}^* = [p_1^1, p_2^1, \ldots, p_\kappa^1]$. The congestion of $\mathbf{p}^*$ is $C(\mathbf{p}^*) = 1$, since edge is used by at most one player. Consider now routing $\mathbf{p} = [p_1^2, p_2^2, \ldots, p_\kappa^2]$, consisting of the second strategy of each player. Routing $\mathbf{p}$ has congestion $C(\mathbf{p}) = \kappa - 1$, since all players except π_i use edge $e_i \in Z$ and all the path segments that use e_i belong to the same channel A_0 for unit length segments. The routing $\mathbf{p}$ is a Nash equilibrium, since if any user π_i attempts to switch to alternate strategy p_i^1, the congestion of the becomes $\kappa + 1 > C(\mathbf{p})$. Therefore we have that: $PoA \geq C(\mathbf{p})/C(\mathbf{p}^*) = \kappa - 1 = |Z| - 1 = \Omega(\beta)$.

Using similar zig-zag paths for the split model by adjusting appropriately the bend distances (see right of Figure 2) we can obtain the following lower bound:

Theorem 8. *In the d-dimensional grid with N nodes, given any $\beta \leq c''N$, for a specific constant c'', there is a split bottleneck game instance R with at most β bends, such that $PoA(R) = \Omega(\beta)$.*

7 Conclusions

We presented new bottleneck games on multidimensional grids whose price of anarchy is analyzed in terms of the number of bends that the paths are allowed to follow. We found that the price of anarchy is proportional to the number of bends. We also provided game instances that show that the price of anarchy results are tight within poly-log factors. A natural question that remains open is whether we can obtain tighter bounds by removing the poly-log factors. Another interesting problem is to study other network topologies and examine how the notion of bends is generalized in them.

References

1. Banner, R., Orda, A.: Bottleneck routing games in communication networks. IEEE Journal on Selected Areas in Communications 25(6), 1173–1179 (2007); also appears in INFOCOM 2006
2. Busch, C., Kannan, R., Vasilakos, A.V.: Quality of routing congestion games in wireless sensor networks. In: Proc. 4th International Wireless Internet Conference (WICON), Maui, Hawaii (November 2008)
3. Busch, C., Magdon-Ismail, M.: Atomic routing games on maximum congestion. Theoretical Computer Science 410(36), 3337–3347 (2009)
4. Busch, C., Magdon-Ismail, M., Xi, J.: Optimal oblivious path selection on the mesh. IEEE Trans. Computers 57(5), 660–671 (2008)

5. Christodoulou, G., Koutsoupias, E.: The price of anarchy of finite congestion games. In: Proceedings of the 37th Annual ACM Symposium on Theory of Computing (STOC), Baltimore, MD, USA, pp. 67–73. ACM (May 2005)
6. Harks, T., Klimm, M., Möhring, R.H.: Strong Nash Equilibria in Games with the Lexicographical Improvement Property. In: Leonardi, S. (ed.) WINE 2009. LNCS, vol. 5929, pp. 463–470. Springer, Heidelberg (2009)
7. Kannan, R., Busch, C.: Bottleneck Congestion Games with Logarithmic Price of Anarchy. In: Kontogiannis, S., Koutsoupias, E., Spirakis, P.G. (eds.) SAGT 2010. LNCS, vol. 6386, pp. 222–233. Springer, Heidelberg (2010)
8. Koutsoupias, E., Papadimitriou, C.: Worst-Case Equilibria. In: Meinel, C., Tison, S. (eds.) STACS 1999. LNCS, vol. 1563, pp. 404–413. Springer, Heidelberg (1999)
9. Leighton, F.T., Maggs, B.M., Rao, S.B.: Packet routing and job-scheduling in $O(congestion + dilation)$ steps. Combinatorica 14, 167–186 (1994)
10. Thomson Leighton, F.: Introduction to Parallel Algorithms and Architectures: Arrays - Trees - Hypercubes. Morgan Kaufmann, San Mateo (1992)
11. Libman, L., Orda, A.: Atomic resource sharing in noncooperative networks. Telecomunication Systems 17(4), 385–409 (2001)
12. Monderer, D., Shapely, L.S.: Potential games. Games and Economic Behavior 14, 124–143 (1996)
13. Papadimitriou, C.: Algorithms, games, and the Internet. In: Proceedings of the 33rd Annual ACM Symposium on Theory of Computing (STOC), Hersonissos, Crete, Greece, pp. 749–753 (July 2001)
14. Rosenthal, R.W.: A class of games possesing pure-strategy Nash equilibria. International Journal of Game Theory 2, 65–67 (1973)
15. Roughgarden, T.: The maximum latency of selfish routing. In: Proceedings of the Fifteenth Annual ACM-SIAM Symposium on Discrete Algorithms (SODA), New Orleans, Louisiana, (USA), pp. 980–981 (January 2004)
16. Roughgarden, T.: Selfish routing with atomic players. In: Proc. 16th Symp. on Discrete Algorithms (SODA), pp. 1184–1185. ACM/SIAM (2005)
17. Roughgarden, T., Tardos, É.: How bad is selfish routing. Journal of the ACM 49(2), 236–259 (2002)
18. Roughgarden, T., Tardos, É.: Bounding the inefficiency of equilibria in nonatomic congestion games. Games and Economic Behavior 47(2), 389–403 (2004)
19. Suri, S., Toth, C.D., Zhou, Y.: Selfish load balancing and atomic congestion games. Algorithmica 47(1), 79–96 (2007)

Optimal Price of Anarchy of Polynomial and Super-Polynomial Bottleneck Congestion Games

Rajgopal Kannan[1], Costas Busch[1], and Athanasios V. Vasilakos[2]

[1] Dept. of Computer Science, Louisiana State University, Baton Rouge, LA 70803
{rkannan,busch}@csc.lsu.edu
[2] Dept. of Telecomm. Engg., Univ. of Western Macedonia, Greece
vasilako@ath.forthnet.gr

Abstract. We introduce *(super) polynomial bottleneck games*, where the utility costs of the players are (super) polynomial functions of the congestion of the resources that they use, and the social cost is determined by the worst congestion of any resource. In particular, the delay function for any resource r is of the form $C_r^{\mathcal{M}_r}$, where C_r is the congestion measured as the number of players that use r, and the degree of the delay function is bounded as $1 \leq \mathcal{M}_r \leq \log C_r$. The utility cost of a player is the sum of the individual delays of the resources that it uses. The social cost of the game is the worst bottleneck resource congestion: $\max_{r \in R} C_r$, where R is the set of resources. We show that for super-polynomial bottleneck games with $\mathcal{M}_r = \log C_r$, the price of anarchy is $o(\sqrt{|R|})$, specifically $O(2^{\sqrt{\log |R|}})$. We also consider general polynomial bottleneck games where each resource can have a distinct monomial latency function but the degree is bounded i.e $\mathcal{M}_r = O(1)$ with constants $\alpha \leq \mathcal{M}_r \leq \beta$ and derive the price of anarchy as $\min\left(|R|, \max\left(\frac{2\beta}{C^*}, (2|R|)^{\frac{1}{\alpha+1}} \cdot (\frac{2\beta}{C^*})^{\frac{\alpha}{\alpha+1}} \cdot (2\beta)^{\frac{\beta-\alpha}{\alpha+1}}\right)\right)$, where C^* is the bottleneck congestion in the socially optimal state. We then demonstrate matching lower bounds for both games showing that this price of anarchy is tight.

1 Introduction

We consider non-cooperative congestion games with n players, where each player has a *pure strategy profile* from which it selfishly selects a strategy that minimizes the player's utility cost function (such games are also known as *atomic* or *unsplittable-flow* games). We focus on *bottleneck congestion games* where the objective for the social outcome is to minimize C, the maximum congestion on any resource. Typically, the congestion on a resource is a non-decreasing function on the number of players that use the resource; here, we consider the congestion to be simply the number of players that use the resource.

Bottleneck congestion games have been studied in the literature [2,4,3] in the context of routing games, where each player's utility cost is the worst resource

R. Jain and R. Kannan (Eds.): GameNets 2011, LNICST 75, pp. 308–320, 2012.

congestion on its strategy. For any resource r, we denote by C_r the number of users that use r in their strategies. In typical bottleneck congestion games, each player i has utility cost function $C_i = \max_{r \in S_i} C_r$, where S_i is the strategy of the player. The social cost is worst congested resource: $C = \max_i C_i = \max_r C_r$.

In [2] the authors observe that bottleneck games are important in networks for various practical reasons. In networks, each resource corresponds to a network link, each player corresponds to a packet, and a strategy represents a path for the packet. In wireless networks, the maximum congested link is related to the lifetime of the network since the nodes adjacent to high congestion links transmit large number of packets which results to higher energy utilization. High congestion links also result in congestion hot-spots which may slow-down the network throughput. Hot spots also increase the vulnerability of the network to malicious attacks which aim to to increase the congestion of links in the hope to bring down the network. Thus, minimizing the maximum congested edge results to hot-spot avoidance and more load-balanced and secure networks.

In networks, bottleneck games are also important from a theoretical point of view since the maximum resource congestion is immediately related to the optimal packet scheduling. In a seminal result, Leighton *et al.* [16] showed that there exist packet scheduling algorithms that can deliver the packets along their chosen paths in time very close to $C + D$, where D is the maximum chosen path length. When $C \gg D$, the congestion becomes the dominant factor in the packet scheduling performance. Thus, smaller C immediately implies faster packet delivery time.

A natural problem that arises in games concerns the effect of the players' selfishness on the welfare of the whole system measured with the *social cost* C. We examine the consequence of the selfish behavior in pure *Nash equilibria* which are stable states of the game in which no player can unilaterally improve her situation. We quantify the effect of selfishness with the *price of anarchy* (*PoA*) [15,21], which expresses how much larger is the worst social cost in a Nash equilibrium compared to the social cost in the optimal coordinated solution. The price of anarchy provides a measure for estimating how closely do Nash equilibria of bottleneck congestion games approximate the optimal C^* of the respective coordinated optimization problem.

Ideally, the price of anarchy should be small. However, the current literature results have only provided weak bounds for bottleneck games. In [2] it is shown that if the resource congestion delay function is bounded by some polynomial with degree k (with respect to the packets that use the resource) then $PoA = O(|R|^k)$, where R is the set of links (resources) in the graph. In [4] the authors consider bottleneck routing games for the case $k = 1$ and they show that $PoA = O(L + \log |V|)$, where L is the maximum path length (maximum number of resources) in the players' strategies and V is the set of nodes in the network. This bound is asymptotically tight (within poly-log factors) since it is shown in [4] that there are game instances with $PoA = \Omega(L)$. Since $L = |R|$, the price of anarchy has to be large, $PoA = \Omega(|R|)$.

1.1 Contributions

The lower bound in [4] suggests that in order to obtain better price of anarchy in bottleneck congestion games (where the social cost is the bottleneck resource C), we need to consider alternative player utility cost functions. Towards this goal, we introduce *(super) polynomial bottleneck games* where the player cost functions are (super) polynomial expressions of the congestions along the resources. In particular, the player utility cost function for player i is: $C_i' = \sum_{r \in S_i} C_r^{\mathcal{M}_r}$, where for each r, the *degree* $\mathcal{M}_r$ ranges between $1 \leq \mathcal{M}_r \leq \log C_r$. Note that the new utility cost is a sum of polynomial or super-polynomial terms on the congestion of the resources in the chosen strategy (instead of the max that we described earlier). The social cost remains the maximum bottleneck congestion C, the same as in typical congestion games.

The new player utility costs have significant benefits in improving both the upper and lower bounds on the price of anarchy for the bottleneck social cost C. Of specific interest are instantiations of the game where the degree $\mathcal{M}_r$ is either a logarithmic function on the congestion, or simply a constant. For *super-polynomial games* with $\mathcal{M}_r = \log C_r$ for each $r \in R$, we prove that the price of anarchy is $o(\sqrt{|R|})$:

$$PoA_{super} = O(2^{\sqrt{\log |R|}}) \tag{1}$$

For general *polynomial games* with $\alpha \leq \mathcal{M}_r \leq \beta$ and constants $1 \leq \alpha \leq \beta$, the price of anarchy is:

$$PoA_{poly} = \min\left(|R|, \max\left(\frac{2\beta}{C^*}, (2|R|)^{\frac{1}{\alpha+1}} \cdot (\frac{2\beta}{C^*})^{\frac{\alpha}{\alpha+1}} \cdot (2\beta)^{\frac{\beta-\alpha}{\alpha+1}}\right)\right) \tag{2}$$

Note that in polynomial games, latency costs are monomials with constant degrees between α and β and thus different resources can have different latency costs even with same congestion. We show that the bounds in Equations 1 and 2 are asymptotically tight by providing specific instances of super-polynomial and polynomial bottleneck games. The price of anarchy bounds above are significant improvements over the price of anarchy from the typical bottleneck games described above.

Polynomial and super-polynomial congestion games are interesting variations of bottleneck games not only because they provide good price of anarchy but also because they represent interesting and important real-life problems. In networks, the overall delay that a packet experiences is directly related with the link congestions along the path and hence the polynomial utility cost function reflects the total delivery delay. In wireless networks, polynomial and super-polynomial player utilities correspond to the total energy that a packet consumes while it traverses the network, and the social cost reflects to the worst energy utilization in any node in the network. Similar benefits from polynomial congestion games appear in the context of job-shop scheduling, where computational tasks require resources to execute. In this context, the social bottleneck cost function C represents the task load-balancing efficiency of the resources, and the player utility costs relate to the makespan of the task schedule. In all the above problems, the degrees $\mathcal{M}_r$ are chosen appropriately to model precisely the involved costs of the resource utilization in each computational environment.

In our analysis, we obtain the price of anarchy upper bound by using two techniques: *transformation* and *expansion*. Consider a game G with a Nash equilibrium S and congestion C. We identify two kinds of players in S: type-A players which use only one resource in their strategies, and type-B players which use two or more resources. In our first technique, transformation, we convert G to a simpler game $\tilde{G}$, having a Nash equilibrium $\tilde{S}$ with congestion $\tilde{C}$, such that $\tilde{C} = O(C)$, and all players in $\tilde{S}$ with congestion above a threshold τ are of type-A; that is, we transform type-B players to type-A players. Having type-A players is easier to bound the price of anarchy. Then, we use a second technique, expansion, which is used to give an upper bound on the price of anarchy of game $\tilde{G}$, which implies an upper bound on the price of anarchy of the original game G.

In [12], we have derived upper bounds for the price of anarchy of games with exponential utility cost functions using similar techniques (transformation and expansion). While exponential cost games have a unique substructure which makes the analysis of Price of Anarchy simpler, we believe these techniques are general enough to adapt in a non-trivial manner for a large class of utility cost functions. For the case of exponential cost games, we obtained logarithmic price of anarchy upper bounds, which was related to the problem structure. Here we obtain tight (optimal) price of anarchy bounds for polynomial and super-polynomial bottleneck games, using a non-trivial application of the general transformation and expansion techniques.

1.2 Related Work

Congestion games were introduced and studied in [20,22], mainly in the context of networks. Typically the social cost is considered to be an aggregate expression on the delay costs of the network edges and the flow that goes through them [5,24,25,26,27]. Specifically, the social cost is $SC = \sum_r C_r \cdot l(C_r)$, where $l(C_r)$ is the delay cost function for resource r, while the player cost is the same as our polynomial game model. Any price of anarchy bounds using social cost SC can be translated to a price of anarchy bound on bottleneck congestion C by amortizing appropriately with the number of resources R. For example, when the latency function is a monomial of the same constant degree d on all resources, the same bounds can be obtained using this method as well as ours (using the $d^{\Theta(d)}$ social cost bound in [1]). However, for atomic congestion games, price of anarchy bounds for SC are not known for super-polynomial delay functions, or for polynomial delay functions of different degrees for the resources, as we consider in this paper. Thus, the techniques that we propose here are useful in providing novel bounds in bottleneck congestion games for a broader range of delay functions.

In [22], Rosenthal proves that congestion games have always pure Nash equilibria. Koutsoupias and Papadimitriou [15] introduced the notion of price of anarchy in the specific *parallel link networks* model in which they provide the price of anarchy bound 3/2. Roughgarden and Tardos [25] provided the first result for splittable flows in general networks in which they showed that the price of anarchy is bounded by 4/3 for a player cost which reflects to the sum of congestions of the resources of a path. Pure equilibria with atomic flow have been studied in [4,5,17,27] (our work fits into this category), and with splittable flow in [23,24,25,26]. Mixed equilibria with atomic flow have been studied in [7,10,14,15,18,19,21], and with splittable flow in [6,9]. The vast majority of the work on congestion games has been performed for parallel link networks, with only a few exceptions on general network topologies [4,5,6,23]. Our work immediately applies to network topologies.

Basic bottleneck routing games have been studied in [2,4] which consider the maximum congestion metric in general networks, and the player cost is equal to the worst congested edge in the chosen routing path. In [2] the authors show the existence and non-uniqueness of equilibria in both the splittable and atomic flow models. They show that finding the best Nash equilibrium that minimizes the social cost is a NP-hard problem. Further, they show that the price of anarchy may be unbounded for specific resource congestion functions. In [3] the $C + D$ social metric is considered. In [11], the authors prove the existence of strong Nash equilibria, which concern coalitions of players, for games with the lexicographic improvement property. Other variations of basic bottleneck games with player coalitions are studied in [8].

Outline of Paper

In Section 2 we give basic definitions. In Section 3 we convert games with type-B players to games with type-A players. In Section 4 we give a bound on the price of anarchy. We finish with providing a lower bound in Section 5.

2 Definitions

A *congestion game* is a strategic game $G = (\Pi_G, R, \mathbb{S}, (l_r)_{r \in R}, (pc_\pi)_{\pi \in \Pi_G})$ where:

- $\Pi_G = \{\pi_1, \ldots, \pi_n\}$ is a non-empty and finite set of players.
- $R = \{r_1, \ldots, r_z\}$ is a non-empty and finite set of resources.
- $\mathbb{S} = \mathbb{S}_{\pi_1} \times \mathbb{S}_{\pi_2} \times \cdots \times \mathbb{S}_{\pi_n}$, where $\mathbb{S}_{\pi_i}$ is a strategy set for player π_i, such that $\mathbb{S}_{\pi_i} \subseteq powerset(R)$; namely, each strategy $S_{\pi_i} \in \mathbb{S}_{\pi_i}$ is pure, and it is a collection of resources. A *game state* (or *pure strategy profile*) is any $S \in \mathbb{S}$. We consider *finite games* which have finite $\mathbb{S}$ (finite number of states).
- In any game state S, each resource $r \in R$ has a *latency cost* denoted $l_r(S)$.
- In any game state S, each player $\pi \in \Pi_G$ has a player cost $pc_\pi(S) = \sum_{r \in S_\pi} l_r(S)$.

Consider a game G with a state $S = (S_{\pi_1}, \ldots, S_{\pi_n})$. The *(congestion) of a resource* r is defined as $C_r(S) = |\{\pi_i : r \in S_{\pi_i}\}|$, which is the number of players that use r in state S. The *(bottleneck) congestion of a set of resources* $Q \subseteq R$ is defined as $C_Q(S) = \max_{r \in Q} C_r(S)$, which is the maximum congestion over all resources in Q. The *(bottleneck) congestion of state* S is denoted $C(S) = C_R(S)$, which is the maximum congestion over all resources in R. When the context is clear, we will drop the dependence on S. We examine polynomial congestion games:

- *Polynomial games:* The latency cost function for any resource r is $l_r = C_r^{\mathcal{M}_r}$, for some integer constants $\mathcal{M}_l \leq \mathcal{M}_r \leq \mathcal{M}_h$.
- *Super-polynomial games:* The delay cost function for any resource r is $d_r = C_r^{\mathcal{M}_r}$, where $\mathcal{M}_r = \log C_r$.

For any state S, we use the standard notation $S = (S_{\pi_i}, S_{-\pi_i})$ to emphasize the dependence on player π_i. Player π_i is *locally optimal* (or *stable*) in state S if $pc_{\pi_i}(S) \leq pc_{\pi_i}((S'_{\pi_i}, S_{-\pi_i},))$ for all strategies $S'_{\pi_i} \in \mathbb{S}_{\pi_i}$. A greedy move by a player π_i is any change of its strategy from S'_{π_i} to S_{π_i} which improves the player's cost, that is, $pc_{\pi_i}((S_{\pi_i}, S_{-\pi_i})) < pc_{\pi_i}((S'_{\pi_i}, S_{-\pi_i}))$. *Best response dynamics* are sequences of greedy moves by players. A state S is in a *Nash Equilibrium* if every player is locally optimal. Nash Equilibria quantify the notion of a stable selfish outcome. In the games that we study there could exist multiple Nash Equilibria.

For any game G and state S, we will consider a *social cost* (or *global cost*) which is simply the bottleneck congestion $C(S)$. A state S^* is called *optimal* if it has minimum attainable social cost: for any other state S, $C(S^*) \leq C(S)$. We will denote $C^* = C(S^*)$. We quantify the quality of the states which are Nash Equilibria with the *price of anarchy* (PoA) (sometimes referred to as the coordination ratio). Let $\mathcal{P}$ denote the set of distinct Nash Equilibria. Then the price of anarchy of game G is:

$$PoA(G) = \sup_{S \in \mathcal{P}} \frac{C(S)}{C^*},$$

We continue with some more special definitions that we use in the proofs. Consider a game G with a socially optimal state $S^* = (S^*_{\pi_1}, \ldots, S^*_{\pi_n})$, and let $S = (S_{\pi_1}, \ldots, S_{\pi_n})$ denote the equilibrium state.

For any resource $r \in R$, we will let Π_r and Π^*_r denote the set of players with r in their equilibrium and socially optimal strategies respectively, i.e $\Pi_r = \{\pi_i \in \Pi_G | r \in S_{\pi_i}\}$ and $\Pi^*_r = \{\pi_i \in \Pi_G | r \in S^*_{\pi_i}\}$.

Let $G = (\Pi_G, R, \mathbb{S}, l, (pc_\pi)_{\pi \in \Pi_G})$ and $\tilde{G} = (\Pi_{\tilde{G}}, \tilde{R}, \tilde{\mathbb{S}}, \tilde{l}, (\tilde{pc}_\pi)_{\pi \in \Pi_{\tilde{G}}})$ be two games.

Definition 1. G **η-dominates** $\tilde{G}$ *if the following conditions hold between them for the highest cost Nash equilibrium and optimal states:* $|\tilde{R}| \leq |R|$, $l = \tilde{l}$, $\tilde{C} \geq C$, $\tilde{C}^* = O(\eta C^*)$, *where* η *is any parameter independent of congestion* C. *Here* C, C^*, $\tilde{C}$ *and* $\tilde{C}^*$ *represent the bottleneck congestions in the highest cost Nash equilibrium and optimal states of* G *and* $\tilde{G}$, *respectively.*

Corollary 1. $PoA(G) \leq \eta \cdot PoA(\tilde{G})$ *for an arbitrary game* G *and dominated game* $\tilde{G}$.

3 Type-B to Type-A Game Transformation

Our approach for obtaining the PoA of an arbitrary game G is to first transform it to a simplified game $\tilde{G}$ with restricted player strategies, obtain the PoA of the restricted game (which should be easier to evaluate than the generic game G) and relate this to the PoA of the unrestricted version G. Transformed game $\tilde{G}$ will consist of players with drastically limited strategies in the equilibrium state. Specifically, for a given game G in equilibrium state S, we consider two special kinds of players with respect to state S:

- *Type-A players:* any player π_i with $|S_{\pi_i}| = 1$.
- *Type-B players:* any player π_i with $|S_{\pi_i}| \geq 2$.

We define type-B games as those containing an arbitrary mix of type-A and type-B players in state S. Thus the type-B label refers to any generic monotonic-bounded congestion game. We define type-A games as those in which highly congested resources (beyond a specific latency-cost dependent threshold that we will define subsequently) are occupied only by type-A players in equilibrium state S. Intuitively, type-A games should be easier to analyze since the equilibrium strategy of players are highly restricted.

Let $\eta > 0$ be a network-related constant (i.e independent of bottleneck congestion). Let τ be an arbitrary congestion threshold such that $\forall r \in R, \forall C_r \geq \tau : \frac{l_r(C_r+1)}{l_r(C_r)} \leq \eta$. For super-polynomial games, where $l_r(C_r) = (C_r)^{\log C_r}$, we can choose τ as any small constant with $\eta = e^2$. For general polynomial games where $l_r(C_r) = (C_r)^{\mathcal{M}_r}$, we can choose $\tau = \max_r \mathcal{M}_r$ with $\eta = e$.

Consider a game $G(S, C, S^*, C^*)$ where S denotes the Nash equilibrium state with the highest social cost (the one having the highest bottleneck congestion) C, and S^* is the socially optimal state with corresponding bottleneck congestion C^*.

Then we have,

Theorem 1. *Every type-B game $G(S, C, S^*, C^*)$ with polynomial or super-polynomial latency costs on resources can be transformed into a type-A game $\tilde{G}(\tilde{S}, \tilde{C}, \tilde{S}^*, \tilde{C}^*)$ in which all resources r with congestion $C_r \geq \tau$ in equilibrium state $\tilde{S}$ are utilized exclusively by type-A players.*

Theorem 2. *Transformed game $\tilde{G}(\tilde{S}, \tilde{C}, \tilde{S}^*, \tilde{C}^*)$ is 7-dominated by $G(S, C, S^*, C^*)$. Specifically, bottleneck congestion in optimal states S^* and $\tilde{S}^*$ of G and $\tilde{G}$ satisfies $C^* \leq \tilde{C}^* \leq 7C^*$ while bottleneck congestion in Nash equilibrium states S and $\tilde{S}$ are the same $C = \tilde{C}$. The Price of Anarchy of G is bounded by $PoA(G) \leq \max(\frac{\tau}{C^*}, 7 \cdot PoA(\tilde{G}))$.*

Proof Sketch of Theorems 1 and 2: We describe a constructive proof of the theorems by iteratively transforming type-B players in G to type-A players in $\tilde{G}$.

We initialize $\tilde{G}$, the input to our transformation algorithm as a restricted version of game G with exactly two strategies per player: $\tilde{S}_\pi = S_\pi$ and $\tilde{S}^*_\pi = S^*_\pi$. We will iteratively transform $\tilde{G}$ into a type-A game by converting all type-B players of cost $\geq l(\tau) + 1$ into type-A players, in phases in decreasing order of player costs. We add and delete players/resources from $\tilde{G}$ iteratively and have a working set of players. However $\tilde{G}$ will always remain in equilibrium state $\tilde{S}$ at every step of the transformation process. When we add a new player π_k to

$\tilde{\Pi}$ we will assign two strategy sets to π_k: an 'equilibrium' strategy $\tilde{S}_{\pi_k}$ and an optimal strategy $\tilde{S}^*_{\pi_k}$. Thus $\tilde{S} = \tilde{S} \bigcup \tilde{S}_{\pi_k}$ and $\tilde{S}^* = \tilde{S}^* \bigcup \tilde{S}^*_{\pi_k}$.

First we convert $\tilde{G}$ into a 'clean' version in which every type-B player $\pi \in \tilde{\Pi}$ has *distinct* resources in its equilibrium and optimal strategies i.e $\tilde{S}_\pi \bigcap \tilde{S}^*_\pi = \emptyset$. If not already true, this can be achieved by creating $|\tilde{S}_\pi \bigcap \tilde{S}^*_\pi|$ new type-A players with identical and one type-B player with disjoint equilibrium and optimal strategies for each original player π. The new type-B player has $\tilde{S}_\pi - \tilde{S}^*_\pi$ and $\tilde{S}^*_\pi - \tilde{S}_\pi$ as its equilibrium and optimal strategy respectively while the new type-A players each use one resource from $|\tilde{S}_\pi \bigcap \tilde{S}^*_\pi|$ as their identical equilibrium and optimal strategies. Note that the new players are also in equilibrium in $\tilde{S}$.

Let π_i be an arbitrary type-B-player using k resources $r_1, r_2, \ldots, r_k$ in its equilibrium strategy $\tilde{S}_{\pi_i}$ that are distinct from the m resources $r^*_1, \ldots, r^*_m$ in its optimal strategy $\tilde{S}^*_{\pi_i}$. Let C_{r_j}, $C_{r^*_j}$ denote the congestion on these resources in equilibrium state $\tilde{S}$. Without loss of generality, assume the resources in $\tilde{S}_{\pi_i}$ and $\tilde{S}^*_{\pi_i}$ have been sorted in non-increasing order of congestion i.e $C_{r_1} \geq C_{r_2} \ldots \geq \ldots C_{r_m}$ and $C_{r^*_1} \leq C_{r^*_2} \ldots \leq \ldots C_{r^*_m}$. Then we have the following:

Lemma 1. *$\tilde{S}_{\pi_i}$ and $\tilde{S}^*_{\pi_i}$ can be partitioned into t pairs $(L_1, L^*_1), (L_2, L^*_2), \ldots, (L_t, L^*_t)$ where*

$$\sum_{r \in L^*_j} l(C_r + 1) \geq \sum_{r \in L_j} l(C_r), \quad 1 \leq j \leq t \tag{3}$$

and further

1. *The L_j's form a disjoint resource partition of $\tilde{S}_{\pi_i}$ i.e $L_j \bigcap L_k = \emptyset$ with $\bigcup_{j=1}^t L_j = \tilde{S}_{\pi_i}$.*
2. *$|L^*_j \bigcap L^*_{j+1}| \leq 1$, for $1 \leq j \leq t$. If $|L^*_j \bigcap L^*_{j+1}| = 1$ then the last resource in L^*_j is the first resource in L^*_{j+1}.*
3. *$\forall j : 1 \leq j \leq t$, either $|L_j| = 1$ or $|L^*_j| = 1$ or both. If $|L_j| > 1$ and $|L^*_j| = 1$ with $L^*_j = \{r^*_p\}$ we must have $C_{r^*_p} \geq \max\{C_r | r \in L_j\}$.*
4. *r^*_m appears at most once in a partition (specifically L^*_t) while r^*_1 appears in at most two partitions. If r^*_1 appears in two partitions then at least one of the partitions contains only one resource (i.e r^*_1). Every other resource $r^*_p \in \tilde{S}^*_{\pi_i}$, $2 \leq p \leq m-1$ appears in at most three partitions. If r^*_p appears in three partitions then two of the partitions contains only r^*_p. If r^*_p appears in two partitions then it is the last resource in the first partition and the first resource in the second partition.*

We label the procedure implementing lemma 1 as Procedure PMS−Partition(). This procedure is used to create new players and forms the basic step in our transformation algorithm. We ensure the equilibrium of these new players in $\tilde{G}$ using the key constructs of *exact matching sets* and *potential matching sets*.

A set of resources $\tilde{R}$ in $\tilde{G}$ forms an exact matching set for a newly created player π_k with newly assigned equilibrium strategy $\tilde{S}_{\pi_k}$ if $\sum_{r \in \tilde{R}} (C_r + 1)^{\mathcal{M}} \geq pc_{\pi_k}(\tilde{S}_{\pi_k}, \tilde{S}_{-\pi_k}) = \sum_{r \in \tilde{S}_{\pi_k}} C_r^{\mathcal{M}}$. Clearly, $\tilde{R}$ can be assigned as the new optimal strategy $\tilde{S}^*_{\pi_k}$ in game $\tilde{G}$ without violating the equilibrium of π_k.

Potential matching sets are defined for newly created type-B players. A potential matching set $\tilde{R}$ is an exact matching set that can 'potentially' be added to the optimal set of resources $\tilde{S}^*_{\pi_k}$ of a type-B player $\pi_k \in \tilde{G}$ without increasing the optimal bottleneck congestion in $\tilde{G}$ from original game G by an η factor i.e $\tilde{C^*} = O(\eta C^*)$.

Now consider a type-B player π_i to be transformed. We partition the resources in its equilibrium and optimal strategies $\tilde{S}_{\pi_i}$ and $\tilde{S}^*_{\pi_i}$ according to PMS−Partition(π_i) and remove it from $\tilde{G}$, i.e $\tilde{S} = \tilde{S} - \tilde{S}_{\pi_i}$ and $\tilde{S}^* = \tilde{S}^* - \tilde{S}^*_{\pi_i}$.

Consider those partition-pairs (L_j, L_j^*) with $|L_j| = 1$. We can create a new type-A player π_k and add it to to $\tilde{G}$ with an equilibrium strategy $\tilde{S}_{\pi_k}$ that is the singleton resource in L_j. Due to the condition in Eq. 3, the set of resources in L_j^* forms an exact matching set for π_k and can therefore be assigned to $\tilde{S}^*_{\pi_k}$. π_k is in equilibrium in $\tilde{G}$ and the equilibrium and optimal congestion on resources in $\tilde{S}_{\pi_k}$ and $\tilde{S}^*_{\pi_k}$ are now the same as before. This forms the 'easy' part of the transformation process.

Consider however, those partitions (L_j, L_j^*) with $1 < |L_j| \le |R|$ and $L_j^* = \{r_l^*\}$. Similar to the above, we can create $|L_j|$ new type-A players and assign a distinct resource in L_j to each such players equilibrium strategy. However if, as above, we assign r_l^*, the single resource in L_j^*, to each players optimal strategy, we might increase the socially optimal congestion $\tilde{C}^*$ of $\tilde{G}$ to as much as $C^* + |R|$, thereby violating the domination of G over $\tilde{G}$. Thus we need to find an appropriate potential matching set from among existing resources and assign them to these players, without increasing the optimal congestion beyond $O(\eta C^*)$. Finding such a set is the 'hard' part of the transformation process.

We form game $\tilde{G}$ by transforming type-B players in distinct phases corresponding to decreasing values of player cost functions $l_r(C_r)$ from $max_r l_r(C)$ down to $min_r l_r(\tau)$. During phase i with congestion index C_i, we transform all type-B players with costs in the range $l_{r_j}(C_i)$ down to $l_{r_k}(C_{i-1})+1$ into type-A players, where $l_{r_j}()$ and $l_{r_k}()$ are respective maximizers in initial game G. To find new potential matching sets for all type-B players without increasing the optimal congestion in $\tilde{G}$ beyond a constant factor, we utilize the set of resources with congestion exactly C_i to generate For details of the proof, please see [13].

4 Price of Anarchy

4.1 Price of Anarchy for Type-A Player Games

By Theorem 2 and Corollary 1, for every type-B game G, we can find a dominated type-A game $\tilde{G}$ such that

$$PoA(\hat{G}) \le 7 \cdot PoA(\tilde{G}) \tag{4}$$

Thus we only need to find an upper bound on the PoA of type-A games. Consider a generic type-A game G with optimal solution $S^* = (S^*_{\pi_1}, \ldots, S^*_{\pi_n})$, optimal congestion C^*, and a Nash equilibrium state $S = (S_{\pi_1}, \ldots, S_{\pi_n})$ with the highest congestion C among all Nash equilibria of G. Define a congestion threshold ψ such that $\forall r \forall C_r : C_r \geq \psi$, we have $C_r \geq \frac{l_r(C_r+1)}{l_r(C_r)} \cdot C^*$. For general polynomial games with $l_r(C_r) = C_r^{\mathcal{M}_r}$, we can choose $\psi = 2\max_r(\mathcal{M}_r, C^*)$. For superpolynomial games with $l_r(C_r) = C_r^{\log C_r}$, we can choose $\psi = e^2 \cdot C^*$. We specify G as a type-A game in which all resources r with $C_r > \psi$ are utilized only by type-A players in in state S. As a consequence of Eq. 4, we can bound the *PoA* of arbitrary type-B games by bounding the ratio C/C^*.

We first define a resource graph $\mathcal{N}$ for state S. There are $V = V_1 \bigcup V_2$ nodes in $\mathcal{N}$. Each resource $r \in R$ with $C_r > \psi$ ($C_r \leq \psi$, resp.) corresponds to the equivalent node $r \in V_1$ ($r \in V_2$). Henceforth we will use the term resource and node interchangeably. For every player π using a resource $x \in V_1$ in equilibrium, there is a directed edge (x, y) between node x and all nodes $y \in V$, where $y \neq x$ is in the optimal strategy set of π i.e $S_\pi = x$ and $y \in S_\pi^*$. We use the notation $\mathrm{Ch}(x)$ to denote the set $\bigcup_{\pi : S_\pi = x} S_\pi^*$. Note that there could be multiple incoming links to a node x from the same node, however there are no self-loops and x can be the child of at most C^* nodes. Also note that nodes in V_2 are terminal nodes that have no outgoing links.

Recursively counting the number of descendants of the root node in $\mathcal{T}$ will help us relate the number of resources $|R|$ with the parameters C and C^* and obtain our *PoA* bound. However since $\mathcal{N}$ can have cycles, we instead modify $\mathcal{N}$ to remove cycles and replace it with a Directed Acyclic Graph (DAG) $\mathcal{T}$ (without increasing the size of $\mathcal{N}$).

Lemma 2. *Resource graph $\mathcal{N}$ can be transformed into expansion DAG $\mathcal{T}$ without affecting the equilibrium state S and optimal congestion C^*, where $|\mathcal{T}| \leq |\mathcal{N}|$.*

Since $\mathcal{T}$ is a DAG we know that it has sink nodes (with outdegree 0). Every node in V_1 is an internal node (with non-zero indegree and outdegree) since it has congestion $> C^*$ and hence the sink nodes in $\mathcal{T}$ are nodes from V_2. Consider the DAG starting at the root node with congestion C. Let κ denote the *PoA* of G.

Lemma 3. *For DAG $\mathcal{T}$ with root node x and congestion $C_x = C$, it holds that*

$$\sum_{r \in \mathrm{Descendants}(x) \bigcap V_2} l_r(\psi) \geq (\kappa - 1) l_x(C)$$

Using the fact that $|V_2| \leq |R|$, and denoting $l_{max}()$ and $l_{min}()$ as the respective maximizers/minimizers among the latency cost functions, we get

$$\kappa l_{min}(C) \leq 2|R| l_{max}(\psi) \tag{5}$$

Note that the the *PoA* of superpolynomial and general polynomial games is at least ψ/C^*.

For superpolynomial games using the fact that $\forall r \in R : l_r(C_r) = C_r^{\log C_r}$ and $\psi = e^2 \cdot C^*$ and simplifying, we get that $\log \kappa + (\log \kappa)^2 \leq \log |R|$. Hence for type-$A$ superpolynomial games the Price of Anarchy is bounded by

$$PoA_{super-poly} = O(2^{\sqrt{\log |R|}})$$

For general polynomial games with $l_r(C_r) = C_r^{\mathcal{M}_r}$, let β and α respectively, denote the largest and smallest degrees in $\mathcal{M}_r$. Substituting in Eq. 5, we get $\kappa C^\alpha \leq 2|R| \cdot \psi^\beta$ or $\kappa^{\alpha+1} \leq 2|R| \cdot (\frac{\psi}{C^*})^\alpha \cdot \psi^{\beta-\alpha}$. Hence for type-$A$ general polynomial games the Price of Anarchy is bounded by

$$PoA_{poly} = \min\left(|R|, \max\left(\frac{\psi}{C^*}, (2|R|)^{\frac{1}{\alpha+1}} \cdot (\frac{\psi}{C^*})^{\frac{\alpha}{\alpha+1}} \cdot \psi^{\frac{\beta-\alpha}{\alpha+1}}\right)\right)$$

Substituting $\psi = 2\max(\beta, C^*)$, we get the result.

5 Lower Bounds on Price of Anarchy

We demonstrate a simple lower bound to show the tightness of the above upper bound. Consider a type-A superpolynomial game G in which C players in Nash equilibrium state utilize the same resource r, i.e $C_r = C$. No other resource is being utilized in this state. The optimal strategy of each player has $C^{\log C}$ unique resources, i.e the congestion in optimal state S^* is $C^* = 1$. Clearly G is in equilibrium and the PoA is $\kappa = C$. The total number of resources R is given by $|R| = C \cdot C^{\log C} + 1$ and hence we have

$$\kappa = 2^{\sqrt{\log(|R|-1)}}$$

Similarly consider a type-A general polynomial game G in which C players in Nash equilibrium state are utilizing only one resource x, where $l_x(C_x) = (C_x)^\alpha$. Let N be any integer, $2 \leq N \leq C$. Each distinct subset of N out of these C players are sharing $\lceil \frac{C^\alpha}{N^\beta} \rceil$ unique resources in optimal state S^*. These resources are used by $N-1$ other players in equilibrium as well as optimal states S and S^*. Hence the equilibrium congestion of these resources is $N-1$ while congestion in optimal state S^* is $C^* = 2N-1$. Note thate there are $\lceil \frac{C}{N} \rceil$ sets of these resources. The latency cost on each of these resources r is $l_r(C_r) = (C_r)^\beta$.

Clearly G is in equilibrium and the PoA is $\kappa = \frac{C}{2N-1}$. The total number of resources $|R| = \frac{C}{N} \cdot \frac{C^\alpha}{N^\beta} + 1$ and hence we have

$$\kappa = \min\left(|R|, \max\left(\frac{C}{2N-1}, (|R|-1)^{\frac{1}{\alpha+1}} \cdot N^{\frac{\beta-\alpha}{\alpha+1}} \cdot \frac{N}{2N-1}\right)\right), 2 \leq N \leq C.$$

Note that this has the same form as the upper bound above.

6 Conclusions

We have considered bottleneck congestion games with polynomial and superpolynomial resource delay cost functions. The price of anarchy result for superpolynomial functions is $o(\sqrt{|R|})$ with respect to the number of resources. We also demonstrate two novel techniques, B to A player conversion and expansion which help us obtain this result. These techniques which enable us to simplify games for analysis are sufficiently general. In future work, we plan to use these techniques to analyze the PoA of games with arbitrary player cost functions.

References

1. Awerbuch, B., Azar, Y., Epstein, A.: The price of routing unsplittable flow. In: Proceedings of the 37th Annual ACM Symposium on Theory of Computing (STOC), Baltimore, MD, USA, pp. 57–66. ACM (May 2005)
2. Banner, R., Orda, A.: Bottleneck routing games in communication networks. IEEE Journal on Selected Areas in Communications 25(6), 1173–1179 (2007); also appears in INFOCOM 2006
3. Busch, C., Kannan, R., Vasilakos, A.V.: Quality of routing congestion games in wireless sensor networks. In: Proc. 4th International Wireless Internet Conference (WICON), Maui, Hawaii (November 2008)
4. Busch, C., Magdon-Ismail, M.: Atomic routing games on maximum congestion. Theoretical Computer Science 410(36), 3337–3347 (2009)
5. Christodoulou, G., Koutsoupias, E.: The price of anarchy of finite congestion games. In: Proceedings of the 37th Annual ACM Symposium on Theory of Computing (STOC), Baltimore, MD, USA, pp. 67–73. ACM (May 2005)
6. Correa, J.R., Schulz, A.S., Stier Moses, N.E.: Computational Complexity, Fairness, and the Price of Anarchy of the Maximum Latency Problem. In: Bienstock, D., Nemhauser, G.L. (eds.) IPCO 2004. LNCS, vol. 3064, pp. 59–73. Springer, Heidelberg (2004)
7. Czumaj, Vocking: Tight bounds for worst-case equilibria. ACM Transactions on Algorithms (TALG) 3 (2007)
8. de Keijzer, B., Schäfer, G., Telelis, O.A.: On the Inefficiency of Equilibria in Linear Bottleneck Congestion Games. In: Kontogiannis, S., Koutsoupias, E., Spirakis, P.G. (eds.) SAGT 2010. LNCS, vol. 6386, pp. 335–346. Springer, Heidelberg (2010)
9. Fotakis, D., Kontogiannis, S.C., Spirakis, P.G.: Selfish unsplittable flows. Theoretical Computer Science 348(2-3), 226–239 (2005)
10. Gairing, M., Lücking, T., Mavronicolas, M., Monien, B.: Computing Nash equilibria for scheduling on restricted parallel links. In: Proceedings of the 36th Annual ACM Symposium on the Theory of Computing (STOC), Chicago, Illinois, USA, pp. 613–622 (June 2004)
11. Harks, T., Klimm, M., Möhring, R.H.: Strong Nash Equilibria in Games with the Lexicographical Improvement Property. In: Leonardi, S. (ed.) WINE 2009. LNCS, vol. 5929, pp. 463–470. Springer, Heidelberg (2009)
12. Kannan, R., Busch, C.: Bottleneck Congestion Games with Logarithmic Price of Anarchy. In: Kontogiannis, S., Koutsoupias, E., Spirakis, P.G. (eds.) SAGT 2010. LNCS, vol. 6386, pp. 222–233. Springer, Heidelberg (2010)
13. Kannan, R., Busch, C.: Optimal price of anarchy for polynomial and super-polynomial bottleneck congestion games. LSU-CSC Tech Report 11-03 (2011)
14. Koutsoupias, E., Mavronicolas, M., Spirakis, P.G.: Approximate equilibria and ball fusion. Theory Comput. Syst. 36(6), 683–693 (2003)
15. Koutsoupias, E., Papadimitriou, C.: Worst-Case Equilibria. In: Meinel, C., Tison, S. (eds.) STACS 1999. LNCS, vol. 1563, pp. 404–413. Springer, Heidelberg (1999)

16. Leighton, F.T., Maggs, B.M., Rao, S.B.: Packet routing and job-scheduling in $O(congestion + dilation)$ steps. Combinatorica 14, 167–186 (1994)
17. Libman, L., Orda, A.: Atomic resource sharing in noncooperative networks. Telecomunication Systems 17(4), 385–409 (2001)
18. Lücking, T., Mavronicolas, M., Monien, B., Rode, M.: A new model for selfish routing. Theoretical Computer Science 406(3), 187–206 (2008)
19. Mavronicolas, Spirakis: The price of selfish routing. Algorithmica 48 (2007)
20. Monderer, D., Shapely, L.S.: Potential games. Games and Economic Behavior 14, 124–143 (1996)
21. Papadimitriou, C.: Algorithms, games, and the Internet. In: ACM (ed.) Proceedings of the 33rd Annual ACM Symposium on Theory of Computing (STOC), Hersonissos, Crete, Greece, pp. 749–753 (July 2001)
22. Rosenthal, R.W.: A class of games possesing pure-strategy Nash equilibria. International Journal of Game Theory 2, 65–67 (1973)
23. Roughgarden, T.: The maximum latency of selfish routing. In: Proceedings of the Fifteenth Annual ACM-SIAM Symposium on Discrete Algorithms (SODA), New Orleans, Louisiana (USA), pp. 980–981 (January 2004)
24. Roughgarden, T.: Selfish routing with atomic players. In: Proc. 16th Symp. on Discrete Algorithms (SODA), pp. 1184–1185. ACM/SIAM (2005)
25. Roughgarden, T., Tardos, É.: How bad is selfish routing. Journal of the ACM 49(2), 236–259 (2002)
26. Roughgarden, T., Tardos, É.: Bounding the inefficiency of equilibria in nonatomic congestion games. Games and Economic Behavior 47(2), 389–403 (2004)
27. Suri, S., Toth, C.D., Zhou, Y.: Selfish load balancing and atomic congestion games. Algorithmica 47(1), 79–96 (2007)

Performance and Convergence of Multi-user Online Learning

Cem Tekin and Mingyan Liu

Department of Electrical Engineering and Computer Science
University of Michigan, Ann Arbor, Michigan, 48109-2122
{cmtkn,mingyan}@umich.edu

Abstract. We study the problem of allocating multiple users to a set of wireless channels in a decentralized manner when the channel qualities are time-varying and unknown to the users, and accessing the same channel by multiple users leads to reduced quality due to interference. In such a setting the users not only need to learn the inherent channel quality and at the same time the best allocations of users to channels so as to maximize the social welfare. Assuming that the users adopt a certain online learning algorithm, we investigate under what conditions the socially optimal allocation is achievable. In particular we examine the effect of different levels of knowledge the users may have and the amount of communications and cooperation. The general conclusion is that when the cooperation of users decreases and the uncertainty about channel payoffs increases it becomes harder to achieve the socially optimal allocation.

Keywords: multi-user learning, multi-armed bandits, spectrum sharing, congestion games.

1 Introduction

In this paper we study the dynamic spectrum access and spectrum sharing problem in a learning context. Specifically, we consider a set of N common channels shared by a set of M users. A channel has time varying rate $r(t)$, and its statistics are not completely known by the users. Thus each user needs to employ some type of learning to figure out which channels are of better quality, e.g., in terms of their average achievable rates. At the same time, simultaneous use of the same channel by multiple users will result in reduced rate due to interference or collision. The precise form of this performance degradation may or may not be known to the user. Thus the users also need to use learning to avoid excess interference or congestion. Furthermore, each user may have private information that is not shared, e.g., users may perceive channel quality differently due to difference in location as well as individual modulation/coding schemes.

Without a central agent, and in the presence of information decentralization described above, we are interested in the following questions: (1) for a given common learning algorithm, does the multiuser learning process converge, and

R. Jain and R. Kannan (Eds.): GameNets 2011, LNICST 75, pp. 321–336, 2012.

(2) if it does, what is the quality of the equilibrium point with respect to a globally optimal spectrum allocation scheme, one that could be computed for a global objective function with full knowledge of channel statistics as well as the users' private information.

A few recent studies have addressed these questions in some special cases. For instance, in [3] it was shown that learning using a sample-mean based index policy leads to a socially optimal (sum of individual utilities) allocation when channels evolve as iid processes and colliding players get zero reward provided that this optimal allocation is such that each user occupies one of the M best channels (in terms of average rates). This precludes the possibility that not all users may have the same set of M best channels, and that in some cases the best option is for multiple users to share a common channel, e.g., when $N < M$.

In this study we investigate under what conditions the socially optimal allocation is achievable by considering different levels of communication (or cooperation) allowed among users, and different levels of uncertainty on the channel statistics. The general conclusion, as intuition would suggest, is that when the cooperation of users increases and the channel uncertainty decreases it becomes easier to achieve the socially optimal welfare. Specifically, we assume that the rate (or reward) user i gets from channel j at time t is of the form $r_j(t)g_j(n_j(t))$ where $r_j(t)$ is the rate of channel j at time t, $n_j(t)$ is the number of users using channel j at time t, and g_j is the user independent interference function (IF) for channel j. This model is richer than the previously used models [3,14,16] since $r_j(t)$ can represent environmental effects such as fading or primary user activity, while g_j captures interactions between users. We consider the following three cases.

In the first case (C1), each channel evolves as an iid random process in time, the users do not know the channel statistics, nor the form of the interference, nor the total number of users present in the system, and no direct communication is allowed among users. A user can measure the overall rate it gets from using a channel but cannot tell how much of it is due to the dynamically changing channel quality (i.e., what it would get if it were the only user) vs. interference from other users. In this case, we show that if all users follow the Exp3 algorithm [7] then the channel allocation converges to a set of pure Nash equilibria (PNE) of a congestion game defined by the IFs and mean channel rates. In this case a socially optimal allocation cannot be ensured, as the set of PNE are of different quality, and in some cases the socially optimal allocation may not be a PNE.

In the second case (C2), each channel again evolves as an iid random process in time, whose statistics are unknown to the user. However, the users now know the total number of users in the system, as well as the fact that the quantitative impact of interference is common to all users (i.e., user independent), though the actual form of the interference function is unknown. In other words the rate of channel j at time t is perceived by user i as $h_j(t, n_j(t))$ so user i cannot distinguish between components $r_j(t)$ and $g_j(n_j(t))$. Furthermore, users are now allowed minimal amount of communication when they happen to be in the same channel, specifically to find out the total number of simultaneous users of that channel. In this case we present a sample-mean based randomized learning policy that achieves socially optimal allocation as time goes to infinity, with a sub-linear regret over the time horizon with respect to the socially optimal allocation.

In the third case (C3), as in case (C2) the users know the total number of users in the system, as well as the fact that the IF is user independent and decreasing without knowing the actual form of the IF. However, the channels are assumed to have constant, albeit unknown, rates. We show that even without any communication among users, there is a randomized learning algorithm that achieves the socially optimal allocation in finite time.

It's worth pointing out that in the settings outlined above, the users are *non-strategic*, i.e., each user simply follow a pre-set learning rule rather than playing a game. In this context it is reasonable to introduce minimal amount of communication among users and assume they may cooperate. It is possible that even in this case the users may not know their IF but only the total rate they get for lack of better detecting capabilities (e.g., they may only be able to detect the total received SNR as a result of channel rate and user interference).

Online learning by a single user was studied by [1,4,6,15], in which sample-mean based index policies were shown to achieve logarithmic regret with respect to the best single-action policy without a priori knowledge of the statistics, and are order-optimal, when the rewards are given by an iid process. In [5,21,22] Markovian rewards are considered, with [22] focusing on *restless* reward processes, where a process continues to evolve according to a Markov chain regardless of the users' actions. In all these studies learning algorithms were developed to achieve logarithmic regret. Multi-user learning with iid reward processes have been studied in a dynamic spectrum context by [3,11,16], with a combinatorial structure adopted in [11], and with collision and random access models in [3,16]. In [13], convergence of multi-user learning with Exp3 algorithm to pure Nash equilibrium is investigated under the collision and fair sharing models. In the collision model, when there is more than one user on a channel all get zero reward, whereas in the random access model one of them, selected randomly, gets all the reward while others get zero reward. In the fair sharing model, a user's utility is inversely proportional to the number of users who are on the same channel with the user. Note that these models do not capture more sophisticated communication schemes where the rate a user gets is a function of the received SNR of the form $g_j(n) = f_j(\frac{P_t}{N_0+(n-1)P_t}) =$ where P_t is the nominal transmit power of all users and N_0 the noise. Moreover, in the above studies the socially optimal allocation is a rather simple one: it is the orthogonal allocation of users to the first M channels with the highest mean rewards. By contrast, we model a more general interference relationship among users, in which an allocation with users sharing the same channel may be the socially optimal one. The socially optimal allocations is not trivial in this case and additional mechanisms may be needed for the learning algorithms to converge.

All of the above mentioned work assumes some level of communication between the users either at the beginning or during the learning. If we assume no communication between the users, achieving the socially optimal allocation seems very challenging in general. Then one may ask if it is possible to achieve some kind of equilibrium allocation. Kleinberg et. al. [14] showed that it is possible for the case when the channel rates are constant and the users do not know the IFs. They show that when the users use *aggregate monotonic selection dynamics*, a variant of Hedge algorithm [10], the allocation converges to *weakly stable equilibria* which is a subset of Nash equilibria (NE) of the congestion game defined by the IFs. They show that for almost all congestion games weakly stable equilibria is the same as PNE.

Other than the work described above [2] considers *spatial congestion games*, a generalization of congestion games and gives conditions under which there exists a PNE and best-response play converges to PNE. A mechanism design approach for socially optimal power allocation when users are strategic is considered in [12].

The organization of the remainder of this paper is as follows. In Sect. 2 we present the notations and definitions that will be used throughout the paper. In Sects. 3, 4, 5 we analyze the cases stated in (C1), (C2), (C3) and derive the results respectively. Conclusion and future research is given in Sect. 6.

2 Preliminaries

Denote the set of users by $\mathcal{M} = \{1, 2, \ldots, M\}$, and the set of channels $\mathcal{N} = \{1, 2, \ldots, N\}$. Time is slotted and indexed by $t = 1, 2, \ldots$ and a user can select a single channel at each time step t. Without loss of generality let $r_j(t) \in [0, 1]$ be the rate of channel j at time t such that $\{r_j(t)\}_{t=1,2,\ldots}$ is generated by an iid process with support $[0, 1]$ and mean $\mu_j \in [0, 1]$. Let $g_j : \mathbb{N} \to [0, 1]$ be the interference function (IF) on channel j where $g_j(n)$ represents the interference when there are n users on channel j. We express the rate of channel j seen by a user as $h_j(t) = r_j(t) g_j(n_j(t))$ when a user does not know the total number of users $n_j(t)$ using channel j at time t as in cases (C1) and (C3). When a user knows $n_j(t)$, we express the rate of channel j at time t as $h_{j,n_j(t)}(t) = r_j(t) g_j(n_j(t))$ as in case (C2). Let $\mathcal{S}_i = \mathcal{N}$ be the set of feasible actions of user i and $\sigma_i \in \mathcal{S}_i$ be the action, i.e., channel selected by user i. Let $\mathcal{S} = \mathcal{S}_1 \times \mathcal{S}_2 \times \ldots \times \mathcal{S}_N = \mathcal{N}^M$ be the set of feasible action profiles and $\sigma = \{\sigma_1, \sigma_2, \ldots, \sigma_M\} \in \mathcal{S}$ be the action profile of the users. Throughout the discussion we assume that the action of player i at time t, i.e., $\sigma_i^{\pi_i}(t)$ is determined by the policy π_i. When π_i is deterministic, $\pi_i(t)$ is in general a function from all past observations and decisions of user i to the set of actions $\mathcal{S}_i$. When π_i is randomized, $\pi_i(t)$ generates a probability distribution over the set of actions $\mathcal{S}_i$ according to all past observations and decisions of user i from which the action at time t is sampled. Since the dependence of actions to the policy is trivial we use $\sigma_i(t)$ to denote the action of user i at time t, dropping the superscript π_i.

Let $K_j(\sigma)$ be the set of users on channel j when the action profile is σ. Let $\mathcal{A}^* = \arg\max_{\sigma \in \mathcal{S}} \sum_{i=1}^{M} \mu_{\sigma_i} g_{\sigma_i}(K_{\sigma_i}(\sigma)) = \arg\max_{\sigma \in \mathcal{S}} \sum_{j=1}^{N} \mu_j K_j(\sigma) g_j(K_j(\sigma))$ be the set of socially optimal allocations and denote by σ^* any action profile that is in the set $\mathcal{A}^*$. Let v^* denote the socially optimal welfare, i.e., $v^* = \sum_{i=1}^{M} \mu_{\sigma_i^*} g_{\sigma_i^*}(K_{\sigma_i^*}(\sigma^*))$ and v_j^* denote the payoff a user gets from channel j under the socially optimal allocation, i.e., $v_j^* = \mu_j g_j(K_j(\sigma^*))$ if $K_j(\sigma^*) \neq 0$. Note that any permutation of actions in σ^* is also a socially optimal allocation since IFs are user-independent.

For any policy π, the regret at time n is

$$R(n) = nv^* - E\left[\sum_{t=1}^{n} \sum_{i=1}^{M} r_{\sigma_i(t)}(t) g_{\sigma_i(t)}(K_{\sigma_i(t)}(\sigma(t)))\right],$$

where expectation is taken with respect to the random nature of the rates and the randomization of the policy. Note that for a deterministic policy expectation is only taken with respect to the random nature of the rates. For any randomized policy π_i, let $p_i(t) = (p_{i1}(t), p_{i2}(t), \ldots, p_{iN}(t))$ be the mixed strategy of user i at time t, i.e., a probability distribution on $\{1, 2, \ldots, N\}$. For a profile of policies $\pi = [\pi_1, \pi_2, \ldots, \pi_M]$ for the users let $p(t) = (p_1(t)^T, p_2(t)^T, \ldots p_M(t)^T)^T$ be the profile of mixed strategies at time t, where $p_i(t)^T$ is the transpose of $p_i(t)$. Then $\sigma_i(t)$ is the action sampled from the probability distribution $p_i(t)$. The dependence of p to π is trivial and not shown in the notation.

3 Allocations Achievable with Exp3 Algorithm (Case 1)

We start by defining a congestion game. A congestion game [17,18] is given by the tuple $(\mathcal{M}, \mathcal{N}, (\Sigma_i)_{i \in \mathcal{M}}, (h_j)_{j \in \mathcal{N}})$, where $\mathcal{M}$ denotes a set of players (users), $\mathcal{N}$ a set of resources (channels), $\Sigma_i \subset 2^{\mathcal{N}}$ the strategy space of player i, and $h_j : \mathbb{N} \to \mathbb{R}$ a payoff function associated with resource j, which is a function of the number of players using that resource. It is well known that a congestion game has a potential function and the local maxima of the potential function corresponds to PNE, and every sequence of asynchronous improvement steps is finite and converges to PNE.

In this section we relate the strategy update rule of Exp3 [7] under assumptions (C1) to a congestion game. Exp3 as given in Fig. 1 is a randomized algorithm consisting of an exploration parameter γ and weights w_{ij} that depend exponentially on the past observations where i denotes the user and j denotes the channel. Each user runs Exp3 independently but we explicitly note the user dependence because a user's action affects other users' updates.

At any time step before the channel rate and user actions are drawn from the corresponding distributions, let R_j denote the random variable corresponding to the reward of the jth channel. Let $G_{ij} = g_j(1 + K_j'(i))$ be the random variable representing the payoff user i gets from channel j where $K_j'(i)$ is the random variable representing the number of users on channel j other than user i. Let $U_{ij} = R_j G_{ij}$ and $\bar{u}_{ij} = E_j[E_{-i}[U_{ij}]]$ be the expected payoff to user i by

Exp3 (for user i)

Initialize: $\gamma \in (0,1)$, $w_{ij}(t) = 1, \forall j \in \mathcal{N}$, $t = 1$
while $t > 0$ **do**

$$p_{ij}(t) = (1-\gamma)\frac{w_{ij}(t)}{\sum_{l=1}^{N} w_{il}(t)} + \frac{\gamma}{N}$$

  Sample $\sigma_i(t)$ from the distribution on $p_i(t) = [p_{i1}(t), p_{i2}(t), \ldots, p_{iN}(t)]$
  Play channel $\sigma_i(t)$ and receive reward $h_{\sigma_i(t)}(t)$
  for $j = 1, 2, \ldots, N$ **do**
    if $j = \sigma_i(t)$ **then**
      Set $w_{ij}(t+1) = w_{ij}(t)\exp\left(\frac{\gamma h_{\sigma_i(t)}(t)}{p_{ij}(t)N}\right)$
    else
      Set $w_{ij}(t+1) = w_{ij}(t)$
    end if
  end for
  $t = t + 1$
end while

Fig. 1. pseudocode of Exp3

using channel j where E_{-i} represents the expectation taken with respect to the randomization of players other than i, E_j represents the expectation taken with respect to the randomization of the rate of channel j. Since the channel rate is independent of users' actions $\bar{u}_{ij} = \mu_j \bar{g}_{ij}$ where $\bar{g}_{ij} = E_{-i}[G_{ij}]$.

Lemma 1. *Under (C1) when all players use Exp3, the derivative of the continuous-time limit of Exp3 is the replicator equation given by*

$$\xi_{ij} = \frac{1}{N}(\mu_j p_{ij}) \sum_{l=1}^{N} p_{il}(\bar{g}_{ij} - \bar{g}_{il}) .$$

Proof. Note that

$$(1-\gamma) w_{ij}(t) = \sum_{l=1}^{N} w_{il}(t)\left(p_{ij}(t) - \frac{\gamma}{N}\right) . \tag{1}$$

We consider the effect of user i's action $\sigma_i(t)$ on his probability update on channel j. We have two cases: $\sigma_i(t) = j$ and $\sigma_i(t) \neq j$. Let $A_{i,j}^{\gamma,t} = \exp\left(\frac{\gamma U_{ij}(t)}{p_{ij}(t)N}\right)$.

Consider the case $\sigma_i(t) = j$.

$$p_{ij}(t+1) = \frac{(1-\gamma) w_{ij}(t) A_{i,j}^{\gamma,t}}{\sum_{l=1}^{N} w_{il}(t) + w_{ij}(t)\left(A_{i,j}^{\gamma,t} - 1\right)} + \frac{\gamma}{N} . \tag{2}$$

Substituting (1) into (2)

$$p_{ij}(t+1) = \frac{\sum_{l=1}^{N} w_{il}(t)\left(p_{ij}(t) - \frac{\gamma}{N}\right) A_{i,j}^{\gamma,t}}{\sum_{l=1}^{N} w_{il}(t)\left(1 + \frac{p_{ij}(t)-\frac{\gamma}{N}}{1-\gamma}\left(A_{i,j}^{\gamma,t} - 1\right)\right)} + \frac{\gamma}{N}$$

$$= \frac{\left(p_{ij}(t) - \frac{\gamma}{N}\right) A_{i,j}^{\gamma,t}}{1 + \frac{p_{ij}(t)-\frac{\gamma}{N}}{1-\gamma}\left(A_{i,j}^{\gamma,t} - 1\right)} + \frac{\gamma}{N} .$$

The continuous time process is obtained by taking the limit $\gamma \to 0$, i.e., the rate of change in p_{ij} with respect to γ as $\gamma \to 0$. Then, dropping the discrete time script t,

$$\dot{p_{ij}} = \lim_{\gamma \to 0} \frac{dp_{ij}}{d\gamma}$$

$$= \lim_{\gamma \to 0} \frac{\left(\frac{-1}{N} A_{i,j}^{\gamma,t} + \left(p_{ij} - \frac{\gamma}{N}\right) \frac{U_{ij}}{p_{ij} N} A_{i,j}^{\gamma,t}\right)\left(1 + \frac{p_{ij}-\frac{\gamma}{N}}{1-\gamma}\left(A_{i,j}^{\gamma,t} - 1\right)\right)}{\left(1 + \frac{p_{ij}-\frac{\gamma}{N}}{1-\gamma}\left(A_{i,j}^{\gamma,t} - 1\right)\right)^2}$$

$$+ \frac{\left(p_{ij} - \frac{\gamma}{N}\right) A_{i,j}^{\gamma,t}\left(\frac{p_{ij}-\frac{1}{N}}{(1-\gamma)^2} A_{i,j}^{\gamma,t} + \frac{p_{ij}-\frac{1}{N}}{1-\gamma}\left(\frac{\gamma}{N} A_{i,j}^{\gamma,t}\right)\right)}{\left(1 + \frac{p_{ij}-\frac{\gamma}{N}}{1-\gamma}\left(A_{i,j}^{\gamma,t} - 1\right)\right)^2} + \frac{1}{N}$$

$$= \frac{U_{ij}(1 - p_{ij})}{N} . \tag{3}$$

Consider the case $\sigma_i(t) = k \neq j$. Then,

$$p_{ij}(t+1) = \frac{(1-\gamma) w_{ij}(t)}{\sum_{l=1}^{N} w_{il}(t) + w_{ik}(t)\left(A_{i,k}^{\gamma,t} - 1\right)} + \frac{\gamma}{N}$$

$$= \frac{p_{ij}(t) - \frac{\gamma}{N}}{1 + \frac{p_{ik}(t)-\frac{\gamma}{N}}{1-\gamma}\left(A_{i,k}^{\gamma,t} - 1\right)} + \frac{\gamma}{N}.$$

Thus

$$\dot{p_{ij}} = \lim_{\gamma \to 0} \frac{\frac{-1}{N}\left(1 + \frac{p_{ik}-\frac{\gamma}{N}}{1-\gamma}\left(A_{i,k}^{\gamma,t} - 1\right)\right)}{\left(1 + \frac{p_{ik}-\frac{\gamma}{N}}{1-\gamma}\left(A_{i,k}^{\gamma,t} - 1\right)\right)^2}$$

$$+ \frac{\left(p_{ij} - \frac{\gamma}{N}\right)\left(\frac{p_{ik}-\frac{1}{N}}{(1-\gamma)^2} A_{i,k}^{\gamma,t} + \frac{p_{ik}-\frac{1}{N}}{1-\gamma}\left(\frac{\gamma}{N} A_{i,k}^{\gamma,t}\right)\right)}{\left(1 + \frac{p_{ik}-\frac{\gamma}{N}}{1-\gamma}\left(A_{i,k}^{\gamma,t} - 1\right)\right)^2} + \frac{1}{N}$$

$$= -\frac{p_{ik} U_{ik}}{N} . \tag{4}$$

Then from (3) and (4), the expected change in p_{ij} with respect to the probability distribution p_i of user i over the channels is

$$\bar{p}_{ij} = E_i[\dot{p}_{ij}] = \frac{1}{N} p_{ij} \sum_{l \in \mathcal{N} - \{j\}} p_{il}(U_{ij} - U_{il}).$$

Taking the expectation with respect to the randomization of channel rates and other users' actions we have

$$\begin{aligned}\xi_{ij} &= E_j[E_{-i}[\bar{p}_{ij}]] \\ &= \frac{1}{N} p_{ij} \sum_{l \in \mathcal{N}-\{j\}} p_{il} \left(E_j[E_{-i}[U_{ij}]] - E_j[E_{-i}[U_{il}]]\right) \\ &= \frac{1}{N}(\mu_j p_{ij}) \sum_{l=1}^{N} p_{il}(\bar{g}_{ij} - \bar{g}_{il}) \ .\end{aligned}$$

□

Lemma 1 shows that the dynamics of a user's probability distribution over the actions is given by a replicator equation which is commonly studied in evolutionary game theory [19,20]. With this lemma we can establish the following theorem.

Theorem 1. *For all but a measure zero subset of* $[0,1]^{2N}$ *from which the* μ_j*'s and* g_j*'s are selected, when* γ *in Exp3 is arbitrarily small, the action profile converges to the set of PNE of the congestion game* $(\mathcal{M}, \mathcal{N}, (\mathcal{S}_i)_{i \in \mathcal{M}}, (\mu_j g_j)_{j \in \mathcal{N}})$.

Proof. Because the replicator equation in Lemma 1 is identical to the replicator equation in [14], the proof of converge to PNE follows from [14]. Here, we briefly explain the steps in the proof. Defining the expected potential function to be the expected value of the potential function ϕ where expectation is taken with respect to the user's randomization one can show that the solutions of the replicator equation converges to the set of fixed points. Then the stability analysis using the Jacobian matrix yields that every stable fixed point corresponds to a Nash equilibrium. Then one can prove that for any stable fixed point the eigenvalues of the Jacobian must be zero. This implies that every stable fixed point corresponds to a *weakly stable* Nash equilibrium strategy in the game theoretic sense. Then using tools from algebraic geometry one can show that almost every weakly stable Nash equilibrium is a pure Nash equilibrium of the congestion game.

We also need to investigate the error introduced by treating the discrete time update rule as a continuous time process. However, by taking γ infinitesimal we can approximate the discrete time process by the continuous time process. For a discussion when γ is not infinitesimal one can define *approximately stable equilibria* [14]. □

The main difference between Exp3 and Hedge [14] is that in Exp3 users do not need to observe the payoffs from the channels that they do not select, whereas Hedge assumes complete observation. In addition to that, we considered the dynamic channel rates which is not considered in [14].

4 An Algorithm for Socially Optimal Allocation with Sub-linear Regret (Case 2)

In this section we propose an algorithm whose regret with respect to the socially optimal allocation is $O(n^{\frac{2M-1+2\gamma}{2M}})$ for $\gamma > 0$ arbitrarily small. Clearly this regret is sublinear and approaches linear as the number of users M increases. This means that the time average of the sum of the utilities of the players converges to the socially optimal welfare. Let $\mathcal{K} = \{k = (k_1, k_2, \ldots, k_N) : k_j \geq 0, \forall j \in \mathcal{N}, k_1 + k_2 + \ldots + k_N = M\}$ denote an allocation of M users to N channels. Note that this allocation gives only the number of users on each channel. It does not say anything about which user uses which channel. We assume that the socially optimal allocation is unique up to permutations so $k^* = \arg\max_{k \in \mathcal{K}} \sum_{j=1}^{N} \mu_j k_j g_j(k_j)$ is unique. We also assume the following stability condition of the socially optimal allocation. Let $v_j(k_j) = \mu_j g_j(k_j)$. Then the stability condition says that $\arg\max_{k \in \mathcal{K}} \sum_{j=1}^{N} k_j \hat{v}_j(k_j) = k^*$ if $|\hat{v}_j(k) - v_j(k)| \leq \epsilon, \forall k \in \{1, 2, \ldots, M\}, \forall j \in \mathcal{N}$, for some $\epsilon > 0$, where $\hat{v}_j : \mathbb{N} \to \mathbb{R}$ is an arbitrary function. Let $T^i_{j,k}(t)$ be the number of times user i used channel j and observed k users on it up to time t. We refer to the tuple (j, k) as an arm. Let $n^i_{j,k}(t)$ be the time of the tth observation of user i from arm (j, k). Let $u^i_{j,k}(t)$ be the sample mean of the rewards from arm (j, k) seen by user i at the end of the tth play of arm (j, k) by user i, i.e., $u^i_{j,k}(t) = (h_{j,k}(n^i_{j,k}(1)) + \ldots + h_{j,k}(n^i_{j,k}(t)))/t$. Then the socially optimal allocation estimated by user i at time t is $k^{i*}(t) = \arg\max_{k \in \mathcal{K}} \sum_{j=1}^{N} k_j u^i_{j,k}(t)$. The pseudocode of the Randomized Learning Algorithm (RLA) is given in Fig. 2. At time t RLA explores with probability $1/(t^{\frac{1}{2M} - \frac{\gamma}{M}})$ by randomly choosing one of the channels and exploits with probability $1 - 1/(t^{\frac{1}{2M} - \frac{\gamma}{M}})$ by choosing a channel which is occupied by a user in the estimated socially optimal allocation.

The following will be useful in the proof of the main theorem of this section.

Lemma 2. *Let $X_i, i = 1, 2, \ldots$ be a sequence of independent Bernoulli random variables such that X_i has mean q_i with $0 \leq q_i \leq 1$. Let $\bar{X}_k = \frac{1}{k} \sum_{i=1}^{k} X_i$, $\bar{q}_k = \frac{1}{k} \sum_{i=1}^{k} q_i$. Then for any constant $\epsilon \geq 0$ and any integer $n \geq 0$,*

$$P\left(\bar{X}_n - \bar{q}_n \leq -\epsilon\right) \leq e^{-2n\epsilon^2}. \tag{5}$$

Proof. The result follows from symmetry and [9]. □

Lemma 3. *For $p > 0, p \neq 1$*

$$\frac{(n+1)^{1-p} - 1}{1-p} < \sum_{t=1}^{n} \frac{1}{t^p} < 1 + \frac{n^{1-p} - 1}{1-p} \tag{6}$$

Proof. See [8]. □

Theorem 2. *When all players use RLA the regret with respect to the socially optimal allocation is $O(n^{\frac{2M-1+2\gamma}{2M}})$ where γ can be arbitrarily small.*

RLA (for user i)

Initialize: $0 < \gamma << 1$, $u^i_{j,k}(1) = 0, T^i_{j,k}(1) = 0, \forall j \in \mathcal{N}, k \in \mathcal{M}$, $t = 1$, sample $\sigma_i(1)$ uniformly from $\mathcal{N}$.
while $t > 0$ **do**
  play channel $\sigma_i(t)$, observe $l(t)$ the total number of players using channel $\sigma_i(t)$ and reward $h_{\sigma_i(t),l(t)}(t)$.
  Set $T^i_{\sigma_i(t),l(t)}(t+1) = T^i_{\sigma_i(t),l(t)}(t) + 1$.
  Set $T^i_{j,l}(t+1) = T^i_{j,l}(t)$ for $(j,l) \neq (\sigma_i(t), l(t))$.
  Set $u^i_{\sigma_i(t),l(t)}(t+1) = \frac{T^i_{\sigma_i(t),l(t)}(t) u^i_{\sigma_i(t),l(t)}(t) + h_{\sigma_i(t),l(t)}(t)}{T^i_{\sigma_i(t),l(t)}(t+1)}$.
  Set $u^i_{j,l}(t+1) = u^i_{j,l}(t)$ for $(j,l) \neq (\sigma_i(t), l(t))$.
  Set $k^{i*}(t+1) = \arg\max_{k \in \mathcal{K}} \sum_{j=1}^N k_j u^i_{j,k_j}(t+1)$.
  Set $\theta^{*i}(t+1)$ to be the set of channels used by at least one user in $k^{*i}(t+1)$.
  Draw i_t randomly from Bernoulli distribution with $P(i_t = 1) = \frac{1}{t^{(1/2M)-\gamma/M}}$
  if $i_t = 0$ **then**
    if $\sigma_i(t) \in \theta^*(t+1)$ and $l(t) = k^{i*}_j(t+1)$ **then**
      $\sigma_i(t+1) = \sigma_i(t)$
    else
      $\sigma_i(t+1)$ is selected uniformly at random from the channels in $\theta^*(t+1)$.
    end if
  else
    Draw $\sigma_i(t+1)$ uniformly at random from $\mathcal{N}$.
  end if
  $t = t + 1$
end while

Fig. 2. pseudocode of RLA

Proof. Let $H(t)$ be the event that at time t there exists at least one user that computed the socially optimal allocation incorrectly. Let ω be a sample path. Then

$$\sum_{t=1}^{n} I(\omega \in H(t)) \leq \sum_{t=1}^{n} \sum_{i=1}^{M} I(k^{*i}(t) \neq k^*)$$

$$\leq \sum_{(t,i,j,l)=(1,1,1,1)}^{(n,M,N,M)} I(|u^i_{j,l}(T^i_{j,l}(t)) - v_j(l)| \geq \epsilon)$$

$$= \sum_{(t,i,j,l)=(1,1,1,1)}^{(n,M,N,M)} I\left(|u^i_{j,l}(T^i_{j,l}(t)) - v_j(l)| \geq \epsilon, T^i_{j,l}(t) \geq \frac{a \ln t}{\epsilon^2}\right)$$

$$+ \sum_{(t,i,j,l)=(1,1,1,1)}^{(n,M,N,M)} I\left(|u^i_{j,l}(T^i_{j,l}(t)) - v_j(l)| \geq \epsilon, T^i_{j,l}(t) < \frac{a \ln t}{\epsilon^2}\right) \quad (7)$$

Let $\epsilon^i_{j,k}(t) = \sqrt{\frac{a \ln t}{T^i_{j,k}(t)}}$. Then $T^i_{j,k}(t) \geq \frac{a \ln t}{\epsilon^2} \Rightarrow \epsilon \geq \sqrt{\frac{a \ln t}{T^i_{j,k}(t)}} = \epsilon^i_{j,k}(t)$. Therefore,

$$I\left(|u^i_{j,l}(T^i_{j,l}(t)) - v_j(l)| \geq \epsilon, T^i_{j,l}(t) \geq \frac{a \ln t}{\epsilon^2}\right) \leq I\left(|u^i_{j,l}(T^i_{j,l}(t)) - v_j(l)| \geq \epsilon^i_{j,l}(t)\right)$$
$$I\left(|u^i_{j,l}(T^i_{j,l}(t)) - v_j(l)| \geq \epsilon, T^i_{j,l}(t) < \frac{a \ln t}{\epsilon^2}\right) \leq I\left(T^i_{j,l}(t) < \frac{a \ln t}{\epsilon^2}\right)$$

Then, continuing from (7),

$$\begin{aligned} &\sum_{t=1}^{n} I(\omega \in H(t)) \\ &\leq \sum_{(t,i,j,l)=(1,1,1,1)}^{(n,M,N,M)} \left(I\left(|u^i_{j,l}(T^i_{j,l}(t)) - v_j(l)| \geq \epsilon^i_{j,l}(t)\right) + I\left(T^i_{j,l}(t) < \frac{a \ln t}{\epsilon^2}\right)\right) \end{aligned} \quad (8)$$

Taking the expectation over (8),

$$\begin{aligned} &E\left[\sum_{t=1}^{n} I(\omega \in H(t))\right] \\ &\leq \sum_{(t,i,j,l)=(1,1,1,1)}^{(n,M,N,M)} P\left(|u^i_{j,l}(T^i_{j,l}(t)) - v_j(l)| \geq \epsilon^i_{j,l}(t)\right) \\ &+ \sum_{(t,i,j,l)=(1,1,1,1)}^{(n,M,N,M)} P\left(T^i_{j,l}(t) < \frac{a \ln t}{\epsilon^2}\right). \end{aligned} \quad (9)$$

We have

$$\begin{aligned} &P\left(|u^i_{j,l}(T^i_{j,l}(t)) - v_j(l)| \geq \epsilon^i_{j,l}(t)\right) \\ &= P\left(u^i_{j,l}(T^i_{j,l}(t)) - v_j(l) \geq \epsilon^i_{j,l}(t)\right) + P\left(u^i_{j,l}(T^i_{j,l}(t)) - v_j(l) \leq -\epsilon^i_{j,l}(t)\right) \\ &= P\left(\frac{S^i_{j,l}(T^i_{j,l}(t))}{T^i_{j,l}(t)} - v_j(l) \geq \epsilon^i_{j,l}(t)\right) + P\left(\frac{S^i_{j,l}(T^i_{j,l}(t))}{T^i_{j,l}(t)} - v_j(l) \leq -\epsilon^i_{j,l}(t)\right) \\ &\leq 2\exp\left(-\frac{2(T^i_{j,l}(t))^2(\epsilon^i_{j,l}(t))^2}{T^i_{j,l}(t)}\right) = 2\exp\left(-\frac{2T^i_{j,l}(t) a \ln t}{T^i_{j,l}(t)}\right) = \frac{2}{t^{2a}}, \end{aligned} \quad (10)$$

where (10) follows from the Chernoff-Hoeffding inequality.

Now we will bound $P\left(T^i_{j,l}(t) < \frac{a \ln t}{\epsilon^2}\right)$. Let $TR^i_{j,l}(t)$ be the number of time steps in which player i played channel j and observed l users on channel j in the time steps where all players randomized up to time t. Then

$$\{\omega : T^i_{j,l}(t) < \frac{a \ln t}{\epsilon^2}\} \subset \{\omega : TR^i_{j,l}(t) < \frac{a \ln t}{\epsilon^2}\}, \quad (11)$$

Thus

$$P\left(T^i_{j,l}(t) < \frac{a\ln t}{\epsilon^2}\right) \leq P\left(TR^i_{j,l}(t) < \frac{a\ln t}{\epsilon^2}\right) . \tag{12}$$

Now we define new Bernoulli random variables $X^i_{j,l}(s)$ as follows: $X^i_{j,l}(s) = 1$ if all players randomize at time s and player i selects channel j and observes l players on it according to the random draw. $X^i_{j,l}(s) = 0$ else. Then $TR^i_{j,l}(t) = \sum_{s=1}^t X^i_{j,l}(s)$. $P(X^i_{j,l}(s) = 1) = \rho_s p_l$ where $p_l = \frac{\binom{M-1}{l-1}\binom{M+N-l-2}{N-2}}{\binom{M+N-1}{N-1}}$ and $\rho_s = \frac{1}{s^{(1/2)-\gamma}}$. Let $s_t = \sum_{s=1}^t \frac{1}{s^{(1/2)-\gamma}}$ Then

$$\begin{aligned} &P\left(TR^i_{j,l}(t) < \frac{a\ln t}{\epsilon^2}\right) \\ &= P\left(\frac{TR^i_{j,l}(t)}{t} - \frac{p_k s_t}{t} < \frac{a\ln t}{t\epsilon^2} - \frac{p_k s_t}{t}\right) \\ &\leq P\left(\frac{TR^i_{j,l}(t)}{t} - \frac{p_k s_t}{t} < \frac{a\ln t}{t\epsilon^2} - \frac{p_k (t+1)^{(1/2)+\gamma} - 1}{t((1/2)+\gamma)}\right), \end{aligned} \tag{13}$$

where (13) follows from Lemma 3. Let $\tau(M, N, \epsilon, \gamma, \gamma', a)$ be the time that for all $k \in \{1, 2, \ldots, M\}$.

$$\frac{p_k (t+1)^{(1/2)+\gamma} - 1}{t((1/2)+\gamma)} - \frac{a\ln t}{t\epsilon^2} \geq t^{(1/2)+\gamma'}, \tag{14}$$

where $0 < \gamma' < \gamma$. Then for all $t \geq \tau(M, N, \epsilon, \gamma, \gamma', a)$ (14) will hold since RHS increases faster than LHS. Thus we have for $t \geq \tau(M, N, \epsilon, \gamma, \gamma', a)$

$$\begin{aligned} &P\left(\frac{TR^i_{j,l}(t)}{t} - \frac{p_k s_t}{t} < \frac{a\ln t}{t\epsilon^2} - \frac{p_k (t+1)^{(1/2)+\gamma} - 1}{t((1/2)+\gamma)}\right) \\ &\leq P\left(\frac{TR^i_{j,l}(t)}{t} - \frac{p_k s_t}{t} < t^{-(1/2)+\gamma'}\right) \\ &\leq e^{-2tt^{2\gamma'-1}} = e^{-2t^{2\gamma'}} \leq e^{-2\ln t} = \frac{1}{t^2}. \end{aligned} \tag{15}$$

Let $a = 1$. Then continuing from (9) by substituting (10) and (15) we have

$$E\left[\sum_{t=1}^n I(\omega \in H(t))\right] \leq M^2 N \left(\tau(M, N, \epsilon, \gamma, \gamma', 1) + 3\sum_{t=1}^n \frac{1}{t^2}\right) . \tag{16}$$

Thus we proved that the expected number of time steps in which there exists at least one user that computed the socially optimal allocation incorrectly is finite. Note that because RLA explores with probability $\frac{1}{t^{1/2M-\gamma/M}}$, the expected number of time steps in which all the players are not randomizing up to time n is

$$\sum_{t=1}^{n}\left(1-\left(1-\frac{1}{t^{(1/2M)-\gamma/M}}\right)^{M}\right)\leq\sum_{t=1}^{n}\frac{M}{t^{1/2M-\gamma/M}}=O(n^{\frac{2M-1+2\gamma}{2M}}). \quad (17)$$

Note that players can choose γ arbitrarily small, increasing the finite regret due to $\tau(M,N,\epsilon,\gamma,\gamma',1)$. Thus if we are interested in the asymptotic performance then $\gamma>0$ can be arbitrarily small.

Now we do the worst case analysis. We classify the time steps into two. *Good* time steps in which all the players know the socially optimal allocation correctly and none of the players randomize excluding the randomizations done for settling down to the socially optimal allocation. *Bad* time steps in which there exists a player that does not know the socially optimal allocation correctly or there is a player that randomizes excluding the randomizations done for settling down to the socially optimal allocation. The number of *Bad* time steps in which there exists a player that does not know the socially optimal allocation correctly is finite while the number of time steps in which there is a player that randomizes excluding the randomizations done for settling down to the socially optimal allocation is $O(n^{\frac{2M-1+2\gamma}{2M}})$. The worst case is when each bad step is followed by a good step. Then from this good step the expected number of times to settle down to the socially optimal allocation is $\left(1-\frac{1}{\binom{M+z^*-1}{z^*-1}}\right)/\left(\frac{1}{\binom{M+z^*-1}{z^*-1}}\right)$ where z^* is the number of channels which has at least one user in the socially optimal allocation. Assuming in the worst case the sum of the utilities of the players is 0 when they are not playing the socially optimal allocation we have

$$\begin{aligned}R(n)&\leq\frac{1-\frac{1}{\binom{M+z^*-1}{z^*-1}}}{\frac{1}{\binom{M+z^*-1}{z^*-1}}}\left(M^2N\left(\tau(M,N,\epsilon,\gamma,\gamma',1)+3\sum_{t=1}^{n}\frac{1}{t^2}\right)+O(n^{\frac{2M-1+2\gamma}{2M}})\right)\\&=O(n^{\frac{2M-1+2\gamma}{2M}})\end{aligned}$$

□

Note that we mentioned earlier, under a classical multi-armed bandit problem approach as cited before [3,4,5,15,16,21,22], a logarithmic regret $O(\log n)$ is achievable. The fundamental difference between these studies and the problem in the present paper is the following: Assume that at time t user i selects channel j. This means that i selects to observe an arm from the set $\{(j,k):k\in\mathcal{M}\}$ but the arm assigned to i is selected from this set depending on the choices of other players.

Also note that in RLA a user computes the socially optimal allocation according to its estimates at each time step. This could pose significant computational effort since integer programming is NP-hard in general. However, by exploiting the stability condition on the socially optimal allocation a user may reduce the number of computations; this is a subject of future research.

5 An Algorithm for Socially Optimal Allocation (Case 3)

In this section we assume that $g_j(n)$ is decreasing in n for all $j \in \mathcal{N}$. For simplicity we assume that the socially optimal allocation is unique up to the permutations of σ^*. When this uniqueness assumption does not hold we need a more complicated algorithm to achieve the socially optimal allocation. All users use the Random Selection (RS) algorithm defined in Fig. 3. RS consists of two phases. Phase 1 is the learning phase where the user randomizes to learn the interference functions. Let $B_j(t)$ be the set of distinct payoffs observed from channel j up to time t. Then the payoffs in set $B_j(t)$ can be ordered in a decreasing way with the associated indices $\{1, 2, \ldots, |B_j(t)|\}$. Let $O(B_j(t))$ denote this ordering. Since the IFs are decreasing, at the time $|B_j(t)| = M$, the user has learned g_j. At the time $|\cup_{j=1}^N B_j(t)| = MN$, the user has learned all IFs. Then, the user computes $\mathcal{A}^*$ and phase 2 of RS starts where the user randomizes to converge to the socially optimal allocation.

Random Selection (RS)

Initialize: $t = 1$, $b = 0$, $B_j(1) = \emptyset, \forall j \in \mathcal{N}$, sample $\sigma_i(1)$ from the uniform distribution on $\mathcal{N}$
Phase 1
while $b < MN$ **do**
  if $h_{\sigma_i(t)}(t) \notin B_{\sigma_i(t)}(t)$ **then**
    $B_{\sigma_i(t+1)}(t+1) \leftarrow O(B_{\sigma_i(t)}(t) \cup h_{\sigma_i(t)}(t))$
    $b = b + 1$
  end if
  Sample $\sigma_i(t+1)$ from the uniform distribution on $\mathcal{N}$
  $t = t + 1$
end while
find the socially optimal allocation σ^*
Phase 2
while $b \geq MN$ **do**
  if $h_{\sigma_i(t)}(t) < v^*_{\sigma_i(t)}$ **then**
    Sample $\sigma_i(t+1)$ from the uniform distribution on $\mathcal{N}$
  else
    $\sigma_i(t+1) = \sigma_i(t)$
  end if
  $t = t + 1$
end while

Fig. 3. pseudocode of RS

Theorem 3. *Under the assumptions of (C3) if all players use RS algorithm to choose their actions, then the expected time to converge to the socially optimal allocation is finite.*

Proof. Let T_{OPT} denote the time the socially optimal allocation is achieved, T_L be the time when all users learn all the IFs, T_F be the time it takes to reach the socially optimal allocation after all users learn all the IFs. Then $T_{OPT} = T_L + T_F$ and $E[T_{OPT}] = E[T_L] + E[T_F]$. We will bound $E[T_L]$ and $E[T_F]$. Let T_i be the first time that i users have learned the IFs. Let $\tau_i = T_i - T_{i-1}, i = 1, 2, \ldots, M$ and $T_0 = 0$. Then $T_L = \tau_1 + \ldots + \tau_M$. Define a Markov chain over all N^M possible configurations of M users over N channels based on the randomization of the algorithm. This Markov chain has a time dependent stochastic matrix which changes at times $T_1, T_2, \ldots, T_M$. Let $P_{T_0}, P_{T_1}, \ldots, P_{T_M}$ denote the stochastic matrices after the times $T_0, T_1, \ldots, T_M$ respectively. This Markov chain is irreducible at all times up to T_M and is reducible with absorbing states corresponding to the socially optimal allocations after T_M. Let $\hat{T}_1, \hat{T}_2, \ldots \hat{T}_M$ be the times that all configurations are visited when the Markov chain has stochastic matrices $P_{T_0}, P_{T_1}, \ldots, P_{T_{M-1}}$ respectively. Then because of irreducibility and finite states $E[\hat{T}_i] < z_1, i = 1, \ldots, M$ for some constant $z_1 > 0$. Since $\tau_i \leq \hat{T}_i, i = 1, \ldots, M$ a.s. we have $E[T_L] < M z_1$. For the Markov chain with stochastic matrix P_{T_M} all the configurations that do not correspond to the socially optimal allocation are transient states. Since starting from any transient state the mean time to absorption is finite $E[T_F] < z_2$, for some constant $z_2 > 0$. □

6 Conclusion

In this paper we studied the decentralized multiuser resource allocation problem with various levels of communication and cooperation between the users. Under three different scenarios we proposed three algorithms with reasonable performance. Our future reserach will include characterization of achievable performance regions for these scenarios. For example, in case 2 we are interested in finding an optimal algorithm and a lower bound on the performance.

References

1. Agrawal, R.: Sample Mean Based Index Policies with $O(\log(n))$ Regret for the Multi-armed Bandit Problem. Advances in Applied Probability 27(4), 1054–1078 (1995)
2. Ahmad, S., Tekin, C., Liu, M., Southwell, R., Huang, J.: Spectrum Sharing as Spatial Congestion Games (2010), `http://arxiv.org/abs/1011.5384`
3. Anandkumar, A., Michael, N., Tang, A.: Opportunistic Spectrum Access with Multiple Players: Learning under Competition. In: Proc. of IEEE INFOCOM (March 2010)
4. Anantharam, V., Varaiya, P., Walrand, J.: Asymptotically Efficient Allocation Rules for the Multiarmed Bandit Problem with Multiple Plays-Part I: IID Rewards. IEEE Trans. Automat. Contr., 968–975 (November 1987)
5. Anantharam, V., Varaiya, P., Walrand, J.: Asymptotically Efficient Allocation Rules for the Multiarmed Bandit Problem with Multiple Plays-Part II: Markovian Rewards. IEEE Trans. Automat. Contr., 977–982 (November 1987)

6. Auer, P., Cesa-Bianchi, N., Fischer, P.: Finite-time Analysis of the Multiarmed Bandit Problem. Machine Learning 47, 235–256 (2002)
7. Auer, P., Cesa-Bianchi, N., Freund, Y., Schapire, R.: The Nonstochastic Multi-armed Bandit Problem. SIAM Journal on Computing 32, 48–77 (2002)
8. Chlebus, E.: An Approximate Formula for a Partial Sum of the Divergent p-series. Applied Mathematics Letters 22, 732–737 (2009)
9. Turner, D.W., Young, D.M., Seaman, J.: A Kolmogorov Inequality for the Sum of Independent Bernoulli Random Variables with Unequal Means. Statistics and Probability Letters 23, 243–245 (1995)
10. Freund, Y., Schapire, R.: Adaptive Game Playing Using Multiplicative Weights. Games and Economic Behaviour 29, 79–103 (1999)
11. Gai, Y., Krishnamachari, B., Jain, R.: Learning Multiuser Channel Allocations in Cognitive Radio Networks: a Combinatorial Multi-armed Bandit Formulation. In: IEEE Symp. on Dynamic Spectrum Access Networks (DySPAN) (April 2010)
12. Kakhbod, A., Teneketzis, D.: Power Allocation and Spectrum Sharing in Cognitive Radio Networks With Strategic Users. In: 49th IEEE Conference on Decision and Control (CDC) (December 2010)
13. Kasbekar, G., Proutiere, A.: Opportunustic Medium Access in Multi-channel Wireless Systems: A Learning Approach. In: Proceedings of the 48th Annual Allerton Conference on Communication, Control, and Computation (September 2010)
14. Kleinberg, R., Piliouras, G., Tardos, E.: Multiplicative Updates Outperform Generic No-Regret Learning in Congestion Games. In: Annual ACM Symposium on Theory of Computing, STOC (2009)
15. Lai, T., Robbins, H.: Asymptotically Efficient Adaptive Allocation Rules. Advances in Applied Mathematics 6, 4–22 (1985)
16. Liu, K., Zhao, Q.: Distributed Learning in Multi-Armed Bandit with Multiple Players. IEEE Transactions on Signal Processing 58(11), 5667–5681 (2010)
17. Monderer, D., Shapley, L.S.: Potential Games. Games and Economic Behavior 14(1), 124–143 (1996)
18. Rosenthal, R.: A Class of Games Possessing Pure-strategy Nash Equilibria. International Journal of Game Theory 2, 65–67 (1973)
19. Sandholm, W.H.: Population Games and Evolutionary Dynamics (2008) (manuscript)
20. Smith, J.M.: Evolution and the Theory of Games. Cambridge University Press (1982)
21. Tekin, C., Liu, M.: Online Algorithms for the Multi-armed Bandit Problem with Markovian Rewards. In: Proceedings of the 48th Annual Allerton Conference on Communication, Control, and Computation (September 2010)
22. Tekin, C., Liu, M.: Online Learning in Opportunistic Spectrum Access: A Restless Bandit Approach. In: 30th IEEE International Conference on Computer Communications (INFOCOM) (April 2011)

Incentivizing Upload Capacity in P2P-VoD Systems: A Game Theoretic Analysis

Weijie Wu[1], John C.S. Lui[1], and Richard T.B. Ma[2]

[1] The Chinese University of Hong Kong
{wjwu,cslui}@cse.cuhk.edu.hk
[2] Advanced Digital Science Center, UIUC and National University of Singapore
tbma@comp.nus.edu.sg

Abstract. Free riding has long been a serious problem in peer-to-peer (P2P) systems due to the selfish behavior of individual users. To conquer this problem, a key design issue of the P2P systems is to appropriately incentivize users to contribute resources. In P2P Video-on-Demand (VoD) applications, content providers need to incentivize the peers to dedicate bandwidth and upload data to one another so as to alleviate the upload workload of their content servers. In this paper, we design a simple yet practical incentive mechanism that rewards each peer based on its dedicated upload bandwidth. We use a *mean field* interaction model to characterize the distribution of number of peers in different video segments, based on which we characterize the content providers' uploading cost as a function of the peers' contribution. By using a game theoretic framework, we analyze the interaction between a content provider's rewarding strategy and the peers' contributing behaviors and derive a unique Stackelberg equilibrium. We further analyze the system efficiency in terms of *the price of anarchy*. Via extensive simulations, we validate the stability and efficiency of our incentive scheme.

Keywords: P2P-VoD, incentive, mean field, Stackelberg game.

1 Introduction

In recent years, we have witnessed the rapid growth of Peer-to-Peer (P2P) systems, many of which have large population bases, e.g., file sharing systems like BitTorrent [4], and Video-on-Demand (VoD) systems like PPLive [1] and PPStream [2]. The key advantage of the P2P architecture is that by utilizing the distributed resources at the peers, the system can be more scalable and fault-tolerant than traditional client-server architectures. Nevertheless, due to the selfish nature of the peers, free-riding [5] often happens where peers do not have incentives to contribute resource for other peers. Thus, designing an effective and practical incentive scheme becomes critical in encouraging the peers to contribute to the system, and thereby improving the system performance. Plenty of work has been done for systems of traditional P2P applications, for example, the tit-for-tat [4] protocol has been well-adopted for file sharing applications.

R. Jain and R. Kannan (Eds.): GameNets 2011, LNICST 75, pp. 337–352, 2012.

However, very limited work has been focusing on the incentive mechanisms for P2P-VoD applications.

What makes it challenging to design incentive schemes for P2P-VoD applications? Compared to file sharing, VoD applications need to satisfy more stringent temporal and spacious constraints for data delivery. To share files, peers exchange segments of files that have not been received. Segments might be received in different orders, therefore, there is hardly a temporal constraint under which a particular segment has to be received. On the contrary, when a user watches a particular video segment, this segment has to be received by the user within a short period of time, while nearby segments would not satisfy the user's instantaneous demand. Even worse, predicting the data demand is difficult because users might fast-forward and/or rewind among the video segments. The tit-for-tat scheme does not work for VoD applications, because the data demand and supply among the peers are highly volatile.

Instead of using a punishment-based scheme like tit-for-tat, we propose and analyze a *reward-based scheme* that incentivizes peers to contribute upload capacity for VoD systems. Our contributions are as follows.

- We model the stochastic operations of the peers and derive the system state by using the *mean field approximation.*
- We propose a practical reward-based incentive scheme based on the dedicated upload capacity of the peers.
- We model the interaction between the content provider and the peers by using a *Stackelberg game.* We derive the unique Stackelberg equilibrium and analyze the efficiency of the equilibrium in terms of *the price of anarchy.*
- We validate the effectiveness of our scheme and the theoretic results via extensive simulations.

Our paper is organized as follows. In Section 2, we present the system model and the reward-based incentive scheme for P2P-VoD systems. In Section 3, we study the system dynamics and characterize the content provider's cost as a function of the peers' dedicated upload bandwidth. In Section 4, we model the strategic behavior between the content provider and the peers, derive a unique system equilibrium, and analyze the efficiency of that. We evaluate the performance of our incentive scheme in Section 5. Section 6 states the related work and Section 7 concludes.

2 System Model and Reward-Based Scheme

In a P2P-VoD system, each peer can support other peers by (1) caching data that would be needed for other peers and (2) uploading data to other peers. Both aspects are equally important because a peer cannot contribute if it either does not have the needed content, or does not have upload capacity. Thus, the design space for an incentive mechanism includes both incentivizing peers to cache the right content as well as to devote upload capacity. Due to the complicated viewing operations supported by VoD systems, data demand and supply can be volatile. Therefore, the optimal data caching policy for a peer might be difficult to predict given its local knowledge of the system. Although the content provider's global knowledge might help guide the data caching policy, collecting this global knowledge and coordinating with peers bring overhead to the system, even if the peers are willing to comply. Our study focuses on incentivizing the peers to dedicate upload capacity, while making a minimum assumption on the data caching policy. Naturally, each peer caches the video segment it recently watched/requested and can contribute to other peers that need the video segment. Under this default policy, a peer cannot contribute much either due to its little upload capacity or the low demand for its recently watched segment.

Since peers only cache the segments of a video they are watching, without loss of generality, we focus on a particular video content provided in the system. We denote N as the long-term average or steady-state number of peers that watch this video. We denote r as the required playback rate, i.e., bits per second, for serving the video. We consider homogeneous peers and explore the symmetric strategies of them. We denote u as the upload capacity of each peer, which is the maximum bandwidth a peer is willing to contribute to the system. In Section 4, the peers will choose u as their strategy to maximize utility.

2.1 Peers' Viewing Behavior

We assume that the system organizes the video as K consecutive data segments. We denote S_i as the ith data segment. We model the user behavior of the system by specifying a set of rate transition probabilities $\{p_{ij} : i, j = 0, 1, \ldots, K\}$. Each p_{ij} denotes the transition probability of a typical user watching S_j after finishing segment S_i. In particular, p_{i0} denotes the probability that a user quits watching after finishing S_i. We denote λ as the aggregate external arrival rate of users that start watching the video. However, users might not start from the very first segment S_1, because they might have watched part of the video before. Thus, p_{0i} denotes the probability that an external arrival will start with S_i. To keep consistency, we define $p_{00} = 0$ and require $\sum_{j=0}^{K} p_{ij} = 1$ for all $i = 0, \ldots, K$. In practice, the probability p_{ij} represents a state transition where a peer performs a play, fast-forward or rewind operation when $j = i+1$, $j > i+1$ or $0 < j < i+1$, respectively.

Under our default caching policy, we assume that after transitioning from segment i to segment j, a peer only provides S_i to other peers if needed when it is viewing S_j. In practice, a peer can cache and upload multiple data segments at any time. However, we will show that even by using such a restricted caching policy, one can still design a simple yet effective scheme to incentivize peers to dedicate upload capacity for other peers, which consequently reduces the workload of the content provider as well as improves the system performance. When a peer downloads data segment i, we assume that the system will direct the demand to the peers that can contribute S_i first, which upload S_i at an aggregate rate that equals the playback rate r. If the supply capacity of S_i is lower than aggregate demand rate, the content server will support the remaining data rate by using content servers' capacity.

2.2 Reward-Based Incentive Scheme

We design an incentive mechanism under which the content provider rewards the peers based on the amount of upload capacity they contribute. The reward can be in various forms, e.g., real money rebate for the service fee and virtual credits or reputation record for advanced services. Notice that any reward scheme can be represented by the currency flow from the content provider to the peers. Even for rewards in virtual currency or reputation, they imply that the P2P-VoD operator needs to invest money for developing advanced/prioritized services for users. We do not restrict the form of implementing the rewards in our paper; however, we use an abstract model to describe the reward in terms of monetary value.

We define the reward W to a peer to be a function of its dedicated maximum upload bandwidth capacity u as

$$W(u) = \int_0^u w(x)dx, \tag{1}$$

where $w(x)$ denotes the marginal reward at the contribution capacity level x. Notice that our incentive scheme is based on the maximal upload bandwidth that a peer is willing to dedicate to the system; in practice, whether a peer will upload data at the maximum capacity u depends on the demand for the data segment the peer has cached and the peer might not upload at the rate of u. We assume that the system will maximize the utilization of upload capacity of the peers whenever their data segments are requested by other peers.

A simple reward scheme is to use a linear reward function

$$W(u) = wu, \tag{2}$$

where the marginal reward $w(x) = w$ is the same for all levels of contribution x. w can also be interpreted as the reward per unit capacity dedicated by a peer. The linear reward scheme can be easily understood by the peers and implemented by the content provider in practice. In what follows, we will start with the linear reward model and extend our results for general reward functions W later.

In summary, under our reward based incentive scheme, the content provider decides the reward function W, and then each peer decides its upload capacity u dedicated to the P2P-VoD system. In Section 4, we present a game theoretic framework to analyze the interaction between the content provider and peers under this reward-based incentive scheme. Before we present this analysis, let us first investigate the distribution of peers in different video segments so as to understand the impact of peers' upload contribution on the content provider's upload cost.

3 Peers' Contribution and Content Provider's Cost

In this section, we derive the distribution of number of peers watching different segments based on the user behavior described in Section 2.1. We further characterize the content provider's upload cost as a function of both the distribution of peers and their upload contribution.

3.1 Distribution of Peers in Different Video Segments

Based on a typical user's viewing behavior described in Section 2.1, we say that a peer is in state i when it is watching segment S_i. Thus, each peer's viewing behavior becomes a random process. For a system with N peers, the state space of the P2P-VoD system is $[1,\ldots,K]^N$. To overcome the large dimensionality of the state space, we tackle the problem from a macro perspective, i.e., instead of observing each peer's individual state, we are only interested in the *fraction* of peers in each of the states or the distribution of the peers in the states. In particular, we use the *mean field interaction model* [3] to calculate the steady state distribution of these peers in the P2P-VoD system. We denote $q_i \in [0,1]$ as the fraction of peers in state i. The system state can now be specified by the vector $(q_1, q_2, \ldots, q_K)$, where $\sum_{i=1}^{K} q_i = 1$.

Theorem 1. *If the P2P-VoD system does not support the rewind operation, i.e.,* $p_{ij} = 0$ *for all* $1 \leq j \leq i$*, then*

$$q_i = \frac{P_{0i}}{\sum_{j=1}^{K} P_{0j}} \quad \forall\, i = 1,\ldots,K, \tag{3}$$

where P_{ij} *denotes the aggregated probability of transitioning from state* i *to* j*, which can be defined recursively as follows.*

$$P_{ij} = \begin{cases} p_{ij} & \text{if } j = i+1, \\ \sum_{k=i+1}^{j-1} p_{ik} P_{kj} + p_{ij} & \text{otherwise.} \end{cases} \tag{4}$$

Proof. Recall λ is the external arrival rate of new peers that start watching the video. In the steady-state, we can express the external arrival rate to state i as λp_{0i}. We denote μ as the rate at which users change states, which includes the events of (1) moving to the next segment, (2) fast-forwarding or rewinding and (3) quitting the system. Because the departure rate of a particular state is proportional to the number of peers in the state, the departure rate from state i is μq_i. After finishing watching S_i, a peer has different probabilities to start watching another segment; and therefore, the transition rate from state i to state j is $\mu q_i p_{ij}$. Each system state i satisfies the following ordinary differential equation (ODE):

$$N \cdot \frac{dq_i}{dt} = \lambda p_{0i} + \sum_{j=1, j\neq i}^{K} \mu q_j p_{ji} - \sum_{k=1, k\neq i}^{K} \mu q_i p_{ik} - \mu q_i p_{i0}, \quad \forall i. \tag{5}$$

The left hand side represents the change in the number of peers in state i. The right hand side counts all the possible cases where changes can incur. The increase of number of peers in state i can occur when external arrivals start watching the video at segment S_i (with a rate of λp_{0i}), or any peer which just finished watching S_j continues (or performs fast-forward or rewind operation) to watch S_i (with a rate of $\mu q_i p_{ji}$). The decrease of number of peers in state i can occur when a peer finishes watching S_i and starts watching another segment S_k (with rate $\mu q_i p_{ik}$), or this peer quits viewing the video (with rate $\mu q_i p_{i0}$).

In steady-state, the external arrival rate to the P2P-VoD system should be equal to the departure rate, i.e.,

$$\lambda = \sum_{i=0}^{K} \mu q_i p_{i0}, \tag{6}$$

and the rate of change in each state should be zero, i.e.,

$$N \cdot \frac{dq_i}{dt} = 0. \tag{7}$$

Given $p_{ij} = 0$ for all $1 \leq j \leq i$, we have $q_1 = \frac{\lambda}{\mu} p_{01}$ and $q_i = \frac{\lambda}{\mu} p_{0i} + \sum_{j=1}^{i-1} p_{ji} q_j$. By recursively solving q_i and requiring $\sum_{i=1}^{K} q_i = 1$, we can derive the above formula. □

3.2 Content Provider's Upload Cost

Since content providers are often charged by their transit providers (ISPs) based on the traffic volume going through them, we assume that the content provider's cost is proportional to the upload capacity needed to support all peers. In steady-state, Nq_i peers watch segment S_i. Given a required playback rate of r, the aggregate required upload capacity for S_i should be $Nq_i r$. In the proof of Theorem 1, we have derived $q_i = \frac{\lambda}{\mu} p_{0i} + \sum_{j=1}^{i-1} p_{ji} q_j$, or equivalently, $Nq_i = N\frac{\lambda}{\mu} p_{0i} + \sum_{j=1}^{i-1} N p_{ji} q_j$. Among the peers watching S_i, $N\frac{\lambda}{\mu} p_{0i}$ peers are external arrivals that start watching at S_i and $Np_{ji}q_j$ peers have viewed S_j before transitioning to watch S_i. Note that we can only assure that peers performing continuous play, i.e., moving from S_i to S_{i+1}, have watched and therefore cached the whole segment of S_i. Under our simplistic caching policy, only the peers that have transitioned from S_i to S_{i+1} and currently watching S_{i+1} can upload S_i to other peers. Suppose each peer contributes u amount of capacity for uploading available video segments, the total available upload capacity for S_i would be $Nq_i p_{i,i+1} u$. If the dedicated peer contribution $Nq_i p_{i,i+1} u$ is less than the required download capacity $Nq_i r$, the content provider needs to upload segment S_i to support the difference in capacity. In particular, the playback requirement of the last segment, $Nq_K r$, must be supported by the content provider. Assume the content provider incurs a cost c_s ($c_s > 0$) per unit bandwidth capacity. Then, the content provider's total upload cost is:

$$C_s(u) = c_s N \times \left[\sum_{i=1}^{K-1} q_i (r - p_{i,i+1} u)^+ + q_K r \right]. \tag{8}$$

Proposition 1. *$C_s(u)$ is a convex and non-increasing function in u.*

Proof. Because c_s, N, q_i and $p_{i,i+1}$ are all positive, the cost $C_s(u)$ is non-increasing in u. Since $(r - p_{i,i+1}u)$ is linear (and therefore, convex) in u, $(r - p_{i,i+1}u)^+ = \max(r - p_{i,i+1}u, 0)$ is convex in u. Given that the convex property keeps under summation operation, we reach the above conclusion. □

Notice that the maximum cost is $C_s(0) = c_s N r$ when the peers do not contribute any capacity and the minimum cost is $C_s(u) = c_s N q_K r$ when u is large enough. In particular, when $u \geq r$, the system might not be able to utilize all peers' upload resource, i.e., fewer peers will participate in data uploading when u increases. This implies the sub-linearity of cost saving of the content provider with respect to the increase of peers' capacity contribution u. We will show an example that validates the convexity feature in Section 5.

4 Game Theoretic Analysis on Incentive Scheme

In this section, we present a game-theoretic model to study the strategies of the content provider and the peers in a P2P-VoD system under the reward-based incentive scheme and analyze the stability and efficiency of the incentive scheme. We define w, the per capacity reward to the peers, as the strategy of the content provider and u, the amount of dedicated capacity, as the strategy of the peers. We assume that peers are homogenous and use the same u strategy in the game. We denote $[0, \bar{w}]$ and $[0, \bar{u}]$ as the strategy space of the content provider and the peers, where $\bar{w}$ and $\bar{u}$ are the upper-bound of the content provider's and peers' strategy respectively.

4.1 Stackelberg Game Model

From the content provider's perspective, it aims at minimizing its total cost, i.e., the cost of uploading and the cost of rewarding the peers. We define the utility of the content provider as the following:

$$\pi_s(w, u) = -C_s(u) - wuN. \tag{9}$$

Similarly, we define the utility of a peer as the reward it receives, minus its cost of upload contribution as the following:

$$\pi_p(u, w) = wu - C_p(u), \tag{10}$$

where $C_p(u)$ denotes the cost of dedicating u amount of capacity. To maximize their utilities, the content provider solves the optimization problem $\max_w \pi_s(w, u)$, and the peers solves $\max_u \pi_p(u, w)$. Here, we do not specify the form of the peer's upload cost function $C_p(u)$. Rather, we assume the cost function satisfies the following property:

(1) $C_p(u)$ is continuous and twice differentiable in u.
(2) $C_p(0) = 0$, $C_p'(u) > 0$, $C_p''(u) > 0$.

$C'_p(u) > 0$ means that a peer's cost increases with its dedicated capacity. $C''_p(u) > 0$ means the marginal cost also increases with the dedicated capacity. The above assumption reflects the fact that a peer's viewing performance would not be affected too much if it contributes a small amount of upload capacity; however, when a peer dedicates much upload capacity, its download rate as well as the performance of video might be substantially reduced.

Proposition 2. *$\pi_p(u, w)$ is a strictly concave function in u.*

Proof. Noting that $C''_p(u) > 0$ implies $-C_p(u)$ is strictly concave in u, and that wu is linear and hence concave in u, we have $\pi_p(u, w)$ strictly concave in u. □

We consider a Stackelberg game [15] where the content provider decides w first, and after that, the peers decide u. It is natural to assume the content provider as the first-mover whereas the peers response to the reward w accordingly, because once u is determined, the content provider would have no incentives to provide any reward for the peers. To obtain the Stackelberg equilibrium of the game, we can use the backward induction [15]. In particular, the peers solve the problem $u^*(w) = \text{argmax}_u\, \pi_p(u, w)$ given any w. By knowing the peers' best responses, the content provider solves the problem $w^* = \text{argmax}_w\, \pi_s(w, u^*(w))$. In what follows, we analyze the existence, uniqueness and efficiency of the Stackelberg equilibrium.

4.2 Existence and Uniqueness of Stackelberg Equilibrium

We start with the following lemma, which establishes the connection between the Stackelberg equilibrium and an optimization problem:

Lemma 1. *If u^* is a solution to the following problem:*

$$\min_u C_s(u) + NuC'_p(u), \tag{11}$$

then there exists a Stackelberg equilibrium $(u^*, u^*C'_p(u^*))$; further, if (u^*, w^*) is a Stackelberg equilibrium, then u^* is the solution to problem (11).

Proof. We start by showing the first half of the statement. Denote $u^* = \text{argmin}_u [C_s(u) + NuC'_p(u)]$ and $w^* = u^*C'_p(u^*)$. We show that (u^*, w^*) is a Stackelberg equilibrium. Since $\pi_p(u, w)$ is strictly concave in u, so for any given w^*, if u^* satisfies $u^*C'_p(u^*) = w^*$, then u^* maximizes the peers' utility $\pi_p(u, w^*)$. Hence, the peers do not have incentives to deviate from u^*. Suppose the content provider has an incentive to deviate from w^* and can obtain higher utility by setting $w = w_0$, where the peers' response is to set $u = u_0$ so that u_0 maximizes $\pi_p(u, w_0)$. Because of the strict concavity of $\pi_p(u, w)$, there are only three possible cases:

(1) $C'_p(u_0) = w_0$ if $C'_p(0) \le w_0 \le C'_p(\bar{u})$; or
(2) $u_0 = 0$ if $C'_p(0) > w_0$; or
(3) $u_0 = \bar{u}$ if $C'_p(\bar{u}) < w_0$.

For any of the above cases, we have

$$\begin{aligned} C_s(u_0) &+ N u_0 C_p'(u_0) \leq C_s(u_0) + N u_0 w_0 \\ < C_s(u^*) &+ N u^* w^* = C_s(u^*) + N u^* C_p'(u^*). \end{aligned} \tag{12}$$

The first inequality holds for the above three cases. The second inequality holds because we assume the content provider can have higher utility by setting $u = u_0$ instead of $u = u^*$. However, $C_s(u_0) + N u_0 C_p'(u_0) < C_s(u^*) + N u^* C_p'(u^*)$ contradicts the fact that u^* is a solution of (11). This implies that the content provider has no incentive to deviate from w^*. Given that we have shown the peers do not have any incentive to deviate from u^* given any w^*, we conclude (u^*, w^*) is a Stackelberg equilibrium[1].

To show the second half of the statement, suppose there exists a Stackelberg equilibrium (u^*, w^*), but u^* is not a solution to (11), i.e., there exists $u_0 \neq u^*$ such that $C_s(u_0) + N u C_p'(u_0) < C_s(u^*) + N u C_p'(u^*)$. Assume the content provider sets $w_0 = u_0 C_p'(u_0)$. Taking the derivative in (10) and noting the strict concavity of $\pi_p(u, w)$, we have the peers' unique best response is $u = u_0$ for given $w_0 = u_0 C_p'(u_0)$. Therefore, $\pi_s(w_0, u_0) = -C_s(u_0) - N u_0 C_p'(u_0) > -C_s(u^*) - N u^* C_p'(u^*) = \pi_s(w^*, u^*)$, which contradicts to the fact that (u^*, w^*) is a Stackelberg equilibrium. This implies u^* must be a solution to (11). □

Theorem 2. *The Stackelberg equilibrium always exists. If $uC_p'(u)$ is strictly convex in u, then the peers' solution u^* at the Stackelberg equilibrium is unique.*

Proof. We first show the existence. The peers solve $\max_u \pi_p(u, w) = wu - C_p(u)$. For any given w, π_p is continuous and strictly concave in u over the compact set $[0, \bar{u}]$. Hence, the optimal solution $u^*(w) = \text{argmax}_u \, \pi_p(u, w)$ exists and is unique. Substituting u by $u^*(w)$ in $\pi_s(w, u)$, the provider's utility $\pi_s(w, u^*(w))$ is continuous in w over the compact set $[0, \bar{w}]$, so $w^* = \text{argmax}_w \, \pi_s(w, u^*(w))$ exists.

Next we show the uniqueness of u^* when $uC_p'(u)$ is strictly convex in u. Since $C_s(u)$ is convex in u (Proposition 1), and $uC_p'(u)$ is strictly convex in u, we can observe that the problem (11) is a strictly convex minimization over a compact set, which has a unique solution. According to Lemma 1, any Stackelberg equilibrium (u^*, w^*) satisfies that u^* is a solution to (11). Therefore, we conclude that the peers' solution in the Stackelberg equilibrium is unique[2]. □

In the proof, we assume $uC_p'(u)$ to be strictly convex. In fact, if the marginal cost $C_p'(u)$ is super-linearly increasing in u, then by multiplying a linear function $f(u) = u$, the term $uC_p'(u)$ can be guaranteed to be strictly convex.

---

[1] Noting the above three cases and that the content provider aims at maximizing its utility, if $u^* > 0$, then the corresponding Stackelberg equilibrium is unique where $w^* = u^* C_p'(u^*)$. If $u^* = 0$, then any (u^*, w^*) where $0 \leq w^* \leq C_p'(0)$ is a Stackelberg equilibrium.

[2] We do not claim the Stackelberg equilibrium is unique. The only chance of having multiple Stackelberg equilibria is $u^* = 0$, where any (u^*, w^*) with $0 \leq w^* \leq C_p'(0)$ is a Stackelberg equilibrium. When $u^* > 0$, the Stackelberg equilibrium is unique, where the content provider sets $w^* = u^* C_p'(u^*)$.

4.3 Efficiency of Stackelberg Equilibrium

Now we discuss the efficiency of the Stackelberg equilibrium. For mathematical simplicity, in this subsection, we assume $C_s(u)$ is differentiable in u.

We define the social welfare, π_w, as the sum of the content provider's and all peers' utilities:

$$\pi_w(u) = \pi_s + N\pi_p = -C_s(u) - NC_p(u). \tag{13}$$

Because of the convexity of $C_s(u)$ and the strict convexity of $C_p(u)$, we immediately have

Proposition 3. *$\pi_w(u)$ is strictly concave in u.*

We define the *price of anarchy* (PoA) [10] to be the ratio of the social welfare at the worst Stackelberg equilibrium to the maximal social welfare one can achieve when varying $u \in [0, \bar{u}]$. In particular, when the Stackelberg equilibrium (u^*, w^*) is unique, we have

$$\text{PoA} = \frac{\pi_w(u^*)}{\pi_w(u_w)}, \tag{14}$$

where $u_w = \text{argmax}_u \pi_w(u)$ and u^* is the peers' solution at the Stackelberg equilibrium. In our model, the social welfare is non-positive, so PoA is in general no less than 1. When PoA is close to 1, it implies the system is in an efficient state. We first state the following lemma:

Lemma 2. *The peers' upload contribution at the Stackelberg equilibrium is no larger than the upload capacity that maximizes the social welfare, i.e., $u^* \leq u_w$.*

Proof. Denote $C_{SW}(u) = -\pi_w(u) = C_s(u) + NC_p(u)$, and $C_{SE}(u) = C_s(u) + uC_p'(u)$. Maximizing the social welfare is equivalent to solving $\min_u C_{SW}(u)$, $u \in [0, \bar{u}]$. According to Lemma 1, u^* can be obtained by solving $\min_u C_{SE}(u)$, $u \in [0, \bar{u}]$. Therefore, u_w and u^* are the minimizers to $C_{SW}(u)$ and $C_{SE}(u)$, respectively. By taking the first order derivative, we have

$$C'_{SW}(u) = C'_s(u) + NC'_p(u), \tag{15}$$
$$C'_{SE}(u) = C'_s(u) + NC'_p(u) + NuC''_p(u). \tag{16}$$

There are only two possible cases regarding $C'_{SW}(u)$:

(1) If $C'_{SW}(u) > 0$, $\forall u \in [0, \infty)$, then $u_w = 0$. Since $NuC''_p(u) \geq 0$, we have $C'_{SE}(u) = C'_{SW}(u) + NuC''_p(u) > 0$, $\forall u \in [0, \infty)$, so $u^* = 0 = u_w$.

(2) If there exists a $u_{SW} \in [0, \infty)$ such that $C'_{SW}(u_{SW}) = 0$, then u_{SW} must be unique due to the strict convexity of $C_{SW}(u)$. We have $u_w = \max(u_{SW}, \bar{u})$. By the concavity assumption on $C_p(u)$ and Proposition 1, $C'_s(u)$ and $NC'_p(u)$ are both non-decreasing in u and $NuC''_p(u) > 0$. Hence, for any $u > u_{SW}$, we have $C'_{SE}(u) > C'_{SW}(u) > C'_{SW}(u_{SW}) = 0$. This implies any $u > u_w = \max(u_{SW}, \bar{u})$ cannot be the minimizer of $C_{SE}(u)$, $u \in [0, \bar{u}]$. Therefore, $u^* \leq u_w$.

Combining the results in the above two cases, we have $u^* \leq u_w$. □

Theorem 3. *Denote u^* as the peers' solution at any Stackelberg equilibrium. If $u^* = 0$ or $u^* = \bar{u}$, the system obtains the maximal social welfare, i.e., PoA=1.*

Proof. If $u^* = 0$, then $C'_{SE}(0) \geq 0$. Suppose $u_w \neq 0$, then by the strict convexity of $C_{SW}(u)$, we have $C'_{SW}(u_w) = 0$ and $C'_{SW}(0) < 0$. From Eq. 15 and Eq. 16, we have $C'_{SE}(0) = C'_{SW}(0) < 0$, which contradicts to $C'_{SE}(0) \geq 0$. Hence, $u_w = 0 = u^*$, so PoA=1.

If $u^* = \bar{u}$, then by Lemma 2, we have $u_w \geq u^* = \bar{u}$. In the meanwhile, $u_w \leq \bar{u}$, so $u_w = \bar{u} = u^*$, and hence PoA=1. □

In general, $\pi_w(u_w)$ and $\pi_w(u^*)$ may not be equal. Given the concavity property in Proposition 3, $\pi_w(u)$ is strictly increasing in $[0, u_w]$. Therefore, the gap between $\pi_w(u_w)$ and $\pi_w(u^*)$ is impacted by the gap between u_w and u^*. In particular, we have $\pi_w(u_w) - \pi_w(u^*) \leq \pi'_w(u^*)(u_w - u^*)$, and PoA$\leq 1 - \frac{\pi'_w(u^*)}{\pi_w(u_w)}(u_w - u^*)$. Given a general form of $C_p(u)$, it is mathematically hard to further quantify PoA for $0 < u^* < \bar{u}$. In the next section, we will use simulations to show the efficiency of the Stackelberg equilibrium in general cases.

4.4 General Reward Scheme

The linear reward model is a simplification of the general reward model in Section 2. If we use the general model where $w(x)$ denotes the marginal reward per upload capacity x, the content provider's problem is

$$\max_{w(x)} \pi_s(w(x), u) = -C_s(u) - Nu \int_0^u w(x)dx, \tag{17}$$

and the peers' problem is

$$\max_{u} \pi_p(u, w(x)) = u \int_0^u w(x)dx - C_p(u). \tag{18}$$

We discuss the following question: among all possible reward models, which reward model can make the content provider obtain the maximal utility?

Theorem 4. *If the content provider can find $u^* = \text{argmax}_u -C_s(u) - NC_p(u)$, then any reward scheme satisfying the following property can make the content provider's utility arbitrarily close to the maximal:*

$$W(u) = \int_0^u w(x)dx \begin{cases} \leq C_p(u) & \text{if } u \neq u^*, \\ = C_p(u) + \epsilon & \text{if } u = u^*, \end{cases} \tag{19}$$

where ϵ is a positive real number and is arbitrarily small.

Proof. For a given reward scheme $w(x)$, assume the peers set $u = u^*(w(x))$ to maximize their utility. We have $\pi_p(u^*, w(x)) = u^* \int_0^{u^*} w(x)dx - C_p(u^*) \geq 0$; otherwise, the peers can obtain $\pi_p(u, w) = 0 > \pi_p(u^*, w(x))$ by setting $u = 0$. Therefore, the content provider's utility $\pi_s(w(x), u)$ in (17) is upper-bounded by $\text{argmax}_u -C_s(u) - NC_p(u)$. The content provider's utility achieved in Eq. 19 is $\text{argmax}_u -C_s(u) - NC_p(u) - \epsilon$, so it can be arbitrarily close to the upper-bound when ϵ is arbitrarily small. □

In this subsection, we relax the requirement of continuity on the reward function $W(u)$. An interesting implication is that the theorem provides us the insight in designing such reward schemes that maximize the content provider's utility. In fact, $\max_u -C_s(u) - NC_p(u)$, or $\min_u C_s(u) + NC_p(u)$ is a standard convex optimization and can be easily solved. After obtaining u^*, we can easily design a reward scheme satisfying (19). For example, we can design

$$w(x) = C_p'(x) + \epsilon\delta(u^*), \tag{20}$$

where $\delta(x)$ is the unit impulse function.

It is also interesting to note, using the general reward scheme in Theorem 4, the procedure of determining u^* is exactly maximizing the social welfare. Therefore, we have PoA=1, i.e., the social welfare is maximized when the content provider maximizes its own utility.

We briefly conclude the result in our game theoretic analysis. The interaction of the content provider and the peers can be viewed as a Stackelberg game where the content provider takes the first action and the peers follow. The existence and uniqueness of Stackelberg equilibrium shows the stability of the reward scheme, while efficiency is quantified by *price of anarchy*. We also point out the content provider's best strategy in designing reward in the general form.

5 Performance Evaluation

In this section, we validate the stability and efficiency of our incentive scheme via extensive simulations. We have the following settings:

- The system consists of $N = 10,000$ peers, one server and one video with $K = 50$ segments and playback rate $r = 500$ Kbps.
- The peers' external arrival probability to segment i is $p_{0i} = \frac{2^{i-1}}{(i-1)!}e^{-2}$ $(1 \le i \le K)$; the probability of doing play operation is $p_{i,i+1} = 1 - 0.6e^{-0.25\times i}$ $(1 \le i \le K-1)$; and the probability of quit operation is $p_{i0} = 1 - p_{i,i+1}$ $(1 \le i \le K-1)$, $p_{K0} = 1$. Assume the fast-forward and rewind operations are rare and can be omitted.
- The content provider's cost per unit capacity $c_s = 1$. The peers' upload cost $C_p(u) = c_p u^\beta$, where c_p and β are parameters we will vary in simulation.
- The content provider chooses $w \in [0, 1]$, and the peers choose $u \in [0, 1000]$ Kbps.

In Fig. 1, we plot the arrival probability p_{0i} and departure probability p_{i0} of each segment i. Observe that when a peer starts watching a video, it has probability $p_{01} = 0.135$ to start from the first segment, and $p_{02} = p_{03} = 0.271$ to start from the second or third segment, but the probability of watching from S_4 and onwards decreases rapidly. This corresponds to the reality where some people start from the beginning, but more people would like to skip the first few segments like advertisement. We can also observe that peers watching S_1 will quit the viewing course with probability 0.6, but the probability of quit operation decreases for

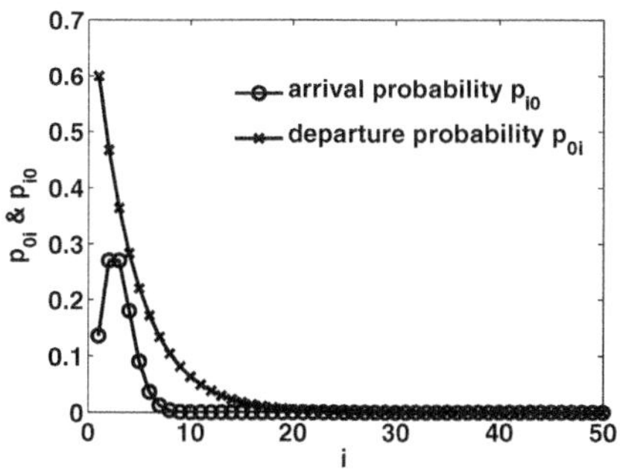

Fig. 1. Arrival and departure probability for each video segment

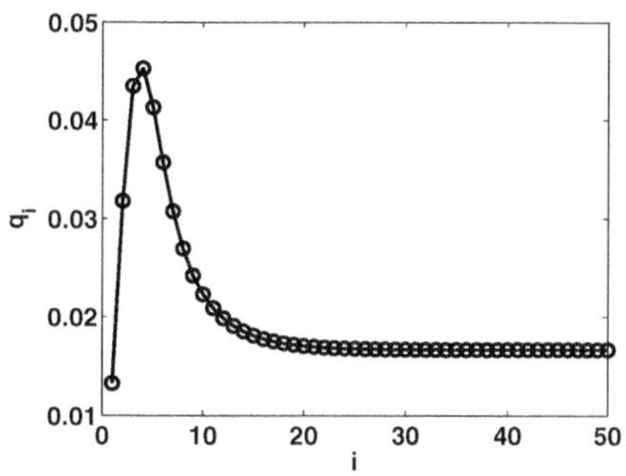

Fig. 2. Fraction of peers in each video segments

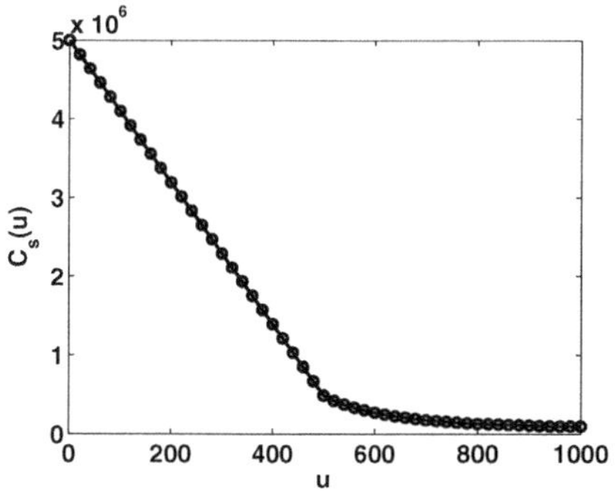

Fig. 3. Content provider's upload cost

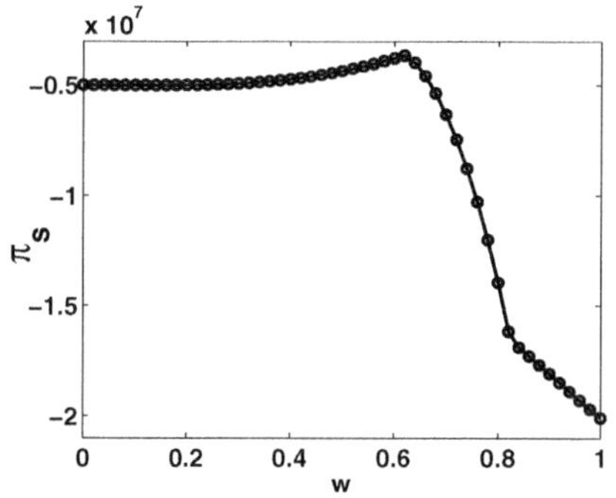

Fig. 4. Content provider's utility

peers watching later segments of the video. In Fig. 2, we plot the fraction of peers q_i for each video segment i. We can see from the figure that there is an increasing trend of popularity from S_1 to S_4, and a decreasing trend thereafter. This trend is due to the peers' viewing behaviors described above.

In Fig. 3, we plot the content provider's upload cost $C_s(u)$ (refer to Eq. 8) when we vary $u \in [0, 1000]$ Kbps. From this figure, we can observe that $C_s(u)$ is convex and non-increasing in u, which validates Proposition 1.

In Fig. 4, we investigate the content provider's utility $\pi_s(w, u)$ (refer to Eq. 9) when varying the unit reward w to peers. Assume the peers decide their upload capacity u to maximize their utility for given w. In this simulation, we fix $c_p = 0.15$ and $\beta = 1.2$ in the peers' upload cost function $C_p = c_p u^\beta$. We plot the content provider's utility π_s when varying $w \in [0, 1]$. When $w = 0$, peers do not contribute any upload bandwidth and thus the content provider's utility equals the negative value of the cost for supporting all peers' viewing requirement. When w increases from 0, the content provider can utilize part of peers' upload capacity so as to increase its utility. However, when w is very large, peers' decision u is also large. The content provider's utility decreases due to the huge amount of reward it has to pay to the peers. There exists an optimal unit reward $w = 0.62$ where the content provider's utility is maximized with value -3.63×10^6.

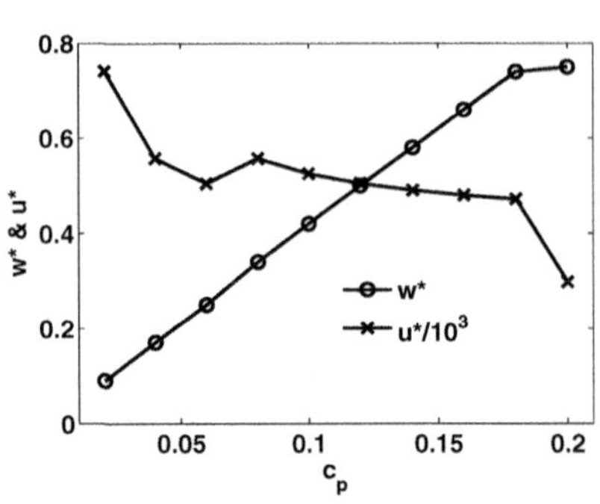

Fig. 5. Stackelberg equilibrium

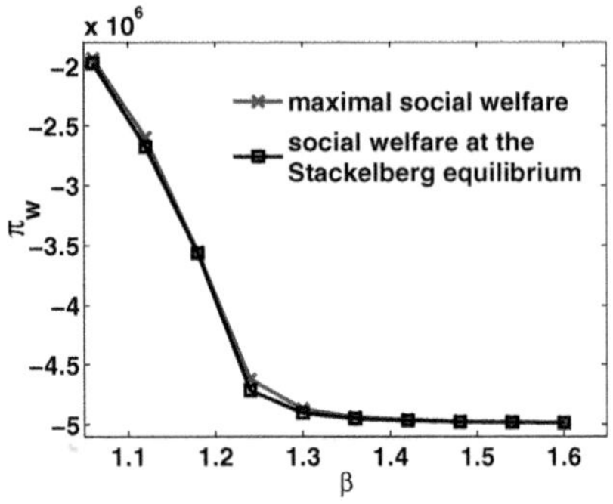

Fig. 6. Social welfare

In Fig. 5, we investigate the Stackelberg equilibrium under different parameters. In particular, we plot the value of w^* and u^* at the Stackelberg equilibrium when fixing $\beta = 1.2$ and varying $c_p \in [0.02, 0.20]$. The figure shows that when c_p increases, i.e., the upload cost of peers increases, the content provider rewards more to peers, and the peers tend to contribute less. However, we also note u^* is not monotonous in c_p, and there is a local minimum of u^* when $c_p = 0.06$.

We compare the maximal social welfare and the social welfare at Stackelberg equilibrium in Fig. 6. We fix $c_p = 0.2$ and vary $\beta \in [1.06, 1.60]$. We can observe from the figure that the social welfare at the Stackelberg equilibrium is always very near to the maximal value, i.e., PoA≈1 for any $\beta \in [1.06, 1.60]$. This simulation result illustrates the efficiency of our incentive scheme.

To summarize, all these simulation results validate our theoretic analysis and show the stability and efficiency of our incentive scheme.

6 Related Work

Incentive issue has received plenty of attentions in P2P applications. Zhao *et al.* [19] proposed a general framework to evaluate the expected performance gain and system robustness for a class of incentive protocols wherein peers can distributively learn to adapt their actions. In [16], the authors used game model to analyze the content production and sharing in P2P networks and compare the performance of different existing incentive schemes. There are also some existing works on designing particular incentive schemes. The first incentive scheme proposed for P2P system is the micropayment in [6]. Misra *et al.* [13] proposed a Shapley value approach in incentive design using a cooperative game model. Reputation [7,9] is another well-known approach where a peer's reputation represents its history of contribution in the system. Ma *et al.* [12] proposed a service differentiation approach in P2P network based on the amount of contribution each node has provided to the network community.

All these existing works are based on general P2P settings or are specifically designed for P2P file sharing systems. However, P2P-VoD systems have special features. Wu *et al.* [18] analyzed how to efficiently utilize the peers' resources, but did not address how to incentivize peers to contribute their resources. Habib *et al.* [8] proposed a service differentiation approach for incentive scheme in P2P multimedia systems, where peers with high contribution have flexibility in peer selection so that they receive better quality of service. Mol *et al.* [14] designed a free-riding-resilient P2P-VoD system where peers favor uploading to other peers who have proven to be good uploaders. These two works are similar because these incentive schemes they proposed are both variants of the tit-for-tat mechanism in file sharing applications. Similar approaches were also proposed for live streaming systems, e.g., in [17], the authors presented a modified tit-for-tat mechanism; in [11], a multi-layered live streaming system punishes the peers with low contribution by providing them with low quality of service. Instead of using the punishment-based approach, we propose a general reward-based incentive scheme where we incentivize the peers to contribute their upload capacity. In practice, peers' upload capacity is constrained by their Internet access types and hence varies a lot. Using our approach, peers with low upload capability can still receive good quality of service provided that they accept a low level of reward.

7 Conclusion

Incentive scheme is a key design issue in P2P applications in order to encourage peers' resource contribution. However, due to the complex and stochastic nature of peers' behavior, it is challenging to design an effective incentive scheme in P2P-VoD systems. In this paper, we propose a simple yet effective reward-based incentive scheme. We model the peers' distribution in movie segments using *mean field approximation*, characterize the content provider's cost in terms of peers' contribution and develop a game-theoretic framework to analyze the interactions of the content provider and peers under our reward-based incentive mechanism. We also show the stability and efficiency of our incentive scheme via extensive simulations.

References

1. pplive, `www.PPLive.com`
2. ppstream, `www.PPStream.com`
3. Benaïm, M., Le Boudec, J.-Y.: A class of mean field interaction models for computer and communication systems. Performance Evaluation 65, 823–838 (2008)
4. Cohen, B.: Incentives build robustness in bittorrent. In: Proceedings of the 1st Workshop on Economics of Peer-to-Peer Systems (2003)
5. Feldman, M., Papadimitriou, C., Chuang, J., Stoica, I.: Free-riding and whitewashing in peer-to-peer systems. IEEE Journal on Selected Areas in Communications 24(5), 1010–1019 (2006)
6. Golle, P., Leyton-Brown, K., Mironov, I.: Incentives for sharing in peer-to-peer networks. In: Proceedings of the 3rd ACM conference on Electronic Commerce, EC (2001)
7. Gupta, M., Judge, P., Ammar, M.: A reputation system for peer-to-peer networks. In: Proceedings of the 13th International Workshop on Network and Operating Systems Support for Digital Audio and Video (NOSSDAV), pp. 144–152 (2003)

8. Habib, A., Chuang, J.: Service differentiated peer selection: an incentive mechanism for peer-to-peer media streaming. IEEE Transactions on Multimedia 8(3), 610–621 (2006)
9. Kamvar, S.D., Schlosser, M.T., Garcia-Molina, H.: The eigentrust algorithm for reputation management in p2p networks. In: Proceedings of the 12th International Conference on World Wide Web (WWW), pp. 640–651 (2003)
10. Koutsoupias, E., Papadimitriou, C.H.: Worst-case equilibria. Computer Science Review 3(2), 65–69 (2009)
11. Liu, Z., Shen, Y., Panwar, S.S., Ross, K.W., Wang, Y.: Using layered video to provide incentives in p2p live streaming. In: Proceedings of the 2007 Workshop on Peer-to-peer Streaming and IP-TV (P2P-TV), pp. 311–316 (2007)
12. Ma, R.T.B., Lee, S.C.M., Lui, J.C.S., Yau, D.K.Y.: Incentive and service differentiation in p2p networks: a game theoretic approach. IEEE/ACM Transactions on Networking 14, 978–991 (2006)
13. Misra, V., Ioannidis, S., Chaintreau, A., Massoulié, L.: Incentivizing peer-assisted services: a fluid shapley value approach. In: Proceedings of the ACM International Conference on Measurement and Modeling of Computer Systems (SIGMETRICS), pp. 215–226 (2010)
14. Mol, J., Pouwelse, J., Meulpolder, M., Epema, D., Sips, H.: Give-to-get: Free-riding-resilient video-on-demand in p2p systems. In: Proceedings of SPIE, Multimedia Computing and Networking Conference, MMCN (2008)
15. Osborne, M.J.: An Introduction to Game Theory. Oxford University Press, USA (2003)
16. Park, J., van der Schaar, M.: A game theoretic analysis of incentives in content production and sharing over peer-to-peer networks. IEEE Journal of Selected Topics in Signal Processing 4(4), 704–717 (2010)
17. Pianese, F., Perino, D., Keller, J., Biersack, E.W.: Pulse: An adaptive, incentive-based, unstructured p2p live streaming system. IEEE Transactions on Multimedia 9(8), 1645–1660 (2007)
18. Wu, W., Lui, J.C.S.: Exploring the optimal replication strategy in p2p-vod systems: Characterization and evaluation. In: Proceedings of the 30th IEEE International Conference on Computer Communications, INFOCOM (2011)
19. Zhao, B.Q., Lui, J.C.S., Chiu, D.-M.: Analysis of adaptive incentive protocols for p2p networks. In: Proceedings of the 28th IEEE International Conference on Computer Communications, INFOCOM (2009)

Service Routing in Multi-ISP Peer-to-Peer Content Distribution: Local or Remote?

Parimal Parag[1], Srinivas Shakkottai[1], and Ishai Menache[2]

[1] Department of Electrical and Computer Engineering
Texas A & M University, College Station, TX 77843, USA
{parimal,sshakkot}@tamu.edu
[2] Department of Electrical and Computer Science
Massachusetts Institute of Technology, Cambridge, MA 02139, USA
t-ismena@microsoft.com

Abstract. The popularity of Peer-to-Peer (P2P) file sharing has resulted in large flows between different ISPs, which imposes significant transit fees on the ISPs in whose domains the communicating peers are located. The fundamental tradeoff faced by a peer-swarm is between free, yet delayed content exchange between intra-domain peers, and inter-domain communication of content, which results in transit fees. This dilemma is complex, since peers who possess the content dynamically increase the content capacity of the ISP domain to which they belong.

In this paper, we study the decision problem faced by peer swarms as a *routing-in-time* problem with *time-varying capacity.* We begin with a system of two swarms, each belonging to a different ISP: One swarm that has excess service capacity (a *steady-state* swarm) and one that does not (a *transient* swarm). We propose an asymptotically accurate fluid-approximation for the stochastic system, and explicitly obtain the optimal policy for the transient swarm in the fluid regime.

We then consider the more complex case where multiple transient swarms compete for service from a single steady-state swarm. We utilize a proportional-fairness mechanism for allocating capacity between swarms, and study its performance as a non-cooperative game. We characterize the resulting Nash equilibrium, and study its efficiency both analytically and numerically. Our results indicate that while efficiency loss incurs due to selfish decision-making, the actual Price of Anarchy (PoA) remains bounded even for a large number of competing swarms.

Keywords: peer-to-peer, overlay network, game theory, price of anarchy.

1 Introduction

Recent trends suggest estimate that 35-90% of bandwidth is consumed by peer-to-peer (P2P) file-sharing applications [1–3]. While there has been some drop in the fraction of P2P traffic for file distribution [4], there has been increased use of P2P for video streaming in systems such as PPLive [5] and QQLive [6]. Thus, P2P applications are likely to increase in number as they prove to be a relatively cheap means of media distribution from a content distributor's perspective.

R. Jain and R. Kannan (Eds.): GameNets 2011, LNICST 75, pp. 353–368, 2012.

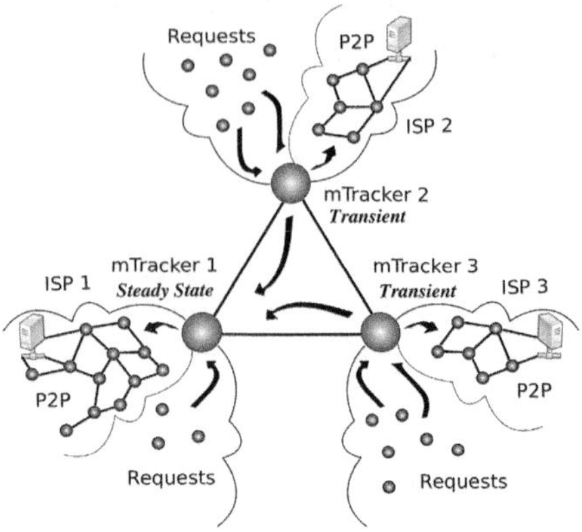

Fig. 1. The MultiTrack architecture described in [7]. The system uses multiple BitTorrent Trackers to achieve a desired delay-tariff tradeoff appropriate to the system state at that time instant. In this paper, we explicitly account for the dynamics of the P2P swarm capacities as a function of time.

P2P networks attempt to keep delays small by leveraging as much end-user bandwidth and storage as possible. However, they are often oblivious of the transit tariff that they impose on the hosting Internet Service Providers (ISPs) due to such optimizations. The pricing architecture of the Internet is tiered, wherein a lower tier ISP must pay a higher tier ISP (from which it obtains service) a tariff for all traffic entering or exiting its domain [8]. Since such lower tier ISPs are usually the ones that provide Internet connectivity to end-users, P2P communication between end-users across ISP domains causes significant tariffs for both of the terminal ISPs. Thus, there appears to be an implicit conflict between P2P applications that seek to find appropriate (lowest delay) peers regardless of the ISP domain in which such a peer might be located, and ISPs that seek to keep traffic localized within their domains. Such conflict has led to efforts by some ISPs to restrict P2P traffic [9].

The most popular P2P system nowadays is BitTorrent [10], which uses a system of *Trackers* to enable peers to find each other. When a peer without the content (called a *leech*) enters the system, it obtains a list of peers that it can communicate with from such a Tracker. The set of peers that is controlled by the Tracker in this fashion is known as the *peer swarm* (or P2P swarm) associated with that Tracker. While the original BitTorrent Tracker is ISP agnostic, Figure 1 depicts a system called MultiTrack described in [7] in which each ISP domain has a different P2P swarm for the same piece of content, with admission to each one being controlled by a separate mTracker. While a single seed is enough for whole swarm to receive the piece of content using P2P methods, it may take a long time to do so. Hence, the mTracker controlling an overloaded swarm (transient swarm) could request service from an mTracker controlling a swarm that has spare capacity (steady-state swarm). This leads to a natural tradeoff between minimization of delay versus transit tariff. However, [7] performs an instantaneous optimization and does not model the phenomenon that when a peer obtains the content it becomes a *seed server*, effectively increasing the peer-swarm capacity with time.

The dynamic evolution of the peer-swarm capacity results in a problem of "routing-in-time", where it is required to route traffic in a system of *time-varying capacity* so as to minimize costs. However, unlike a general routing-in-time problem, there are predictive models that describe the evolution of the swarm capacity [11–13], which are employed in the present paper. Since each peer is hosted within a particular ISP domain, once served, that peer could become a seed server in its host domain. Hence, each mTracker must take a decision on whether to keep a request local (and potentially incur a delay cost) or to forward it to a different ISP domain (and incur a transit tariff) with the knowledge that a request forwarded to a different domain at a particular time instant could result in a new local seed once that request has been satisfied. In this paper, we will consider two important questions pertaining to routing in P2P swarms:

- What is the appropriate routing in time profile that would minimize delay plus tariff costs in a transient swarm?
- Assuming there are multiple transient swarms competing for capacity available at a steady-state swarm, how should be capacity divided, and what are the consequences of non-cooperative competition for capacity?

1.1 Related Work

While the original BitTorrent Tracker was ISP agnostic, there have been several attempts to enable Trackers to become ISP-aware. Papers such as [7, 14–16] seek for the right tradeoff between delay costs and transit tariffs by attempting to keep traffic local whenever possible. For example, [15] suggests to optimize for minimum tariffs, and then develops heuristics for allowing a certain fraction of peers to be non-local so as to ensure that delays are not excessively large. In [7] the objective is to design a distributed control scheme called MultiTrack, that would achieve a desired delay-tariff tradeoff in a distributed fashion. However, as stated above, none of these references considers the evolution of peer-swarm capacity as a function of time.

We implicitly make the assumption that the newly created seeds would be willing to serve content to the leeches in their ISP domain. Incentives for such a seeds to share content may be provided by trading a local currency in exchange for files, see, e.g., [17, 18]. The objective in this paper is to optimize over the predicted capacity of P2P swarms, assuming that a mechanism for such trade is in place.

1.2 Contribution and Content

In this paper, we study a P2P model consisting of multiple ISP domains, each associated with a Tracker and a peer swarm that has N peers. Each swarm consists of peers that possess the content (seeds) and leeches, who may become seeds over time. We assume that one of the ISPs has a peer swarm that is in steady-state, meaning that its offered load is lower than the available content-capacity. Such peer swarm can act as a content server to boost the performance of those swarms that have not yet reached steady state. However, a transit tariff must be paid for access to remote swarms, and Trackers must take decisions on whether to request remote service or keep traffic local.

In Section 2, we develop a general stochastic model for a system with a single transient swarm. We propose a deterministic fluid model which is amenable for analysis, and has a similar behavior as the stochastic model as the swarm size N becomes large. In Section 3, we use the fluid model to show the optimality of an intuitive remote service profile, which is to request the entire capacity from the steady state swarm until a "stopping time", after which no more service should be requested. We obtain an explicit expression for this stopping time. We demonstrate through simulations (Section 4) that the optimal routing policy obtained for the fluid model is near-optimal with respect to the original stochastic model. In Section 5, we adopt a proportional-fairness like mechanism for dividing the capacity of a steady state swarm between multiple transient swarm. Such a mechanism naturally gives rise to a non-cooperative game between the swarms, each of which selfishly decides on a bid and a stopping time. We analyze the resulting game and provide bounds on the "Price of Anarchy" under symmetry assumptions. We supplement the analysis of the non-cooperative game with simulations in Section 6, which indicate that the experienced efficiency loss in terms of overall user cost is not more than 30%, and much less for a small number of swarms.

2 Single Transient Swarm

We first consider the case of a single transient swarm with N peers interested in a certain piece of content, and a single steady-state swarm that has a total upload capacity of NC distributed among N seeds. Thus, C is the maximum "per-requester capacity" available from the steady-state swarm. The transient swarm can make a time-varying request for service $C(t)$ from the steady state swarm. There are two ways in which the leeches in the transient swarm can obtain the content of interest. First, they contact other peers in the transient swarm uniformly and at random, and if the contacted peer has the file (i.e., it is a seed) the content may be downloaded. Alternatively, based on the choice of $C(t)$, seeds in the steady-state swarm are directed by the Tracker controlling their swarm to contact the Tracker controlling the transient swarm. This Tracker selects a peer that does not possess the content (i.e., a leech) from the transient swarm, and the seed from the steady-state swarm uploads the content to this peer. Below, we will develop a stochastic model corresponding to above dynamics, and then simplify it to a deterministic fluid model.

2.1 Stochastic System Model

We can think of the transient swarm as a graph $G = (V, E)$, with vertex set V corresponding to the set of peers, and edge set E corresponding to a communication link between two peers. By assumption, we have $N = |V|$. For simplicity of presentation, we consider a fully-connected graph G. However, our conclusions can easily be generalized to ensemble of randomly chosen K-regular graphs. Let $\mathcal{P}(t) \subseteq V$ be the set of seeds in the transient swarm at some time t, such that there are $P(t) = |\mathcal{P}(t)|$ seeds.

Assumption 1. *We assume that the capacity requested by the transient-swarm is piece-wise constant with time. Thus, time is divided into* phases $j \in \mathbb{N}_0$ *with*

$$C(t) = C_j \quad \forall t \in [T_{j-1}, T_j), \ \textit{where } T_{-1} = 0 \textit{ and } C_j \leq C \textit{ for all } j.$$

We denote the time spent in phase i as $\tau_i = T_i - T_{i-1}$. We will see later that our results hold for a general $C(t)$ as well. Leeches in the transient swarm contact each other uniformly and at random and download the file at rate η if it is available with the contacted peer. Also, in the jth phase the seeds in the steady state swarm are randomly directed to serve one of the leeches in the transient swarm with an upload rate C_j each. We create a tractable stochastic model by making the following simplifying assumptions.

Assumption 2. *We assume that each leech in the transient swarm is equipped with a clock, which ticks at a random interval that is an independent and exponentially distributed random variable X_k, with mean η^{-1} for peer $k \notin \mathcal{P}(t)$ in the transient swarm. Similarly each seed l in the steady-state swarm $l \in \{1, 2, \ldots, N\}$ has a clock that ticks at times that are denoted by exponential i.i.d. random variables Y_l with mean C_j^{-1} in the jth phase.*

When its clock ticks, each leech in the transient swarm contacts a neighbor uniformly at random. If the contacted peer k happens to be a seed (i.e., $k \in P(t)$), the leech downloads the content. When the clock of a seed in the steady state swarm ticks, it contacts its Tracker that directs it (via the Tracker controlling the transient swarm) to one of the leeches in the transient swarm, and uploads the content. A final assumption completes the model.

Assumption 3. *We assume that content is downloaded instantaneously.*

The assumption implies that an if a leech contacts a peer without the file of interest, it has to wait for a random interval of time before trying again. This allows for load balancing at seeds, since the number of leeches contacting each seed would be finite with high probability. Our approach provides a lower bound on system performance. The following lemma characterizes the evolution of the number of seeds as a continuous-time Markov chain (CTMC).

Proposition 1. *Let $P(t)$ be the number of seeds at any time t, in the single transient swarm model described above, that satisfies Assumptions 1, 2, and 3. If we assume that phase changes can occur only at the instants when $P(t)$ increases, then the number of seeds $P(t)$ evolves according to a CTMC as depicted in Figure 2, where the state-dependent jump-rate in phase j is*

$$\lambda_{P(t)} = NC_j + \frac{P(t)(N - P(t))}{N}\eta, \qquad P(t) \in \{1, \ldots, N\}.$$

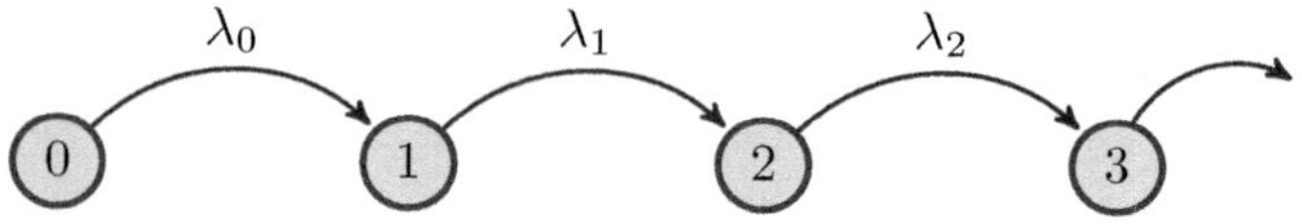

Fig. 2. Continuous time Markov chain governing the evolution of the number of seeds in the transient swarm

We show that as the peer-population N grows large, the evolution of the number of seeds $P(t)$ in the transient swarm can be modeled deterministically by the following differential equation

$$\frac{dP(t)}{dt} = \lambda_{P(t)}. \tag{1}$$

To this end, we show that for large N, the time-interval to increase the number of seeds from m to n in the deterministic model is identical to the corresponding time-interval in the stochastic model with probability one.

Theorem 1. *For a user population N, we denote the time-interval $T_{mn}(N) = t_{n+1} - t_m$ for number of seeds to increase from m to $n+1$ under the stochastic model under consideration. As user population N grows large*

$$\lim_{N\to\infty} T_{mn}(N) = \int_{\frac{m}{N}}^{\frac{n}{N}} \frac{dy}{\eta y(1-y) + C_j} \quad \textit{with probability } 1. \tag{2}$$

Notice (2) *is the integral form of the expected differential equation in* (1).

2.2 Delay Calculation Using the Fluid Model

As before, let $P(t)$ be the number of seeds in the transient swarm at t. From Theorem 1, the evolution of $P(t)$ is given by (1), restated below:

$$\frac{dP}{dt}(t) = \eta\frac{P(t)}{N}(N - P(t)) + NC(t). \tag{3}$$

We consider the evolution of $P(t)$ until $P(t) = N - 1$, at which point we say that the remaining one leech will be served in constant time. Let $y(t) = P(t)/N$ be the fraction of seeds in the system at time t. We denote the fraction of seeds at the beginning of phase i by $y_{i-1} = y(T_{i-1})$, with the convention $T_{-1} = 0$. We further define fraction $\alpha_i \triangleq (\theta_i - \eta y_{i-1})/(\eta y_{i-1} - \theta'_i)$, where $\theta_i \geq \theta'_i$ are the solutions to the quadratic equation $\theta^2 - \eta\theta - \eta C_i = 0$. The explicit evolution of $y(t)$ in time is presented in the following lemma.

Proposition 2. *We can write $y(t)$ in terms of positive difference $\Delta\theta_i = \theta_i - \theta'_i$ as*

$$y(t) = \frac{\theta'_i}{\eta} + \frac{\Delta\theta_i/\eta}{1 + \alpha_i e^{-\Delta\theta_i(t-T_{i-1})}} \qquad t \in [T_{i-1}, T_i). \tag{4}$$

For a finite number of $K + 1$ phases, we can compute the total delay seen by all the leeches as the difference in area between curves N and $P(t)$ for the interval $[0, T_K]$, where $P(T_K) = N - 1$[1]. That is, we have $T_0 \leq T_j \leq T_K$ such that $y(T_K) = 1 - 1/N$. The average per-requester delay D seen by leeches is the area between the curve $d(t) = 1$ (which is the demand curve, since the all leeches demand service at time zero) and $y(t)$, (the service curve) which can be expressed in terms of rates η, $\{(\theta_j, \theta'_j)\}$ and fractions $\{\alpha_j\}$.

Proposition 3. *The average per-requester delay D_i in phase i can be expressed as*

$$D_i = \frac{\theta'_i \tau_i}{\eta} - \frac{1}{\eta} \ln \left(\frac{\eta y_{i-1} - \theta'_i}{\eta y_i - \theta'_i} \right). \tag{5}$$

The aggregate average per-requester delay in $K+1$ finite phases is $D = \sum_{i=0}^{K} D_i$.

Now that we have expressions for the delay experienced in each phase, we optimize over delay and transit tariff in the following section.

3 Single Swarm Optimization

Let the transit tariff per unit traffic be denoted by p. Since this value is fixed, we can equivalently assume that the Tracker controlling the transient-swarm asks for a per-requester capacity $C(t)$ from the steady-state swarm, at a rate of p per unit capacity. The value of the per-requester capacity must be chosen such that a linear combination f of average per-requester delay D and transit tariff per unit traffic is minimized. That is, we wish to minimize

$$f = D + p \int_{t=0}^{\infty} C(t) dt. \tag{6}$$

Consider the case of piece-wise constant $C(t)$ with two non-zero phases. The problem reduces to the following optimization problem

$$\begin{aligned} &\text{minimize } f(\tau_0, \tau_1) \triangleq D + p\,(C_0 \tau_0 + C_1 \tau_1) \\ &\text{such that } C_0, C_1 \leq C. \end{aligned} \tag{7}$$

We now present two results that lead us to the intuitively appealing conclusion that remote capacity usage is necessary only in the first phase, and the amount of capacity used during that one phase to should be the maximum possible. We present these results without the proof due to space limitations.

Lemma 1. *The cost function f associated with the ISP is minimized for problem (7) when phase 1 is stopped at time T_1 such that $y_1 = y^*$, where*

$$y^* = \min \left\{ \frac{1}{\eta p}, \frac{1}{\eta p'} \right\} \text{ where } p'^{-1} = \eta \left(1 - \frac{1}{N} \right). \tag{8}$$

[1] Such a computation of total delay is valid for any work conserving policy.

Lemma 2. *The average per-requester delay D is minimized for problem (7) when the phase-interval with the smaller remote service-rate is zero.*

We next characterize the stopping time after which remote capacity usage is detrimental. Proof is omitted due to space constraints.

Theorem 2. *The cost function f associated with supporting a P2P swarm is minimized for the problem (7), if the maximum remote service capacity C available is utilized till an optimum stopping time τ^*, such that $y(\tau^*) = y^*$ defined in (8). In other words, setting remote service-rates as $C_0 = C, C_1 = 0$, and phase-change times as $\tau_0 = \tau^*$ and $\tau_1 = 0$ minimizes f. The optimal stopping-time τ^* in terms of ϕ, ϕ', such that $\phi + \phi' = \eta$ and $\phi\phi' = -\eta C$, is*

$$\tau^* = \frac{1}{\Delta\phi} \ln \left(\alpha \frac{(\eta y^* - \phi')}{(\phi - \eta y^*)} \right), \text{ where } \Delta\phi = \phi - \phi'. \tag{9}$$

We can also characterize the optimal per-user delay

$$D = \frac{\phi' \tau^*}{\eta} + \frac{1}{\eta} \ln \left(1 - \frac{\eta y^*}{\phi'} \right) - \frac{1}{\eta} \ln \left(\eta y^* p' \right), \tag{10}$$

and the associated cost function

$$f = D + pC\tau^*. \tag{11}$$

Finally, we have the following corollary that shows that our restriction to piecewise constant functions is actually not binding.

Corollary 1. *Let T be such that $y(T) = 1 - 1/N$. Then, the minimizer function in $\mathcal{C} = \{C(t), t \in [0,T] : C(t) \text{ simple }, 0 \le C(t) \le C, \forall t \in [0,T]\}$ for the following optimization problem*

$$\textit{minimize } f = D + p \int_0^T C(t)dt \textit{ such that } C(t) \in \mathcal{C},$$

is the following function $C(t) = C\chi_{[0,\tau^*]}(t)$ where τ^* is defined in (9).

Proof. It follows from induction using Lemma 1, Lemma 2, and Theorem 2.

Given a price p for service from remote swarm, one can find the optimal stopping time τ^*. It can quickly be seen that this stopping time is a non-increasing function of price p. In fact, it stays constant for $p \le p'$ and starts decreasing when $p > p'$. We are also interested in finding how the total cost per-user for the transient swarm increases with the price p for service from the remote swarm. It follows from equation (8), (10), and (11), that

$$\frac{df}{dp} = \begin{cases} C\tau^*(p') & p \le p' \\ \frac{Cy^*}{\eta y^*(1-y^*)+C} + C\tau^*(p) & p > p' \end{cases}. \tag{12}$$

Therefore, it is clear that the total cost per-user in the transient swarm is concave increasing in the remote service usage price p.

In conclusion, we have shown in this section that in order to minimize the total cost, a transient swarm should utilize both local P2P dissemination as well as all the capacity available from a remote steady state swarm up to a stopping time, after which the transient swarm has enough seeds that the correct decision is to not utilize the remote capacity. We found an explicit characterization of this stopping time, whose value has the intuitive property of being non increasing in transit tariff.

4 Stochastic Simulations of the Single Swarm

We perform simulations of the Markov chain described in Section 2.1 to illustrate the nature of the analytical results derived above. For the optimal case, we calculate the optimal stopping time based on the derivations relating to the fluid model of the previous section, but apply the resulting policy to the stochastic system described in Section 2.1. In our simulations, we took the population of peers in ISP 1 to be $N = 10^4$. The number is realistic for many P2P scenarios. Similar results hold for smaller values of N. However, the differential equations approximate the stochastic system model better with increasing number of peers. The capacity per-seed at the steady-state swarm is assumed to be $C = 1$, and the upload rate for peers in the transient swarm is $\eta = 0.01$ units. We consider the case where the transit tariff per-unit traffic follows $p = 4p'$, where $p' = \frac{N}{\eta(N-1)}$ as defined in Lemma 1 and its numerical value is 100.01 in current setting.

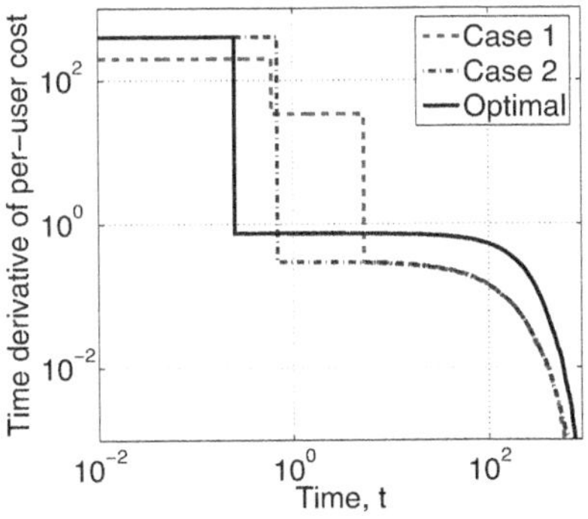

(a) Cost evolution for different policies

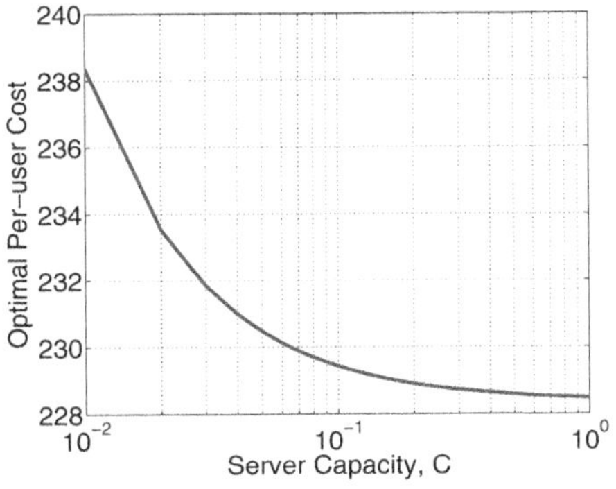

(b) Per-user optimal costs

Fig. 3. Left figure shows cost evolution for optimal vs. arbitrary stopping time. The area between curves is the total system cost. Right figure shows the optimal cost for different server capacities.

Our first objective is to illustrate that our optimal stopping time policy yields a lower cost than other policies. Hence, we create two arbitrary policies that are described below. We run the simulation until the transient swarm has $N-1$ seeds. We can compute the time derivative $\frac{df}{dt}(t)$ of the per-user cost function f from the definition of per-user delay and cost function (11). Therefore, we have

$$\frac{df}{dt}(t) = 1 - y(t) + pC(t). \tag{13}$$

We plot this time derivative $\frac{df}{dt}(t)$ in Figure 3(a) for three different cases. In all cases, the area enclosed between the curves and the axes is the total cost of the system. The dashed curve denotes the first case, where requests are served in two phases by the steady-state swarm. In the zeroth phase, the remote service capacity is $C_0 = C/2$ and the phase ends when a three-tenth of the population has the desired content, i.e. $y_0 = 0.3$. In the first phase, server-rate is $C_1 = C/12$ and the phase ends when seven-tenth of the user population has the desired content, i.e. $y_1 = 0.7$. The dotted curve denotes the case when there is a single phase where maximum available server-capacity C is used by the requesting swarm. However, in this case, the stopping time is not optimal, and the phase ends when seven-tenth of the user population has the desired content. The solid curve denotes the time-derivative of cost for the optimal case. Note that since the scale is log-log to illustrate the differences in the curves, the large difference between the optimal area and suboptimal area is not prominently visible.

In Figure 3(b), we plot the minimum per-user cost f for ISP 1 as a function of available per-requester server-capacity C, for the same parameters for user-population, server-capacity, upload rate, and tariff per-unit traffic. Essentially, we plot the equivalent of the fluid equation (1) in the stochastic setting when y^* has been chosen optimally as a function of constant C. The available per-requester server-capacity C takes values from 0 to 1. Clearly, the minimum per-user cost decreases and the rate of decay decreases with C as expected from the proof of Theorem 2.

We conclude that the optimal policy in the fluid system is indeed optimal in the original stochastic system, and that the parameters τ^*, y^* that can be calculated from the optimal deterministic policy are essentially optimal in the stochastic case.

5 Multiple Swarms: Collaboration or Competition?

We now consider the case of Q transient P2P swarms, each controlled by a distinct Tracker $i \in \{1, 2, \ldots, Q\}$. There is also a single steady state swarm indexed by 0 from which all the transient swarms attempt to obtain service. Thus, we have swarm 0 in steady state with total available capacity NC, while every other swarm $i \in \{1, 2, \ldots, Q\}$ starts with 0 initial service capacity and number of leeches N. Suppose that the transit tariff to reach the steady state swarm is p for all the transient swarms (this value could also be 0). In addition to this tariff, each transient swarm i bids a value p_i indicating its willingness to pay the steady state swarm for service. The steady state swarm must use some mechanism to decide how much of its capacity to allocate to each transient swarm.

5.1 Proportional-Fairness Mechanism

We propose to use the proportional-fairness mechanism for the steady state swarm to allocate capacity amongst the transient swarms. Under this mechanism the allocation to each transient swarm i is given by

$$\textbf{Mechanism:} \quad C_i = \frac{p_i C}{\sum_{j=1}^{k} p_j}. \tag{14}$$

The mechanism is very simple to implement, and has been successfully used in communication networks for apportioning bandwidth to competing flows [19]. Further, it has been shown to have a bounded inefficiency even with strategic users that optimize against the mechanism [20]. It is therefore, a good candidate for our system of competing transient swarms.

The Tracker associated with a transient swarm i can utilize the capacity allocated to it for any duration that it chooses, and pays $p_i + p$ per unit traffic during that time. We assume that once a certain amount of capacity has been allocated to a transient swarm i, it cannot be withdrawn and reallocated to some other swarm. Such a scheme is consistent with the idea developed in the previous section that the capacity from the transient swarm is most useful during the initial stage, and also simplifies our analysis.

5.2 Definition of the Game

We utilize the same fluid approximation developed in Section 2.1 to describe the dynamics of each swarm. The cost function f_i associated with swarm i for the multiple ISP scenario can be expressed in terms of the fixed transit tariff p, the bid p_i, the per-requester delay D_i, and allocated capacity C_i as

$$f_i(p_i, p_{-i}) = D_i + (p + p_i)C_i\tau_i. \tag{15}$$

Note that in general f_i is also a function of the upload capacity η_i of peers in swarm i. We can then define a strategic game $\mathcal{G} =< \mathcal{Q}, \mathcal{P}, \mathcal{F} >$, where $\mathcal{Q}$ is the set of Trackers (players), $\mathcal{P}$ is the set of bid profiles (action sets) and $\mathcal{F} = \{f_1, f_2, \cdots, f_Q\}$.

Our first objective is to find the socially optimal way of bidding when all Trackers collaborate. The objective here is to minimize the sum of the costs incurred by all swarms. Secondly, we also wish to compute what the cost is if Trackers are selfish and act individually and rationally to arrive at a bid decision. In the following, we follow the notation $p_{-i} = \sum_{j \neq i} p_j$. We will only analyze the symmetric case where $\eta_i = \eta$ for all swarms $i \in \{1, 2, \ldots, Q\}$, and consider the scenario where the transit tariff p is lower bounded by $p \geq p' = N/(\eta(N-1))$. Therefore, by Theorem 2, the total available capacity is used only till each ISP reaches the fraction $y_i = \frac{1}{\eta(p+p_i)}$.

5.3 Collaborative Scenario

We consider cooperating Trackers, who wish to jointly minimize the aggregate cost. Consider a set of bids $P = \{p_i : i = 1, 2, \ldots, Q\}$. Then the problem that the Trackers wish to solve is

$$\textbf{Opt:} \quad \min_{p_i \geq 0} \sum_{i=1}^{Q} f_i(P). \tag{16}$$

Theorem 3. *For collaborative scenario, optimal set of bids is* $P^* = \{0, 0, \ldots, 0\}$.

5.4 Multiplayer Game

We now consider the non-cooperative situation, where every Tracker acts according to its own self interest. We assume each Tracker makes a rational decision, assuming every other Tracker does the same. We also assume that each bid is made without knowledge of any other Tracker's bid. In this setting, we wish to find the Nash equilibrium (if it exists) of the bid p_i made by each Tracker i. Hence, each Tracker wishes to solve the following problem.

$$\textbf{Game:} \quad \min_{p_i \geq 0} f_i(p) \qquad \forall i \in \{1, 2, \ldots, Q\}. \tag{17}$$

The following theorem provides the necessary conditions for the existence of a symmetric Nash equilibrium of bids for this non-cooperative strategic game. We omit the proof in the interest of space.

Theorem 4. *If the number of competing swarms Q is such that $\eta < 2C/Q$ then the strategic game $\mathcal{G}$ has a pure strategy Nash Equilibrium of set of bids $P = \{\beta, \beta, \ldots, \beta\}$.*

Since the NE exists, we identify a pure strategy NE in terms of a common bid β made by all the Trackers. Note, that $h(\beta) = 0$ and τ_i is a logarithmic function of p_i (see (9)). We identify upper and lower bounds on the pure, symmetric bid in the following theorem stated without the proof.

Theorem 5. *The optimal bid β for the strategic game $\mathcal{G}$ is bounded above and below by the following values*

$$\frac{Q-1}{\eta Q + 4C} \leq \beta \leq \frac{Q-1}{\eta Q + 4C} \left(\frac{1 + \frac{p\eta(\eta+2C_i)}{4C_i}}{1 - \frac{(Q-1)\eta(\eta+2C_i)}{4C(\eta+4C_i)}} \right), \tag{18}$$

under the condition $\eta < 2C/Q$.

It is clear that it is difficult to analytically compute Nash equilibrium β. Therefore, we make an approximation to get some insight into how the optimal bid changes with the number of transient swarms competing for the available capacity at the steady-state swarm. In the regime $\eta \ll C_i$, we have $y_i \approx C_i\tau_i$. Under this approximation, from Theorem 5 we obtain

$$\beta \approx \frac{Q-1}{\eta Q + 4C} \qquad \forall i \in \{1,\dots,Q\}. \tag{19}$$

Notice that as the number of competing Trackers go up, so does the bid at Nash equilibrium. Such an increased bid is the price paid for uncoordinated behavior by the different Trackers. Our approximation is a lower bound on the Nash equilibrium bid. Also, it is clear from (18) that the approximation error is small when $\eta \ll C_i$ and $p(\eta + 2C_i) \ll 1$.

5.5 The Price of Anarchy

In most work on selfish decision making, it is found that individual optimization has a negative impact on the total value of a system. We observed that the lack of coordination results in a bid that is linearly increasing in number of ISPs Q. How different would the system cost be in such a scenario? Note that since are dealing with costs, a larger PoA is worse. Due to symmetry in the problem, each Tracker i bids the same value and receives a total service-rate NC/Q from the steady-state swarm. For the optimal case, when the players collaborate, this bid is 0 and the per-user cost for each transient swarm is f_i^{opt}. When the players are selfish, they bid value β additional to the base price p as tariff per-unit traffic.

Following terminology from [21], we define the "price of anarchy" as

$$\text{PoA} \triangleq \frac{\sum_{j=1}^{Q} f_i^{\text{game}}}{\sum_{j=1}^{Q} f_i^{\text{opt}}}. \tag{20}$$

However, unlike most work on the price of anarchy, we are less interested in the regime where the number of players is large. In other words, $Q \to \infty$ is less interesting to us since the number of peer swarms simultaneously competing for capacity is likely to be fairly small, although each swarm might have many thousand peers. Thus, our primary focus will be on obtaining good bounds on the PoA for relatively small values of Q, and in this regime we have the following theorem. Proof is omitted due to space constraints.

Theorem 6. *The price of anarchy (PoA) for strategic game $\mathcal{G}$ is bounded above by the following*

$$1 + \frac{2\ln\left(1+\frac{\beta}{p}\right)}{\eta f_i(p') + \frac{8C_i}{(4C_i+\eta)}\ln\left(\frac{p}{p'}\right)}$$

where β is the Nash Equilibrium bid for the game.

Therefore, if number of Trackers Q is small enough, such that $\eta \ll C_i$, then we can approximate PoA to be

$$\text{PoA} \approx 1 + \frac{2\ln\left(1+\frac{\beta}{p}\right)}{\eta f_i(p') + 2\ln\left(\frac{p}{p'}\right)}. \tag{21}$$

Notice that if we use the approximate value of β given in (19), the upper bound above is a function of known parameters of the system. Such a form is appealing since it a simple upper bound on the PoA. In the next section, we will numerically solve the game, and compare the actual price of anarchy to the bound derived above.

6 Numerical Studies of the Game

For numerical studies, we considered the symmetric case where each transient swarm has the same user population $N = 10^4$, and upload rate of seeds $\eta = .01$. The available per-user capacity at the steady-state swarm was taken to be 1. We considered the base price $p = p'$ where $p' = 100.01$ as in Section 4. We varied the number of transient swarms from 2 to 50.We plot the bid at Nash equilibrium along with the upper and lower bounds in the Figure 4. Since $\eta < 2C_i$ for the chosen values of Q, the bounds are reasonable. It must be noted that the bounds get worse with increase in number of swarms and they do not hold when number of transient swarms become large. Figure 5(a) shows the actual

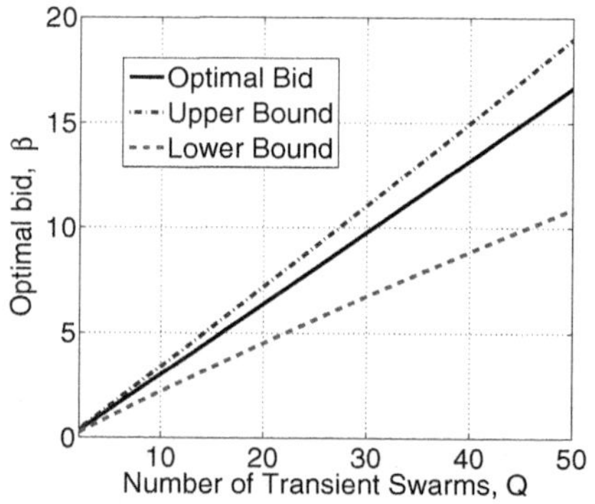

Fig. 4. Optimal bid in the symmetric case, and its upper and lower bounds

PoA computed numerically, and its approximation computed analytically. As expected, the PoA grows with number of transient swarms. However, when Q becomes large, we are no longer in the regime $\eta < 2C_i$. We see in Figure 5(b) that PoA remains bounded even when number of swarms become large. Further, our approximation of the PoA remains bounded as well. While we do not expect the number of competing swarms to be this large in reality, it is interesting the PoA is at most about 30% even for large Q.

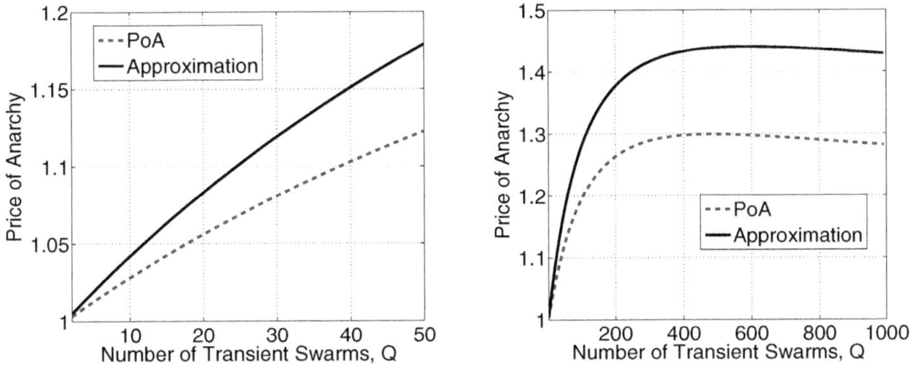

(a) PoA and approximation (b) PoA with increasing swarm size

Fig. 5. Illustration of the price of anarchy for different swarm sizes. As the swarm size increases, it remains bounded.

7 Conclusion

We studied in this paper the basic dilemma faced by any content distributor that wishes to utilize the inherent capacity scaling effects of P2P networks, but also does not want to impose excessive transit tariffs on the ISPs hosting the peers. We showed that since a P2P network has a capacity that scales as the number of users served, the greatest gain for usage of the steady-state swarm is in the initial phase, with the duration of usage that depends on the transit tariff. We also considered the case of multiple ISPs competing for capacity, and showed that while the resulting equilibrium is suboptimal, performance is adequate. We believe that besides the specific results, the model proposed in this paper can be used for more complicated P2P interactions that we will explore in the future..

Acknowledgments. Research was funded in part by NSF grants CNS 0904520 and CNS 0963818, the Google Research Awards program, and Qatar Telecom, Doha, Qatar.

References

1. Fraleigh, C., Moon, S., Lyle, B., Cotton, C., Khan, M., Moll, D., Rockell, R., Seely, T., Diot, C.: Packet-level traffic measurements from the Sprint IP backbone. IEEE Network Magazine 17(6), 6–16 (2003)
2. Gummadi, K.P., Dunn, R.J., Saroiu, S., Gribble, S.D., Levy, H.M., Zahorjan, J.: Measurement, modeling, and analysis of a peer-to-peer file-sharing workload. In: Proc. SOSP (October 2003)
3. Bangeman, E.: P2P responsible for as much as 90 percent of all Net traffic. ArsTechnica, September 3 (2007)
4. Labovitz, C., McPherson, D., Iekel-Johnson, S.: 2009 Internet Observatory report. In: NANOG-47 (October 2009)

5. PPLive (2009), http://www.pplive.com/
6. QQLive (2009), http://www.qqlive.com/
7. Reddy, V., Kim, Y., Shakkottai, S., Reddy, A.L.N.: MultiTrack: A Delay and Cost Aware P2P Overlay Architecture. In: ACM SIGMETRICS as a poster, Seattle, WA (June 2009)
8. Huston, G.: ISP Survival Guide: Stratagies for Running a Competitive ISP. John Wiley and Sons, New York (1999)
9. Broache, A.: FCC chief grills Comcast on BitTorrent blocking. C|Net News.com, February 25 (2008)
10. BitTorrent (2005), http://www.bittorrent.com/
11. Yang, X., de Veciana, G.: Performance of Peer-to-Peer Networks: Service Capacity and Role of Resource Sharing Policies. In: Performance Evaluation: Special Issue on Performance Modeling and Evaluation of P2P Computing Systems, vol. 63 (2006)
12. Qiu, D., Srikant, R.: Modeling and performance analysis of BitTorrent-like peer-to-peer networks. In: Proc. ACM SIGCOMM, Portland, Oregon, USA (August 2004)
13. Shakkottai, S., Johari, R.: Demand Aware Content Distribution on the Internet. IEEE/ACM Transactions on Networking 18(2), 476–489 (2010)
14. Aggarwal, V., Feldmann, A., Scheideler, C.: Can ISPs and P2P users cooperate for improved performance? ACM Computer Communication Review 37(3) (July 2007)
15. Xie, H., Yang, Y.R., Krishnamurthy, A., Liu, Y., Silberschatz, A.: P4P: Portal for P2P applications. In: Proc. ACM SIGCOMM (August 2008)
16. Choffnes, D.R., Bustamante, F.: Taming the torrent: A practical approach to reducing cross-ISP traffic in p2p systems. In: Proc. ACM SIGCOMM, Seattle, WA (August 2008)
17. Aperjis, C., Freedman, M.J., Johari, R.: Peer-assisted content distribution with prices. In: Proc. 4th ACM SIGCOMM Conference on emerging Networking EXperiments and Technologies, CoNext 2008 (December 2008)
18. Menasché, D.S., Massoulié, L., Towsley, D.F.: Reciprocity and barter in peer-to-peer systems. In: INFOCOM 2010, San Diego, CA (2010)
19. Kelly, F.P.: Charging and rate control for elastic traffic. European Transactions on Telecommunications 8, 33–37 (1997)
20. Johari, R., Tsitsiklis, J.N.: Efficiency loss in a network resource allocation game. Mathematics of Operations Research 29, 407–435 (2004)
21. Roughgarden, T., Tardos, E.: How bad is selfish routing? In: IEEE Symposium on Foundations of Computer Science, pp. 93–102 (2000)

Bargaining and Peering between Network Content/Coverage Providers

Guosen Feng, Jianwei Huang, and Dah Ming Chiu*

Department of Information Engineering
The Chinese University of Hong Kong
{fgs009,jwhuang,dmchiu}@ie.cuhk.edu.hk

Abstract. Both content quality and market coverage have significant impacts on a network content provider's revenue. In this paper, we present a preliminary study on how providers' cooperation and adoption of special content can affect the content quality and market coverage. We first consider a baseline case, where providers have static contents and do not cooperate. We derive the providers' coverages based on the quality of the contents and user subscription fees. Then we consider how cooperation and content sharing can help providers to improve their revenues. The key insight is that cooperation will be desirable when the providers' total revenue is increased and properly shared by an inter-provider financial transfer. In the case of linear advertisement functions, cooperation will happen when providers have different abilities in generating advertisement revenue and have proper subscription fees. We further consider the dynamic content case, where a provider can introduce some high quality special content for a short amount of time to attract users to switch from one provider to the other. We show that the switching cost, the valuation of content, and time discount factor all play important roles in deciding the benefit of special content.

Keywords: Network Content Providers, Content Coverage, Peering, Bargaining, Cooperation.

1 Introduction

High quality contents attract great interest from users, and can significantly increase the market coverage of *network content providers*. This in turn will increase the providers' advertisement and subscription incomes. One way for a network content provider to obtain high quality contents is to cooperate (or peer) with other providers and share contents. However, peering agreements are not always easy to reach. For example, Google TV, a new Internet-connected television platform, aims at providing users with new experiences of enjoying both traditional TV and web contents [1]. But some content providers (*e.g.*,

* This work is supported by the General Research Fund (Project Number 412710) established under the University Grant Committee of the Hong Kong Special Administrative Region, China.

R. Jain and R. Kannan (Eds.): GameNets 2011, LNICST 75, pp. 369–382, 2012.

Fig. 1. Network architecture among advertisers, network content providers, and end users

NBC and ABS in the U.S.) choose to block Google TV from accessing their TV programs. These content providers are afraid that this new technology may influence advertisers' choice of advertisement platforms (*e.g.*, Google TV vs. NBC's website) and reduce their advertisement incomes. A proper financial agreement between Google and the content providers may resolve this issue. Besides regular contents, there are some special contents (*e.g.*, world cup programs) that can attract many users during a certain period of time, and can be used as a powerful tool for providers to gain additional market share. The 2010 world cup broadcasting right issue in Hong Kong showed how fiercely content providers bargain over the special content delivering right [2]. The official broadcast right holder (iCable in Hong Kong) wanted the content to reach a bigger coverage together with its own advertisement. While other content providers (*e.g.*, TVB and ATV) wished to purchase the broadcasting right without the advertisement from iCable. A final agreement was reached which led to a win-win situation of both sides. In a third example, mobile TV program providers want to reach a large audience. However, indoor users typically have difficulty in accessing the mobile TV programs with a high quality due to a poor cellular signal receptions. Owners of large office buildings and shopping malls may help to "amplify" the signals through special equipments, and receive payments from the mobile TV providers for providing the extra coverage.

In this paper, we are motivated by the above three examples and want to study the interactions among multiple network content providers over content, coverage, and the possible strategies of cooperation. We will assume that network content providers obtain revenue through two approaches: advertisement income based on the agreement with advertisers and the market coverage, and subscription income based on the content quality and the subscription fees. The network architecture is illustrated in Fig. 1.

Many papers have studied network providers' strategies on maximizing revenues. Some studies (*e.g.*, [3]-[5]) focused on analyzing how to choose contents and determine advertisement lengths to attract users. Other results (*e.g.*, [6]-[7]) focused on how to increase revenue through either an advertisement-sponsored only approach or a subscription-and-advertisement-sponsored approach. However, none of the prior results have considered the cooperation among providers. Another line of research investigated the cooperation issues among providers. Reference [8,9] examined the incentive for ISPs to interconnect and developed Shapley value based revenue distribution mechanisms. Reference [10] examined the optimal pricing strategies for content delivery. Reference [11] considered rate allocation of ISPs with content providers' participation. However, most of these results assumed that each user has a fixed subscription to one provider and can not switch to different providers. Also, none of them have taken the advertising income into account during the cooperation. In our paper, we consider the interactions of advertisers, network providers, and users, where users may switch between providers depending on the contents and subscription fees.

In our paper, we will focus on the interactions of two network content providers in three cases. In the first baseline "static" case, both providers have fixed quality contents over time and they do not cooperate with each other. We will examine the users' subscription choices and the corresponding market share. Then we look at the second "cooperation" case, and study how cooperation can increase the providers' revenues with static contents. In the third "dynamic" case, we will consider how the introduction of a special content will impact the users' subscription choices and the providers' revenues.

Our main results and contributions are as follows:

- *General Network Model:* We present a model that captures the interactions among advertisers, network content providers, and users interact, and explain how users choices influence providers' content strategies and revenues.
- *Win-win Cooperation Agreement:* We propose a Nash bargaining based cooperation framework between providers, by considering the changes of content, advertisement, and coverage with the cooperation. We characterize the necessary condition for cooperation to happen, and show that a provider's bargaining power depends on its capability of generating advertisement revenue.
- *Impact of Dynamic Content:* One-time special content induces users to switch providers. We show how the switching cost, content evaluation, and time discount factor together determine a user's subscription decision and the providers' revenues.

The rest of the paper is organized as follows. Section 2 presents the static baseline model and Section 3 shows the cooperation strategy of providers through bargaining. The impact of dynamic content without cooperation between providers is given in Section 4. We conclude in Section 5 and summarize the future work directions.

2 A Static Baseline Model

We consider a duopoly market of two network providers: A and B. Each provider has a dedicated advertiser, who pays the provider advertising fees based on the provider's coverage. A provider's coverage depends on the number of users subscribing to its service. A provider can attract subscribers by high quality contents or a low monthly subscription fee.

2.1 Content Qualities and Subscribing Fees

We consider a period of T time slots, where each time slot has a unit length (*e.g.*, representing one month of time). The content quality of a provider $i \in \{A, B\}$

in time slot $t \in \{1, \ldots, T\}$ is q_{it}. In this section, we assume that both providers have *static* contents, *i.e.*, $q_{it} = q_{it'} = q_i$ for any $t, t' \in \{1, \ldots, T\}$ and both $i = A$ and $i = B$. Without loss of generality, we assume that provider A has a more popular content, *i.e.*, $q_A \geq q_B$. This may reflect the fact that provider A has a larger budget and can purchase higher qualities contents than provider B. We will come back to the budget issue in Section 4.

During each time slot, provider $i \in \{A, B\}$ charges each of its subscriber p_i. As provider A has a better content, it can charge a higher subscribing fee, *i.e.*, $p_A > p_B$.[1] *We further assume that both p_A and p_B are fixed throughout this paper.* This allows us to focus our study on the impact of content choices and provider competitions. In our future work, we will further discuss how providers optimize their subscription fees in a game theoretical setting.

2.2 Users' Utilities

Users may achieve different satisfaction levels by consuming the same contents. We characterize a user with two parameters: θ representing the user's valuation of the content quality, and δ representing the user's time discount factor over future contents. A user's total utility of subscribing and consuming contents from provider $i \in \{A, B\}$ over T time slot is

$$U(\theta, \delta) = \theta \sum_{t=1}^{T} \delta^{t-1} q_{it} - p_i T. \tag{1}$$

For a user who is indifferent of choosing either provider, we have the following relationship between θ and δ:

$$\theta \sum_{t=1}^{T} \delta^{t-1} q_{At} - p_A T = \theta \sum_{t=1}^{T} \delta^{t-1} q_{Bt} - p_B T. \tag{2}$$

Based on (2), we can compute the boundary evaluation $\theta^*(\delta)$ as a function of δ, which is illustrated in Fig. 2. Users with parameters (θ, δ) below the boundary will choose to subscribe to provider B, while the users above the boundary will subscribe to provider A.

2.3 Providers' Coverages and Revenues

For the rest of the analysis, we assume that both θ and δ are uniformly distributed in $[0, 1]$. Without loss of generality, we normalize the total users population to be 1. Then the area under the boundary represents the market share

[1] Assume this is not true and $p_A \leq p_B$. Then all users will choose provider A, who offers a better content with a lower fee. Provider B will have no subscribers and will be out of the market. This is apparently not an interesting case and will not be further discussed in this paper.

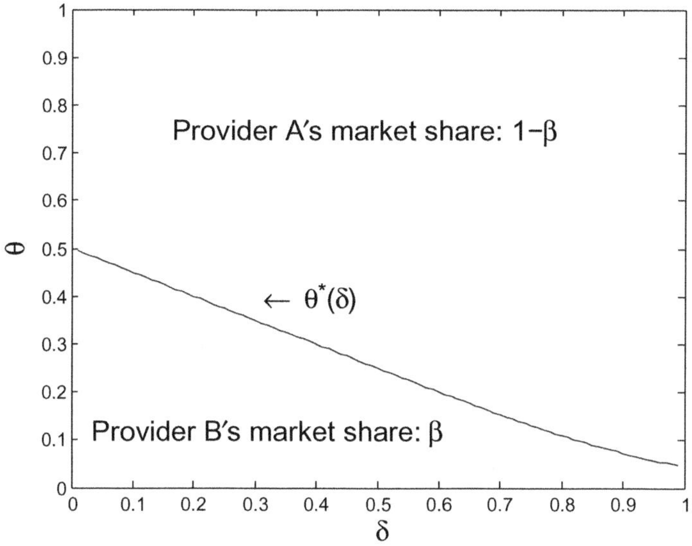

Fig. 2. Two providers' market shares

of provider B (denoted as β), and the provider A has a market share of $1-\beta$:

$$\text{Provider } A\text{'s coverage} : 1-\beta = 1-\int_0^1 \frac{(p_A-p_B)T}{\sum_{t=1}^T \delta^{t-1} q_{At} - \sum_{t=1}^T \delta^{t-1} q_{Bt}} d\delta, \quad (3)$$

$$\text{Provider } B\text{'s coverage} : \beta = \int_0^1 \frac{(p_A-p_B)T}{\sum_{t=1}^T \delta^{t-1} q_{At} - \sum_{t=1}^T \delta^{t-1} q_{Bt}} d\delta. \quad (4)$$

A provider's revenue includes both the users' subscription fee and the advertisement fee. We assume that provider $i \in \{A, B\}$ has a advertisement revenue function $f_i(\cdot)$ per time slot. Here $f_i(\cdot)$ is an increasing function of its market share. If there are no users accessing the contents, no advertisers will pay for the advertisement. Thus $f_i(0) = 0$. The two providers' revenues over T time slots are

$$\pi_A = (f_A(1-\beta) + p_A \cdot (1-\beta)) \cdot T, \quad (5)$$

$$\pi_B = (f_B(\beta) + p_B \cdot \beta) \cdot T. \quad (6)$$

2.4 Content Procurement Strategies

Each provider may change its revenue through contents procurement.[2] We consider two possibilities in the next two sections: peering between providers to share contents and increase coverage (Section 3) and introducing special content to attract users to switch providers (Section 4).

[2] Recall that we have assumed the subscribing fees p_A and p_B are fixed in this paper.

3 The Peering and Bargaining of Providers

3.1 Peering Agreement

When two providers cooperate (or peer) with each other, we assume that one provider will purchase the *whole* content from the other provider. Since provider A has the better content (*i.e.*, $q_A \geq q_B$ as assumed in Section 2), provider A will be the seller and provider B will be the buyer.

However, two providers have different concerns when peering. From A's point of view, it wishes to deliver both the content and its advertisement in B's network, so as to increase the advertisement revenue from its dedicated advertiser. A also wishes B to pay for the usage of the content. From B's point of view, it wishes to carry its own advertisement in order to get payment from its own dedicated advertiser. B also wishes A to pay for the additional coverage after peering with B.

Next we describe a general peering agreement. When peering, provider A and B will deliver the same content (*i.e.*, the original content of provider A). As for advertisement, provider A will deliver its own advertisement. Provider B delivers α portion of A's advertisement and (1-α) portion of its own advertisement. Finally, B pays provider A a one-time payment c for peering over T time slots, where c can be either positive or negative. Figures 3 and 4 illustrate two providers' contents and advertisements with and without peering. The bargaining variables are the advertisement ratio α and payment c. Figure 5 illustrates this bargaining process.

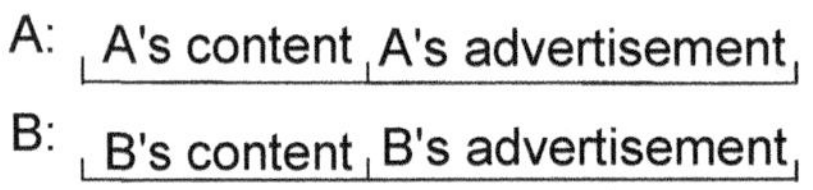

Fig. 3. Two providers' contents and advertisements without peering

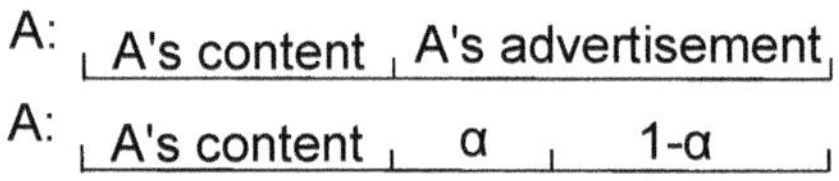

Fig. 4. Two providers' contents and advertisements with peering

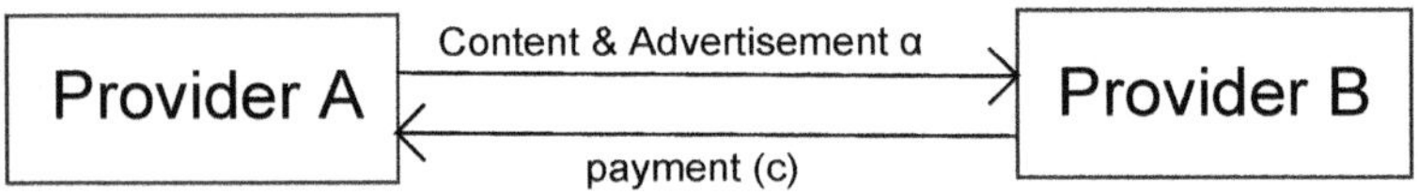

Fig. 5. Bargaining model

3.2 Change of Coverage

Now let us consider how users change their subscriptions when providers peer. Since now both providers have the same contents and provider B charges a lower price $p_B < p_A$, then all users will choose to subscribe to provider B.[3] Provider

[3] Here we assume that all users are freely to choose. This may not be the case where users already sign contracts with their providers. The additional switching cost because of this will be further discussed in Section 4.

A will get zero subscriber. However, notice that since α part of provider A's advertisement is delivered through B, then provider B's coverage also contributes to the advertisement revenue of A.

3.3 Providers' Revenues

Now let us compute the providers' revenues with peering. For provider A, its advertisement can reach all users with α fraction of the time. Hence, provider A's revenue with cooperation is

$$\pi_A^c(\alpha, c) = \alpha f_A(1)T + c. \tag{7}$$

For provider B, it can only deliver its own advertisement to its own users with $(1 - \alpha)$ fraction of the time. Thus, its revenue with cooperation is

$$\pi_B^c(\alpha, c) = (1 - \alpha) f_B(1)T + p_B T - c. \tag{8}$$

3.4 Nash Bargaining Problem

Next we model the bargaining problem based on the Nash bargaining solution [13], which is the unique bargaining solution that satisfies the axioms of *Pareto efficiency, symmetry, invariance*, and *independence of irrelevant alternatives.*

Definition 1. *A cooperation strategy (α^*, c^*) is a Nash bargaining solution if it solves the following problem:*

$$maximize_{\alpha \in [0,1], c} \quad (\pi_A^c(\alpha, c) - \pi_A) \cdot (\pi_B^c(\alpha, c) - \pi_B), \tag{9}$$

where π_A and π_B are the revenue obtained without cooperation as in (5) and (6).

It is clear that both providers should achieve revenues no worse than their non-cooperative revenues (*i.e.*, π_A and π_B) at the Nash bargaining solution. Otherwise, at least one provider does not have the incentive to bargain. This means that a cooperation agreement can be achieved if and only if the following condition holds:

$$\alpha f_A(1) + (1 - \alpha) f_B(1) > f_A(1 - \beta) + f_B(\beta) + (p_A - p_B)(1 - \beta). \tag{10}$$

With a proper choice of c, condition (10) can ensure that both providers get better payoffs through cooperation.

The optimal solution of (9) depends on the revenue functions $f_A(\cdot)$ and $f_B(\cdot)$. As an illustrative example, we consider linear advertisement revenue functions $f_A(x) = k_A \cdot x$ and $f_B(x) = k_B \cdot x$. Higher values of k_A and k_B lead to higher values of advertisement revenue with the same user coverage. Next we summarize the optimal solution of (9) depending on three possible relationships between k_A and k_B, with detailed proofs given in [14].

Scenario 1. $k_A = k_B$.

In this case, both providers have the same advertisement revenue function. The advertisements from both advertisers are equally important. If we plug $k_A = k_B$ into condition (10), then the left hand side (LHS) equals k_A and the right hand side (RHS) equals $k_A + (p_A - p_B)(1-\beta)$. Since $p_A > p_B$, we know that the LHS actually is less than RHS, and thus condition (10) does not hold. This means that providers will not choose to cooperate in this case.

Scenario 2. $k_A > k_B$.

In this case, provider A has a stronger ability in generating advertisement revenue than B. We can show that the optimal advertising strategy is $\alpha^* = 1$, in which both providers deliver the same advertisement originally belonging to provider A. With $\alpha^* = 1$, condition (10) becomes

$$k_A > k_A(1-\beta) + k_B\beta + (p_A - p_B)(1-\beta),$$

which means that the subscription fees p_A and p_B need to satisfy

$$p_A - p_B < \frac{(k_A - k_B)\beta}{1-\beta}$$

so that the providers want to cooperate. When the providers want to cooperate, we can further show that the optimal payment strategy c^* from provider B to provider A is

$$c^* = \frac{1}{2}\big((p_A + p_B)(1-\beta) - (k_A + k_B)\beta\big) \cdot T, \tag{11}$$

which can be either positive or negative. For example, if the revenue income is much larger than the user subscription fee, *i.e.*, $(k_A + k_B)\beta > (p_A + p_B)(1-\beta)$, then $c^* < 0$. This is because provider A's increase in advertisement income with cooperation is much larger than provider B's revenue increase by getting more subscribers. Then provider A should share the additional income with B.

Scenario 3. $k_A < k_B$.

In this case, provider A has a weaker capability in generating advertisement revenue than B. We can show that the optimal advertising strategy is $\alpha^* = 0$, in which the two providers deliver their own advertisements. With $\alpha^* = 0$, condition (10) becomes

$$k_B > k_A(1-\beta) + k_B\beta + (p_A - p_B)(1-\beta),$$

which is equivalent to

$$p_A - p_B < k_B - k_A,$$

which can be satisfied under proper values of p_A and p_B. When the providers want to cooperate, we can further show that the optimal payment strategy c^* from provider B to provider A is

$$c^* = \frac{1}{2} \cdot (k_A + k_B + p_1 + p_2)(1-\beta) \cdot T > 0.$$

In this case, provider A loses all the subscribers and can not get any advertisement revenue. As a result, provider B should compensate A's loss.

4 Impact of Dynamic Content

In this section, we consider how a provider can change its coverage (and thus the revenue) by introducing some special content (*i.e.*, content with a very high quality) over a short time period.

For a fair comparison, we assume that each provider has a finite budget. Introducing (purchasing) a special content in one time slot will decrease the content quality of the remaining $T-1$ time slots. This means that the content will become "dynamic" over the entire T time slots.

Without loss of generality, we assume that provider B introduces a special content with a high quality $q_s > q_A$ in the first time slot. As a result, the content quality for the remaining $t \in \{2, \ldots, T\}$ slots reduces to q'_B. Assuming a linear relationship between the budget and the content quality, then the finite budget constraint means that

$$q_s + (T-1)q'_B = Tq_B. \tag{12}$$

Here q_B is the static content quality introduced in Section 2. Figure 6 illustrates the change of provider B's content qualities with and without the special content.

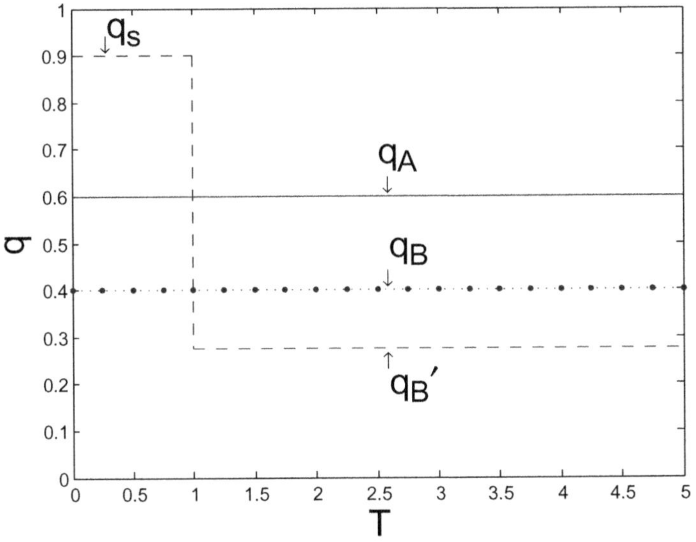

Fig. 6. Dynamic change of Content

Now let us consider if any user wants to switch from provider B to provider A after the special content is introduced at provider B. Recall that the utility of a user subscribing to provider B is

$$U_{B,static} = \theta \sum_{t=1}^{T} \delta^{t-1} q_B - p_B T$$

before introducing the special content and

$$U_{B,dynamic} = \theta q_s + \theta \sum_{t=2}^{T} \delta^{t-1} q'_B - p_B T$$

after introducing the special content. The utility change is

$$\begin{aligned} U_{B,dynamic} - U_{B,static} &= \theta q_s + \theta \sum_{t=2}^{T} \delta^{t-1} q'_B - \theta \sum_{t=1}^{T} \delta^{t-1} q_B \\ &= \theta \left(q_s - q_B + \sum_{t=2}^{T} \delta^{t-1} \left(q'_B - q_B \right) \right) \\ &\geq \theta \left(q_s - q_B + (T-1) \left(q'_B - q_B \right) \right) \\ &= 0. \end{aligned}$$

The inequality follows from $\delta < 1$ and $q'_B < q_B$, and the last equality follows from (12). This means that no user will switch from provider B to A after introducing the special content.

On the other hand, some original subscribers of provider A might want to switch to provider B due to the special content. However, these users will incur a switching cost z as they break the original contract.[4] Then a user will only switch from provider A to provider B if his utility improves after the switching, *i.e.*,

$$\theta(q_s + \sum_{t=2}^{T} \delta^{t-1} q'_B) - p_B \cdot T - z \geq \theta \sum_{t=1}^{T} \delta^{t-1} q_A - p_A \cdot T. \tag{13}$$

Let

$$Q(\delta) = (q_s + \sum_{t=2}^{T} \delta^{t-1} q'_B) - \sum_{t=1}^{T} \delta^{t-1} q_A.$$

Then (13) is equivalent to

$$\theta \cdot Q(\delta) \geq z - (p_A - p_B) \cdot T. \tag{14}$$

Function $Q(\delta)$ is a strictly decreasing in δ. It is positive when δ is close to 0 and negative when δ is close to 1. We can denote δ^{th} as the unique value such that $Q(\delta^{th}) = 0$.

[4] In a more general model, different users might have different switching costs, depending on when they signed the contract with their existing provider.

Next, we will discuss three cases based on the switching cost z.

Scenario 1. *Large switching cost:* $z > (p_A - p_B)T$.

In this case, users have to pay a high switching cost to switch. Users who are indifferent in terms of switching to provider B or stay with provider A have a parameter $\theta_{large}(\delta)$ that satisfies

$$\theta_{large}(\delta) = \frac{z - (p_A - p_B)T}{Q(\delta)}, \quad \forall \delta < \delta^{th}, \tag{15}$$

which is shown in Fig. 7. Here we need to have $\delta < \delta^{th}$ (*i.e.*, $Q(\delta) > 0$), otherwise no user will switch from A to B. Users with parameter (θ, δ) on the left hand side of the boundary will choose provider B. In fact, all users who are below curve $\theta^*(\delta)$ choose provider B even without the special content. Only users who are above the curve $\theta^*(\delta)$ and on the left of curve $\theta_{large}(\delta)$ are the switching users. We further notice that the threshold $\theta_{large}(\delta)$ increases with δ. This means that when δ increases, users need to have a larger evaluation θ in order to switch to provider B. Users on the right hand side of boundary $\theta_{large}(\delta)$ will stick to the original providers (either A or B).

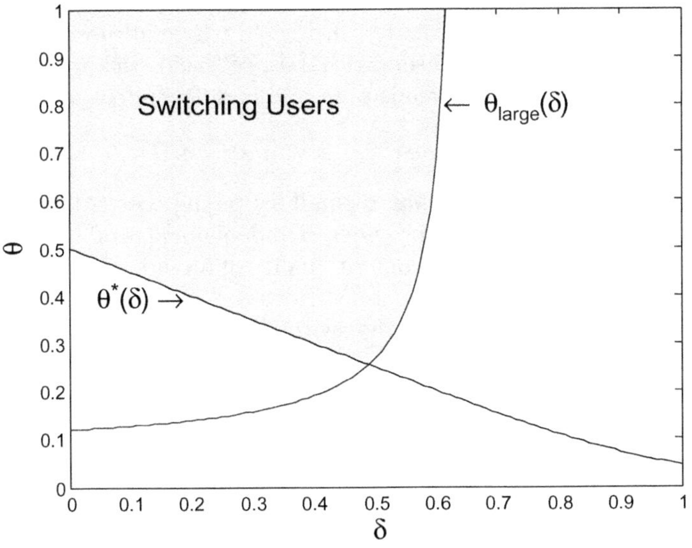

Fig. 7. Switching users under the high switching cost

Scenario 2. *Intermediate Switching Cost:* $z = (p_A - p_B)T$

In this case, users originally with provider A will switch to provider B if

$$\theta \cdot Q(\delta) \geq 0. \tag{16}$$

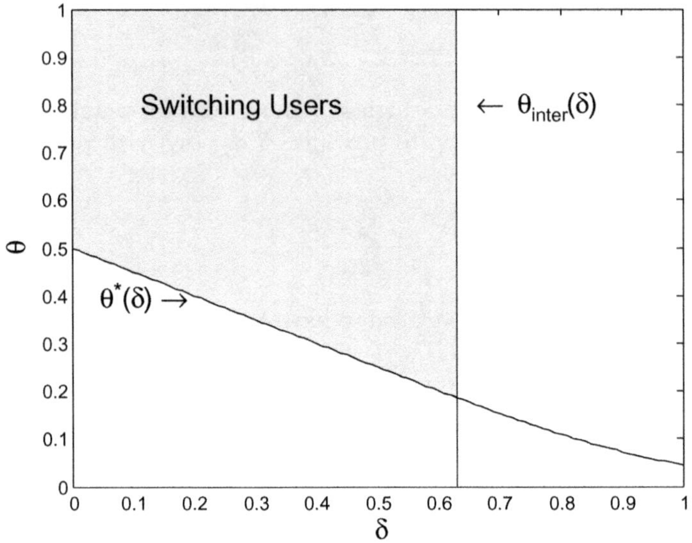

Fig. 8. Intermediate Switching Cost

The boundary value of θ is illustrated in Fig. 8, where all users with a $\delta \leq \delta^{th}$ will choose provider B, and all users with a $\delta > \delta^{th}$ will stick with their original choices. The boundary only depends on δ and is independent of the value of θ.

Scenario 3. *Small Switching Cost:* $0 \leq z < (p_A - p_B)T$.

In this case, users only need to pay a small switching cost. All users who have a $\delta < \delta^{th}$ will definitely choose provider B independent of θ. For users with a $\delta > \delta^{th}$ (*i.e.*, $Q(\delta) < 0$), the total content quality of provider B is less than A even after the introduction of special content. However, some users may still choose to switch from provider A to B, if the switch can bring significant reduction in terms of subscription fees (*i.e.*, $(p_A - p_B)T$). Thus the boundary value of θ can be characterized as

$$\theta_{small}(\delta) = \frac{z - (p_A - p_B)T}{Q}, \quad \forall \delta > \delta^{th}, \tag{17}$$

which is shown in Fig. 9. All users below the boundary and above the curve $\theta^*(\delta)$ will switch from provider A to B. Users who are under the curve $\theta^*(\delta)$ always stay with provider B. Notice that the boundary $\theta_{small}(\delta)$ actually decreases with θ.

The above three scenarios illustrate the importance of special content. When the switching cost z is large, only users with high valuations of the current content and content qualities (*i.e.*, small δ and large θ) will switch from provider A to B. As the switching cost decreases, the switching threshold moves towards the right and more users want to switch. This means that the provider B's revenue will increase due to more coverage and provider A's revenue will decrease.

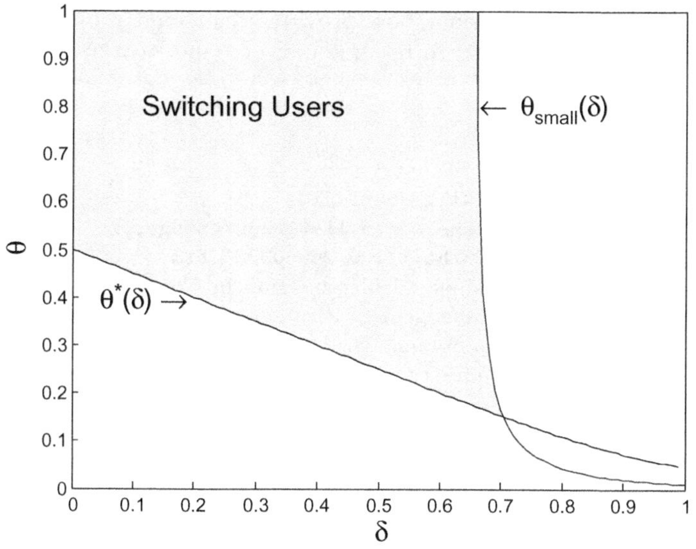

Fig. 9. Small Switching Cost

5 Conclusion and Future Work

Both content quality and market coverage have significant impacts on a network content provider's revenue. In this paper, we present a preliminary study on how providers' cooperation and adoption of special content can affect providers' content quality and change the market coverage. We first consider a baseline case, where providers have static content and do not cooperate. In this case, we derive the coverage of the providers based on the quality of the contents and user subscription fees. Then we consider how cooperation and content sharing can help providers to improve their revenues. The key insight is that cooperation will be desirable when the providers' total revenue is increased and properly shared by inter-provider financial transfers. In the case of linear advertisement functions, cooperation will happen when providers have different abilities in generating advertisement revenue and set subscription fees properly. We further consider the dynamic content case, where a provider can introduce some high quality special content for a short amount of time to attract users. We show that the switching cost, the valuation of content, and time discount factor all play important roles in deciding the benefit of special content.

There are several ways to extend this work. One direction is to consider the case where both providers can purchase special contents, and then it is possible for a user to switch more than once during T time slots. The two providers will engage in a game theoretical interaction in terms of the timing, quality, and length of the special contents. The other direction is to consider the strategic interactions between advertisers and content providers, *e.g.*, when an advertiser has the choice to work with more than one content provider to maximize its revenue. Finally, we will consider how providers can jointly optimize the subscription fees with the contents to become more competitive in the market.

References

1. http://en.wikipedia.org/wiki/Google_TV
2. http://archive.news.gov.hk/isd/ebulletin/en/category/businessandfinance/100426/html/100426en03005.htm
3. Gal-Or, E., Dukes, A.: Minimum Differentiation in Commercial Media Markets. Journal of Economics and Management Strategy 12, 291–325 (2003)
4. Gabszewicz, J., Laussel, D., Sonnac, N.: Programming and Advertising Competition in the Broadcasting Industry. Journal of Economics and Management Strategy 13, 657–669 (2004)
5. Anderso, S., Coate, S.: Market Provision of Broadcasting: A Welfare Analysis Review of Economic Studies. Review of Economic Studies 72, 947–972 (2005)
6. Peitza, M., Valletti, T.: Content and advertising in the media: Pay-tv versus free-to-air. International Journal of Industrial Organization 26(4), 949–965 (2008)
7. Kind, H., Nilssen, T., Sorgard, L.: Business Models for Media Firms: Does Competition Matter for How They Raise Revenue? Marketing Science 28(6), 1112–1128 (2009)
8. Richard, T.B.M., Chiu, D.M., Lui, C.S.: Interconnecting eyeballs to content: a shapley value perspective on isp peering and settlement. In: Proceedings of the 3rd international Workshop on Economics of Networked Systems. ACM, New York (2008)
9. Cheung, Y., Chiu, D.M., Huang, J.: Can bilateral ISP peering lead to network-wide cooperative? In: IEEE International Conference on Communications and Networks (August 2008)
10. Hosanagar, K., Krishnan, R., Smith, M., Chuang, J.: Optimal pricing of content delivery network (CDN) services. In: Proceedings of the 37th Annual Hawaii International Conference on System Sciences (2004)
11. Hande, P., Chiang, M., Calderbank, R., Rangan, S.: Network pricing and rate allocation with content provider participation. In: IEEE INFOCOM, pp. 990–998 (2009)
12. Fisher, F., McGowan, J., Evans, S.: The Audience-Revenue Relationship for local Television Stations. The Bell Journal of Economics 11(2), 694–708 (1980)
13. Binmore, K., Rubinstein, A., Wolinsky, A.: The Nash Bargaining Solution in Economic Modelling. RAND Journal of Economics 17, 176–188 (1986)
14. Feng, G., Huang, J., Chiu, D.M.: Bargaining and Peering between Network Content/Coverage Providers, Technical Report, The Chinese University of Hong Kong, http://home.ie.cuhk.edu.hk/~jwhuang/publication/ContentTechReport.pdf

Coalition Stability under QoS Based-Market Segmentation

Dominique Barth[1], Johanne Cohen[1],
Loubna Echabbi[2], and Hélène Le Cadre[1]

[1] PRiSM-CNRS, université de Versailles Saint-Quentin, 45, avenue des Etats-Unis, 78035 Versailles, France
firstname.lastname@prism.uvsq.fr
[2] INPT, 2, avenue Allal El Fassi, Rabat, Maroc
echabbi@inpt.ac.ma

Abstract. N independent sources choose their provider depending on the perceived costs associated with each provider. The perceived cost is the sum of the price and quality of service proposed by the provider coefficiented by the source sensitivity to the quality of service. The source chooses the smallest cost provider or refuses to subscribe if all the perceived costs are above her maximum admissible opportunity cost. First, we detail the market segmentation between the providers as function of the quality of service sensitivity. Then, we prove that in case where coalitions emerged and under defensive equilibria, the game characteristic function would be submodular meaning that the Shapley value would be a fair and stable way to share the grand coalition revenue.

Keywords: cooperative game, Shapley value.

1 Introduction

The inter-carrier network of Internet that interconnects different operator domains is a wide and diversified network in constant growth in terms of number of domain operators. These operators have to manage complex technical and economic interactions. In this context, the different operators first provide (and sell) network services needed to ensure the level of performance required by the end-user applications. The inter-carrier network is thus today a technico-economic system in which competition and inter-dependencies prevail. The study presented in this paper is a contribution to the FP7 project ETICS that aims at creating a new ecosystem of innovative QoS-enabled interconnection models between Network Service Providers allowing for a fair distribution of revenue shares among all the actors of the service delivery value-chain.

The objective of this paper is to evaluate the benefit of network providers acting on a same market when they have together a privileged partnership in terms of economic alliance. As an example of network scenario in this context, we briefly describe here the Game as a Service scenario studied in the context of ETICS (see Figure 1). We consider a set of network providers, each one proposing a catalogue of on-line games to final users (such providers act as service retailers). The games they propose are developed and managed by cloud providers and require network transport services. In

R. Jain and R. Kannan (Eds.): GameNets 2011, LNICST 75, pp. 383–396, 2012.

the context of the present article, cloud providers and network services are considered as cost impacts for service retailers, and these costs can be shared by some of these retailers if they choose to be member of a same alliance.

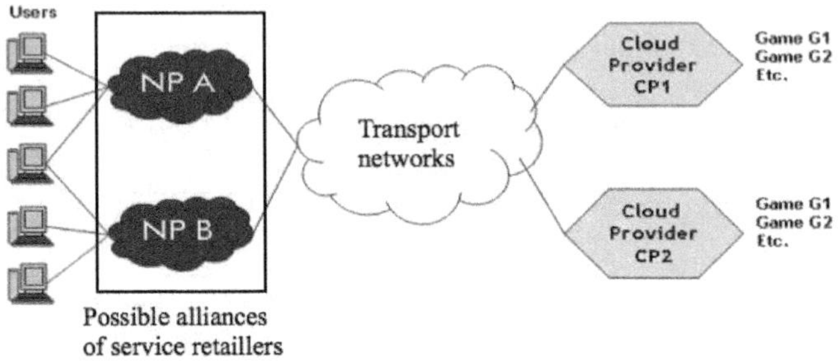

Fig. 1. "Game as a service" scenario

Most articles in economics assume that the providers have a fixed consumer basis i.e., that no churn is possible, or that demand is simply a linear function of price. In the rapid growing literature on revenue management, one of the most important issues is how to model provider demand forecasting. Demand is usually represented as an explicit/implicit function of price and the root tactic upon which revenue management is based is to change prices dynamically to maximize immediate or short-run revenue [9]. Kwon et al. propose a novel approach in [5] where non-cooperative providers learn dynamically their demand which is governed by a continuous time differential equation based on an evolutionary game theory perspective. However, the learning process requires data to efficiently update its forecastings. The problem is that in most real systems, data are missing or even altered by noise or measurement errors. The idea of our article is to take into account the source perception of both the prices and quality of service of the providers, and also their capacity to churn from one provider to the others or even, to refuse to subscribe. Moreover, uncertainty on the providers' knowledge about the sources' preferences is incorporated by assuming that their sensitivity to the quality of service is distributed according to a density function.

Taking into account the individual source preferences, the article shows that providers have always interest to cooperate through a grand coalition, provided the coalition revenue is shared according to the Shapley value, and that this is a stable organization of the market. Besides, most articles in the economic literature are restricted to monopolistic or duopolistic cases of competition [4]; on the contrary, our article is extended to an arbitrary large number of interacting providers using specific game properties.

The paper is organized as follow. The game is described in Section 2: first the relations between two interacting providers are analyzed in Section 2.1, then a two provider game is solved in Section 2.2 and generalized to the case of three providers in Section 2.3. Finally, the game is extended to the n provider case in Section 3: the market segmentation is determined in Section 3.1 and we prove that the characteristic function associated with the cooperative game is submodular, implying that the grand coalition should remain stable with the Shapley value as revenue sharing mechanism, in Section 3.2. We conclude in Section 4.

2 Game Description

In this work, n providers in competition want to interconnect with N sources. On one hand, each provider i's quality of service (QoS) level q_i, is fixed and known publicly. It proposes an access price p_i according to its QoS level q_i. On the other hand, the sources have the choice either to connect or not. Besides, they select their provider depending on the QoS they perceive and also on the proposed access prices. The provider selection process of a generic source k depends its opportunity cost. *Opportunity cost* is the cost related to the best choice available to someone who has picked among several mutually exclusive choices.

In our article, the sources have indeed the choice between buying an access to one of the n providers or refusing to subscribe. Opportunity cost is a key concept in economics [1]; it has already been applied to the telecommunications bundle offer market [12] and to the pricing of virtual mobile network operators' services [6]. In our model, each source k ($k = 1, 2, ..., N$) has an opportunity cost towards each provider i ($i = 1, 2, ..., n$). It is defined as $c_k(i) = p_i - \beta_k q_i$ where p_i and q_i are the access price and QoS level for provider i respectively, while $\beta_k \in [0; 1]$ captures source k's sensitivity to the QoS.

Besides, we make the assumption that all source have a same maximum opportunity cost $c_{\max} > 0$, above which they will refuse to buy access. The need to introduce a maximum admissible opportunity cost results from the following observation: a source will refuse to buy access or delay the subscription process either if the access price is too high, or if the QoS is not good enough.

We make the fundamental assumption that the source chooses the provider having the smallest opportunity cost or refuses to subscribe if this latter is larger than $c_{\max}$.

n_i measures provider i's market share. Provider i's utility is the difference between the revenue generated by the source subscriptions and his fixed cost I_i. Let n_i be the pourcentage of sources subcribing to provider i

$$\pi_i = n_i N p_i - I_i \tag{1}$$

In a non-cooperative setting, each provider aims at maximizing his own utility by determining the optimal access price. The two-level game between the providers can be described as follows

(1) Providers determine simultaneously and independently their access prices.

(2) Depending on the perceived opportunity costs, source k chooses the provider having the smallest opportunity cost or refuses to connect.

To cope with the uncertainty on the sources' preferences, we assume that the QoS sensitivity parameter is distributed according to the uniform density on the interval $[0; 1]$ i.e., $\beta_k \sim \mathcal{U}[0; 1]$. Besides, the QoS levels being fixed a priori, we make the assumption that $0 < q_n < q_{n-1} < q_{n-2} < ... < q_1 < +\infty$.

2.1 Relationship between Two Providers

We focus on two providers i, j. Without loosing generality, we assume that $i < j$. As already stated, the opportunity cost associated by source k to provider i is

$$c_k(i) = p_i - \beta_k q_i \tag{2}$$

and for provider j, we get

$$c_k(j) = p_j - \beta_k q_j \tag{3}$$

According to the assumptions introduced previously in this section, we know that the QoS are initially fixed so that $0 < q_j < q_i < +\infty$.

Lemma 1. *Source k prefers provider i over provider j if, and only if, $0 < p_i < p_j$.*

Proof of Lemma 1. Suppose that $0 < p_i < p_j$. It implies that $c_k(i) < c_k(j)$ by definition of both providers' opportunity costs as described in Equations (2) and (3). But, it means that provider j would not have any client i.e., his market share, as defined in Section 1, would vanish ($n_j = 0$) since by definition, the sources subscribe to the provider having the smallest opportunity cost. □

An immediate consequence of Lemma 1 is that it is mandatory to impose $0 < p_j < p_i$ to guarantee that both providers i and j might have non-negative market shares.

We now aim at determining the values of the indifference bounds when studying the interactions between two providers only.

Lemma 2. *For any provider i, source k prefers provider i to no subscription if, and only if, $B_i < \beta_k$ where $B_i \equiv \frac{p_i - c_{\max}}{q_i}$.*

Proof of Lemma 2. Source k prefers provider i to no subscription when $c_k(i) < c_{\max}$ (by definition of opportunity cost). So, we have

$$c_k(i) < c_{\max} \Leftrightarrow p_i - \beta_k q_i < c_{\max}.$$

This last inequality can be re-ordered to give a lower bound for β_k

$$p_i - \beta_k q_i < c_{\max} \Leftrightarrow \frac{p_i - c_{\max}}{q_i} < \beta_k$$

□

Note that if $B_i = \beta_k$, source k is indifferent between provider i and no subscription.

Lemma 3. *For any providers $i < j$, source k prefers provider i to provider j if, and only if, $B_{i,j} < \beta_k$ where $B_{i,j} \equiv \frac{p_i - p_j}{q_i - q_j}$.*

Proof of Lemma 3. Source k prefers provider i to provider j if, and only if, $c_k(i) < c_k(j)$. The last inequality can be rewritted as follow:$p_i - \beta_k q_i < p_j - \beta_k q_j$. So, this last inequality gives a lower bound for β_k.

$$\begin{aligned} p_i - \beta_k q_i < p_j - \beta_k q_j &\Leftrightarrow p_i - p_j < \beta_k (q_i - q_j) \\ &\Leftrightarrow \frac{p_i - p_j}{q_i - q_j} < \beta_k. \end{aligned}$$

□

Presently, we aim at ordering these bound values on the interval $[0;1]$ which will enable us to determine the analytical expressions of both providers' market shares. For the remaining of this paper, we will introduce some notations : for any providers i, j, we define $B_i \equiv \frac{p_i - c_{\max}}{q_i}$. and $B_{i,j} \equiv \frac{p_i - p_j}{q_i - q_j}$. Moreover, an additional proposition is required:

Proposition 1. *Let two providers i, j such that $i < j$.*

$$B_j < B_{i,j} \Leftrightarrow B_j < B_i < B_{i,j}$$

Due to the lack of place, the proof of this proposition is in [2].

2.2 Case of Two Providers

In this section, we assume that provider 1 and 2 only, propose access services to N independent sources. We want to determine the complete ordering of the bounds B_1, B_2, $B_{1,2}$. Using Proposition 1, two cases might arise.

Case (1): $B_2 < B_{1,2} \Leftrightarrow B_2 < B_1 < B_{1,2}$.
Case (2) : $B_{1,2} < B_2 \Leftrightarrow B_{1,2} < B_1 < B_2$.

We detail these two cases below.

Case (1): $B_2 < B_{1,2} \Leftrightarrow B_2 < B_1 < B_{1,2}$.

Figure 2 represents the source preferences ordered according to their β_k values. We have placed the indifference bounds on the β_k-axis. Using Lemmas 2 and 3, we infer the source prefences as functions of the β_k values. If $\beta_k \in [0; B_2[$, then the source prefers to not subscribe; if $\beta_k \in [B_2; B_{1,2}[$, the source prefers provider 2; if $\beta_k \in [B_{1,2}; 1]$, the source prefers provider 1.

Fig. 2. Case where $B_2 < B_1 < B_{1,2}$

Now, we will focus on Case (2) : $B_{1,2} < B_2 \Leftrightarrow B_{1,2} < B_1 < B_2$. Figure 3 represents the source preferences ordered according to their β. We have placed the indifference bounds on the β_k-axis. Using the same way as Case (1) (applying Lemmas 2 and 3), we infer the source prefences as functions of the β_k values. If $\beta_k \in [0; B_1[$, then the source prefers to not subscribe; if $\beta_k \in [B_1; 1]$, the source prefers provider 1.

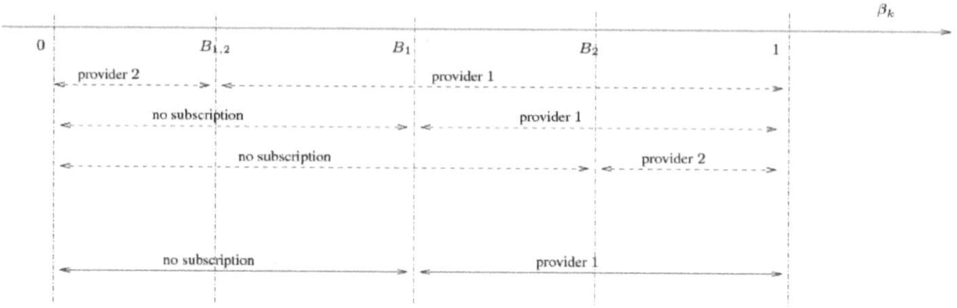

Fig. 3. Case where $B_{1,2} < B_1 < B_2$

Note that in Case (1), both providers might have non-negative market shares. Under such assumptions, both providers would have the opportunity to survive and to contract with sources. As a consequence, a duopoly might emerge [7], [11]. Recall that a duopoly is the simplest case of oligopoly. In economics, duopoly models are shared between cooperative equilibrium models like cartel and non-cooperative equilibrium models like Cournot, Stackelberg or Bowley duopoly models [7]. On the contrary, in Case (2), solely provider 1 can survive on the market leading to a monopol [7], [11]. This remark is summarized in the following corollary.

Corollary 1. *In case of two providers, a duopoly might emerge if, and only if* $B_2 < B_1 < B_{1,2}$.

We want to determine conditions on the game parameters q_1, q_2 and the access prices upper bound denoted by $p_{\max}$, so that the game belongs to Case (1). We let $\alpha_{1,2} = p_1 - p_2$ be providers 1 and 2's price margin and suppose that there exists a real $\varepsilon > 0$ such that $\varepsilon < \alpha_{1,2}$.

Proposition 2. *If* ε *and* $p_{\max}$ *are chosen so that* $\frac{p_{\max}}{\varepsilon} < \frac{q_2}{q_1 - q_2}$, *then a duopoly might emerge.*

Proof of Proposition 2. The game belongs to Case (1) if, and only if

$$B_{1,2} > B_2 \Leftrightarrow p_1 q_2 - p_2 q_1 > \underbrace{-c_{\max}(q_1 - q_2)}_{<0}$$

A sufficient condition to guarantee the emergence of a duopoly is to have

$$p_1 q_2 - p_2 q_1 > 0 \Leftrightarrow \frac{\alpha_{1,2}}{p_2} > \frac{q_1}{q_2} - 1$$
$$\Leftrightarrow p_2 < \alpha_{1,2} \frac{q_2}{q_1 - q_2}$$

If $p_{\max} < \varepsilon \frac{q_2}{q_1 - q_2}$, the last inequality is automatically satisfied. This condition is equivalent with $\frac{p_{\max}}{\varepsilon} < \frac{q_2}{q_1 - q_2}$. □

As already stated, the sources' QoS sensitivity being distributed according to the uniform density on the interval $[0; 1]$, it is easy to infer both providers' market shares using Figure 2:

$$n_1 = 1 - B_{1,2} = 1 - \frac{p_1 - p_2}{q_1 - q_2},$$
$$n_2 = B_{1,2} - B_2 = \frac{p_1 - p_2}{q_1 - q_2} - \frac{p_2 - c_{\max}}{q_2}.$$

This short computation enables us to obtain the market segmentation as a function of the game parameters.

In a non-cooperative setting, each provider determines selfishly the access price maximizing his utility using the market segmentation. This is summarized in the following proposition.

Proposition 3. *In a non-cooperative game, the prices maximizing the providers' utilities are* $p_1^* = (q_1 - q_2)\Big[\frac{1}{2} + \frac{1}{4q_1 - q_2}(\frac{q_2}{2} + c_{\max})\Big]$ *and* $p_2^* = 2\frac{q_1 - q_2}{4q_1 - q_2}\Big[\frac{q_2}{2} + c_{\max}\Big]$.

Proof of Proposition 3. Provider i ($i = 1, 2$)'s utility has been introduced in Equation 1: $\pi_i = n_i N p_i - I_i$ where I_i is provider i's fixed cost. By differentiation of π_1 with respect to p_1, we obtain p_1^* as a linear function of p_2^* i.e., $p_1^* = \frac{(q_1 - q_2) + p_2^*}{2}$, provider 1's cost ($I_1$) being fixed. Identically, by differentiation of π_2 with respect to p_2, we obtain p_2^* as a linear function of p_1^* i.e., $p_2^* = \frac{q_2}{2q_1} p_1^* + \frac{c_{\max}(q_1 - q_2)}{2q_1}$. Note that π_1 and π_2 are second order polynomial equations in p_1 and p_2 respectively whose highest order coefficient is negative; hence the extremum obtained by differentiation of their utilities coincides with a global maximum. Solving a linear system of two equations with two unknown variables, we obtain $p_1^* = (q_1 - q_2)\Big[\frac{1}{2} + \frac{1}{4q_1 - q_2}(\frac{q_2}{2} + c_{\max})\Big]$ and $p_2^* = 2\frac{q_1 - q_2}{4q_1 - q_2}\Big[\frac{q_2}{2} + c_{\max}\Big]$. □

To end this two provider game section, we compare the profit resulting from cooperation for the providers against the selfish maximization of their utilities and try to answer the following question: in case of a duopoly, do the providers have incentives to cooperate?

We suppose that when the access providers become allied, they share the alliance revenue according to the Shapley value. There exists many other sharing mechanisms like the nucleolus, proportional allocation, supply chain contract mechanisms, etc. [11], [13]. But, their study is out of the scope of the present article.

We prove that in this case, the providers always prefer to form an alliance than to maximize their utility independently. This result is summarized in the following lemma.

Lemma 4. *In the case of a duopoly, the providers always prefer to become allied than to selfishly maximize their own utility provided the alliance revenue is shared according to the Shapley value.*

Note that Lemma 4 is simply a by-product of the Shapley value definition [11].

2.3 Case of Three Providers

In this section, we focus on the case of three interacting providers. As previously, we assume that the QoS levels are ordered so that $0 < q_3 < q_2 < q_1 < +\infty$. Generalizing Propostion 2 to the case of three interacting providers, we obtain the following price ordering: $0 < p_3 < p_2 < p_1$.

Applying Proposition 2 to all the coalitions containing two providers and assuming that there exists a real $\varepsilon > 0$ such that $\varepsilon < \min\{\alpha_{i,j} | \forall (i,j) \in \{1,2,3\}^2, i \neq j\}$, where $\alpha_{i,j}$ is the difference between provider i's price and provider j's price (i.e : $\alpha_{i,j} = p_i - p_j$). We obtain three relations on the bound ordering:

- If $\frac{p_{\max}}{\varepsilon} < \frac{q_2}{q_1 - q_2}$ then $B_2 < B_1 < B_{1,2}$.
- If $\frac{p_{\max}}{\varepsilon} < \frac{q_3}{q_1 - q_3}$ then $B_3 < B_1 < B_{1,3}$.
- If $\frac{p_{\max}}{\varepsilon} < \frac{q_3}{q_2 - q_3}$ then $B_3 < B_2 < B_{2,3}$.

If these three relations are simultaneously satisfied then $B_3 < B_2 < B_1$.

Presently, we want to determine the total ordering of $B_{2,3}$, $B_{1,3}$, $B_{1,2}$ on the interval $[0; 1]$.

Lemma 5.

$$B_1 < B_{1,2} \Leftrightarrow B_1 < B_{2,3} < B_{1,2}$$

Due to the lack of place, the proof of Lemma 5 is in [2].

As in the case with two providers, we want to determine conditions on the game parameters q_1 , q_2, q_3, $c_{\max}$, $p_{\max}$ guaranteeing that competition is possible on the market i.e., that the three providers might have non-negative market shares. Note that in this case, the competition would be total (i.e : all providers are in competition).

Proposition 4. *If ε and $p_{\max}$, $c_{\max}$ are chosen so that $\frac{p_{\max} - c_{\max}}{\varepsilon} < \frac{q_1}{q_1 - q_2}$, then the competition would be total.*

Proof of Proposition 4.

$$B_1 < B_{1,2} \Leftrightarrow \frac{p_1 - c_{\max}}{q_1} < \frac{p_1 - p_2}{q_1 - q_2}$$
$$\Leftrightarrow p_1 - c_{\max} < \alpha_{1,2} \frac{q_1}{q_1 - q_2}$$

A sufficient condition to satisfy this last inequality is to have $p_{\max} - c_{\max} < \varepsilon \frac{q_1}{q_1 - q_2}$. □

Under the conditions introduced in Propositions 5 and 4, we obtain the following indifference bound ordering in case of three providers in competition: $0 < B_3 < B_{2,3} < B_{1,2} < 1$ as depicted in Figure 4.

This reinforces the economic intuition behind the problem. Indeed, the market should enable competition between providers which might improve the sources' welfare. Competition is possible if each provider's market share remains positive. Besides, it is quite logical to note that on the β_k-axis, the provider with the smallest QoS (q_3) captures the sources having small β_k values; the provider with intermediate QoS (q_2) captures the sources having intermediate β_k values; the provider with the highest QoS (q_1) captures the sources having hight sensitivites in the QoS.

Note that according to Lemma 5, if $B_1 > B_{1,2}$ then $B_1 > B_{2,3} > B_{1,2}$ and provider 2 market share equals zero since provider 2 is always dominated by the other provider in the source preferences. This point is summarized in the corollary below.

Corollary 2. *In case of three providers, the competition is total if, and only if* $B_1 < B_{2,3} < B_{1,2}$.

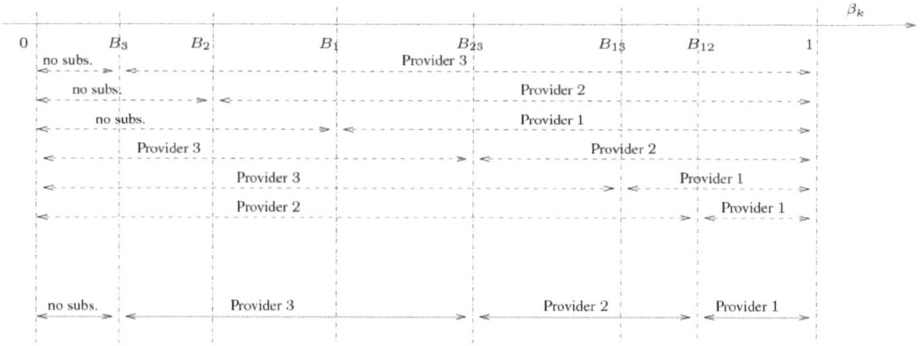

Fig. 4. Indifference bound ordering in case of three providers in competition

3 Can Cooperation Emerge in Case of n Providers?

In this section, we aim at generalizing the results obtained in Section 2 to the case of n interacting providers.

3.1 Generalization of the Game Resolution to n Providers

The indifference bound ordering derived in Sections 2.1 and 2.3, can be generalized recursively, to give the following ordering for n providers.

Theorem 1. *In case of* $n \geq 2$ *providers, the competition is total if, and only if,* $0 < B_n < B_{j-1,j} < B_{j-2,j-1} < 1$ *for any integer* j *such that* $3 \leq j \leq n$. *Moreover, in this case each source* k *prefers* j *if* $B_{j,j+1} \leq \beta_k \leq B_{j-1,j}$ *where* $2 \leq j \leq n-1$, 1 *if* $B_{1,2} \leq \beta_k \leq 1$, *and* n *if* $B_n \leq \beta_k \leq B_{n-1,n}$.

Proof of Theorem 1. We proceed by recurrence on the number n, of interacting providers. For $n = 3$, the result has been proved in Corollary 2.

Suppose that at rank n, $B_n < B_{n-1,n} < B_{n-2,n-1} < ... < B_{1,2}$ and each source k prefers j if $B_{j,j+1} \leq \beta_k \leq B_{j-1,j}$ where $2 \leq j \leq n-1$, 1 if $B_{1,2} \leq \beta_k \leq B_1$, and n if $B_n \leq \beta_k \leq B_{n-1,n}$.

At rank $n+1$, a new provider enters the market. Using the assumption introduced in Section 2, provider $n+1$ chooses his QoS so that $0 < q_{n+1} < q_n < q_j$ for any integer j such that $3 \leq j < n$. Using Lemma 1, we obtain the following ordering on the providers' prices: $p_{n+1} < p_n < p_j$ for any integer j such that $3 \leq j < n$.

Now, we want to order B_n, $B_{n,n+1}$, B_{n+1} and $B_{n-1,n}$. We need to compare $B_{n,n+1}$ to B_n. Two cases are possible: $B_{n,n+1} < B_n$ and $B_{n,n+1} > B_n$.

If $B_{n,n+1} < B_n$ then from Proposition 1, we get $B_{n,n+1} < B_{n+1} < B_n$. But, from Lemma 3, source k prefers provider $n+1$ to provider n if $B_{n,n+1} > \beta_k$. Moreover, from Lemma 2 source k prefers no subcription to provider n if $B_{n,n+1} < \beta_k < B_n$. And by recurrence hypothesis, each source k prefers another provider to provider n if $\beta_k > B_n$. So, in this case, provider n market share would vanish since each source prefers $n+1$ to n. Therefore, the competition cannot be total.

If $B_n < B_{n,n+1}$ then from Proposition 1, we infer that $B_{n,n+1} > B_n > B_{n+1}$. In this case, we need to compare $B_{n-1,n}$ to $B_{n,n+1}$.

By absurd reasoning, assume that $B_{n-1,n} < B_{n,n+1}$. From Lemma 2 source k prefers no subcription to provider n if $\beta_k < B_n$. Moreover, source k prefers provider $n+1$ to provider n if $B_n < \beta_k < B_{n,n+1}$ and also if $B_n < \beta_k < B_{n-1,n}$. Moreover, by recurrence hypothesis, each source k prefers another provider to provider n if $B_{n,n+1} < \beta_k$. So, on any sub-interval on $[0;1]$, provider n would be dominated by an other provider meaning that none of the sources would agree to subscribe to provider n's service. Thus, provider n market share would vanish and the competition cannot be total.

Finaly, we focus on the fact that $B_{n+1} < B_{n,n+1} < B_{n-1,n}$. By recurrence hypothesis, we have each source k prefers j if $B_{j,j+1} \leq \beta_k \leq B_{j-1,j}$ where $2 \leq j \leq n-1$, 1 if $B_{1,2} \leq \beta_k \leq 1$. Moreover each source k prefers n to provider ℓ if $B_n \leq \beta_k \leq B_{n-1,n}$ and if $1 \leq \ell < n$. It remains to focus on interval $[B_{n+1}, B_{n,n+1}]$. From Lemma 3, each source k prefers $n+1$ to provider n if $\beta_k \leq B_{n,n+1}$ and n to provider $n+1$ otherwise. So, it this case, the competition is total is total and the reccurence hypothesis holds.

Hence, the competition is total if, and only if $B_{n-1,n} > B_{n,n+1} > B_n > B_{n+1}$.

Note that all the other indifference bounds ($B_{j-1,j}$ forall integer j such that $2 \leq j < n$) remain identical to the n provider case. □

Using the same principles as in Sections 2.1 and 2.3, the sources are shared between the providers according to the following rule.

For provider $k = 2, ..., n-1$, the market share is $n_k = B_{k-1,k} - B_{k,k+1}$ (from Theorem 1). On the boundaries, provider 1's market share takes the form $n_1 = 1 - B_{1,2}$ while provider n's market share is $n_n = B_{n-1,n} - B_n$. The market segmentation is perfectly determined as a function of the game parameters i.e., $c_{\max}, q_1, q_2, ..., q_n$.

To determine the prices maximizing each provider's utility, we substitute the analytical expressions of the providers' market shares in their utility as defined in Equation (1) and derive the providers' utilities with respect to the prices. It gives us a linear system on n equations in the n unknown prices

$$
\begin{aligned}
p_1^* &= \frac{1}{2}p_2^* + \frac{1}{2}(q_1 - q_2),\\
p_k^* &= \frac{1}{2}\frac{q_k - q_{k+1}}{q_{k-1} - q_{k+1}}p_{k-1}^* + \frac{1}{2}\frac{q_{k-1} - q_k}{q_{k-1} - q_{k+1}}p_{k+1}^*, \text{ for } k = 2, ..., n-1,\\
p_n^* &= \frac{1}{2}\frac{q_n}{q_{n-1}}p_{n-1}^* + \frac{1}{2}\frac{q_{n-1} - q_n}{q_{n-1}}.
\end{aligned}
$$

The problem can be written under a matricial form. To simplify the expressions, we let

$$A_n = \begin{pmatrix} 0 & -\frac{1}{2} & 0 & 0 & \cdots & 0 & 0 & 0 \\ \frac{1}{2}\frac{q_2-q_1}{q_1-q_3} & 0 & \frac{1}{2}\frac{q_1-q_2}{q_1-q_3} & 0 & \cdots & 0 & 0 & 0 \\ \ddots & \ddots & \ddots & 0 & 0 & 0 & 0 & 0 \\ 0 & \cdots & \frac{1}{2}\frac{q_k-q_{k+1}}{q_{k-1}-q_{k+1}} & 0 & \frac{1}{2}\frac{q_{k-1}-q_k}{q_{k-1}-q_{k+1}} & 0 & \cdots & 0 \\ 0 & 0 & \ddots & \ddots & \ddots & 0 & \cdots & 0 \\ 0 & 0 & \cdots & 0 & 0 & \frac{1}{2}\frac{q_{n-1}-q_n}{q_{n-2}-q_n} & 0 & \frac{1}{2}\frac{q_{n-2}-q_{n-1}}{q_{n-2}-q_n} \\ 0 & 0 & 0 & \cdots & 0 & 0 & \frac{1}{2}\frac{q_n}{q_{n-1}} & 0 \end{pmatrix}.$$

Note that A_n is a tri-diagonal matrix. Using this notation, the linear system of equations in the prices can be arranged to give

$$\begin{pmatrix} p_1^* \\ p_2^* \\ \vdots \\ p_n^* \end{pmatrix} \underbrace{(I - A_n)}_{B_n} = \begin{pmatrix} \frac{q_1-q_2}{2} \\ \vdots \\ 0 \\ \vdots \\ \frac{c_{\max}(q_{n-1}-q_n)}{2q_{n-1}} \end{pmatrix} \tag{4}$$

Lemma 6. *The prices maximizing the providers' utilities are uniquely defined as solutions of the matricial Equation (4).*

Proof of Lemma 6. For the sake of simplicity, we let $B_n = I - A_n$. The prices maximizing the providers' utilities are uniquely defined if, and only if, B_n is invertible i.e., its determinant does not vanish. We proceed by recursion.

At rank $n = 1$, we have $B_1 = 1$ and hence $\det B_1 = 1 > 0$.

At rank $n = 2$, $\det B_2 = 1 - \frac{1}{4}\frac{q_2-q_3}{q_2-q_3}$. We check that $\det B_2 > \frac{3}{4}$. A fortiori, we get: $\det B_2 - \frac{1}{4}\det B_1 > 0$.

At rank $n-1$, we make the recursive hypothesis that: $\det B_{n-1} - \frac{1}{4}\det B_{n-2} > 0$.

At rank n, B_n being a tri-diagonal matrix, we have the following well-known relation between its determinant and the minor determinants: $\det B_n = \det B_{n-1} - \frac{1}{4}\frac{q_n}{q_{n-1}}\frac{q_{n-2}-q_{n-1}}{q_{n-2}-q_n}\det B_{n-2}$. We check easily that: $-\frac{1}{4}\frac{q_n}{q_{n-1}}\frac{q_{n-2}-q_{n-1}}{q_{n-2}-q_{n-1}} > -\frac{1}{4}$. Then, $\det B_n - \frac{1}{4}\det B_{n-1} > \det B_{n-1} - \frac{2}{4}\det B_{n-2} > 0$ using rank $n-1$ recursive hypothesis. □

3.2 Stability of the Shapley Value as a Revenue Sharing Mechanism

Suppose that $\mathcal{S}$ is the set of all the possible coalitions of providers. It is well-known that the cardinal of $\mathcal{S}$ equals $2^n - 1$ provided n providers are interacting on the market. The providers cooperate in order to maximize their joint utility by increasing the alliance total market share. Formally, it can be described as follows: for any coalition $s \in \mathcal{S}$, π_s is the utility of coalition s, and we have

$$\max_{n_s} \pi_s = \sum_{k \in s} \pi_k$$

$$\max_{n_{\mathcal{S}-s}} \pi_{\mathcal{S}-s} = \sum_{k \in \mathcal{S}-s} \pi_k \tag{5}$$

where n_s (resp. $n_{\mathcal{S}-s}$) contains coalition s (resp. $\mathcal{S} - s$) total market share. Going back to the analytical expressions of the providers' market shares, we note that coalition s market share relies solely on the prices proposed by the providers having the highest and the worst QoS levels respectively, in the coalition.

Proposition 5. *The utility $\pi(.)$ is submodular in the n providers cooperative game described by Equations (5).*

Proof of Proposition 5. Consider two coalitions s, s' such that $s \subset s' \subset \{1, 2, ..., n\}$ and a provider $j \in \{1, 2, ..., n\} - s'$. We want to show that $\pi_{s\cup j} - \pi_s \geq \pi_{s'\cup j} - \pi_s' \Leftrightarrow \pi_{s\cup j} - \pi_{s'\cup j} \geq \pi_s - \pi_s'$.

We observe that
$[\pi_{s\cup j} - \pi_{s'\cup j}] - [\pi_s - \pi_s'] = (p_s' - p_{s',j})n_{s'} - (p_s - p_{s,j})n_s + \underbrace{(p_{s,j} - p_{s',j})n_j}_{\geq 0}$.

Now, we note that $(p_s' - p_{s',j})n_{s'} - (p_s - p_{s,j})n_s \geq (p_s' - p_{s,j})n_{s'} - (p_s - p_{s,j})n_s$ since $s \subset s'$ implies that $p_s \geq p_s'$ and in turn that $p_{s,j} \geq p_{s',j}$.

We let l be the provider belonging to coalition s which proposes the smallest QoS in coalition s i.e., $q_l = \min_{i \in s} q_i$ and k be the provider belonging to coalition s' which proposes the smallest QoS in coalition s' i.e., $q_k = \min_{i \in s'} q_i$.

Then $n_s = B - \frac{p_k - p_l}{q_k - q_l}$ and $n_{s'} = B - \frac{p_j - p_k}{q_j - q_k}$ with $B_{j,k} \leq B_{k,l}$ and B the upper indifference bound delimiting coalitions s and s' market shares. By definition and using the inclusion property $(p_{s'} - p_{s,j})(B_{j,k} - B) \geq (p_s - p_{s,j})(B_{k,l} - B)$. Multiplying each term of the inequality by -1 we obtain the proposition result. □

The Shapley value is then the center of gravity of the core of the n provider cooperative game as described in Equations (5). Therefore, it is still a fair and stable revenue sharing mechanism.

4 Conclusion

We have proved that the Shapley value is the center of gravity of the n provider cooperative game taking into account the sources' individual preferences. Therefore, it is always a stable mechanism to share the grand coalition total revenue, meaning that none of the providers has incentives to deviate from it or leave the grand coalition. It is therefore most likely that tacit alliances emerge, to the disadvantage of the sources.

However, in practice, alliances do not emerge on every market since for instance, collusion may be forbidden due to competition policy. Indeed, courts punish explicit accords whose objectives are clearly to decrease the competition. Heavy sanctions have been applied to the international accords on the vitamin market (855 millions of euros), on lysine and citric acid (200 millions of dollars and emprisonment years) [?] and more recently, on the memory chip market (331 millions of euros).

In the present article, we have assumed that the providers' QoS levels were fixed. Extensions should be envisaged by adding another level in the game description. The resulting three level game resolution might be tackled using numerical approaches. In practice i.e., in the telecommunication business area, when information about the providers' QoS is known publicly, it is possible for some providers to set a price that kick other providers out of the market. However, QoS is difficult to estimate with accuracy and performance measures are usually very costly to perform. Consequently, it might be interesting to assume that in the game with the other providers, each provider ignores his rivals' true QoS levels but try to infer them using side observations.

Finally, hierarchical relations between the providers should be taken into account. Indeed, some of them might lack the infrastructure and buy their QoS to some others, owning a network. The QoS fixation/negotiation market will add another level of complexity to the game and the revenue sharing mechanism might be more complex to design [10]. Generally speaking, it might be possible to design contract mechanisms which would force the providers to cooperate. Analogies with the supply chain theory is possible [3] but it would require to evaluate complex power relations between all the involved providers. This task is not straightforward compared to the Shapley value which provides a direct measure of the providers' inter-relations.

References

1. Acemoglu, D., Ozdagalr, A.: Competition and Efficiency in Congested Market. Mathematics of Operations Research 32, 1–31 (2007)
2. Bourreau, M.: Coalition, Industrial Organization Course, master IREN (2010)
3. Cachon, G.: Supply chain coordination with contracts. In: Graves, S., de Kok, T. (eds.) Handbooks in Operations Research and Management Science: Supply Chain Management (2003)
4. Fershtman, C., Kamien, M.I.: Dynamic Duopolistic Competition with Sticky Prices. Econometrica 55, 1151–1164 (1987)
5. Kwon, C., Friesz, T.L., Mookherjee, R., Yao, T., Baichun, F.: Non-cooperative competition among revenue maximizing service providers with demand learning. European Journal of Operational Research 197, 981–996 (2008)
6. Le Cadre, H., Bouhtou, M.: Modelling Dynamic Interconnection Relations between MNO and MVNO: Is Cooperative Content Investment Profitable for Both Providers? To appear in Telecommunication Systems Journal 52(1-2)
7. Mas-Colell, A., Whinston, M.D., Green, J.R.: Microeconomic Theory. Oxford University Press (1995)
8. Maxima, a free downloadable computer algebra system, `http://maxima.sourceforge.net/`
9. McGill, J., Van Ryzin, G.: Revenue management: Research overview and prospects. Transportation Science 33, 233–256 (1999)

10. Musacchio, J., Schartz, G., Walrand, J.: A Two-sided Market Analysis of Provider Investment Incentives and Application to the Net-Neutrality Issue. Review of Network Economics 8, 22–39 (2009)
11. Myerson, R.: Game Theory, Analysis of conflict, 6th edn. Harvard University Press (2004)
12. Pohjola, O.-P., Kilkki, K.: Value-based methodology to analyze communication services. Netnomics 8, 135–151 (2007)
13. Saad, W., Han, Z., Debbah, M., Hjorungnes, A., Basar, T.: Coalitional Game Theory for Communication Networks: A Tutorial. IEEE Signal Processing Magazine, Special Issue on Game Theory 26, 77–97 (2009)

On the Shapley-Like Payoff Mechanisms in Peer-Assisted Services with Multiple Content Providers*

Jeong-woo Cho and Yung Yi

Dept. of Electrical Engineering, KAIST, South Korea
{ggumdol,yiyung}@kaist.ac.kr

Abstract. This paper studies an incentive structure for cooperation and its stability in peer-assisted services when there exist multiple content providers, using a coalition game theoretic approach. We first consider a generalized coalition structure consisting of multiple providers with many assisting peers, where peers assist providers to reduce the operational cost in content distribution. To distribute the profit from cost reduction to players (*i.e.*, providers and peers), we then establish a generalized formula for individual payoffs when a "Shapley-like" payoff mechanism is adopted. We show that the grand coalition is *unstable*, even when the operational cost functions are concave, which is in sharp contrast to the recently studied case of a single provider where the grand coalition is stable. We also show that irrespective of stability of the grand coalition, there always exist coalition structures which are not convergent to the grand coalition. Our results give us an important insight that a provider does not tend to cooperate with other providers in peer-assisted services, and be separated from them. To further study the case of the separated providers, three examples are presented; *(i)* underpaid peers, *(ii)* service monopoly, and *(iii)* oscillatory coalition structure. Our study opens many new questions such as realistic and efficient incentive structures and the tradeoffs between fairness and individual providers' competition in peer-assisted services.

1 Introduction

The Internet is becoming more content-oriented, and the need of cost-effective and scalable distribution of contents has become the central role of the Internet. Uncoordinated peer-to-peer (P2P) systems, *e.g.*, BitTorrent, has been successful in distributing contents, but the rights of the content owners are not protected well, and most of the P2P contents are in fact illegal. In its response, a new type of service, called *peer-assisted services,* has received significant attentions these days. In peer-assisted services, users commit a part of their resources to assist content providers in content distribution with objective of enjoying both

* This work was supported by Brain Korea 21 Project, BK Electronics and Communications Technology Division, KAIST in 2011, and KCC (Korea Communications Commission), Korea, under the R&D program supervised by the KCA (Korea Communications Agency)" (KCA-2012-11913-05004).

R. Jain and R. Kannan (Eds.): GameNets 2011, LNICST 75, pp. 397–412, 2012.

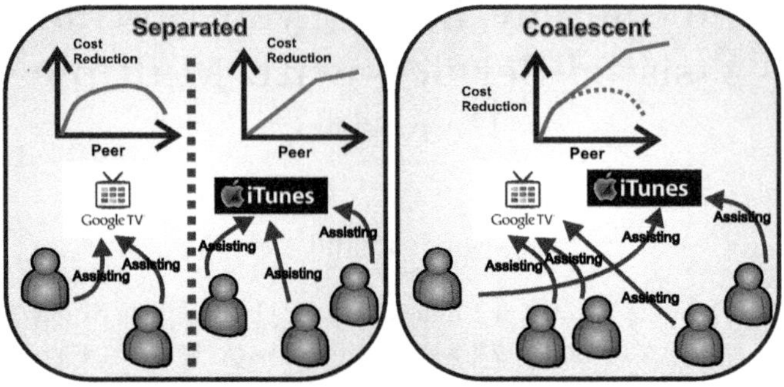

Fig. 1. Coalition Structures for a Dual-Provider Network

scalability/efficiency in P2P systems and controllability in client-server systems. Examples of peer-assisted services include nano data center [1] and IPTV [2], where high potential of operational cost reduction was observed. For instance, there are now 1.8 million IPTV subscribers in South Korea, and the financial sectors forecast that by 2014 the IPTV subscribers is expected to be 106 million, see, *e.g.*, [3]. However, it is clear that most users will not just "donate" their resources to content providers. Thus, the key factor to the success of peer-assisted services is how to (economically) incentivize users to commit their valuable resources and participate in the service.

One of nice mathematical tools to study incentive-compatibility of peer-assisted services is the coalition game theory which covers how payoffs should be distributed and whether such a payoff scheme can be executed by rational individuals or not. In peer-assisted services, the "symbiosis" between providers and peers are sustained when *(i)* the offered payoff scheme guarantees fair assessment of players' contribution under a provider-peer coalition and *(ii)* each individual has no incentive to exit from the coalition. In the coalition game theory, the notions of Shapley value and the core have been popularly applied to address *(i)* and *(ii)*, respectively, when the entire players cooperate, referred to as the grand coalition. A recent paper by Misra *et al.* [4] demonstrates that the Shapley value approach is a promising payoff mechanism to provide right incentives for cooperation in a *single-provider* peer-assisted service.

However, in practice, the Internet consists of multiple content providers, even if only giant providers are counted. The focus of our paper is to study the cooperation incentives for *multiple* providers. In the multi-provider case, the model clearly becomes more complex, thus even classical analysis adopted in the single-provider case becomes much more challenging, and moreover the results and their implications may experience drastic changes. To motivate further, see an example in Fig. 1 with two providers (Google TV and iTunes) and consider two cases of cooperation: *(i) separated,* where there exists a fixed partition of peers for each provider, and *(ii) coalescent,* where each peer is possible to assist any provider. In the separated case, a candidate payoff scheme is based on the Shapley value in each separated coalition. Similarly, in the coalescent case, the Shapley value is also a candidate payoff scheme after the worth function of the grand coalition N (the player set) is defined appropriately. A reasonable definition of the worth function[1] can be the total cost reduction, maximized over all combinations of peer partitions to each provider[2]. Then, it is not hard to see that the cost reduction for the coalescent case exceeds that for the separated case, unless the two partitions are equivalent in both cases. This implies that at least one individual in the separated case is *underpaid* than in the coalescent case under the Shapley-value based payoff mechanism. Thus, providers and users are recommended to form the grand coalition and be paid off based on the Shapley value, *i.e.*, the due desert.

However, it is still questionable whether peers will stay in the grand coalition and thus the consequent Shapley-value based payoff mechanism is desirable in the multi-provider setting. In this paper, we anatomize incentive structures in peer-assisted services with multiple content providers and focus on stability issues from two different angles: stability at equilibrium of Shapley value and convergence to the equilibrium.

Our main contributions are summarized as follows:

1) We first provide a closed-form formula of the Shapley value for a general case of multiple providers and peers. To that end, we define a worth function to be the maximum total cost reduction over all possible peer partitions to each provider. Due to the intractability of analytical computation of the Shapley value, we take a fluid-limit approximation that assumes a large number of peers and re-scales the system with the number of peers. This is a non-trivial generalization of the Shapley value for the single-provider case in [4]. In fact, our formula in Theorem 1 establishes the general Shapley value for distinguished *multiple* atomic players and infinitesimal players in the context of the Aumann-Shapley (A-S) prices [5] in coalition game theory.
2) We prove in Theorem 2 that the Shapley value for the multiple-provider case is not in the core under mild conditions, *e.g.*, each provider's cost function is concave. This is in stark contrast to the single-provider case where the concave cost function stabilizes the equilibrium.
3) We study, for the first time, the endogenous formation of coalitions in peer-assisted services by introducing the stability notion defined by the seminal work of Hart and Kurz [6]. We show that, if we adopt a Shapley-like payoff mechanism, called Aumann-Drèze value, irrespective of stability of the grand coalition, there always exist initial states which are not convergent to the grand coalition. An interesting fact from this part of study is that peers and providers have opposite cooperative preferences, *i.e.*, peers prefer to cooperate with more providers, whereas providers prefer to be separated from other providers.

---

[1] We establish in Section 3.1 that this definition is *derived* directly from an essential property of coalition.

[2] The notion of peer partitions implicitly assumes that a peer assists only one provider. However, our model is not restricted in the sense that we will study the regime of a large number of peers for mathematical tractability, in which case a peer assisting two providers can be regarded as two distinct peers assisting distinct providers.

In short, the Shapley payoff regime cannot incentivize rational players to form the grand coalition, implying that *fair* profit-sharing and *opportunism* of players cannot stand together. If the grand coalition is broken up, the Shapley payoff scheme cannot be executed because the total profit of a coalition *differs* from the sum of Shapley payoffs in the coalition. Only payoff mechanisms for general coalition structure may be used. In conjunction with this point, we present three examples for non-cooperation among providers who adopt the Shapley-like payoff scheme: *(i)* the peers are underpaid than their Shapley payoffs, *(ii)* a provider with more "advantageous" cost function monopolizes all peers, and *(iii)* Shapley value for each coalition gives rise to an oscillatory behavior of coalition structures. These examples suggest that the system with the separated providers may be even unstable as well as unfairness in a peer-assisted service market.

The rest of the paper is organized as follows. In Section 2, we define Shapley and Aumann-Drèze values with minimal formalism. After formulating the fluid Aumann-Drèze formula for multiple providers, we establish results on the stability-related concepts in Section 4 to substantiate that it is very unlikely that the grand coalition occurs. Then we point out main drawbacks of Aumann-Drèze value in Section 5 and conclude this paper.

2 Preliminaries

Since this paper investigates a multi-provider case, where a peer can choose any provider to assist, we start this section by defining a coalition game with a peer partition (*i.e.*, a coalition structure) and introducing the payoff mechanism thereof.

2.1 Game with Coalition Structure

A game with coalition structure is a triple $(N, v, \mathcal{P})$ where N is a player set and $v : 2^N \to \mathbb{R}$ (2^N is the set of all subsets of N) is a worth function, $v(\emptyset) = 0$. $v(K)$ is called the worth of a coalition $K \subseteq N$. $\mathcal{P}$ is called a *coalition structure* for (N, v); it is a partition of N where $C(i) \in \mathcal{P}$ denotes the coalition containing player i. For your reference, a coalition structure $\mathcal{P}$ can be regarded as a set of disjoint coalitions. The *grand coalition* is the partition $\mathcal{P} = \{N\}$. For instance[3], a partition of $N = \{1, 2, 3, 4, 5\}$ is $\mathcal{P} = \{\{1, 2\}, \{3, 4, 5\}\}$, $C(4) = \{3, 4, 5\}$, and the grand coalition is $\mathcal{P} = \{\{1, 2, 3, 4, 5\}\}$. $\mathcal{P}(N)$ is the set of all partitions of N. For notational simplicity, a game *without* coalition structure $(N, v, \{N\})$ is denoted by (N, v). A value of player i is an operator $\phi_i(N, v, \mathcal{P})$ that assigns a payoff to player i. We define $\phi_K = \sum_{i \in K} \phi_i$ for all $K \subseteq N$.

To conduct the equilibrium analysis of coalition games, the notion of *core* has been extensively used to study the stability of grand coalition $\mathcal{P} = \{N\}$:

Definition 1 (Core). *The core is defined by* $\{\phi(N, v) \mid \sum_{i \in N} \phi_i(N, v) = v(N)$ *and* $\sum_{i \in K} \phi_i(N, v) \geq v(K), \forall K \subseteq N\}$.

[3] A player i is an *element* of a coalition $C = C(i)$, which is in turn an *element* of a partition $\mathcal{P}$. $\mathcal{P}$ is an element of $\mathcal{P}(N)$ *while* a subset of 2^N.

If a payoff vector $\phi(N, v)$ lies in the core, no player in N has an incentive to split off to form another coalition K because the worth of the coalition K, $v(K)$, is no more than the payoff sum $\sum_{i \in K} \phi_i(N, v)$. Note that the definition of the core hypothesizes that the grand coalition is already formed *ex-ante*. We can see the core as an analog of Nash equilibrium from noncooperative games. Precisely speaking, it should be viewed as an analog of *strong Nash equilibrium* where no arbitrary coalition of players can create worth which is larger than what they receive in the grand coalition. If a payoff vector $\phi(N, v)$ lies in the core, then the grand coalition is stable with respect to any collusion to break the grand coalition.

2.2 Shapley Value and Aumann-Drèze Value

On the premise that the player set is not partitioned, *i.e.*, $\mathcal{P} = \{N\}$, the Shapley value, denoted by φ (not ϕ), is popularly used as a fair distribution of the grand coalition's worth to individual players, defined by:

$$\varphi_i(N, v) = \sum_{S \subseteq N \setminus \{i\}} \frac{|S|!(|N| - |S| - 1)!}{|N|!} \left(v(S \cup \{i\}) - v(S)\right). \quad (1)$$

Shapley [7] gives the following interpretation: "*(i)* Starting with a single member, the coalition adds one player at a time until everybody has been admitted. *(ii)* The order in which players are to join is determined by chance, with all arrangements equally probable. *(iii)* Each player, on his admission, demands and is promised the amount which his adherence contributes to the value of the coalition." The Shapley value quantifies the above that is axiomatized (see Section 2.3) and has been treated as a worth distribution scheme. The beauty of the Shapley value lies in that the payoff "summarizes" in *one* number all the possibilities of each player's contribution in every coalition structure.

Given a coalition structure $\mathcal{P} \neq \{N\}$, one can obtain the Aumann-Drèze value (A-D value) [8] of player i, also denoted by φ, by taking $C(i)$, which is the coalition containing player i, to be the player set and by computing the Shapley value of player i of the *reduced* game $(C(i), v)$. It is easy to see that the A-D value can be construed as a direct extension of the Shapley value to a game with coalition structure. Note that both Shapley value and A-D value are denoted by φ because the only difference is the underlying coalition structure $\mathcal{P}$.

2.3 Axiomatic Characterizations of Values

We provide here the original version [7] of the axiomatic characterization of the Shapley value.

Axiom 1 (Coalition Efficiency, CE). $\sum_{j \in C} \phi_j(N, v, \mathcal{P}) = v(C), \ \forall C \in \mathcal{P}$.

Axiom 2 (Coalition Restricted Symmetry, CS). *If* $j \in C(i)$ *and* $v(K \cup \{i\}) = v(K \cup \{j\})$ *for all* $K \subseteq N \setminus \{i, j\}$, *then* $\phi_i(N, v, \mathcal{P}) = \phi_j(N, v, \mathcal{P})$.

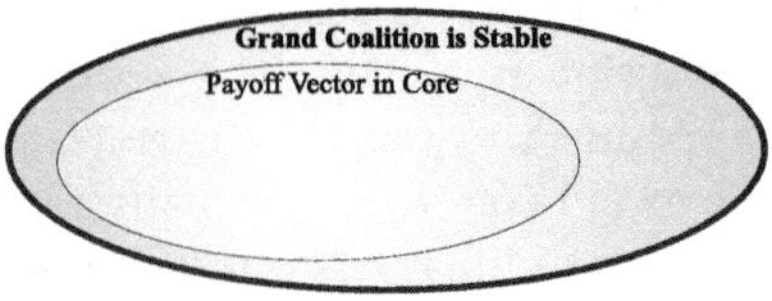

Fig. 2. If a payoff vector lies in the core, the grand coalition is stable [6].

Axiom 3 (Additivity, ADD). $\phi_i(N, v + v', \mathcal{P}) = \phi_i(N, v, \mathcal{P}) + \phi_i(N, v', \mathcal{P})$ *for all coalition functions* v, v' *and* $i \in N$.

Axiom 4 (Null Player, NP). *If* $v(K \cup \{i\}) = v(K)$ *for all* $K \subseteq N$, *then* $\phi_i(N, v, \mathcal{P}) = 0$.

Recall that the basic premise of the Shapley value is that the player set is not partitioned, *i.e.*, $\mathcal{P} = \{N\}$. Also, the Shapley value, defined in (1), is *uniquely* characterized by **CE**, **CS**, **ADD** and **NP** for $\mathcal{P} = \{N\}$ [7]. The A-D value is also *uniquely* characterized by **CE**, **CS**, **ADD** and **NP** (Axioms 1-4), but in this case for arbitrary coalition structure $\mathcal{P}$ [8]. In the literature, *e.g.*, [9,10], the A-D value has been used to analyze the *static* games where a coalition structure is *exogenously* given.

Definition 2 (Coalition Independent, CI). *If* $i \in C \subseteq N$, $C \in \mathcal{P}$ *and* $C \in \mathcal{P}'$, *then* $\phi_i(N, v, \mathcal{P}) = \phi_i(N, v, \mathcal{P}')$.

From the definition of the A-D value, the payoff of player i in coalition $C(i)$ is affected neither by the player set N nor by coalitions $C \in \mathcal{P}$, $C \neq C(i)$. Note that only $C(i)$ contains the player i. Thus, it is easy to prove that the A-D value is coalition independent. From **CI** of the A-D value, in order to decide the payoffs of a game with general coalition structure $\mathcal{P}$, it suffices to decide the payoffs of players within each coalition, say $C \in \mathcal{P}$, without considering other coalitions $C \in \mathcal{P}$, $C \neq C(i)$. In other words, once we decide the payoffs of a coalition $C \in \mathcal{P}$, the payoffs remains unchanged even though other coalitions, $C' \in \mathcal{P}$, $C' \neq C$, vary. Thus, for any given coalition structure $\mathcal{P}$, any coalition $C \in \mathcal{P}$ is just two-fold in terms of the number of providers in C: *(i)* one provider or *(ii)* two or more providers, as depicted in Fig. 1.

The notion of **CI** also enables us to define the stability of a game with coalition structure in the following simplistic way:

Definition 3 (Stable Coalition Structure [6]). *We say that a coalition structure* $\mathcal{P}'$ *blocks* $\mathcal{P}$, *where* $\mathcal{P}', \mathcal{P} \in \mathcal{P}(N)$, *with respect to* ϕ *if and only if there exists some* $C \in \mathcal{P}'$ *such that* $\phi_i(N, v, \{C, \cdots\}) > \phi_i(N, v, \mathcal{P})$ *for all* $i \in C$. *In this case, we also say that* C *blocks* $\mathcal{P}$. *If there does not exist any* $\mathcal{P}'$ *which blocks* $\mathcal{P}$, $\mathcal{P}$ *is called stable.*

Due to **CI** of the A-D value, all stability notions defined by the seminal work of Hart and Kurz [6] coincide with the above simplistic definition, as discussed by Tutic [11]. Definition 3 can be intuitively interpreted that, if there exists any subset of players C who improve their payoffs away from the current coalition structure, they *will* form a new coalition C. In other words, if a coalition structure $\mathcal{P}$ has any blocking coalition C, some rational players will break $\mathcal{P}$ to increase their payoffs. The basic premise here is that players are not clairvoyant, *i.e.*, they are interested only in improving their instant payoffs. If a payoff vector lies in the core, the grand coalition is stable in the sense of Definition 3, but the converse is not necessarily true (see Fig. 2).

3 Coalition Game in Peer-Assisted Services

In this section, we first define a coalition game in a peer-assisted service with multiple content providers by classifying the types of coalition structures as *separated*, where a coalition includes only one provider, and *coalescent*, where a coalition is allowed to include more than one providers (see Fig. 1). To define the coalition game, we will define a worth function of an arbitrary coalition $S \subseteq N$ for such two cases.

3.1 Worth Function in Peer-Assisted Services

Assume that players N are divided into two sets, the set of content providers $Z := \{p_1, \cdots, p_\zeta\}$, and the set of peers $H := \{n_1, \cdots, n_\eta\}$, *i.e.*, $N = Z \cup H$. We also assume that the peers are homogeneous, *e.g.*, the same computing powers, disk cache sizes, and upload bandwidths. Later, we discuss that our results can be readily extended to nonhomogeneous peers. The set of peers assisting providers is denoted by $\bar{H} := \{n_1, \cdots, n_{x \cdot \eta}\}$ where $x = |\bar{H}|/\eta$, *i.e.*, the fraction of assisting peers. We define the worth of a coalition S to be the amount of cost reduction due to cooperative distribution of the contents by the players in S in both separated and coalescent cases.

<u>**Separated case**</u>: Denote by $\Omega_p^\eta(x(S))$ the operational cost of a provider p when the coalition S consists of provider p and $x(S) \cdot \eta$ assisting peers. Since the operational cost cannot be negative, we assume $\Omega_p^\eta(x(S)) > 0$. Note that from the homogeneity assumption of peers, the cost function depends only on the fraction of assisting peers. Then, we define the worth function $\hat{v}(S)$ for the coalition S as:

$$\hat{v}(S) := \Omega_p^\eta(0) - \Omega_p^\eta(x(S)) \tag{2}$$

where $\Omega_p^\eta(0)$ corresponds to the cost when there are no assisting peers. For notational simplicity, in what follows, $x(S)$ is denoted by x from now on.

<u>**Coalescent case**</u>: In contrast to the separated case, where a coalition includes a single provider, the worth for the coalescent case is not clear yet, since depending on which peers assist which providers the amount of cost reduction may differ. One of reasonable definitions would be the maximum worth out of all peer partitions, *i.e.*, the worth for the coalescent case is defined by:

$$v(S) = \max\left\{\textstyle\sum_{C \in \mathcal{P}} \hat{v}(C) \;\middle|\; \mathcal{P} \in \mathcal{P}(S) \text{ such that } |Z \cap C| = 1,\ \forall C \in \mathcal{P}\right\}. \tag{3}$$

The definition above implies that we *view* a coalition containing more than one provider as the most productive coalition whose worth is *maximized* by choosing the optimal partition $\mathcal{P}^*$ among all possible partitions of S. Note that (3) is consistent with the definition (2) for $|Z \cap S| \leq 1$, *i.e.*, $v(S) = \hat{v}(S)$ for $|Z \cap S| \leq 1$.

Four remarks are in order. First, as opposed to [4] where $\hat{v}(\{p\}) = \eta R - \Omega_p^\eta(0)$ (R is the subscription fee paid by any peer), we simply assume that $\hat{v}(\{p\}) = 0$. Note that, as discussed in [10, Chapter 2.2.1], it is no loss of generality to assume that, initially, each provider has earned no money. In our context, this means that it does not matter how much fraction of peers is subscribing to each provider because each peer has already paid the subscription fee to providers *ex-ante.*

Second, it is also important to note that we cannot always assume that $\Omega_p^\eta(x)$ is monotonically decreasing because providers have to pay the electricity expense of the computers and the maintenance cost of the hard disks of assisting peers. For example, a recent study [12] found that Annualized Failure Rate (AFR) of hard disk drives is over 8.6% for three-year old ones. We discuss in Appendix of [13] that, if we consider a more *intelligent* coalition, the cost function $\Omega_p^\eta(x)$ is always non-increasing. However, we assume the following to simplify the exposition:

Assumption 1. *$\Omega_p^\eta(x)$ is non-increasing in x for all $p \in Z$.*

Third, the worth function in peer-assisted services can reflect the diversity of peers. It is not difficult to extend our result to the case where peers belong to distinct classes. For example, peers may be distinguished by different upload bandwidths and different hard disk cache sizes. A point at issue for the multiple provider case is whether peers who are *not* subscribing to the content of a provider may be allowed to assist the provider or not. On the assumption that the content is ciphered and not decipherable by the peers who do not know its password which is given only to the subscribers, providers will allow those peers to assist the content distribution. Otherwise, we can easily reflect this issue by dividing the peers into a number of classes where each class is a set of peers subscribing to a certain content.

Lastly, it should be pointed out that the worth function in (3) is *rigorously* selected in order to satisfy a basic property:

Definition 4 (Superadditivity). *A worth function v is superadditive if the following holds: $(S, T \subseteq N$ and $S \cap T = \emptyset) \Rightarrow v(S \cup T) \geq v(S) + v(T)$.*

Suppose we have a superadditive worth function v'. We require $v'(S) = \hat{v}(S)$ if S includes one provider. It follows from the definition of v in (3) that $v'(\cdot)$ is no greater than $v(\cdot)$, *i.e.*, $v(\cdot) \geq v'(\cdot)$ because v is the total cost reduction that is maximized over all possible peer partitions to each provider. In the meantime, since v' is superadditive, it must satisfy $v'(S \cup T) \geq v'(S) + v'(T)$ for all disjoint S, $T \subseteq N$, implying that $v'(\cdot) \geq v(\cdot)$. This completes the proof of the following lemma.

Lemma 1. *When the worth is given by (2), there exists a superadditive worth function, uniquely given by (3).*

Superadditivity is one of the most elementary properties, which ensures that the core is nonempty by appealing to Bondareva-Shapley Theorem [10, Theorem 3.1.4]. In light of this lemma, we can clearly restate that our objective in this paper is to analyze the incentive structure of peer-assisted services when the worth of coalition is superadditive. This objective then necessarily implies the form of worth function in (3).

3.2 Fluid Aumann-Drèze Value for Multiple-Provider Coalitions

So far we have defined the worth of coalitions. Now let us *distribute* the worth to the players for a given coalition structure $\mathcal{P}$. Recall that the payoffs of players in a coalition are independent from other coalitions by the definition of A-D payoff. Pick a coalition C without loss of generality, and denote the set of providers in C by $\bar{Z} \subseteq Z$. With slight notational abuse, the set of peers assisting $\bar{Z}$ is denoted by $\bar{H}$. Once we find the A-D payoff for a coalition consisting of arbitrary provider set $\bar{Z} \subseteq Z$ and assisting peer set $\bar{H} \subseteq H$, the payoffs for the separated and coalescent cases in Fig. 1 follow from the substitutions $\bar{Z} = Z$ and $\bar{Z} = \{p\}$, respectively. In light of our discussion in Section 2.2, it is more reasonable to call a Shapley-like payoff mechanism 'A-D payoff' and 'Shapley payoff' respectively for the partitioned and non-partitioned games $(N, v, \{\bar{Z} \cup \bar{H}, \cdots\})$ and $(N, v, \{Z \cup H\})$[4].

Fluid Limit: We adopt the limit axioms for a large population of users to overcome the computational hardness of the A-D payoffs:

$$\lim_{\eta \to \infty} \widetilde{\Omega}_p^\eta(\cdot) = \widetilde{\Omega}_p(\cdot) \text{ where } \widetilde{\Omega}_p^\eta(\cdot) = \tfrac{1}{\eta}\Omega_p^\eta(\cdot) \tag{4}$$

which is the asymptotic operational cost per peer in the system with a large number of peers. We drop superscript η from notations to denote their limits as $\eta \to \infty$. From the assumption $\Omega_p^\eta(x) > 0$, we have $\widetilde{\Omega}_p(x) \geq 0$. To avoid trivial cases, we also assume $\widetilde{\Omega}_p(x)$ is not constant in the interval $x \in [0, 1]$ for any $p \in Z$. We also introduce the payoff of each provider per user, defined as $\widetilde{\varphi}_p^\eta := \frac{1}{\eta}\varphi_p^\eta$. We now derive the fluid limit equations of the payoffs which can be obtained as $\eta \to \infty$. The proof of the following theorem is given in Appendix of [13].

Theorem 1 (A-D Payoff for Multiple Providers). *As η tends to ∞, the A-D payoffs of providers and peers under an arbitrary coalition $C = \bar{Z} \cup \bar{H}$ converge to the following equation:*

$$\begin{cases} \widetilde{\varphi}_p^{\bar{Z}}(x) = \widetilde{\Omega}_p(0) - \sum_{S \subseteq \bar{Z} \setminus \{p\}} \int_0^1 u^{|S|}(1-u)^{|\bar{Z}|-1-|S|} \left(M_\Omega^{S \cup \{p\}}(ux) - M_\Omega^S(ux) \right) \mathrm{d}u, \\ \widetilde{\varphi}_n^{\bar{Z}}(x) = -\sum_{S \subseteq \bar{Z}} \int_0^1 u^{|S|}(1-u)^{|\bar{Z}|-|S|} \frac{\mathrm{d}M_\Omega^S}{\mathrm{d}x}(ux)\mathrm{d}u, \end{cases} \tag{5}$$

[4] On the contrary, the term 'Shapley payoff' was used in [4] to refer to the payoff for the game $(N, v, \{\bar{Z} \cup \bar{H}, \cdots\})$ where a proper subset of the peer set assists the content distribution.

where $p \in \bar{Z}$ *and* $n \in \bar{H}$. *Here* $M_{\Omega}^{S}(x) := \min\left\{\sum_{i\in S} \widetilde{\Omega}_i(y_i) \mid \sum_{i\in S} y_i \le x,\ y_i \ge 0\right\}$ *and* $M_{\Omega}^{\emptyset}(x) := 0$. *Note that* $M_{\Omega}^{\{p\}}(x) = \widetilde{\Omega}_p(x)$.

The following corollaries are immediate as special cases of Theorem 1, which we will use in Section 5.

Corollary 1 (A-D Payoff for Single Provider). *As* η *tends to* ∞, *the A-D payoffs of providers and peers who belong to a single-provider coalition,* i.e., $\bar{Z} = \{p\}$, *converge to the following equation:*

$$\begin{cases} \widetilde{\varphi}_p^{\{p\}}(x) = \widetilde{\Omega}_p(0) - \int_0^1 M_{\Omega}^{\{p\}}(ux)\mathrm{d}u, \\ \widetilde{\varphi}_n^{\{p\}}(x) = -\int_0^1 u \frac{\mathrm{d}M_{\Omega}^{\{p\}}}{\mathrm{d}x}(ux)\mathrm{d}u, \qquad \text{for } n \in \bar{H}. \end{cases} \tag{6}$$

Corollary 2 (A-D Payoff for Dual Providers). *As* η *tends to* ∞, *the A-D payoffs of providers and peers who belong to a dual-provider coalition,* i.e., $\bar{Z} = \{p, q\}$, *converge to the following equation:*

$$\begin{cases} \widetilde{\varphi}_p^{\{p,q\}}(x) = \widetilde{\Omega}_p(0) - \int_0^1 uM_{\Omega}^{\{p,q\}}(ux)\mathrm{d}u - \int_0^1 (1-u)M_{\Omega}^{\{p\}}(ux)\mathrm{d}u + \int_0^1 uM_{\Omega}^{\{q\}}(ux)\mathrm{d}u, \\ \widetilde{\varphi}_n^{\{p,q\}}(x) = -\int_0^1 u^2 \frac{\mathrm{d}M_{\Omega}^{\{p,q\}}}{\mathrm{d}x}(ux)\mathrm{d}u - \sum_{i\in\{p,q\}} \int_0^1 u(1-u)\frac{\mathrm{d}M_{\Omega}^{\{i\}}}{\mathrm{d}x}(ux)\mathrm{d}u, \end{cases} \tag{7}$$

where p *and* q *are interchangeable and* $n \in \bar{H}$.

Note that our A-D payoff formula in Theorem 1 generalizes the formula in Misra *et al.* [4, Theorem 4.3] (*i.e.*, $|Z| = 1$). It also establishes the A-D values for distinguished *multiple* atomic players (the providers) and infinitesimal players (the peers), in the context of the Aumann-Shapley (A-S) prices [5] in coalition game theory.

Our formula for the peers can be interpreted as follows. Take the second line of (7) as an example. Recall the definition of the Shapley value (1). The payoff of peer n is the *marginal* cost reduction $v(S \cup \{n\}) - v(S)$ that is *averaged* over all equally probable arrangements, *i.e.*, the orders of players. It is also implied by (1) that the *expectation* of the marginal cost is computed under the assumption that the events $|S| = y$ and $|S| = y'$ for $y \neq y'$ are *equally probable*, *i.e.*, $\mathsf{P}(|S| = y) = \mathsf{P}(|S| = y')$. Therefore, in our context of infinite player game in Theorem 1, for every values of ux along the interval $[0, x]$, the subset $S \subseteq \bar{Z} \cup \bar{H}$ contains ux fraction of the peers. More importantly, the probability that each provider is a member of S is simply u because the size of peers in S, ηux, is infinite as $\eta \to \infty$ so that the *size* of S is not affected by whether a provider belongs to S or not. Therefore, the marginal cost reduction of each peer on the condition that both providers are contained in S becomes $-u^2\frac{\mathrm{d}M_{\Omega}^{\{p,q\}}}{\mathrm{d}x}(ux)$. Likewise, the marginal cost reduction of each peer on the condition that only one provider is in the coalition is $-u(1-u)\frac{\mathrm{d}M_{\Omega}^{\{p\}}}{\mathrm{d}x}(ux)$.

4 Instability of the Grand Coalition

In this section, we study the stability of the grand coalition to see if *rational* players are willing to form the grand coalition, only under which they can be paid their respective *fair* Shapley payoffs. The key message of this section is that the rational behavior of the providers makes the Shapley value approach *unworkable* because the major premise of the Shapley value, the grand coalition, is not formed in the multi-provider games.

4.1 Stability of the Grand Coalition

Guaranteeing the stability of a payoff vector has been an important topic in coalition game theory. For the single-provider case, $|Z| = 1$, it was shown in [4, Theorem 4.2] that, if the cost function is decreasing and concave, the Shapley incentive structure lies in the core of the game. What if for $|Z| \geq 2$? Is the grand coalition stable for the multi-provider case? Prior to addressing this question, we first define the following:

Definition 5 (Noncontributing Provider). *A provider $p \in Z$ is called noncontributing if $M_{\Omega}^{Z}(1) - M_{\Omega}^{Z\setminus\{p\}}(1) = \widetilde{\Omega}_p(0)$.*

To understand this better, note that the above expression is equivalent to the following:

$$\Big(\sum_{i\in Z} \widetilde{\Omega}_i(0) - M_{\Omega}^{Z}(1)\Big) - \Big(\sum_{i\in Z\setminus\{p\}} \widetilde{\Omega}_i(0) - M_{\Omega}^{Z\setminus\{p\}}(1)\Big) = 0 \qquad (8)$$

which implies that there is no difference in the total cost reduction, irrespective of whether the provider p is in the provider set or not. Interestingly, if all cost functions are concave, there exists at least one noncontributing provider.

Lemma 2. *Suppose $|Z| \geq 2$. If $\widetilde{\Omega}_p(\cdot)$ is concave for all $p \in Z$, there exist $|Z| - 1$ noncontributing providers.*

To prove this, recall the definition of $M_{\Omega}^{Z}(\cdot)$:

$$M_{\Omega}^{Z}(x) = \min_{y\in Y(x)} \sum_{i\in Z} \widetilde{\Omega}_i(y_i)$$

$$\text{where } Y(x) := \{(y_1, \cdots, y_{|Z|}) \mid \sum_{i\in Z} y_i \leq x,\ y_i \geq 0\}.$$

Since the summation of concave functions is concave and the minimum of a concave function over a convex feasible region $Y(x)$ is an *extreme* point of $Y(x)$ as shown in [14, Theorem 3.4.7], we can see that the solutions of the above minimization are the extreme points of $\{(y_1, \cdots, y_{|Z|}) \mid \sum_{i\in Z} y_i \leq x,\ y_i \geq 0\}$, which in turn imply $y_i = 0$ for $|Z| - 1$ providers in Z. Note that the condition $|Z| \geq 2$ is *necessary* here.

We are ready to state the following theorem, a direct consequence of Theorem 1. The proof is given in Appendix of [13].

Theorem 2 (Shapley Payoff Not in the Core). *If there exists a noncontributing provider, the Shapley payoff for the game $(Z \cup H, v)$ does not lie in the core.*

It follows from Lemma 2 that, if all operational cost functions are concave and $|Z| \geq 2$, the Shapley payoff does not lie in the core. This result appears to be in best agreement with our usual intuition. If there is a provider who does not contribute to the coalition at all in the sense of (8) and is still being paid due to her potential for imaginary contribution assessed by the Shapley formula (1), which is not actually exploited in the current coalition, other players may improve their payoff sum by expelling the noncontributing provider.

The condition $|Z| \geq 2$ plays an essential role in the theorem. For $|Z| \geq 2$, the concavity of the cost functions leads to the Shapley value not lying in the core, whereas, for the case $|Z| = 1$, the concavity of the cost function is proven to make the Shapley incentive structure lie in the core [4, Theorem 4.2].

4.2 Convergence to the Grand Coalition

The notion of the core lends itself to the stability analysis of the grand coalition *on the assumption* that the players are already in the equilibrium, *i.e.*, the grand coalition. However, Theorem 2 still lets further questions unanswered. In particular, for the non-concave cost functions, it is unclear if the Shapley value is not in the core, which is still an open problem. We rather argue here that, whether the Shapley value lies in the core or not, the grand coalition is unlikely to occur by showing that the grand coalition is not a global attractor under some conditions.

To study the convergence of a game with coalition structure to the grand coalition, let us recall Definition 3. It is interesting that, though the notion of stability was not used in [4], one main argument of this work was that the system with one provider would converge to the grand coalition, hinting the importance of the following convergence result with multiple providers. The proof of the following theorem is given in Appendix of [13].

Theorem 3 (A-D Payoff Does Not Lead to the Grand Coalition). *Suppose* $|Z| \geq 2$ *and* $\widetilde{\Omega}_p(y)$ *is not constant in the interval* $y \in [0, x]$ *for any* $p \in Z$ *where* $x = |\bar{H}|/|H|$*. The followings hold for all* $p \in Z$ *and* $n \in \bar{H}$*.*

- *The A-D payoff of provider* p *in coalition* $\{p\} \cup \bar{H}$ *is larger than that in all coalition* $T \cup \bar{H}$ *for* $\{p\} \subsetneq T \subseteq Z$*.*
- *The A-D payoff of peer* n *in coalition* $\{p\} \cup \bar{H}$ *is smaller than that in all coalition* $T \cup \bar{H}$ *for* $\{p\} \subsetneq T \subseteq Z$*.*

In plain words, a provider, who is in cooperation with a peer set, will receive the highest dividend when she cooperates only with the peers excluding other providers whereas each peer wants to cooperate with as many as possible providers. It is surprising that, for the multiple provider case, *i.e.*, $|Z| \geq 2$, each provider benefits from forming a single-provider coalition *whether* the cost function is concave *or not*. There is no *positive* incentives for providers to cooperate with each other under the implementation of A-D payoffs. On the contrary, a peer always looses by leaving the grand coalition.

Upon the condition that each provider begins with a single-provider coalition with a sufficiently large number of peers, one cannot reach the grand coalition

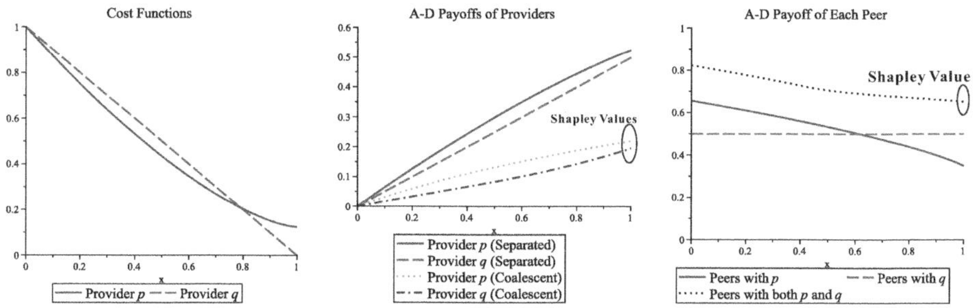

Fig. 3. Example 1: A-D Payoffs of Two Providers and Peers for Convex Costs

because some single-provider coalitions are already *stable* in the sense of the stability in Definition 3. That is, the grand coalition is not the global attractor. For instance, take $\mathcal{P} = \{\{p\} \cup H, \cdots\}$ as the current coalition structure where all peers are possessed by the provider p. Then it follows from Theorem 3 that players cannot make any transition from $\mathcal{P}$ to $\{\Phi \cup H, \cdots\}$ where $\Phi \subseteq Z$ is any superset of $\{p\}$ because provider p will not agree to do so.

5 A Critique of the A-D Payoff for Separate Providers

The discussion so far has focused on the stability of the grand coalition. The result in Theorem 2 suggests that if there is a noncontributing (free-riding) provider, which is true even for concave cost functions for multiple providers, the grand coalition will not be formed. The situation is aggravated by Theorem 3, stating that *single-provider coalitions* (*i.e.*, the separated case) will persist if providers are rational. We now illustrate the weak points of the A-D payoff under the single-provider coalitions with a few representative examples.

Example 1 (Unfairness). *Suppose that there are two providers,* i.e., $Z = \{p, q\}$, *with* $\widetilde{\Omega}_p(x) = 7(x-1)^{1.5}/8 + 1/8$ *and* $\widetilde{\Omega}_q(x) = 1 - x$, *both of which are decreasing and convex. All values are shown in Fig. 3 as functions of* x. *In line with Theorem 3, providers are paid more than their Shapley values, whereas peers are paid less than theirs.*

We can see that each peer n *will be paid* $21/32$ $(\widetilde{\varphi}_n^{\{p\}}(0))$ *when he is contained by the coalition* $\{p, n\}$ *and the payoff decreases with the number of peers in this coalition. On the other hand, provider* p *wants to be assisted by as many peers as possible because* $\widetilde{\varphi}_p^{\{p\}}(x)$ *is increasing in* x. *If it is possible for* n *to prevent other peers from joining the coalition, he can get* $21/32$. *However, it is more likely in real systems that no peer can kick out other peers, as discussed in [4, Section 5.1] as well. Thus,* p *will be assisted by* $x = 0.6163$ *fraction of peers, which is the unique solution of* $\widetilde{\varphi}_n^{\{p\}}(x) = \widetilde{\varphi}_n^{\{q\}}(x)$ *while* q *will be assisted by* $1 - x = 0.3837$ *fraction of peers.*

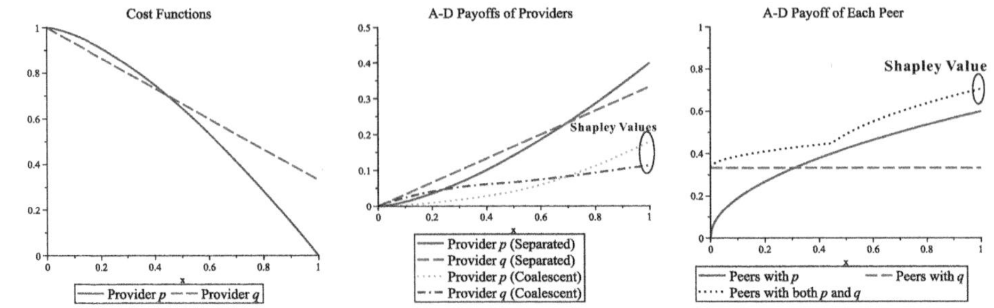

Fig. 4. Example 2: A-D Payoffs of Two Providers and Peers for Concave Costs

Example 2 (Monopoly). *Consider a two-provider system* $Z = \{p, q\}$ *with* $\widetilde{\Omega}_p(x) = 1 - x^{3/2}$ *and* $\widetilde{\Omega}_q(x) = 1 - 2x/3$, *both of which are decreasing and concave. All values including the Shapley values are shown in Fig. 4. Not to mention unfairness in line with Example 1 and Theorem 3, provider p monopolizes the whole peer-assisted services. No provider has an incentive to cooperate with other provider. It can be seen that all peers will assist provider p because* $\widetilde{\varphi}_n^{\{p\}}(x) > \widetilde{\varphi}_n^{\{q\}}(x)$ *for* $x > 25/81$. *Appealing to Definition 3, if the providers are initially separated, the coalition structure will converge to the service monopoly by p. In line with Lemma 2 and Theorem 2, even if the grand coalition is supposed to be the initial condition, it is not stable in the sense of the core. The noncontributing provider (Definition 5) in this example is q.*

The last example illustrates the A-D payoff can even induce an analog of the limit cycle in nonlinear control theory.

Example 3 (Oscillation). *Consider a game with two providers and two peers where* $N = \{p_1, p_2, n_1, n_2\}$. *If* $\{n_1\}$, $\{n_2\}$ *and* $\{n_1, n_2\}$ *assist the content distribution of* p_1, *the reduction of the distribution cost is respectively 10\$, 9\$ and 11\$ per month. However, the hard disk maintenance cost incurred from a peer is 5\$. In the meantime, if* $\{n_1\}$, $\{n_2\}$ *and* $\{n_1, n_2\}$ *assist the content distribution of* p_2, *the reduction of the distribution cost is respectively 6\$, 3\$ and 13\$ per month. In this case, the hard disk maintenance cost incurred from a peer is supposed to be 2\$ due to smaller contents of* p_2 *as opposed to those of* p_1. *We refer to Appendix of [13] for a detailed analysis.*

Table 1 contains A-D payoffs (and Shapley payoffs for the grand coalition) and blocking coalitions $C \subseteq N$ for any coalition structure where, for notational simplicity, we adopt a simplified expression for coalitional structure $\mathcal{P}$: a coalition $\{a, b, c\} \in \mathcal{P}$ is denoted by abc and each singleton set $\{i\}$ is denoted by i. We first observe that the Shapley payoff of this example does not lie in the core.

As time tends to infinity, the A-D payoff exhibits an oscillation of the partition $\mathcal{P}$ consisting of the four recurrent coalition structures as shown in Fig. 5. As of now, from the-state-of-the-art in the literature on this behavior [11], it is not yet clear how this behavior will be developed in large-scale systems.

Table 1. Example 3: A-D Payoff and Blocking Coalition C

	$\{p_1p_2, n_1n_2\}$	$\{p_1p_2, n_1, n_2\}$	$\{p_1, p_2, n_1n_2\}$	$\{p_1, p_2, n_1, n_2\}$	$\{p_1n_1, p_2n_2\}$
φ_{p_1}	0	0	0	0	5/2=2.5
φ_{p_2}	0	0	0	0	1/2=0.5
φ_{n_1}	0	0	0	0	5/2=2.5
φ_{n_2}	0	0	0	0	1/2=0.5
C	p_1n_1,p_1n_2,p_2n_1,p_2n_2,$p_1p_2n_1n_2$,$p_2n_1n_2$				$p_2n_1n_2$
recurrent	**X**	**X**	**X**	**X**	**O**

	$\{p_1p_2n_1n_2\}$	$\{p_1p_2n_1, n_2\}$	$\{p_2n_1n_2, p_1\}$	$\{p_1p_2n_2, n_1\}$	$\{p_1n_2, p_2, n_1\}$
φ_{p_1}	7/6 = 1.17	7/6=1.17	0	5/3=1.67	2
φ_{p_2}	19/6 = 3.17	2/3=0.67	23/6=3.83	1/6=0.17	0
φ_{n_1}	17/6 = 2.83	19/6=3.17	10/3=3.33	0	0
φ_{n_2}	11/6 = 1.83	0	11/6=1.83	13/6=2.17	2
C		p_1n_2		p_1n_1,p_2n_1	
recurrent	**X**	**X**	**O**	**X**	**X**

	$\{p_1n_1n_2, p_2\}$	$\{p_1n_1, p_2, n_2\}$	$\{p_1, n_1, p_2n_2\}$	$\{p_1n_2, p_2n_1\}$	$\{p_1, n_2, p_2n_1\}$
φ_{p_1}	11/6=1.83	5/2=2.5	0	2	0
φ_{p_2}	0	0	1/2=0.5	2	2
φ_{n_1}	-1/6=-0.17	5/2=2.5	0	2	2
φ_{n_2}	-2/3=-0.67	0	1/2=0.5	2	0
C	p_1n_1,p_1n_2,p_2n_1, p_2n_2,$p_2n_1n_2$, n_1, n_2, n_1n_2	$p_2n_1n_2$,p_2n_2	p_1n_1,p_1n_2,p_2n_1, $p_1p_2n_1n_2$, $p_2n_1n_2$	p_1n_1	p_1n_1,p_1n_2, $p_1p_2n_1n_2$ $p_2n_1n_2$
recurrent	**X**	**O**	**X**	**O**	**X**

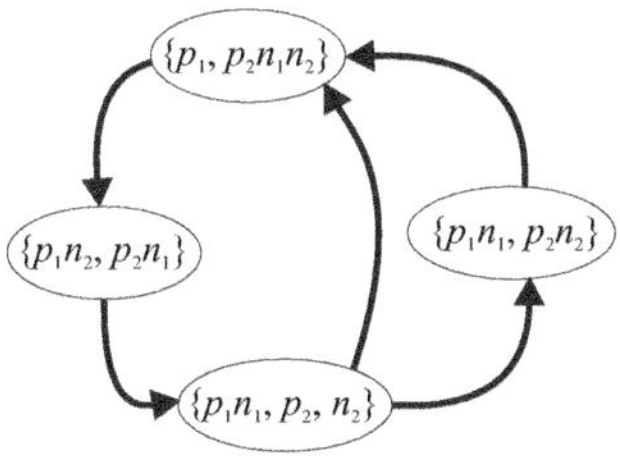

Fig. 5. Example 3: A-D Payoff Mechanism Leads to Oscillatory Coalition Structure

6 Concluding Remarks

A quote from an interview of BBC iPlayer with CNET UK [15]: "*Some people didn't like their upload bandwidth being used. It was clearly a concern for us, and we want to make sure that everyone is happy, unequivocally, using iPlayer.*"

In this paper, we have studied whether the Shapley incentive structure in peer-assisted services would be in conflict with the pursuit of profits by rational content providers and peers. A lesson from our analysis is summarized as: even though it is righteous to pay peers more because they become relatively more useful as the number of peer-assisted services increases, the content providers will not admit that peers should receive their due deserts. The providers tend to persist in single-provider coalitions. In the sense of the classical stability notion, called 'core', the cooperation would have been broken even if we had begun with the grand coalition as the initial condition. Secondly, we have illustrated yet another problems when we use the Shapley-like incentive for the exclusive single-provider coalitions. These results suggest that the profit-sharing system, Shapley value, and hence its fairness axioms, are not compatible with the selfishness of the content providers.

References

1. Valancius, V., Laoutaris, N., Massoulié, L., Diot, C., Rodriguez, P.: Greening the Internet with Nano Data Centers. In: Proc. ACM CoNEXT (December 2009)
2. Cha, M., Rodriguez, P., Moon, S., Crowcroft, J.: On next-generation telco-managed P2P TV architectures. In: Proc. USENIX IPTPS (February 2008)
3. RNCOS, Global IPTV market forecast to 2014, Market Research Report (February 2011)
4. Misra, V., Ioannidis, S., Chaintreau, A., Massoulié, L.: Incentivizing peer-assisted services: A fluid Shapley value approach. In: Proc. ACM Sigmetrics (June 2010)
5. Aumann, R., Shapley, L.: Values of Non-Atomic Games. Princeton University Press (1974)
6. Hart, S., Kurz, M.: Endogenous formation of coalitions. Econometrica 51, 1047–1064 (1983)
7. Shapley, L.: A Value for n-Person Games. In: Kuhn, H.W., Tucker, A.W. (eds.) Contribution to the Theory of Games II. Annals of Mathematics Studies, vol. 28. Princeton University Press (1953)
8. Aumann, R., Drèze, J.: Cooperative games with coalition structures. International Journal of Game Theory 3, 217–237 (1974)
9. Saad, W., Han, Z., Debbah, M., Hjørungnes, A., Başar, T.: Coalitional game theory for communication networks. IEEE Signal Processing Mag 26(5), 77–97 (2009)
10. Peleg, B., Sudhölter, P.: Introduction to the Theory of Cooperative Games, 2nd edn. Springer, Heidelberg (2007)
11. Tutic, A.: The Aumann-Drèze value, the Wiese value, and stability: A note. International Game Theory Review 12(2), 189–195 (2010)
12. Pinheiro, E., Weber, W., Barroso, L.A.: Failure trends in a large disk drive population. In: Proc. USENIX FAST (February 2007)
13. Cho, J., Yi, Y.: On the Shapley-like payoff mechanisms in peer-assisted services with multiple content providers (March 2011), `http://arxiv.org/abs/1012.2332`
14. Bazaraa, M.S., Sherali, H.D., Shetty, C.M.: Nonlinear Programming: Theory and Algorithms, 2nd edn. John Wiley & Sons Inc. (1993)
15. Lanxon, N.: iPlayer uncovered: What powers the BBC's epic creation? In: CNET UK (May 2009)

Economic Viability of Femtocell Service Provision

Lingjie Duan and Jianwei Huang*

Department of Information Engineering, The Chinese University of Hong Kong
{dlj008,jwhuang}@ie.cuhk.edu.hk

Abstract. Femtocells can effectively resolve the poor connectivity issue of indoor cellular users. This paper investigates the economic incentive for the cellular operator to add femtocell service on top of its existing macrocell service. We model the interactions between a cellular operator and users as a Stackelberg game: in Stage I the operator determines spectrum allocations and pricing decisions of femtocell and macrocell services, and in Stage II the users with heterogeneous macrocell channel qualities and spectrum efficiencies choose between the two services and decide their bandwidth usages. We show that the operator will choose to only provide femtocell service if femtocell service has full spatial coverage as macrocell service. In this case, the operator can serve more users at a higher price and thus obtain a higher profit. However, with the additional requirement that users need to achieve payoffs no worse than using the original macrocell service, we show that the operator will always provide macrocell service (with or without the femtocell service). Finally, we study the impact of operational cost on femtocell service provision, where we show that the operator will always provide both services. We also show that as such cost increases, fewer users are served by femtocell service and the operator's profit decreases.

Keywords: Femtocells, Stackelberg game, spectrum allocations, pricing.

1 Introduction

The next generation 4G cellular systems aim at providing end users with high data rates and reliable services by operating at wider and higher frequency bands (*e.g.*, 2496MHz-to-2690MHz for TD-LTE in U.S.). However, severe signal attenuation at these high frequencies often causes poor signal receptions for indoor users, who are far away and separated by walls from outdoor cellular base station in local macrocells.[1]

* This work is supported by the General Research Funds (Project Number 412509 and 412710) established under the University Grant Committee of the Hong Kong Special Administrative Region, China.

[1] A macrocell is the typical cell in a cellular network that provides radio coverage served by a power cellular base station [1].

R. Jain and R. Kannan (Eds.): GameNets 2011, LNICST 75, pp. 413–428, 2012.

To solve the poor signal reception problem for indoor users, researchers have proposed the idea of femtocell (*e.g.,* [7–10]). Compared to macrocells, femtocells are short-range, low deployment cost, and low power user-deployed tiny base stations. A user can deploy a femtocell at home and connect it to the wireline broadband Internet connection, *e.g.,* the digital subscriber line (DSL). Femtocells are often managed by the same operator that controls the macrocells, and they can provide better quality of services to indoor users as they are very close to users' cell phones. Despite of the obvious motivation to deploy femtocell service, the operator needs to carefully consider several issues that will affect the economic return of the femtocell service.

First, the femtocell service needs to share the limited licensed bands with the macocell service. There are two types of sharing schemes. The first scheme is "separate carriers", where the femtocells and macrocells occupy non-overlapping spectrum bands (*e.g.,* [11, 13, 14]). The second scheme is "shared carrier" (or "partially shared carrier"), where macrocells and femtocells operate on (partially) overlapping bands (*e.g.,* [10, 15, 19]). The first scheme is easy to manage but reduces the available spectrum for both services. The second scheme requires efficient distributed interference management mechanisms that are still open research problems. In this paper, our analysis will focus on the first "separate carriers" scheme.

Second, when an operator introduces the femtocell service and charges a higher price, some users who originally experience good macrocell service quality may actually experience a decrease in payoff. It is important to ensure the satisfaction of these users by keeping the original macrocell service available at the original price. This will limit the resource allocation to femtocell service.

Third, although femtocells are low in *deployment* costs, the femtocell service may incur additional *operational* cost compared to macrocells. Femtocell users' traffic needs to go through wireline broadband Internet connections. The wireline Internet Service Providers (ISPs) may impose additional charges on the femtocell related traffics [16]. Also, since the femtocell users' traffics will go through the ISP's network before reaching the cellular operator's own network, issues such as synchronization with macrocells become more challenging to resolve [17, 18].

In this paper, we will discuss the economic incentive of the operator's femtocell service provision, by considering three issues discussed above. We want to understand when and how the operator should offer the femtocell service, and the impacts on the original macrocell service. Our main results and contributions include:

- *A Dynamic Decision Model:* We model and analyze the interactions between an cellular operator and users as a two-stage Stackelberg game. Users experience different spectrum efficiencies with the macrocell service, but achieve the same maximum spectrum efficiency with the femtocell service. Thus users have different preferences between macrocells and femtocells. The operator makes spectrum allocations and pricing decisions for both macrocell and femtocell services to maximize its profit.

- *Profit-Maximizing with Femtocell Service Only:* If femtocell service has the same maximum coverage as macrocell service, then a profit-maximizing operator will choose to only offer femtocell service to all its users.
- *Dual Service Provision Considering Users' Reservation Payoffs:* If we consider users' reservation payoffs as what they can achieve with the original macrocell service, then offering femtocell service only may force some users to leave and thus may not be optimal to the operator. In this case, we characterize when and how the operator should provide the femtocell service together with the macrocell service (*i.e.*, dual services) so that all users achieve payoffs no worse than their reservation payoffs.
- *Impact of Femtocell Operational Cost:* When femtocell service incurs operational cost to the operator, the operator will always serve users by dual services. As the cost decreases, more users are served by the femtocell service and the operator obtains a higher profit.

Most prior work on femtocells focused on various technical issues in service provision (*e.g.*, access control and resource management [10,11,14,15,19]). Only few papers discussed the economic issues of femtocells (*e.g.*, [13, 20, 21]). The key difference between our paper and the existing literature is that we study the operator's provision of dual services in terms of both spectrum allocations and pricing decisions. We also characterize the impacts of users' reservation payoffs and the femtocell operational costs.

The rest of the paper is organized as follows. We introduce the network model of macrocell service in Section 2, which services as a benchmark for later analysis. In Section 3, we introduce the network model of femtocell service and analyze how the operator provides dual services in terms of spectrum allocations and pricing. Then we extend the results in Section 3 to Sections 4 and 5, by examining the impacts of users' reservation payoffs and femtocell operational costs. We conclude our work in Section 6. **Due to space limit, all proof details are included in the online technical report [23].**

2 Benchmark Scenario: Macrocell Service Only

As a benchmark case, we first look at how the operator prices the macrocell service to maximize its profit before introducing the femtocell service. When we consider the introduction of femtocell service in Sections 3, 4, and 5, the operator should achieve a profit no worse than this benchmark case. Also, what users get in this benchmark case will serve as their reservation payoffs in Section 4.

We consider an operator who owns a single macrocell without frequency reuse.[2] As shown in Fig. 1, we model the interactions between the operator and the users as a two-stage Stackerberg game. In Stage I, the operator determines the macrocell price p_M per unit bandwidth to maximize its profit. Here, subscript M denotes macrocells. In Stage II, each user decides how much bandwidth

[2] The results of this paper can be extended to a multiple macrocell scenario, where frequency reuse is allowed over macrocells.

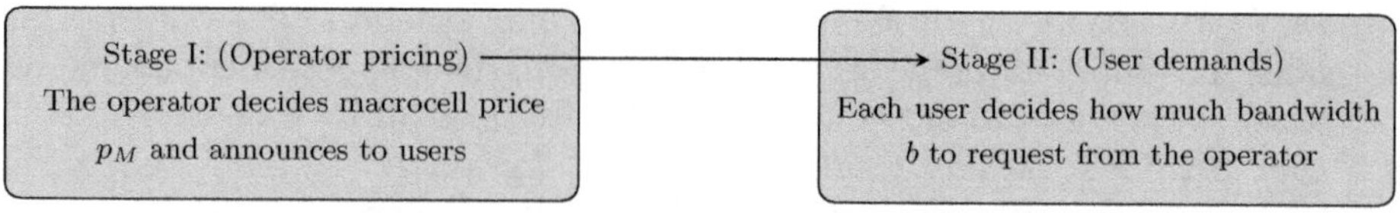

Fig. 1. Two-stage Stackelberg game between the operator and users. Stage I: the operator decides macrocell price per unit bandwidth to maximize its profit. Stage II: users choose how much resource to request in order to maximize their payoffs.

to purchase to maximize its payoff. This usage-based pricing scheme is widely used in practice [22]. We solve this two-stage Stackelberg game by backward induction [3].

2.1 Users' Requests in Service and Bandwidth in Stage II

Different users experience different channel conditions to the macrocell base stations due to different locations, and thus achieve different data rates when using the same amount of bandwidth. We model the users' channel heterogeneity by a *macrocell spectrum efficiency* θ, which is assumed to be uniformly distributed in $[0, 1]$ (see Fig. 2).[3] A larger θ means a better channel condition and a higher spectrum efficiency *when using the macrocell service*. In Section 3, we will show that all users achieve the same maximum spectrum efficiency with femtocell service. We also normalize the total user population to be 1.

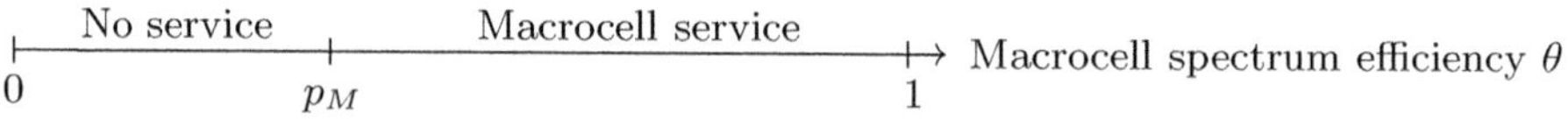

Fig. 2. Distribution of users' macrocell spectrum efficiencies

For a user with a macrocell spectrum efficiency θ, it obtains a utility $u(\theta, b)$ (*e.g.*, data rate) when using a bandwidth b [2],

$$u(\theta, b) = \ln(1 + \theta b).$$

The user needs to pay a linear payment $p_M b$ to the operator, where the price p_M is announced by the operator in Stage I. The user's *payoff* is the difference of its utility and payment, *i.e.*,

$$\pi_M(\theta, b, p_M) = \ln(1 + \theta b) - p_M b. \tag{1}$$

The optimal value of bandwidth (demand) that maximizes the user's payoff with the macrocell service is

[3] The uniform distribution is assumed for analytical tractability. However, a more complicated distribution will not change the main engineering insights obtained in this paper.

$$b^*(\theta, p_M) = \begin{cases} \frac{1}{p_M} - \frac{1}{\theta}, & \text{if } p_M \leq \theta, \\ 0, & \text{otherwise,} \end{cases} \tag{2}$$

which is decreasing in p_M and increasing in θ (if $p_M \leq \theta$). The user's maximum payoff with macrocell service is

$$\pi_M(\theta, b^*(\theta, p_M), p_M) = \begin{cases} \ln\left(\frac{\theta}{p_M}\right) - 1 + \frac{p_M}{\theta}, & \text{if } p_M \leq \theta, \\ 0, & \text{otherwise,} \end{cases} \tag{3}$$

which is always nonnegative.

2.2 Operator's Pricing in Stage I

Next we consider the operator's optimal choice of price p_M in Stage I. To achieve a positive profit, the operator needs to set $p_M \leq \max_{\theta \in [0,1]} \theta = 1$, so that at least some user purchases some positive bandwidth in Stage II. The fraction of users choosing macrocell service is $1 - p_M$ as shown in Fig. 2. The total user demand is

$$Q_M(p_M) = \int_{p_M}^{1} \left(\frac{1}{p_M} - \frac{1}{\theta}\right) d\theta = \frac{1}{p_M} - 1 + \ln p_M, \tag{4}$$

which is a decreasing function of p_M. On the other hand, the operator has a limited bandwidth supply B, and thus can only satisfy the demand no larger than B.

The operator chooses price p_M to maximize its profit, *i.e.*,

$$\max_{0 < p_M \leq 1} \pi^{operator}(p_M) = p_M \min\left(B, \frac{1}{p_M} - 1 + \ln p_M\right). \tag{5}$$

Theorem 1 characterizes the unique optimal solution of Problem (5).

Theorem 1. *The equilibrium macrocell price p_M^* is the unique solution of the following equation:*

$$B = \frac{1}{p_M^*} - 1 + \ln p_M^*. \tag{6}$$

Furthermore, the total user demand $Q(p_M^*) = B$. Finally, the equilibrium price p_M^* decreases with B, and the operator's equilibrium profit $\pi^{operator}(p_M^*)$ increases with B.

Notice that all users with a macrocell spectrum efficiency θ less than p_M^* will not receive macrocell service. When the total bandwidth B is small, the equilibrium macrocell price p_M^* is close to 1 and thus most users will not get service. This motivates the operator to adopt the femtocell service so that it can serve these users and generate additional profits.

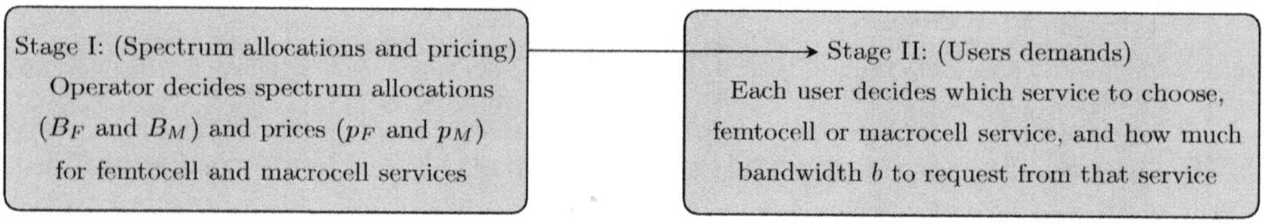

Fig. 3. Two-stage Stackelberg game between the operator and users. Stage I: the operator decides spectrum allocations and prices of macrocell and femtocell services; Stage II: users choose which service to access and how much resource usage to request from the operator.

3 Provision of Femtocell Service

We now consider how femtocell service can improve the operator's profit. The analysis in this section is based on several simplified assumptions:

- All users have a zero reservation payoff, and thus will accept the femtocell service as long as its payoff is positive. This assumption will be relaxed in Section 4.
- The femtocell service does not incur any additional operational cost compared to the macrocell service. This assumption will be relaxed in Section 5.
- The femtocells have the same maximum coverage as the macrocell service, and each user can access to both macrocell and femtocell services. This assumption will be relaxed in our technical report [23].

We are interested in answering the following two questions:

- Is it economically viable for the the operator to introduce the femtocell service?
- If so, how should the operator allocate bandwidth for and price macrocell and femtocell services?

Under the assumptions in this section, we will show that the operator will choose to only provide femtocell service (*i.e.*, no macrocell service) to all users and charge a higher price p_F^* than the optimal macrocell price p_M^* derived in Section 2.

More specifically, we will look at a two-stage Stackelberg game as in Fig. 3. In Stage I, the operator determines bandwidth allocated to femtocell service (femtocell band B_F) and to macrocell (macrocell band B_M), with $B_F + B_M = B$. The operator also determines the femtocell price p_F and macrocell price p_M. In Stage II, each user decides which service to choose and how much bandwidth to purchase. If a user's demand cannot be satisfied by its preferred service, it will switch and purchase bandwidth from the other service.[4] We will again analyze this two-stage Stackelberg game by using backward induction.

[4] Here we focus on a large group of users, where a user's demand is infinitesimal compared to the total demand. Thus we can ignore the case where a user purchases bandwidth from both services.

3.1 Users' Requests in Service and Bandwidth in Stage II

If a user has a macrocell spectrum efficiency θ, its optimal payoff by using the macrocell service is given in (3). Next we consider users' payoffs by using the femtocell service.

Since femtocell base stations are deployed indoors and are very close to the users' cell phones, we assume that all users using the femtocell service have equal good channel conditions and achieve the same *maximum* spectrum efficiency. This means that independent of the macrocell spectrum efficiency θ, each user achieves the same payoff $\pi_F(b)$ when using a bandwidth of b,

$$\pi_F(b, p_F) = \ln(1+b) - p_F b. \tag{7}$$

The user's optimal demand in femtocells is

$$b^*(p_F) = \begin{cases} \frac{1}{p_F} - 1, & \text{if } p_F \leq 1, \\ 0, & \text{otherwise.} \end{cases} \tag{8}$$

A user's maximum payoff with the femtocell service is

$$\pi_F(b^*(p_F), p_F) = \begin{cases} \ln\left(\frac{1}{p_F}\right) - 1 + p_F, & \text{if } p_F \leq 1, \\ 0, & \text{otherwise,} \end{cases} \tag{9}$$

which is always nonnegative.

It is clear that a user with small macrocell spectrum efficiency θ can get a better payoff by using the femtocell service instead of the macrocell service. We can imagine a threshold of θ that separates the users of two services. Next we define two different types of thresholds.

Definition 1 (Users' *preferred* partition threshold θ_{th}^{pr}). *Users with $\theta \in [0, \theta_{th}^{pr})$ prefer to use the femtocell service, and users with $\theta \in [\theta_{th}^{pr}, 1]$ prefer to use the macrocell service.*

Definition 2 (Users' partition threshold θ_{th}). *The partition threshold θ_{th} is the minimum macrocell spectrum efficiency among all the users served (may not prefer to be served) by the macrocell service. Users with $\theta \in [\theta_{th}, 1]$ receive the macrocell service finally, while users with $\theta \in [0, \theta_{th})$ receive either the femtocell service or no service.*

If all users' demands from their preferred services are satisfied, then users' preferred partition threshold equals users' partition threshold (*i.e.*, $\theta_{th}^{pr} = \theta_{th}$). However, in general θ_{th} may be different from θ_{th}^{pr}, depending on the operator's choice of B_F and B_M in the first stage.

By comparing a user's optimal payoff with macrocell and femtocell services in (3) and (9), we have the following result.

Lemma 1. *Users' preferred partition threshold $\theta_{th}^{pr} = p_M/p_F$. Users with a small macrocell spectrum efficiency $\theta < p_M/p_F$ prefer the femtocell service, and users with a large macrocell spectrum efficiency $\theta > p_M/p_F$ prefer the macrocell service.*

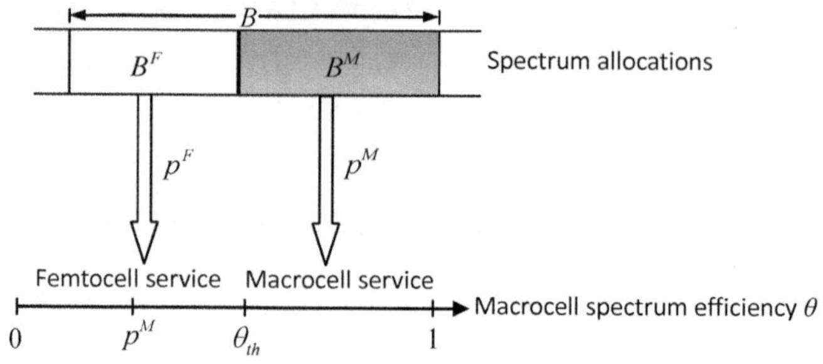

Fig. 4. Operations of macrocell and femtocell services

Now we introduce the concept of *finalized demand.*

Definition 3 (User's Finalized Demand). *If a user's demand from its preferred service is satisfied, then its finalized demand is its prefered demand. If a user's demand from its preferred service is not satisfied, then the user may switch to the alternative service and the new demand becomes the finalized demand.*

3.2 Operator's Spectrum Allocations and Pricing in Stage I

Now we are ready to study Stage I, where the operator determines B_F, B_M, p_F, and p_M to maximize its profit (see Fig. 4). Let us denote the operator's equilibrium decisions as B_F^*, B_M^*, p_F^*, and p_M^*, which lead to the users' equilibrium partition threshold (Definition 2) equal to θ_{th}^*. It is clear that the femtocell price p_F^* is larger than the macrocell price p_M^*, otherwise all users will choose the femtocell service.

Lemma 2. *At the equilibrium, the operator's total bandwidth B equals users' total finalized demand.*

Based on Lemma 2, we can further show that bandwidth allocated to each service equals users' total finalized demand in that service. That is,

$$B_F^* = \int_0^{\theta_{th}^*} \left(\frac{1}{p_F^*} - 1\right) d\theta = \theta_{th}^* \left(\frac{1}{p_F^*} - 1\right), \tag{10}$$

$$B_M^* = \int_{\theta_{th}^*}^{1} \left(\frac{1}{p_M^*} - \frac{1}{\theta}\right) d\theta = \frac{1-\theta_{th}^*}{p_M^*} + \ln\theta_{th}^*, \tag{11}$$

and $B_F^* + B_M^* = B$. This means that we only need to solve for the equilibrium decisions of θ_{th}^*, p_M^*, and p_F^*. The operator's profit-maximization problem is

$$\max_{p_M, p_F, \theta_{th}} \pi^{operator}(p_M, p_F, \theta_{th}) = p_F\theta_{th}\left(\frac{1}{p_F} - 1\right) + p_M\left(\frac{1-\theta_{th}}{p_M} + \ln\theta_{th}\right)$$

$$\text{subject to} \quad p_M \le \theta_{th} \le 1,$$

$$\theta_{th}\left(\frac{1}{p_F} - 1\right) + \frac{1-\theta_{th}}{p_M} + \ln\theta_{th} = B. \tag{12}$$

By examining (12), we have the following result.

Theorem 2. *At the equilibrium, the operator will only provide femtocell service, i.e.,* $B_F^* = B$ *and* $B_M^* = 0$*. All users will use femtocell service, i.e., users' equilibrium partition threshold* $\theta_{th}^* = 1$*. The equilibrium femtocell price is*

$$p_F^* = \frac{1}{1+B}, \tag{13}$$

and the operator's equilibrium profit is

$$\pi^{operator*} = \frac{B}{1+B}. \tag{14}$$

Theorem 2 is easy to understand. As the femtocell service provides a higher QoS to all users,[5] the operator can attract the users with small macrocell spectrum efficiency θ, and sell out the whole bandwidth B at a price $p_F^* = 1/(1+B)$ *higher* than the equilibrium macrocell price p_M^* in Theorem 1. This means that the operator achieves a higher profit by only providing femtocell service.

However, a user who has a large θ (*e.g.*, $\theta = 1$) will achieve a smaller payoff $\pi_F(b^*(p_F^*), p_F^*)$ with femtocell service than the payoff $\pi_M(\theta, b^*(\theta, p_M^*), p_M^*)$ with macrocell service. If we treat $\pi_M(\theta, b^*(\theta, p_M^*), p_M^*)$ as a user's *reservation payoff* below which the user will not accept the femtocell service,[6] then the operator can no longer only provide femtocell service. Next section studies this case in details.

4 Impact of Users' Reservation Payoffs

In this section, we will consider the operator's decisions by assuming that each user with a macrocell spectrum efficiency θ receives a payoff no less than $\pi_M(\theta, b^*, p_M^*)$ as calculated in (3). This means that the operator always needs to provide macrocell service at the same price as p_M^* derived based on (6). Also, all users' preferred demands in macrocell service should be satisfied. Next we consider a two-stage decision process similar to Fig. 3. The only difference here is that the operator needs to satisfy users' reservation payoffs.

In this section only, we assume that the operator has a priority to serve the users with the smallest θ first in femtocells. This is reasonable since femtocell service aims at improving QoS of indoor users especially for those with a small spectrum efficiency. These users cannot use macrocell service and will be happy to pay a high price for the femtocell service. For users with a high macrocell spectrum efficiency, they have the additional choice of macrocell service and will not use femtocell service if p_F is high.

---

[5] The only exception will be users with $\theta = 1$, who have a zero size support under the uniform distribution assumption of θ.

[6] For example, the user may switch to a different operator who provides macrocell service.

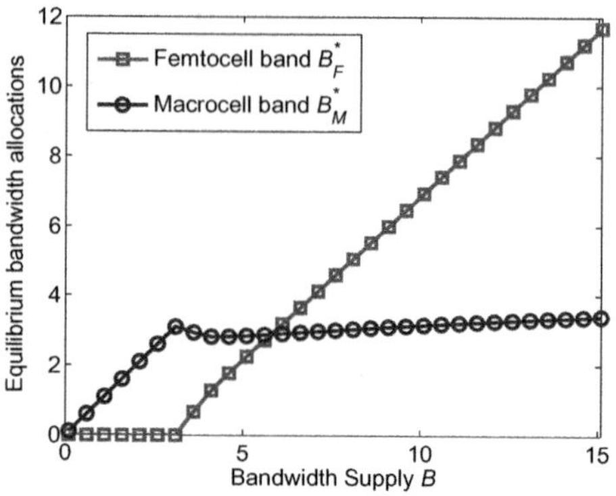

Fig. 5. The operator's equilibrium femtocell band B_F^* and macrocell band B_M^* as functions of supply B considering users' reservation payoffs

We will again use backward induction to analyze the problem. As Stage II is the same as Section 3.1, we will focus on the operator's decisions on B_M, B_F, and p_F in Stage I.

Lemma 3. *At the equilibrium, one of the following is true:*

- *Only users with $\theta \in [p_M^*, 1]$ are served with the macrocell service, and no users are served with the femtocell service.*
- *All users with $\theta \in [0, 1]$ are served, by either the macrocell service or the femtocell service.*

Lemma 3 shows that the equilibrium femtocell band is either $B_F^* = 0$ or $B_F^* \geq \int_0^{p_M^*} (\frac{1}{p_F^*} - 1) d\theta$. This implies that when B is small, the operator needs to allocate all its bandwidth supply B for macrocell service to reach users' reservation payoffs. Only when B is large, the operator can serve all users by dual services (*i.e.*, macrocell and femtocell services).

The operator's profit-maximization problem can be simplified as

$$\max_{p_F, \theta_{th}} \pi^{operator}(p_F, \theta_{th}) = p_F \int_0^{\theta_{th}} \left(\frac{1}{p_F} - 1\right) d\theta + p_M^* \int_{\theta_{th}}^1 \left(\frac{1}{p_M^*} - \frac{1}{\theta}\right) d\theta$$

$$\text{subject to} \quad p_M^* \leq \theta_{th} \leq p_M^*/p_F,$$

$$\int_0^{\theta_{th}} \left(\frac{1}{p_F} - 1\right) d\theta + \int_{\theta_{th}}^1 \left(\frac{1}{p_M^*} - \frac{1}{\theta}\right) d\theta \leq B, \tag{15}$$

where p_M^* is computed from (6), and the right inequality of the first constraint means that the operator cannot violate users' preferences in macrocell service. In the second constraint, the first and second terms on the left hand side are the users' finalized total demands in femtocells and macrocells, respectively.

Problem (15) is difficult to solve in closed form, so we use numerical results to illustrate some interesting insights.

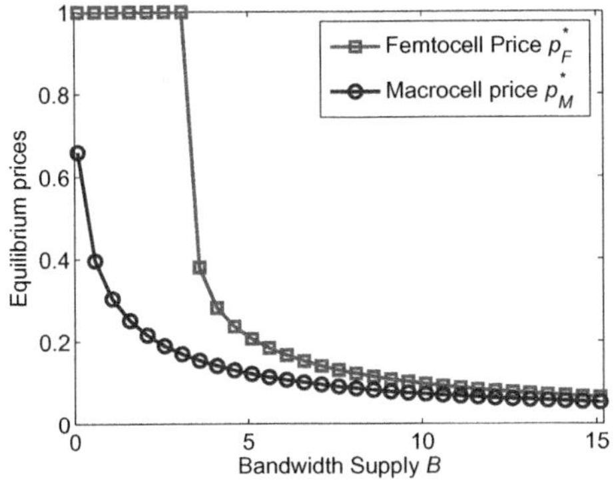

Fig. 6. The operator's equilibrium femtocell price p_F^* and macrocell price p_M^* as functions of bandwidth supply B considering users' reservation payoffs

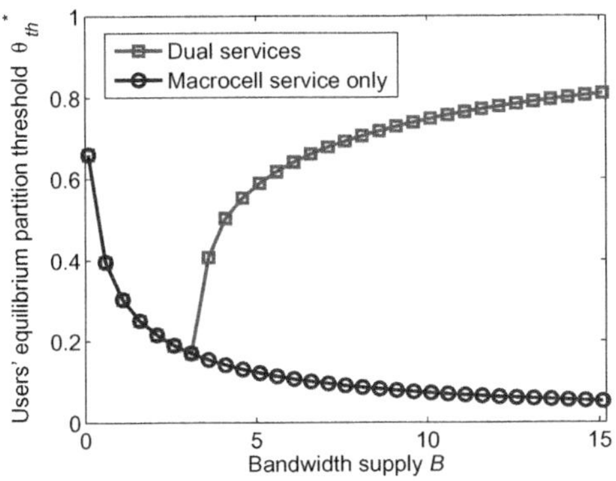

Fig. 7. Users' equilibrium partition thresholds θ_{th}^* as functions of bandwidth supply B in two cases: dual services considering users' reservation payoffs and macrocell service only as in Section 2

Figure 5 shows the operator's equilibrium bandwidth allocations to dual services (*i.e.*, B_F^* and B_M^*) as functions of the total bandwidth supply B. Figure 5 is consistent with Lemma 3, where only macrocell service is available (*i.e.*, $B_F^* = 0$) in the low supply regime (*i.e.*, $B < 3.5$), and both services are available in the high supply regime (*i.e.*, $B \geq 3.5$). In the high supply regime, femtocell band B_F^* increases faster than the macrocell band B_M^* when B increases. This is because the operator can obtain a higher profit by providing femtocell service, which charges users a higher price compared with macrocell service.

Figure 6 shows the operator's equilibrium femtocell price p_F^* and macrocell price p_M^* as functions of total bandwidth B. We can observe that in high supply regime p_F^* decreases faster than p_M^* as B increases, which means that the operator wants to attract more users to femtocell service.

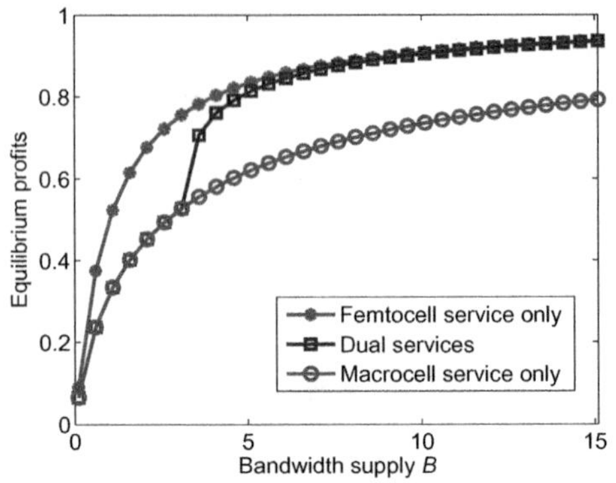

Fig. 8. The operator's equilibrium profits in three cases: femtocell service only as in Section 3, dual services considering users' reservation payoffs, and macrocell service only as in Section 2

Figure 7 shows the users' equilibrium partition threshold θ_{th}^* with dual services in this section, comparing to the partition threshold of the macrocell service only case in Section 2. In the low supply regime (*i.e.*, $B < 3.5$), both curves overlap with each other, as the dual services degenerate to the macrocell service only in this regime. However, as the total bandwidth B becomes very large in high supply regime, the operator will announce similar femtocell and macrocell prices, and most users will choose to use femtocell service. Comparing with the femtocell service only provision in Section 3 (without considering users' reservation payoffs), here users with a large θ will choose to stay with the macrocell service and are not affected by the introduction of femtocell service.

Figure 8 compares the operator's profits in three different cases: femtocell service only as in Section 3, dual services as in this section, and macrocell service only as in Section 2. In the low supply regime, dual services degenerate to the macrocell service case. In the high supply regime, the profit of the dual services becomes closer to the femtocell service only case as B increases. This means that considering users' reservation payoffs will not lead to significant profit loss when the total resource is abundant. In this case, only users with a θ very close to 1 will stay with macrocell service and all other users will choose the femtocell service.

5 Impact of Femtocell Operational Cost

In Section 3, we have assumed that there is no additional operational cost of the femtocell service. The data from the femtocells will be delivered through the wireline Internet connection of an ISP back to the operator's cellular network, free of charge. However, this is only reasonable when the operator and the ISP belong to the same entity or the ISP is sharing-friendly as in [4,5]. In this section, we consider the case where the ISP will charge the operator usage-based fees for

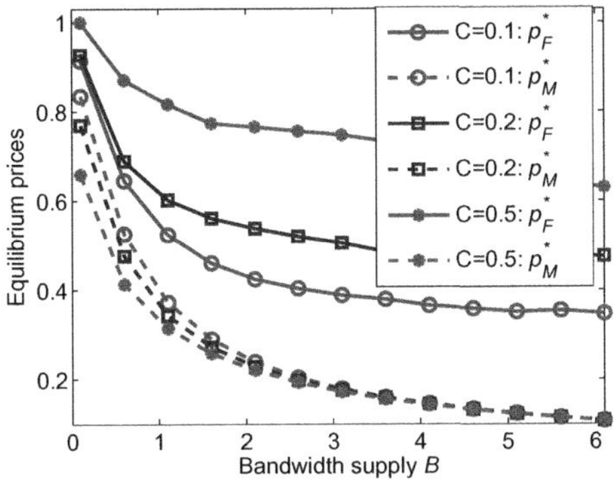

Fig. 9. The equilibrium femtocell price p_F^* and macrocell price p_M^* as functions of supply B and cost C in dual services considering femtocell operational cost

using the wireline Internet connection. We are interested in understanding how this operational cost affects the provision of femtocell service.

For simplicity, we assume that the operational cost is linearly proportional to femtocell bandwidth with the coefficient C. We focus on the case of $C \in (0, 1)$. It is easy to show that if $C \geq 1$, then the operator will charge a femtocell price $p_F > 1$, and no user will choose the femtocell service based on (8).

We consider a two-stage decision process similar as Fig. 3. The analysis of Stage II is the same as in Section 3.1. Here we will focus on the operator's decisions on B_M, B_F, p_M, and p_F in Stage I. Following a similar analysis as in Section 3, we can show that the total bandwidth B will be fully utilized at the equilibrium (*i.e.*, Lemma 2). Then we can formulate the operator's profit-maximization problem as

$$\begin{aligned} \max_{p_M, p_F, \theta_{th}} \quad & \pi^{operator}(p_M, p_F, \theta_{th}) = (p_F - C)\theta_{th}\left(\frac{1}{p_F} - 1\right) + p_M\left(\frac{1-\theta_{th}}{p_M} + \ln\theta_{th}\right) \\ \text{subject to} \quad & p_M \leq \theta_{th} \leq 1, \\ & \theta_{th}\left(\frac{1}{p_F} - 1\right) + \frac{1-\theta_{th}}{p_M} + \ln\theta_{th} = B. \end{aligned} \tag{16}$$

Then we have the following result.

Theorem 3. *With a femtocell operational cost $C \in (0, 1)$, the operator always provides both femtocell service and macrocell service at the equilibrium, and $p_M^* \leq \theta_{th}^* < 1$.*

Note that p_M^* is the equilibrium macrocell price, and θ_{th}^* is the users' equilibrium partition threshold with dual services. Intuitively, a positive operational cost C forces the operator to charge a higher femtocell price p_F^* than the value in (13). However, the small payment from users with a large value of θ (who only experience a little QoS improvement) in femtocell service cannot cover the increased operation cost to the operator. As a result, the operator will serve these users by macrocell service.

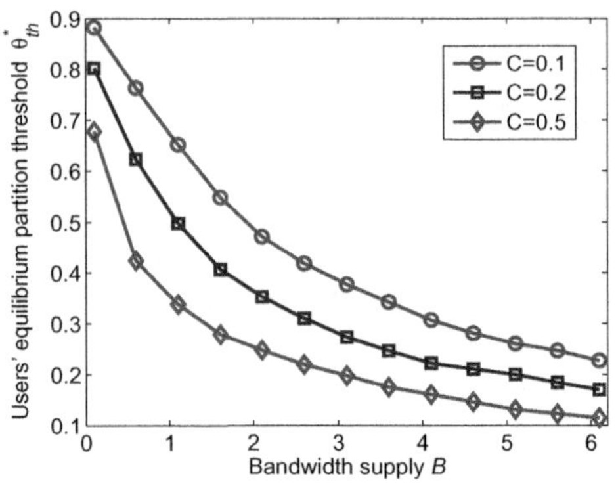

Fig. 10. The users' equilibrium partition threshold θ_{th}^{*} as a function of supply B and cost C in dual services considering femtocell operational cost

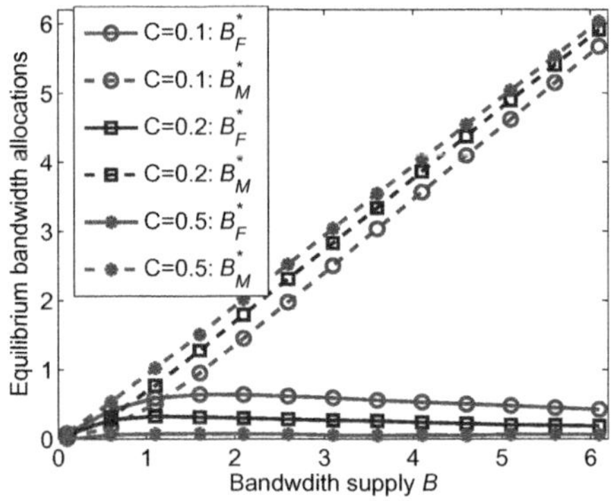

Fig. 11. The equilibrium femtocell band B_F^* and macrocell band B_M^* as functions of supply B and cost C in dual services considering femtocell operational cost

Problem (16) is difficult to solve in closed form, so we use numerical results to illustrate some interesting insights.

Figure 9 shows the operator's equilibrium femtocell price p_F^* and macrocell price p_M^* as functions of bandwidth supply B and femtocell operational cost C. The femtocell price p_F^* is always larger than C in order to be profitable. The macrocell price p_M^* does not need to compensate any cost. When B increases, the operator can set p_M^* as low as needed to maximize its profit, while p_F^* is lower-bounded by C. This explains why the $p_F^* - p_M^*$ widens as B increases, and such gap becomes even bigger with a larger C.

Figure 10 shows users' equilibrium partition threshold θ_{th}^* as a function of B and C. The threshold θ_{th}^* decreases in both B and C, which means that more users will choose to use the macrocell service due to the increase of $p_F^* - p_M^*$.

Figure 11 shows the operator's equilibrium bandwidth allocations to dual services as functions of B and C. When B is small, femtocell band B_F^* increases with B since the operator wants to serve more users at higher femtocell price.

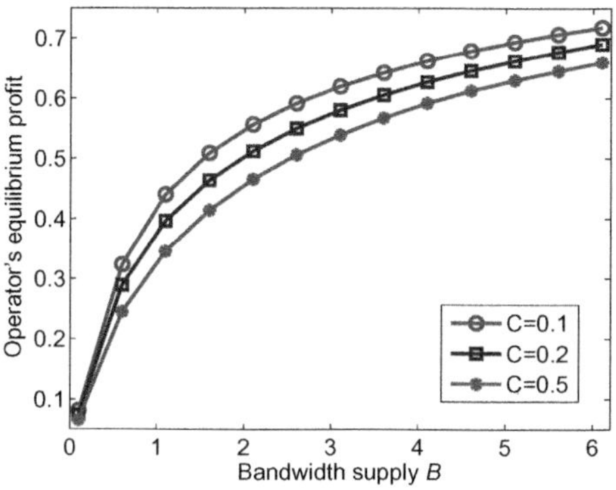

Fig. 12. The operator's equilibrium profit $\pi^{operator*}$ as a function of supply B and cost C in dual services considering femtocell operational cost

When B is large, B_F^* decreases with B since the high femtocell price (relative to macrocell price) makes more users choose the macrocell service.

Figure 12) shows that the operator's equilibrium profit increases in B and decreases in C.

6 Conclusion

This paper studies the economic incentive for a cellular operator to add the femtocell service on top of its existing macrocell service. We analyze the operator's equilibrium decisions in terms of spectrum allocations and pricing of both types of services. We show that compared to macrocell service, femtocell service can attract more users at a higher price and increase the operator's profit. However, the requirement of satisfying users' reservation payoffs (*i.e.*, what they can achieve with the original macrocell service) prevents the operator from only providing femtocell service. In the case of small total bandwidth B, the operator actually cannot even provide any femtocell service. Also, in the case where femtocell service has an additional operational cost, it is always a good idea for the operator to provide both femtocell and macrocell services to maximize its profit. In our technical report [23], we further look at the realistic case where the femtocell service has a smaller coverage than the macrocell service.

References

1. Schiller, J.H.: Mobile communications. Addison Wesley (2003)
2. Shen, H., Basar, T.: Optimal nonlinear pricing for a monopolistic network service provider with complete and incomplete information. IEEE Journal on Selected Areas in Communications 25(6), 1216–1223 (2007)
3. Myerson, R.B.: Game theory: analysis of conflict. Harvard University Press, Cambridge (2002)

4. Electronic Frontier Foundation (EFF) Wireless friendly ISP list, http://www.eff.org/Infrastructure/Wireless_cellular_radio/wireless_friendly_isp_list.html
5. Speakeasy NetShare, http://www.speakeasy.net/netshare/
6. Presentations by ABI Research, Picochip, Airvana, IP.access, Gartner, Telefonica Espana. In: 2nd Int'l. Conf. Home Access Points and Femtocells, http://www.avrenevents.com/dallasfemto2007/purchase_presentations.htm
7. picoChip Designs Ltd., The case for home base stations, White Paper (April 2007), http://www.femtoforum.org/femto/Files/File/picoChipFemtocellWhitePaper1.1.pdf
8. Claussen, H., Ho, L.T.W., Samuel, L.G.: An overview of the femtocell concept. Bell Labs Technical Journal 13(1), 221–245 (2008)
9. Chandrasekhar, V., Andrews, J., Gatherer, A.: Femtocell networks: a survey. IEEE Commun. Mag. 46(9), 59–67 (2008)
10. Yeh, S.-P., Talwar, S., Lee, S.-C., Kim, H.: WiMAX femtocells: a perspective on network architecture, capacity, and coverage. IEEE Commun. Mag. 46(10), 58–65 (2008)
11. Hobby, J.D., Claussen, H.: Deployment options for Femtocells and their impact on existing macrocellular networks. Bell Labs Technical Journal 13(4), 145–160 (2009)
12. Spectrum policy task force report, Federal Communications Commission, US (November 2002)
13. Shetty, N., Parekh, S., Walrand, J.: Economics of femtocells. In: IEEE GLOBECOM 2009 (2009)
14. Wu, J.-S., Chung, J.-K., Sze, M.-T.: Analysis of uplink and downlink capacities for two-tier cellular system. IEE Proceedings- Communications 144(6), 405–411 (1997)
15. Chandrasekhar, V., Andrews, J.: Uplink capacity and interference avoidance for two-tier cellular networks. In: IEEE GLOBECOM 2007, pp. 3322–3326 (November 2007)
16. McKnight, L.W., Bailey, J.P.: Internet economics. MIT Press, Cambridge (1998)
17. Yoon, J., Lee, J., Lee, H.S.: Multi-hop based network synchronization scheme for femtocell systems. In: IEEE PIMRC 2010 (2010)
18. Kim, R.Y., Kwak, J.S., Etemad, K.: WiMax femtocell: requirements, challenges, and solutions. IEEE Communications Magazine (September 2009)
19. Ho, L., Claussen, H.: Effects of user-deployed, co-channel femtocells on the call drop probability in a residential scenario. In: Proc. PIMRC 2007, pp. 1–5 (September 2007)
20. Claussen, H., Ho, L., Samuel, L.: Financial analysis of a pico-cellular home network deployment. In: IEEE ICC 2007, pp. 5603–5609 (June 2007)
21. Yun, S., Yi, Y., Cho, D., Mo, J.: Open or close: on the sharing of femtocells. To appear at IEEE INFOCOM Mini-Conference (April 2011)
22. Courcoubetis, C., Weber, R.: Pricing communication networks- economics, technology and modeling. Wiley, New York (2003)
23. Duan, L., Huang, J.: Economic Incentives of Femtocell Service Provision. Technical Report, http://home.ie.cuhk.edu.hk/~jwhuang/publication/FemtocellTechReport.pdf

Jamming Game in a Dynamic Slotted ALOHA Network

Andrey Garnaev[1], Yezekael Hayel[2],
Eitan Altman[3], and Konstantin Avrachenkov[4]

[1] Saint Petersburg State University
St Petersburg, Russia
garnaev@yahoo.com
[2] University of Avignon
Avignon, France
yezekael.hayel@univ-avignon.fr
[3] INRIA Sophia Antipolis
Sophia Antipolis, France
altman@sophia.inria.fr
[4] INRIA Sophia Antipolis
Sophia Antipolis, France
k.avrachenkov@sophia.inria.fr

Abstract. In this paper we suggest a development of the channel capacity concept for a dynamic slotted ALOHA network. Our object is to find maxmin successful transmissions of an information over a dynamic communication channel. To do so, we analyze an ALOHA-type medium access control protocol performance in the presence of a jammer. The time is slotted and the system is described as a zero-sum multistage matrix game. Each player, the sender and the jammer, have different costs for respectively sending their packets and jamming, and the jammer wants to minimize the payoff function of the sender. For this case, we found explicit expression of the equilibrium strategies depending on the costs for sending packets and jamming. Properties of the equilibrium are investigated. In particular we have found a simple linear correlation between the probabilities to act for both players in different channel states which are independent on the number of packets left to send. This relation implies that increasing activity of the jammer leads to reducing activity of the user at each channel state. The obtained results are generalized for the case where the channel can be in different states and change according to a Markov rule. Numerical illustrations are performed.

1 Introduction

The first work related Game theory and Information theory through a max-min problem was proposed in 1952 by Mandelbrot in his PhD Thesis *Contribution a la theorie mathematique des jeux de communication*. He has studied the problem of communication through a noisy channel as a two-player zero-sum game where the sender maximizes mutual information and the noise minimizes it, subject to

R. Jain and R. Kannan (Eds.): GameNets 2011, LNICST 75, pp. 429–443, 2012.

average power constraints. It has been shown that an i.i.d Gaussian signaling scheme and an i.i.d. Gaussian noise distribution are robust, in that any deviation of either the signal or noise distribution reduces or increases (respectively) the mutual information. Hence, the solution to this game-theoretic problem yields a rate of $\log(1+\frac{P}{N_0})$ which is defined now as the Shannon capacity.

Recall that channel capacity is the tightest upper bound on the amount of information that can be reliably transmitted over a communication channel with noise. In this paper we suggest a development of this channel capacity concept for a dynamic slotted ALOHA network. Our object is to find maxmin successful transmissions of an information over a dynamic communication channel. Therefore, our work extends in a simple way the concept of Shannon capacity in a ALOHA network.

The ALOHA protocol proposed in[4], is a totally decentralized mechanism for defining a medium access protocol without carrier sense in a multi-user environment. The slotted-ALOHA has been proposed in [5] by introducing the synchronization between the devices. This distributed mechanism leads several extensions and is the base of several cellular networks protocols like GSM. There are several works on the the study of non-cooperation between users in an ALOHA network. For example in [6], the authors consider an ALOHA game which the users decides and advertises their transmission probability but keeps their desired throughput private. They study the existence of equilibrium points that could possibly be reached by the users for given user throughput demands. The users' convergence to equilibrium points is analyzed using a specified potential function that governs their dynamics. We can cite also the papers [8] and [9] in which the authors extend the previous model by incorporating channel state information as affecting the transmission policy. They have also shown that there exists particular configurations with several Nash equilibrium. Another model with partial information is proposed in [7] in which mobiles do not know the number of backlogged packets at other nodes. A Markov chain analysis is used to obtain optimal and equilibrium retransmission probabilities and throughput. Jamming in an ALOHA network has been first study in [2]. The performance of the system is defined as the minmax of a two-person constant sum game. The author considers the expected forward progress by taking into account geometrical considerations and routing protocols. In [3], the authors consider an ALOHA non-cooperative game in which one player is a jammer. The authors consider only probability of sending packet or jamming without an energy cost. In [10] jamming and transmission costs were employed for the plot of one step jamming game. Note that besides ALOHA network the jamming problem has been studied for a variety of wireless network settings including sensor networks [11] and other general wireless network models [12,13,14,15,16].

In this paper we assume that a user wants to transmit a message of N packets in a time smaller or equal to T. In fact, we assume that T is an exponentially distributed random variable with the mean $1/\lambda$. Why exponentially distributed? Delay-tolerant networks (DTN) are complex distributed systems that are composed of wireless mobile/fixed nodes, and they are typically assumed to experience frequent, long-duration partitioning, and intermittent node connection [18]. There have been various research works on the characteristics of the intercontact time between nodes [19]. Initial works typically assumed that the CCDF (complementary cumulative distribution function) of the inter-contact time decays exponentially over time and it is generally modeled using an exponential random variable [20]. This assumption is supported by numerical simulations conducted under most existing mobility models in the literature [21].

Note that as T is a duration, we should have $\lambda < 1$. This parameter represents the average time between two transmission attempts. We consider a slotted model. In each slot, the user sends a packet with probability p and the jammer tries to jam with probability q. The user obtains one as the reward only if he sends successfully the file of N packets within the time frame T, otherwise the user gets the zero reward. We describe this plot using a multistage zero-sum game. The value of the game and the optimal strategies are found in closed form. In particular we show that if the transmission cost is too big then the game has a saddle point. At this equilibrium, for both players there is no sense to act i.e. to transmit and to jam. If the jamming cost is too big and the transmission cost is not too big then there is no sense for the jammer to jam. Of course, since there is no jamming threat and the transmission is not too costly then the user transmits packets safely. If both jamming and transmission costs are not too big, then mixed equilibrium arises where both players act equalizing chances for the opponent. We have established a conservation law for the activities of the user and the jammer. In particular, an increase of the jammer activity results in a decrease of the user activity. Furthermore, the conservation law is invariant with respect to the amount of data to send.

1.1 Organization of the Paper

The rest of this paper is organized as follows. In Section 2 and its subsection formulation and solution of the ALOHA game is given. Numerical modelling is performed in Section 3. In Section 4 the obtained results are generalized for the case where the channel can be in different states and change according to a Markov rule. Discussion of the obtain result and also a possible generalization of the game can be found in Section 5.

2 Model

We analyze an ALOHA-type medium access control protocol performance in the presence of a jammer with static channel state. We assume that a user wants to transmit a message of N packets in a time smaller or equal to T. In each slot the user sends a packet with probability p and the jammer tries to jam with probability q. The user obtains R as the reward only if he sends successfully the file of N packets within the time frame T, otherwise the user gets the zero reward.

For each transmission attempt, the sender will pay a cost C_T, and respectively, for each jamming attack, the jammer will pay C_J. Let V_i be the expected reward for total successful transmission when there are still i packets needed to be sent.

$$V_i = \max_p \min_q \left(-C_T p + C_J q + p(1-q)\lambda V_{i-1} + (1-p(1-q))\lambda V_i\right)$$

with $V_0 = R$, where R is the reward for successful transmission of all the packets.

Then the problem can be reformulated in the following multistage form:

$$V_i = \frac{1}{1-\lambda}\mathrm{val}\begin{matrix} \\ \mathrm{T} \\ \mathrm{N}\end{matrix}\begin{matrix} \mathrm{J} & \mathrm{N} \\ \left(\begin{matrix} -C_T + C_J \\ C_J \end{matrix}\right. & \left.\begin{matrix} \lambda(V_{i-1} - V_i) - C_T \\ 0 \end{matrix}\right)\end{matrix}, \tag{1}$$

where *val* means either maxmin or the value of the game if maxmin coincides with minmax.

We study now the optimal strategies of the players and the value of the game. The results are collected in Theorems 1– 3 and their proofs are supplied in Appendix.

First, we will show that if the transmission cost is too big then the game has a saddle point telling that for both players there is no sense to act (to transmit and to jam).

Theorem 1. *Let there be still i packets needed to be sent. Then (N, N) is a saddle point if and only if*

$$\lambda V_{i-1} \le C_T,$$

then $V_i = 0$.
In particular, if the transmission cost C_T is too big, namely,

$$\lambda R \le C_T \tag{2}$$

then $V_i = 0$, $i \ge 1$ and for both players there is no sense to act (to transmit and to jam).

Second, we will show that if jamming cost C_J is too big then there is no sense for the jammer to jam. Of course, since there is no jamming thread and the transmission is not too costly then the user transmit packets safely.

Theorem 2. *Let there be still i packets needed to be sent. Let assume that*

$$C_T < \lambda R \ \mathit{and}\ R\lambda(1-\lambda) + \lambda C_T \le C_J. \tag{3}$$

Then (T, N) is a saddle point for $i \le i_*$, (N, N) is a saddle point for $i > i_*$ and

$$V_i = \begin{cases} R\lambda^i - \dfrac{1-\lambda^i}{1-\lambda} C_T, & i \le i_* \\ 0, & i > i_*, \end{cases}$$

where i_* is given as follows:

$$i_* = \left\lceil \frac{\ln\left(\frac{C_T}{\lambda(R(1-\lambda)+C_T)}\right)}{\ln(\lambda)} \right\rceil. \tag{4}$$

Finally we will consider the case where jamming and transmission costs are not too big. Then mixed equilibrium arises in which both players act with some probabilities.

Theorem 3. *Let there be still i packets needed to be sent. Let*

$$C_T < \lambda R \text{ and } C_J < R\lambda(1-\lambda) + \lambda C_T. \tag{5}$$

Then the game has mixed equilibrium for $i < i_*$ where i_* is the minimal integer such that

$$V_{i-1} - \frac{C_J}{1-\lambda} + \sqrt{\left(V_{i-1} - \frac{C_J}{1-\lambda}\right)^2 + 4\frac{C_J C_T}{(1-\lambda)\lambda}} \leq \frac{2\max\{C_T, C_J\}}{\lambda}.$$

The value of the game for $i < i_*$ is given by

$$V_i = \frac{C_J}{1-\lambda}\left(1 - \frac{C_T}{\lambda}\frac{1}{V_{i-1} - V_i}\right). \tag{6}$$

The equilibrium mixed strategies $(p_i, 1-p_i)$ and $(q_i, 1-q_i)$ are given as follows:

$$p_i = \frac{C_J}{\lambda(V_{i-1} - V_i)}, \quad q_i = 1 - \frac{C_T}{\lambda(V_{i-1} - V_i)}.$$

For $i \geq i_*$ $V_i = 0$ and (N, N) is a saddle point.

It is interesting to note that there is a simple linear correlation between the probabilities to act for both players, namely

$$C_T p_i + C_J q_i = C_J. \tag{7}$$

This relation is independent of the number of number of packets left to send and moreover, this relation establishes a conservation law for the total activities of the user and the jammer in the regime of mixed strategies. In particular, an increase of the jammer activity results in a decrease of the user activity. Furthermore, this conservation law is invariant with respect to the amount of data to send.

3 Numerical Illustrations

As a numerical example consider situation with $\lambda = 0.9$ and $R = 1$. The value of the game in Figure 1 for transmission cost $C_T \in [0.001, 0.02]$, 2, 3 and 4 packets left to send and jamming cost $C_J = 0.005$ and $C_J = 0.01$. Also the optimal user strategy for transmission cost $C_T \in [0.001, 0.02]$, 2 and 4 packets left to send and jamming cost $C_J = 0.005$ and $C_J = 0.01$. The optimal jammer strategy and 2 and 4 packets left to send and jamming cost $C_J = 0.01$. We can restrict ourself mainly to the optimal use's strategies because a strong linear correlation between them and the jammer's strategies (7).

One can see that the value of the game and the optimal use's strategies are very sensitive to the changing of the environment. We observe that the activity of the user is decreasing with the cost of transmission, which is an intuitive result. Moreover, if jamming cost is decreasing then jammer activity arises (because the activity of the user decreases) and the value of the game goes down.

Finally, when transmission cost C_T increases, difference in user's payoff is increasing under different environment conditions.

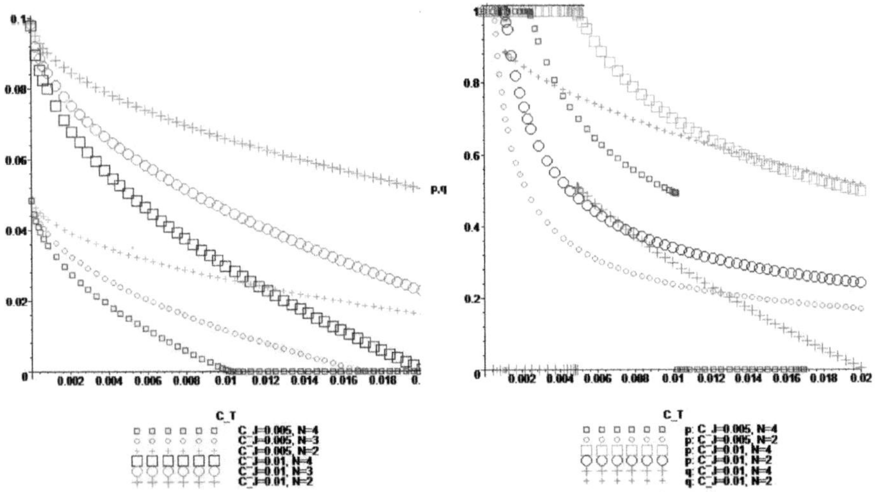

Fig. 1. The value of the game and the user's equilibrium strategy

4 Markov ALOHA Game

In this section we consider a variation of the game for the case where the channel can be in two states: good (1) and bad (0), and it can change its state according to a Markov rule. We denote by X_t the state of the channel at time slot t. Namely, with probability α_{xy}, $x, y = 0, 1$ the channel switches from state x to state y, i.e. $\text{Prob}(X_{t+1} = y | X_t = x)$. So, $\alpha_{x0} + \alpha_{x1} = 1$ and $\alpha_{0y} + \alpha_{1y} = 1$. We also assume that the probability of successful transmission in state x, if there is no jamming, is γ_x, where $\gamma_1 = 1$ and $\gamma_0 = \gamma$. So, if there is jamming then transmission is blocked with sure. If there is no jamming in the good channel state, then the transmission performs with sure and in the bad channel state it carries on with probability γ. Let $V_{i,x}$ be the expected reward for total successful transmission when there are still i packets needed to be sent and the channel is in state x. The action of the sender and the jammer depends on the state x of the channel. We define now by p_x (resp. q_x) the probability of transmission (resp. of jamming) when the channel is in state x. Then, in general case for $V_{i,x}$ the following maxmin equations hold:

$$\begin{aligned}V_{i,x} &= \max_{p_x}\min_{q_x}\Big\{-C_T p_x + C_J q_x \\ &+ \lambda\Big(p_x q_x(\alpha_{xx}V_{i,x} + \alpha_{xy}V_{i,y}) \\ &+ p_x(1-q_x)(\gamma_x\alpha_{xx}V_{i-1,x} + \gamma_x\alpha_{xy}V_{i-1,y}) \\ &+ p_x(1-q_x)((1-\gamma_x)\alpha_{xx}V_{i,x} + (1-\gamma_x)\alpha_{xy}V_{i,y}) \\ &+ (1-p_x)(\alpha_{xx}V_{i,x} + \alpha_{xy}V_{i,y})\Big)\Big\}\end{aligned} \tag{8}$$

with

$$V_{0,x} = R \tag{9}$$

and $\{x, y\} = \{0, 1\}$.

Then the problem can be reformulated in the following multistage form:

$$\begin{aligned}&V_{i,1} - \lambda(\alpha_{11}V_{i,1} + \alpha_{10}V_{i,0}) \\ &= \mathrm{val}\begin{matrix} \\ \mathrm{T} \\ \mathrm{N}\end{matrix}\begin{matrix}\mathrm{J} & \mathrm{N} \\ \left(\begin{matrix} -C_T + C_J \\ C_J\end{matrix}\right. & \left.\begin{matrix}\lambda(\alpha_{11}W_{i,1} + \alpha_{10}W_{i,0}) - C_T \\ 0\end{matrix}\right)\end{matrix},\end{aligned} \tag{10}$$

and

$$\begin{aligned}&V_{i,0} - \lambda(\alpha_{00}V_{i,0} + \alpha_{01}V_{i,1}) \\ &= \mathrm{val}\begin{matrix} \\ \mathrm{T} \\ \mathrm{N}\end{matrix}\begin{matrix}\mathrm{J} & \mathrm{N} \\ \left(\begin{matrix} -C_T + C_J \\ C_J\end{matrix}\right. & \left.\begin{matrix}\lambda\gamma(\alpha_{00}W_{i,0} + \alpha_{01}W_{i,1}) - C_T \\ 0\end{matrix}\right)\end{matrix},\end{aligned} \tag{11}$$

with

$$W_{i,x} = V_{i-1,x} - V_{i,x} \text{ for } x = 1, 2,$$

where *val* means either maxmin or the value of the game if maxmin coincides with minmax.

4.1 Solution of Markov ALOHA Game

In this Section we will find solution of the Markov ALOHA game. First note that straightforward from (10) and (11) and Theorem 8 we have the following result.

Theorem 4. *(N, N) is a saddle point for both state if and only if*

$$\alpha_{11}V_{i-1,1} + \alpha_{10}V_{i-1,0} \leq C_T/\lambda,$$
$$\alpha_{01}V_{i-1,1} + \alpha_{00}V_{i-1,0} \leq C_T/(\gamma\lambda).$$

In particular, if $V_{i,1} = 0$ then $V_{i,0} = 0$. Also, if the transmission cost C_T is too big, namely,

$$C_T \geq \lambda R$$

then there is no sense in transmission at all and so in jamming, then

$$V_{i0} = V_{i1} = 0, \quad i \geq 1.$$

So, we can assume now that

$$C_T < \lambda R$$

Then we have only for situation left to deal with:

(a) the jamming cost is too big that jammer does not jam in both state, so users can send packets safely,
(b) the jamming cost is too big for bad channel state and not to big for good channel state, so in bad channel state users stick to pure equilibrium strategies (T, N) meanwhile in the bad channel state users employ mixed equilibrium strategies,
(c) the jamming cost is not big and then the users acts according to mixed equilibrium strategies.

These three situations are described in the following theorems.

Theorem 5. *(T, N) is the saddle point for both states if and only if*

$$\begin{aligned} V_{i,1} &= \lambda(\alpha_1 V_{i,1} + \alpha_{10}V_{i,0}) - C_T, \\ V_{i,0} &= \lambda\frac{\gamma\alpha_{00} + \lambda(1-\gamma)\alpha_{01}\alpha_{10}}{1-\lambda(1-\gamma)\alpha_{00}}V_{i-1,0} \\ &\quad + \lambda\frac{\gamma\alpha_{01} + \lambda(1-\gamma)\alpha_{01}\alpha_{11}}{1-\lambda(1-\gamma)\alpha_{00}}V_{i-1,1} \\ &\quad - \frac{1+\lambda(1-\gamma)\alpha_{01}}{1-\lambda(1-\gamma)\alpha_{00}}C_T \end{aligned} \tag{12}$$

and

$$\frac{C_T}{\lambda} \leq \alpha_{11}(V_{i-1,1} - V_{i,1}) + \alpha_{10}(V_{i-1,0} - V_{i,0}) \leq \frac{C_J}{\lambda},$$
$$\frac{C_T}{\lambda\gamma} \leq \alpha_{01}(V_{i-1,1} - V_{i,1}) + \alpha_{00}(V_{i-1,0} - V_{i,0}) \leq \frac{C_J}{\lambda\gamma}$$

In particular for $i = 1$:

$$\begin{aligned} V_{1,1} &= \lambda R - C_T, \\ V_{1,0} &= \lambda R\frac{\gamma + \alpha_{01}\lambda(1-\gamma)}{1-\alpha_{00}\lambda(1-\gamma)} - \frac{1+\alpha_{01}\lambda(1-\gamma)}{1-\alpha_{00}\lambda(1-\gamma)}C_T \end{aligned} \tag{13}$$

and

$$R - \frac{C_J}{\lambda} \le \alpha_{11} V_{1,1} + \alpha_{10} V_{1,0} \le R - \frac{C_T}{\lambda},$$
$$R - \frac{C_J}{\lambda\gamma} \le \alpha_{00} V_{1,0} + \alpha_{01} V_{1,1} \le R - \frac{C_T}{\lambda\gamma}.$$

Now we consider the situation where both players in both states according to equilibrium apply mixed strategies.

Theorem 6. $(p_{i,x}, q_{i,x})$, $x = 0, 1$ *be the equilibrium in mixed strategy if and only if*

$$\begin{aligned}
p_{i,1} &= \frac{C_J}{\lambda(\alpha_{11} V_{i-1,1} + \alpha_{10} V_{i-1,0} - \alpha_{11} V_{i,1} - \alpha_{10} V_{i,0})},\\
q_{i,1} &= 1 - \frac{C_T}{\lambda(\alpha_{11} V_{i-1,1} + \alpha_{10} V_{i-1,0} - \alpha_{11} V_{i,1} - \alpha_{10} V_{i,0})},\\
p_{i,0} &= \frac{C_J}{\lambda\gamma(\alpha_{01} V_{i-1,1} + \alpha_{00} V_{i-1,0} - \alpha_{01} V_{i,1} - \alpha_{00} V_{i,0})},\\
q_{i,0} &= 1 - \frac{C_T}{\lambda\gamma(\alpha_{01} V_{i-1,1} + \alpha_{00} V_{i-1,0} - \alpha_{01} V_{i,1} - \alpha_{00} V_{i,0})},
\end{aligned}$$

and $V_{i,1}$ *and* $V_{i,0}$ *are solutions of equations*

$$\begin{aligned}
&V_{i,1} - \lambda(\alpha_{i1} V_{i,1} + \alpha_{10} V_{i,0})\\
&= C_J - \frac{C_T C_J}{\lambda(\alpha_{11} V_{i-1,1} + \alpha_{10} V_{i-1,0} - \alpha_{11} V_{i,1} - \alpha_{10} V_{i,0})},\\
&V_{i,0} - \lambda(\alpha_{01} V_{i,1} + \alpha_{00} V_{i,0})\\
&= C_J - \frac{C_T C_J}{\lambda\gamma(\alpha_{01} V_{i-1,1} + \alpha_{00} V_{i-1,0} - \alpha_{01} V_{i,1}) - \alpha_{00} V_{i,0})},
\end{aligned} \tag{14}$$

where the following conditions have to hold:

$$\max\{C_T, C_J\} \le \lambda[\alpha_{11}(V_{i-1,1} - V_{i,1}) + \alpha_{10}(V_{i-1,0} - V_{i,0})],$$

$$\max\{C_T, C_J\} \le \lambda\gamma[\alpha_{01}(V_{i-1,1} - V_{i,1}) + \alpha_{00}(V_{i-1,0} - V_{i,0})].$$

It is interesting to note that there is a simple linear correlation independent on the number packets left to send between the probabilities to act for both players in different channel states, namely

$$C_T p_{i,x} + C_J q_{i,x} = C_J, \quad x = 0, 1$$

which implies the fact that increasing activity of the jammer leads to reducing activity of the user at each channel state.

Finally we consider the situation with jamming cost which is to high to jam in the good channel state and at the same time it allows to jam in the good channel state.

Theorem 7. *$(p_{i,1}, q_{i,1})$ and (T, N) be the equilibrium for good and bad channel states if and only if*

$$V_{i,0} = \frac{\lambda\gamma(\alpha_{00}V_{i-1,0} + \alpha_{01}V_{i-1,1}) - C_T}{1 - \lambda(1-\gamma)\alpha_{00}} + \frac{\lambda(1-\gamma)\alpha_{01}V_{i,01}}{1 - \lambda(1-\gamma)\alpha_{00}}$$

and

$$\begin{aligned} &V_{i,1} - \lambda(\alpha_{i1}V_{i,1} + \alpha_{10}V_{i,0}) \\ &= C_J - \frac{C_T C_J}{\lambda(\alpha_{11}V_{i-1,1} + \alpha_{10}V_{i-1,0} - \alpha_{11}V_{i,1} - \alpha_{10}V_{i,0})}, \end{aligned}$$

where the following conditions have to hold:

$$\begin{aligned} &\frac{C_T}{\lambda} \le \alpha_{11}(V_{i-1,1} - V_{i,1}) + \alpha_{10}(V_{i-1,0} - V_{i,0}) \le \frac{C_J}{\lambda}, \\ &\frac{\max\{C_T, C_J\}}{\lambda} \le \alpha_{11}(V_{i-1,1} - V_{i,1}) + \alpha_{10}(V_{i-1,0} - V_{i,0}). \end{aligned}$$

Then, we have obtained, in a general framework, where the channel can be in good or bad state, the existence of different equilibrium even in pure or in mixed strategy. In the next section, we explore a particular asymmetric case for the transition probabilities.

4.2 A Particular Case: The Asymmetric Case

In this Section we consider in detail the asymmetric case $\alpha_{11} = \alpha_{01} = \alpha$ and $\alpha_{00} = \alpha_{01} = 1 - \alpha$. Then in the situation with mixed strategies in both states by Theorem 6 we have that

$$\begin{aligned} &V_{i,1} - \lambda(\alpha V_{i,1} + (1-\alpha)V_{i,0}) = C_J \\ &- \frac{C_T C_J}{\lambda(\alpha V_{i-1,1} + (1-\alpha)V_{i-1,0} - \alpha V_{i,1} - (1-\alpha)V_{i,0})}, \\ &V_{i,0} - \lambda(\alpha V_{i,1} + (1-\alpha)V_{i,0}) = C_J \\ &- \frac{C_T C_J}{\lambda\gamma(\alpha V_{i-1,1} + (1-\alpha)V_{i-1,0} - \alpha V_{i,1} - (1-\alpha)V_{i,0})}, \end{aligned}$$

Summing up the last two equations multiply by α and $1 - \alpha$ respectively, and substracting from the first equation the second one multiplied by γ we obtain the following two relations first of them give a recurrent formula for finding the expected value of payoff $\alpha V_{i,1} + (1-\alpha)V_{i,0}$ at different states, and the second one gives a strong linear correlation between payoffs:

$$(1-\lambda)(\alpha V_{i,1} + (1-\alpha)V_{i,0}) = C_J - \frac{C_T C_J(\alpha + (1-\alpha)/\gamma)}{\lambda(\alpha V_{i-1,1} + (1-\alpha)V_{i-1,0} - \alpha V_{i,1} - (1-\alpha)V_{i,0})}, \tag{15}$$

$$(1-\lambda(1-\gamma)\alpha)V_{i,1} = (\gamma + \lambda(1-\gamma)(1-\alpha))V_{i,0} + C_J(1-\gamma). \tag{16}$$

Then, subtracting (16) from (15) implies:

$$\alpha V_{i,1} + (1-\alpha)V_{i,0} = B$$

with

$$B := \frac{\alpha V_{i-1,1} + (1-\alpha)V_{i-1,0} + \dfrac{C_J}{1-\lambda}}{2} - \frac{1}{2}\left[\left(\alpha V_{i-1,1} + (1-\alpha)V_{i-1,0} - \frac{C_J}{1-\lambda}\right)^2 + 4\frac{(\gamma + \lambda(1-\gamma)(1-\alpha))C_J C_T}{(1-\lambda)\lambda}\right]^{1/2}$$

Thus, the optimal payoffs are given as follows:

$$V_{i,0} = \frac{1 - \lambda\alpha(1-\gamma)B - \alpha(1-\gamma)C_J}{1-\alpha(1-\gamma)},$$

$$V_{i,1} = \frac{(\gamma + \lambda(1-\gamma(1-\alpha)))B + (1-\alpha)(1-\gamma)C_J}{1-\alpha(1-\gamma)}$$

5 Discussion and Extensions

In this paper we suggested a development of the channel capacity concept for a dynamic slotted ALOHA network. We found maxmin successful transmission of an information over a dynamic communication channel. To do so, we analyzed a simple ALOHA-type medium access control protocol performance in the presence of a jammer as a zero-sum dynamic game. The obtained results are generalized for the case where the channel can be in different states and change according to a Markov rule. We considered only the simplest case the channel can be in two states: good (1) and bad (0). If there is jamming then transmission is blocked with sure. If there is no jamming in the good channel state, then the transmission performs with sure and in the bad channel state it carries on with a probability γ. The probabilities with which the channel switches from one state to the other are known and fixed. For this game also the recurrent formulas for finding the optimal solution are obtained. As the other direction of the investigation we are planning to deal with the uncomplete information case, say, when jamming cost and transmission costs are unknown to the rival correspondingly.

6 Appendix

Before solving our game (1) let us remind the following result [1] which supplies all the equilibrium for 2 × 2 matrix zero-sum game.

Theorem 8. *Let A be the zero-sum game with the following matrix:*

$$A = \begin{pmatrix} A_{11} & A_{12} \\ A_{21} & A_{22} \end{pmatrix}.$$

This game

(a) *either has a saddle point (each saddle point can be found as the an element of this matrix which is the minimal one in its row and it is the maximal one in its column),*
(b) *or a couple of mixed equilibrium strategies $(x, 1-x)$, $(y, 1-y)$ where*

$$x = \frac{A_{22} - A_{21}}{A_{11} - A_{12} + A_{22} - A_{21}},$$
$$y = \frac{A_{22} - A_{12}}{A_{11} - A_{12} + A_{22} - A_{21}},$$
$$v = \frac{A_{11}A_{22} - A_{12}A_{21}}{A_{11} - A_{12} + A_{22} - A_{21}}.$$

Note that the mixed equilibrium exists if and only if either

$$A_{11} > A_{12}, A_{12} < A_{22}, A_{22} > A_{21}, A_{21} < A_{11} \tag{17}$$

or

$$A_{11} < A_{12}, A_{12} > A_{22}, A_{22} < A_{21}, A_{21} > A_{11}. \tag{18}$$

In our case

$$\begin{aligned} A_{11} &= -C_T + C_J, \;\; A_{12} = \lambda(V(i-1) - V(i)) - C_T, \\ A_{21} &= C_J, \;\; A_{22} = 0. \end{aligned} \tag{19}$$

Then, only two pairs of strategies (N, N) and (T, N) could be saddle points in our game under some circumstance. Theorems 1 and 2 supply the condition under which either (N, N) or (T, N) is saddle point. Theorem 3 deals with the rest case, namely, where the mixed equilibrium arises.

Proof of Theorem 1: By (19) (N, N) presents a saddle point if and only if

$$\lambda(V_{i-1} - V_i) - C_T \le 0 \text{ for any } i.$$

and the result follows.

Proof of Theorem 2: By (19) (T, N) presents a saddle point if and only if

$$\lambda(V_{i-1} - V_i) - C_T > 0$$

and

$$C_J - C_T \geq \lambda(V_{i-1} - V_i) - C_T$$

which is equivalent to

$$C_T \leq \lambda(V_{i-1} - V_i) \leq C_J. \tag{20}$$

Thus, by (20), (T, N) is a saddle point if the jamming cost has to be bigger than the transmission one, namely

$$C_T \leq C_J. \tag{21}$$

Also, since (T, N) is a saddle point, by (1), we have that

$$V_i = \frac{1}{1-\lambda}\left(\lambda(V_{i-1} - V_i) - C_T\right).$$

Thus,

$$V_i = \lambda V_{i-1} - C_T. \tag{22}$$

Substituting (22) into (20) turns (20) into the following equivalent form:

$$\frac{C_T}{\lambda} \leq V_{i-1} \leq \frac{1}{\lambda(1-\lambda)}(C_J - \lambda C_T). \tag{23}$$

Now let have a look at (23) for $i = 1$. Since $V_0 = R$ then the left part of (23) is clear. The right part of (23) holds if

$$R\lambda(1-\lambda) + \lambda C_T \leq C_J. \tag{24}$$

Then by induction from (22) we can obtain that

$$V_i = R\lambda^i - \frac{1-\lambda^i}{1-\lambda} C_T \text{ while (23) holds.}$$

Also, since $\lambda C_T > R$ then (24) implies (21).
It is clear that V_i is decreasing function from $V_0 = R$ and $V_\infty = -C_T/(1-\lambda)$ and (4) holds, where i_* is the root of the equation

$$R\lambda^i - \frac{1-\lambda^i}{1-\lambda} C_T = \frac{C_T}{\lambda}$$

Finally note that by (3)

$$\max\{C_T, R\lambda(1-\lambda) + \lambda C_T\} = R\lambda(1-\lambda) + \lambda C_T.$$

This completes proof of Theorem 2.

Proof of Theorem 3: In this Theorem we want to find mixed strategies and the condition where they take place. Since by (19 $A_{21} = C_J > C_J - C_T = A_{11}$ then the situation (17) cannot hold. Also, $A_{22} = 0 < C_J = A_{21}$. Then conditions (18) are equivalent to the following two inequalities:

$$-C_T + C_J < \lambda(V_{i-1} - V_i) - C_T \text{ and } 0 < \lambda(V_{i-1} - V_i) - C_T.$$

Thus, we have the following condition for existence of equilibrium in mixed strategies:

$$\max\{C_T, C_J\} \le \lambda(V_{i-1} - V_i). \tag{25}$$

Then, by Theorem 8, we have that (6) holds. Introduce the following notation:

$$W_i = V_{i-1} - V_i.$$

In the new notation, (6) can be presented in the following way:

$$-W_i + V_{i-1} = \frac{C_J}{1-\lambda}\left(1 - \frac{C_T}{\lambda W_i}\right). \tag{26}$$

So,

$$W_i = \frac{V_{i-1} - \frac{C_J}{1-\lambda} \pm \sqrt{(V_{i-1} - \frac{C_J}{1-\lambda})^2 + 4\frac{C_J C_T}{(1-\lambda)\lambda}}}{2}.$$

Since, by (25), $W_i > 0$ from the last relation we have that

$$W_i = \frac{V_{i-1} - \frac{C_J}{1-\lambda} + \sqrt{(V_{i-1} - \frac{C_J}{1-\lambda})^2 + 4\frac{C_J C_T}{(1-\lambda)\lambda}}}{2}. \tag{27}$$

Then, substituting (27) into (25) implies the following equivalent presentation for (25) just in terms of V_{i-1}:

$$\frac{2C_T}{\lambda} \le V_{i-1} - \frac{C_J}{1-\lambda} + \sqrt{(V_{i-1} - \frac{C_J}{1-\lambda})^2 + 4\frac{C_J C_T}{(1-\lambda)\lambda}}. \tag{28}$$

Also, (27) yields that $V(i)$ has the form

$$V_i = \frac{V_{i-1} + \frac{C_J}{1-\lambda} - \sqrt{\left(V_{i-1} - \frac{C_J}{1-\lambda}\right)^2 + 4\frac{C_J C_T}{(1-\lambda)\lambda}}}{2} \tag{29}$$

and

$$V_{i-1} - V_i = \frac{V_{i-1} + \sqrt{\left(\frac{C_J}{1-\lambda} - V_{i-1}\right)^2 + 4\frac{C_J C_T}{(1-\lambda)\lambda}} - \frac{C_J}{1-\lambda}}{2} > 0. \tag{30}$$

Thus, V_i given by (29) is decreasing on i. Then by (27) W_i is also decreasing. Finally, we have to check whether (25) holds for $i = 1$. By (30) it is equivalent to

$$\sqrt{\left(\frac{C_J}{1-\lambda} - R\right)^2 + 4\frac{C_J C_T}{(1-\lambda)\lambda}} \ge \frac{C_J}{1-\lambda} - R + \frac{2\max\{C_T, C_J\}}{\lambda}. \tag{31}$$

Since for $C_T > C_J$ the inequality (31) is equivalent to $C_T \le \lambda R$, and for $C_T < C_J$ the inequality (31) is equivalent to $C_J \le \lambda C_T + \lambda R(1-\lambda)$ we have the following result supplying the value of the game. This completes proof of Theorem 3.

References

1. Owen, G.: Game Theory. W.B.Sanders, Philadelphia (1982)
2. Zander, J.: Jamming in Slotted ALOHA Multihop Packet Radio Networks. IEEE Trans. on Comm. 39(10) (1991)
3. Ma, R., Misra, V., Rubenstein, D.: An Analysis of Generalized Slotted-Aloha Protocols. IEEE/ACM Trans. on Networking 17(3) (2009)
4. Abramson, N.: The Aloha-system-another alternative for computer communications. In: AFIPS Conf. Proc., vol. 36, pp. 295–298 (1970)
5. Roberts, L.: Aloha Packet System with and without slots and capture. ACM SIGCOMM Comput. Comm. Rev. 5(2) (1975)
6. Jin, Y., Kesidis, G.: Equilibria of a Noncooperative Game for Heterogeneous Users of an ALOHA Network. IEEE Communications Letters 6(7) (2002)
7. Altman, E., ELAzouzi, R., Jimenez, T.: Slotted Aloha as a game with partial information. Computer Networks 45(6) (2004)
8. Menache, I., Shimkin, N.: Efficient Rate-Constrained Nash Equilibrium in Collision Channels with State Information. In: Proc. of INFOCOM 2008 (2008)
9. Menache, I., Shimkin, N.: Fixed-Rate Equilibrium in Wireless Collision Channels. In: Chahed, T., Tuffin, B. (eds.) NET-COOP 2007. LNCS, vol. 4465, pp. 23–32. Springer, Heidelberg (2007)
10. Sagduyu, Y.E., Ephremides, A.: A game-theoretic analysis of denial of service attacks in wireless random access. Journal of Wireless Networks 15, 651–666 (2009)
11. Xu, X., Trappe, W., Zhang, Y., Wood, T.: The Feasibility of Launching and Detecting Jamming Attacks in Wireless Networks. In: Proc. MobiHoc 2005, Urbana-Champaign, IL (May 2005)
12. Sagduyu, Y.E., Berry, R.A., Ephremides, A.: Wireless Jamming Attacks under Dynamic Traffic Uncertainty. In: Proc. of WiOpt 2010 (2010)
13. Sagduyu, Y.E., Berry, R.A., Ephremides, A.: Jamming Games for Power Controlled Medium Access with Dynamic Traffic. In: Proc. of ISIT 2010 (2010)
14. Altman, E., Avrachenkov, K., Garnaev, A.: A Jamming Game in Wireless Networks with Transmission Cost. In: Chahed, T., Tuffin, B. (eds.) NET-COOP 2007. LNCS, vol. 4465, pp. 1–12. Springer, Heidelberg (2007)
15. Altman, E., Avratchenkov, K., Garnaev, A.: Jamming Game with Incomplete Information about the Jammer. In: Proc. of GameComm 2009 (2009)
16. Alpcan, T., Basar, T.: A Game Theoretic Analysis of Intrusion Detection in Access Control Systems. In: Proc. of IEEE Conference on Decision and Control, Paradise Island, Bahamas (2004)
17. Altman, E., Hordijk, A.: Zero-sum Markov games and worst-case optimal control of queueing systems. QUESTA 21, 415–447 (1995), Special issue on optimization of queueing systems
18. Fall, K.: A delay-tolerant network architecture for challenged internets. In: Proc. of SIGCOMM 2003, pp. 27–34 (2003)
19. Hui, P., Chaintreau, A., Scott, J., Gass, R., Crowcroft, J., Diot, C.: Pocket switched networks and the consequences of human mobility in conference environments. In: Proceedings of ACM SIGCOMM First Workshop on Delay Tolerant Networking and Related Topics (2005)
20. Grossglauser, M., Tse, D.N.C.: Mobility increases the capacity of ad hoc wireless networks. IEEE/ACM Trans. Netw. 10(4), 477–486 (2002)
21. Sharma, G., Mazumdar, R., Shroff, N.B.: Delay and capacity trade-offs in mobile ad hoc networks: A global perspective. In: Proc. of INFOCOM (2006)

Nash Equilibria for Weakest Target Security Games with Heterogeneous Agents*

Benjamin Johnson[1,4], Jens Grossklags[2], Nicolas Christin[1,3], and John Chuang[4]

[1] CyLab, Carnegie Mellon University
[2] College of Information Sciences and Technology, Pennsylvania State University
[3] Information Networking Institute, Carnegie Mellon University
[4] School of Information, University of California, Berkeley

Abstract. Motivated attackers cannot always be blocked or deterred. In the physical-world security context, examples include suicide bombers and sexual predators. In computer networks, zero-day exploits unpredictably threaten the information economy and end users. In this paper, we study the conflicting incentives of individuals to act in the light of such threats.

More specifically, in the weakest target game an attacker will *always* be able to compromise the agent (or agents) with the lowest protection level, but will leave all others unscathed. We find the game to exhibit a number of complex phenomena. It does not admit pure Nash equilibria, and when players are heterogeneous in some cases the game does not even admit mixed-strategy equilibria.

Most outcomes from the weakest-target game are far from ideal. In fact, payoffs for most players in any Nash equilibrium are far worse than in the game's social optimum. However, under the rule of a social planner, average security investments are extremely low. The game thus leads to a conflict between pure economic interests, and common social norms that imply that higher levels of security are always desirable.

Keywords: Security, Economics, Game Theory, Heterogeneity.

1 Introduction

Motivated by observations about widespread and frequent security failures, Hal Varian started a conversation on the role of public goods dilemmas in the reliability and security context [19]. We continued this investigation by analyzing three canonical interdependency scenarios (i.e., weakest-link, average effort and best-shot) in the presence of two investment strategies [7]. Under the assumption of these particularly strong interdependencies, a failure to achieve a common

* This research was partially supported by CyLab at Carnegie Mellon under grant DAAD19-02-1-0389 from the Army Research Office, and by the National Science Foundation under ITR award CCF-0424422 (TRUST).

R. Jain and R. Kannan (Eds.): GameNets 2011, LNICST 75, pp. 444–458, 2012.

protection goal leads to a compromise of the entire network of agents. For example, in the weakest-link game the lack of protection effort by a single agent will immediately be exploited by an attacker to harm all agents.[1]

However, such strong interdependency effects are not always present in practice or attackers will not be able to exploit them efficiently under all circumstances. Similarly, even with significant protection investments a highly motivated attacker can rarely be fully blocked and deterred.[2] Rather, many situations result in asymmetric security consequences, i.e., some agents have their security violated while others remain unharmed.[3] We considered this case by proposing a novel strategic security interaction: the weakest-target game [7].[4] Here, an attacker will *always* be able to compromise the agent (or agents) with the lowest protection level, but will leave all others unscathed. Many financially motivated attacks can be explained by considering the weakest target game. For example, botnet herders need to compromise a large number of computing resources at low cost to implement several of their small margin business concepts (e.g., spam distribution [9]). For such purposes, miscreants frequently utilize relatively complex malware that supports a number of attack strategies [18]. A comprehensive defense against such malware becomes increasingly cumbersome because of the inclusion of zero-day exploits, the delivery via different channels (automated scans, email, peer-to-peer networks, ...) and the targeting of different operating systems.[5]

In this paper, we conduct an in-depth investigation of the weakest-target game considering homogeneous and heterogeneous agents. We add to our previous work [7,8] by deriving more general results and related proofs about properties of the game. In particular, we provide a non-existence proof for pure Nash equilibria, and exact conditions for mixed Nash equilibria for 2-player and N-player games under different parameter conditions. We also discuss important effects that result from the group dynamics inherent in the game.

The weakest target game is not anticipated to be a "bearer of good news" to the security community because it further exacerbates the conflicting incentives of defensive actors. On the one hand, agents, in their avoidance efforts to become the weakest target, may want to engage in significant security investments. On

---

[1] Variations can be considered with less strict contribution and associated failure conditions. See, for example, the literature on better-shot and weaker-link games [4].

[2] See, for example, the recent successful attacks against Google, Visa/Mastercard, and the US government.

[3] In complementary work, we also discussed this observation and distinguished more generally between tightly and loosely coupled networks [6].

[4] We discussed a simplified version of this game initially in the context of network economics [3].

[5] Consider, for example, the Stuxnet worm that carried four zero-day exploits. Further, it initially infected targets via USB drives, while newer information suggested that Stuxnet also replicated via computer networks. See:
`http://www.eweek.com/c/a/Security/Sophisticated-Stuxnet-Worm-Uses-4-Microsoft-Zeroday-Bugs-629672/`

the other hand, the availability of a cheap mitigation alternative weakens the incentives of all agents to invest in prevention. A similar effect occurs when the population of agents includes at least one subject that has little of value to lose. The result is a game with a particularly perverse set of incentives yielding very ineffective defensive strategies.

In the remainder of this paper, we first conduct a brief discussion on relevant concepts in classical game-theory and security economics. We then present the mathematical model and analysis before offering concluding remarks.

2 Related Work

The strategic aspects apparent in the weakest-target game are complementary to incentive structures analyzed in diverse games in the area of conflict studies. For example, in the Game of Chicken, two agents are driving a car towards a cliff. Both agents can agree to partake in a low-payoff safe choice by stopping early. Alternatively, one of them might dare to deviate by jumping from the car late causing an increase in her own reputation and a reduction of the other's social status. But if both players opt for the daring choice, the result is of disastrous proportions: aiming for an edge, both agents fall off the cliff [16]. In this prominent example for an anti-coordination game, there are two pure Nash equilibria with asymmetric appeal to the players. However, drivers can agree to disagree by selecting a mixed Nash to moderate between the agents' desires (e.g., [5]).

In the Prisoner's Dilemma, rational agents fail to cooperate when facing the enticement of individually beneficial defection from the socially desirable outcome [15]. Following these incentives, the players have to settle for a low-payoff Nash equilibrium. The Prisoner's Dilemma has been motivation for recent research work on interdependent security in which an agent can invest in protection effort against attacks directly targeting her, but is helpless if an attack is unknowingly spread by her peers [11]. Interdependent security games are helpful to better understand large group effects in the presence of misaligned or conflicting incentives. For example, these games exhibit strong tipping effects that can shift the economy of agents from full prevention efforts to passivity, and vice versa [10]. (See also [12] for an analysis with a similar focus.)

Recent work on linear influence networks also introduce complexity in the decision-making process. Linear influence networks allow for the fine-grained modeling of asset and risk interdependencies (see [13] and [14]). These studies offer an alternative approach to capturing diversity and heterogeneity of incentives for security decision-making. For further research considering the importance of network interdependencies, we also refer to a number of recent review efforts in the area of security economics [1,2,17].

3 Model

Each of $N \in \mathbb{N}$ players is responsible for choosing security investments for a single computer that is connected to other computers through a network. The network is subject to the risk of an external breach that occurs with exogenous probability $p \in (0, 1]$. If the network is breached, the attacker finds the player (or players) with the least amount of protection investment and obliterates her (their) computer(s).[6] All may not be lost though, as players may choose among two types of security investments to mitigate against damages of a successful breach. They may choose a protection investment, which benefits the public network, and is exemplified by investments such as installing antivirus software or firewalls; or they can choose a *self-insurance* investment, which benefits only the contributing user and is exemplified by an investment such as maintaining extensive private data backups [7]. A full protection investment costs $b_i \in \mathbb{R}^+$ to player i, and a full self-insurance investment costs $c_i \in \mathbb{R}^+$. Players may also choose a partial investment. The choice variables are thus a protection investment level $e_i \in [0, 1]$ and a self-insurance investment level $s_i \in [0, 1]$. Player i begins the game with an initial endowment $M_i \in \mathbb{R}^+$, and suffers a maximum loss of $L_i \in \mathbb{R}^+$ if a security breach occurs. The utility for player i as a result of the investment choice (e_i, s_i) is given by

$$U_i(e_i, s_i) = M_i - pL_i \cdot 1_{e_i \leq \min_{j \neq i} e_j} \cdot (1 - s_i) - be_i - cs_i \ , \qquad (1)$$

where $1_{e_i \leq \min_{j \neq i} e_j} = 1$ if $e_i \leq \min_{j \neq i} e_j$ and 0 otherwise.

4 Analysis

We begin by proving that the game does not admit a pure strategy Nash equilibrium. The result holds for any number of players, and assumes only that the cost and risk parameters (pL_i, c_i, b_i) are all positive.

We next provide a complete characterization of mixed-strategy Nash equilibria in the weakest target game with two players. In brevity, we show that, when either the parameters are symmetric, or when the maximum payoff of players is determined by protection costs rather than self-insurance costs, then exists a well-defined mixed strategy equilibrium. If parameters are not symmetric and self-insurance costs for one player are low, we show that a mixed strategy equilibria does not exist.

Finally, we address the case of N players. We begin by exhibiting a mixed strategy equilibrium in the homogeneous version. Then we derive a framework for addressing the full heterogeneous version.

4.1 No Pure Strategies

Theorem 1. *The weakest target game does not admit a pure strategy Nash equilibrium.*

[6] In previous work, we referred to this scenario as the weakest target game without mitigation. A slightly different version (with mitigation) allows agents to invest in full protection with the benefit of immunity from attacks [7].

Proof. We will divide into three parts the set of configurations in which each player plays a pure strategy. For each part we will then show that a strategy configuration of the prescribed type fails to be a Nash equilibrium.

First suppose that the strategy configuration has $e_i = 0$ for every i. In this case, the utility of each player i is $M_i - \min\{c_i, pL_i\}$. In such a configuration, if any one player i were to play $(e_i, s_i) = (\epsilon, 0)$ with $\epsilon < \min\{c_i, pL_i\}$, then she would no longer be a weakest target, and her payoff would improve to $M_i - \epsilon > M_i - \min\{c_i, pL_i\}$. Thus, a strategy configuration of this type is not a Nash equilibrium.

Next suppose that the strategy configuration has $e_i = x$ for every i, with $x > 0$. In this case, every player is a weakest target, the utility of player i is $M_i - b_i x - pL_i$. Player i could now improve her utility by playing $(e_i, s_i) = (0, 0)$, and reaping $M_i - pL_i > M_i - b_i x - pL_i$. Thus, such a strategy configuration cannot be a Nash equilibrium.

Finally, suppose the strategy configuration is such that some two players i and j have different protection investment levels (say $e_i < e_j$). In such a configuration, player j is not the weakest target, and thus has a utility of $M - b_j e_j$. If player j were to select a slightly lower investment level, say $x = \frac{e_j + e_i}{2}$, then the corresponding utility would become $M_j - b_j x > M_j - b_j e_j$. Thus, a configuration of this type cannot be a Nash equilibrium.

This exhausts all cases of pure strategies. We have shown that none of the cases is a Nash equilibrium strategy. Thus, no pure strategy Nash equilibrium can exist.

4.2 Mixed Strategies

Mixed Strategy Descriptions. Best response pure strategies for agent i always have one of the two forms: $(e_i, s_i) = (0, 1)$ or $(e_i, s_i) = (x, 0)$ for some $x \in [0, 1]$. So to describe a mixed strategy over this set of pure strategies, it suffices to specify the probability of playing $s_i = 1$ given $e_i = 0$, and a cumulative distribution function $F_i : \mathbb{R} \to [0, 1]$ defined such that $F_i[x]$ is the probability that $e_i < x$.[7]

We will dispense with giving the first part of the equilibrium conditions (involving self-insurance versus passivity in the case of no protection investment) because that part of the strategy is trivial to determine[8] and it does not affect whether the strategy is part of an equilibrium.[9] So, to describe a mixed strategy for player i, it suffices to define a non-decreasing left-continuous function $F_i : \mathbb{R} \to [0, 1]$ satisfying $F_i(x) = 0$ for $x \leq 0$ and $F_i(x) = 1$ for $x > 1$.

[7] The use of $<$ in our definition of F_i differs from the standard treatment of cumulative distributions, which uses $\leq$ instead. We adopt the former convention so that the probabilities we care about are easy to describe in terms of F_i.

[8] The structure of the game dictates that player i would self-insure in this instance if and only if $c_i \leq pL_i$

[9] The choice between self-insuring and remaining passive poses no externalities.

Mixed Strategy Equilibria for 2 Players

Theorem 2. *In a two-player weakest target game with parameter conditions satisfying* $\frac{\min\{b_1,c_1,pL_1\}}{b_1} = \frac{\min\{b_2,c_2,pL_2\}}{b_2}$, *the following mixed strategy is a Nash equilibrium.*

$$F_1(x) = \begin{cases} 0 & \text{for } x \le 0 \\ \frac{b_2 x}{pL_2} + 1 - \frac{\min\{c_2,pL_2\}}{pL_2} & \text{for } x \in \left(0, \frac{\min\{b_2,c_2,pL_2\}}{b_2}\right] \\ 1 & \text{for } x > \frac{\min\{b_2,c_2,pL_2\}}{b_2} \end{cases}$$

$$F_2(x) = \begin{cases} 0 & \text{for } x \le 0 \\ \frac{b_1 x}{pL_1} + 1 - \frac{\min\{c_1,pL_1\}}{pL_1} & \text{for } x \in \left(0, \frac{\min\{b_1,c_1,pL_1\}}{b_1}\right] \\ 1 & \text{for } x > \frac{\min\{b_1,c_1,pL_1\}}{b_1} \end{cases}$$

Proof. See Appendix 6.1

Theorem 3. *In a two-player weakest target game with parameter conditions satisfying* $\frac{\min\{b_1,c_1,pL_1\}}{b_1} < \frac{\min\{b_2,c_2,pL_2\}}{b_2}$ *and* $pL_1 \le c_1$, *the following mixed strategy is a Nash equilibrium.*

$$F_1(x) = \begin{cases} 0 & \text{for } x \le 0 \\ \frac{b_2 x}{pL_2} + 1 - \frac{b_2 L_1}{b_1 L_2} & \text{for } x \in \left(0, \frac{pL_1}{b_1}\right] \\ 1 & \text{for } x > \frac{pL_1}{b_1} \end{cases}$$

$$F_2(x) = \begin{cases} 0 & \text{for } x \le 0 \\ \frac{b_1 x}{pL_1} & \text{for } x \in \left(0, \frac{pL_1}{b_1}\right] \\ 1 & \text{for } x > \frac{pL_1}{b_1} \end{cases}$$

Proof. See Appendix 6.2

Theorem 4. *In the remaining parameter condition,* $\frac{\min\{b_1,c_1,pL_1\}}{b_1} < \frac{\min\{b_2,c_2,pL_2\}}{b_2}$ *and* $c_1 < pL_1$, *there does not exist a mixed-strategy Nash equilibrium.*

Proof. See Appendix 6.3

Mixed Strategy Equilibria for *N* Players. First we consider what happens in a game where the parameters are homogeneous – that is, $c_i = c$, $b_i = b$, and $L_i = L$. This scenario was addressed for a limited parameter range in [7].

Theorem 5. *In the case of homogeneous parameters, a Nash equilibrium occurs when each player plays the following mixed strategy.*

$$F_i(x) = \begin{cases} 0 & \text{for } x \le 0 \\ 1 - \sqrt[N-1]{\frac{\min\{c,pL\}-bx}{pL}} & \text{for } x \in \left(0, \frac{\min\{b,c,pL\}}{b}\right] \\ 1 & \text{for } x > \frac{\min\{b,c,pL\}}{b} \end{cases}$$

Proof. See appendix 6.4

With fully heterogeneous parameters, the analysis becomes significantly more complex. We derive a basic framework for a mixed equilibrium strategy involving N heterogeneous agents, but omit the exact specification of the necessary conditions for space reasons and will include them in the extended version.

Theorem 6. *The following mixed strategy describes an equilibrium in which each player j receives an expected utility of $M_j - \min\{c_j, pL_j\}$ for a continuous range of plays. (Assume WLOG that $\frac{\min\{b_1,c_1,pL_1\}}{b_1} \leq \frac{\min\{b_j,c_j,pL_j\}}{b_j}$ for $j > 1$).*

$$F_i(x) = \begin{cases} 0 & \text{for } x \leq 0 \\ 1 - \dfrac{\sqrt[N-1]{\prod_{j=1}^{N} \frac{\min\{c_j,pL_j\}-b_j x}{pL_j}}}{\frac{\min\{c_i,pL_i\}-b_i x}{pL_i}} & \text{for } x \in \left(0, \frac{\min\{b_1,c_1,pL_1\}}{b_1}\right] \\ 1 & \text{for } x > \frac{\min\{b_1,c_1,pL_1\}}{b_1} \end{cases}$$

Proof. See Appendix 6.5

4.3 Social Optimum: The Sacrificial Lamb

In the weakest target game, the least protected players always bear the brunt of the attack. The socially desirable outcome then ensures a minimization of the aggregate loss for all players. Typically, a planner will elect a specific agent, i.e., the sacrificial lamb, to invest in the smallest possible security effort to attract the attacker. If self-insurance is less costly than the loss from a security compromise, then the planner will invest in mitigation for the designated agent. Under heterogeneity assumptions, the social planner must undertake this comparison across all agents, i.e., she typically needs to identify the two agents who have the least to lose, $c_{min} = \min\{c_i\}$, and the lowest cost self-insurance option, $pL_{min} = \min\{pL_j\}$, respectively.[10] She will then select the agent with the lower of the two values. All other agents will merely invest in a token security effort, ϵ, to escape the attack.

With this strategy configuration, the total cost for all players is approximately $\min\{c_{min}, pL_{min}\} + \epsilon$. This is the best strategy possible up to a factor of epsilon.

5 Discussion

In the absence of the mediating presence of a social planner, agents are struggling to find a cheap way to ameliorate the threat from the attacker and the behavior of the agents can become quite complex. In the following, we illustrate two main behavioral trends related to the mixed strategy play identified in the analysis section.

First, when for all agents the cost of protection is lower than the cost of self-insurance, $b_i < c_i$, and preventive efforts are considered worthwhile, $b_i < pL_i$,

[10] It is, of course, possible that a single agent acts in both roles, $i = j$.

then agents' behavior can be described as a "race against the wall," i.e, they try to avoid a security compromise by selecting a very high protection effort. But according to the rules of the game an escape from the attacker's wrath is not possible and agents understand that collusion on the highest protection level would only benefit the attacker and would be wasteful from a cost perspective. Therefore, all agents probabilistically lower their security efforts to a certain degree with smaller probabilities assigned to lower protection levels. That is, agents engage in an implicit and tacit process of risk sharing. Theorem 2, when $b_i < min\{c_i, pL_i\}$, is an example for this type of behavior.

Second, a more nuanced behavior can be observed when the population includes at least one player with a low self-insurance cost, $c_i < b_i$ and $c_i < pL_i$, or a low potential loss, $pL_i < b_i$. We term this agent an implicit leader in the game. It is easy to see that she would not invest more than a certain threshold amount, $e_{max} = \frac{\min\{c_i, pL_i\}}{b_i}$, in protection efforts. And other players can infer that they will lose at most $b_j \cdot \left(\frac{\min\{c_i, pL_i\}}{b_i} + \epsilon\right)$ by investing a small amount more than e_{max} in preventive efforts. However, to achieve an equilibrium outcome close to this scenario the other players need to motivate the leader to invest in e_{max}, otherwise, the game play would unravel. That is, they need to ensure that for a whole range of parameters, the leader has an incentive to invest in protection efforts. In order to achieve this goal they need to "support the leader and share the burden" by engaging in probabilistic protection efforts below e_{max}. An example for this scenario is put forward in Theorem 3.

Adding more players and more heterogeneous preferences to the game play increases the complexity of the outcome. For example, if an agent is endowed with a extremely low cost of self-insurance it is very difficult to persuade her to act as a leader. In future work we will more thoroughly cover the nuances in the N-player version of the game, and illustrate the important behaviors with graphical representations. In the future, we also plan to engage in experimental validations of the predicted outcomes.

6 Conclusion

The weakest target game is interesting and well-motivated, but difficult to analyze. It does not admit pure Nash equilibria, and when players are heterogeneous in some cases the game does not even admit mixed-strategy equilibria. When mixed strategy equilibria do occur they are dominated by phenomena which we identified in the discussion section.

Most outcomes are far from ideal. In fact, *payoffs* for most players in any Nash equilibrium are far worse than in the game's social optimum. However, under the social planner rule *average security investments* are extremely low. This leads to a conflict between budgetary interests and a desire for increased security readiness.

References

1. Anderson, R., Moore, T.: The economics of information security. Science 314(5799), 610–613 (2006)
2. Böhme, R., Schwartz, G.: Modeling cyber-insurance: Towards a unifying framework. In: Proceedings of the Ninth Workshop on the Economics of Information Security (WEIS 2010), Cambridge, MA (June 2010)
3. Christin, N., Grossklags, J., Chuang, J.: Near rationality and competitive equilibria in networked systems. In: Proceedings of ACM SIGCOMM 2004 Workshop on Practice and Theory of Incentives in Networked Systems (PINS), Portland, OR, pp. 213–219 (August 2004)
4. Cornes, R., Sandler, T.: The theory of externalities, public goods, and club goods. Cambridge University Press, Cambridge (1986)
5. Dixit, A., Skeath, S.: Games of Strategy. Norton & Company, New York (1999)
6. Fultz, N., Grossklags, J.: Blue versus Red: Towards a Model of Distributed Security Attacks. In: Dingledine, R., Golle, P. (eds.) FC 2009. LNCS, vol. 5628, pp. 167–183. Springer, Heidelberg (2009)
7. Grossklags, J., Christin, N., Chuang, J.: Secure or insure? A game-theoretic analysis of information security games. In: Proceedings of the 2008 World Wide Web Conference (WWW 2008), Beijing, China, pp. 209–218 (April 2008)
8. Grossklags, J., Christin, N., Chuang, J.: Security and insurance management in networks with heterogeneous agents. In: Proceedings of the 9th ACM Conference on Electronic Commerce (EC 2008), Chicago, IL, pp. 160–169 (July 2008)
9. Kanich, C., Kreibich, C., Levchenko, K., Enright, B., Voelker, G., Paxson, V., Savage, S.: Spamalytics: An empirical analysis of spam marketing conversion. In: Proceedings of the Conference on Computer and Communications Security (CCS), Alexandria, VA, pp. 3–14 (October 2008)
10. Kearns, M., Ortiz, L.: Algorithms for interdependent security games. In: Thrun, S., Saul, L., Schölkopf, B. (eds.) Advances in Neural Information Processing Systems 16, pp. 561–568. MIT Press, Cambridge (2004)
11. Kunreuther, H., Heal, G.: Interdependent security. Journal of Risk and Uncertainty 26(2-3), 231–249 (2003)
12. Lelarge, M., Bolot, J.: Network externalities and the deployment of security features and protocols in the Internet. In: Proceedings of the ACM International Conference on Measurement and Modeling of Computer Systems (SIGMETRICS 2008), Annapolis, MA, pp. 37–48 (June 2008)
13. Miura-Ko, A., Yolken, B., Mitchell, J., Bambos, N.: Security decision-making among interdependent organizations. In: Proceedings of the 21st IEEE Computer Security Foundations Symposium (CSF 2008), Pittsburgh, PA, pp. 66–80 (June 2008)
14. Nguyen, K., Alpcan, T., Basar, T.: Stochastic games for security in networks with interdependent nodes. In: Proceedings of the International Conference on Game Theory for Networks (GameNets 2009), Istanbul, Turkey, pp. 697–703 (May 2009)
15. Rapoport, A., Chammah, A.: Prisoner's Dilemma: A Study in Conflict and Cooperation. Ann Arbor Paperbacks, University of Michigan Press (1965)
16. Rapoport, A., Chammah, A.: The game of chicken. American Behavioral Scientist 10(3), 10–28 (1966)

17. Roy, S., Ellis, C., Shiva, S., Dasgupta, D., Shandilya, V., Wu, Q.: A survey of game theory as applied to network security. In: Proceedings of the 43rd Hawaii International Conference on System Sciences (HICSS 2010), Koloa, HI, pp. 1–10 (January 2010)
18. Skoudis, E.: Malware: Fighting malicious code. Prentice Hall, Upper Saddle River (2004)
19. Varian, H.: System reliability and free riding. In: Camp, J., Lewis, S. (eds.) Economics of Information Security. Advances in Information Security, vol. 12, pp. 1–15. Kluwer Academic Publishers, Dordrecht (2004)

Appendix

6.1 Proof of Theorem 2

Proof. First note that under the specified parameter conditions, each F_i is a left-continuous increasing function on $\mathbb{R}$, that $F_i(x) = 0$ for $x \leq 0$ and that $F_i(x) = 1$ for $x \geq 1$. Thus F_i describes a valid mixed strategy for player i. To show that the mixed-strategy configuration is a Nash equilibrium, we will consider two parameter sub-cases separately.

Case 1: $b_i \leq \min\{c_i, pL_i\}$ for $i = 1, 2$
In this case, we have $\frac{\min\{b_1, c_1, pL_1\}}{b_1} = 1$. So the mixed strategies simplify to:

$$F_1(x) = \begin{cases} 0 & \text{for } x \leq 0 \\ \frac{b_2 x}{pL_2} + 1 - \frac{\min\{c_2, pL_2\}}{pL_2} & \text{for } x \in (0, 1] \\ 1 & \text{for } x > 1 \end{cases}$$

$$F_2(x) = \begin{cases} 0 & \text{for } x \leq 0 \\ \frac{b_1 x}{pL_1} + 1 - \frac{\min\{c_1, pL_1\}}{pL_1} & \text{for } x \in (0, 1] \\ 1 & \text{for } x > 1 \end{cases}$$

As the strategies are now symmetric it suffices to assume that player 1 is playing F_1 and show that player 2's response strategy is optimal. So assume player 1 is playing F_1.

– If player 2 plays $e_2 = 0$, her resulting payoff will be $M_2 - \min\{c_2, pL_2\}$.
– If player 2 plays $e_2 = x \in (0, 1]$, her resulting payoff will be $M_2 - b_2 x - pL_2(1 - F_1(x)) = M_2 - b_2 x - pL_2 \left(\frac{\min\{c_2, pL_2\}}{pL_2} - \frac{b_2 x}{pL_2} \right) = M_2 - \min\{c_2, pL_2\}$

We see that player 2 receives the same payoff regardless of her choice of strategies. Thus, playing a mixed strategy distribution over all possible strategies is an optimal response strategy, and hence the strategy configuration is a mixed-strategy equilibrium.

Case 2: $\min\{c_1, pL_1\} < b_1$ and $\frac{\min\{c_1, pL_1\}}{b_1} = \frac{\min\{c_2, pL_2\}}{b_2}$
In this case the mixed strategies simplify to:

$$F_1(x) = \begin{cases} 0 & \text{for } x \leq 0 \\ \frac{b_2 x}{pL_2} + 1 - \frac{\min\{c_2, pL_2\}}{pL_2} & \text{for } x \in \left(0, \frac{\min\{c_2, pL_2\}}{b_2}\right] \\ 1 & \text{for } x > \frac{\min\{c_2, pL_2\}}{b_2} \end{cases}$$

$$F_2(x) = \begin{cases} 0 & \text{for } x \le 0 \\ \frac{b_1 x}{pL_1} + 1 - \frac{\min\{c_1, pL_1\}}{pL_1} & \text{for } x \in \left(0, \frac{\min\{c_1, pL_1\}}{b_1}\right] \\ 1 & \text{for } x > \frac{\min\{c_1, pL_1\}}{b_1} \end{cases}$$

Again the strategies are symmetric so it suffices to assume player 1 is playing F_1 and consider the best response of player 2.

- Again, if player 2 plays $e_2 = 0$, her resulting payoff will be $M_2 - \min\{c_2, pL_2\}$.
- If player 2 plays $e_2 = x \in \left(0, \frac{\min\{c_2, pL_2\}}{b_2}\right]$, her resulting payoff will be $M_2 - b_2 x - pL_2(1 - F_1(x)) = M_2 - b_2 x - pL_2 \left(\frac{\min\{c_2, pL_2\}}{pL_2} - \frac{b_2 x}{pL_2}\right) = M_2 - \min\{c_2, pL_2\}$.
- If player 2 were to play $e_2 = x > \frac{\min\{c_2, pL_2\}}{b_2}$, her resulting payoff would be $M_2 - b_2 x - pL_2(1 - F_1(x)) = M_2 - b_2 x < M_2 - b_2 \cdot \frac{\min\{c_2, pL_2\}}{b_2} = M - \min\{c_2, pL_2\}$.

We see that the first two options yield the same payoff, and the third option yields a suboptimal payoff. Since response strategy for player 2 described by F_2 is a mixed strategy over pure strategies of only the first two forms, it is a best response strategy. We see again in this case that the mixed strategy configuration is a Nash equilibrium.

6.2 Proof of Theorem 3

Proof. Again F_i describes a valid mixed strategy for player i. Assume that player 1 is playing F_1 and consider the utility of player 2's response strategy.

- If player 2 were to play $e_2 = 0$, she would reap $M_2 - \min\{c_2, pL_2\}$. (Note that according to F_2 she plays this strategy with probability zero.)
- If player 2 plays $e_2 = x \in \left(0, \frac{pL_1}{b_1}\right]$, her payoff is $M_2 - b_2 x - pL_2(1 - F_1(x)) = M_2 - b_2 x - pL_2 \left(\frac{b_2 L_1}{b_1 L_2} - \frac{b_2 x}{pL_2}\right) = M_2 - \frac{b_2 L_1}{b_1} > M_2 - b_2 \cdot \frac{\min\{c_2, pL_2\}}{b_2} = M - \min\{c_2, pL_2\}$.
- Finally, if player 2 were to play $e_2 = x > \frac{pL_1}{b_1}$, her payoff would be $M_2 - b_2 x < M_2 - b_2 \cdot \frac{pL_1}{b_1}$.

We see that the optimal payoff player 2 can achieve is $M_2 - \frac{b_2 p L_1}{b_1}$. She achieves this utility exactly when she plays $x \in \left(0, \frac{pL_1}{b_1}\right]$; and this is exactly the set of strategies that she plays according to her mixed-strategy specification F_2.

Next assume that player 2 is playing F_2 and consider the utility of player 1's response strategy.

- If player 1 plays $e_1 = 0$, she reaps $M - pL_1$.
- If player 1 plays $e_1 = x \in \left(0, \frac{pL_1}{b_1}\right]$, she reaps $M_1 - b_1 x - pL_1(1 - F_2(x)) = M_1 - b_1 x - pL_1 \left(1 - \frac{b_1 x}{pL_1}\right) = M - pL_1$.

– Lastly, if player 1 plays $e_1 = x > \frac{pL_1}{b_1}$, she reaps $M_1 - b_1 x < M_1 - b_1 \cdot \frac{pL_1}{b_1} = M_1 - pL_1$.

We see that player 1 maximizes her utility by playing any of the first two strategy conditions, and this conforms to the prescription of F_1. So player 1 is playing an optimal response strategy.

This completes the proof that this strategy configuration is a Nash equilibrium.

6.3 Proof of Theorem 4

Proof. First note that the parameter conditions imply that $\frac{\min\{b_1, c_1, pL_1\}}{b_1} \neq 1$ and hence $c_1 < \min\{pL_1, b_1\}$.

Suppose that there does exist a mixed strategy Nash equilibrium under these parameter conditions. For $i = 1, 2$, let X_i be the set of pure strategies that occur in player i's mixed strategy; and let F_i be the cumulative distribution function for e_i in player i's mixed strategy, defined so that $F_i(x) = Pr[e_i < x]$. Note that each F_j in monotone non-decreasing and left-continuous.[11]

We next prove a sequence of lemmas that give more structure to the functions F_j. Ultimately, these lemmas will result in a contradiction involving the behavior of F_2 near the point $x = 0$, demonstrating that functions satisfying the prescribed properties cannot exist.

Lemma 1. *There exists a real number β with $0 < \beta \leq \frac{c_1}{b_1}$ such that for each j, $F_j(\beta) = 1$, but for every real number $\alpha < \beta$, $F_j(\alpha) < 1$.*

Proof. The utility of player 1's mixed strategy is at least $M_1 - c_1$. Hence any choice of $e_1 > \frac{c_1}{b_1}$ is deterministically suboptimal. Let β be the least upper bound on e_1 in player 1's mixed strategy. (More formally, we could define β to be the maximum element in $\overline{X_1}$). Then $\beta \leq \frac{c_1}{b_1}$. Interpreting the definition of least upper bound into the language of F_1, we also have $F_1(\alpha) < 1$ for every $\alpha < \beta$, and $F_1(\gamma) = 1$ for every $\gamma > \beta$.

Now, player 2 must have elements in her mixed strategy that take e_2 arbitrarily close to β from below. Otherwise, player 1 would have chosen an upper bound lower than β to obtain a better utility. Thus we have $F_2(\alpha) < 1$ for every $\alpha < \beta$. Also player 2 cannot have any part of her mixed strategy include a protection level strictly higher than β, for otherwise she would have preferred to reduce this expenditure by a small amount to be closer to β and obtain an improved utility. Thus $F_2(\gamma) = 1$ for every $\gamma > \beta$.

Next we see that player 1 cannot play the pure strategy $e_1 = \beta$ with positive probability. Otherwise, the discontinuity of F_1 at β would cause player 2 to

[11] Note that our use of $<$ as opposed to $\leq$ differs from the standard treatment of cumulative distribution functions. Our notation yields left-continuous as opposed to right continuous. The reason we use the formulation with $<$ is that we need to know when $e_i \geq x$; this is easily expressed algebraically in terms of the predicate $e_i < x$, but using $\leq$ would make it cumbersome.

receive a strictly higher payoff from playing $\beta + \epsilon$ (for sufficiently small ϵ) than from playing $\beta - \epsilon$. This contradicts the presumed optimality of player 2's mixed strategy, which contains plays greater than $\beta - \epsilon$ for every ϵ but no plays of $\beta + \epsilon$. Similarly $F_2(\beta) = 1$ by a completely analogous argument to the one above.

This completes the proof of the lemma.

Lemma 2. $\lim_{x \to 0^+} F_2(x) = 0$

Proof. In words, this lemma says that player 2's mixed strategy cannot contain a pure strategy component of the form $e_2 = 0$ with positive probability. To see this, observe that the maximum benefit player 2 can achieve from playing $e_2 = 0$ is $M_2 - \min\{c_2, pL_2\}$. However, using the result from the previous lemma, if player 2 were to play $e_2 = \frac{c_1}{b_1}$, she would fail to be the weakest target with probability 1, and would thus receive a utility of $M - b_2 \cdot \frac{c_1}{b_1}$. Since our parameter conditions imply $\frac{c_1}{b_1} < \frac{\min\{c_2, pL_2\}}{b_2}$, we have $M - b_2 \cdot \frac{c_1}{b_1} > M - b_2 \cdot \frac{\min\{c_2, pL_2\}}{b_2} = M - \min\{c_2, pL_2\}$. Thus playing $e_2 = 0$ is a suboptimal strategy for player 2. I.e., $\lim_{x \to 0^+} F_2(x) = 0$.

Lemma 3. *For $j \neq i$, and for $w, x \in X_j$, we have $F_j(w) = F_j(x) - \frac{b_i}{pL_i}(x - w)$.*

Proof. From the weakest target game definition, player j loses pL_j whenever $0 < e_j \leq e_i$, and this happens with probability $1 - F_i(e_j)$. We see that for $x > 0$ and for $j \neq i$ the utility of player j is directly related F_i via

$$U_j(x, 0) = M_j - b_j x - pL_j(1 - F_i(x)).$$

Now in a mixed strategy equilibrium all pure strategy components yield the same utility, hence for each $w, x \in X_i$ we have $U_i(w) = U_i(x)$. By rewriting the expression in terms of F_j, we obtain the result

$$F_j(w) = F_j(x) - \frac{b_i}{pL_i}(x - w).$$

Lemma 4. *For $j \neq i$, and for $w \in X_j$, we have $F_j(w) = 1 - \frac{b_i}{pL_i}(\beta - w)$.*

Proof. Since each F_j is left continuous, we have $\lim_{\alpha \to \beta^-} F_j(\alpha) = F_j(\beta) = 1$. Let $w \in X_j$ and let $\langle \alpha_n \rangle_{n \in \mathbb{N}}$ be a sequence from X_j that converges to β. (Such a sequence exists in X_j from the arguments given in Lemma 1.) Then from Lemma 3 we have $F_j(w) = F_j(\alpha_n) - \frac{b_i}{pL_i}(\alpha_n - w)$ for each n. Taking the limit of both sides yields

$$F_j(w) = 1 - \frac{b_i}{pL_i}(\beta - w).$$

Lemma 5. *For $j \neq i$, and for $z \in (0, \beta)$, we have $F_j(z) \geq 1 - \frac{b_i}{pL_i}(\beta - z)$.*

Proof. If $z \notin X_i$, then we cannot use equality of utilities, but using Lemma 4 and the properties of F_j as a cumulative distribution, we have for $z \in (0, \beta)$:

$$\begin{aligned} F_j(z) &= \inf\{F_j(w) : w \in X_j \text{ and } w > z\} \\ &= \inf\left\{1 - \frac{b_i}{L_i}(\beta - w) : w \in X \text{ and } w > z\right\} \\ &\geq 1 - \frac{b_i}{pL_i}(\beta - z) \end{aligned}$$

Finally, for the punchline,

Using Lemma 5, for every $\epsilon \in (0, \beta)$ we have

$$\begin{aligned} F_2(\epsilon) &\geq 1 - \frac{b_1}{pL_1}(\beta - \epsilon) \\ &= 1 + \frac{b_1}{pL_1}\epsilon - \frac{b_1}{pL_1}\beta \\ &> 1 - \frac{b_1}{pL_1} \cdot \frac{c_1}{b_1} \\ &= 1 - \frac{c_1}{pL_1} \end{aligned}$$

In particular, $\lim_{\epsilon \to 0^+} F_2(\epsilon) \geq 1 - \frac{c_1}{pL_1} > 0$. This contradicts the conclusion of Lemma 2.

We conclude that no mixed strategy equilibrium can exist.

6.4 Proof of Theorem 5

Proof. F_i describes a valid mixed strategy for player i. Assume that all players $j \neq i$ are playing F_j and consider the best response of player i.

- If player i plays $e_i = 0$, she reaps $M - \min\{c, pL\}$.
- If player i plays $e_i = x \in \left(0, \frac{\min\{b,c,pL\}}{b}\right]$, her payoff is $M - bx - pL \cdot \prod_{j \neq i}(1 - F_j(x)) = M - bx - pL \cdot \prod_{j \neq i}\left(\sqrt[N-1]{\frac{\min\{c,pL\}-bx}{pL}}\right) = M - bx - pL \cdot \frac{\min\{c,pL\}-bx}{pL} = M - \min\{c, pL\}$.
- Finally, if player i were to play $e_i = x > \frac{\min\{b,c,pL\}}{b}$, then this is possible only if $\min\{b, c, pL\} = \min\{c, pL\}$, and in this case, her payoff would be at most $M - bx < M - b \cdot \frac{\min\{c,pL\}}{b} = M - \min\{c, pL\}$.

We see that the optimal payoff player i can achieve is $M - \min\{c, pL\}$. She achieves this utility exactly when she plays $x \in \left[0, \frac{\min\{b,c,pL\}}{b}\right]$; and this is exactly the set of strategies that she plays according to her mixed-strategy specification F_i.

6.5 Proof of Theorem 6

Proof. The algebraic part of the formula is derived as follows:

$$M_i - \min\{c_i, pL_i\} = M_i - b_i x - pL_i \prod_{j\neq i}(1 - F_j(x))$$

$$\frac{\min\{c_i, pL_i\} - b_i x}{pL_i} = \prod_{j\neq i}(1 - F_j(x))$$

$$\prod_{k=1}^{N} \frac{\min\{c_k, pL_k\} - b_k x}{pL_k} = \prod_{k=1}^{N} (1 - F_k(x))^{N-1}$$

$$\frac{\prod_{k=1}^{N} \frac{\min\{c_k, pL_k\} - b_k x}{pL_k}}{\left(\frac{\min\{c_i, pL_i\} - b_i x}{pL_i}\right)^{N-1}} = \frac{\prod_{k=1}^{N} (1 - F_k(x))^{N-1}}{\left(\prod_{j\neq i}(1 - F_j(x))\right)^{N-1}}$$

$$\frac{\prod_{k=1}^{N} \frac{\min\{c_k, pL_k\} - b_k x}{pL_k}}{\left(\frac{\min\{c_i, pL_i\} - b_i x}{pL_i}\right)^{N-1}} = (1 - F_i(x))^{N-1}$$

$$\sqrt[N-1]{\frac{\prod_{k=1}^{N} \frac{\min\{c_k, pL_k\} - b_k x}{pL_k}}{\left(\frac{\min\{c_i, pL_i\} - b_i x}{pL_i}\right)^{N-1}}} = (1 - F_i(x))$$

$$1 - \sqrt[N-1]{\frac{\prod_{k=1}^{N} \frac{\min\{c_k, pL_k\} - b_k x}{pL_k}}{\left(\frac{\min\{c_i, pL_i\} - b_i x}{pL_i}\right)^{N-1}}} = F_i(x)$$

Noisy Mean Field Game Model for Malware Propagation in Opportunistic Networks

Hamidou Tembine, Pedro Vilanova, and Mérouane Debbah

Ecole Supérieure d'Electricité,
Chaire Alcatel-Lucent en radio-flexible,
Supelec, France
firstname.familyname@supelec.fr
AMCS, KAUST, Saudi Arabia

Abstract. In this paper we present analytical mean field techniques that can be used to better understand the behavior of malware propagation in opportunistic large networks. We develop a modeling methodology based on stochastic mean field optimal control that is able to capture many aspects of the problem, especially the impact of the control and heterogeneity of the system on the spreading characteristics of malware. The stochastic large process characterizing the evolution of the total number of infected nodes is examined with a noisy mean field limit and compared to a deterministic one. The stochastic nature of the wireless environment make stochastic approaches more realistic for such types of networks. By introducing control strategies, we show that the fraction of infected nodes can be maintained below some threshold. In contrast to most of the existing results on mean field propagation models which focus on deterministic equations, we show that the mean field limit is stochastic if the second moment of the number of object transitions per time slot is unbounded with the size of the system. This allows us to compare one path of the fraction of infected nodes with the stochastic trajectory of its mean field limit. In order to take into account the heterogeneity of opportunistic networks, the analysis is extended to multiple types of nodes. Our numerical results show that the heterogeneity can help to stabilize the system. We verify the results through simulation showing how to obtain useful approximations in the case of very large systems.

1 Introduction

In modern times, the massive use of information formed an interconnected global society of billions in which communication systems are vital infrastructures. Among them, wireless communications is without doubt one of the most explosive developments ever to have taken place in the telecommunications industry. At the same time, over the last years, we have seen the power of microprocessors double about every 18 months, becoming considerably smaller, cheaper and abundant; indeed, they are ubiquitous and are even finding their way into everyday objects. Those technology trends - tiny, cheap processors with integrated

R. Jain and R. Kannan (Eds.): GameNets 2011, LNICST 75, pp. 459–474, 2012.

sensors and wireless communications formed the technological basis for a new era of vast number of smart objects that communicate via wireless links.

Those incredibly diverse and complex wireless communication networks brought several fundamental technical issues: networks must be designed to carry out the intended functions in an efficient and predictable way; they must be manageable and upgradeable; and the most of all: reliable. Today, one of the major emerging threat against reliability is malware, that is, malicious self-replicating code. Threats posed by malware ranges from attacks against the confidentiality of the communication to attacks that actually alter the information traffic, hence destroying the integrity of the network.

Recently, malware outbreaks designed for personal computer environments on wired networks like those of Slammer and Code Red worms over the Internet have already inflicted severe economic damages, infecting thousands of hosts in short periods of time. Then, it is of vital importance to predict the limits of the damages that the attackers can inflict in large wireless networks, subject to its fundamental limitations, such as limited energy, unreliable communications and topology changes due mobility.

In this work we illustrate how mean field approaches can be used to reduce the complexity in the analysis of the damage that can be inflicted in a large network with opportunistic interaction of the objects. Opportunistic networking is one of the emerging communication paradigms in wireless mobile communications, in which the communication opportunities are based on sporadic and intermittent contacts. In contrast with the extensive work of malware propagation in wired networks ([1]), large wireless networks have not yet received similar attention, with a few exceptions ([2,3]).

Our Contribution

Contributions in this paper are twofold. First, from the theoretical point of view, we present a novel mean field approach, which tries to overcome one of the limitations of the "classical" mean field, that is, the approximation of an inherently stochastic system with a *deterministic* representation (ordinary differential equation) [4]. We propose a new, more general approach in which we preserve the main advantage of the classic mean field, that is, the reduction of the number of parameters in the analysis of large systems, but adding a random or "noisy" component. This new addition could lead to a more realistic mathematical model of the original situation, for example when many local object transitions occurs at the same time making the second moment unbounded. In this context the work [5] is not applicable anymore because the second moment of the number of object transitions per time slot may not vanish when the number of objects goes to infinity. A typical scenario is when many players do parallel transitions. The idea of the proposed analysis is that, if the third order in the Taylor approximation of the regular function of the mean field is bounded, then the noise may not be negligible but a convergence to a stochastic mean field limit can be established. Inspired from the work of [6] based on multidimensional diffusion processes, we were able to establish a mean field convergence to *non-deterministic* differential equations and extend the previous works in mean field interaction models (with and without controls). This new mean field limit which is stochastic is called *"noisy" mean field limit* and applied in this work to malware propagation in opportunistic networks.

Second, from the malware propagation modelling point of view, we extend the model developed in [7] in which the types are not used and the impact of the control parameters are not specially studied. This leads to a limitation in the results obtained, because different types of systems could lead for example to slower rates of propagation. They can represent, for example, different operating systems, different versions of the operating systems or patched/unpatched version of the same operating system. To the best of our knowledge, in most of the related work about malware spreading in large networks authors do not model the heterogeneity of the network.

Most of the mean field studies do not examine consider control framework. We observe that control parameters are important in the mean field limit since they give new insights to uncontrolled mean field framework which may be constrained (energy limitation). This helps in controlling the proportion of infected nodes.

Organization

The rest of the paper is structured as follows. In next section we present basics on stochastic games with random set of interacting players. In section **??**, we overview existing mean field models in discrete time. We develop a controlled mean field framework in section 2 and in section 3 we provide a general convergence to mean field which is characterized by a stochastic differential equation and the payoff evolution are solution of partial differential equations. Finally, we apply the *noisy mean field* framework to opportunistic wireless large networks.

2 Controlled Mean Field Interaction Model

In this section, we introduce a controlled mean field interaction model. The finite version of this model is a particular case of stochastic games with individual states. We restrict our attention into a particular class of behavioral strategies within we are able to establish the mean field convergence. This restriction is due to the fact that when the number goes to infinity, the dimension of the set of stationary strategies goes to infinity as well as. By letting the size of the system go to infinity, the discrete stochastic game problem is replaced by a limit of a system of Hamilton-Jacobi-Bellman equations coupled with a mean field limit ODE or coupled system of Bellman-Shapley optimality and discrete mean field evolution, that are deterministic and where the dimensionality of the original system has been transformed in the mass-behavior of the system.

Time $t \in \mathbb{N}$ is discrete. The *global state* of the system at time t is $(S(t), X^n(t)) = (S(t), X_1^n(t), ..., X_n^n(t))$. Denote by $A^n(t) = (A_1^n(t), \ldots, A_n^n(t))$ the action profile at time t. The system $(S(t), X^n(t))$ is Markovian once the action profile $A^n(t)$ are drawn under Markovian strategies. We denote the set of Markovian strategies by $\mathcal{U}$. $M_x^n(t)$ is the fraction of players who belong to the population of individual state x. Similarly, we associate the process $U_a^n(t) = \frac{1}{n}\sum_{j=1}^n \mathbb{1}_{\{A_j^n(t)=a\}}$ to the fraction of actions.

Strategies and Random Set of Interacting Players

At time slot t, an ordered list $\mathcal{B}_t^n$, of players in $\{1, 2, \ldots, n\}$, is randomly selected without repetition as follows: First we draw a random number of players k_t such that $\mathbb{P}(|\mathcal{B}_t^n| = k \mid M^n(t) = m) =: J_k^n(m)$ where the distribution $J_k^n(m)$ is given for any n, $m \in \{0, \frac{1}{n}, \frac{2}{n}, \ldots, 1\}^{|\mathcal{X}|}$. Second, we set $\mathcal{B}_t^n$ to an ordered list of k_t players drawn uniformly at random among the $n(n-1)...(n-k_t+1)$ possible ones.

Each player such that $j \in \mathcal{B}_t^n$ takes part in a one-shot interaction at time t, as follows. First, each selected player $j \in \mathcal{B}_t^n$ chooses an action $a_{j,t} \in \mathcal{A}(s, x_j)$ with probability $u(a_j \mid s, x_j)$ where (s, x_j) is the current player state. The stochastic array u is the strategy profile of the population.

Denote the current set of interacting players $\mathcal{B}_t^n = \{j_1, \ldots, j_k\}$. Given the actions $a_{j_1}, ..., a_{j_k}$ drawn by the k players, we draw a new set of individual states $(x'_{j_1}, ..., x'_{j_k})$ and resource state s' with probability $L_{s;s'}^n(k, m, a)$, where a is the vector of the selected actions by the interacting players.

We assume that for any given Markovian strategy, the transition kernel L^n is invariant by any permutation of the index of the players within the same type. This implies in particular that the players are only distinguishable through their individual state. Moreover, this means that the process $M^n(t)$ is also Markovian once the sequence of strategy is given. Denote by $w_{s,s'}^n(u, m)$ the marginal transition probability between the resource states. Given any Markov strategy and any vector m of Δ, the resource state generates an irreducible Markov decision process with limiting invariant measure $w_s(u, m)$. Then, we can simplify the analysis by fixing the resource state $S(t) = s$ without losing generality. The model is entirely determined by the probability distributions J^n, the transition kernels L^n and the strategy profile u.

3 Noisy Mean Field Approach

We provide a general convergence result of the mean field to a stochastic differential equation and a martingale problem is formulated for the the law of the process M_t^n. We are able to establish a mean field convergence to non-deterministic differential equations, thus, extending the previous works in mean field interaction, in mean field Markov decision teams, in mean field Markov games [5,8,9,10]. We show that even if the expected number of players that do a transition in one time slot is not bounded, one can have a mean field limit, in such a case a stochastic one. This mean field limit is referred as *noisy mean field*.

Before presenting the main theoretical results of this paper, we first introduce some preliminary notions. The evolution of the system depends on the decision of the interacting players. Given a history $h_t = (S(0), X^n(0), A^n(0), \ldots, S(t) = s, X^n(t), A^n(t))$. $X^n(t+1)$ evolves according to the transition probability

$$L^n(x'; x, u, s) = \mathbb{P}\left(X^n(t+1) = x' \mid h_t\right)$$

The term $L^n(x'; x, u, s)$ is the transition kernel on $\mathcal{X}^n$ under the strategy U^n. Let $x^n = (x_1^n, \ldots, x_n^n)$ such that $\frac{1}{n}\sum_{j=1}^n \delta_{x_j^n} = m$ and define

$$\mathcal{L}^n(m'; m, u, s) := \sum_{\substack{(x'_1,\ldots,x'_n)\\ \frac{1}{n}\sum_{j=1}^n \delta_{x'_j}=m'}} L^n(x'; x, u, s)$$

$$= \mathbb{P}(M^n(t+1)=m' \mid M^n(t)=m, U^n(t)=u, S(t)=s) = \mathbb{P}(M^n(t+1)=m' \mid \tilde{h}_t)$$

where $\tilde{h}_t = (S(t'), X^n(t'), A^n(t'), t' \leq t, S(t)=s, X^n(t)=x^n)$.The term $\mathcal{L}^n(m'; m, u, s)$ corresponds to the projected kernel of L^n. Below we give sufficient conditions on the transition kernels to get a weak convergence of the process M_t^n under the strategy $U^n(t)$.

We now present the main assumptions of this paper.

A1: $w_s(u, m)$ is continuously differentiable in m and u. Note that this assumption was already implicit in the approach of [7] via the smoothness of the drift.
A2: There exists a continuous mapping $f : \mathbb{R}^{|\mathcal{X}|} \times \mathcal{U} \times \mathcal{S} \longrightarrow \mathbb{R}^{|\mathcal{X}|}$ such that $\forall s \in \mathcal{S}$;

$$\lim_n \sup_{u\in\mathcal{U}} \sup_{\|m\|\leq 1} \left\| \frac{f^n(m, u, s)}{\delta_n} - f(m, u, s) \right\| = 0$$

where

$$f_x^n(m, u, s) = \int_{m'\in\mathbb{R}^{|\mathcal{X}|}} 1\!\!1_{\{\|m'-m\|\leq 2\}}(m'_x - m_x)\mathcal{L}^n(dm'; m, u, s),$$

for $x \in \mathcal{X}$, and $s \in \mathcal{S}$. This assumption is analogous to H2 in which control parameters are added.
A3: There exists $\delta_n \searrow 0$ and a continuous mapping $a : \mathbb{R}^{|\mathcal{X}|} \times \mathcal{U} \times \mathcal{S} \longrightarrow \mathbb{R}^{|\mathcal{X}|\times|\mathcal{X}|}$ such that $\forall s \in \mathcal{S}$;

$$\lim_n \sup_{u\in\mathcal{U}} \sup_{\|m\|\leq 1} \left\| \frac{a^n(m, u, s)}{\delta_n} - a(m, u, s) \right\| = 0$$

where

$$a_{x,x',s}^n(m, u, s) = \int_{m'\in\mathbb{R}^{|\mathcal{X}|}} 1\!\!1_{\{\|m'-m\|\leq 2\}}(m'_x - m_x)(m'_{x'} - m_{x'})\mathcal{L}^n(dm'; m, u, s),$$

for $(x, x', s) \in \mathcal{X}^2 \times \mathcal{S}$, where the third moment is finite. Note that under H3, $a \equiv 0$.

A4: For all $\epsilon > 0$ and $\forall s \in \mathcal{S}$;

$$\lim_{n} \sup_{u \in \mathcal{U}} \frac{1}{\delta_n} \int_{m' \in \mathbb{R}^{|\mathcal{X}|}} 1\!\!1_{\{\|m'-m\|>\epsilon\}} \mathcal{L}^n(dm'; m, u, s) = 0$$

Note that assumption H3 and lemma 3 from [7], implies $A4$.
A4': $\forall s \in \mathcal{S}$;

$$\sup_{u \in \mathcal{U}} \sup_{m \in \mathbb{R}^{|\mathcal{X}|}} \sup_{n \geq 1} \left[\| \frac{a^n(m,u,s)}{\delta_n} \| + \| \frac{f^n(m,u,s)}{\delta_n} \| \right] < \infty$$

The smoothness assumptions H4-H5 in [7] imply $A4'$ which implies $A4$ too.

Theorem 1. *Let $M_0^n \longrightarrow \pi$ in law where π is a probability measure. Under A1-A4 [resp. A1-A4'], the process M_t^n converges in law to a weak [resp. strong] solution of the Itô stochastic differential equation given by*

$$dm = \tilde{f}(m,u)dt + \tilde{\sigma}(m,u)dW_t$$

where $\tilde{\sigma}\tilde{\sigma}^t = \tilde{a}$, $\tilde{f}(m,u) = \sum_s w_s(m,u) f(m,u,s)$, $\tilde{a}(m,u) = \sum_s w_s(m,u)$ $a(m,u,s)$ and $w(m,u)$ is the invariant probability of the transition matrix $L(m,u)$

A proof can be found in [11].

Remark. This result generalizes the deterministic mean field limit conditions established in [7] for $\mathcal{U}$ equal to a singleton. It generalizes also the deterministic controlled mean field dynamics obtained in [5] for stationary strategies.

Remark. The conditions $A1 - A4, A4'$ are weaker than those given in [7]. Under the conditions in [7], the noise term $\frac{1}{\delta_n} a^n \longrightarrow 0$ when n goes the infinity. Moreover, the continuity assumption on the drift limit is not needed. If $\tilde{f}$ admits a unique integral curve, and $\tilde{a}$ is bounded and continuous then, the result applies as well. This allows us to apply it in wide range of networking scenarios with discontinuous drift limit but lower semi-continuity properties. Note that the uniformity in u may not be satisfied. In that case, a local mean field solution is derived. An example of such discontinuity is provided in [5]. Our result extends also the convergence theorem in [5] which was restricted to stationary strategies. Here u_t is an admissible strategy at time t.

Definition 1 (Individual optimization framework). *To a game as defined above, we associate a macroscopic population game, defined as follows. Each member j of the population, with state $X_j(t)$ and a population profile $m[u](t)$. The initial condition of the game is $S(0) = s, X_j(0) = x$, $m[u](0) = m_0$. The population profile is solution to the noisy mean field evolution and $S(t), X_j(t)$ evolves as a jump process given by the marginal of q which depends on $m(t)$, $u(t)$ and the strategy $u'(t)$ of j. Further, let $V_T(s,x,u',u,m)$ be the $T-$stage payoff of player j in this game, given that $S(0) = s, X_j(0) = x$ and $m(0) = m_0$, i.e. $V_T(s,x,u',u,m) =$*

$$\mathbb{E}\left(g(m(T)) + \int_0^T r(S(t), X_j(t), u'(t), u(t), m(t))dt \middle| s, x, m_0 \right)$$

where m is solution of the Itô SDE

$$dm(t) = \tilde{f}(m(t), u(t))dt + \tilde{\sigma}(m(t), u(t))dW_t \tag{1}$$

Using Itô's formula [12] the expected payoff evolution for a fixed horizon T is then given by

$$\begin{cases} v_{u',u}(T,m) = g(m) \\ -\partial_t v_{u',u}(t,m) = \bar{r}(t,m,u,u') + \sum_{x\in\mathcal{X}} \tilde{f}_x(m,u)\partial_{m_x} v_{u',u}(t,m) \\ \qquad + \frac{1}{2} \sum_{(x,x')\in\mathcal{X}^2} \tilde{a}_{x,x'}(m,u)\partial^2_{m_x m_{x'}} v_{u',u}(t,m) \end{cases}$$

where $v_{u',u}(t,m) = \mathbb{E}_{S,X_j} V_T(s,x,u',u,m)$ starting at time t and $\bar{r}(t,m,u,u') = \mathbb{E}_{S,X_j} r(S,X_j,u',u,m)$ is the global expected mean field payoff.

Centralized Mean Field Control. We now provide the feedback optimality principle for the global expected mean field payoff $\bar{r}(t,m,u,u)$. Please note that in this case, each member of the population uses the same strategy. The $T-$stage mean field optimization problem writes

$$\sup_u \ \mathbb{E}g(m(T)) + \int_0^T \bar{r}(t, m(t), u(t), u(t)) \ dt$$

subject to the stochastic differential equation (1).

A strategy $u_*(t) = \phi(t,m)$ constitutes an optimal mean field solution if there exist a continuously differentiable function $\bar{v}(t,m)$ satisfying the following Hamilton-Jacobi-Bellman equation combined with the mean field SDE. By combining the system, one gets the mean field optimality:

Proposition 1. *The mean field optimality for horizon T is given by*

$$\begin{cases} \bar{v}(T,m) = g(m) \\ -\partial_t \bar{v} = \sup_{u_t\in\mathcal{U}_t} \left\{ \bar{r}(t,m_t,u_t,u_t) + \sum_{x\in\mathcal{X}} \tilde{f}_x(m_t,u_t)\partial_{m_x}\bar{v} \right. \\ \qquad \left. + \frac{1}{2} \sum_{(x,x')\in\mathcal{X}^2} \tilde{a}_{x,x'}(m_t,u_t)\partial^2_{m_x m_{x'}}\bar{v} \right\} \\ dm_t = \tilde{f}(m_t,u_t)dt + \tilde{\sigma}(m_t,u_t)dW_t, \ t > 0 \\ m_0 = m \end{cases}$$

where $\bar{v} = \bar{v}(t,m)$.

Please note that this kind of systems may not have a solution. Counterexamples can be found in [11].

Discrete Time Mean Field Limit

If the step-size δ_n is not vanishing when n goes to infinity, then the deterministic mean field limit is in discrete time and driven by the probability transition $L_t(u,m)$. Given an initial population profile m_0 and a terminal payoff, the sequence of population profiles $\{m_t\}_t$ is driven by the transition probabilities $\{L_{t,s,x,s',x'}(u,m_t)\}_t$.

$$m_{t+1}(x) = \sum_{x' \in \mathcal{X}} m_t(x') L_{t,x,x'}(u_t, m_t), \tag{2}$$

where

$$L_{t,x,x'}(u,m) = \sum_{k \geq 0} \sum_{s} w_s(u,m) L_{t,s,x,x'}(u,m;k) J_k(m),$$

$L_{t,s,x,x'}(u,m;k)$ is the limiting probability transition from x to x' when the resource state is s and the number of interacting players is k. Combining with the Bellman-Shapley optimality criterion, one gets the following system in the finite horizon case:

$$\begin{cases} v_{j,t}(s,x,m) = \max_{u_j} \Bigg\{ r(s,x,u,m_t) \\ \qquad + \sum_{s',x'} L(s',x'|s,x,u,m_t) v_{j,t+1}(s',x') \Bigg\} \\ m_{t+1}(x) = \sum_{x' \in \mathcal{X}} m_t(x') L_{t,x',x}(u,m_t) \end{cases} \tag{3}$$

4 Application to Malware Propagation

In this section, we apply the mean field approach to a controlled malware propagation in opportunistic networks. The malware propagation model is based on [7] in which the impact of the control parameters is not examined and the player types are not used. The types can represent, for example, different operating systems, different versions of the operating systems, or patched/unpatched version of the same operating system. In most of the related work about malware spreading in large networks authors do not model the heterogeneity of systems which forms the network, as far as we know. This leads to a limitation in the results obtained, because different types of systems could lead for example to slower rates of propagation.

In this example we have mobile nodes that can be infected by a malicious code. There are two infected states: passive and active. Non infected nodes are susceptible. Then, the set of possible state of a node is $\{P, A, S\}$ (for passive, active and susceptible) and the set of possible types is $\{\theta_1, \theta_2\}$. The state of the system at time t is $X^n(t) = (P_1(t), P_2(t), A_1(t), A_2(t), S(t))$, where $P(t) + A(t) + S(t) = n$, $\sum_j P_j(t) = P(t)$, $\sum_j A_j(t) = A(t)$, $\forall t$ and n is the total number of mobiles in the system. In this example, there is no resource. The occupancy measure is $M_\theta^n(t) = (P_\theta(t)/n, A_\theta(t)/n, S(t)/n) = (P_\theta^n(t), A_\theta^n(t), S^n(t))$. At every time step we want to control the proportion of infected nodes, which is $I^n(t) := A^n(t) + P^n(t)$. There are two fundamental ways to get infected:

1. Caused by a system flaw. (e.g: an exploit that could allow arbitrary code execution).
2. Caused by human flaw. (e.g: the user is deceived and executes a dangerous piece of code).

We can model this system as a controlled mean field interaction model. The interaction is simulated using the following rules:

1. A passive node may become susceptible (inoculation) with probability δ_P.
2. A passive node with type θ may opportunistically encounter another passive node of type θ', and both become active. This occurs with probability proportional to the frequency of other passive nodes at time t. For type θ, the probability is $\lambda(P_{\theta'}^n(t) - \frac{1}{n}1\!\!1_{\{\theta=\theta'\}})$. Note that the passive node can decide to contact the other passive node or not, so there are two possible actions: $\{m, \bar{m}\}$ (for meet and not meet). Those events will be modeled as a Bernoulli random variable with success (meeting) probability δ_m, which represents $u(m|P, \theta)$. Here we model the possibility of getting infected by a system flaw.
3. An active node may become susceptible (inoculation) with probability δ_A.
4. An active node of type θ may become passive with probability $\beta \frac{P_\theta^n(t)}{h_\theta + P_\theta^n(t)}$ at time t. Here is assumed that, at high concentrations of passive nodes, each active node infects some maximum number of passive ones per time step. This reflects finite total bandwidth. The parameter $0 \leq \beta \leq 1$ has the interpretation of the maximum infection rate. The parameter $0 \leq h_\theta \leq 1$ is the passive node density at which the infection proceeds at half of its maximum rate. Here we model the possibility of getting infected by a system flaw.
5. A susceptible node may become active with probability δ_S
6. A susceptible node may become passive via two ways. First, δ_{Sm} is the probability of getting infected by a human flaw. In this case, the susceptible node can "decide" to get deceived or not, so there are two possible actions: $\{o, \bar{o}\}$. The stationary strategy in this case will be modeled as a coin toss with probability δ_e. Second, $\eta(P_\theta^n(t) + P_{\theta'}^n(t))$ models the probability of encountering a passive node. In this case, the passive node can decide to contact the susceptible node or not, and it is modeled analogously to the other two cases.

Table 1. Probabilities, effects, actions and payoffs

Case	Transition proba. ($\theta, \theta' \in \{1,2\}$).	$M_\theta^n(t+1) - M_\theta^n(t)$	Actions	Payoff contrib.
1	$P_\theta^n(t)\delta_P$	$(-1,0,1)/n$	singleton set	$-1/n$
2	$P_\theta^n(t)\delta_m^2\lambda(P_\theta^n(t) - \frac{1}{n})$	$(-2,2,0)/n$	$\{m, \bar{m}\}$	0
3	$A_\theta^n(t)\delta_A$	$(0,-1,1)/n$	singleton set	$-1/n$
4	$A_\theta^n(t)\beta\frac{P_\theta^n(t)}{h_\theta + P_\theta^n(t)}$	$(-1,1,0)/n$	singleton set	0
5	$S^n(t)\delta_S$	$(0,1,-1)/n$	singleton set	$1/n$
6	$S^n(t)(\delta_e\delta_{Sm} + \delta_m\eta P^n(t))$	$(1,0,-1)/n$	$\{o, \bar{o}, m, \bar{m}\}$	$1/n$

At every time step, one of the transitions is randomly selected and performed. The number of nodes that do a transition in one time slot is always 0, 1 or 2. In order to control the infected population, each transition has a certain payoff contribution which could be 0 if no infected node is inoculated, $1/n$ if there is a node which is inoculated and $-1/n$ if one node is infected. In Table 1 are the transition probabilities, the contribution to $M^n(t+1) - M^n(t)$, the set of actions, and the contribution to the total payoff.

The intensity, that is, the probability that one arbitrary object does a transition in one time slot is of the order of $1/n$. The drift, that is, the expected change of M^n in one time step, given the current state of the system is:

$$f_\theta^n(m) = n\mathbb{E}(M_\theta^n(t+1) - M_\theta^n(t)|M^n(t) = m) =$$

$$\begin{pmatrix} -p_\theta\delta_P - 2p_\theta\delta_m^2\lambda\frac{np_\theta - 1}{n} - a_\theta\beta\frac{p_\theta}{h_\theta + p_\theta} + s(\delta_e\delta_{Sm} + \delta_m\eta(p_\theta + p_{\theta'})) \\ 2p_\theta\delta_m^2\lambda\frac{np_\theta - 1}{n} - a_\theta\delta_A + a_\theta\beta\frac{p_\theta}{h_\theta + p_\theta} + s\delta_S \\ p_\theta\delta_P + a_\theta\delta_A - s\delta_S - s(\delta_e\delta_{Sm} + \delta_m\eta(p_\theta + p_{\theta'})) \end{pmatrix}$$

where $m = (p_\theta, p_{\theta'}, a_\theta, a_{\theta'}, s)$. Then the limit $f(m)$ is

$$\begin{pmatrix} -p_\theta\delta_P - 2\lambda p_\theta^2\delta_m^2 - a_\theta\beta\frac{p_\theta}{h_\theta + p_\theta} + s(\delta_e\delta_{Sm} + \delta_m\eta(p_\theta + p_{\theta'})) \\ -p_{\theta'}\delta_P - 2\lambda p_{\theta'}^2\delta_m^2 - a_{\theta'}\beta\frac{p_{\theta'}}{h_{\theta'} + p_{\theta'}} + s(\delta_e\delta_{Sm} + \delta_m\eta(p_\theta + p_{\theta'})) \\ 2\lambda p_\theta^2\delta_m^2 - a_\theta\delta_A + a_\theta\beta\frac{p_\theta}{h_\theta + p_\theta} + s\delta_S \\ 2\lambda p_{\theta'}^2\delta_m^2 - a_{\theta'}\delta_A + a_{\theta'}\beta\frac{p_{\theta'}}{h_{\theta'} + p_{\theta'}} + s\delta_S \\ (p_\theta + p_{\theta'})\delta_P + (a_\theta + a_{\theta'})\delta_A - 2s\delta_S - 2s(\delta_e\delta_{Sm} + \delta_m\eta(p_\theta + p_{\theta'})) \end{pmatrix}$$

4.1 Homogeneous System

We briefly mention the homogeneous mean field. The drift is obtained by computing the expected changes in one time slot: $f^n(m) = \mathbb{E}(M^n(t+1) - M^n(t)|M^n(t) = m) =$

$$\frac{1}{n}\begin{pmatrix} -p\delta_P - 2p\delta_m^2\lambda\frac{np-1}{n} - a\beta\frac{p}{h+p} + s(\delta_e\delta_{Sm} + \delta_m\eta p) \\ 2p\delta_m^2\lambda\frac{np-1}{n} - a\delta_A + a\beta\frac{p}{h+p} + s\delta_S \\ p\delta_P + a\delta_A - s\delta_S - s(\delta_e\delta_{Sm} + \delta_m\eta p) \end{pmatrix}$$

where $m = (p, a, s)$. Then the limit is

$$f(m) = \begin{pmatrix} -p\delta_P - 2p^2\delta_m^2\lambda - a\beta\frac{p}{h+p} + s(\delta_e\delta_{Sm} + \delta_m\eta p) \\ 2p^2\delta_m^2\lambda - a\delta_A + a\beta\frac{p}{h+p} + s\delta_S \\ p\delta_P + a\delta_A - s\delta_S - s(\delta_e\delta_{Sm} + \delta_m\eta p) \end{pmatrix}$$

In all the simulations, we kept this parameters unchanged: $\beta = 10^{-2}$, $\delta_A = 5 \times 10^{-3}$ and $\delta_P = \delta_S = \delta_{Sm} = 10^{-4}$. On parameter h (the passive node density at which the infection of active nodes proceeds at half of it maximum value) depends the stability of the system. Here we set $h = 10^{-2}$ in order to obtain an unstable behaviour. Regarding the control parameters, $\delta_m = 1$, $\delta_e = 1$ means no control.

We investigate the evolution of the system in the following scenarios:

- Trajectory of one run of the simulation (figure 1),
- Mean trajectory of multiple simulations (figure 2),
- Mean field limit trajectory (figure 3),
- Trajectory of the payoff function,
- Controlled mean field limit (figure 4),
- Trajectory of the heterogeneous malware propagation (figure 6),
- Optimal control under the mean field limit (figure 7),
- Noisy mean field (stochastic path), (figure 8).

These configurations are analyzed with control and without control parameters. We observe that the time mean $\frac{1}{T}\int_0^T m(s)ds$ is asymptotically close to the stationary point inside the limit cycle.

Uncontrolled Behaviour. In figures 1,2,3 we can see the simulation results, using the well-known algorithm for exact simulation of a discrete time Markov chain. The initial configuration is $(0.2, 0, 0.8)$. An oscillating behaviour of the total reward can be seen.

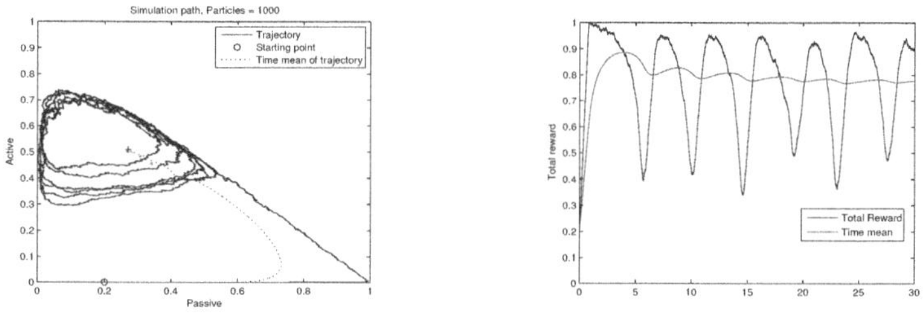

Fig. 1. Left: Path of one simulation without control. Right: Total reward.

Controlled Behaviour. In order to illustrate the control parameters, suppose we want to keep the proportion of infected nodes below 0.9 for all times. One simple way to achieve this is to reduce the contact tendency of a passive node. Then, we set $\delta_m = 0.075$, which is the more relevant control parameter. The results can be seen in figures 4 and 5.

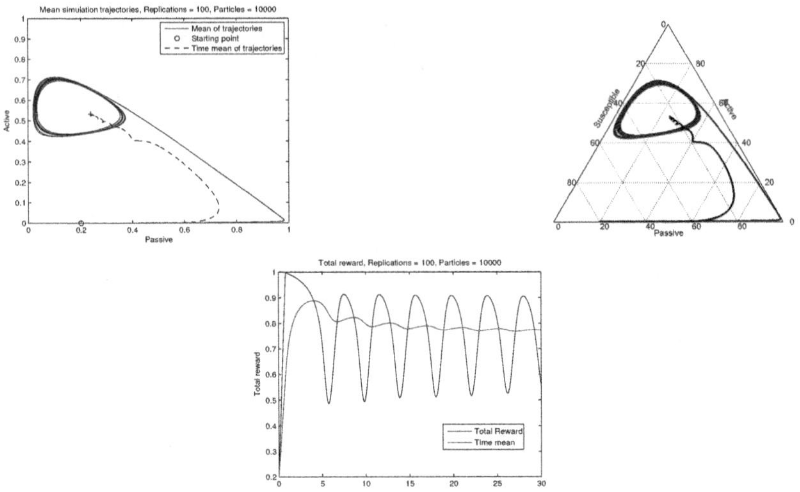

Fig. 2. Leftmost: Mean of simulation trajectories without control. Rightmost: Total reward.

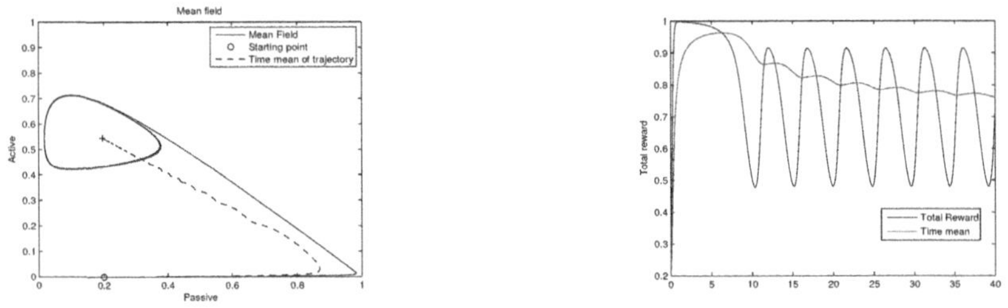

Fig. 3. Left: Deterministic mean field without control. Right: Total reward.

4.2 Heterogeneous System

We investigate numerically the behavior of the mean field limit for two types θ and θ'. In figure 6 we can see that it is possible to stabilize the homogeneous system using classes.

4.3 Optimal Strategy for the Homogeneous System

Since the payoff function is the same for all the players i.e. $r_j(\cdot) = r(\cdot)$ the discounted stochastic game with common payoff can be transformed in a team problem. Moreover, the set of actions is the same for all the players. In figure 7, we plot the optimal strategy obtained by solving the system in the finite horizon case (equation (3)). The existence of a dominant strategy can be observed in the plot.

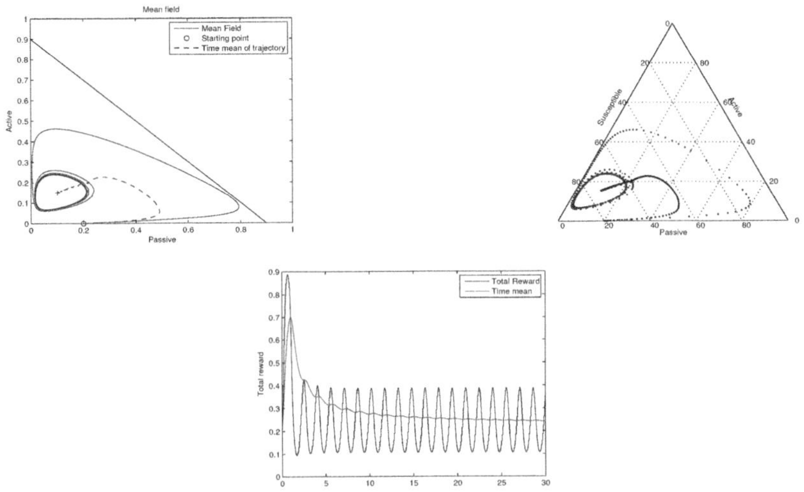

Fig. 4. Leftmost: Deterministic mean field with controlled passive node contact tendency. Rightmost: Total reward.

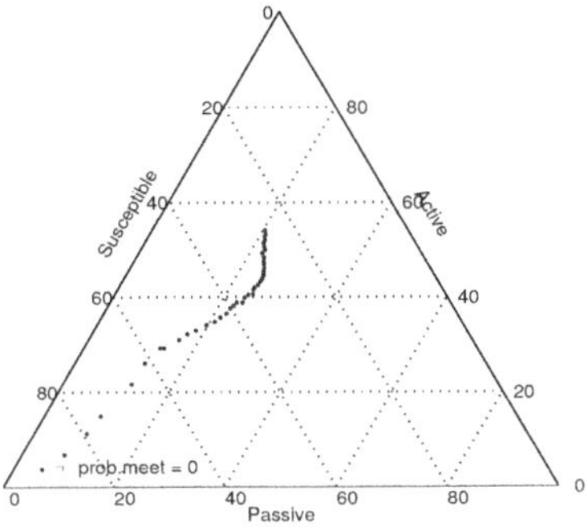

Fig. 5. Evolution of the limit point of the time mean, for $\delta_m = 0, 0.025, .., 0.925, 1$

4.4 Noisy Mean Field

In order to give some feeling on how the noisy mean field evolves with time, we show in figure 8 two different realizations, for $n = 1000$ and $n = 2000$. The variance of the noise is obtained from $\tilde{\sigma}^n = \frac{1}{\delta_n} f^n - f$ which its norm sup is bounded by $\frac{2D\lambda}{n}$, then we have $d\tilde{m} = f(\tilde{m})dt + \tilde{\sigma}^n dW_t$. Note that the smooth version of the last equation can be $\dot{m}^n = \frac{f^n(m^n)}{\delta_n}$. It is worth to mention that, in this case, the simulation algorithm is only exact when the numerical time step vanishes. In figure 9 we compare the deterministic mean field versus the mean trajectory of the noisy mean field.

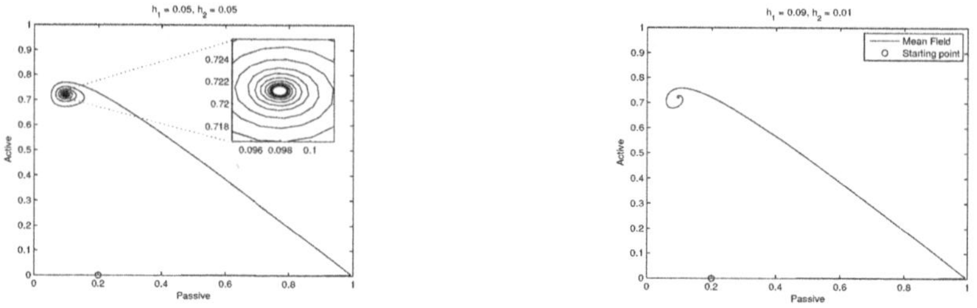

Fig. 6. Left: Unstable deterministic mean field, for $h_1 = 0.05$, $h_2 = 0.05$. Right: Stable mean field $h_1 = 0.09$, $h_2 = 0.01$.

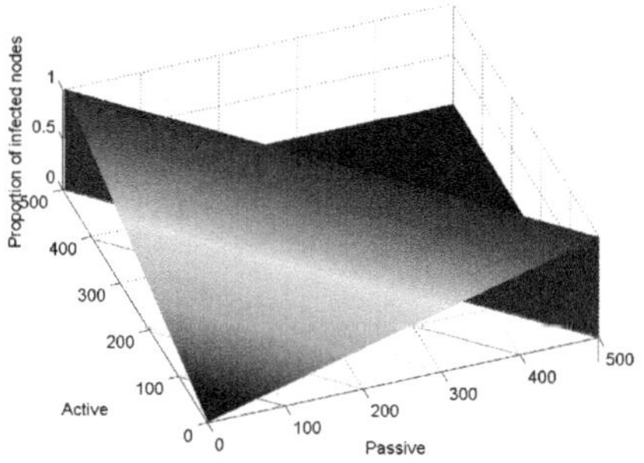

Fig. 7. Optimal strategy using backward induction for the homogeneous system

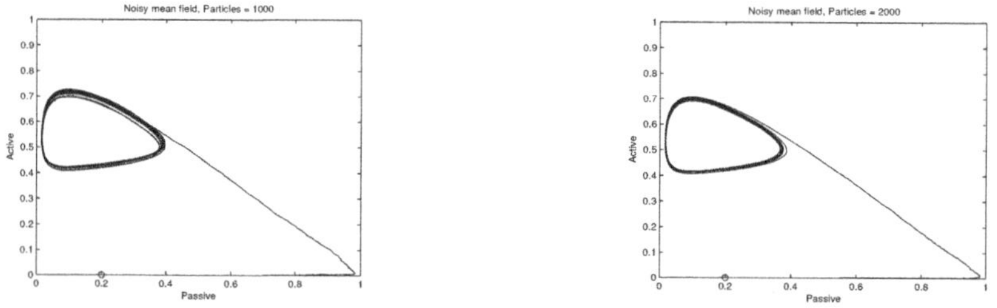

Fig. 8. Left: Path of one simulation of the noisy mean field, for $n = 1000$, Right: $n = 2000$

4.5 Effect of the System Size

In this subsection we examine the effect of n of the mean field object M^n. We represent the evolution (in time) for different value of n and same state-component (see figures 10).

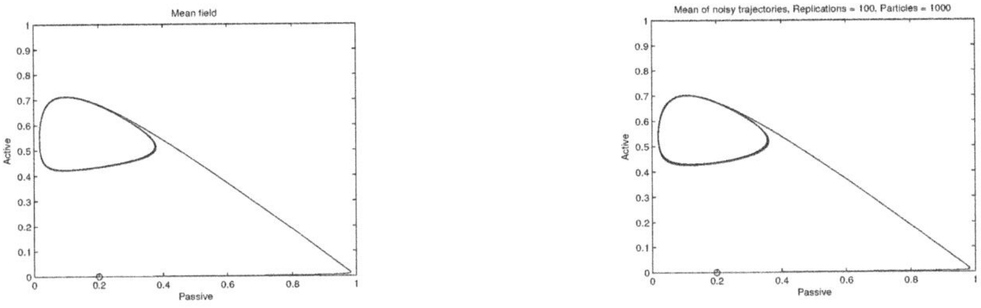

Fig. 9. Left: Deterministic mean field, Right: Mean of noisy trajectories

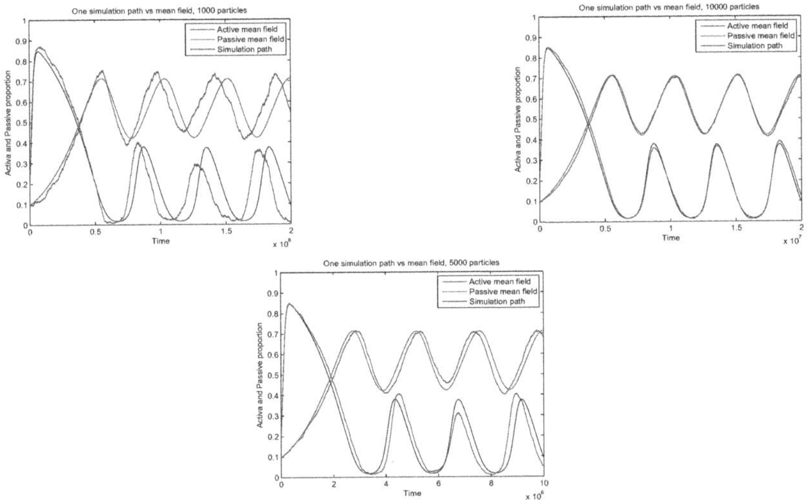

Fig. 10. Mean field vs Simulation: Impact of n

5 Concluding Remarks

We have studied mean field stochastic games and established a mean field convergence to Itô stochastic differential equations. Our convergence result opens new questions for discontinuous mapping (namely the drift and the variance). In that case one may expect to have a stochastic differential inclusion as in [13] for which the existence of solution needs more attention. This leads to ask what will happen when the discontinuities are coming from the strategies as it is the case for the best response correspondence? We do not have answers to this question and postpone it to future work. In practical scenarios, when the second moment conditions are not satisfied, the conditions for mean field convergence given in [5,14] are not applicable. In these cases, the weak conditions that we provided here cover a wide range of networking scenarios. The cycling behavior of the fraction of infected is sensitive to the control parameters. It would be useful to understand this relationship better by taking into account the network topology.

References

1. Zou, C.C., Gong, W., Towsley, D.F., Gao, L.: The monitoring and early detection of internet worms. IEEE/ACM Trans. Netw. 13(5), 961–974 (2005)
2. Karyotis, V., Kakalis, A., Papavassiliou, S.: Malware-propagative mobile ad hoc networks: Asymptotic behavior analysis. J. Comput. Sci. Technol. 23(3), 389–399 (2008)
3. Xu, Y., Wang, W.: The speed of information propagation in large wireless networks. In: IEEE Infocom (2008)
4. Tembine, H., Altman, E., ElAzouzi, R., Hayel, Y.: Evolutionary games in wireless networks. IEEE Trans. on Systems, Man, and Cybernetics, Part B, Special Issue on Game Theory 40, 634–646 (2010)
5. Tembine, H., Le Boudec, J.Y., ElAzouzi, R., Altman, E.: Mean field asymptotic of markov decision evolutionary games and teams. In: The Proc. of GameNets (May 2009)
6. Stroock, D.W., Varadhan, S.R.S.: Multidimensional diffusion processes. Springer, New York (1979)
7. Benaïm, M., Le Boudec, J.-Y.: A class of mean field interaction models for computer and communication systems. Perform. Eval. 65(11-12), 823–838 (2008)
8. Weintraub, G.Y., Benkard, L., Van Roy, B.: Oblivious equilibrium: A mean field approximation for large-scale dynamic games. In: Advances in Neural Information Processing Systems, vol. 18 (2006)
9. Tembine, H.: Mean field stochastic games: convergence, q/h learning, optimality. In: American Control Conference, ACC 2011, San Francisco, California, US (2011)
10. Tembine, H.: Hybrid mean field game dynamics in large populations. In: American Control Conference, ACC 2011, San Francisco, California, US (2011)
11. Tembine, H.: Mean field stochastic games: Simulation, dynamics and applications. Suplec (2010)
12. Oksendal, B.: Stochastic Differential Equations: An Introduction with Applications (Universitext), 6th edn. Springer, Heidelberg (2003)
13. Benaim, M., Hofbauer, J., Sorin, S.: Stochastic approximation and differential inclusions. SIAM J. Control and Optimization 44, 328–348 (2005)
14. Tembine, H.: Population games in large-scale networks: time delays, mean field and applications, 250 pages. LAP Lambert Academic Publishing (December 2010) ISBN 978-3-8383-6392-9

The Impact of Incomplete Information on Games in Parallel Relay Networks

Hongda Xiao and Edmund M. Yeh

Electrical Engineering, Yale University
{hongda.xiao,edmund.yeh}@yale.edu

Abstract. This work considers the impact of incomplete information on incentives for node cooperation in parallel relay networks with one source node and multiple relay nodes. All nodes are selfish and strategic, interested in maximizing their own profit instead of the social welfare. We consider the practical situation where a node cannot observe the state of links adjacent to other nodes. We examine a general game setting where the source has full bargaining power, and propose a framework for analyzing the efficiency loss induced by incomplete information.

1 Introduction

There is now widespread awareness of the importance of incentives in the management of communication networks. Network nodes often cannot be relied upon to cooperatively implement network algorithms in the service of the social good. Instead, selfish nodes will behave in a given manner only if it is profitable for them to do so. Of clear interest is the impact of such selfish actions on the social good. From the network point of view, it is important to design incentives such as pricing schemes, which induce selfish behavior aligned with the social good.

In single-hop and multi-hop networks, the incentive issue and its impact on social efficiency have been extensively studied[1,2,3,4,5,6,7,8,9]. All these papers, however, assume a complete information setting where players in the network game have complete knowledge about quantities such as the state of network links. In practice, this assumption is often too strong. Information regarding network quantities is typically incomplete and imperfect. In a multi-hop network such as the Internet, a source does not typically have perfect information on the congestion state of links a few hops away [10]. In wireless networks, the source usually cannot observe or test the channel state from a relay to the destination. Neither can a relay observe the channel state from other relays to the destination. Given the above, it is clear that in analyzing selfish behavior in network settings, the role of incomplete information must be emphasized.

One approach to network design problems with incomplete information is through dominant implementable mechanisms [11]. This idea has been used in the context of spectrum auctions [12] and communication networks [13]. These mechanisms, however, require a centralized authority and extra funding from an outsider. This makes the extension to general multi-hop networks difficult.

R. Jain and R. Kannan (Eds.): GameNets 2011, LNICST 75, pp. 475–489, 2012.

Another approach, based on the idea of Bayesian Nash Equilibrium, a generalization of the Nash Equilibrium concept, is advocated in [14]. Here, the authors consider selfish routing in a single-hop network, where every source node knows only its own traffic requirement, but has knowledge of the traffic distribution of other sources. While the results in [14] are appealing, it remains unclear how they might extend to the multi-hop network situation.

In this work, we investigate the impact of incomplete information on the problem of incentives in a two-hop parallel relay network. In our setting, the state of the links adjacent to a given relay is not observable by the source or the other relays, although the prior distribution of the link state is known. We consider a game with full source bargaining power, where the source offers a general contract mapping signals from the relays to traffic allocations and transfer payments. Given the proposed contract, the relays decide whether to accept or not. Those relays which choose to accept the contract then send link-state-dependent signal functions to the source, which then allocates traffic and transfer payments according to the proposed contract. This general setting includes as a specific case of pricing games where the signals from the relays consist of nonlinear traffic-dependent charging functions. We study the Bayesian Nash equilibria corresponding to the general game. To provide a benchmark, we first show that in the game with complete information, (Bayesian) Nash equilibria exist and are all efficient. Next, we investigate the game with incomplete information. To deal with the difficulty of characterizing the Bayesian Nash Equilibria in this case, we first show that if a resource allocation outcome can be realized by a Bayesian Nash equilibrium, then there exists a "truth telling" Bayesian Nash equilibrium that realizes the outcome. We then show that the outcome of the "truth telling" Bayesian Nash equilibrium coincides with that of the Nash equilibrium for a *complete information game*, in which the link cost functions are replaced by a specified "virtual cost functions." Using this approach, we obtain for a symmetric network scenario a bound on the amount of inefficiency which may result from incomplete information.

2 Network Model and Problem Formulation

Consider a parallel relay network modeled by a directed graph $G = (V, E)$, with a single source s, destination d, and a set of relays I, where $|I| = n$. We assume that there is no direct link between s and d. Instead, The relays in I are used to forward traffic in a two-hop fashion from s to d.

We shall consider two scenarios. In the first *inelastic* scenario, the source has a fixed rate r_s of transmission. This rate must be carried by the relays in I, where the traffic rate forwarded by relay i is r_i, and $\sum_{i=1}^{n} r_i = r_s$. In the second *elastic* scenario, the source may be willing to withhold some of its transmission rate, according to how the cost of sending traffic affects it overall utility. Let r_0 denote the amount of traffic withheld or rejected. Then $r_s - r_0$ is the total admitted traffic from the source. A traffic vector $\mathbf{r} \triangleq (r_0, r_1, \ldots, r_n) \in \mathbb{R}_+^{n+1}$ is a feasible routing of the source traffic if it satisfies $r_0 + \sum_{i=1}^{n} r_i = r_s$.

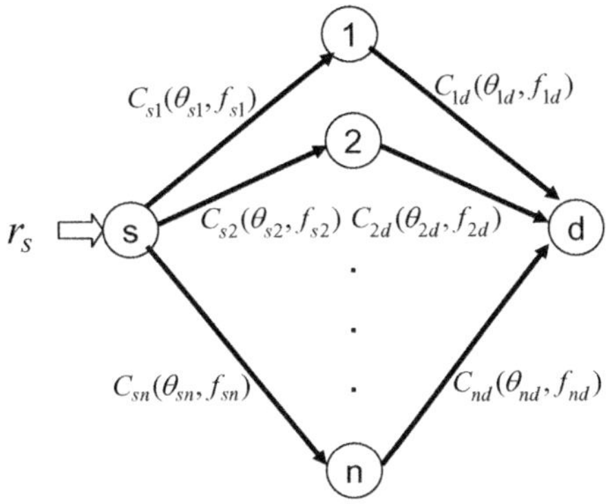

Fig. 1. Relay channel

2.1 Cost Function and Utility Function

In general, for any relay node i, there is a cost involved in forwarding traffic for source s. This cost typically depends both on the properties of the links adjacent on relay i and the amount of traffic flowing through those links. Denote the traffic flow on link $(i, j) \in E$ by f_{ij}. We assume that link (i, j) has a cost function $C_{ij}(\theta_{ij}, f_{ij})$ with $C_{ij}(\theta_{ij}, 0) = 0$, where θ_{ij} is a measure of the quality of link (i, j). This quality may have different physical meanings in different contexts. For example, if the cost function reflects the queuing delay on (i, j), then using the M/M/1 approximation, $C_{ij}(\theta_{ij}, f_{ij}) = \frac{f_{ij}}{k_{ij} - f_{ij}}$. Here, θ_{ij} denotes the link capacity k_{ij}. For another example, consider the cost of power assumption required for transmitting traffic of rate f_{ij} over a wireless link with channel gain g_{ij}, bandwidth W, and receiver noise power N. Using the Shannon capacity formula, we have $f_{ij} = W \log(1 + g_{ij} P_{ij}/N)$, where P_{ij} is transmission power required on link (i, j). Thus, the link cost is

$$C_{ij}(\theta_{ij}, f_{ij}) = \frac{N}{g_{ij}}(2^{f_{ij}/W} - 1).$$

Here, θ_{ij} denotes the channel gain g_{ij}.

Now consider the overall cost $C_i(\theta_i, r_i)$ for relay node i to forward traffic of rate r_i from source s to destination d, where θ_i measures the quality or *type* of the *path* from s to d through i. We assume that $C_i(\theta_i, r_i) = C_{si}(\theta_{si}, r_i) + C_{id}(\theta_{id}, r_i)$. The costs $C_i(\theta_i, r_i)$ are particularly amenable to analysis if θ_i can be expressed as a simple scalar function of θ_{si} and θ_{id}: $\theta_i = h(\theta_{si}, \theta_{id})$. This is true in the example of the power consumption cost function given above, where $\theta_{ij} = g_{ij}$ is the channel gain on link (i, j). Normalizing the bandwidth and receiver noise power to 1, we have

$$C_i(\theta_i, r_i) = P_{si} + P_{id} = (2^{r_i} - 1)(g_{si}^{-1} + g_{id}^{-1}) = (2^{r_i} - 1)\theta_i^{-1}, \tag{1}$$

where $\theta_i \triangleq (g_{si}^{-1} + g_{id}^{-1})^{-1} = (\theta_{si}^{-1} + \theta_{id}^{-1})^{-1}$. In this paper, we focus on situations where the path quality θ_i can be expressed as a scalar function of θ_{si} and θ_{id}. We further assume that θ_i belongs to a compact interval $[\underline{\theta}_i, \overline{\theta}_i]$.

Motivated by the power consumption example, we assume that $C_i(\theta_i, r_i)$ is twice continuously differentiable on $[\underline{\theta}_i, \overline{\theta}_i] \times [0, r_s]$, and strictly increasing and strictly convex in r_i: $\partial C_i(\theta_i, r_i)/\partial r_i > 0$ and $\partial^2 C_i(\theta_i, r_i)/\partial r_i^2 > 0$. Also, assume that $C_i(\theta_i, r_i)$ is strictly decreasing in θ_i: $\partial C_i(\theta_i, r_i)/\partial \theta_i < 0$. Furthermore, assume $\partial^2 C_i(\theta_i, r_i)/\partial \theta_i \partial r_i \leq 0$.

Now consider the source s. Let the utility function of the source be given by $W_s(\theta_s, r)$, where $\theta_s \in [\underline{\theta}_s, \overline{\theta}_s]$ parameterizes the utility for the source, and r is the source rate admitted into the network. For example, the source utility may be $W_s(\theta_s, r) = \theta_s \log(1 + r)$. Assume that $W_s(r) = W_s(r_s)$ for all $r \geq r_s$, i.e. r_s is the maximum desired source rate. $W_s(\theta_s, r)$ is assumed to be continuously differentiable, strictly increasing and strictly concave in r on $[0, r_s]$. Let $C_s(\theta_s, r_0) \triangleq W_s(r_s) - W_s(r_s - r_0)$ denote the source's *utility loss* from having traffic of rate r_0 withheld from the network. Since $W_s(r_s)$ is a constant, it can be seen that $C_s(\theta_s, r_0)$ is continuously differentiable on $[\underline{\theta}_s, \overline{\theta}_s] \times [0, r_s]$, strictly increasing and strictly convex in r_0: $\partial C_s(\theta_s, r_0)/\partial r_0 > 0$ and $\partial^2 C_s(\theta_s, r_0)/\partial r_0^2 > 0$. Furthermore, we assume that $C_s(\theta_s, r_0)$ is strictly decreasing in θ_s: $\partial C_s(\theta_s, r_0)/\partial \theta_s < 0$. Finally, it can be seen that $C_s(\theta_s, 0) = 0$ for all θ_s.

2.2 Socially Optimal Allocation

A socially optimal traffic allocation in a parallel relay network is an allocation which minimizes the total network cost, assumed to be the sum of the link costs. Let $R \triangleq \{(r_0, r_1, ..., r_n) : r_j \geq 0 \; \forall j = 0, \ldots, n, \sum_{j=0}^{n} r_j = r_s\}$ be the set of feasible traffic allocations, and let $\mathbf{r} = (r_0, r_1, ..., r_n) \in R$ denote the vector of traffic rates in the network.

Definition 1. *A traffic allocation vector* $\mathbf{r}^*$ *is called socially optimal if*

$$\mathbf{r}^* \in \arg\min_{\mathbf{r} \in R} C_s(\theta_s, r_0) + \sum_{i=1}^{n} C_i(\theta_i, r_i). \tag{2}$$

Since the link cost functions $C_i(\theta_i, r_i)$ as well as $C_s(\theta_s, r_0)$ are all strictly increasing and strictly convex, the socially optimal allocation $\mathbf{r}^*$ exists and is unique. The conditions for specifying $\mathbf{r}^*$ can be obtained using the Kuhn-Tucker conditions. Let $c_i(\theta_i, r_i) \triangleq \partial C_i(\theta_i, r_i)/\partial r_i$ and $c_s(\theta_s, r_0) \triangleq \partial C_s(\theta_s, r_0)/\partial r_0$ denote the marginal cost function of link i and the marginal cost function of the overflow link for source s, respectively.

For the case of an inelastic source, $\mathbf{r}^* = (0, r_1^*, \ldots, r_n^*)$ is the socially optimal allocation if and only if for each $i = 1, \ldots, n$,

$$c_i(\theta_i, r_i^*) = c^* \text{ if } r_i^* > 0, \quad c_i(\theta_i, r_i^*) > c^* \text{ if } r_i^* = 0. \tag{3}$$

For the case of an elastic source, $\mathbf{r}^* = (r_0^*, r_1^*, \ldots, r_n^*)$ is the socially optimal allocation if and only if (3) holds and furthermore,

$$c_s(\theta_s, r_0^*) = c^* \text{ if } r_0^* > 0, \quad c_s(\theta_s, r_0^*) > c^* \text{ if } r_0^* = 0.$$

2.3 Game Structure

We assume that the (maximum) source input rate r_s and the parameter θ_s are known to all nodes. We also assume that θ_i is randomly distributed according to distribution function $F_i(\theta_i)$. In practical network scenarios, the exact realization of θ_i is typically known only to relay i, and not to the source or to the relays other than i. Thus, θ_i is *private information* to relay i. Nevertheless, the source and other relays may still have knowledge of the distribution $F_i(\theta_i)$. For instance, a wireless source or a relay $j \neq i$ may know the distribution of the channel gains for relay i, but typically does not know the realization of those channel gains.

In order for the source node to allocate its traffic intelligently in the presence of incomplete information regarding the θ_i's, it needs to observe some "signal" from the relay nodes. This can be realized by having the relay node send a signal according to the realization of its type to the source.[1] Let M_i be the set of signals for relay i, where M_i is a subset of the set of differentiable functions on $[0, r_s]$. The signal map for relay i is

$$s_i : \Theta_i \to M_i,$$

where $\Theta_i \triangleq [\underline{\theta}_i, \overline{\theta}_i]$ and $s_i(\theta_i) = m_i(\cdot)$.

Given the signals $m_i(\cdot), i = 1, \ldots, n$, the source decides on an allocation of its traffic as well as a vector of transfer payments to the relays. This allocation is called a *contract*. Let $\mathbf{r} = (r_0, r_1, ..., r_n) \in R$ denote the vector of traffic rates in the network. Note that for the inelastic case, $r_0 = 0$. Now let $\mathbf{t} = (t_1, t_2, ..., t_n) \in \mathbb{R}^n_+$ be the vector of transfer payments, where t_i is the transfer payment to relay i. Let $M \triangleq M_1 \times \cdots \times M_n$ and $T \triangleq \mathbb{R}^n_+$. Then the allocation map of the source node is

$$g : M \to R \times T,$$

where $g(m_1(\cdot), \ldots, m_n(\cdot)) = (\mathbf{r}, \mathbf{t})$.

The above framework encompasses many forms of pricing games explored in previous literature. For instance, in [9], the relay signals are simply charging functions $P_i(\cdot)$, and the transfer payments are required to equal the charges demanded by the relays, i.e. $t_i = P_i(r_i)$.

The signal maps of the relays along with the allocation map of the source realize a corresponding network allocation map

$$f : \Theta \to R \times T,$$

where $f(\theta_1, \ldots, \theta_n) = g(s_1(\theta_1), \ldots, s_n(\theta_n)) = (\mathbf{r}, \mathbf{t})$.

In the game with incomplete information corresponding to the above setting, the utility of the source is given by

$$U_s(\theta_s, g(s_1(\theta_1), \ldots, s_n(\theta_n)) = W_s(r_s) - C_s(\theta_s, r_0) - \sum_{i=1}^{n} t_i.$$

[1] One can also consider the possibility of the source sending a signal according to its type θ_s. However, since we assume θ_s is known to all network nodes, we do not consider this possibility here.

The utility of relay i is given by

$$U_i(\theta_i, g(s_1(\theta_1), \ldots, s_n(\theta_n))) = t_i - C_i(\theta_i, r_i).$$

The game with incomplete information proceeds as follows. First, each relay i observes its own private information θ_i. Second, the source provides a contract for the relay nodes. The contract announces the source allocation rule $g : M \to R \times T$. Third, the relays simultaneously decide to either accept or reject the contract. If a given relay accepts the contract, then it will participate in the game which follows. Otherwise, the relay quits and receives zero utility.[2] Fourth and finally, the relay nodes simultaneously send their signals to the source, and the source allocates rates and transfer payments according to the announced g.

In the following, we give the formal definition of the Bayesian Nash equilibrium corresponding to the game with incomplete information described above. Let $\theta \triangleq (\theta_1, \ldots, \theta_n)$, $\theta_{-i} \triangleq (\theta_j)_{j \neq i}$, and $s_{-i}(\theta_{-i}) \triangleq (s_j(\theta_j))_{j \neq i}$.

Definition 2. *A Bayesian Nash Equilibrium of the above game is a set of strategies $\{s_1, \ldots, s_n, g\}$ satisfying*

1. *for each relay node i and for every feasible $\widetilde{s}_i : \Theta_i \to M_i$,*

$$E_{\theta_{-i}} \{U_i(\theta_i, g(s_i(\theta_i), s_{-i}(\theta_{-i})))\} \geq E_{\theta_{-i}} \{U_i(\theta_i, g(\widetilde{s}_i(\theta_i), s_{-i}(\theta_{-i})))\}\,; \quad (4)$$

2. *for every feasible $\widetilde{g} : M \to R \times T$,*

$$E_\theta \{U_s(\theta_s, g(s(\theta)))\} \geq E_\theta \{U_s(\theta_s, \widetilde{g}(s(\theta)))\}\,. \quad (5)$$

3 Games with Complete Information

We examine games where the source has full bargaining power, in the sense that source can offer any contract $g \in R \times T$.[3] We first investigate the (Bayesian) Nash equilibria which can result from games with source bargaining power in the case of complete information. Here, we show that all (Bayesian) Nash equilibria are efficient. Then, we proceed to the case of incomplete information, and characterize the potential inefficiencies associated with that case.

In the complete information game, the source can observe the type vector $\theta = (\theta_1, \ldots, \theta_n)$ of the relays, and then design the allocation map g according to θ. Since the type θ_i is not private to relay i, relay i cannot manipulate this information in designing its signalling strategy s_i. Since the source can observe θ, it can effectively ignore the strategies of the relays in designing g. Nevertheless, the source needs to ensure that the relays will accept its proposed contract and

[2] Note that the relays which quit can simply be left out of the game formulation. Thus, without loss of generality, we assume for the rest of the paper that the source plays the game in a manner which gives non-negative expected utility to all relays, so that all relays stay in the game.

[3] In games with partial bargaining power, such as auctions where the transfers are determined by bids, the source can propose only a subset of $R \times T$.

stay in the game. The latter will hold as long as $E_{\theta_{-i}}\{U_i(\theta_i, g(\theta))\} = E_{\theta_{-i}}\{t_i - C_i(\theta_i, r_i)\} \geq 0$ for all i. That is, all relays receive non-negative expected utility by accepting the contract proposed by the source, and therefore are willing to participate in the game.

Lemma 1. *In any (Bayesian) Nash Equilibrium of the complete information game with source bargaining power, all relays receive zero utility.*

Proof. Suppose that there exists a (Bayesian) Nash Equilibrium where the source allocation rule

$$g(m_1(\cdot), ..., m_n(\cdot)) = (\mathbf{r}, \mathbf{t})$$

is such that $U_i(\theta_i, r_i, t_i) = t_i - C_i(\theta_i, r_i) > 0$ for some i. Since the source can observe θ, it could select another allocation rule $g'(m_1(\cdot), ..., m_n(\cdot)) = (\mathbf{r}', \mathbf{t}')$ such that

$$r_i' = r_i, i = 1, \ldots, n; \quad t_i' = t_i - \epsilon, \;\; t_j' = t_j \text{ for all } j \neq i$$

where ϵ is small enough so that $t_i' - C_i(\theta_i, r_i') > 0$. Note that the set of relays which would opt to accept contract g and stay in the game is the same as the set for contract g'. On the other hand, by shifting its allocation rule from g to g', the source has strictly decreased its total transfer payment, while keeping the same traffic allocation. Thus, the source's utility is strictly increased. This contradicts our assumption of being at a Nash equilibrium.

Theorem 1. *In the complete information game with source bargaining power, all (Bayesian) Nash equilibria are efficient.*

Proof. At any Nash equilibrium, the source maximizes its utility

$$U_s(\theta_s, g(s_1(\theta_1), \ldots, s_n(\theta_n)) = W_s(\theta_s, r_s) - C_s(\theta_s, r_0) - \sum_{i=1}^{n} t_i.$$

By Lemma 1, at the equilibrium, we have $t_i = C_i(\theta_i, r_i)$ for all i. Thus, the traffic allocation by the source minimizes $C_s(\theta_s, r_0) + \sum_{i=1}^{n} C_i(\theta_i, r_i)$, and therefore the equilibrium is efficient.

Using Lemma 1 and Theorem 1, we can easily solve for the Nash equilibrium of the complete information game with source bargaining power. By Theorem 1, the source allocation rule at the equilibrium may be obtained by solving for the socially optimal traffic allocation $\mathbf{r}^*$, where $\mathbf{r}^* = \arg\min_{\mathbf{r} \in R} C_s(\theta_s, r_0) + \sum_{i=1}^{n} C_i(\theta_i, r_i)$. As noted in Section 2.2, due to the strict convexity of the optimization problem, $\mathbf{r}^*$ exists and is unique. By Lemma 1, at the equilibrium, the transfer payment $t_i = C_i(\theta_i, r_i^*)$ for every $i = 1, \ldots, n$.[4] For the relays, any feasible signal map s_i may be chosen for the equilibrium.

[4] Recall that $C_i(\theta_i, 0) = 0$.

4 Games with Incomplete Information

We now turn to the case where the source cannot observe the types of the relay. In this case, the relay nodes can manipulate their types in order to obtain more utility, and the source can no longer design the allocation map according to θ. Instead, the source must maximize its expected utility according to the signals sent by the relays. The characterization of Bayesian Nash Equilibria for this case is very difficult due to the complexity of the strategy set and the possible behaviors of source and relays. Nevertheless, we devise a method for characterizing outcomes corresponding to the Bayesian Nash Equilibria which avoids the difficulty of calculating the the equilibria explicitly. We shall do this in two steps. First, we show that if a resource allocation outcome can be realized by a Bayesian Nash equilibrium for a game with source bargaining in which every relay receives non-negative expected utility, then there exists a "truth telling" Bayesian Nash equilibrium that realizes the outcome. Second, we show that the outcome of the "truth telling" Bayesian Nash equilibrium coincides with that of the Nash equilibrium for a *complete information game*, in which the link cost functions are replaced by a specified "virtual cost functions."

Definition 3. *A Bayesian Nash Equilibrium of the game with source bargaining power is* truth telling *if $M = \Theta$ and every relay node is willing to report its true type to the source node.*

Theorem 2. *If a resource allocation outcome f can be realized by a Bayesian Nash Equilibrium of the game with source bargaining power, in which every relay receives non-negative expected utility, then there exists a truth telling Bayesian Nash Equilibrium which realizes f.*

Proof. Suppose there is a Bayesian Nash Equilibrium which realizes the allocation outcome $f(\theta)$. By the definition of the Bayesian Nash Equilibrium, we have (4) and (5). Now observe that by (4), we must have

$$\theta_i \in \arg\max_{\widetilde{\theta}_i} E_{\theta_{-i}} \left\{U_i(\theta_i, g(s_i(\widetilde{\theta}_i), s_{-i}(\theta_{-i})))\right\} \quad \text{for all } i.$$

Otherwise, if there exists some θ' such that $E_{\theta_{-i}} \{U_i(\theta_i, g(s_i(\theta'_i), s_{-i}(\theta_{-i})))\} > E_{\theta_{-i}} \{U_i(\theta_i, g(s_i(\theta_i), s_{-i}(\theta_{-i})))\}$, then there is another strategy $s'_i(\theta)$ satisfying $s'_i(\theta_i) = s_i(\theta'_i)$ and $s'_i(\theta) = s_i(\theta)$ for all $\theta \neq \theta_i$, such that $E_{\theta_{-i}} \{U_i(\theta_i, g(s'_i(\theta_i), s_{-i}(\theta_{-i})))\} > E_{\theta_{-i}} \{U_i(\theta_i, g(s_i(\theta_i), s_{-i}(\theta_{-i})))\}$, violating (4). Therefore, since $g(s_1(\theta_1), \ldots, s_n(\theta_n)) = f(\theta)$, we have

$$\theta_i \in \arg\max_{\widetilde{\theta}_i \in \Theta_i} E_{\theta_{-i}} \left\{U_i(\theta_i, f(\widetilde{\theta}_i, \theta_{-i}))\right\} \quad \text{for all } i, \tag{6}$$

$$f \in \arg\max_{\widetilde{f}} E_{\theta} \left\{U_s(\theta_s, \widetilde{f}(\theta))\right\}. \tag{7}$$

Thus, there exists a direct truth telling Bayesian Nash Equilibrium with the outcome $f(\theta)$.

Theorem 2 says that the set of outcomes corresponding to Bayesian Nash Equilibria for the game with source bargaining power and incomplete information is a subset of the outcomes corresponding to truth telling Bayesian Nash Equilibria, in which each relay proposes its type truthfully to the source, and the source optimally allocates rates according to the relays' types. This finding simplifies our analysis considerably, since we can now focus on the truth telling Bayesian Nash Equilibria in order to bound the efficiency loss introduced by incomplete information in games with source bargaining power.

We now investigate the outcomes which can be realized by truth telling Bayesian Nash Equilibria. Notice that these equilibria correspond to the solutions of the optimization problem given by (6) and (7), in addition to the non-negative expected utility constraint

$$E_{\theta_{-i}}\{U_i(\theta_i, r_i)\} = E_{\theta_{-i}}\{t_i(\theta, \mathbf{r}) - C_i(\theta_i, r_i(\theta))\} \geq 0 \quad \text{for all } i. \tag{8}$$

and feasibility constraint $\mathbf{r} \in R$.

Theorem 3. *The set of solutions for the optimization problem defined by* (6)-(8) *is the same as the set of outcomes corresponding to the Nash equilibria for the complete information game in which the link cost functions* $C_i(\theta_i, r_i)$ *are replaced by*

$$J_i(\theta_i, r_i) = C_i(\theta_i, r_i) - \frac{1 - F_i(\theta_i)}{f_i(\theta_i)} \frac{\partial C_i(\theta_i, r_i)}{\partial \theta_i}. \tag{9}$$

Proof. Please see the Appendix.

We refer to the functions J_i as *virtual cost functions.* The proof involves examining the first and second order optimality conditions, using the envelope theorem, and expressing the expected utility of the source in an equivalent form. We skip the proof here due to space constraints. Note that by Theorem 1, all Nash equilibria corresponding to games with complete information are efficient. Thus, the set of outcomes corresponding to the Nash equilibria for the complete information game with virtual link cost functions $J_i(\theta_i, r_i)$ is given by

$$\mathbf{r}' = \arg\min_{\mathbf{r} \in R} C_s(\theta_s, r_0) + \sum_{i=1}^{n} J_i(\theta_i, r_i).$$

5 Efficiency Analysis

In this section, we bound the amount of inefficiency in the outcomes for games with incomplete information. We focus on the inelastic scenario where $r_0 = 0$. Following [1], define the price of anarchy for type θ as:

$$\rho(\theta) = \frac{\max_{\mathbf{r} \in R^E} \sum_i C_i(\theta_i, r_i)}{\min_{\mathbf{r} \in R} \sum_i C_i(\theta_i, r_i)} \tag{10}$$

where R^E is the set of all Bayesian Nash Equilibria for the game with incomplete information, and $R^J \equiv \arg\min_{\mathbf{r}\in R}\sum_i J_i(\theta_i, r_i)$. By Theorems 2 and 3, we have $R^E \subseteq R^J$. Therefore,

$$\rho(\theta) \le \frac{\max_{\mathbf{r}\in R^J}\sum_i C_i(\theta_i, r_i)}{\min_{\mathbf{r}\in R}\sum_i C_i(\theta_i, r_i)} \tag{11}$$

Since the link cost functions are strictly convex, the socially optimal allocation $\mathbf{r}^*$ are given by the necessary and sufficient conditions in (3). An allocation $\mathbf{r}'$ in R^J must satisfy the following necessary conditions: for all $i \in \{1, \ldots, n\}$ such that $r_i' > 0$,

$$\frac{\partial C(\theta_i, r_i')}{\partial r_i} - \frac{1 - F_i(\theta_i)}{f_i(\theta_i)}\frac{\partial^2 C(\theta_i, r_i')}{\partial\theta_i \partial r_i} \le \frac{\partial C(\theta_j, r_j')}{\partial r_j} - \frac{1 - F_j(\theta_j)}{f_j(\theta_j)}\frac{\partial^2 C(\theta_i, r_j')}{\partial\theta_i \partial r_j} \quad \text{for all } j \tag{12}$$

We now bound the price of anarchy in the symmetric case.

Theorem 4. *Consider the symmetric case where the link cost functions $C_i(\theta_i, r_i)$ and the type distributions $F_i(\theta_i)$ are the same for all relays. If (i) $J(\theta_i, r_i)$ is convex in r_i and decreasing in θ_i, (ii) $X(\theta_i, r_i) \equiv J(\theta_i, r_i) - C(\theta_i, r_i)$ is concave in θ_i, (iii) $\frac{\partial X(\theta_i, r_i)}{\partial\theta_i \partial r_i} \le 0$, then the price of anarchy $\rho(\theta)$ can be upper bounded as follows. If the marginal cost function $c(\theta_i, r_i) = \frac{\partial C(\theta_i, r_i)}{\partial r_i}$ is concave, then $\rho(\theta) \le n$, where n is the number of relays. If $\frac{c(\underline{\theta}, r_s)}{c(\overline{\theta}, 0)} \le k$ for some constant k, then $\rho(\theta) \le k$.*

Proof. Let $(r_i') \in \arg\min_i \sum_i J(\theta_i, r_i)$, and r_i^* be the efficient allocation. Let $x(\theta_i, r_i) = \frac{\partial X(\theta_i, r_i)}{\partial r_i}$. Suppose $\theta_i > \theta_j$, and both carry traffic in optimal point, thus for the optimal allocation, $r_i^* > r_j^* > 0$. As $X(\theta_i, r_i)$ is concave in r_i, thus $x(\theta_i, r_i^*) < x(\theta_i, r_j^*)$. As the cost is decreasing in θ_i, we have $x(\theta_i, r_j^*) < x(\theta_j, r_j^*)$ Thus $x(\theta_i, r_i^*) < x(\theta_j, r_j^*)$. As $c(\theta_i, r_i^*) = c(\theta_j, r_j^*)$, $c(\theta_i, r_i^*) + x(\theta_i, r_i^*) < c(\theta_j, r_j^*) + x(\theta_j, r_j^*)$. But to maximize the virtual cost, we have $c(\theta_i, r_i') + x(\theta_i, r_i') \ge c(\theta_j, r_j') + x(\theta_j, r_j')$. As the virtual cost function is convex in r_i, we get $r_i' > r_i^*$. Thus $c(\theta_i, r_i') > c(\theta_i, r_i^*) = c(\theta_j, r_j^*) > c(\theta_j, r_j')$. As the cost function is convex in r_i and decreasing in θ_i, $c(\max_i \theta_i, r_s) \ge c(\theta_i, r_i')$. Thus $C(\max_i \theta_i, r_s) \ge \sum_i C(\theta_i, r_i')$. Thus the price of anarchy has the same bound with game 2.

As an example of a link cost function/type distribution pair which satisfies the assumptions in Theorem 4, consider the uniform type distribution over $[0, 1]$ and the cost function $C_i(\theta_i, r_i) = \frac{1}{\theta_i}(2^{r_i} - 1)$.

If either the virtual cost functions $J_i(\theta_i, r_i)$ are not convex, or the link cost functions and type distributions are not the same across relays, then higher prices of anarchy may result. Consider the situation in Figure 2. Here, there are only two relays. $(r_1^*, r_s - r_1^*)$ is the efficient allocation. Since the type distributions are not the same, the marginal virtual costs are as indicated in the figure. To minimize the sum of the virtual costs, the source allocates all traffic to relay 2, while this allocation is clearly the worst outcome for minimizing the sum of the link costs.

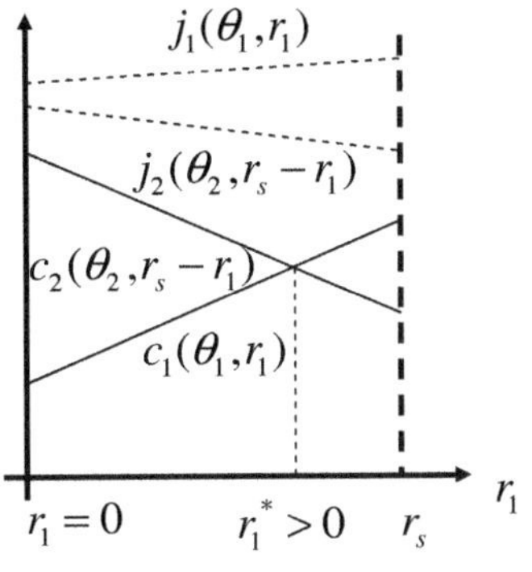

Fig. 2. Efficiency Loss in Asymmetric Case

6 Conclusion

This work investigated the impact of incomplete information on incentives for node cooperation in parallel relay networks. We considered a general game setting where the source has full bargaining power. We first showed that in the game with complete information, (Bayesian) Nash equilibria exist and are all efficient. Next, we investigated the game with incomplete information. To deal with the difficulty of characterizing the Bayesian Nash Equilibria in this case, we first showed that if a resource allocation outcome can be realized by a Bayesian Nash equilibrium, then there exists a "truth telling" Bayesian Nash equilibrium that realizes the outcome. We then showed that the outcome of the "truth telling" Bayesian Nash equilibrium coincides with that of the Nash equilibrium for a complete information game, in which the link cost functions are replaced by a specified "virtual cost functions." Using this approach, we obtained for a symmetric network scenario a bound on the amount of inefficiency which may result from incomplete information.

References

1. Roughgarden, T., Tardos, E.: How bad is selfish routing? In: Proceedings of the 41st Annual Symposium on Foundations of Computer Science (2000)
2. He, L., Walrand, J.: Pricing and revenue sharing strategies for internet service providers. In: Proceedings of the 24th Annual Joint Conference of the IEEE Computer and Communications Societies, vol. 1, pp. 205–216 (2005)
3. Shakkottai, S., Srikant, R.: Economics of network pricing with multiple isps. IEEE/ACM Transactions on Networking 14(6), 1233–1245 (2006)
4. Acemoglu, D., Ozdaglar, A.: Competition and efficiency in congested markets. Math. Oper. Res. 32(1), 1–31 (2007)
5. Neely, M.: Optimal pricing in a free market wireless network. In: IEEE 26th IEEE International Conference on Computer Communications, INFOCOM 2007, pp. 213–221 (May 2007)

6. Zhong, S., Chen, J., Yang, Y.: Sprite: a simple, cheat-proof, credit-based system for mobile ad-hoc networks. In: Proceedings of IEEE INFOCOM, vol. 3, pp. 1987–1997 (March 2003)
7. Ileri, O., Mau, S.-C., Mandayam, N.: Pricing for enabling forwarding in self-configuring ad hoc networks. IEEE Journal on Selected Areas in Communications 23(1), 151–162 (2005)
8. Wang, B., Han, Z., Liu, K.R.: Distributed relay selection and power control for multiuser cooperative communication networks using stackelberg game. IEEE Transactions on Mobile Computing 8(7), 975–990 (2009)
9. Xi, Y., Yeh, E.: Pricing, competition, and routing for selfish and strategic nodes in multi-hop relay networks. In: Proceedings of the 27th Annual Joint Conference of the IEEE Computer and Communications Societies, pp. 1463–1471 (2008)
10. Bertsekas, D.P., Gallager, R.: Data Networks, 2nd edn. Prentice Hall (1992)
11. Nisan, N., Ronen, A.: Algorithmic mechanism design (extended abstract). In: Proceedings of the Thirty-First Annual ACM Symposium on Theory of Computing, STOC 1999, pp. 129–140. ACM, New York (1999)
12. Archer, A., Papadimitriou, C., Talwar, K.: An approximate truthful mechanism for combinatorial auctions with single parameter agents. Internet Mathematics 1(2), 129–150 (2006)
13. Dobson, I., Carreras, B., Lynch, V., Newman, D.: Communication requirements of vcg-like mechanisms in convex environments. In: Proceedings of the Allerton Conference on Communication, Control, and Computing (2005)
14. Gairing, M., Monien, B., Tiemann, K.: Selfish routing with incomplete information. Theory of Computing Systems 42, 91–130 (2008), doi:10.1007/s00224-007-9015-8

7 Appendix

Proof of Theorem 3: The first and second-order conditions for (6) are:

$$\left.\frac{dE_{\theta_{-i}}\left\{U_i(\theta_i, f(\widetilde{\theta}_i, \theta_{-i}))\right\}}{d\widetilde{\theta}_i}\right|_{\widetilde{\theta}_i=\theta_i} = 0 \tag{13}$$

and

$$\left.\frac{d^2E_{\theta_{-i}}\left\{U_i(\theta_i, f(\widetilde{\theta}_i, \theta_{-i}))\right\}}{d\widetilde{\theta}_i^{\,2}}\right|_{\widetilde{\theta}_i=\theta_i} \leq 0. \tag{14}$$

The first-order condition is equivalent to

$$E_{\theta_{-i}} \left.\frac{dt_i(\widetilde{\theta}_i, \theta_{-i})}{d\widetilde{\theta}_i}\right|_{\widetilde{\theta}_i=\theta_i} \tag{15}$$

$$= E_{\theta_{-i}} \left.\left\{\frac{\partial C_i(\theta_i, r_i(\widetilde{\theta}_i, \theta_{-i}))}{\partial r_i} \frac{dr_i(\widetilde{\theta}_i, \theta_{-i})}{d\widetilde{\theta}_i}\right\}\right|_{\widetilde{\theta}_i=\theta_i}. \tag{16}$$

The second-order condition is equivalent to

$$E_{\theta_{-i}} \frac{d^2 t_i(\widetilde{\theta}_i, \theta_{-i})}{d\widetilde{\theta}_i^{\,2}}\Bigg|_{\widetilde{\theta}_i=\theta_i}$$

$$\leq E_{\theta_{-i}} \left\{ \frac{\partial^2 C_i(\theta_i, r_i(\widetilde{\theta}_i, \theta_{-i}))}{\partial r_i^2} \left[\frac{dr_i(\widetilde{\theta}_i, \theta_{-i})}{d\widetilde{\theta}_i} \right]^2 \right. \tag{17}$$

$$\left. + \frac{\partial C_i(\theta_i, r_i(\widetilde{\theta}_i, \theta_{-i}))}{\partial r_i} \frac{d^2 r_i(\widetilde{\theta}_i, \theta_{-i})}{d\widetilde{\theta}_i^{\,2}} \right\}\Bigg|_{\widetilde{\theta}_i=\theta_i}.$$

By evaluating the first-order condition at θ_i differentiating with respect to θ_i, we get:

$$E_{\theta_{-i}} \left\{ \frac{d^2 \{t_i(\theta_i, \theta_{-i})\}}{d\theta_i^2} \right\}$$

$$= E_{\theta_{-i}} \left\{ \frac{\partial^2 C_i(\theta_i, r_i(\theta_i, \theta_{-i}))}{\partial r_i^2} \left[\frac{dr_i(\theta_i, \theta_{-i})}{d\theta_i} \right]^2 \right. \tag{18}$$

$$+ \frac{\partial C_i(\theta_i, r_i(\theta_i, \theta_{-i}))}{\partial r_i} \frac{d^2 r_i(\theta_i, \theta_{-i})}{d\theta_i^2}$$

$$\left. + \frac{\partial^2 C_i(\theta_i, r_i(\theta_i, \theta_{-i}))}{\partial r_i \partial \theta_i} \frac{dr_i(\theta_i, \theta_{-i})}{d\theta_i} \right\}.$$

Comparing with the second-order condition, we get

$$E_{\theta_{-i}} \frac{\partial^2 C_i(\theta_i, r_i(\theta_i, \theta_{-i}))}{\partial r_i \partial \theta_i} \frac{dr_i(\theta_i, \theta_{-i})}{d\theta_i} \leq 0. \tag{19}$$

We have already assumed that

$$\frac{\partial^2 C_i(\theta_i, r_i(\theta_i, \theta_{-i}))}{\partial r_i \partial \theta_i} \leq 0 \text{ for each } \theta_{-i}. \tag{20}$$

Notice that when an outcome can be realized by a Bayesian Nash Equilibrium, the following condition must hold:

$$\frac{\partial r_i(\theta_i, \theta_{-i})}{\partial \theta_i} \geq 0 \text{ given any } \theta_{-i} \tag{21}$$

Otherwise, the source would allocate a higher rate to a lower type relay, which is not optimal. Notice that by (20) and (21), (19) automatically holds.

Thus, the following conditions are necessary for the first and second-order conditions to hold.

$$\frac{dE_{\theta_{-i}} \left\{ t_i(\widetilde{\theta}_i, \theta_{-i}) \right\}}{d\widetilde{\theta}_i}$$

$$= E_{\theta_{-i}} \frac{\partial C_i(\theta_i, r_i(\widetilde{\theta}_i, \theta_{-i}))}{\partial r_i} \frac{dr_i(\widetilde{\theta}_i, \theta_{-i})}{d\widetilde{\theta}_i}\Bigg|_{\widetilde{\theta}_i=\theta_i}$$

$$\frac{\partial r_i(\theta_i, \theta_{-i})}{\partial \theta_i} \geq 0 \text{ given any } \theta_{-i}$$

Let $V_i(\theta_i, \theta_{-i}) = \max_{\widetilde{\theta_i}} U_i(\theta_i, r_i(\widetilde{\theta_i}, \theta_{-i}), t_i(\widetilde{\theta_i}, \theta_{-i}))$. We use the envelope theorem just as we did in the previous sections:

$$\begin{aligned} \frac{dE_{\theta_{-i}} V_i(\theta_i, \theta_{-i})}{d\theta_i} &= \left.\frac{\partial E_{\theta_{-i}} U_i(\theta_i, r_i(\widetilde{\theta_i}, \theta_{-i}), t_i(\widetilde{\theta_i}, \theta_{-i}))}{\partial \theta_i}\right|_{\widetilde{\theta_i}=\theta_i} \\ &= -\left.\frac{\partial E_{\theta_{-i}} C_i(\theta_i, r_i(\widetilde{\theta_i}, \theta_{-i}))}{\partial \theta_i}\right|_{\widetilde{\theta_i}=\theta_i} \end{aligned} \tag{22}$$

Let $\overline{\theta_i}$ and $\underline{\theta_i}$ be the upper and lower bounds on relay node i's type, then

$$\begin{aligned} & E_{\theta_{-i}} V_i(\theta_i, \theta_{-i}) \\ &= E_{\theta_{-i}} V_i(\underline{\theta_i}, \theta_{-i}) - \int_{\underline{\theta_i}}^{\theta_i} \frac{\partial E_{\theta_{-i}} C_i(\theta_i, r_i(\theta_i, \theta_{-i}))}{\partial \theta_i} d\theta_i \end{aligned} \tag{23}$$

We see from the above equation that, as we already assumed $\frac{\partial C_i(\theta_i, r_i)}{\partial \theta_i} < 0$, the expected utility of relay i is non-decreasing with respect to θ_i. Thus, to guarantee that constraints (8) holds, the lowest type must receive non-negative profit. On the other hand, the relay with the lowest type can never receive a positive profit, otherwise the source will reduce its profit by some small amount and still guarantee that the contract is acceptable to all, which contradicts the definition of Bayesian Nash Equilibrium. Thus, the lowest type relay should receive zero profit.

$$E_{\theta_{-i}} V_i(\underline{\theta_i}, \theta_{-i}) = 0 \tag{24}$$

Plugging in, we get

$$E_{\theta_{-i}} V_i(\theta_i, \theta_{-i}) = -\int_{\underline{\theta_i}}^{\theta_i} \frac{\partial E_{\theta_{-i}} C_i(\theta_i, r_i(\theta_i, \theta_{-i}))}{\partial \theta_i} d\theta_i \tag{25}$$

Suppose the type distribution function of relay i is $F_i(\theta_i)$ and the density is $f_i(\theta_i)$. Let R be the expected revenue of the source node. Then,

$$
\begin{aligned}
R &= E_\theta \left\{ W_s(r_s) - C_s(\theta_s, r_0) - \sum_i t_i(\theta) \right\} \\
&= E_\theta \left\{ W_s(r_s) - C_s(\theta_s, r_0) - \sum_i V_i(\theta) - \sum_i C_i(\theta) \right\} \\
&= E_\theta \left\{ W_s(r_s) - C_s(\theta_s, r_0) - \sum_i C_i(\theta) \right\} + \sum_i \int_{\underline{\theta_i}}^{\overline{\theta_i}} f_i(\theta_i) E_{\theta_{-i}} \left[\int_{\underline{\theta_i}}^{\theta_i} \frac{\partial C_i(\theta_i', r_i(\theta_i', \theta_{-i}))}{\partial \theta_i'} d\theta_i' \right] d\theta_i \\
&= E_\theta \left\{ W_s(r_s) - C_s(\theta_s, r_0) - \sum_i C_i(\theta) \right\} - \sum_i \int_{\underline{\theta_i}}^{\overline{\theta_i}} E_{\theta_{-i}} \left[\int_{\underline{\theta_i}}^{\theta_i} \frac{\partial C_i(\theta_i', r_i(\theta_i', \theta_{-i}))}{\partial \theta_i'} d\theta_i' \right] \times d(1 - F_i(\theta_i)) \\
&= E_\theta \left\{ W_s(r_s) - C_s(\theta_s, r_s) - \sum_i C_i(\theta) \right\} - \sum_i E_{\theta_{-i}} \left[\int_{\underline{\theta_i}}^{\theta_i} \frac{\partial C_i(\theta_i', r_i(\theta_i', \theta_{-i}))}{\partial \theta_i'} d\theta_i' \right] \times (1 - F_i(\theta_i))|_{\underline{\theta_i}}^{\overline{\theta_i}} \\
&\quad + \sum_i E_{\theta_{-i}} \int_{\underline{\theta_i}}^{\overline{\theta_i}} (1 - F_i(\theta_i)) \times d \left[\int_{\underline{\theta_i}}^{\theta_i} \frac{\partial C_i(\theta_i', r_i(\theta_i', \theta_{-i}))}{\partial \theta_i'} d\theta_i' \right] \\
&= E_\theta \left\{ W_s(r_s) - C_s(\theta_s, r_0) - \sum_i C_i(\theta) \right\} + \sum_i E_{\theta_{-i}} \int_{\underline{\theta_i}}^{\overline{\theta_i}} (1 - F_i(\theta_i)) \times d \left[\int_{\underline{\theta_i}}^{\theta_i} \frac{\partial C_i(\theta_i', r_i(\theta_i', \theta_{-i}))}{\partial \theta_i'} d\theta_i' \right] \\
&= E_\theta \{ W_s(r_s) - C_s(\theta_s, r_0) \} - \sum_i E_{\theta_{-i}} \int_{\underline{\theta_i}}^{\overline{\theta_i}} C_i(\theta_i, r_i(\theta_i, \theta_{-i})) - \frac{1 - F_i(\theta_i)}{f_i(\theta_i)} \frac{\partial C_i(\theta_i, r_i(\theta))}{\partial \theta_i} f_i(\theta_i) d\theta_i \\
&= E_\theta [W_s(r_s) - C_s(\theta_s, r_0)] - E_\theta \sum_i \left(C_i(\theta_i, r_i(\theta)) - \frac{1 - F_i(\theta_i)}{f_i(\theta_i)} \frac{\partial C_i(\theta_i, r_i(\theta))}{\partial \theta_i} \right)
\end{aligned}
$$

Thus, we obtain a game with complete information and full source bargaining power where the revenue function is changed to $J_i(\theta_i, r_i)$ rather than $C_i(\theta_i, r_i)$.

Hierarchical Coalition Formation Game of Relay Transmission in IEEE 802.16m

Dusit Niyato[1], Xiangyun Zhou[2], Are Hjørungnes[2], Ping Wang[1], and Yifan Li[1]

[1] School of Computer Engineering, Nanyang Technological University (NTU), Singapore
[2] UNIK - University Graduate Center, University of Oslo, Kjeller, Norway

Abstract. One of the main features of IEEE 802.16m is the relay transmission which could not only extend the service coverage, but also improve the quality-of-service (QoS) to the mobile stations. In this paper, we consider the *cooperation* among relay stations and mobile stations to improve the performance of relay transmission in IEEE 802.16m network. In particular, the relay stations and mobile stations are rational to maximize their own benefit by forming coalitions. A hierarchical coalition formation game is introduced which is similar to the Stackelberg game. At the upper level, relay stations are considered to be the leaders to cooperate with each other to relay the data from base station to the mobile station. At the lower level, mobile stations are considered to be the followers cooperating with each other to relay the data from base and relay stations. Given the coalition formed by the relay station (i.e., leaders), mobile stations (i.e., followers) form the coalitions such that their individual payoffs are maximized. Knowing this behavior of mobile station, relay stations form their coalitions to maximize their individual payoffs. The analysis based on Markov model is introduced to obtain the stable coalitional structures of both leaders and followers. This hierarchical coalition formation game model will be useful for jointly investigating the self-interest behaviors of relay stations and mobile stations in IEEE 802.16m relay network.

Keywords: Cooperative communications and networking, coalitional game theory, Markov model, IEEE 802.16m.

1 Introduction

IEEE 802.16m is proposed for mobile broadband access as an enhancement of the existing WiMAX systems. IEEE 802.16m will meet all requirements of 4G wireless system under the ITUs International Mobile Telecommunications Advanced (IMT-Advanced program) [1]. IEEE 802.16m will incorporate the relay transmission as a cost-effective approach to extend the coverage area and improve the capacity of the network as specified in the standard [2,3]. Relay transmission will be based on the cooperative communications in which the relay station (RS) forwards the data between base station (BS) and mobile station (MS). Radio resource management will remain the important and open issues in IEEE 802.16m

R. Jain and R. Kannan (Eds.): GameNets 2011, LNICST 75, pp. 490–505, 2012.

relay network. One of them is the relay selection, especially, in the environment where RSs and MSs are rational to maximize their own benefits. In this case, RSs and MSs can cooperate to perform relay transmission to enhance the performance, but this will incur the cost (e.g., due to energy consumption). Therefore, with rational behavior, the analysis of the decision making for cooperation in IEEE 802.16m relay network will be required.

In this paper, we apply coalitional game theory to study the cooperation among the RSs and MSs in relay transmission. RSs in the same coalition will perform relay transmission for their subscribed MSs. In addition, if MSs form coalition, they will also perform relay transmission for each other to improve the transmission rate further. However, relay transmission incurs the cost (e.g., energy consumption), RSs and MSs have to form coalition such that their individual payoffs are maximized. As the main contribution, a hierarchical coalition formation game model is proposed to analyze this decision making process of RSs and MSs jointly. This game model is similar to the Stackelberg game, where RSs are considered to be the leaders forming coalition before MSs which are considered to be the followers. The stable coalitional structures (i.e., set of coalitions of all leaders and followers) are obtained using hierarchical Markov model. The performance evaluation clearly shows the impact of various parameters to the network performances (e.g., channel quality and cost of relay transmission). This hierarchical coalition formation game model will be useful for the implementation of the IEEE 802.16m relay network.

The rest of this paper is organized as follows: Related works are reviewed in Section 2. Section 3 describes the system model and assumptions. Section 4 presents the formulation of hierarchical coalition formation game for relay transmission. Section 5 presents the numerical results. The summary is given in Section 6.

2 Related Works

2.1 IEEE 802.16 Relay Networks

IEEE 802.16m is an enhancement of IEEE 802.16 standards which will support mobile multihop relay (MMR) networks [4]. Various issues were addressed for such networks. For example, in [5], a cooperative relay selection algorithm was proposed. In this algorithm, the signal intensity is used by base station to select the best relay station. In [6], a resource allocation algorithm for multicast service in IEEE 802.16j relay network was proposed. The objective is to maximize the total number of recipients constrained by the transmission budget. The resource is allocated for the base station and relay nodes to achieve this objective. In [7], the problem of relay station placement in IEEE 802.16 relay network was addressed. The objective is to minimize the number of relay stations required to meet all demand of users. An efficient heuristic algorithm was proposed to obtain the solution of placement.

2.2 Game Theory and Cooperative Communications

Due to the nature of the cooperative communications, game theory has been adopted to analyze various issues [8]. In [9], a game model was presented to investigate the cooperation among nodes using decode-and-forward (DF) cooperative transmission with Rayleigh fading channels. In [10], a bargaining game was formulated to study the bandwidth allocation problem between a source node and a number of relay nodes. Also, the conditions under which the source and relay nodes will cooperate were analyzed. The relay selection and power control problems in cooperative relay network were addressed in [11] as a two-level Stackelberg game. In this game, the source node, considered as a buyer, pays to the relay nodes to provide them with an incentive to cooperate and forward the signal to the destination. In [12], coalitional game theory was used, in combination with cooperative transmission, to solve the boundary node problem in an ad hoc packet forwarding network. A grouping algorithm for relay selection was proposed in [13] to minimize transmit power. Also, an optimal rate allocation scheme among the relay nodes was studied.

Although existing literature addressed various issues in cooperative communications using game theory, none of these works considered the problem of performing coalition formation among rational nodes (i.e., nodes are self-interested to maximize their individual payoffs). Especially, when a hierarchy exists in the relay network in which relay stations and mobile stations can form coalitions to achieve their goals. The closest work is [14] where a coalitional game framework was proposed for cooperative communications. However, no analysis on the stability of the resulting coalitional structure was considered. Also, the joint formations of relay stations and mobile stations were ignored.

3 System Model and Assumptions

In this section, the network model of IEEE 802.16m relay network considered in this paper is presented (Section 3.1). Then, the details of the relay transmission in such a network are given (Section 3.2). After that, an overview of the hierarchical coalition formation game model, which is the main contribution of this paper is presented (Section 3.3).

3.1 Network Model and Coalition Formation

We consider IEEE 802.16m relay network. Without loss of generality, a service area with one base station is considered. There are R relay stations (RSs) providing the relay service in this service area in which the set of RSs is denoted by $\mathcal{R}$. There are N mobile stations (MSs) in this service area, each of which subscribes to one of the RSs. A set of MSs subscribed to RS j is denoted by $\mathcal{N}_j$ and the set of all MSs is denoted by $\mathcal{N} = \bigcup_{j \in \mathcal{R}} \mathcal{N}_j$. For MS i, the corresponding RS is denoted by $j = J(i)$. When BS transmits data to the target MS i, RS

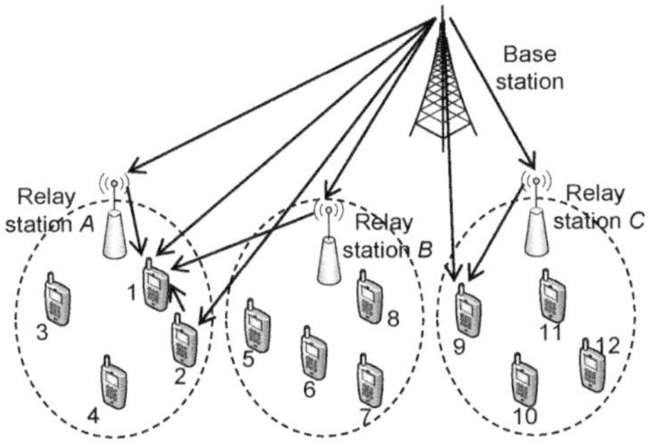

Fig. 1. System model for relay transmission with coalitions $\{A, B\}$ and $\{C\}$

$j = J(i)$ performs relay transmission to improve the transmission rate in the downlink direction of target MS i.[1]

RSs can cooperate by forming the coalitions (i.e., groups) to perform relay transmission for the MSs subscribed to the RSs in the same coalition. The coalition of RSs is denoted by $\mathcal{J}$.[2] In addition, we also assume that MSs can help each other by performing relay transmission. The coalition of MSs is denoted by $\mathcal{I}$.

Fig. 1 shows example of network model with three RSs (i.e., $\mathcal{R} = \{A, B, C\}$) and twelve MSs (i.e., $\mathcal{N} = \{1, \ldots, 12\}$) where $\mathcal{N}_1 = \{1, \ldots, 4\}$, $\mathcal{N}_2 = \{5, \ldots, 8\}$, and $\mathcal{N}_3 = \{9, \ldots, 12\}$. RSs A and B form a coalition denoted by $\mathcal{J} = \{A, B\}$, while RS C acts alone. Also, MSs 1 and 2 form a coalition denoted by $\mathcal{I} = \{1, 2\}$. In this case, when BS transmits data to MS 1 which subscribes to RS A, RSs A and B as well as MS 2 will perform the relay transmission for MS 1 to improve the performance (Fig. 1). Similarly, when BS transmits data to MS 2, RSs A and B as well as MS 1 will perform the relay transmission. However, BS transmits data to MS 9 which subscribes to RS C. Since RS C does not form coalition with other RSs and MS 9 does not form coalition with other MSs, only RS C will perform the relay transmission for MS 9.

3.2 Relay Transmission

The frame structure for IEEE 802.16m cooperative relay transmission is divided into uplink and downlink subframes. For a subframe (e.g., downlink), it consists of access zone and relay zone to support transmission from base station to mobile station and from relay station to mobile station. We assume that the transmissions from BS to MS and from RS to MS use cooperative diversity based on a decode-and-forward strategy as in [15]. However, the approach that we propose in this paper can accommodate other strategies (e.g., amplify-and-forward). In the first phase of the cooperative diversity scheme (i.e., access zone

[1] The model is also applicable to the uplink transmission.

[2] In this paper, to simplify the presentation, notations of coalition and set are assumed to be the same.

in the frame structure), BS transmits using a particular adaptive modulation and coding (AMC) mode to a target MS. In this phase, RSs and MSs receive the signals. In the second phase, the RSs which are in the same coalition as the RS which the target MS is subscribed to as well as the MSs which are in the same coalition as the target MS repeat the transmission from BS with the same AMC mode. At the end of the second phase, the target MS achieves a gain in SNR by combining the space-time decoded signals with those received in the first phase.

Let $\mathcal{J}$ and $\mathcal{I}$ denote the coalitions of RSs and MSs to perform relay transmission for MS i, respectively. Let γ_i, γ_j, and $\gamma_{i'}$ denote the instantaneous SNR from BS, from RS j and MS i' to the target MS i, respectively. If there are more than one relay transmission from RS (or MS) with the perfect channel between BS to RS (or MS), the post-processing SNR at the target MS i can be expressed as follows:

$$\gamma_i^{\text{post}} = \gamma_i + \sum_{j \in \mathcal{J}} \gamma_j + \sum_{i' \in \mathcal{I} \setminus \{i\}} \gamma_{i'}. \tag{1}$$

When considering Rayleigh fading channels, the cumulative distribution function (CDF) of the post-processing SNR at the MS with single RS j (i.e., no coalition of RS and MS) is given by [15].

$$F(\gamma) = \frac{\overline{\gamma}_i}{\overline{\gamma}_i - \overline{\gamma}_j}(1 - e^{-\frac{\gamma}{\overline{\gamma}_i}}) + \frac{\overline{\gamma}_j}{\overline{\gamma}_j - \overline{\gamma}_i}(1 - e^{-\frac{\gamma}{\overline{\gamma}_j}}), \tag{2}$$

where $\overline{\gamma}_i$ and $\overline{\gamma}_j$ are, respectively, the corresponding average SNRs from BS and RS j to the target MS i. For the multiple relay transmission case, (2) can be extended as in (3) [15], where $|\mathcal{J} \cup \mathcal{I}| \geq 3$, $I = 1 - e^{-\frac{\gamma}{\overline{\gamma}_i}}$, $B_j = 1 - e^{-\frac{\gamma}{\overline{\gamma}_j}}$,

$$F(\gamma) = \left(\prod_{j \in \mathcal{J}} \frac{\overline{\gamma}_i}{\overline{\gamma}_i - \overline{\gamma}_j}\right) \left(\prod_{i' \in \mathcal{I} \setminus \{i\}} \frac{\overline{\gamma}_i}{\overline{\gamma}_i - \overline{\gamma}_{i'}}\right) I + \sum_{j \in \mathcal{J}} \left(\frac{\overline{\gamma}_j}{\overline{\gamma}_j - \overline{\gamma}_i} \prod_{j' \in \mathcal{J} \setminus \{j\}} \frac{\overline{\gamma}_j}{\overline{\gamma}_j - \overline{\gamma}_{j'}}\right) B_j \times \\ \sum_{i' \in \mathcal{I} \setminus \{i\}} \left(\frac{\overline{\gamma}_{i'}}{\overline{\gamma}_{i'} - \overline{\gamma}_i} \prod_{i'' \in \mathcal{I} \setminus \{i, i'\}} \frac{\overline{\gamma}_{i'}}{\overline{\gamma}_{i'} - \overline{\gamma}_{i''}}\right) B_{i'}. \tag{3}$$

and $B_{i'} = 1 - e^{-\frac{\gamma}{\overline{\gamma}_{i'}}}$. Note that for DF based relay transmission, the choice of an AMC mode only depends on the post-processing SNR at the corresponding destination. Given the available AMC modes as well as the minimum required SNR threshold Γ_c for each mode c, the probability of using mode c for target MS i can be calculated as $\alpha_{i,c}(\mathcal{J}, \mathcal{I}) = F(\Gamma_{c+1}) - F(\Gamma_c)$. The transmission rate of the target MS i can then be obtained from

$$R_i(\mathcal{J}, \mathcal{I}) = \sum_{c \in \mathcal{C}} \rho_c \alpha_{i,c}(\mathcal{J}, \mathcal{I}), \tag{4}$$

where $\mathcal{C}$ is a set of AMC modes, and ρ_c is the transmission rate of AMC mode c in packets/frame.

3.3 Hierarchical Coalition Formation Game

If RSs and MSs, in a service area of IEEE 802.16m relay network, are rational, they can cooperate by forming coalitions such that their individual payoffs are maximized. For MS i, the payoff is defined as a function of transmission rate and cost of relay transmission as follows:

$$U_i(\mathcal{J},\mathcal{I}) = R_i(\mathcal{J},\mathcal{I}) - \beta_i(|\mathcal{I}| - 1), \tag{5}$$

for $i \in \mathcal{I}$ and $J(i) \in \mathcal{J}$, where β_i is the cost factor of MS i. Note that the cost of relay transmission increases as the number of members in the same coalition increases. Similarly, for RS j, the payoff is defined as the total transmission rate of all subscribed MSs and cost of relay transmission as follows:

$$V_j(\mathcal{J}) = \sum_{i\in\mathcal{N}_j} \overline{R}_i(\mathcal{J}) - \beta_j \sum_{j'\in\mathcal{J}\setminus j} |\mathcal{N}_{j'}|, \tag{6}$$

where $\overline{R}_i(\mathcal{J})$ is the average transmission rate of MS i given coalition $\mathcal{J}$ of RSs[3]. Given the payoff functions of MSs and RSs defined in (5) and (6), respectively, there is a tradeoff for them as the rational entities to form the coalition. In particular, forming coalition can help the target MS to gain higher transmission rate from relay transmission. However, it may incur too high cost if the coalition is composed of many members. In addition, MSs and RSs form different coalitions. To address this coalition formation in IEEE 802.16m relay network, the hierarchical coalition formation game model is proposed to analyze the stability of cooperation behavior of RSs and MSs. The hierarchical coalition formation game model is composed of two coalitional games, i.e., coalition formation of RSs and MSs at the upper and lower levels, respectively. These two coalitional games are interrelated in which the coalition of RSs will affect the performance of MSs, and, hence, the coalition formation of MSs. Also, the coalition formation of MSs affects the transmission rate of each other in which the RSs have to take this effect into account. The details of the hierarchical coalition formation game model is presented in the next section.

4 Hierarchical Coalition Formation Game

In this section, first the definition of the coalition game of RSs and MSs is given (Section 4.1). The stable coalitional structure is analyzed (Section 4.2).

4.1 Game Definition

RSs and MSs will form coalitions separately, but their resulting coalitions can influence the decision of each other. We assume that RSs and MSs are rational

[3] This average transmission rate $\overline{R}_i(\mathcal{J})$ will be obtained later in this paper, specifically in (20).

to maximize their individual payoffs. Also, RSs can form coalition before MSs, and MSs can fully observe the coalition of RSs (i.e., by checking the relayed signal). By adopting the concept of Stackelberg game, in the hierarchical coalition formation game model, the leaders are RSs while the followers are MSs. In this case, first, MSs will form the coalition according to the coalition of RSs such that their individual payoffs are maximized. The RSs (i.e., the leaders) having this knowledge will form their coalition accordingly. Note that the proposed hierarchical coalitional game has a *non-transferable utility (NTU)*, since the value (i.e., transmission rate minus cost) of any coalition of RSs and MSs cannot be transferred (divided) arbitrarily among the members of a given coalition. Let $\mathcal{J}_x$ and $\mathcal{I}_y$ denote the coalitions of RSs and MSs, respectively, where x and y are the indexes of coalitions. We can define the coalitional structures of RSs and MSs as follows: r_u denotes the coalitional structure of RSs defined as $r_u = \{\dots, \mathcal{J}_x, \dots\}$, such that $\mathcal{R} = \bigcup_{\mathcal{J}_x \in r_u}$, where u is an index of coalitional structure. Similarly, m_v denotes the coalitional structure of MSs defined as $m_v = \{\dots, \mathcal{I}_y, \dots\}$, such that $\mathcal{N} = \bigcup_{\mathcal{I}_y \in m_v} \mathcal{I}_y$, where v is the index.

Consider example in Fig. 1, the coalitional structures of three RSs (i.e., A, B, and C) are defined as follows: $r_1 = \{\{A\},\{B\},\{C\}\}$, $r_2 = \{\{A,B\},\{C\}\}$, $r_3 = \{\{A,C\},\{B\}\}$, $r_4 = \{\{A\},\{B,C\}\}$, and $r_5 = \{\{A,B,C\}\}$. The coalitional structures of four MSs (i.e., 1, 2, 3, and 4) subscribed to RS A are defined as follows: $m_1 = \{\{1\},\{2\},\{3\},\{4\}\}, m_2 = \{\{1,2\},\{3\},\{4\}\}, m_3 = \{\{1\},\{2\},\{3,4\}\}$, $m_4 = \{\{1,3\},\{2\},\{4\}\}$, $m_5 = \{\{1\},\{3\},\{2,4\}\}$, $m_6 = \{\{1,4\},\{2\},\{3\}\}$, $m_7 = \{\{1\},\{4\},\{2,3\}\}$, $m_8 = \{\{1,2\},\{3,4\}\}$, $m_9 = \{\{1,3\},\{2,4\}\}$, $m_{10} = \{\{1,4\}, \{2,3\}\}$, $m_{11} = \{\{1,2,3\},\{4\}\}$, $m_{1,2} = \{\{1,2,4\},\{3\}\}$, $m_{13} = \{\{1,3,4\},\{2\}\}$, $m_{14} = \{\{1\},\{2,3,4\}\}$, and $m_{15} = \{\{1,2,3,4\}\}$.

Given coalition $\mathcal{J}$ of RS $j \in \mathcal{J}$ such that $j = J(i)$, the actions of MS i in forming coalition are as follows [16]:

- *Joining:* Let $\mathbb{M}_{\text{jn}}$ denote the set of candidate coalitions of MSs that can join together to form a new single coalition $\mathcal{I}_{y'}$. If all MSs $i \in \mathcal{I}_y \in \mathbb{M}_{\text{jn}}$ can gain higher individual payoffs, i.e.,

$$U_i(\mathcal{J}, \mathcal{I}_{y'}) \geq U_i(\mathcal{J}, \mathcal{I}_y), \quad \forall i \in \mathcal{I}_y, \tag{7}$$

  where $\mathcal{I}_{y'} = \bigcup_{\mathcal{I}_y \in \mathbb{M}_{\text{jn}}} \mathcal{I}_y$, then the coalitions can decide to join together.
- *Splitting:* Given a coalition $\mathcal{I}_y$, the MSs in this coalition can split (i.e., be partitioned) into multiple new coalitions $\mathcal{I}_{y'}$, if all the MSs $i \in \mathcal{I}_y$ can gain higher individual payoffs, i.e.,

$$U_i(\mathcal{J}, \mathcal{I}_{y'}) \geq U_i(\mathcal{J}, \mathcal{I}_y), \quad \forall i \in \mathcal{I}_y, \tag{8}$$

  where $\mathcal{I}_y = \bigcup_{\mathcal{I}_{y'} \in \mathbb{M}_{\text{sp}}} \mathcal{I}_{y'}$ and $\mathbb{M}_{\text{sp}}$ is the set of new coalitions of MSs.

If RSs observe the coalition formation of MSs at the steady state and then perform their own coalition formation, the payoff function of RS (i.e., previously defined in (6)) can be expressed as a function of its own coalition and a set of coalitional structure of MSs, i.e., $V_j(\mathcal{J}, \Xi^\dagger(r))$. $\Xi^\dagger(r) = \{\dots, m^\dagger(r), \dots\}$ is a set of coalitional structure $m^{\dagger}(r)$ of MSs at the steady state given coalitional structure r of RSs. The actions of RS j are as follows:

- *Joining:* Let $\mathbb{R}_{\text{jn}}$ denote the set of candidate coalitions of RSs that can join together to form a new single coalition $\mathcal{J}_{x'}$. If all RSs $j \in \mathcal{J}_x \in \mathbb{R}_{\text{jn}}$ can gain higher individual payoffs, i.e.,

$$V_j(\mathcal{J}_{x'} \in r', \Xi^{\dagger}(r')) \geq V_j(\mathcal{J}_x \in r, \Xi^{\dagger}(r)), \quad \forall j \in \mathcal{J}_x, \tag{9}$$

  where $\mathcal{J}_{x'} = \bigcup_{\mathcal{J}_x \in \mathbb{R}_{\text{jn}}} \mathcal{J}_x$, then the coalitions can decide to join together.
- *Splitting:* Given a coalition $\mathcal{J}_x$, the RSs in this coalition can split (i.e., be partitioned) into multiple new coalitions $\mathcal{J}_{x'}$, if all the RSs $j \in \mathcal{J}_x$ can gain higher individual payoffs, i.e.,

$$V_j(\mathcal{J}_{x'} \in r', \Xi^{\dagger}(r')) \geq V_j(\mathcal{J}_x \in r, \Xi^{\dagger}(r)), \quad \forall j \in \mathcal{J}_x, \tag{10}$$

  where $\mathcal{J}_x = \bigcup_{\mathcal{J}_{x'} \in \mathbb{R}_{\text{sp}}} \mathcal{J}_{x'}$ and $\mathbb{R}_{\text{sp}}$ is the set of new coalitions of RSs.

The stable coalitional structure (i.e., a set of coalitions) is considered to be the solution. For MSs as the followers, given the coalitional structure r of RSs, the stable coalitional structure $m^*(r)$ of MSs can be defined based on the following condition:

$$U_i(\mathcal{J} \in r, \mathcal{I}^*) \geq U_i(\mathcal{J} \in r, \mathcal{I}), \quad \forall i \in \mathcal{N}, \tag{11}$$

for all $\mathcal{I}^* \in m^*(r)$ and $\mathcal{I} \in m(r)$ where $m^*(r) \neq m(r)$. In this case, coalitional structures $m^*(r)$ and $m(r)$ of MSs are defined as the function of coalitional structure r of RSs. From (11), at the stable coalitional structure $m^*(r)$ of MSs given coalitional structure r of RSs, none of MS can join or split and result in the new coalitions which improve the payoff if other MSs keep the coalition unchanged.

Let $\Xi^*(r) = \{\ldots, m^*(r), \ldots\}$ denote the set of stable coalitional structure $m^*(r)$ of MSs at the steady state given coalitional structure r of RS. For RSs as the leaders, the stable coalitional structure r^* can be defined based on the following condition:

$$V_j(\mathcal{J}^* \in r^*, \Xi^*(r^*)) \geq V_j(\mathcal{J} \in r, \Xi^*(r)), \quad \forall j \in \mathcal{R}, \tag{12}$$

where $r^* \neq r$. From (12), at the stable coalitional structure r^* of RSs, none of RS can join or split and result in the new coalitions which improve the payoff if other RSs keep the coalition unchanged given that the MSs are at their stable coalitional structure.

To obtain this stable coalitional structures of RSs and MSs, the Markov model will be developed.

4.2 Stable Coalitional Structure

To analyze the stability of the coalition formation game, a Markov model can be used [17]. Since RSs and MSs can form coalitions separately (but can influence each other), we can develop the hierarchical Markov model which is composed of two models for MSs and RSs at the lower and upper levels, respectively. The state of each Markov is defined as the coalitional structure[4]. In this case, the MSs form coalition and reach the stable coalitional structure. Then, the RSs will observe the stable coalitional structure of MSs and form coalition accordingly until the stable coalitional structure of RSs is reached.

Markov Model of Coalition Formation of Mobile Stations. We first analyze the dynamics of coalition formation of MSs. The state space of Markov model for the coalition formation of MSs is defined as a function of coalitional structure r of RSs as follows:

$$\Omega(r) = \{m_v \mid v = \{1, \ldots, D_{|\mathcal{N}|}\}\}, \tag{13}$$

where m_v represents a coalitional structure (spanning all MSs). $D_{|\mathcal{N}|}$ is the Bell number obtained from

$$D_i = \sum_{j=0}^{i-1} \binom{i-1}{j} D_j, \text{ for } i \geq 1, \text{ and } D_0 = 1. \tag{14}$$

The transition probability matrix of the Markov model for the coalition formation of MSs is expressed again as a function of coalitional structure r of RSs as follows:

$$\mathbf{P}(r) = \begin{bmatrix} P_{m_1,m_1}(r) & \cdots & P_{m_1,m_{D_{|\mathcal{N}|}}}(r) \\ P_{m_2,m_1}(r) & \cdots & P_{m_2,m_{D_{|\mathcal{N}|}}}(r) \\ \vdots & \ddots & \vdots \\ P_{m_{D_{|\mathcal{N}|}},m_1}(r) & \cdots & P_{m_{D_{|\mathcal{N}|}},m_{D_{|\mathcal{N}|}}}(r) \end{bmatrix}, \tag{15}$$

where $P_{m_v,m_{v'}}(r)$ is the probability of transition from state m_v to $m_{v'}$. Let $\mathbb{P}_{m_v,m_{v'}}$ denote the set of candidate MSs which are bound to make a coalition formation decision which will result in the change of the coalitional structure from m_v to $m_{v'}$. This transition probability $P_{m_v,m_{v'}}(r)$ can be obtained from

$$P_{m_v,m_{v'}}(r) = \begin{cases} \prod_{i \in \mathbb{P}_{m_v,m_{v'}}} \delta_{\text{ms}} \theta_i(m_{v'} | m_v, r), & \text{if } m_v \twoheadrightarrow m_{v'}, \\ 0, & \text{otherwise}, \end{cases} \tag{16}$$

where $m_v \twoheadrightarrow m'_v$ is a feasibility condition. In particular, if a coalitional structure $m_{v'}$ is reachable from m_v given the decision of all MSs, then the condition $m_v \twoheadrightarrow m_{v'}$ is true. Otherwise, condition $m_v \twoheadrightarrow m_{v'}$ becomes false. δ_{ms} is the probability that an MS makes a decision (e.g., $\delta = 0.5$). $\theta_i(m_{v'}|m_v, r)$ is the best-reply rule of MS i. That is, $\theta_i(m_{v'}|m_v, r)$ is the probability that MS i changes decision, and then the coalitional structure changes from m_v to $m_{v'}$

[4] In the rest of this paper, the terms "state" of Markov chain and "coalitional structure" of coalition formation game are used interchangeably.

given coalitional structure r of RSs. This best-reply rule of MS i is defined as follows:

$$\theta_i(m_{v'}|m_v, r) = \begin{cases} \hat{\theta}, \text{ if } U_i(\mathcal{J}, \mathcal{I}_y \in m_{v'}) \geq U_i(\mathcal{J}, \mathcal{I}_y \in m_v), \\ \epsilon, \text{ otherwise}, \end{cases} \tag{17}$$

where $0 < \hat{\theta} \leq 1$ is a constant (e.g., $\hat{\theta} = 0.1$), and ϵ is a small probability that the MS makes an irrational decision. It is assumed that the MS can make an irrational coalition formation decision due to lack of information or the need for "exploration" in the learning process.

The diagonal element of matrix $\mathbf{P}(r)$ defined in (15) is obtained from

$$P_{m_v,m_v}(r) = 1 - \sum_{m_{v'} \in \Omega(r) \backslash \{m_v\}} P_{m_v,m_{v'}}(r). \tag{18}$$

Given the coalitional structure r of RSs, the stable coalitional structure $m^*(r)$ of MSs can be obtained. This stable coalitional structure exhibits internal and external stability [17]. Internal stability implies that, given a coalition, no MS in this coalition has an incentive to leave this coalition and act alone (non-cooperatively as a singleton), since the payoff any MS receives in the coalition is higher than that received when acting non-cooperatively. External stability implies that, in a given partition, no MS can improve its payoff by switching its current coalition and join another one. In particular, a coalitional structure $m^*(r)$ is said to be *stable*, if the conditions for internal and external stability are verified for all the coalitions in $m^*(r)$. A stable coalitional structure $m^*(r)$ of MSs can be identified from the stationary probability of the Markov model defined with state space in (13) and transition probability in (15). The stationary probability of the Markov model for the coalition formation of MSs can be obtained by solving

$$\boldsymbol{\pi}^T(r)\mathbf{P}(r) = \boldsymbol{\pi}^T(r), \quad \text{and} \quad \boldsymbol{\pi}^T(r)\mathbf{1} = 1, \tag{19}$$

where $\boldsymbol{\pi}(r) = \left[\pi_{m_1}(r) \cdots \pi_{m_v}(r) \cdots \pi_{m_{D_{|\mathcal{N}|}}}\right]^T$ is a vector of stationary probabilities and $\pi_{m_v}(r)$ is the probability that the coalitional structure m_v will be reached given the coalitional structure r of RSs. $\mathbf{1}$ is a vector of ones.

If the probability of irrational decisions approaches zero (i.e., $\epsilon \to 0^+$), there could be an ergodic set $\mathbb{E}_{\mathrm{ms}}(r) \subseteq \Omega(r)$ of states m_v in the Markov model for coalition formation of MSs defined by the state space in (13) and the transition probability in (15). This ergodic set $\mathbb{E}_{\mathrm{ms}}(r)$ exists if $P_{m_v,m_{v'}}(r) = 0$ for $m_v \in \mathbb{E}_{\mathrm{ms}}(r)$ and $m_{v'} \notin \mathbb{E}_{\mathrm{ms}}(r)$, and no nonempty proper subset of $\mathbb{E}_{\mathrm{ms}}(r)$ has this property. In this regard, the singleton ergodic set is the set of absorbing states.

Once all MSs reach the state in an ergodic set, they will remain in this ergodic set forever. Therefore, the absorbing state is referred to as the stable coalitional structure $m^*(r)$, and $\mathbb{E}_{\mathrm{ms}}(r)$ is the set of stable coalitional structures of MSs where $m^*(r) \in \mathbb{E}_{\mathrm{ms}}(r)$. With this stable coalitional structure, no MS has an incentive to change the decision given the prevailing coalitional structure.

Given the coalitional structure r of RSs, the average transmission rate of MS i (i.e., used to calculate the payoff of RS i as defined in (6)) can be obtained from

$$\overline{R}_i(\mathcal{J}) = \sum_{m_v \in \Omega(r)} \pi_{m_v}(r) \left(\sum_{\mathcal{I}_y \in m_v} R_i(\mathcal{J}, \mathcal{I}_y) \right), \tag{20}$$

for $\mathcal{J} \in r$ where $R_i(\mathcal{J}, \mathcal{I})$ is the transmission rate of MS i which can be obtained from (4) given coalition $\mathcal{J}$ of RSs and coalition $\mathcal{I}_y$ of MSs.

Markov Model of Coalition Formation of Relay Stations. We then analyze the dynamics of coalition formation of RSs. The state space of Markov model for the coalition formation of RSs is defined as follows:

$$\Psi = \{r_u \mid u = \{1, \ldots, D_{|\mathcal{R}|}\}\}, \tag{21}$$

where r_u represents a coalitional structure (spanning all RSs), and $D_{|\mathcal{R}|}$ is the Bell number obtained from (14).

The transition probability matrix of the Markov model for the coalition formation of RSs is expressed as follows:

$$\mathbf{Q} = \begin{bmatrix} Q_{r_1,r_1} & Q_{r_1,r_2} & \cdots & Q_{r_1,r_{D_{|\mathcal{R}|}}} \\ Q_{r_2,r_1} & Q_{r_2,r_2} & \cdots & Q_{r_2,r_{D_{|\mathcal{R}|}}} \\ \vdots & \ddots & \vdots & \\ Q_{r_{D_{|\mathcal{R}|}},r_1} & Q_{r_{D_{|\mathcal{R}|}},r_2} & \cdots & Q_{r_{D_{|\mathcal{R}|}},r_{D_{|\mathcal{R}|}}} \end{bmatrix}, \tag{22}$$

where $Q_{r_u,r_{u'}}$ is the probability of transition from state r_u to $r_{u'}$. Let $\mathbb{Q}_{r_u,r_{u'}}$ denote the set of candidate RSs which can make coalition formation decision and result in the change of the coalitional structure from r_u to $r_{u'}$. Similar to (16), this transition probability $Q_{r_u,r_{u'}}$ can be obtained from

$$Q_{r_u,r_{u'}} = \begin{cases} \prod_{j \in \mathbb{Q}_{r_u,r_{u'}}} \delta_{\text{rs}} \phi_j(r_{u'}|r_u), \text{ if } r_u \twoheadrightarrow r_{u'}, \\ 0, \qquad \text{otherwise,} \end{cases} \tag{23}$$

where again $r_u \twoheadrightarrow r'_u$ is a feasibility condition. δ_{rs} is the probability that an RS makes a decision. $\phi_j(r_{u'}|r_u)$ is the best-reply rule of RS j. That is, $\phi_j(r_{u'}|r_u)$ is the probability that RS j changes decision, and, hence, the coalitional structure changes from r_u to $r_{u'}$. This best-reply rule of RS j is defined as follows:

$$\phi_j(r_{u'}|r_u) = \begin{cases} \hat{\phi}, \text{ if } V_j(\mathcal{J} \in r_{u'}) \geq V_j(\mathcal{J} \in r_u), \\ \epsilon, \text{ otherwise,} \end{cases} \tag{24}$$

where $0 < \hat{\phi} \leq 1$ is a constant (e.g., $\hat{\phi} = 0.1$), and ϵ is the probability of RS to make an irrational decision. Diagonal element of matrix $\mathbf{Q}$ defined in (22) can be obtained from

$$Q_{r_u,r_u} = 1 - \sum_{r_{u'} \in \Psi \setminus \{r_u\}} Q_{r_u,r_{u'}}. \tag{25}$$

Similar to that of MSs, the stable coalitional structure of RSs can be analyzed. Specifically, a coalitional structure r^* is said to be *stable*, if the conditions for internal and external stability are verified for all the coalitions in r^*. To obtain the stable coalitional structure r^* of RSs, the transition probability in (22) is used. The stationary probability of the coalitional structure of RSs can be obtained by solving

$$\boldsymbol{\mu}^T \mathbf{P} = \boldsymbol{\mu}^T, \quad \text{and} \quad \boldsymbol{\mu}^T \mathbf{1} = 1, \tag{26}$$

where $\boldsymbol{\mu} = \begin{bmatrix} \mu_{r_1} & \cdots & \mu_{r_u} & \cdots & \mu_{r_{D_{|\mathcal{R}|}}} \end{bmatrix}^T$ is a vector of stationary probabilities and μ_{r_u} is the probability that the coalitional structure r_u will be reached.

For $\epsilon \to 0^+$, there could be an ergodic set $\mathbb{E}_{\text{rs}} \subseteq \Psi$ of states r_u in the Markov model defined by the state space in (21) and the transition probability in (22). This ergodic set $\mathbb{E}_{\text{rs}}$ exists if $Q_{r_u, r_{u'}} = 0$ for $r_u \in \mathbb{E}_{\text{rs}}$ and $r_{u'} \notin \mathbb{E}_{\text{rs}}$, and no nonempty proper subset of $\mathbb{E}_{\text{rs}}$ has this property. In this regard, the singleton ergodic set is the set of absorbing states. The absorbing state is referred to as the stable coalitional structure r^*, and $\mathbb{E}_{\text{rs}}$ is a set of stable coalitional structures of RSs, where $r^* \in \mathbb{E}_{\text{rs}}$. With this stable coalitional structure, no RS has an incentive to change the decision given the prevailing coalitional structure.

While (19) and (26) are used to obtain the stationary probabilities for the coalition formations of MSs and RSs, respectively, the joint stationary probability for coalitional structures m_v and r_u for MSs and RSs, respectively, can be obtained from

$$\psi_{r_u, m_v} = \pi_{m_v}(r_u)\mu_{r_u}. \tag{27}$$

The average payoff of MS i can be obtained

$$\overline{U}_i = \sum_{r_u \in \Psi} \sum_{m_v \in \Omega(r_u)} \psi_{r_u, m_v} U_i(\mathcal{J} \in r_u, \mathcal{I} \in m_v) \tag{28}$$

and the average payoff of RS j can be obtained from

$$\overline{V}_j = \sum_{r_u \in \Psi} \mu_{r_u} V_j(\mathcal{J} \in r_u). \tag{29}$$

5 Performance Evaluation

5.1 Parameter Setting

The relay transmission based on IEEE 802.16m is considered. There is single base station (BS) and three relay stations (RSs) whose set is denoted by $\mathcal{R} = \{A, B, C\}$. There are twelve advanced mobile stations (MSs) whose set is denoted by $\mathcal{N} = \{1, \ldots, 12\}$. We consider the case that only the MSs subscribed to the same RS can perform relay transmission for each other. The SNR from BS to all MSs is 5dB. The SNR from RSs to their subscribed MSs is 5dB. The SNR from RSs A, B, and C to MSs 5-8, 9-12, and 1-4 is 3dB, respectively. The SNR from RSs A, B, and C to MSs 9-12, 1-4, and 5-8 is 1dB, respectively. The SNR among MSs subscribed to the same RS is 3dB. The cost factor of MSs is 0.2, and that of RSs is 0.05.

5.2 Numerical Results

Payoff under Different Coalition Fig. 2 shows the payoff of MS 1 under different coalitions of RSs and MSs for which the cost factor of MS is 0.3. It is observed that when RS A which MS 1 is subscribed to does not form any coalition, the payoff is the lowest, and as there are more members in the coalition with MS 1, the payoff increases. This result is from the fact that without relay transmission from other RSs, MS 1 can gain higher transmission rate and higher payoff only by forming the coalition with other MSs. However, if RS A forms coalition $\{A, B, C\}$, the transmission rate of MS 1 is high (the top curve in Fig. 2). In this case, MS 1 may not achieve the highest payoff if it forms coalition $\{1, 2, 3, 4\}$ since the cost of relay transmission for MSs 2, 3, and 4 is higher than the transmission rate gained from them performing relay transmission for MS 1. As a result, MS 1 will split from coalition $\{1, 2, 3, 4\}$ to $\{1, 2, 3\}$ which yields higher payoff due to lower cost. To analyze this complex decision making process of RSs and MSs, an analytical model would be required.

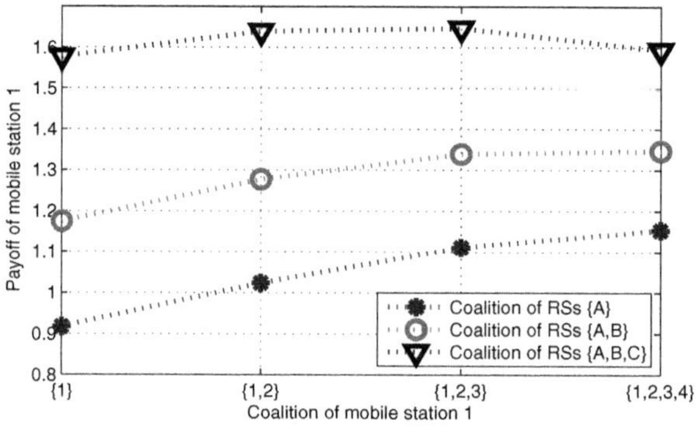

Fig. 2. Payoff of MS 1 under different coalitions of RSs and MSs

Performance under Varied SNR Fig. 3 shows the stable coalitional structure of RS under different SNR from BS to MSs. As expected, when the SNR from BS increases, the transmission rate from BS to the MS increases. Therefore, the cooperation among MSs and RSs is not necessary, and they split into smaller coalitions. The similar result is also observed when the SNR from RS and MS changes. For example, if the SNR from RS B to MSs subscribed to RS A increases, it is likely that RS A will form coalition with RS B. We omit this similar result for brevity of the paper.

Fig. 4 shows the minimum SNR between MS 4 and MSs 1, 2, 3 such that MS 4 will join coalition with MSs 1, 2, and 3. As the SNR among MSs 1, 2, and 3 increases, the minimum SNR between MS 4 and MSs 1, 2, 3 to maintain the stable coalition increases. Since the SNR among MSs 1, 2, and 3 increases, they gain high transmission rate without relay transmission from MS 4. Therefore, for MS 4 to join the coalition, the SNR between MS 4 and other MSs has to be

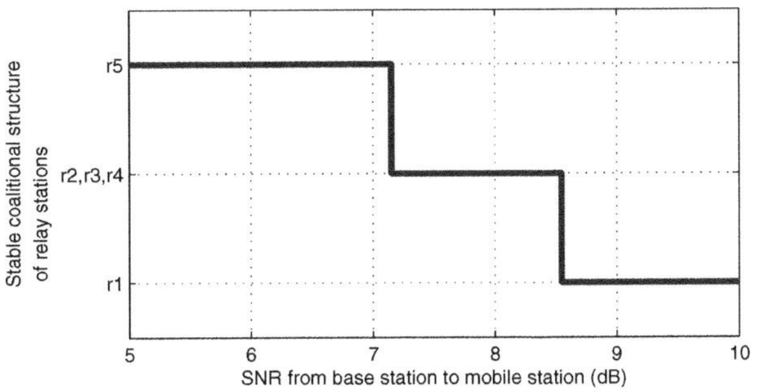

Fig. 3. Stable coalitional structure under SNR from base station

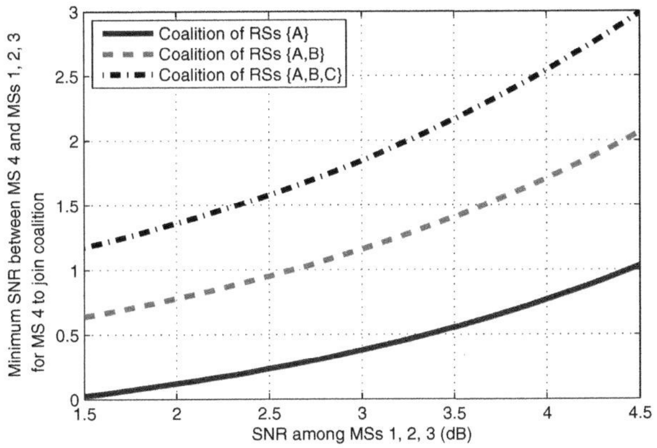

Fig. 4. Maximum SNR between MS 4 and MSs 1, 2, 3 such that the MS 4 will join coalition

high enough. Also, given different coalition of RSs (i.e., the bigger coalition of RSs, the higher minimum SNR between MS 4 and other MSs to maintain stable coalition).

Based on the above results, the coalition formation among RSs and MSs can be affected by various factors. The proposed hierarchical coalition formation game will be useful to analyze and obtain the stable coalition formation under complex environment of IEEE 802.16m relay network.

6 Summary

In this paper, the novel hierarchical coalition formation game has been proposed to model the cooperation among rational relay stations and mobile stations jointly. The cooperation is performed through the relay transmission which can improve the performance, but also incurs the cost. This hierarchical coalition formation game is similar to the Stackelberg game in which relay stations are considered to be the leaders making cooperation decision before mobile stations (i.e., the followers). To analyze the stable coalitional structure of the coalition formation, the hierarchical Markov model has been developed. The extensive performance evaluation has been performed. The numerical results show that cost of relay transmission and channel quality have the effect to the coalition formation of both RSs and MSs.

For the future work, the queueing dynamics of the mobile stations due to the relay transmission will be considered in the coalition formation.

Acknowledgment. This work was done in the Centre for Multimedia and Network Technology (CeMNet) of the School of Computer Engineering, Nanyang Technological University, Singapore. This work was supported by the Research Council of Norway through projects 176773/S10, 197565/V30, and 183311/S10.

References

1. Papapanagiotou, I., Toumpakaris, D., Lee, J., Devetsikiotis, M.: A survey on next generation mobile WiMAX networks: objectives, features and technical challenges. IEEE Communications Surveys & Tutorials 11(4), 3–18 (2009)
2. Yang, Y., Hu, H., Xu, J., Mao, G.: Relay technologies for WiMAX and LTE-advanced mobile systems. IEEE Communications Magazine 47(10), 100–105 (2009)
3. Loa, K., Wu, C.C., Sheu, S.T., Yuan, Y., Chion, M., Huo, D., Xu, L.: IMT-advanced relay standards. IEEE Communications Magazine 48(8), 40–48 (2010)
4. Genc, V., Murphy, S., Yu, Y., Murphy, J.: IEEE 802.16J relay-based wireless access networks: an overview. IEEE Wireless Communications 15(5), 56–63 (2008)
5. Zhanjun, L., Chunlin, C., Yun, L., Qilie, L., Hongcheng, Z.: Cooperative relay selection strategies in two-hop IEEE 802.16 relay networks. In: Proceedings of International Conference on Future Computer and Communication (ICFCC), vol. 2, V2-504–V2-508 (May 2010)
6. Kuo, W.-H., Lee, J.-F.: Multicast recipient maximization in IEEE 802.16j WiMAX relay networks. IEEE Transactions on Vehicular Technology 59(1), 335–343 (2010)
7. Yang, D., Fang, X., Xue, G., Tang, J.: Relay station placement for cooperative communications in WiMAX networks. In: Proceedings of IEEE Global Telecommunications Conference (GLOBECOM), pp. 1–5 (December 2010)
8. Han, Z., Niyato, D., Saad, W., Başar, T., Hjørungnes, A.: Game Theory in Wireless and Communication Networks: Theory, Models, and Applications. Cambridge University Press (2011)
9. Chen, Y., Kishore, S.: A game-theoretic analysis of decode-and-forward user cooperation. IEEE Transactions on Wireless Communications 7(5), 1941–1951 (2008)
10. Zhang, Z., Shi, J., Chen, H.-H., Guizani, M., Qiu, P.: A cooperation strategy based on Nash bargaining solution in cooperative relay networks. IEEE Transactions on Vehicular Technology 57(4), 2570–2577 (2008)
11. Wang, B., Han, Z., Liu, K.J.R.: Distributed relay selection and power control for multiuser cooperative communication networks using Stackelberg fame. IEEE Transactions on Mobile Computing 8(7), 975–990 (2009)
12. Han, Z., Poor, H.V.: Coalition games with cooperative transmission: A cure for the curse of boundary nodes in selfish packet-forwarding wireless networks. IEEE Transactions on Communications 57(1), 203–213 (2009)

13. Ng, C.Y., Lok, T.M.: Grouping algorithm for partner selection in cooperative transmission. In: Proceedings of International Conference on Ubiquitous and Future Networks (ICUFN), pp. 133–138 (June 2010)
14. Mukherjee, A., Kwon, H.M.: A coalition game framework for decode-and-forward relay networks. In: Proceedings of IEEE Vehicular Technology Conference Fall (VTC) (September 2009)
15. Can, B., Yanikomeroglu, H., Onat, F.A., de Carvalho, E., Yomo, H.: Efficient cooperative diversity schemes and radio resource allocation for IEEE 802.16j. In: Proceedings of IEEE Wireless Communications and Networking Conference (WCNC), pp. 36–41 (March-April 2008)
16. Saad, W., Han, Z., Debbah, M., Hjørungnes, A., Basar, T.: Coalitional game theory for communication networks: A tutorial. IEEE Signal Processing Magazine 26(5), 77–97 (2009)
17. Arnold, T., Schwalbe, U.: Dynamic coalition formation and the core. Journal of Economic Behavior & Organization 49(3), 363–380 (2002)

A Game-Theoretic Framework for Resource Allocation in IEEE 802.16j Transparent Relay Networks

Hui-Tang Lin[1,2] and Ying-You Lin[1]

[1] Institute of Computer and Communication Engineering, National Cheng Kung University, Taiwan (R.O.C.)
[2] Department of Electrical Engineering, National Cheng Kung University, Taiwan (R.O.C.)
{htlin,q3897107}@mail.ncku.edu.tw

Abstract. This paper focuses on downlink resource allocation in IEEE 802.16j transparent relay networks. A game theoretical formulation is derived where a resource allocation problem is represented as a two-stage bargaining game. Based on game formulation and solutions, the proposed approach not only provides improved performance but also supports fairness among the inter-class and intra-class users according to their heterogeneities in terms of the rate requirement, channel conditions, and link types. The simulation results confirm that the proposed scheme achieves a tradeoff between effective data rate and proportional fairness while also outperforming the static scheme and guaranteeing the rate requirements.

Keywords: OFDMA, 802.16j networks, game theory, resource allocation, transparent relay.

1 Introduction

Next generation (i.e.,4G) broadband wireless access networks are designed to provide high data rate to support fixed and mobile users having different quality of service (QoS) requirements. The system based on IEEE 802.16e standards is among leading candidates for 4G wireless networks. In such networks, a single base station covers a cellular area with radius of one mile. However, its initial field trials have limited coverage and poor service for indoor as well as cell boundary users. To address drawbacks, the IEEE 802.16j standard [1] has been developed to extend 802.16e to achieve coverage extension and capacity enhancement with full backward compatibility to 802.16e Mobile Stations (MSs).

The IEEE 802.16j standard [1] specifies that a relay station which supports multi-hop relay operations can operate in two different modes: transparent and non-transparent modes. In the transparent relay mode, a transparent RS is not allowed to transmit control messages such as preamble, FCH, UL/DL-MAP and DCD/UCD. Instead, the BS directly sends control messages to MSs. Hence, the scheduling for both uplink and downlink transmission are performed at the BS centrally. In doing this, the control overhead is reduced and the end-to-end throughput is improved. However, the transparent mode only supports when all MSs are within the coverage range of the BS. In contrast, a non-transparent RS is designed to support MSs which are out of the BS

R. Jain and R. Kannan (Eds.): GameNets 2011, LNICST 75, pp. 506–515, 2012.

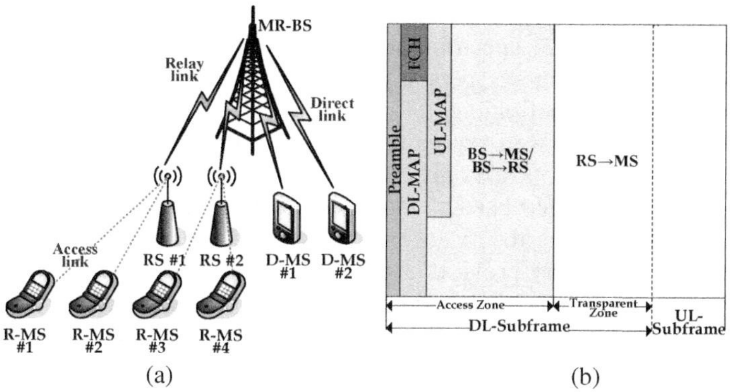

Fig. 1. IEEE 802.16j transparent relay networks: (a)topology and (b) frame structure

coverage range by transmitting/receiving control messages as well as data packets to/from those MSs. In the non-transparent mode, the transmission scheduling can be done either in a centralized or a distributed way. For both operating modes, resource allocation is one of the most important issues concerns the system throughput. Therefore, this study focuses on the resource allocation issue of 802.16j transparent relay systems and leaves the problem of non-transparent relays as the future work.

Figure 1 shows the topology and the frame structure of an 802.16j transparent relay network. In Fig. 1(a), the one-hop links between the multi-hop relay base station (MR-BS) and the transparent relay stations (T-RSs) are referred to as *relay links*, while the links between the T-RSs and the relay mobile stations (R-MSs) are referred to as *access links*. The links between the MR-BS and the directed mobile stations (D-MSs) are *direct links*. As shown in Fig. 1(b), the 802.16j transparent relay frame is comprised of the downlink and uplink subframes. Each downlink subframe is further partitioned into access and transparent zones. The MR-BS uses the access zone to transmit packets to D-MSs and T-RSs. The packets received in the access zone by a T-RS are subsequently relayed to R-MSs in the transparent zone.

Although the resource allocation problem of relay networks has been extensively investigated in [2-4], their solutions are under a fixed zone partitioning structure which ignores the 802.16j specifications. Applying these solutions under a heavy network load will result in poor system utilization due to their inability to adapt to varying network dynamics (e.g., varying traffic load to D-MSs and R-MSs, various MS channel conditions, and differing minimum reserved rates of service flows). Several resource allocation schemes based on dynamic zone partitioning (DZP) have also been proposed for relay networks [5-7]. However, the scheme proposed in [5] does not take into account the various minimum reserved traffic rates (MRTR) of different service flows in the network. Those presented in [6-7] are distributed approaches, and are unsuitable for centrally-oriented networks such as 802.16j transparent relay networks.

Recently, game theory has been widely used for resource management weighing different network constraints such as power, QoS requirements, channel conditions, etc. In particular, many studies have focused on using a non-cooperative game approach to model the bandwidth allocation problem in wireless networks. In [8], a game-theoretic based scheduler for bandwidth provisioning in 4G BWA technologies is presented. This approach simultaneously controls network congestion and fairness while providing differentiated QoS guarantees. In [9], the hierarchical bandwidth sharing problem in dynamic spectrum access is formulated as an interrelated market model from a microeconomics perspective. A similar approach has been used in [10]-[12] with varying definitions of the utility function and different network constraints. However, it is well known that Nash equilibrium is Pareto inefficient and the pricing mechanisms proposed only provide some Pareto improvements [13].

In recently years, there has been an increasing trend to use the axiomatic bargaining theory of cooperative game theory to examine resource allocation problems [13-15]. Axiomatic bargaining theory provides a good analytical framework to derive a unique desirable operation point that is fair and Pareto optimal. Accordingly, this study presents a DZP resource allocation scheme based on a two-stage bargaining game for accomplishing downlink transmission under varying network dynamics in a 802.16j transparent relay network. The game formulation is based on a bottom-up model where player utility functions in each stage reflect the corresponding bargaining solutions. According to game formulation and solutions, a tradeoff between effective data rate and proportional fairness for relay systems is provided. Simulation results show the resulting resource allocation is both fair and efficient among inter-class and intra-class users. To the best of the authors' knowledge, the proposed scheme represents the first attempt to present a game formulation for the DZP problem in 802.16j transparent relay systems.

2 Dynamic Zone Partitioning Game

In developing the DZP resource allocation scheme, this study assumes the 802.16j transparent relay network utilizes partial usage of sub-carriers (PUSC) mode in orthogonal frequency division multiple access (OFDMA) systems and operates under heavy traffic load environment. It is further assumed that the downlink subframe contains N resource slots, comprising N_{az} access zone resource slots and N_{tz} transparent zone resource slots. Finally, an assumption is made that the system consists of K service classes, indexed by the set $\Omega=\{1,2,\ldots,K\}$. For each class $k\in\Omega$, the corresponding service flows are denoted by the set $\Phi_k=\{1,2,\ldots, I_k\}$.

2.1 Two-Stage Bargaining Game

The proposed DZP resource allocation scheme is based on a two-stage bargaining game. In the first stage, resource allocation between different service classes is modeled as an *inter-class* bargaining game, in which each player represents a particular service class, i.e., a class of service flows with the same minimum reserved rate. The strategy for a player k, where all $k\in\Omega$, is the number of resource slots allocated for transmission in a downlink subframe. The payoff for service class k is evaluated using a utility parameter u_k, defined as a function of the corresponding class-level performance.

To ensure an efficient and fair distribution of the resource slots amongst all the service classes, the solution of the bargaining game is obtained using Nash Bargaining Solution (NBS). The NBS is defined as follows:

Definition 1 Nash Bargaining Solution (NBS): *Let* $C = \{k\in\Omega \mid \exists u\in U,\ u_k > u_{dis,k}\}$ *be the set of players(i.e., classes) which can achieve a utility strictly greater than the* *disagreement utility $u_{dis,k}$. Under the five general axioms* [15], *Nash showed that the unique NBS (i.e., $\phi(u^*, u_{dis,k})$) is that which maximizes the Nash Product, i.e.,*

$$\phi(u^*, u_{dis,k}) \in \arg\max_{u > u_{dis}} \prod_{k=1}^{K} (u_k(N_k) - u_{dis,k}) , \tag{1}$$

where $\sum_{k=1}^{K} N_k \leq N$, $u_k(.)$ is the utility function for class k, and N_k is the total number of resource slots allocated to class k.

Once each class has obtained its solution (i.e., N^*_k) by performing a local search method, the second stage of the bargaining game is performed to distribute the resource slots allocated to each class amongst all service flows within the class. In other words, the second stage of the bargaining game is an *intra-class* bargaining problem. For a service class *k*, each of the service flows within the class represents a player in the game. The strategy of each player *i* (where $i \in \Phi_k$) is the number of allocated resource slots received for its downlink transmission. Importantly, in obtaining the solution to the game, each player must receive an equal payoff in every frame since each service flow has the same minimum reserved rate. To achieve this goal, Egalitarian Bargaining Solution (EBS) [16] is used to obtain the bargaining solution. The EBS is defined as follows.

Definition 2 (Egalitarian Bargaining Solution (EBS)): *The total number of allocated resource slots for class k is N^*_k. Thus, $N^*_{k,i}$ is the EBS for a service flow i in the intra-class bargaining game if and only if the numbers of allocated resource slots received by any two players p and v satisfy*

$$\xi_{k,p}(N^*_{k,P}) = \xi_{k,v}(N^*_{k,v}),\ \forall p,\ v \in \Phi_k,\ p \neq v , \tag{2}$$

where $\sum_{i=1}^{I_k} N^*_{k,i} = N^*_k$ and $\xi_{k,i}(.)$ is the utility function of service flow i in class k.

2.2 Utility Formulation for Two-Stage Bargaining

When conducting a bargaining process in two stages, a critical problem is that of modeling the player utility functions in each stage in such a way as to reflect the corresponding bargaining solutions. In the proposed DZP scheme, the aim is to provide an effective throughput-fairness allocation for all service flows under varying network dynamics. Thus, it is reasonable to take a bottom-up approach to model the utility functions in the two-stage bargaining game by first deriving the bargaining power of each service flow within a class and then deriving the overall bargaining power of the class.

1) Intra-class Bargaining Utility Formulation: Let $R_k = \{r_{k,1,s}, r_{k,2,s}, \ldots, r_{k,I,s}\}$ for all $s \in$ {D-MS, R-MS} be the set of per slot transmission rates for the service flows in class *k*. To obtain EBS for the different service flows within the class, the utility function (i.e., the payoff) for service flow *i* in the current frame is assumed to be its amount of data which can be sent in this frame, i.e.,

$$\xi_{k,i}(N_{k,i,s}) = r_{k,i,s} \cdot N_{k,i,s},\ i \in \Phi_k , \tag{3}$$

where $N_{k,i,s}$ is the number of resource slots allocated to the service flow i. If flow i is destined for a D-MS, the per slot transmission rate $r_{k,i,D}$ is easily obtained from the feedback received at the MR-BS from the D-MS. However, if flow i is destined for an R-MS (denoted as R-MS i), the per slot transmission rate is given by combining the transmission rate on the relay and access links, i.e.,

$$r_{k,i,R} = 1/((r_{k,i,B\_R})^{-1} + (r_{k,i,R\_M})^{-1}) \ , \tag{4}$$

where r_{k,i,B_R} is the per slot transmission rate from the MR-BS to the T-RS associated with R-MS i, and r_{k,i,R_M} is the per slot transmission rate from T-RS to R-MS i. This utility formulation applied in Eq. (2) guarantees EBS fairness amongst the different service flows within a class.

2) Inter-class Bargaining Utility Formulation: In formulating the utility function of each class in the inter-class bargaining stage of the game, the transmission rate of a class is defined as follows.

Definition 3 (Class Transmission Rate (CTR)): *$r_{CTR,k}$ is the class transmission rate for class k and is equal to the summation of the amount of data sent in this frame of all service flows within the class divided by the total number of resource slots θ_k allocated to the class, i.e.,*

$$r_{CTR,k} = \sum_{i=1}^{I_K} \frac{(\xi_{k,i}(N_{k,i,s}))}{\theta_k}, \tag{5}$$

where $\theta_k = \sum_{i=1}^{I_K} (\xi_{k,i}(N_{k,i,s})/(r_{k,i,s}))$.

Since the EBS of the intra-class bargaining game guarantees that each flow within class k has the same payoff, the class transmission rate is obtained as follows:

$$\begin{aligned} r_{CTR,k} &= \frac{(\xi_{k,i}(N_{k,i,s}))\times I_k}{\theta_k} \\ &= \frac{(\xi_{k,i}(N_{k,i,s}))\times I_k}{\sum_{i=1}^{I_k}(\xi_{k,i}(N_{k,i,s})/(r_{k,i,s}))} = \frac{I_k}{\sum_{i=1}^{I_k} 1/(r_{k,i,s})} . \end{aligned} \tag{6}$$

Therefore, to measure the achievable utility of a particular class in each frame, the following logarithmic utility function is introduced:

$$u_k(N_k) = \log(r_{CTR} \cdot N_k - \sum_{i=1}^{I_k} \eta_{\min,k,i} - \sum_{i=1}^{I_k} c_{k,i}), \tag{7}$$

where $\eta_{min,k,i}$ and $c_{k,i}$ are the minimum guaranteed amount of data and the control message overhead (i.e., Preamble and DL/UL MAP) should be sent per frame for service flow i in class k in the current frame, respectively. Note that $\eta_{min,k,i}$ can be derived from its minimum reserved traffic rate.

2.3 Zone Partition

Once the inter-class bargaining stage has determined the allocated resources N^*_k for each class k, the number of transmission slots $N^*_{k,i,s}$ required for service flow i associated with class k can be determined by Eq. (2) and Eq. (3) as follows:

$$N^*_{k,i,s} = \left\lfloor N'_k \cdot \left(r^{-1}_{k,i,s} / \sum_{i=1}^{I_k} r^{-1}_{k,i,s} \right) \right\rfloor , \tag{8}$$

where $N_k' = N_k^* - \sum_{i=1}^{I_k} (c_{k,i} / r_{k,i,s})$. If flow i is destined to an R-MS, $N^*_{k,i,R}$ is further divided into $N^*_{k,i,R,az}$ and $N^*_{k,i,R,tz}$, i.e., the number of resource slots required for R-MS i downlink transmission in the access zone and the transparent zone, respectively. $N^*_{k,i,R\text{-},az}$ and $N^*_{k,i,R,tz}$ are expressed as

$$N^*_{k,i,R,az} = \left\lfloor N^*_{k,i,s} \cdot \left(r_{k,i,R\_M} / \left(r_{k,i,B\_R} + r_{k,i,R\_M} \right) \right) \right\rfloor \tag{9}$$

and

$$N^*_{k,i,R,tz} = \left\lceil N^*_{k,i,R,az} \cdot \left(r_{k,i,B\_R} / r_{k,i,R\_M} \right) \right\rceil . \tag{10}$$

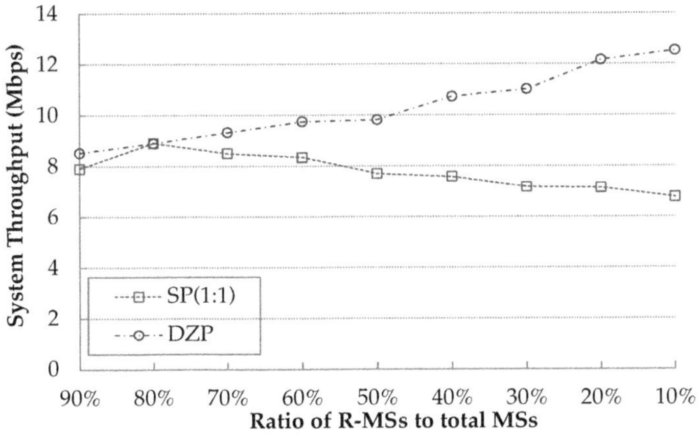

Fig. 2. System throughput for various R-MSs to total MSs ratios

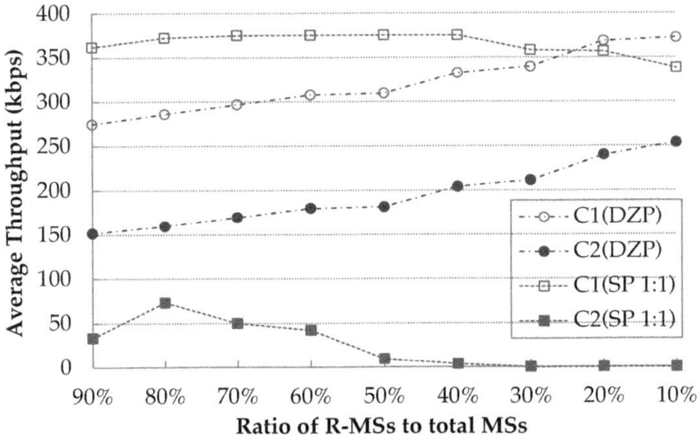

Fig. 3. Average Throughput for various R-MSs to total MSs ratios

As a result, in each downlink sub-frame, the exact number of resource slots required for data transmission in the access zone and transparent zone are given respectively by

$$N^{*}_{az} = \sum_{k=1}^{K} \sum_{i=1}^{I_k} \left(N^{*}_{k,i,D} + N^{*}_{k,i,R,az} + \left(c_{k,i} / r_{k,i,s} \right) \right) \tag{11}$$

and

$$N^{*}_{tz} = \sum_{k=1}^{K} \sum_{i=1}^{I_k} \left(N^{*}_{k,i,R,tz} \right). \tag{12}$$

Once N^{*}_{az} and N^{*}_{tz} have been derived, the partitioning of the downlink sub-frame into the access zone and the relay zone is easily achieved.

3 Simulation Results and Analysis

The performance of the proposed DZP scheme was evaluated using an 802.16j transparent relay simulation model constructed using QualNet Simulator 4.5 [17]. The simulation model included a single cell comprising a MR-BS, six T-RSs and 40 randomly-distributed MSs. The simulations assumed 20 MHz bandwidth, an OFDMA frame size of 20 ms, and the downlink-to-uplink ratio was specified as 2:1. Two different service classes were simulated, namely class 1 (C1) and class 2 (C2). Each MS has a service flow which has an equal probability of being in either C1 or C2. The packet arrival at MR-BS was modeled as a Poisson process. The traffic rates for C1 and C2 service flows were specified as 375 kbps and 256 kbps, respectively. Finally, the minimum reserved traffic rates for C1 and C2 were respectively set at 256 and 128 kbps. The performance of the proposed bargaining game-based DZP scheme was compared with Strict Priority (SP) scheme [18]. Note that SP scheme is regarded as a static scheme which has a fixed boundary between the access zone and the transparent zone with a ratio of 1:1.

In addition, to measure the user fairness within a class, the fairness metric uses the traditional Jain's Fairness Index (FI) [19].

$$\mathrm{FI} = \frac{\left(\sum_{i=1}^{N} Thr_i \right)^2}{\left(N \sum_{i=1}^{N} Thr_i^2 \right)} \tag{13}$$

where Thr_i is the average throughput of service flow i in a class and $0 \le \mathrm{FI} \le 1$.

Figure 2 shows the variation of the system throughput with respect to the ratio of R-MSs to total MSs. It can be seen that the system throughput of the SP scheme decreases with the ratio of R-MSs to total MSs. In contrast to the SP scheme, the system throughput of the proposed DZP scheme increases with increasing ratio of R-MSs to total MSs. The results show the proposed DZP scheme improves the capacity of the SP approach by up to 33% on average. This is because when the ratio of R-MSs to all MSs increases, the R-MSs need more resource slots to transmit data packets. By employing the proposed DZP scheme, the boundary between the access zone and the transparent zone can be adaptively adjusted to improve resource utilization according to the traffic rate requirements of MSs with different link types.

Figure 3 plots the C1 and C2 average throughputs of the proposed DZP scheme compared with those of SP with varying ratio of R-MSs. It can be observed that the C2 average throughput of the SP scheme is lower than its minimum reserved traffic rates due to the essence of the SP scheme. Since the proposed DZP scheme employs the NBS solutions at the first stage bargaining game, the average throughput of each class in DZP always meets the minimum reserved traffic rate required by each class despite the fact that the service flows are receiving lower allocations than the scenarios with a lower ratio of R-MSs. The SP scheme favors the MSs of C1, where C1 always has higher priority to get requested bandwidth before bandwidth being distributed to the MSs of C2. The average throughput of the proposed DZP is superior to that of the SP scheme when the ratio of R-MSs is greater than 30%. The reason is that the proposed DZP scheme is able to adaptively select a suitable zone boundary between the access zone and the transparent zone.

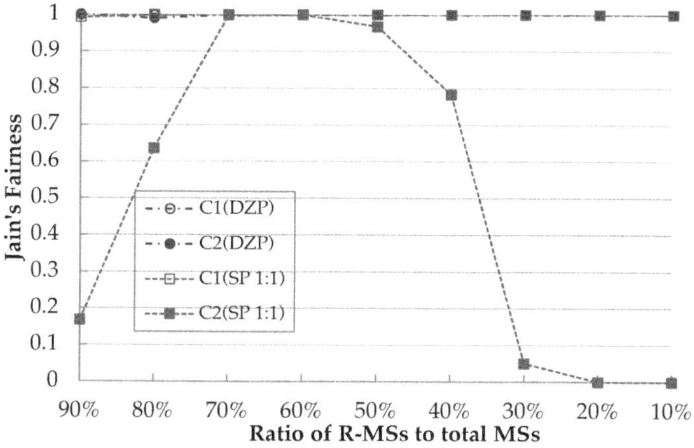

Fig. 4. Jain's Fairness for various R-MSs to total MSs ratios

Figure 4 illustrates the variation of the class fairness with respect to the ratio of R-MSs to total MSs. It shows that both C1 and C2 in the proposed DZP have almost perfect fairness among all service flows in the class while the SP approach is not fair for the C2 service flows. This is a result of the EBS of the second stage game formulation in the proposed DZP.

4 Conclusion

A dynamic zone partitioning game has been presented for solving the resource allocation problem in IEEE 802.16j transparent relay networks under varying network dynamics. The proposed scheme is able to adaptively adjust the boundary between the access zone and the transparent zone in the downlink subframe. It provides a tradeoff between fairness and throughput while meeting rate guarantees to all the service flows within the system irrespective of whether they are direct or relay MSs. The simulation results have confirmed that the proposed game-theoretic-based resource allocation scheme ensures efficient and fair bandwidth distribution amongst the different service flows within the IEEE 802.16j transparent relay network.

Acknowledgements. The authors would like to thank the National Science Council, Taiwan, R.O.C., for supporting this research under grant NSC 97-2221-E-006 -175 -MY3.

References

[1] IEEE Std 802.16j/D7: Air Interface for Fixed and Mobile Broadband Wireless Access Systems, Multihop Relay Specification

[2] Bae, C., Cho, D.: Fairness-Aware Adaptive Resource Allocation Scheme in Multihop OFDMA Systems. IEEE Communication Letters 11(2), 134–136 (2007)

[3] Salem, M., Adinoyi, A., Yanikomeroglu, H., Falconer, D., Kim, Y.-D.: A Fair Radio Resource Allocation Scheme for Unbiquitous High-data-rate Coverage in OFDMA-based Cellular Relay Networks. In: Proceedings of IEEE Global Communications Conference (GLOBECOM), Honolulu, Hawaii, USA (December 2009)

[4] Xiao, L., Cuthbert, L.: Load Based Relay Selection Algorithm for Fairness in Relay Based OFDMA Cellular Systems. In: Proceedings of IEEE Wireless Communications and Networking Conference, WCNC (April 2009)

[5] Chan, I.K., Liao, W.: Adaptive Bandwidth Allocation for TCP Traffic in IEEE 802.16j Wireless Networks with Transparent Relay Stations. In: Proceedings of IEEE Personal, Indoor and Mobile Radio Communications, PIMRC (September 2008)

[6] Wang, L., Ji, Y., Liu, F.: A Semi-distributed Resource Allocation Scheme for OFDMA Relay-Enhanced Downlink Systems. In: Proceedings of IEEE Global Communications Conference (GLOBECOM) Workshops (November 2008)

[7] Kaneko, M., Popovski, P., Hayashi, K.: Throughput-Guaranteed Resource Allocation Algorithms for Relay-Aided Cellular OFDMA System. IEEE Transactions on Vehicular Technology 58(4) (May 2009)

[8] AbuAli, N., Hayajneh, M., Hassanein, H.: Congestion-Based Pricing Resource Management in Broadband Wireless Networks. IEEE Transactions on Wireless Communications 9(8) (August 2010)

[9] Niyato, D., Hossain, E.: A Microeconomic Model for Hierarchical Bandwidth Sharing in Dynamic Spectrum Access Networks. IEEE Transactions on Computers 59(7) (July 2010)

[10] Niyato, D., Hossain, E.: Competitive Pricing in Heterogeneous Wireless Access Networks: Issues and Approaches. IEEE Network (November/December 2008)

[11] Rubaiyat Kibria, M., Jamalipour, A.: Game Theoretic Outage Compensation in Next Generation Mobile Networks. IEEE Transactions on Wireless Communications 8(5) (May 2009)

[12] Jia, J., Zhang, Q.: Competitions and Dynamics of Duopoly Wireless Service Providers in Dynamic Spectrum Market. In: Proceedings of ACM International Symposium on Mobile Ad Hoc Networking and Computing, May 26-30 (2008)

[13] Cao, X.-R., Shen, H.-X., Milito, R., Wirth, P.: Internet Pricing With a Game Theoretical Approach: Concepts and Examples. IEEE/ACM Transactions on Networking 10(2) (April 2002)

[14] Hew, S.-L., White, L.B.: Cooperative Resource Allocation Games in Shared Networks: Symmetric and Asymmetric Fair Bargaining Models. IEEE Transactions on Wireless Communications 7(11) (November 2008)

[15] Suris, J.E., DaSilva, L.A., Han, Z., MacKenzie, A.B., Komali, R.S.: Asymptotic Optimality for Distributed Spectrum Sharing using Bargaining Solutions. IEEE Transactions on Wireless Communications 8(10) (October 2009)

[16] Peters, H.J.M.: Axiomatic Bargaining Game Theory. Kluwer Academic Publishers (1992)

[17] Qualnet Simulator Home, `http://www.scalable-networks.com/`

[18] Kitti, W., Aura, G.: Packet scheduling for QoS support in IEEE 802.16 broadband wireless access systems. International Journal of Communication System 16(1), 81–96 (2003)

[19] Jain, R., Chiu, D., Hawe, W.: A quantitative measure of fairness and discrimination for resource allocation in shared computer systems. DEC Research Report TR-301 (September 1984)

Dynamic Spectrum Negotiation with Asymmetric Information

Yang Yan, Jianwei Huang, Xiaofeng Zhong, and Jing Wang

[1] Department of Electronic Engineering, Tsinghua University, Beijing, China
[2] Department of Information Engineering, CUHK, Hong Kong, China
yanyang07@mails.tsinghua.edu.cn, jwhuang@ie.cuhk.edu.hk
{zhongxf,wangj}@tsinghua.edu.cn

Abstract. Spectrum scarcity is becoming a serious issue due to the rapid development of wireless communication technology. Dynamic spectrum sharing can effectively improve the spectrum usage by allowing secondary unlicensed users (SUs) to dynamically and opportunistically share the spectrum with primary licensed users (PUs). In this paper, we investigate a spectrum negotiation mechanism under incomplete information in a dynamic environment, where both the PU and the SU can obtain rate increases through cooperative communications. Specifically, an SU relays traffics for a PU in exchange for dedicated transmission time for the SU's own communication needs. We model the bargaining process as dynamic Bayesian games and characterize the Perfect Bayesian Equilibria under different system model parameters. Analysis and numerical results indicate that both PU and SU obtain performance improvements compared with no cooperation, and thus achieve a win-win situation via the spectrum negotiation.

Keywords: dynamic spectrum negotiation, incomplete information, game theory, perfect bayesian equilibrium.

1 Introduction

Wireless spectrum is generally regarded as a scarce resource, and has been tightly controlled through licensing. Recent field measurements showed that, however, most licensed spectrum bands are heavily underutilized even in densely populated urban areas [1]. This indicates that the current fixed license-based spectrum allocation policy is not efficient. As one promising technology to address this issue, cognitive radio technology [2] enables efficient dynamic spectrum sharing among secondary *unlicensed* users (SUs) and primary *licensed* users (PUs). One way to achieve dynamic spectrum sharing is through market-driven spectrum negotiation/bargaining[1], which leads to a win-win situation by improving the payoffs of both PUs and SUs. In spectrum negotiation, PUs and SUs jointly decide (i) how PUs allocate resource to SUs, and (ii) how SUs compensate PUs either by offering monetary payments or providing performance improvements.

[1] In this paper, we use "negotiation" and "bargaining" interchangeably.

R. Jain and R. Kannan (Eds.): GameNets 2011, LNICST 75, pp. 516–529, 2012.

There are two key challenges for implementing spectrum negotiation: *asymmetric information* and *dynamics*. Spectrum negotiation often involves incomplete and asymmetric information, such as transmission power, energy cost, and channel state information of the users. This often complicates the decisions. Moreover, the negotiation can involve multiple periods, during which PUs and SUs need to update beliefs about its opponent's information and adjust their own strategies accordingly. In this paper, we present a first step on understanding how to address the above two issues for spectrum negotiation.

There are two types of spectrum negotiation mechanisms: *monetary-exchange* and *resource-exchange*. In a monetary-exchange mechanism, SUs provide (virtual) money to the PUs in exchange of spectrum resource. Reference [3] studied a sequential second-price auction mechanism for sharing a wireless resource (bandwidth or power) among competing transmitters, where a spectrum manager is responsible for allocating spectrum among non-cooperative secondary users. Reference [4] addressed spectrum pricing in a cognitive radio network, where multiple PUs compete to offer spectrum access opportunities to SUs. Reference [5] further investigated price competition with multiple PUs and multiple SUs by taking into account both bandwidth uncertainty and spatial reuse. The monetary-exchange model is suitable when PUs have temporarily idle resource. However, when PUs' own transmission demands cannot be satisfied (e.g., when PUs suffer from poor channel conditions), there will be no resource left for allocating to SUs. In this case, the resource-exchange model can be helpful.

In a resource-exchange model, SUs provide communication resources (e.g., the transmission power) to assist PUs' transmissions in exchange for spectrum usage. In this paper, we consider an resource-exchange model, where PU has a poor channel link to its receiver (e.g., base station). Our study is motivated by [6], where the remuneration of spectrum negotiation is realized by the cooperative transmission of SUs. In [6], a single PU knows utility functions of the SUs, and thus there is no asymmetric or incomplete information. The PU optimizes the resource splitting based on either instantaneous or long-term network channel state information of the whole system. Due to complete information, SUs simply follow the PU's spectrum leasing decision and there is no bargaining. Reference [7] considered a similar cooperative spectrum sharing scheme between multiple PUs and multiple SUs, again based on complete information. The only recent publication dealing with incomplete information of spectrum negotiation is [8], which proposed a contract mechanism between one PU and multiple SUs in a static network environment.

Compared with [6] and [7], we study a *dynamic* spectrum bargaining process with *incomplete information*, where both the PU and the SU have the negotiation power. The incomplete information model better captures the reality of wireless communications, but so far has received little attention in the literature due to its high analysis complexity. The most related research result in terms of methodology is the incomplete information game of wireless power control [9], where the authors studied a completely different application scenario (not related to cognitive radio).

Our paper is the first one that jointly considers incomplete information and dynamic bargaining in spectrum negotiation. The key results and contributions are summarized as follows:

- *Introduction of asymmetric information:* We model the realistic situation where the PU does not know the complete information of the SU. Specifically, we assume that PU is unaware of the exact value of SU's energy cost C, but has a belief about C's distribution. We compute the equilibrium behaviors for both PU and SU with such asymmetric information.
- *Dynamic negotiation process:* We investigate a multi-stage bargaining model, where both the PU and the SU have negotiation power. As the bargaining proceeds, PU and SU must adjust their beliefs about incomplete information and strategies accordingly. This can be modeled as a dynamic Bayesian game which is challenging to analyze. We explicitly compute and characterize multiple equilibria for a two-stage bargaining game. We demonstrate the existence of two types of equilibria, and show that one is more Pareto efficient than the other in the expected sense.

The rest of the paper is organized as follows. We introduce the system model and problem formulation in Section 2. In Section 3, we analyze the one-stage bargaining model, which serves as the basis for the two-stage bargaining model in Section 4. In Section 5, we discuss various insights obtained from the equilibrium analysis through numerical studies. We conclude in Section 6.

2 PU-SU Negotiation and Cooperation Model

Throughout this paper, we denote *primary user* and *secondary user* as PU and SU, respectively. We consider a time-slotted system, where one PU negotiates the spectrum allocation with one SU. Figure 1 introduces the key notations of the system model. Here, TP and RP represent the primary transmitter and receiver, and TS and RS represent the secondary transmitter and receiver. Let h_p, h_s, h_{ps} and h_{sp} denote the channel gains of the links between TP and RP, TS and RS, TP and TS, and TS and RP, respectively. We consider a block fading channel model, where the channel gains are fixed during one time slot and can change across time slots. We further assume that both PU and SU know the channel gains of all the links through a proper feedback mechanism[2]. The fixed transmission powers of the PU and SU are denoted as P_t and P_s, respectively. Our model can describe the situation where a cellular subscriber (PU) wants to send traffic to a base station which is far away, and a laptop user (SU) can help to relay the traffic for the cellular subscriber (see Fig. 1).

This stylized model enables us to analytically study the challenging issues of asymmetric information and dynamics. Such one-to-one model has been common in the economics literatures (e.g., [12–14]). In contrast, some recent economics

[2] In this paper, we focus on the study of incomplete information related to energy cost. The study of incomplete information related to channel conditions can be studied based a similar methodology and will be considered in the future work.

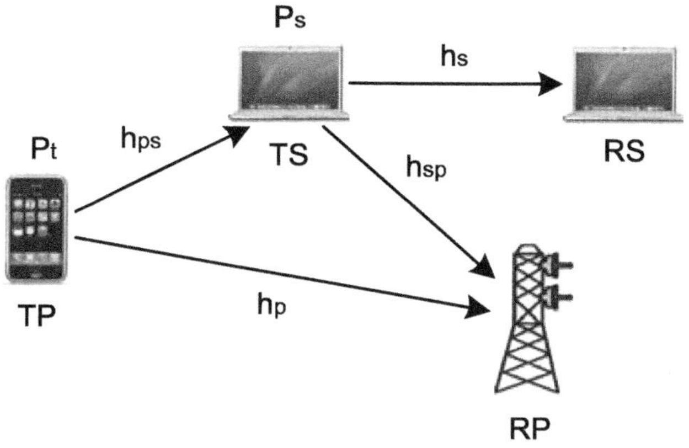

Fig. 1. PU-SU Cooperation Model

literatures (e.g., [15]) discussed more complicated models (e.g., one-to-many bargaining model) in a static or dynamic setting without information incompleteness. However, our focus in this paper is on dynamics in the bargaining with incomplete information, in which case the negotiation among multiple users is difficult to analyze even in the economics literature. We believe that a thorough understanding of the one-to-one model can help us to better tackle the one-to-many and many-to-many spectrum negotiation scenarios in the future.

Next we describe how the system works. Without spectrum negotiation, the PU directly transmits from TP to RP and achieves a data rate R_{dir} (determined by transmission power P_t and direct channel gain h_p). In that case, the SU cannot transmit and achieves a zero rate. However, spectrum negotiation becomes attractive to the PU if the direct channel h_p is poor but the relay channels h_{ps} and h_{sp} are good. In this case, the PU can increase its own data rate by using the SU transmitter TS as a relay. To attract the help from SU, the PU offers the SU α fraction of the transmission time for SU's own transmission (from TS to RS). Apparently, a larger α means a higher data for SU but a lower data rate for PU (due to the reduction of transmission time). PU and SU will bargain with each other to determine the value of α that is acceptable by both sides. Figure 2 illustrates three possibilities of the bargaining result. Without loss of generality, we normalize the time slot length T to 1, and denote $\tau \ll 1$ as the time "wasted" due to bargaining.

- Figure 2(a): At the beginning of the time slot, PU can choose to directly transmit during the whole time slot without bargaining, if it believes that SU cannot provide a performance improvement.
- Figure 2(b): If PU decides to bargain, then the unit time slot is divided into two phases: bargaining and transmission. During the bargaining phase, PU and SU negotiate on the fraction of the remaining time, α, to be allocated for the SU's own transmission. If no agreement is reached, PU proceeds with direct transmission for the remaining time without the cooperation of SU.

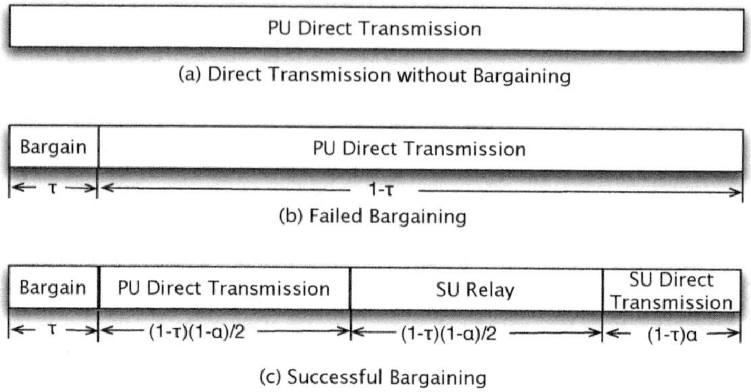

Fig. 2. Three Possibilities in Negotiation Process

- Figure 2(c): If an agreement is reached in the bargaining phase, then PU and SU transmit in an amplified-and-forward relay mode. The transmission can be further divided into three time periods. In the first period, primary transmitter TP broadcasts the data, and both primary receiver RP and secondary transmitter TS receive the information. In the second period, the secondary transmitter TS amplifies the received signal and forwards it to the primary receiver RP. The primary receiver RP combines the signal received from the first and second periods based on the maximum ratio combining scheme and decodes the signal. In the third period, SU utilizes the remaining third period (i.e., α fraction of the $(1-\tau)$ transmission time) for its own transmission.

In this paper, we assume that SU is an *energy-constrained* device (**e.g., wireless sensor or mobile device**). Parameter C is the SU's energy cost, which is related to SU's current battery status. We further assume that the precise value of C is known by SU but not by PU[3], thus the information is asymmetric. However, PU knows the distribution of C, and hence it is incomplete information. To simplify the analysis, we assume that C follows a uniform distribution[4].

Let us consider how both users make their decisions. SU chooses to accept or reject the offer α to maximize its utility function U_s, which is the difference between the achievable data rate and the energy cost. By taking the SU's response into consideration, PU chooses the optimal offer α to maximize its utility function U_p, which is the *expected data rate increase*. Details of U_s and U_p will be given shortly.

Finally, we note that the bargaining period with length τ in Fig. 2(b) and (c) can include multiple stages, i.e., users can bargain multiple times within the same

[3] If PU knows the value of C, it can simply calculate α so that SU gets zero payoff. This is a special case of the general model discussed here.

[4] The analysis for other distributions will be technically more involved but offers essentially the same engineering insights.

time slot. As a preliminary result, we first study the one-stage bargaining case in Section 3. This will help us to study the multi-stage bargaining in Section 4.

3 One-Stage Bargaining Game

In this section, we consider the case where there is *at most* one stage of bargaining in a time slot. The proportion of the time slot after bargaining is $\delta = (1-\tau) < 1$, where τ is the duration of one-stage bargaining. PU needs to decide (i) whether to bargain, and (ii) the optimal offer α if it decides to bargain. SU should decide whether to accept the offer of α (if the PU offers one). **The SU's utility function** U_s is the difference between the achievable data rate and the energy cost,

$$U_s(\alpha) = \delta\left(\alpha R_s - \frac{1+\alpha}{2} P_s C\right),$$

where $R_s = \log(1 + P_s h_s)$ is the SU's own fixed data rate per unit time between its own transmitter and receiver (TS and RS). A good way to understand the utility function is to think C as data rate per *watt* the SU can get if it does not relay for PU. Therefore, $U_s(\alpha)$ is SU's *data rate increase* by accepting the offer α. Given PU's offer α, the optimal decision for SU is obvious: accept α if and only if $U_s(\alpha) > 0$.

Now let us consider PU's rate increase maximization problem. Without bargaining, the PU can always achieve a rate of R_{dir} through direct transmission as in Fig. 2(a). In that case, its *rate increase* is zero. Now let us calculate how much PU can gain by bargaining with SU. Without any prior knowledge, PU assumes that SU's energy cost C follows a uniform distribution in $[K_1, K_2]$, where $0 \leq K_1 < K_2$. Due to such incomplete information, PU does not know whether the SU will accept or reject a particular choice of α before offering it. If the SU rejects the offer as in Fig. 2(b), PU can only directly transmit in the remaining δ time and achieve a *negative* rate improvement $(\delta R_{dir} - R_{dir}) < 0$. If SU accepts the offer as in Fig. 2(c), PU receives a rate increase of $\delta \frac{1-\alpha}{2} R_r - R_{dir}$, which can be either positive or negative. Here R_{dir} and R_r are the data rates achieved by PU's direct transmission and AF relay [10] respectively, i.e., $R_{dir} = \log(1+P_t h_p)$ and $R_r = \log\left(1 + P_t h_p + \frac{P_t P_s h_{ps} h_{sp}}{P_t h_{ps} + P_s h_{sp} + 1}\right)$. Therefore, if the PU decides to bargain with the SU, it will choose α to maximize the **PU's utility (expected rate increase)** defined as

$$(\delta R_{dir} - R_{dir}) \operatorname{Prob}(U_s(\alpha) \leq 0) + \left(\delta \frac{1-\alpha}{2} R_r - R_{dir}\right) \operatorname{Prob}(U_s(\alpha) > 0) \quad (1)$$

where $\operatorname{Prob}(U_s(\alpha) \leq 0)$ and $\operatorname{Prob}(U_s(\alpha) > 0)$ are the probabilities for SU to reject and accept offer, respectively. Denote the optimal time fraction that maximizes (1) as α^*. The PU will choose to bargain if and only if $U_p(\alpha^*) \geq 0$. Otherwise, it will simply choose direct transmission as in Fig. 2(a). The optimal choice α^* that maximizes (1) is given in the following theorem:

Theorem 1. *When $K_1 > R_s/P_s$, $U_p(\alpha) = -(1-\delta)R_{dir} < 0$ for any $\alpha \in [0,1]$. When $K_1 \leq R_s/P_s$, then*

$$\alpha^* = \min\left(\max\left(\bar{\alpha}_p, \frac{K_1}{2R_s/P_s - K_1}\right), \min\left(\left|\frac{K_2}{2R_s/P_s - K_2}\right|, 1\right)\right) \quad (2)$$

where

$$\bar{\alpha}_p = 2\sqrt{\frac{(R_s/P_s)(R_r - R_{dir})}{R_r(2R_s/P_s - K_1)}} - \frac{1}{2}.$$

The proof of Theorem 1 can be found in our online technical report [16]. When K_1 (the minimum possible value of the SU's energy cost C) is larger than R_s/P_s, the SU will not accept any offer α from the PU. In this case, the PU knows that the bargaining will fail and thus will choose direct transmission. When $K_1 \leq R_s/P_s$, PU will choose the optimal offer α^* in (2) to achieve the best tradeoff of rate increase and performance loss. We want to emphasize again that PU will compare $U_p(\alpha^*)$ with zero and decides whether it is worth trying to bargain or not.

The one-stage bargaining game is a subgame for the multi-stage bargaining game in Section 4, and we will use Theorem 1 in the later analysis.

4 Two-Stage Bargaining Game

Now we return to the dynamic bargaining case, where the bargaining within a time slot can happen over more than one stage. For the ease of illustration, we will focus on the two-stage bargaining case. Similar to the one-stage game, here the utility functions are PU's *expected data rate increase* and SU's *date rate increase.* We assume that PU's belief about SU's energy cost C *at the beginning of stage one of bargaining* is a uniform distribution over $[0, K]$. We denote δ_1 and δ_2 as the proportions of the slot after bargaining in the first and second stage. By setting different values of δ_1 and δ_2, we can model different bargaining overhead. For the two-stage bargaining game, PU's strategy involves (1) whether to bargain at the beginning of the first stage, (2) if yes, what the first stage offer α_1 is, and (3) if SU rejects α_1, what the second stage offer α_2 is.

Figure 3 illustrates the sequential decisions and possible scenarios of this two-stage bargaining game. PU and SU make decisions alternatively at the non-leaf nodes. PU first makes the decision on whether to bargain. If it selects direct transmission (D), the game ends. Otherwise, PU offers α_1 to SU. If SU accepts this offer, then the game ends. If SU rejects the offer, then PU makes a second offer α_2 to SU. Finally, SU either accepts or rejects α_2. The game ends in both cases. Every possible ending of the game is denoted by a black solid square together with the corresponding utilities (*data rate increases*) of PU (upper value) and SU (lower value).

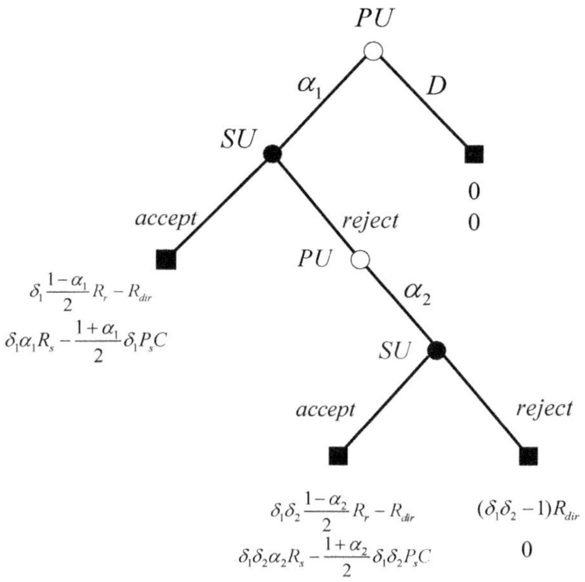

Fig. 3. Game Tree of the Two-stage Bargaining

The two-stage bargaining game is a dynamic Bayesian game (DBG)[5], and is more complex than the one-stage bargaining model for two reasons: (i) The SU may reject the first stage offer α_1 even though its utility is positive, if it believes that the second stage offer α_2 is much better; (ii) The PU needs to update its belief on SU's energy cost C at the end of the first stage by taking the SU's strategic decision process into consideration.

We want to point out the assumption that once the PU decides to bargain, it cannot choose direct transmission before the two-stage process finishes. Such restriction leads to a simplified model. Allowing the PU to choose direct transmission after the first stage will introduce more possibilities and make the analysis much more complicated. On the other hand, our focus of this paper is to consider the impact of dynamic bargaining on the choices of PU and SU. The key insights of PU's belief updates and SU's ideas on the sequential strategy for the second stage are unlikely to change with the more complicated model.

One of the commonly used solution concepts for a DBG is the Perfect Bayesian Equilibrium (PBE), which is a strategy profile and belief system in a dynamic game of incomplete information so that they satisfy the following three requirements [12].

Requirement 1. *The player with the move must have a **belief** (probability distribution) about incomplete information.*

[5] The precise definition of DBG is beyond the scope of this paper. See [11, 12] for details.

Requirement 2. *Given their beliefs, the players'* ***strategies*** *must be sequentially rational. That is, the action taken by the player must be optimal given the player's belief and the other player's subsequent strategies.*

Requirement 3. *A player's belief is determined by the Bayes' rule and the players' equilibrium strategies.*

Back to our model, the strategies and beliefs of the PU and SU are as follows.

- PU's strategy: whether to bargain at the beginning of the game, the first-stage offer α_1 if decides to bargain, and the second-stage offer $\alpha_2(\alpha_1)$ (i.e., as a function of α_1) after SU rejects α_1.
- PU's belief: $\mu_1(C)$ denotes PU's belief on C at the beginning of the first stage, and $\mu_2(C|\alpha_1)$ denotes PU's *updated* belief about C at the beginning of the second stage after SU rejects α_1.
- SU's strategy: $[\mathcal{A}_1(\alpha_1|C), \mathcal{A}_2(\alpha_2|C,\alpha_1)]$. When its energy cost is C, $\mathcal{A}_1(\alpha_1|C) = 1$ if SU accepts α_1, and $\mathcal{A}_1(\alpha_1|C) = 0$ otherwise. $\mathcal{A}_2(\alpha_2|C,\alpha_1) = 1$ if SU accepts α_2 after rejecting α_1, and $\mathcal{A}_2(\alpha_2|C,\alpha_1) = 0$ otherwise.
- SU's belief: since SU knows C, its belief is a singleton set (i.e., no uncertainty).

Before presenting our PBE analysis on the two-stage bargaining game, we want to discuss more about PBE and the method used in this subsection. First, according to Requirement 3, PBE requires the *consistency* between one player's belief and players' equilibrium strategies. Second, such an equilibrium often cannot be constructed by the traditional *backward induction method* through the game tree, as we need to construct a subgame-perfect Nash equilibrium[6]. Thus, the analysis of PBE is typically problem specific.

We will start the PBE analysis with the second stage. Since this is the last stage of the game, the analysis is similar as the one-stage game in Section III. The key difference is that PU does not have the option of direct transmission at the beginning of the second stage, and it needs to optimize the choice of α_2.

Specifically, we can apply Requirement 2 to solve SU's strategy $\mathcal{A}_2(\alpha_2|C,\alpha_1)$. Since this is the *last* move of the game, the optimal strategy for SU in the second stage is to accept α_2 if and only if the SU's utility $\delta_1\delta_2\alpha_2 R_s - \frac{1+\alpha_2}{2}\delta_1\delta_2 P_s C > 0$. Such decision is independent of α_1.

Given SU's optimal strategy in the second stage $\mathcal{A}_2(\alpha_2|C,\alpha_1)$, we can apply Requirement 2 to compute the PU's optimal strategy in the second stage. The PU will calculate the optimal α_2 that maximizes PU's expected utility function U_p, given PU's *updated* belief $\mu_2(C|\alpha_1)$ and SU's *subsequent* strategy $\mathcal{A}_2(\alpha_2|C,\alpha_1)$.

[6] Requirement 2 explains that a player's strategy at a given information set is based in part on its belief. However, the player's belief is from the players' strategies higher up the game tree according to Requirement 3. But, Requirement 2 (*sequential rationality*) means that these strategies higher up the game tree are based in part on the players' subsequent strategies, including the strategies at the given information set. This *circularity* implies that a single pass backward induction through the game tree typically will not suffice to compute a PBE. See [12] for details.

The tricky part is how to compute the belief $\mu_2(C|\alpha_1)$, which depends on the interaction in the first stage (Requirement 3). For it, we need to understand the SU's equilibrium strategy in the first stage in order to update PU's belief in the second stage.

Consider arbitrary first and second stage offers α_1 and α_2[7]. We further assume that K is reasonably large ($K > \frac{R_s}{P_s}$). We further assume that K (the upper bound of C) is reasonably large ($K > \frac{R_s}{P_s}$). Define

$$C^*(\alpha_1, \alpha_2) = \frac{2R_s(\alpha_1 - \delta_2\alpha_2)}{P_s((1+\alpha_1) - \delta_2(1+\alpha_2))}.$$

The following lemma provides SU's equilibrium strategy in the first stage for given α_1 and α_2.

Lemma 1. *SU* rejects α_1 *in the first stage if one of the following is true: (i)* $C \in [\frac{2\alpha_1 R_s}{P_s(1+\alpha_1)}, K]$ *and* $\alpha_1 > \alpha_2$, *(ii)* $C \in [C^*(\alpha_1, \alpha_2), K]$ *and* $\delta_2\alpha_2 < \alpha_1 \leq \alpha_2$, *or (iii)* $C \in [0, K]$ *and* $\alpha_1 \leq \delta_2\alpha_2$. *SU* accepts α_1 *otherwise.*

The proof of Lemma 1 can be found in [16]. With Lemma 1, we can characterize two types of PBEs of the two-stage bargaining game. We want to emphasize that PBEs in Theorem 2 and 3 are just *possible* ones. Whether they exist in a particular game requires further investigation. The proofs of Theorem 2 and 3 can be found in [16]. The physical meanings of the PBE will be further discussed in Section V.

4.1 Type I PBE

Let us look at the first type of PBE, where α_2^* is slight better than α_1^* (i.e., $\delta_2\alpha_2^* < \alpha_1^* \leq \alpha_2^*$). An SU with a small energy cost will accept α_1 in the first stage, so that it can start to benefit immediately. An SU with a medium or large energy cost will wait for the second stage hoping for a better offer. In the second stage, only an SU with a medium energy cost will accept α_2, and an SU with a high energy cost has to reject α_2.

Note that the SU does not know the value of α_2 in the first stage, and thus it needs to make the above decisions by *anticipating* the value of α_2. The PU needs to decide α_1 and α_2 by taking the SU's anticipation into consideration. A PBE exists if the SU's anticipation is *consistent* with what the PU offers. The first type of PBE is summarized in Theorem 2.

Theorem 2. *Given* α_1, *the beliefs and strategies for PU and SU are:*

– $\alpha_2^*(\alpha_1)$: *PU's second stage offer is the solution of the following fixed point equation of* α_2*:*

$$\alpha_2 = \min\left(\max\left(\alpha_p^*(K_1(\alpha_1,\alpha_2)), \frac{K_1(\alpha_1,\alpha_2)}{2R_s/P_s - K_1(\alpha_1,\alpha_2)}\right), \min\left(\left|\frac{K}{2R_s/P_s - K}\right|, 1\right)\right),$$

[7] Later on we will show that the optimal value of α_2 will depend on α_1.

where

$$\alpha_p^*(K_1(\alpha_1,\alpha_2)) = 2\sqrt{\frac{R_s(R_r - R_{dir})}{P_s R_r(2R_s/P_s - K_1(\alpha_1,\alpha_2))}} - \frac{1}{2},$$

and

$$K_1(\alpha_1,\alpha_2) = \frac{2R_s(\alpha_1 - \delta_2\alpha_2)}{P_s((1+\alpha_1) - \delta_2(1+\alpha_2))}.$$

- $\mu_1(C)$: *PU believes* C *is uniformly distributed in* $[0, K]$.
- $\mu_2(C|\alpha_1)$: *PU updates its belief on* C *as uniformly distributed in* $[C^*(\alpha_1, \alpha_2^*(\alpha_1)), K]$.
- $\mathcal{A}_1(\alpha_1|C)$: *SU rejects* α_1 *if* $C \in [C^*(\alpha_1, \alpha_2^*(\alpha_1)), K]$.
- $\mathcal{A}_2(\alpha_2|C,\alpha_1)$: *SU accepts* α_2 *if and only if* $\delta_1\delta_2\alpha_2 R_s - \frac{1+\alpha_2}{2}\delta_1\delta_2 P_s C > 0$.

Finally, PU computes first stage offer α_1^* *to maximize*

$$\left(\delta_1\frac{1-\alpha_1}{2}R_r - R_{dir}\right)P_1 + \left(\delta_1\delta_2\frac{1-\alpha_2^*(\alpha_1)}{2}R_r - R_{dir}\right)P_2 + (\delta_1\delta_2 - 1)R_{dir}P_3,$$

where $P_1 = \frac{K_1(\alpha_1)}{K}$, $P_2 = \frac{\frac{2\alpha_2}{1+\alpha_2}\frac{R_s}{P_s} - K_1(\alpha_1)}{K}$, *and* $P_3 = \frac{K - \frac{2\alpha_2}{1+\alpha_2}\frac{R_s}{P_s}}{K}$. *PU chooses direct transmission if* $U_p(\alpha_1^*) < 0$. *The above beliefs and strategies constitute a PBE if and only if* $\delta_2\alpha_2^*(\alpha_1^*) < \alpha_1^* \leq \alpha_2^*$.

4.2 Type II PBE

Next we examine the second type of PBE, where α_2^* is much larger than α_1^* (i.e., $\alpha_1^* \leq \delta_2\alpha_2^*$). If an SU believes that the second stage offer α_2 is much better than the first stage offer α_1 (even after considering time discount δ_2), it will definitely reject the first stage offer. In the second stage, an SU with a small C will accept the offer and an SU with a high C will reject the offer. This is summarized in Theorem 3.

Theorem 3. *The following beliefs and strategies constitute infinitely many PBEs.*

- α_2^*: *PU's second stage offer equals to a constant independent of* α_1,

$$\min\left(\max\left(\alpha_p^*, 0\right), \min\left(\left|\frac{K}{2R_s/P_s - K}\right|, 1\right)\right),$$

where

$$\alpha_p^* = \sqrt{\frac{2(R_r - R_{dir})}{R_r}} - \frac{1}{2}.$$

- α_1^*: *any value satisfying* $\alpha_1^* \leq \delta_2\alpha_2^*$.
- $\mu_1(C) = \mu_2(C|\alpha_1)$: *PU believes* C *is uniformly distributed in* $[0, K]$ *in the first and second stage.*
- $\mathcal{A}_1(\alpha_1|C)$: *SU never accepts* α_1.
- $\mathcal{A}_2(\alpha_2|C,\alpha_1)$: *SU accepts* α_2 *if and only if* $\delta_1\delta_2\alpha_2 R_s - \frac{1+\alpha_2}{2}\delta_1\delta_2 P_s C > 0$.

5 Numerical Results

In this section, we simulate and compare the PU's and SU's data rate increases at the equilibria in the two-stage game. We set transmission power $P_t = P_s = 1$, SU's maximum energy cost $K = 1.5R_s/P_s$, and $\delta_1 = \delta_2 = \delta = 0.8$. The PU's direct transmission rate $R_{dir} = 1$ and rate achieved via relay $R_r = 500$. In this case, SU's cooperative transmission can bring a significant improvement to PU's data rate. The SU's own data rate $R_s = 10$.

Figure 4 and 5 show the equilibrium data rate increases of PU with different energy cost C. Two figures correspond to two different types of PEBs of the same game. The *mean payoffs* (dotted curves) denote the average value over all possible values of C. Since both PU and SU obtain positive mean payoffs, spectrum leasing leads to a win-win situation.

Figure 4 corresponds to the PBE in Theorem 2. We can classify PU's payoff (circle line) into three regions depending on the SU's energy cost C: (i) Small C (e.g., $C \leq 6$ in Fig. 4): SU accepts α_1 and thus PU and SU receive significant data rate increases. (ii) Medium C (e.g., $C \in [7, 9]$ in Fig. 4): SU rejects α_1 but accepts α_2. Compared to the small C case, PU's payoff dramatically decreases, since α_2 is larger than α_1 and more time is wasted in the bargaining. SU's payoff decreases smoothly between these two regions, since the larger offer α_2 mitigates the negative effect of additional bargaining overhead. (iii) Large C (e.g., $C \geq 10$ in Fig. 4): SU rejects both offers and PU receives a negative payoff.

Figure 5 corresponds to the PBE in Theorem 3. In this PBE, the SU never accepts the first stage offer α_1, as it expects that the second stage offer α_2 is much better. As a result, the two-stage game becomes similar to a one-stage game. We can classify the PU's payoff into two regions based on the value of

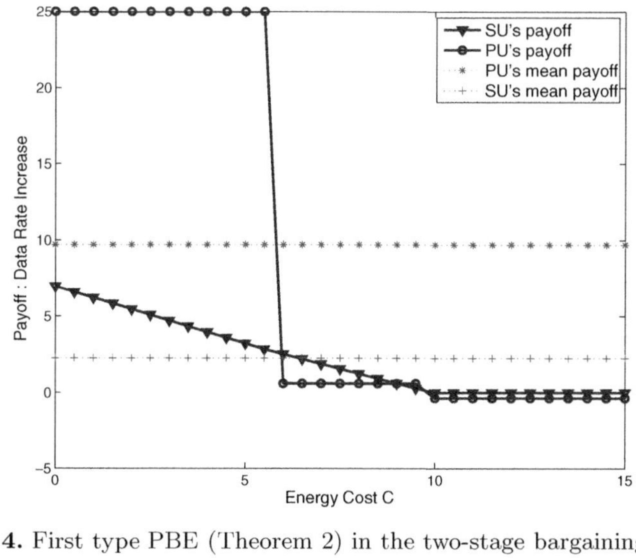

Fig. 4. First type PBE (Theorem 2) in the two-stage bargaining game

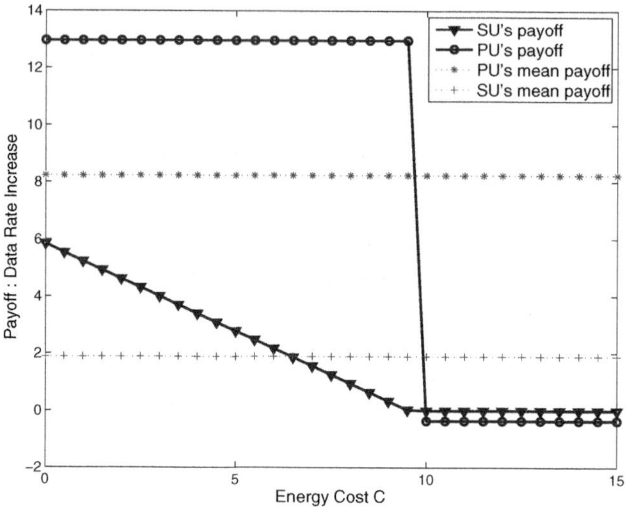

Fig. 5. Second type PBE (Theorem 3) in the two-stage bargaining game

C, following a similar argument as for Fig. 4. By comparing Figure 4 and 5, we notice that the PU's expected utility and SU's utility are both higher in Fig. 4 than in Fig. 5. The key reason is that the PBE in Fig. 5 always wastes the first stage bargaining opportunity. In other words, the PBE in Theorem 2 *Pareto dominates* the PBE in Theorem 3.

6 Conclusion

This paper studies a *dynamic* spectrum negotiation problem with *incomplete information*. We model the interactions between a PU and an SU as a dynamic Bayesian game, and derive two types of Perfect Bayesian Equilibria. Simulations show that both equilibria can coexist in the same game, and one type Pareto dominates the other in the expected sense. This paper represents an initial step of studying the resource allocation in cognitive radio networks with incomplete information and dynamics. Only one PU and one SU bargaining model has been studied, and it is certainly desirable although very challenging to consider dynamic bargaining among many PUs and SUs.

Acknowledgements. This work has been supported by CSC Postgraduate Scholarship, NSF of China No.60832008, National S&T Major Project 2010ZX03003-002-02, TNList Cross-discipline Foundation, China's 973 Project 2007CB310608, China's 863 Project 2009AA011501, PCSIRT, and National S&T Pillar Program 2008BAH30B09. This work is also supported by the General Research Funds (Project No. 412509 and 412710) established under the University Grant Committee of the Hong Kong Special Administrative Region, China. The authors express their gratitude to DoCoMo Beijing Communications Laboratories Co., Ltd for their helpful comments and criticisms.

References

1. McHenry, M.A., McCloskey, D.: Spectrum Occupancy Measurements: Chicago, Illinois, November 16-18, Shared Spectrum Company, Technical report (2005)
2. Haykin, S.: Cognitive Radio: Brain-Empowered Wireless Wommunications. IEEE Journal on Selected Areas in Communications 23(2), 201–220 (2005)
3. Bae, J., Beigman, E., Berry, R., Honig, M., Vohra, R.: Sequential Bandwidth and Power Auctions for Distributed Spectrum Sharing. IEEE Journal on Selected Areas in Communications 26(7), 1193–1203 (2008)
4. Niyato, D., Hossain, E.: Competitive Pricing for Spectrum Sharing in Cognitive Radio Networks: Dynamic Game, Inefficiency of Nash equilibrium, and Collusion. IEEE Journal on Selected Areas in Communications 26(1), 192–202 (2008)
5. Kasbekar, G., Sarkar, S.: Spectrum Pricing Games with Bandwidth Uncertainty and Spatial Reuse in Cognitive Radio Networks. In: ACM MobiHoc (2010)
6. Simeone, O., Stanojev, I., Savazzi, S., Bar-Ness, Y., Spagnolini, U., Pickholtz, R.: Spectrum Leasing to Cooperating Secondary ad hoc Networks. IEEE Journal on Selected Areas in Communications 26(1), 203–213 (2008)
7. Jing, Z., Qian, Z.: Stackelberg Game for Utility-Based Cooperative Cognitive Radio Networks. In: ACM MobiHoc, pp. 23–32 (2009)
8. Lingjie, D., Lin, G., Jianwei, H.: Contract-Based Cooperative Spectrum Sharing. In: IEEE International Dynamic Spectrum Access Networks (DySPAN) Symposium, Aachen, Germany (2011)
9. Adlakha, S., Johari, R., Goldsmith, A.: Competition in Wireless Systems via Bayesian Interference Games. arXiv, vol.cs.GT (2007)
10. Cover, T.M., El Gamaal, A.: Capacity Theorems for the Relay Channel. IEEE Transaction on Information Theory 25(5), 572–584 (1979)
11. Fudenberg, D., Tirole, J.: Game Theory. The MIT Press, Cambridge (1991)
12. Gibbons, R.: Game Theory for Applied Economists. Princeton University Press (1992)
13. Rubinstein, A.: A Bargaining Model with Incomplete Information about Time Preferences. In: Econometrica (1985)
14. Chatterjee, K., Samuelson, W.: Bargaining under Incomplete Information. Operations Research, 835–851 (1983)
15. Rubinstein, A., Osborne, M.: Bargaining and Markets. Academic Press (1990)
16. Yang, Y., Jianwei, H., Xiaofeng, Z., Jing, W.: Dynamic Spectrum Negotiation with Asymmetric Information. The Chinese University of Hong Kong, Technical report (2010), `http://home.ie.cuhk.edu.hk/~jwhuang/publication/DynamicBargainTechReport.pdf`

Evolution of Cooperation: A Case with Interference-Aware Cooperative Spectrum Sensing in Cognitive Radio Networks

You-En Lin[1] and Hung-Yun Hsieh[1,2]

[1] Graduate Institute of Communication Engineering
[2] Department of Electrical Engineering,
National Taiwan University, Taipei, Taiwan 106
hyhsieh@cc.ee.ntu.edu.tw

Abstract. We consider in this work a group of secondary users with backlogged traffic to transmit in the primary network. To avoid interfering with the primary user, each secondary user must perform interference-aware spectrum sensing before transmission. Unlike conventional sensing techniques, interference-aware spectrum sensing allows a secondary user to adjust its sensing parameters for optimal performance depending on the probability of interfering with the primary user. While interference-aware sensing can achieve better performance for individual users, challenges arise when secondary users collaborate with each other for cooperative spectrum sensing due to their unequal interference probabilities that result in a conflict for setting the optimal sensing parameters. To model this problem, we consider an interference-aware cooperative sensing game and analyze player behaviors under such a game. We find that there is a unique pure Nash equilibrium of the game, but it tends to deviate from the desirable solution of social optimum. We then design a repeated game based on evolutionary game theory to address this problem. Players in the repeated game have the chance to revenge "uncooperative" players in ensuing repetitions for driving the equilibrium to the social optimum. We show through numerical results that the proposed game of evolution does achieve the desirable performance for interference-aware cooperative sensing in dynamic spectrum access.

1 Introduction

The concept of dynamic spectrum access allows unlicensed secondary users to opportunistically fill in the white space left by the licensed primary user for better utilization of the spectrum. To enable such opportunistic spectrum access, an important step for the secondary user is to sense the spectrum before transmission to detect the presence of the primary user [17,4]. Due to the stochastic nature of channel fading and/or background noise, however, it is possible that the signal detector at the secondary user makes an inaccurate detection decision in terms of false alarm and missed detection. While occurrences of false alarm and missed detection are equally undesirable that should ideally be minimized

R. Jain and R. Kannan (Eds.): GameNets 2011, LNICST 75, pp. 530–545, 2012.

whenever possible, the fundamental tradeoffs between the two types of errors prevent the detector from operating under arbitrarily small probabilities of false alarm and missed detection [10,14].

To operate the signal detector in the desired operation region, related work has proposed several sensing techniques and optimization frameworks for optimizing the detector performance. For example, the authors in [12] aim to design an optimal detector under non-Gaussian noise that can minimize the probability of false alarm while ensuring that the probability of missed detection is below the target threshold. While it is reasonable to limit the maximum probability of missed detection while minimizing the probability of false alarm for ensuring that the primary user is properly protected without suffering from undesirable interference [12,11,18,9], such a spectrum sensing strategy can become *conservative* and result in sub-optimal performance for opportunistic spectrum access. The reason is that different secondary users could potentially incur different degrees of interference on the primary user when missed detection occurs. If secondary users are required to provide the same level of protection to the primary user, a secondary user with a lower probability of interference should ideally be allowed to have a larger probability of missed detection during spectrum sensing. Probability of spectrum hole discovery and hence spectrum utilization can thus be improved by employing such an interference-aware spectrum sensing technique.

While interference-aware spectrum sensing can potentially improve the performance for individual users in opportunistic spectrum access, challenges arise when secondary users collaborate with each other for cooperative spectrum sensing. In cooperative spectrum sensing, a secondary transmitter cooperates with a group of secondary users to make the transmission decision through fusing the local decisions of individual signal detectors. It has been shown in related work that cooperative sensing can indeed improve the sensing performance for individual detectors under channel artifacts such as fading and noise [8,6,13]. In interference-aware cooperative spectrum sensing, however, since the optimal sensing parameters depend on the interference probability of the concerned transmitter, it is possible that the final decision is not optimal for other cooperating users. In such a case, a secondary user may not have sufficient incentive to perform local spectrum sensing and contribute to the fusion process only to obtain a final decision that does not improve its performance. If secondary users are allowed to make their own decisions without considering the resultant benefit of the whole group, interference-aware cooperative spectrum sensing cannot be guaranteed to achieve the desirable performance gain due to cooperation.

To model this problem and gain more insights on finding the solution, we resort to game theory and focus on the negotiation process for sensing parameters in interference-aware cooperative spectrum sensing. A secondary user acts as a player in the game and it participates in the game to determine its sensing parameters for maximizing its payoff. Under such a game model, as we show in Section 3, there exists a pure Nash equilibrium where no player can change its sensing parameters unilaterally without decreasing its payoff. The Nash equilibrium, however, fails to provide the socially optimal performance for all players due to the selfish actions of individual players. To address the problem of sub-optimality due to non-cooperation, we then design a repeated game based on evolutionary game theory. In the repeated game, a stage game is played in repetitions where each player can learn from history the actions of other players to adapt its action in ensuing repetitions. Four different roles including *Solitary, Leader, Follower, and Glutton* that players in the game can choose to act are proposed. To stimulate cooperation, a special role of *Avenger* is also included so disadvantaged player can "punish" non-cooperative players with the goal of driving the equilibrium of the game to social optimum. We show through numerical results that the proposed repeated game can indeed stimulate cooperation through evolution of player roles and achieve the desirable performance for interference-aware cooperative sensing in opportunistic spectrum access.

The paper is organized as follows. Section 2 presents the network scenario as well as the spectrum sensing and sharing models. Based on the system models, Section 3 formulates a game for interference-aware cooperative spectrum sensing and then presents the best responses of the players and the Nash equilibrium thus obtained. Due to the selfish behavior of individual players, it is also shown in Section 3 that the Nash equilibrium will deviate form the social optimum. To address the problem, a repeated game based on evolutionary game theory is proposed in Section 4 to stimulate cooperation among secondary users and drive the equilibrium to the social optimum. Evaluation results in Section 5 show that the proposed evolutionary game can indeed achieve the desired performance with significant performance gains.

2 System Model

We consider a simple network scenario with one primary transmitter serving as the base station of a cellular network or a TV broadcast station with service range R. To leverage the spectrum holes in the frequency bands used by primary users, secondary users can sense the activity of the primary cell and reuse the spectrum when primary users are not active. Even when primary users are active, secondary users can still reuse the spectrum as long as no undesirable interference (as explained in Section 2.2) is incurred on primary users.

In such a scenario, a challenging issue is to achieve reliable spectrum sensing when secondary users are outside the cell and the signal from the primary transmitter is relatively weak. To mitigate the problem of channel artifacts such as fading and noise, we assume that a group of secondary users $\mathcal{G}$ with backlogged traffic cooperate for spectrum sensing. Secondary users operate on a TDMA-basis by following the frame structure shown in Figure 1. In each frame, secondary users in $\mathcal{G}$ first negotiate the sensing parameters (e.g. detection threshold for an energy detector) to be used during the following sensing period. After negotiation, each secondary user performs local spectrum sensing and then broadcasts the local decision to be fused for the final decision in cooperative sensing. If transmission opportunity (spectrum hole) is identified based on the sensing result, secondary users send in turns during the transmission slots until the next frame starts. In the following, we explain in more details the spectrum sensing and sharing models used for opportunistic spectrum access.

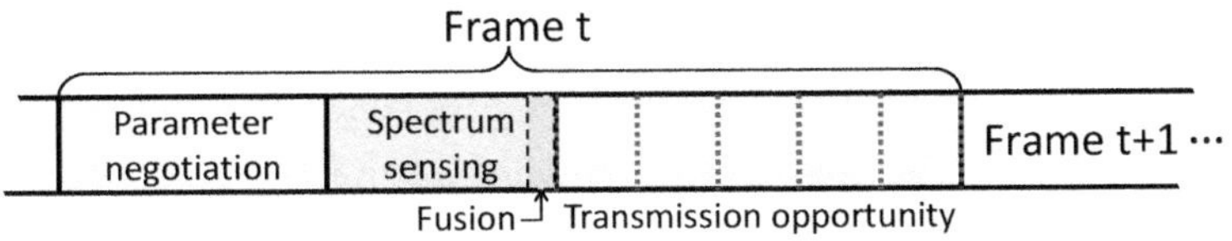

Fig. 1. Frame structure for secondary users

2.1 Spectrum Sensing Model

We consider the use of an energy detector for detecting the presence of the primary user due to its applicability to a wide range of signals and mathematical amenity compared to other detectors [5,7]. A secondary user first senses the local spectrum and takes N samples of the signal measurements. Denote H_0 as the case when the primary transmitter is inactive (no transmission) and H_1 as the case when the primary transmitter is active. Sample $Y_j[n]$ taken at time n by secondary user j can then be written as follows:

$$Y_j[n] = \begin{cases} W_j[n], & H_0, \\ X_j[n] + W_j[n], & H_1, \end{cases} \tag{1}$$

where $W_j[n]$ is the sample of the additive white Gaussian noise (AWGN) with variance σ^2, and $X_j[n]$ is the sample of the primary signal that we model as a Gaussian signal [11]. To decide whether the primary signal is present, secondary user j first computes a test statistic as $Z_j = \frac{1}{N}\sum_{n=1}^{N} |Y_j[n]|^2$. Based on the test statistic Z_j, secondary user j makes a positive decision ($\mathcal{D} = 1$) regarding the presence of the primary user if $Z_j \geq \theta_j$, where θ_j denotes its detection threshold. Otherwise, the spectrum is considered empty ($\mathcal{D} = 0$).

Due to the stochastic nature of the background noise, it is possible that the decision regarding the spectrum usage is not correct. False alarm occurs when the primary user is inactive but determined to be active by the secondary user, whereas missed detection occurs when the primary user is active but determined as being inactive. With sufficiently large number of samples N, the probability of false alarm $\mathbb{P}_{\mathrm{FA}}^{(j)}$ and missed detection $\mathbb{P}_{\mathrm{MD}}^{(j)}$ observed by secondary user j can be derived as follows [11]:

$$\mathbb{P}_{\mathrm{FA}}^{(j)} = Q\left(\frac{\theta_j - \sigma^2}{\sqrt{2/N}\sigma^2}\right), \tag{2}$$

$$\mathbb{P}_{\mathrm{MD}}^{(j)} = 1 - Q\left(\frac{\theta_j - (P_j + \sigma^2)}{\sqrt{2/N}(P_j + \sigma^2)}\right), \tag{3}$$

where P_j is the mean power of the received primary signal at secondary user j, and $Q(\cdot)$ is the complementary CDF of the standard Gaussian distribution.

After the decision of the local spectrum sensing is made by each secondary user, cooperative spectrum sensing is performed using decision fusion. We consider in this paper the use of the OR rule for fusing local decisions from individual detectors. The probabilities of missed detection and false alarm for the cooperative sensing set $\mathcal{G}$ can then be expressed as $\mathbb{P}_{\mathrm{MD}}^{(\mathcal{G})} = \prod_{j\in\mathcal{G}} \mathbb{P}_{\mathrm{MD}}^{(j)}$ and $\mathbb{P}_{\mathrm{FA}}^{(\mathcal{G})} = 1 - \prod_{j\in\mathcal{G}}(1 - \mathbb{P}_{\mathrm{FA}}^{(j)})$. It is clear that $\mathbb{P}_{\mathrm{MD}}^{(\mathcal{G})}$ depends on detection thresholds θ_j of individual detectors and $\mathbb{P}_{\mathrm{MD}}^{(\mathcal{G})}$ is smaller than $\mathbb{P}_{\mathrm{MD}}^{(j)}$ of any user j in the cooperative set.

2.2 Spectrum Sharing Model

As we mentioned earlier, in opportunistic spectrum access, a secondary user is allowed to use the spectrum as long as it does not interfere with the primary user. We thus define for secondary user j the probability of interfering with the primary user as $\mathbb{P}_{\mathrm{I}}^{(j)} = \mathrm{Prob}(\gamma < \gamma_t)$, where γ is the mean SINR during the reception of a packet with M symbols for a primary user at the closest point to the secondary user. Assuming Gaussian primary and secondary signals similar to [11], we can derive the probability of interference for sufficiently large value of M as follows:

$$\mathbb{P}_{\mathrm{I}}^{(j)} = Q\left(\frac{P_p/\gamma_t - P_c - \sigma^2}{\sqrt{2/M\,(P_p/\gamma_t + P_c^2 + \sigma^4)}}\right), \tag{4}$$

where P_p and P_c are the mean powers of primary and secondary signals at the primary receiver respectively.

Note that $\mathbb{P}_{\mathrm{I}}^{(j)}$ is the probability of causing interference when *both the primary and secondary transmitters are active.* That is, it is a conditional probability that applies when missed detection occurs. To allow for better spectrum reuse based on interference-aware sensing, we can formulate the following constraint

$$\mathbb{P}_{\mathrm{MD}}^{(\mathcal{G})}\mathbb{P}_{\mathrm{I}}^{(j)} \le \varepsilon \tag{5}$$

to ensure that the primary receiver at the worst point is not interfered (unable to correctly decode the primary signal) by the secondary user with a probability of at least $1-\varepsilon$. For simplicity, we say that *the secondary user does not interfere with the primary user when Inequality* (5) *holds.* If Inequality (5) does not hold, on the other hand, the secondary user is not allowed to send for spectrum sharing. Since different secondary users may incur unequal degrees of interference on the primary user, there exists a conflict of interests in setting individual detection thresholds for obtaining $\mathbb{P}_{\mathrm{MD}}^{(\mathcal{G})}$ in cooperative spectrum sensing. We elaborate the problem in the following section.

3 Interference-Aware Sensing Game

In this section, we first formulate a game to model the negotiation process for spectrum sensing parameters in interference-aware cooperative spectrum sensing. We then find the best response of each player and derive the unique Nash equilibrium and social optimum for the game.

3.1 Game Formulation

During the negotiation process of sensing parameters, each player (secondary user) negotiates the detection threshold (that affects the final probability of missed detection) to maximize its payoff. To determine the payoff, we note that a spectrum hole exists for a secondary user under the following two cases:

- $\text{Prob}(\mathcal{D} = 0 \mid H_0)$: The primary user is not active, and the secondary user does not make any false alarm about the spectrum usage. The expected throughput for this case can be expressed as $(1-\mathbb{P}_{\text{FA}}^{(\mathcal{G})})(1-\mathbb{P}_{\text{E}})C_0^{(j)}$, where $\mathbb{P}_{\text{E}}$ is the probability that the primary transmitter uses the spectrum, i.e. $\mathbb{P}_{\text{E}} = \text{Prob}(H_1)$, and $C_0^{(j)}$ is the channel capacity at the corresponding receiver of secondary user j when the primary user is inactive.
- $\text{Prob}(\{\mathcal{D} = 0 \mid H_1\} \cap \{\gamma \geq \gamma_t\})$: The secondary user makes missed detection when the primary user is active, but transmission of the secondary user does not interfere with the primary receiver at the worst point ($\gamma \geq \gamma_t$). The expected throughput for this case can be expressed as $\mathbb{P}_{\text{MD}}^{(\mathcal{G})}\mathbb{P}_{\text{E}}(1-\mathbb{P}_{\text{I}}^{(j)})C_1^{(j)}$, where $C_1^{(j)}$ is the channel capacity at the corresponding receiver of secondary user j when the primary user is active.

We can therefore define the payoff for secondary user j as the overall expected throughput by summing the achievable throughput in the two cases:

$$C^{(j)} = \mathbb{P}_{\text{MD}}^{(\mathcal{G})}\mathbb{P}_{\text{E}}(1-\mathbb{P}_{\text{I}}^{(j)})C_1^{(j)} + (1-\mathbb{P}_{\text{FA}}^{(\mathcal{G})})(1-\mathbb{P}_{\text{E}})C_0^{(j)}. \tag{6}$$

A rational secondary user j thus may adjust its detection threshold θ_j for maximizing the expected throughput. To ensure that primary users are properly protected, however, we define the utility function for each player as follows:

$$U^{(j)} = \begin{cases} C^{(j)}, & \mathbb{P}_{\text{MD}}^{(\mathcal{G})}\mathbb{P}_{\text{I}}^{(j)} \leq \varepsilon \\ 0, & \text{Otherwise} \end{cases}, \; j \in \mathcal{G}. \tag{7}$$

That is, a secondary user gets the payoff (expected throughput) only when Inequality (5) holds. Otherwise, the secondary user is not allowed to use the spectrum, and its payoff is zero.

3.2 Best Response

The utility function of secondary user j in Equation (7) is a function of $\mathbb{P}_{\text{MD}}^{(\mathcal{G})}$, and hence utility $U^{(j)}$ depends not only on the detection threshold θ_j of player j but also on those of other players. Note that maximization of the utility function with respect to θ_j as shown in Equation (7) belongs to monotonic optimization [15]. Hence, the optimal value of θ_j that maximizes the utility occurs on the boundaries of the constraints as follows:

– If $\left(\mathbb{P}_{\mathrm{I}}^{(j)} \prod_{k\in\mathcal{G},k\neq j} \mathbb{P}_{\mathrm{MD}}^{(k)}\right) > \varepsilon$, optimal θ_j occurs on the boundary of the interference constraint such that $\mathbb{P}_{\mathrm{MD}}^{(\mathcal{G})}\mathbb{P}_{\mathrm{I}}^{(j)} = \varepsilon$. Using Equation (3), we have

$$\theta_j^* = (\sigma^2 + P_j)\{1 - \sqrt{2/N}Q^{-1}[\varepsilon/(\mathbb{P}_{\mathrm{I}}^{(j)} \prod_{k\in\mathcal{G},k\neq j} \mathbb{P}_{\mathrm{MD}}^{(k)})]\}. \quad (8)$$

– If $\left(\mathbb{P}_{\mathrm{I}}^{(j)} \prod_{k\in\mathcal{G},k\neq j} \mathbb{P}_{\mathrm{MD}}^{(k)}\right) \leq \varepsilon$, optimal θ_j occurs when $\mathbb{P}_{\mathrm{MD}}^{(j)} = 1$ or $\mathbb{P}_{\mathrm{FA}}^{(j)} = 0$. It can be observed from Equations (2) and (3) that $\theta_j^* \to \infty$. This is a special case since the local sensing results contributed by other players are already sufficient for protecting the primary receiver. It is thus not necessary for player j to perform local sensing, and its best strategy to maximize the expected throughput thus is not to perform any local sensing but use the result from cooperative sensing.

A player in the game can therefore determine its detection threshold (best response) for maximizing its payoff given the detection thresholds of other players.

3.3 Nash Equilibrium

As discussed in Section 3.2, the maximum payoff for player j occurs on the constraint boundary, i.e. $\mathbb{P}_{\mathrm{MD}}^{(\mathcal{G})} = \varepsilon/\mathbb{P}_{\mathrm{I}}^{(j)}$. While $\mathbb{P}_{\mathrm{MD}}^{(\mathcal{G})}$ is shared by all members in cooperative sensing, $\mathbb{P}_{\mathrm{I}}^{(j)}$ is different for different players in the group. For a player with a larger $\mathbb{P}_{\mathrm{I}}^{(j)}$, it would desire a smaller $\mathbb{P}_{\mathrm{MD}}^{(\mathcal{G})}$ to maximize its payoff without violating the interference constraint. It therefore needs to reduce the detection threshold θ_j such that $\mathbb{P}_{\mathrm{MD}}^{(j)}$ and $\mathbb{P}_{\mathrm{MD}}^{(\mathcal{G})}$ can be decreased. On the other hand, a player with a smaller $\mathbb{P}_{\mathrm{I}}^{(j)}$ would desire a larger $\mathbb{P}_{\mathrm{MD}}^{(\mathcal{G})}$ by increasing detection threshold θ_j and $\mathbb{P}_{\mathrm{MD}}^{(j)}$. Negotiation of detection thresholds comes to an end when no player wants to change its threshold unilaterally. As we show in the following, there is exactly one pure Nash equilibrium in the game.

Firstly, according to Equation (3), the probability of missed detection $\mathbb{P}_{\mathrm{MD}}^{(j)}$ and the detection threshold θ_j has 1-to-1 mapping. We can then use $\mathbb{P}_{\mathrm{MD}}^{(j)}$ in place of θ_j as the response of player j for notation clarity. Now let player k be the one with the maximum interference probability, i.e. $k = \arg\max_{j\in\mathcal{G}} \mathbb{P}_{\mathrm{I}}^{(j)}$. The unique pure Nash equilibrium of the interference-aware spectrum sensing game can then be written as $(1, \cdots, \varepsilon/\mathbb{P}_{\mathrm{I}}^{(k)}, \cdots, 1)$. Equivalently, in Nash equilibrium the probability of missed detection of all players is 1 except for player k such that $\mathbb{P}_{\mathrm{MD}}^{(\mathcal{G})} = \mathbb{P}_{\mathrm{MD}}^{(k)} = \varepsilon/\mathbb{P}_{\mathrm{I}}^{(k)}$. In such a case, player k takes all responsibility of spectrum sensing because it has the most stringent interference constraint for protecting the primary user. Other players are simply free-riders who can obtain the final sensing decision without contributing any local sensing results. Therefore, under the Nash equilibrium, the cooperative spectrum sensing game turns out to non-cooperative, single-node spectrum sensing. The Nash equilibrium clearly is far from being social optimum in terms of cooperative spectrum sensing. This motivates us to develop rules for the game so that a socially optimal equilibrium can be achieved.

3.4 Social Optimum

To compare the Nash equilibrium against the social optimum, we first define the social utility as the sum of the expected throughput of all secondary users in $\mathcal{G}$. Finding the social optimum is then equal to finding the optimal solution of the following optimization problem:

$$\underset{\theta_j,\, j\in\mathcal{G}}{\text{Maximize}} \sum_{j\in\mathcal{G}} C^{(j)}, \text{ subject to } \mathbb{P}_{\text{MD}}^{(\mathcal{G})}\mathbb{P}_{\text{I}}^{(j)} \leq \varepsilon, \forall j \in \mathcal{G}. \tag{9}$$

Although Problem (9) is not a convex problem, it belongs to monotonic programming and hence the global optimal solution can be found.

To start, note that due to the monotonic increasing property of $\mathbb{P}_{\text{MD}}^{(\mathcal{G})}$ with respect to θ_j, the interference constraint for the user with the largest interference probability, say user $k = \arg\max_{j\in\mathcal{G}} \mathbb{P}_{\text{I}}^{(j)}$, will hit the boundary first. Hence, the $|\mathcal{G}|$ interference constraints can be replaced by $\mathbb{P}_{\text{MD}}^{(\mathcal{G})} = \varepsilon/\mathbb{P}_{\text{I}}^{(k)}$. In addition, since θ_j and $\mathbb{P}_{\text{MD}}^{(j)}$ has 1-to-1 mapping, the optimization problem can then be rewritten as:

$$\underset{\mathbb{P}_{\text{MD}}^{(j)},\, j\in\mathcal{G}}{\text{Maximize}} \prod_{j\in\mathcal{G}} \left(1 - \mathbb{P}_{\text{FA}}^{(j)}\right), \text{ subject to } \prod_{j\in\mathcal{G}} \mathbb{P}_{\text{MD}}^{(j)} = \frac{\varepsilon}{\mathbb{P}_{\text{I}}^{(k)}}. \tag{10}$$

As can be observed from Problem (10), the objective and constraint functions are products of $|\mathcal{G}|$ independent functions *each with only one variable as the parameter*. Such a property makes it amenable to a solution based on dynamic programming [1] that solves a complex problem by recursively breaking into simpler sub-problems. To detail, assume cooperative set $\mathcal{G} = \{1, 2, ..., |\mathcal{G}|\}$ without loss of generality. The original problem can be solved in $|\mathcal{G}|$ stages (sub-problems), and each stage can be solved using the following recursive equation:

$$\begin{aligned} f^{(j)}(S^{(j)}) = \underset{\mathbb{P}_{\text{MD}}^{(i)}}{\text{Maximize}} \; & \left(1 - \mathbb{P}_{\text{FA}}^{(j)}\right) \cdot f^{(j+1)}\left(S^{(j+1)}\right), \\ \text{subject to } & \mathbb{P}_{\text{MD}}^{(j)} \cdot S^{(j+1)} = S^{(j)}, \end{aligned} \tag{11}$$

where state $S^{(j)}$ indicates the value of the constraint function counting only detectors $\{j, ..., |\mathcal{G}|\}$ and $f^{(j)}\left(S^{(j)}\right)$ is the maximum cumulative payoff from stage j to $|\mathcal{G}|$ under state $S^{(j)}$.

Algorithm 1 thus shows the pseudo-code for finding the optimal thresholds in interference-aware cooperative sensing. While Problem (10) has continuous state in [0, 1], it can be solved through proper discretization of the state space [2]. Input to the algorithm includes δ_1 and δ_2 that are used to control the search granularity for state $S^{(j)}$ and variable $\mathbb{P}_{\text{MD}}^{(j)}$ respectively. The algorithm consists

Algorithm 1. Finding optimal detection thresholds

Input: Protection constraint $\varepsilon/\mathbb{P}_{\mathrm{I}}^{(k)}$, granularity δ_1 and δ_2
Output: Detection thresholds θ_j^*, $\forall j \in \mathcal{G} = \{1, 2, ..., |\mathcal{G}|\}$

01: **for** $(j = |\mathcal{G}|;\ j \geq 1;\ j--)$ // iterate for stages
02: **for** $(\widehat{S} = \varepsilon/\mathbb{P}_{\mathrm{I}};\ \widehat{S} \leq 1;\ \widehat{S} = \widehat{S} + \delta_1)$ // iterate for states
03: **if** $(j == |\mathcal{G}|)$ // starting stage
04: $f^{(j)}\left(\widehat{S}\right) = 1 - \mathbb{P}_{\mathrm{FA}}^{(j)}\left(\widehat{S}\right),\ \mathbb{P}_{\mathrm{MD}}^{(j)}\left(\widehat{S}\right) = \widehat{S}$
05: **else**
06: **for** $(\widehat{\mathbb{P}}_{\mathrm{MD}} = \widehat{S};\ \widehat{\mathbb{P}}_{\mathrm{MD}} \leq 1;\ \widehat{\mathbb{P}}_{\mathrm{MD}} = \widehat{\mathbb{P}}_{\mathrm{MD}} + \delta_2)$
07: $\widehat{f} = \left[1 - \mathbb{P}_{\mathrm{FA}}^{(j)}\left(\widehat{\mathbb{P}}_{\mathrm{MD}}\right)\right] \cdot f^{(j+1)}\left(\frac{\widehat{S}}{\widehat{\mathbb{P}}_{\mathrm{MD}}}\right)$
08: **if** $\left(\widehat{f} > f^{(j)}\left(\widehat{S}\right)\right)$
09: $f^{(j)}\left(\widehat{S}\right) = \widehat{f},\ \mathbb{P}_{\mathrm{MD}}^{(j)}\left(\widehat{S}\right) = \widehat{\mathbb{P}}_{\mathrm{MD}}$
10: **end if**
11: **end for** // found optimal $\widehat{\mathbb{P}}_{\mathrm{MD}}$
12: **end if**
13: **if** $(j == 1)$ **break** // only one state for stage 1
14: **end for**
15: **end for**
16: **for** $(j = 1, \widehat{S} = \varepsilon/\mathbb{P}_{\mathrm{I}}^{(k)};\ j \leq |\mathcal{G}|;\ j++)$ // trace back
17: $\theta_j^* = (\sigma^2 + P_j)\left\{1 - \sqrt{\frac{2}{N}}Q^{-1}\left[\mathbb{P}_{\mathrm{MD}}^{(j)}\left(\widehat{S}\right)\right]\right\}$
18: $\widehat{S} = \widehat{S}/\mathbb{P}_{\mathrm{MD}}^{(j)}\left(\widehat{S}\right)$
19: **end for**

of three layers of loops: the first layer from Line 1 to Line 15 is used to iterate from stage $|\mathcal{G}|$ to stage 1. For each stage, each possible state is examined for calculating the cumulative payoff as shown between Line 2 and Line 14. The best payoff for each state is obtained by solving Problem (11) as shown between Line 6 and Line 11. Once the optimal $\mathbb{P}_{\mathrm{MD}}^{(j)}$ leading to the best payoff is obtained for each state, it is recorded to facilitate back-tracing when stage 1 reaches. Line 16 to Line 19 show how back-trace from stage 1 to stage $|\mathcal{G}|$ is done for retrieving the optimal sequence of states and detection thresholds. We note that while the feasible region of $\mathbb{P}_{\mathrm{MD}}^{(j)}$ of each detector j is between 0 and 1, in traversing states from stage $|\mathcal{G}|$ backwards, we can utilize the concept of the cutting plane method [3] to cut the feasible region and reduce the computation complexity. Briefly, since $S^{(1)} = \varepsilon/\mathbb{P}_{\mathrm{I}}^{(k)}$ and $S^{(j+1)} = S^{(j)}/\mathbb{P}_{\mathrm{MD}}^{(j)}$, we have $S^{(j)} \in [\varepsilon/\mathbb{P}_{\mathrm{I}}^{(k)}, 1]$ and $\mathbb{P}_{\mathrm{MD}}^{(j)} \in [S^{(j)}, 1]$. Limiting the search range for $S^{(j)}$ in each stage (Line 2) and $\mathbb{P}_{\mathrm{MD}}^{(j)}$ in each state (Line 6) can significantly reduce the computation complexity.

4 Evolution of Cooperation

The game presented in Section 3 can be regarded as a one-shot game that only models spectrum sensing within a frame in Figure 1, where secondary users pay the cost of spectrum sensing and earn the payoff from spectrum usage. For a group of backlogged secondary users, however, the demand for spectrum access can extend to several frames. Therefore, the one-shot game can be played repeatedly for continuous sensing and access of the spectrum. As we have shown in Section 3, the Nash equilibrium of the one-shot game is not the social optimum due to individual players selfishly maximizing their own utilities. To stimulate cooperation, we design a repeated game in this section with the goal of achieving the socially optimal equilibrium. Each secondary user in the repeated game maintains a list to remember payoff and corresponding actions in the past such that a player can adapt its best response based on the concept of evolution. In addition, we design the mechanism for cooperative players to punish selfish ones through repetitions of the game. In the following, we first analyze and classify possible strategies for a player to adopt in the repeated game, and then we present the evolutionary learning model.

4.1 Strategies in the Repeated Game

In a repeated game, a player may take into account the impact of its current action on future actions of other players. In addition, a player may have its own "personality" in choosing the best strategy of action during repetitions of the game. We classify four different strategies (roles) to model the behavior of players as follows:

Solitary. A solitary, say user j, is a conservative player who sets the detection threshold as in the case of single-node sensing. As a result, even if all other players are free-riders such that $\mathbb{P}_{\mathrm{MD}}^{(i)} = 1$, $i \neq j$, a solitary can still ensure that the final decision $\mathbb{P}_{\mathrm{MD}}^{(\mathcal{G})} = \mathbb{P}_{\mathrm{MD}}^{(j)}$ will meet its interference constraint. Specifically, a solitary sets $\mathbb{P}_{\mathrm{MD}}^{(j)}$ as $\varepsilon/\mathbb{P}_{\mathrm{I}}^{(j)}$ to satisfy the interference constraint $\mathbb{P}_{\mathrm{MD}}^{(\mathcal{G})}\mathbb{P}_{\mathrm{I}}^{(j)} \leq \varepsilon$ while maximizing its utility. Clearly, without relying on the cooperation of others, a solitary can always get non-zero payoff as the case for single-node sensing. However, it cannot benefit from the increased payoff due to cooperative sensing. Hence, if not all player are free-riders, $\mathbb{P}_{\mathrm{MD}}^{(\mathcal{G})}$ is smaller than the expected value, and the utility of a solitary is not maximized.

Leader. In contrast to a solitary, a leader aims to fully utilize the gain of cooperative sensing through active coordination of the negotiation process among players. Effectively, a leader sets its constraint as the most stringent one such that $\mathbb{P}_{\mathrm{MD}}^{(\mathcal{G})} = \varepsilon/\mathbb{P}_{\mathrm{I}}^{(j)}$. It then applies Algorithm 1 to compute the optimal detection thresholds and then broadcasts the results to all players. In this way, its utility is maximized if all other players follow the coordination; if some players do not follow the coordination, however, its utility is not maximized.

Follower. Instead of coming up with the "optimal" threshold for itself, a follower simply uses the suggestion from other players. If there is only one leader in the game, a follower simply follows the command of the former. If there are more than one leaders in the game, a follower adopts the command from the leader with the minimum detection threshold value. Finally, if no leader exists in the game, a follower copies the minimum threshold among other players.

Glutton. A glutton, say player j, at each repetition maximizes its utility by pushing $\mathbb{P}_{\mathrm{MD}}^{(\mathcal{G})}$ to its constraint boundary such that $\mathbb{P}_{\mathrm{MD}}^{(\mathcal{G})} = \varepsilon/\mathbb{P}_{\mathrm{I}}^{(j)}$. To do so, a glutton collects the responses from other players and then makes its decision as $\mathbb{P}_{\mathrm{MD}}^{(j)} = \min\{1, \varepsilon/(\mathbb{P}_{\mathrm{I}}^{(j)} \prod_{i\in\mathcal{G}, i\neq j} \mathbb{P}_{\mathrm{MD}}^{(i)})\}$, where $\mathbb{P}_{\mathrm{MD}}^{(i)}$, $i \neq j$ are responses of other players. It is possible, however, that a glutton fails to collect the correct responses from all other players such that estimation of responses is needed. While over-estimation of responses results in less payoff, under-estimation results in zero payoff due to violation of the interference constraint.

In addition to the aforementioned four strategies in the repeated game, a special strategy called **Avenger** is proposed to stimulate cooperation. An avenger is a special role used for punishing non-cooperation (selfishness) of other players. If the final result of the negotiation fails to satisfy the interference constraint of a player, the "unsatisfied" player deviates from the original role by always claiming a positive decision ($\mathcal{D} = 1$) regardless of its sensing result. Consequently, $\mathbb{P}_{\mathrm{MD}}^{(j)} = 0$ and hence $\mathbb{P}_{\mathrm{MD}}^{(\mathcal{G})} = 0$ for the whole group. Equivalently, the expected payoff is zero for all players in the group, including the avenger itself. Since no player including selfish ones can get non-zero payoff, it is possible that in the next repetition non-cooperative players will be less "selfish" for improving its utility.

We have identified four possible strategies (roles) in the repeated game, but a secondary user may still need some rule for determining a suitable role for maximizing its utility. We present in the next section how the theory of evolution can be used for role learning.

4.2 Evolution of Strategies

As in the theory of evolution, a player selects the fittest strategy to be applied (for survival) in the game. If a strategy can bring more payoff than others to a player, the player has tendency to use it in ensuing game repetitions and hence the probability (time ratio) of choosing that strategy will increase. The replicator equation used in evolutionary game theory [16] for modeling the population of a species under *Nature* selection can then be applied to govern the dynamic increase and decrease of the probability distribution.

To start, Algorithm 2 shows the flow of how a secondary user plays the repeated game and learns its role based on the replicator equation. Let $\mathbb{P}_{\mathrm{S}}^{(j)}$, $\mathbb{P}_{\mathrm{L}}^{(j)}$, $\mathbb{P}_{\mathrm{F}}^{(j)}$ and $\mathbb{P}_{\mathrm{G}}^{(j)}$ denote the probabilities (time ratios) of player j choosing Solitary, Leader, Follower and Glutton respectively as the strategy. Initially, a player assigns equal probability to each strategy (Line 1). After the game starts, in each repetition, a player randomly chooses a strategy as the action based on the current distribution of strategies (Line 3). When a strategy is selected, a player follows the rule of the strategy to determine its $\mathbb{P}_{\mathrm{MD}}$ and detection threshold as presented in Section 4.1 (Line 4). After the negotiation process is complete,

Algorithm 2. Role evolution of player $j \in \mathcal{G}$

Input: Protection constraint $\varepsilon/\mathbb{P}_{\mathrm{I}}^{(j)}$ and learning step size α
Output: $\mathbb{P}_{\mathrm{S}}^{(j)}$, $\mathbb{P}_{\mathrm{L}}^{(j)}$, $\mathbb{P}_{\mathrm{F}}^{(j)}$ and $\mathbb{P}_{\mathrm{G}}^{(j)}$

01: $\mathbb{P}_{\mathrm{S}}^{(j)} = \mathbb{P}_{\mathrm{L}}^{(j)} = \mathbb{P}_{\mathrm{F}}^{(j)} = \mathbb{P}_{\mathrm{G}}^{(j)} = 0.25$ and $t = 1$
02: **while** $(j \in \mathcal{G})$
03: $A[t] = \mathrm{rand}(\mathbb{P}_{\mathrm{S}}^{(j)}, \mathbb{P}_{\mathrm{L}}^{(j)}, \mathbb{P}_{\mathrm{F}}^{(j)}, \mathbb{P}_{\mathrm{G}}^{(j)})$
04: Claim $\mathbb{P}_{\mathrm{MD}}^{(j)}$ according to role $A[t]$
05: **if** $\prod_{i \in \mathcal{G}} \mathbb{P}_{\mathbf{MD}}^{(i)} > \varepsilon/\mathbb{P}_{\mathbf{I}}^{(j)}$
06: Become an Avenger
07: **end if**
08: Apply corresponding θ_j to sense the spectrum
09: Earn the corresponding utility $U^{(j)}[t]$ as shown in Equation (7)
10: Record the best response up to t: $A^* = A[\arg \max_{1 \le \tau \le t} U^{(j)}[\tau]]$
11: Update fitness: $\phi_j(A^*) = \phi_j(A^*) + 1$
12: Update strategy distribution:
$\mathbb{P}_a^{(j)} = \{(1+\alpha)\left(\phi_j(a) - \bar{\phi_j}\right)\}\, \mathbb{P}_a^{(j)}, a \in \{\mathrm{S, L, F, G}\}$
13: $t = t + 1$
14: **end while**

a player checks $\mathbb{P}_{\mathrm{MD}}$ claimed by all other players. If the player fails to access the spectrum due to violation of its interference constraint, it switches to an Avenger; otherwise, it uses the sensing parameter as claimed to sense the spectrum (Line 5 to Line 8), and then evaluates its payoff using the utility function shown in Equation (7). A player maintains a data structure to remember its responses and the correspond payoff in history (Line 10). The player can then update the distribution of strategies using the replicator equation as follows.

In the replicator equation, the increasing rate of strategy a of player j can be modeled as:

$$\dot{\mathbb{P}}_a^{(j)} = \left[\phi_j(a) - \bar{\phi_j}\right] \mathbb{P}_a^{(j)},\ a \in \{\mathrm{S, L, F, G}\}, \tag{12}$$

where $\phi_j(a)$ is the fitness of strategy $a \in \{\mathrm{S, L, F, G}\}$ and $\bar{\phi_j}$ is the mean fitness of strategies of player j that can be written as $\bar{\phi_j} = \sum_{a \in \{\mathrm{S,L,F,G}\}} \mathbb{P}_a^{(j)} \phi_j(a)$. If a strategy has larger fitness than the mean fitness, its increasing rate $\dot{\mathbb{P}}_a^{(j)}$ is positive. Otherwise, the rate decreases. In addition, since the population of the next generation also depends on the current population, the increasing rate $\dot{\mathbb{P}}_a^{(j)}$ is proportional to the current value of $\mathbb{P}_a^{(j)}$. To reflect the objective of a player, the fitness of a strategy is designed as the times of bringing maximum payoff through history. Player j can then update the probability of each strategy in each repetition based on the following equation:

$$\mathbb{P}_a^{(j)} = \mathbb{P}_a^{(j)} + \alpha \dot{\mathbb{P}}_a^{(j)},\ a \in \{\mathrm{S, L, F, G}\}, \tag{13}$$

where α is a positive constant indicating the learning step size. It can be easily shown that the sum of $\mathbb{P}_a^{(j)}$ is equal to 1 after the update because the sum of

$\mathbb{P}_a^{(j)}$ is one initially and the sum of $\dot{\mathbb{P}}_a^{(j)}$ is zero. As repetition proceeds, a role with the largest probability for a player becomes the best role for itself.

5 Evaluation Results

In this section, we first show the convergence of the repeated game based on Algorithm 2 and then show its performance gain compared to the Nash equilibrium of the stage game under different scenarios.

5.1 Evolution of Strategies

We consider a network of three secondary users with different levels of interference to the primary user. $\mathbb{P}_\mathrm{I}$ of three users are 0.99, 0.9 and 0.6 respectively, and the requirement of the protection threshold ε is 0.05. Figure 2 thus shows the evolution of roles for players 1 and 3. Initially, each player assigns equal weight to each role. After 100 repetitions of the sensing game, the weight of each role becomes different, where player 1 has more probability to be a Glutton or a Solitary and player 3 enjoys the benefits of being a Leader. After another 100 repetitions of the game, the distributions exhibit quite different behaviors. Since player 1 has the most stringent constraint, it is okay for it to be a Leader while the other two act as Followers. On the other hand, it would be problematic if player 3 becomes a Leader or player 1 becomes a Glutton. Thus, the probability of (Leader, Follower, Follower) for players 1, 2, and 3 respectively shows clear increase. After 400 repetitions, the optimal roles for individual players become clear where player 1 acts as the Leader to solve the detection thresholds for all players, and other players follow the coordination of the Leader. Since detection thresholds are solved with the tightest constraint by Algorithm 1, it is also a social optimal solution. On the other hand, if the game is played without the Avenger, players quickly realize that greedy and risky strategies bring zero payoff. Consequently, all players become solitary and act conservatively. The results without the Avenger are not presented due to lack of space.

5.2 Gain of Evolution

Figure 3 shows the results when the cooperative set varies from 2 to 6 nodes. The right Y-axis shows the total expected throughput of the secondary users, and the left Y-axis shows the price of anarchy (PoA) defined as the achieved value over the social optimal value. It can be observed from the figure that in all cases the total expected throughput in the repeated game approaches the optimal performance and outperforms the Nash equilibrium. In turns of the PoA, it can be observed that the PoA of the Nash equilibrium decreases from 73% to 65% as the number of players increases. The reason is that under the Nash equilibrium, all players rely on the one with the most stringent constraint to sense the spectrum, and the gain of cooperation due to node diversity is not fully utilized.

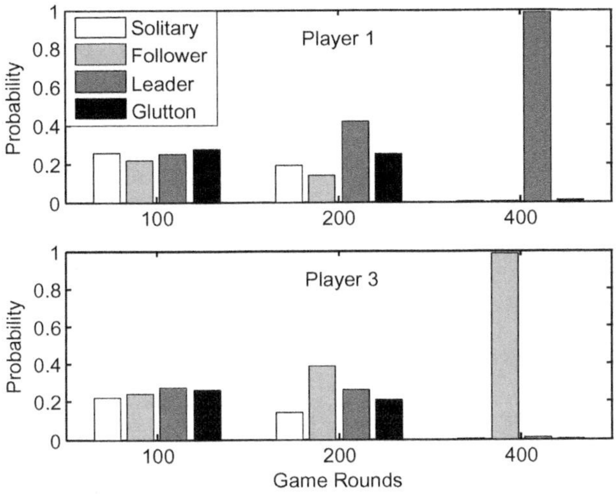

Fig. 2. Evolution of strategies in the repeated game

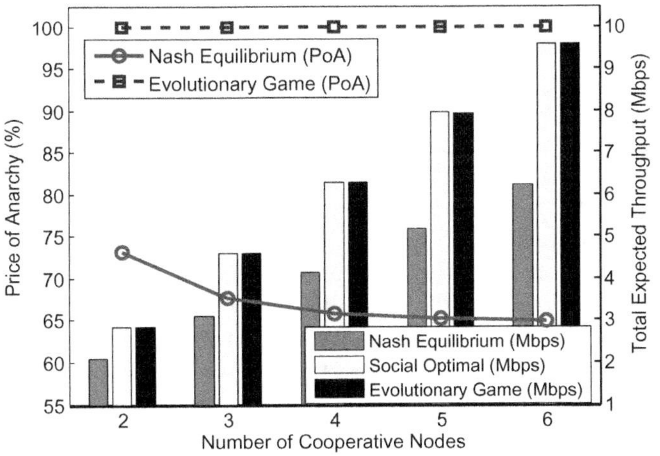

Fig. 3. Performance for different numbers of players

Figure 4 shows the results as the external interference varies from 0 to 1 (1 means that additional interference equal to the background noise is added during spectrum sensing). Note that as the external interference increases, it is more difficult to correctly detect the activity of the primary user and cooperation among nodes becomes more important. As the figure shows, the performance gain (compared to the result of the Nash equilibrium) of the repeated game increases as external interference increases and as the number of nodes increases. This substantiates that the proposed repeated game can indeed address the non-cooperative problem in the original one-shot game.

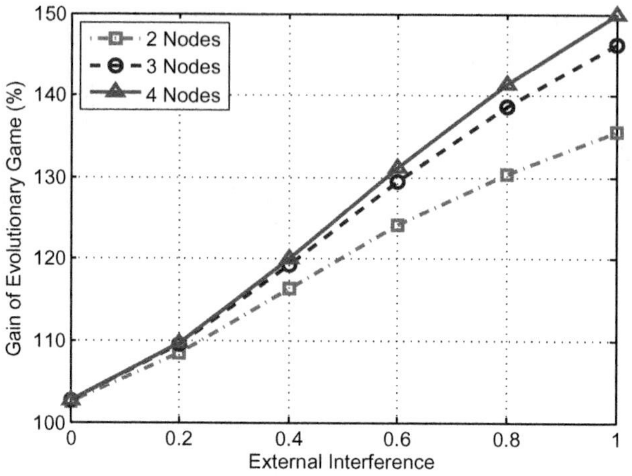

Fig. 4. Performance gains for varying external interference

6 Conclusions

In this work, we investigated the problem of interference-aware spectrum sensing for opportunistic spectrum access in cognitive radio networks. We showed that because different secondary users may have different levels of interference to the primary user, there is a conflict in setting the optimal sensing parameters for cooperative spectrum sensing. An interference-aware sensing game does not solve the problem since its Nash equilibrium will deviate from the social optimum. To address this problem, we designed a repeated game based on evolutionary game theory so players have the chance to revenge "non-cooperative" players in ensuing repetitions for driving the equilibrium to the social optimum. We showed through numerical results that the proposed repeated game does achieve the desirable performance for interference-aware cooperative sensing in opportunistic spectrum access.

Acknowledgements. This work was supported in part by funds from the National Science Council under Grants NSC-99-2219-E-002-020 and NSC-99-2911-I-002-001.

References

1. Bellman, R.: The theory of dynamic programming. Bulletin of the American Mathematical Society 60, 503–516 (1954)
2. Chow, C.S., Tsitsiklis, J.: An optimal multigrid algorithm for continuous state discrete time stochastic control. In: Proceedings of the 27th IEEE Conference on Decision and Control, vol. 3, pp. 1908–1912 (December 1988)
3. Cornuejols, G.: Valid inequalities for mixed integer linear programs. Mathematical Programming 112(1), 3–44 (2008)

4. Datla, D., Wyglinski, A., Minden, G.: A spectrum surveying framework for dynamic spectrum access networks. IEEE Transactions on Vehicular Technology 58(8), 4158–4168 (2009)
5. Digham, F.F., Alouini, M.S., Simon, M.K.: On the energy detection of unknown signals over fading channels. IEEE Transactions on Communications 55(1), 21–24 (2007)
6. Ganesan, G., Li, Y.: Cooperative spectrum sensing in cognitive radio networks. In: Proceeding of IEEE DySPAN, Baltimore, MD, USA, pp. 137–143 (November 2005)
7. Gardner, W.: Signal interception: A unifying theoretical framework for feature detection. IEEE Transactions on Communications 36(8), 897–906 (1988)
8. Ghasemi, A., Sousa, E.: Collaborative spectrum sensing for opportunistic access in fading environments. In: IEEE International Symposium on New Frontiers in Dynamic Spectrum Access Networks, pp. 131–136 (November 2005)
9. Huang, S., Liu, X., Ding, Z.: Optimal transmission strategies for dynamic spectrum access in cognitive radio networks. IEEE Transactions on Mobile Computing 8(12), 1636–1648 (2009)
10. Kay, S.M.: Fundamentals of Statistical Signal Processing: Detection Theory. Prentice Hall (February 1998)
11. Liang, Y.C., Zeng, Y., Peh, E., Hoang, A.T.: Sensing-throughput tradeoff for cognitive radio networks. IEEE Transactions on Wireless Communications 7(4), 1326–1337 (2008)
12. Moghimi, F., Nasri, A., Schober, R.: LP-norm spectrum sensing for cognitive radio networks impaired by non-Gaussian noise. In: Proceedings of IEEE Global Telecommunications Conference (GLOBECOM), pp. 1–6 (November 2009)
13. Peh, E., Liang, Y.C., Guan, Y.L., Zeng, Y.: Optimization of cooperative sensing in cognitive radio networks: A sensing-throughput tradeoff view. IEEE Transactions on Vehicular Technology 58(9), 5294–5299 (2009)
14. Sahai, A., Tandra, R., Mishra, S.M., Hoven, N.: Fundamental design tradeoffs in cognitive radio systems. In: Proceedings of the First International Workshop on Technology and Policy for Accessing Spectrum (TAPAS), New York, NY, USA (2006)
15. Tuy, H.: Monotonic optimization: Problems and solution approaches. SIAM Journal on Optimization 11(2), 464–494 (2000)
16. Weibull, J.W.: Evolutionary Game Theory. MIT Press (1995)
17. Yucek, T., Arslan, H.: A survey of spectrum sensing algorithms for cognitive radio applications. IEEE Communications Surveys and Tutorials 11(1), 116–130 (2009)
18. Zhang, T., Wu, Y., Lang, K., Tsang, D.: Optimal scheduling of cooperative spectrum sensing in cognitive radio networks. IEEE Systems Journal 4(4), 535–549 (2010)

A Game Theoretic Approach for Multi-hop Power Line Communications*

Walid Saad[1], Zhu Han[2], and Harold Vincent Poor[1]

[1] Electrical Engineering Department, Princeton University, Princeton, USA
{saad,poor}@princeton.edu
[2] Electrical and Computer Engineering Department, University of Houston, Houston, USA
zhan2@mail.uh.edu

Abstract. In this paper, a model for multi-hop power line communication is studied in which a number of smart sensors, e.g., smart meters, seek to minimize the delay experienced during the transmission of their data to a common control center through multi-hop power line communications. This problem is modeled as a network formation game and an algorithm is proposed for modeling the dynamics of network formation. The proposed algorithm is based on a myopic best response process in which each smart sensor can autonomously choose the path that connects it to the control center through other smart sensors. Using the proposed algorithm, the smart sensors can choose their transmission path while optimizing a cost that is a function of the overall achieved transmission delay. This transmission delay captures a tradeoff between the improved channel conditions yielded by multi-hop transmission and the increase in the number of hops. It is shown that, using this network formation process, the smart sensors can self-organize into a tree structure which constitutes a Nash network. Simulation results show that the proposed algorithm presents significant gains in terms of reducing the average achieved delay per smart sensor of at least 28.7% and 60.2%, relative to the star network and a nearest neighbor algorithm, respectively.

1 Introduction

The use of power lines as a means for communications has been adopted by utility companies for many decades in order to transmit control and monitoring data in power systems. Recently, power line communication (PLC) has emerged as a key technology that enables the delivery of new applications such as broadband Internet, telephony, automation, remote metering, as well as in-home delivery of a variety of data and multimedia services [1, 2, 3, 4]. While the full potential of PLC is yet to be exploited in the market, it is expected that PLC will play a major role as an enabler for efficient communications in the emerging smart grid and "Internet of things" networks.

In particular, within large-scale networks such as the smart grid, PLC is one of the main candidate technologies that can be used to ensure data communication between the smart sensors that are typically used to collect data (e.g., household loads, monitoring

* This work is supported by NSF Grants CNS-0910461, CNS-0953377, CNS-0905556, and ECCS-1028782, and by DTRA Grant HDTRA1-07-1-0037.

R. Jain and R. Kannan (Eds.): GameNets 2011, LNICST 75, pp. 546–561, 2012.

data, maintenance, prices inquiry, etc.) and transmit it to a control center or other central nodes within the network [2]. In this respect, enabling such PLC-based applications faces a variety of challenges such as channel modeling, medium access, security, efficient data transmission, and advanced network planning. For instance, in [4], a survey of smart grid applications for PLC is provided. The authors discuss a variety of usage scenarios for PLC such as advanced metering infrastructure, vehicle-to-grid communications, and substations interconnection in the medium voltage part of the grid. The work in [5] discusses the use of relaying techniques, such as decode-and-forward, for improving the capacity and coverage of an in-home broadband PLC network. In [6], the authors propose a space-time coding technique for improving retransmissions through repeaters using PLC channels. Performance analysis of a variety of channel models for PLC is done in [7, 8, 9, 10] and the references therein. The use of PLC for demand response in the smart grid is analyzed in [11]. An extensive treatment of communications and other networking issues in the smart grid is found in [2].

In essence, PLC can operate on either broadband or narrowband frequencies [1]. Depending on the mode of operation, the potential PLC applications can vary. While broadband PLC is suitable for Internet services, in-home entertainment, or demand response applications, narrowband PLC is bound to be used for advanced metering, vehicle-to-grid networking, as well as other smart grid applications [1, 2, 4]. In particular, narrowband PLC is a suitable means for interconnecting the smart sensors that are used for control, load monitoring, price inquiry, and other metering purposes in the smart grid. In fact, narrowband PLC has been widely used for advanced metering infrastructure in Europe [4]. One of the key challenges of adopting narrowband PLC as a communication technology between the sensors of the smart grid stems from the limited capacity of PLC channels which decreases quickly with distance as discussed in [1, Chap. 5] and [7]. Overcoming these limitations in existing PLC networks is typically done by dimensioning the network, prior to deployment, so as to ensure a reasonable capacity for every point-to-point PLC communication [1, 7].

However, for large-scale heterogeneous networks, such as the smart grid, a predetermined dimensioning may not be possible. For example, the deployment of smart meters is restricted by the locations of the related homes or businesses, irrespective of the communication technology that will be adopted between these meters. Hence, the ability to control the potential PLC capacity within a large cyber-physical network such as the smart grid faces several constraints and practical restrictions. In such networks, the PLC capacity limitations can lead to large delays or limited coverage, which constitute key quality-of-service (QoS) requirements for most smart grid applications [12]. As a result, it is of central interest to design intelligent and advanced algorithms that enable narrowband PLC communication in networks such as the smart grid or the Internet of things, while maintaining a reasonable QoS (e.g., low delays) given the large-scale, heterogeneous, and decentralized nature of these networks.

The main contribution of this paper is to propose a novel multi-hop protocol for narrowband PLC communication suitable for cyber-physical networks such as the smart grid. Hence, our objective is to develop an algorithm that enables multi-hop PLC communication between a number of smart sensors that need to send their data (e.g., meter readings, load reports, control and monitoring data, power quality, or pricing information) to a common access point (e.g., a control center or a repeater). For this purpose, we formulate a network formation game in which the smart sensors are the players and the strategy of each smart sensor is to select the preferred next hop for transmitting its data, using narrowband PLC. Then, we propose a network formation algorithm which enables the smart sensors to take distributed decisions regarding their PLC transmission path while minimizing their cost function which captures the overall experienced transmission delay. Using the proposed algorithm, the smart sensors can self-organize into a Nash network, i.e., a stable tree structure which connects them to the common control center or repeater. Simulation results show that the proposed algorithm yields a significant reduction of the average delay per smart sensor when compared to the star network or a nearest neighbor approach.

The rest of this paper is organized as follows: Section 2 presents the system model. In Section 3, we formulate the network formation game between the smart sensors while in Section 4, we present the proposed algorithm. Simulation results are presented and analyzed in Section 5. Finally, conclusions are drawn in Section 6.

2 System Model

Consider an area of a smart grid network composed of M physically interconnected smart sensors, e.g., smart meters, and let $\mathcal{M}$ denote the set of all of these smart sensors. Each smart sensor in the set $\mathcal{M}$ needs to communicate different information such as meter readings, load reports, pricing inquiries, event detection data, or network repair information to a common access point (CAP). This access point can be either a control center installed by the grid operator or a repeater that connects this area to other parts of the smart grid. In order to communicate, these smart sensors operate using high data rate (HDR) narrowband power line communication. Hence, all frequencies used throughout this paper are chosen from within the typical narrowband PLC values which range between 3 kHz and 148.5 kHz in Europe and up to 500 kHz in the USA [1].

We consider that the CAP allows each smart sensor $i \in \mathcal{M}$ to communicate using a channel having a bandwidth W, assumed to be the same for all smart sensors, without loss of generality. To do so, the CAP divides its range of frequencies $F_{\text{CAP},1} \leq |f| \leq F_{\text{CAP},2}$ into $K_{\text{CAP}} \triangleq \frac{F_{\text{CAP},2}-F_{\text{CAP},1}}{W}$ channels with equal bandwidth W. Subsequently, the CAP assigns each one of these channels to a requesting smart sensor $i \in \mathcal{M}$. We let $\mathcal{K}_{\text{CAP}}$ denote the set of all K_{CAP} channels that the CAP offers. Each channel $k \in \mathcal{K}_{\text{CAP}}$ is defined by a range of frequencies $F_{k,1} \leq |f| \leq F_{k,2} = F_{k,1} + W$. We consider a general case in which the CAP assigns the channels randomly depending on when a certain smart sensor requests to communicate. Further, hereinafter, we assume that the CAP is able to provide one channel for every smart sensor in the network and, thus, $M \leq K_{\text{CAP}}$. Nonetheless, the analysis can be easily extended to the case in which $M > K_{\text{CAP}}$ by adding extra control centers or by adopting advanced techniques for multiple PLC access which are outside the scope of this paper.

In narrowband PLC communication, the colored background noise has a (double sided) power spectral density $N(f)$ which decreases for increasing frequencies and can be expressed by [1, Chap. 5] (in W/Hz):

$$N(f) = 1/2 \cdot 10^{\gamma - 4 \cdot 10^{-5} f}, \tag{1}$$

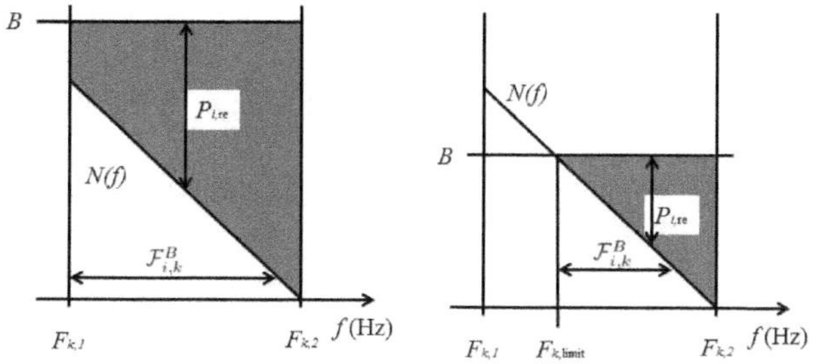

(a) Region 1, $F_{k,1} \leq |f| \leq F_{k,2}$ (b) Region 2, $F_{k,\text{limit}} \leq |f| \leq F_{k,2}$

Fig. 1. Two regions (dependent on the frequency range $\mathcal{F}_{i,k}^{B}$) used for finding the capacity of a PLC transmission, based on (2)-(4), for a smart sensor i using a certain channel k

where γ is normally distributed with mean -8.64 and standard deviation 0.5. As observed in [7], the value of γ is often estimated from measurements and has a worst case (high noise) value of -7.64 and a best case (low noise) value of -9.64.

Each smart sensor $i \in \mathcal{N}$ has a transmit power of P_i and needs to send L_i packets of R bits each to the CAP. In this respect, given the noise in (1) the capacity $C_{i,\text{CAP}}^{k}$of any point-to-point PLC communication link between a smart sensor $i \in \mathcal{N}$ and the CAP, using a channel $k \in \mathcal{K}_{\text{CAP}}$ is given by the so-called *water-filling* solution [7],[1, Chap. 5]:

$$C_{i,\text{CAP}}^{k} = \int_{f \in \mathcal{F}_{i,k}^{B}} 1/2 \log_2 \left[\frac{B}{N(f)} \right] df, \tag{2}$$

where $\mathcal{F}_{i,k}^{B}$ is the range of frequencies for channel k (i.e., within $F_{k,1} \leq |f| \leq F_{k,2} = F_{k,1} + W$) for which we have:

$$N(f) \leq B, \tag{3}$$

where B is the solution to

$$P_{i,\text{re}} = P_i \cdot 10^{-\kappa d_{i,\text{CAP}}} = \int_{f \in \mathcal{F}_{i,k}^{B}} [B - N(f)] \,. \tag{4}$$

Here, κ is the attenuation factor which ranges between 40 dB/km and 100 dB/km and $d_{i,\text{CAP}}$ is the distance between smart sensor i and the CAP. Note that, (2), (3), and (4) are derived based on the assumption that the transfer function $H(f)$ is chosen such that, within the frequency range of channel k, the channel gain is equal to one[1], i.e., $|H(f)|^2 = 1$. However, (2)-(4) can be easily modified to accommodate any other transfer function $H(f)$ as shown in [7].

As discussed in [1, Chap. 5], the solution to (2)-(4) can be found by splitting the problem into the two regions of Figure 1 in order to find the frequency range $\mathcal{F}_{i,k}^{B}$. In the first region shown in Figure 1(a), the level B is larger than the noise level and, thus, this region corresponds to the case in which the received signal power is relatively large,

[1] This consideration on the transfer function is inspired from the channel filter of the well-known CENELEC A-band [7].

i.e., there exists enough power to fill the entire frequency region. In contrast, the second region, shown in Figure 1(b), has a level B smaller than the noise level within a portion of the frequency range. This region corresponds to the case in which the received signal power is relatively small (due to attenuation) and, hence, it is used to fill the portion of the spectrum with a small noise level, i.e., below B.

The expression of the capacity in (2) differs between the two regions. In order to compute the expression within each region, it is of interest to find the critical distance $\tilde{d}^k_{i,\text{CAP}}$ such that for $d_{i,\text{CAP}} \leq \tilde{d}^k_{i,\text{CAP}}$ the operation is in Region 1 and for $d_{i,\text{CAP}} > \tilde{d}^k_{i,\text{CAP}}$ the operation is in Region 2. In other words, the critical distance represents the distance after which we move from the analysis of Region 1, i.e., Figure 1(a), to that of Region 2, i.e., Figure 1(b). For the transmission between a smart sensor $i \in \mathcal{M}$ to the CAP over a channel $k \in \mathcal{K}_{\text{CAP}}$, this critical distance $\tilde{d}^k_{i,\text{CAP}}$ can be given by [7]:

$$\tilde{d}^k_{i,\text{CAP}} = -\kappa^{-1} \cdot \log_{10}\left[\frac{(W-2)N(F_{k,1}) - 2N(F_{k,2})}{4 \cdot 10^{-5} \cdot \ln 10 \cdot P_i}\right], \tag{5}$$

where $N(f)$ is given by (1).

In Region 1 where $d_{i,\text{CAP}} \leq \tilde{d}^k_{i,\text{CAP}}$, as clearly seen from Figure 1(a), $\mathcal{F}^B_{i,k}$ is simply the entire frequency band available at channel k, i.e., $F_{k,1} \leq |f| \leq F_{k,2}$. Therefore, given the critical distance in (5), determining the capacity in Region 1 is straightforward and can be computed from (2) and (4) as follows [7]:

$$C^k_{i,\text{CAP}} = (1 + \log_2 B + 3.3\gamma)W + 6.65 \cdot 10^{-5} \cdot \left(F^2_{k,2} - F^2_{k,1}\right), \tag{6}$$

where B is given by:

$$B = \frac{[N(F_{k,1}) - N(F_{k,2})]}{9.2 \cdot 10^{-5} \cdot W} + \frac{10^{-\kappa d_{i,\text{CAP}}} \cdot P_i}{2W}. \tag{7}$$

In Region 2 where $d_{i,\text{CAP}} > \tilde{d}^k_{i,\text{CAP}}$, $\mathcal{F}^B_{i,k}$ is a region such that $F_{k,\text{limit}} \leq |f| \leq F_{k,2}$, where $F_{k,\text{limit}} > F_{k,1}$ is the point after which we have (3) verified (see Figure 1(b)). In this case, determining the capacity using (2)-(4) requires finding $F_{k,\text{limit}}$ first. While the computation can be analytically complex as discussed in [7], first, one can find an expression that links the distance and $F_{k,\text{limit}}$ as follows [7]:

$$\begin{aligned} P_i \cdot 10^{-\kappa d_{i,\text{CAP}}} - 2.2 \cdot 10^4 N(F_{k,2}) &= -2F_{k,\text{limit}} N(F_{k,\text{limit}}) \\ &\quad + 2N(F_{k,\text{limit}})(F_{k,2} - 1.1 \cdot 10^4) \end{aligned} \tag{8}$$

An analytical solution for (8) in which $F_{k,\text{limit}}$ is a real number expressed as a function of $d_{i,\text{CAP}}$ can be found as a function of the "Lambert W" function, as follows:

$$F_{k,\text{limit}} = (F_{k,2} - 1.1 \cdot 10^4) - \frac{W_L\left(g(d_{i,\text{CAP}})10^{(F_{k,2} - 1.1 \cdot 10^4) 4 \cdot 10^{-5}}\right)}{4 \cdot 10^{-5} \ln 10} \tag{9}$$

where $W_L(\cdot)$ is the "Lambert W" function and $g(d_{i,\text{CAP}})$ is given by

$$g(d_{i,\text{CAP}}) = 9.2 \cdot 10^{-5}(P_i \cdot 10^{-\gamma - \kappa d_{i,\text{CAP}}} - 1.1 \cdot 10^{4 - 4 \cdot 10^{-5} \cdot F_{k,2}}), \tag{10}$$

Then, by using (9) and (2), we can find a closed-form expression for the capacity in Region 2 as follows:

$$C_{i,\text{CAP}}^{k} = (F_{k,2} - F_{k,\text{limit}}) \log_2 (10^{-4 \cdot 10^{-5}} F_{k,\text{limit}}) + 6.65 \cdot 10^{-5} (F_{k,2}^{2} - F_{k,\text{limit}}^{2}) \quad (11)$$

Given any smart sensor $i \in \mathcal{M}$ that needs to send L_i packets of R bits each to the CAP while using a certain channel $k \in \mathcal{K}_{\text{CAP}}$, we define a cost function that captures the transmission delay as follows:

$$r_i(G) = \tau_{i,\text{CAP}} = \frac{R \cdot L_i}{C_{i,\text{CAP}}^{k}}, \quad (12)$$

where G is a star network graph centered at the CAP and connecting it to the smart sensors with direct transmission links (i.e., edges).

For narrowband PLC, it is well known that the capacities, as given in (6) and (11), are large for small distances, however, they decay very fast with distance. Subsequently, the delay in (12) can increase significantly, notably for a large-scale smart grid network in which the smart sensors need to communicate with a relatively far CAP. For many of the emerging applications within smart grid networks, delay and capacity are key QoS requirements [12], and, thus, it is of interest to design an approach that enables the smart sensors to utilize narrowband PLC for sending their data, while maintaining reasonable delays. For instance, by exploiting the fact that the capacities in (6) and (11) can be large for small to medium distances, one can develop a multi-hop scheme that enables the smart sensors to relay each others' data, while optimizing the delay in (12).

3 A Game Theoretic Approach for Multi-hop PLC Transmission

In order to improve their delays while communicating with the CAP, the smart sensors in $\mathcal{M}$ can interact with one another in order to perform multi-hop transmission. By doing so, the smart sensors can exploit the fact that the capacity of a narrowband PLC channel as captured by (2) is large for small distances but decays fast as the communication distances become larger. To perform multi-hop PLC communication, the smart sensors will essentially try to interact with their neighbors and decide on which hop to use given their traffic and potential PLC capacity.

In order to model these interactions between the smart sensors, we use the analytical framework of network formation games [13, 14, 15, 16]. Network formation games involve situations in which a number of players need to interact in order to decide on the formation of a network graph among them. In a network formation game, the outcome is essentially a graph structure that interconnects the various players while capturing their individual objectives. In this respect, to overcome the capacity limitations inherent to narrowband PLC-based networks, we propose a network formation game in which the smart sensors are the players and the objective is to form a multi-hop tree structure that enables each smart sensor to reduce its delay.

Hence, the result of the proposed smart sensors network formation game is a *directed* graph $G(\mathcal{M}, \mathcal{E})$ with $\mathcal{M}$ being the set of vertices of the graph (i.e., the smart sensors) and $\mathcal{E}$ being the set of all edges (links) between pairs of smart sensors. Note that, for the scope of this paper, we limit our attention to *tree structures* in which each smart sensor selects only one parent node for transmission. Each directed link between two smart sensors $i \in \mathcal{M}$ and $j \in \mathcal{M}$, denoted $(i, j) \in \mathcal{E}$, corresponds to a traffic flow over the narrowband PLC channel from smart sensor i to smart sensor j. Prior to delving into the details of the proposed network formation game, we will first define the notion of a path:

Definition 1. A *path* between two smart sensors i and j in a graph structure G is a sequence of smart sensors $i_1, \ldots, i_L$ such that $i_1 = i$, $i_L = j$ and each directed link $(i_l, i_{l+1}) \in G$ for each $l \in \{1, \ldots, L-1\}$.

In the proposed network formation game, each smart sensor will have a single path to the CAP due to the fact that we consider multi-hop tree structures between the smart sensors. As a result, we have a network formation game between the smart sensors in which the strategy of each smart sensor is to select its preferred path to destination. Formally, we can delineate the possible actions or strategies that a smart sensor can take in the proposed PLC network formation game as follows. The strategy space of any smart sensor $i \in \mathcal{M}$ is the set of possible smart sensors (or the CAP) that i can connect to. Consequently, the strategy of the smart sensor i is to select the link that it wants to form out of its available strategy space. Essentially, a smart sensor $i \in \mathcal{M}$ can connect either directly to the CAP or through any other smart sensor $j \in \mathcal{M},\ j \neq i$ as long as j is not, itself, connected to i. In other words, a smart sensor i cannot connect to another smart sensor j which is already connected to i, i.e., if $(j, i) \in G$, then $(i, j) \notin G$.

Hence, for a given network graph G, let $\mathcal{A}_i = \{j \in \mathcal{M} \setminus \{i\} | (j, i) \in G\}$ be the set of smart sensors from which smart sensor i accepted a link (j, i) and $\mathcal{S}_i = \{(i, j) | j \in \mathcal{V} \setminus (\{i\} \bigcup \mathcal{A}_i)\}$ be the set of links corresponding to the nodes (smart sensors or the CAP) that i can connect to, with $\mathcal{V}$ defined as the set of all smart sensors *and* the CAP. In consequence, the strategy of a smart sensor i can be formally defined as the link $s_i \in \mathcal{S}_i$ that it wants to form. As we consider tree structures, the strategy of a smart sensor can be seen as a *replace* operation using which a smart sensor $i \in \mathcal{M}$ replaces its current link with a new link from $\mathcal{S}_i$.

For narrowband PLC transmission, whenever a smart sensor needs to select a strategy, i.e., connect to another smart sensor (or the CAP), it needs to obtain an appropriate channel for transmission. In this respect, we define, for each smart sensor $i \in \mathcal{M}$, a set $\mathcal{K}_i$ that represents the set of *all* channels that i is able to offer to other smart sensor wishing to use i for multi-hop communication. In essence, for any given channel $k \in \mathcal{K}_i$ defined with a range of frequencies $F_{k,1} \leq |f| \leq F_{k,1} + W$, as the frequency $F_{k,1}$ increases, the capacity in (2) increases, thus reducing the delay (given that the bandwidth W is assumed to be the same for all channels). Thus, whenever a smart sensor i connects to a smart sensor j, we assume that j would assign the channel $k_i \in \mathcal{K}_j$ having the largest frequency $F_{k_i,1}$ to i, i.e., $k_i \in \underset{l \in \mathcal{K}_j}{\arg\max}\, F_{l,1}$. This best available channel is, thus, the channel that yields the lowest delay (among other *available* channels at j) for smart sensor i as per (13).

Further, we assume that the number of channels that a smart sensor i is able to offer is limited, due to the resource constrained nature of these smart sensors. Therefore, we have $|\mathcal{K}_i| < K_{\text{CAP}}$, $\forall i \in \mathcal{M}$, where $K_{\text{CAP}} = \mathcal{K}_{\text{CAP}}$ is the number of channels that the CAP can offer and $|\cdot|$ is the cardinality of a set operator. Thus, we can highlight the following property for our proposed smart sensors network formation game:

Property 1. For the proposed smart sensors network formation game, the number of nodes that a smart sensor $i \in \mathcal{M}$ serves within a graph G (i.e., the number of smart sensors in $\mathcal{A}_i$) is limited by the available channels in $\mathcal{K}_i$, and, thus, we have $|\mathcal{A}_i| \leq |\mathcal{K}_i|$.

As a result of Property 1, whenever a smart sensor $i \in \mathcal{M}$ has already accepted its maximum number of connections, i.e., $|\mathcal{A}_i| = |\mathcal{K}_i|$, it can no longer accept additional connections to serve. In this regard, denoting by $G + s_i$ as the graph G modified when a smart sensor i deletes its current link in G and adds a new link $s_i = (i, j)$, we define the concept of a *feasible* strategy as follows:

Definition 2. A strategy $s_i \in \mathcal{S}_i$, i.e., a link $s_i = (i, j)$, is a *feasible strategy* for a smart sensor $i \in \mathcal{V}$ if and only if smart sensor j can still offer a channel for i, i.e., $|\mathcal{A}_j| + 1 \leq |\mathcal{K}_j|$. For any smart sensor $i \in \mathcal{M}$, the set of all feasible strategies is denoted by $\hat{\mathcal{S}}_i \subseteq \mathcal{S}_i$.

Hence, a feasible strategy for any smart sensor i is simply a link $s_i = (i, j)$ in which the receiving smart sensor j has at least one channel available that can be assigned to smart sensor i. Whenever a smart sensor i plays a feasible strategy $s_i \in \hat{\mathcal{S}}_i$ while all the remaining smart sensors maintain a vector of strategies $\boldsymbol{s}_{-i}$, we let $G_{s_i, \boldsymbol{s}_{-i}}$ denote the resulting network graph.

In the proposed network formation game, the objective of each smart sensor $i \in \mathcal{M}$ is to select the path that minimizes its overall transmission delay when sending its data to the CAP, either directly or through multi-hop. Hence, given any tree structure $G_{s_i, \boldsymbol{s}_{-i}}$ resulting from the strategy selections of all the smart sensors in $\mathcal{M}$, the cost function of any smart sensor $i \in \mathcal{M}$ which selected a feasible strategy $s_i = (i, j) \in \hat{\mathcal{S}}_i$ having a corresponding path $q_i = \{i_1, \ldots, i_L\}$, with $i_1 = i$, $i_2 = j$ and i_L being the CAP, is captured by the total delay experienced by i which is given by:

$$c_i(G_{s_i, \boldsymbol{s}_{-i}}) = \tau_{i, s_i} = \sum_{(i_l, i_{l+1}) \in q_i} \tau_{i_l, i_{l+1}}, \tag{13}$$

where $\tau_{i_l, i_{l+1}}$ is the delay experienced during the transmission from smart sensor i_l to smart sensor i_{l+1} which can be given by:

$$\tau_{i_l, i_{l+1}} = \frac{R \cdot L_i}{C^k_{i_l, i_{l+1}}}, \tag{14}$$

where L_i is the number of packets of R bits that i needs to transmit and $C^k_{i_l, i_{l+1}}$ is the capacity for the narrowband PLC transmission between i_l and i_{l+1} over channel $k \in \mathcal{K}_{i_{l+1}}$. The capacity $C^k_{i_l, i_{l+1}}$ is computed using the method developed in Section 2 for the direct transmission to the CAP, i.e., using (2) - (4). Note that, whenever $G_{s_i, \boldsymbol{s}_{-i}}$ is a star network, i.e., all smart sensors are connected directly to the CAP, (13) reduces to (12).

Hence, in the proposed network formation game, the objective of each smart sensor is to interact with its neighbors in order to identify a strategy that can minimize its cost function in (13). These interactions are essentially non-cooperative as each smart sensor is selfish, i.e., interested in optimizing its individual cost as per (13). In this game, finding a suitable path to the CAP is a challenging task for each smart sensor, given the capacity limitations of narrowband PLC as well as the limited number of connections that a smart sensor can actually serve as highlighted in Property 1. Having formulated a network formation game among the smart sensors, our next step is to develop an algorithm that can model the interactions among the smart sensors that seek to form the network tree structure for multi-hop narrowband PLC transmission.

4 Distributed Network Formation Algorithm

Before discussing the details of the algorithm, we highlight that, for any developed algorithm, the resulting network structure will always be a *connected* graph as follows:

Property 2. Any network graph resulting from a network formation algorithm applied to the smart sensors game formulated in Section 3 is a *connected* tree structure rooted at the CAP, as long as $M < K_{\mathrm{CAP}}$.

Proof. Consider any network graph G in which there exists a certain smart sensor i that is disconnected from the CAP, i.e., no path of transmission (direct or multi-hop) exists between i and the CAP. In this case, the disconnected smart sensor i will experience an infinite delay as its data is not being transmitted, and, thus, its cost in (13) is maximized. As a result, no smart sensor has an incentive to disconnect from the CAP since such a disconnection will drastically increase its delay. Hence, as long as each smart sensor can always connect to the CAP, i.e., $M \leq K_{\mathrm{CAP}}$, then any network graph G formed for the proposed game is a connected tree structure rooted at the CAP.

A direct result of this property is that any smart sensor that is unable to connect to other smart sensors for performing multi-hop PLC will eventually use a direct transmission channel to the CAP, as long as such a channel exists, i.e., $M \leq K_{\mathrm{CAP}}$ (which is an assumption maintained throughout this paper). In this regard, we consider that the initial starting point for our network formation game is the star network in which all smart sensors are connected directly to the CAP, prior to interacting for further network formation decisions.

For any smart sensor $i \in \mathcal{M}$, given the set of feasible strategies $\hat{\mathcal{S}}_i$, we define the *best response* strategy as follows [15].

Definition 3. A strategy $s_i^* \in \hat{\mathcal{S}}_i$ is a *best response* for any smart sensor $i \in \mathcal{M}$ if $c_i(G_{s_i^*,\boldsymbol{s}_{-i}}) \leq c_i(G_{s_i,\boldsymbol{s}_{-i}})$, $\forall s_i \in \hat{\mathcal{S}}_i$. Thus, the best response for smart sensor i is to select the *feasible* link that minimizes its cost given that the other smart sensors maintain their vector of feasible strategies $\boldsymbol{s}_{-i}$.

Using the best responses of the smart sensors, we can develop a distributed network formation algorithm. To do so, we consider that the smart sensors are myopic, in the sense that each smart sensor seeks to reduce its delay by observing only the *current state* of the network without taking into account any potential future evolutions of the network graph. Developing algorithms for myopic network formation is a challenging task that has been receiving a significant attention in game theoretical research (e.g., see [13, 15, 16] and references therein). The challenging aspect of this problem stems from the fact that one deals with discrete strategy sets (i.e., forming links) and with the formation of network graphs in which adding or removing a single link can affect the overall network performance. The existing game theoretical literature on network formation games studies various myopic algorithms for different game models with directed and undirected graphs [13, 15, 16]. For our proposed smart sensors network formation algorithm, we construct an algorithm that is based on some of the models in [13] and [15], but modified to accommodate the specifics of the narrowband PLC multi-hop game. Hence, we define an algorithm where each round is mainly composed of three stages: a network discovery stage, a myopic network formation stage and a multi-hop PLC transmission phase.

Initially, the smart sensors start by using direct transmission within a star network. During the first stage of the proposed algorithm, the smart sensors attempt to discover some of their neighboring nodes, either by doing some monitoring of the communication in the star network or by using information downloaded from the network operator itself. Once each smart sensor obtains some information on the current nodes within the initial network, it can start with the second stage of the algorithm in which the main goal is to form the multi-hop tree structure. During the myopic network formation stage, the smart sensors perform pairwise negotiations (e.g., using some kind of dedicated PLC control channel), sequentially, in order to assess potential network formation decisions. In this stage, we consider that the smart sensors can make their decisions in a sequential, yet arbitrary order. In practice, this order can be dictated by which smart sensor requests first to form its link. Thus, in the myopic network formation stage, each smart sensor i selects a certain feasible strategy from its space $\hat{\mathcal{S}}_i$ so as to minimize its cost in (13). Each *iteration* in the network formation stage of the algorithm consists of a single sequence of plays during which all M smart sensors make their strategy choices to myopically react to the choices of the other smart sensors. The myopic network formation phase can consist of one or more iterations. In every iteration t, during its turn, each smart sensor i chooses to play its best response $s_i^* \in \hat{\mathcal{S}}_i$ in order to minimize its cost at each round given the current network graph resulting from the strategies of the other smart sensors. The best response of each smart sensor is a replace operation as the smart sensor disconnects its current link to the CAP while replacing it with another link that minimizes its cost (if such a link is available). Hence, the proposed network formation stage is based on the iterative feasible best responses of the smart sensors. When it converges, the network formation stage is guaranteed to reach a network in which no smart sensor can reduce its delay by changing its current link, i.e., a Nash network, defined as follows for the studied game [15]:

Definition 4. A network graph $G(\mathcal{M}, \mathcal{E})$ in which no smart sensor i can reduce its cost by unilaterally changing its feasible strategy $s_i \in \hat{\mathcal{S}}_i$ is a *Nash network*.

A Nash network is simply the concept of a Nash equilibrium applied to a network formation game. In the proposed game, a Nash network would, thus, be a network

Table 1. Proposed network formation algorithm

Initial State

The starting network is a star network in which each smart sensor is connected directly to the CAP.

The proposed algorithm consists of three stages

Stage 1 - Network Discovery:

Each smart sensor monitors the transmissions in the star network. Given the monitoring results and, possibly, assistance from the operator, each sensor gathers information on the other nodes.

Stage 2 - Myopic Network Formation:

repeat

In an arbitrary but sequential order, the smart sensors perform network formation.

a) In every iteration t of Stage 2, each smart sensor i plays its feasible best response $s_i^* \in \hat{\mathcal{S}}_i$, while minimizing its cost.

b) The best response s_i^* of each smart sensor is a *replace* operation using which a smart sensor i splits from its current parent smart sensor and replaces it with a new smart sensor that improves its cost.

until convergence to a final Nash tree G_{final}.

Stage 3 - Multi-hop PLC Transmission:

During this phase, data transmission from the smart sensors occurs using the assigned channels and hops within the formed network tree structure G_{final}.

---

where no smart sensor can improve its utility, by unilaterally changing its current link, given the current strategies of all other smart sensors.

Once a Nash network G_{final} forms, the last stage of the algorithm begins. This stage represents the actual data transmission phase, whereby the smart sensor can transmit their data using multi-hop PLC communication over the existing tree architecture G_{final}. A summary of the proposed algorithm is given in Table 1.

The proposed algorithm can be implemented in a distributed way within any network requiring PLC communication such as emerging smart grid networks. In essence, the smart sensors (e.g., meters) can perform the algorithm of Table 1 with little reliance on the CAP or other centralized control centers. For instance, the only role that may be required from the CAP is to provide the smart meters with some assistance in the network discovery phase, i.e., Stage 1 of the algorithm in Table 1. Once the sensors are aware of their environment, within every iteration t, during its turn, each smart sensor can engage in pairwise negotiations with the discovered nodes in order to find its feasible best response from the set of feasible strategies. During this process, the smart sensors need to only communicate in pairs and assess their potential cost function as per (13). Subsequently, each smart sensor can select its myopic best response, leading to a new iteration, until reaching the final Nash network. The worst case complexity for selecting the feasible best response (finding a preferred partner) for any smart sensor i is $O(M)$ where M is the total number of smart sensors. In practice, the complexity is much smaller as the smart sensors do not negotiate with the smart sensors that are connected to them, nor with the smart sensors that are too far away.

5 Simulation Results and Analysis

We consider a smart grid network deployed within a square area of 4 km $\times$ 4 km in which the control center (or a repeater) is placed at the middle. The smart sensors are deployed randomly inside this area and utilize narrowband PLC for transmission. Using typical parameters from [1] and [7], we choose a bandwidth of $W = 12.5$ kHz for every channel, we set the transmit power of any smart sensor $i \in \mathcal{M}$ to $P_i = 25$ W, we set the attenuation level to $\kappa = 0.007$ and we set $\gamma = -8.64$. The values of γ and κ are chosen within the typical best and worst cases [7]. For frequencies, we consider the narrowband PLC frequency range from 10 kHz to 235 kHz. Using this range, the CAP is able to offer 18 channels and, hence, it can accommodate up to $M = 18$ smart sensors. Each smart sensor can offer up to 5 channels for the nodes wishing to connect to it. The number of channels are picked randomly from integers between 1 and 5. All channels offered by the smart sensors are within the range of 10 kHz to 235 kHz. The packet size is set to $R = 2048$ bits and the number of packets is set to $L_i = 1$ packet for all $i \in \mathcal{M}$.

In Figure 2, we show a snapshot of a tree structure resulting from the proposed algorithm for a network with $M = 10$ randomly deployed smart sensors. This figure shows how a tree structure can form as a result of the distributed decisions of the smart sensors. In this snapshot, we can see that the smart sensors select their path not only based on distance but also on the offered channels. For instance, although smart sensor 6 is closer to smart sensor 4 than to smart sensor 5, it prefers to connect to 5. This is due to the fact that smart sensor 5 offers smart sensor 6 a communication over a channel k such that $F_{k,1} = 137.5$ kHz while smart sensor 4 offers a channel k such that $F_{k,1} = 37.5$ kHz. Hence, smart sensor 6 prefers to operate at a higher frequency as this ensures a higher capacity, and, eventually a better delay. Further, due to Property 1, smart sensor 7 decides to connect smart sensor 8 instead of smart sensor 10 since the latter can offer only two frequencies and has already assigned these frequencies to smart sensors 2 and 8. The strategies of all other nodes in Figure 2 are chosen by the smart sensors using a somewhat similar reasoning. Moreover, the network in Figure 2 is a Nash network as no smart sensor has an incentive to unilaterally change its current link. For example, consider smart sensor 6 whose feasible strategies are all other smart sensors and the CAP. If smart sensor 6 decides to disconnect from smart sensor 5 and connect to:

- Smart sensors 1, 2, 7, 8, 9, or 10 its delay increase from 102.5 ms to about 382 seconds.
- Smart sensor 3, its delay increase from 10 ms to 153 ms.
- Smart sensor 4, its delay increase from 102.5 ms to 120.1 ms.

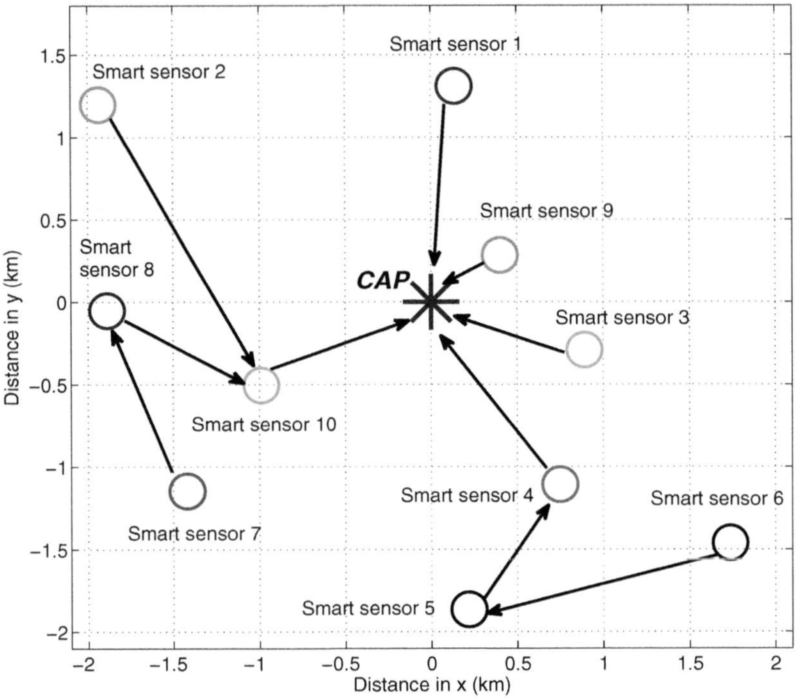

Fig. 2. Snapshot of a tree graph formed using the proposed algorithm for a network with $M = 10$ randomly deployed smart sensors

Hence, clearly, smart sensor 6 has no incentive to change its current strategy. Similar results can be seen for all other smart sensors in the Nash network of Figure 2. In a nutshell, Figure 2 shows how the smart sensors can self-organize into a Nash network while optimizing their delay given the offered frequencies and available potential partners.

Figure 3 shows the average achieved delay per smart sensor as the number of smart sensors M varies. The results are averaged over the random positions and channel selections of the smart sensors. The performance of the proposed network formation algorithm is compared with the direct transmission performance, i.e., the star network, as well as with a nearest neighbor algorithm in which each smart sensor selects the closest partner (in terms of distance) to connect to. In this figure, we can see that, as the number M of smart sensors in the network increases, the average achieved delay per smart sensor increases for the star network and the nearest neighbor algorithm. For the star network, this increase is due to the fact that, as the network size grows, it becomes more likely to have smart sensors that are far away from the CAP, and, thus achieving a poor capacity. Moreover, for the star network, the increase in the network size, constrains the channels that the CAP can offer. Hence, a larger star network will encompass smart sensors that are using channels in the lower part of the frequency band, and, thus, achieving a lower capacity as per (2)-(4).

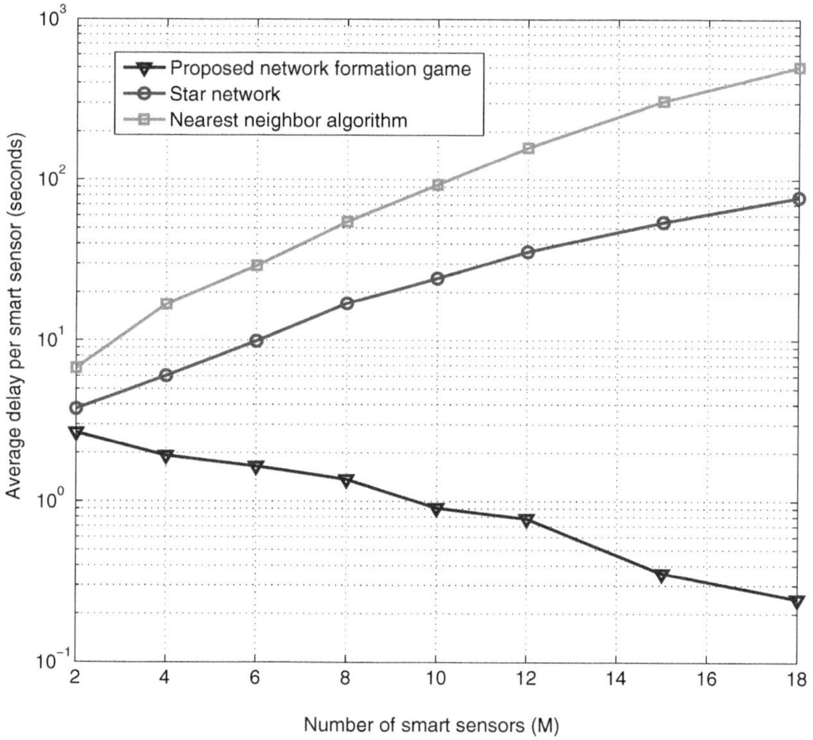

Fig. 3. Performance assessment of the proposed distributed network formation algorithm as the number of smart sensors M in the network (average over random positions and random channel choices of the smart sensors)

In the case of the nearest neighbor algorithm, the increase in the average delay with increasing M is due to the fact that, as more smart sensors are deployed in the network, the average delay resulting from a nearest neighbor-based multi-hop transmission increases due to the additional traffic. Moreover, for the nearest neighbor case, a smart sensor makes its selection solely based on distance and, hence, may connect to another smart sensor that is offering channels in the lower part of the band, hence, decreasing the potential capacity that can be achieved. Consequently, as the nearest neighbor algorithm yields, on the average, longer transmission paths with little capacity gains, as seen in Figure 3, its achieved average delays are larger than the star network.

In contrast, Figure 3 shows that, for the proposed network formation game, the average delay per smart sensor decreases with the network size. This result is interpreted by the fact that as the network size M grows, each smart sensor has a larger pool of partners from whom to select. Moreover, the increase in the number of smart sensors is accompanied by an increase in the number of possible transmission paths and channels that can be used. As a result, as more smart sensors are deployed, each smart sensor is able to exploit further the benefits of the proposed network formation algorithm in order to minimize its delay. In this respect, Figure 3 demonstrates that, at all network sizes M, the proposed network formation algorithm yields significant reductions of at least 28.7% and 60.2% in terms of the average delay per smart sensor, relative to the star network and the nearest neighbor algorithm, respectively.

6 Conclusions

In this paper, we have introduced a novel model for multi-hop communications in cyber-physical networks (such as the smart grid) that are bound to adopt narrowband power line communication for data transmission. In this respect, we have formulated a network formation game among a number of smart sensors (e.g., smart meters) that seek to transmit their data, using multi-hop, to a common control center or repeater. We have shown that the outcome of the formulated game is a tree structure that interconnects the smart sensors. To form this tree structure, we have developed a distributed myopic algorithm based on game theory. Using the proposed network formation algorithm, each smart sensor is able to decide, in a distributed manner, on its preferred path for data transmission in such way as to optimize a cost function that captures the overall transmission delay. We have shown that, using the proposed algorithm, the smart sensors self-organize into a Nash network in which no node has an incentive to change its current data transmission path. Simulation results have demonstrated that the proposed algorithm presents a significant advantage in terms of reducing average achieved delay per smart meter of at least 28.7% and 60.2%, relative to the star network and a nearest neighbor algorithm, respectively. Future extensions to this work can consider interference over the narrowband power line communication channel, advanced channel scheduling techniques, as well as network formation algorithms that can adapt to rapidly changing environments.

References

1. Ferreira, H.C., Lampe, L., Newbury, J., Swart, T.G.: Power Line Communications: Theory and Applications for Narrowband and Broadband Communications over Power Lines. Wiley, New York (2010)
2. Hossain, E., Han, Z., Poor, H.V.: Smart Grid Communications and Networking. Cambridge University Press, Cambridge (2011)
3. Galli, S., Scaglione, A., Dostert, K.: Broadband is power: Internet access through the power line network. IEEE Commun. Mag. 41(5), 82–83 (2003)
4. Galli, S., Scaglione, A., Wang, Z.: Power line communications and the smart grid. In: Proc. International Conference on Smart Grid Communications, Gaithersburg, MD, USA (October 2010)
5. Tonello, A.M., Versolatto, F., D'Alessandro, S.: Opportunistic relaying in In-Home PLC networks. In: Proc. IEEE Global Commun. Conf., Miami, FL, USA (December 2010)
6. Lampe, L., Schober, R., Yiu, S.: Distributed space-time coding for multihop transmission in power line communication networks. IEEE J. Select. Areas Commun. 24(7), 1389–1400 (2006)
7. Galli, S.: On the channel capacity of a European-style residential power circuit. In: Proc. International Symposium on Power Line Communication and its Applications, Tokyo, Japan (March 1998)
8. Canete, F., Cortes, J., Diez, L., Entrambasaguas, J.: Analysis of the cyclic short-term variation of indoor power line channels. IEEE J. Select. Areas Commun. 24(7), 1327–1338 (2006)

9. Galli, S.: A simple two-tap statistical model for the power line channel. In: Proc. International Symposium on Power Line Communication and its Applications, Rio de Janeiro, Brazil (March 2010)
10. Kim, I.H., Varadarajan, B., Dabak, A.: Performance analysis and enhancements of narrowband OFDM powerline communication systems. In: Proc. International Conference on Smart Grid Communications, Gaithersburg, MD, USA (October 2010)
11. Mohagheghi, S., Stoupis, J., Wang, Z., Li, Z., Kazemzadeh, H.: Demand response architecture: Integration into the distribution management system. In: Proc. International Conference on Smart Grid Communications, Gaithersburg, MD, USA (October 2010)
12. Li, H., Zhang, W.: QoS routing in smart grid. In: Proc. IEEE Global Commun. Conf., Miami, FL, USA (December 2010)
13. Demange, G., Wooders, M.: Group formation in economics: networks, clubs and coalitions. Cambridge University Press, Cambridge (2005)
14. Han, Z., Niyato, D., Saad, W., Başar, T., Hjørungnes, A.: Game Theory in Wireless and Communication Networks: Theory, Models and Applications. Cambridge University Press, Cambridge (2011)
15. Derks, J., Kuipers, J., Tennekes, M., Thuijsman, F.: Local dynamics in network formation, Maastricht University, Department of Mathematics, Maastricht, The Netherlands (December 2007), http://www.math.unimaas.nl/PERSONAL/jeand/downlds/derdyn.pdf
16. Johari, R., Mannor, S., Tsitsiklis, J.: A contract based model for directed network formation. Games and Economic Behavior 56, 201–224 (2006)

"To Sense" or "Not to Sense" in Energy-Efficient Power Control Games

Maël Le Treust[1], Yezekael Hayel[2], Samson Lasaulce[1], and Mérouane Debbah[3]

[1] Laboratoire des signaux et systèmes
(CNRS - SUPELEC - Paris 11)
91190 Gif-sur-Yvette, France
{letreust,lasaulce}@lss.supelec.fr
[2] Laboratoire d'information d'Avignon
Université d'Avignon
84911 Avignon, France
yezekael.hayel@univ-avignon.fr
[3] Chaire Alcatel
SUPELEC
91190 Gif-sur-Yvette, France
debbah@supelec.fr

Abstract. A network of cognitive transmitters is considered. Each transmitter has to decide his power control policy in order to maximize energy-efficiency of his transmission. For this, a transmitter has two actions to take. He has to decide whether to sense the power levels of the others or not (which corresponds to a finite sensing game), and to choose his transmit power level for each block (which corresponds to a compact power control game). The sensing game is shown to be a weighted potential game and its set of correlated equilibria is studied. Interestingly, it is shown that the general hybrid game where each transmitter can jointly choose the hybrid pair of actions (to sense or not to sense, transmit power level) leads to an outcome which is worse than the one obtained by playing the sensing game first, and then playing the power control game. This is an interesting Braess-type paradox to be aware of for energy-efficient power control in cognitive networks.

1 Introduction

In fixed communication networks, the paradigm of peer-to-peer communications has known a powerful surge of interest during the the two past decades with applications such as the Internet. Remarkably, this paradigm has also been found to be very useful for wireless networks. Wireless ad hoc and sensor networks are two illustrative examples of this. One important typical feature of these networks is that the terminals have to take some decisions in an autonomous (quasi-autonomous) manner. Typically, they have to choose their power control and resources allocation policy. The corresponding framework, which is the one of this paper, is the one of distributed power control or resources allocation. More specifically, the scenario of interest is the case of power control in cognitive

R. Jain and R. Kannan (Eds.): GameNets 2011, LNICST 75, pp. 562–577, 2012.

networks. Transmitters are assumed to be able to sense the power levels of neighboring transmitters and adapt their power level accordingly. The performance metric for a transmitter is the energy-efficiency of the transmission [5] that is, the number of bits successfully decoded by the receiver per Joule consumed at the transmitter.

The model of multiuser networks which is considered is a multiple access channel with time-selective non-frequency selective links. Therefore, the focus is not on the problem of resources allocation but only on the problem of controlling the transmit power over quasi-static channels. The approach of the paper is related to the one of [8][7] where some hierarchy is present in the network in the sense that some transmitters can observe the others or not; also the problem is modeled by a game where the players are the transmitters and the strategies are the power control policies. One the differences with [8][7] is that *every* transmitter can be cognitive and sense the others but observing/sensing the others has a cost. Additionally, a new type of power control games is introduced (called hybrid power control games) in which an action for a player has a discrete component namely, to sense or not to sense, and a compact component namely, the transmit power level. There are no general results for equilibrium analysis in the game-theoretic literature. This is a reason why some results are given in the 2-player case only, as a starting point for other studies. In particular, it is shown that it is more beneficial for every transmitter to choose his discrete action first and then his power level. The (finite) sensing game is therefore introduced here for the first time and an equilibrium analysis is conducted for it. Correlated equilibria are considered because they allow the network designer to play with fairness, which is not possible with pure or mixed Nash equilibria.

This paper is structured as follows. A review of the previous results regarding the one-shot energy efficient power control game is presented in Sec. 2. The sensing game is formally defined and some equilibrium results are stated in Sec. 3. A detailed analysis of the 2-players sensing is provided in Sec. 4 and the conclusion appears in Sec. 5.

2 Review of Known Results

2.1 Review of the One-Shot Energy-Efficient Power Control Game (Without Sensing)

We review a few key results from [6] concerning the static non-cooperative PC game. In order to define the static PC game some notations need to be introduced. We denote by R_i the transmission information rate (in bps) for user i and f an efficiency function representing the block success rate, which is assumed to be sigmoidal and identical for all the users; the sigmoidness assumption is a reasonable assumption, which is well justified in [11][4]. Recently, [3] has shown that this assumption is also justified from an information-theoretic standpoint. At a given instant, the SINR at receiver $i \in \mathcal{K}$ writes as:

$$\mathrm{SINR}_i = \frac{p_i |g_i|^2}{\sum_{j \neq i} p_j |g_j|^2 + \sigma^2} \tag{1}$$

where p_i is the power level for transmitter i, g_i the channel gain of the link between transmitter i and the receiver, σ^2 the noise level at the receiver, and f is a sigmodial efficiency function corresponding to the block success rate. With these notations, the static PC game, called $\mathcal{G}$, is defined in its normal form as follows.

Definition 2.1 (Static PC game). The static PC game is a triplet $\mathcal{G} = (\mathcal{K}, \{\mathcal{A}_i\}_{i\in\mathcal{K}}, \{u_i\}_{i\in\mathcal{K}})$ where $\mathcal{K}$ is the set of players, $\mathcal{A}_1, ..., \mathcal{A}_K$ are the corresponding sets of actions, $\mathcal{A}_i = [0, P_i^{\max}]$, $P_i^{\max}$ is the maximum transmit power for player i, and $u_1, ..., u_k$ are the utilities of the different players which are defined by:

$$u_i(p_1, ..., p_K) = \frac{R_i f(\mathrm{SINR}_i)}{p_i} \text{ [bit/J]}. \tag{2}$$

In this game with complete information ($\mathcal{G}$ is known to every player) and rational players (every player does the best for himself and knows the others do so and so on), an important game solution concept is the NE (i.e., a point from which no player has interest in unilaterally deviating). When it exists, the non-saturated NE of this game can by obtained by setting $\frac{\partial u_i}{\partial p_i}$ to zero, which gives an equivalent condition on the SINR: the best SINR in terms of energy-efficiency for transmitter i has to be a solution of $xf'(x) - f(x) = 0$ (this solution is independent of the player index since a common efficiency function is assumed, see [4] for more details). This leads to:

$$\forall i \in \{1, ..., K\}, \; p_i^* = \frac{\sigma^2}{|g_i|^2} \frac{\beta^*}{1-(K-1)\beta^*} \tag{3}$$

where β^* is the unique solution of the equation $xf'(x) - f(x) = 0$. By using the term "non-saturated NE" we mean that the maximum transmit power for each user, denoted by $P_i^{\max}$, is assumed to be sufficiently high not to be reached at the equilibrium i.e., each user maximizes his energy-efficiency for a value less than $P_i^{\max}$ (see [8] for more details). An important property of the NE given by (3) is that transmitters only need to know their individual channel gain $|g_i|$ to play their equilibrium strategy. One of the interesting results of this paper is that it is possible to obtain a more efficient equilibrium point by repeating the game $\mathcal{G}$ while keeping this key property.

2.2 Review of the Stackelberg Energy-Efficient Power Control Game (With Sensing)

Here we review a few key results from [7]. The framework addressed in [7] is that the existence of two classes of transmitters are considered: those who can sense and observe the others and those who cannot observe. This establishes a certain hierarchy between the transmitters in terms of observation. A suited model to study this is the Stackelberg game model [13]: some players choose their transmit power level (these are the leaders of the power control game) and the others observe the played action and react accordingly (these are the followers of the game). Note that the leaders know they are observed and take this into account for deciding. This leads to a game outcome (namely a Stackelberg equilibrium) which Pareto-dominates the one-shot game Nash equilibrium (given by (3)) when there is no cost for sensing [8]. However, when the fraction of time to sense is taken to be $\alpha > 0$, the data rate is weighted by $(1 - \alpha)$ and it is not always beneficial for a transmitter to sense [7]. The equilibrium action and utility for player i when he is a game leader (L) are respectively given by

$$p_i^L = \frac{\sigma^2}{|g_i|^2} \frac{\gamma^*(1+\beta^*)}{1-(K-1)\gamma^*\beta^*-(K-2)\beta^*} \tag{4}$$

where γ^* is the unique solution of $x\left[1 - \frac{(K-1)\beta^*}{1-(K-2)\beta^*}x\right] f'(x) - f(x) = 0$ and

$$u_i^L = \frac{|g_i|^2}{\sigma^2} \frac{1-(K-1)\gamma^*\beta^*-(K-2)\beta^*}{\gamma^*(1+\beta^*)} f(\gamma^*). \tag{5}$$

On the other hand, if player i is a follower (F) we have that:

$$p_i^F = \frac{\sigma^2}{|g_i|^2} \frac{\beta^*(1+\gamma^*)}{1-(K-1)\gamma^*\beta^*-(K-2)\beta^*} \tag{6}$$

and

$$u_i^F = (1-\alpha)\frac{|g_i|^2}{\sigma^2} \frac{1-(K-1)\gamma^*\beta^*-(K-2)\beta^*}{\beta^*(1+\gamma^*)} f(\beta^*). \tag{7}$$

3 A New Game: The $K-$Player Sensing Game

3.1 Sensing Game Description

In the two hierarchical power control described above, the transmitter is, by construction, either a cognitive transmitter or a non-cognitive one and the action of a player consists in choosing a power level. Here, we consider that all transmitters can sense, the power level to be the one at the Stackelberg equilibrium, and the action for a player consists in choosing to sense (S) or not to sense (NS). This game is well defined only if at least one player is a follower (i.e., he senses) and one other is the leader (i.e., he does not sense). We assume in the following that the total number of transmitters is $K + 2$, where K transmitters are considered as usual players and the two last are a follower and a leader. Define the $K-$player sensing game as a triplet:

$$G = (K, (\mathcal{S})_{i\in K}, (U_i)_{i\in K}) \tag{8}$$

where the actions set are the same for each player $i \in K$, sense or not sense: $\mathcal{S} = (S, NS)$. The utility function of each player $i \in K$ depends on his own channel state g_i and transmission rate R_i but also on the total number of players F playing the sensing action and the number of players that non sense denoted

L. Denote $U_i^S(F,L)$ the utility of player i when playing action sensing S whereas $F-1$ other players are also sensing and L other players are non-sensing. The total number of player is $F+L=K$.

$$U_i^S(F,L) = \frac{g_i R_i}{\sigma^2} \frac{f(\beta^*)}{N\beta^\star(N+\gamma_{L+1}^\star)}$$
$$\left(N^2 - N\beta^\star - [(N+\beta^\star)L + (F+1)\beta^\star]\,\gamma_{L+1}^\star\right)$$
$$U_i^{NS}(F,L) = \frac{g_i R_i}{\sigma^2} \frac{f(\gamma_L^*)}{N\gamma_{L+1}^\star(N+\beta^\star)}$$
$$\left(N^2 - N\beta^\star - [(N+\beta^\star)L + (F+1)\beta^\star]\,\gamma_{L+1}^\star\right)$$

with γ_L^* solution of $x(1-\epsilon_L x)f'(x) = f(x)$ with:

$$\epsilon_L = \frac{(K+2-L)\beta^\star}{N^2 - N(K+1-L)\beta^\star}. \tag{9}$$

3.2 The Sensing Game Is a Weighted Potential Game

The purpose of this section is to show that the sensing game may be an exact potential game. However, this holds under restrictive assumptions on the channel gains. It is then shown, as a second step, that the game is a weighted potential game. For making this paper sufficiently self-containing we review important definitions to know on potential games.

Definition 3.1 (Monderer and Shapley 1996 [9]). *The normal form game G is a potential game) if there is a potential function $V : S \longrightarrow \mathbb{R}$ such that*

$$U_i(s_i, s_{-i}) - U_i(t_i, s_{-i}) = V(s_i, s_{-i}) - V(t_i, s_{-i}), \tag{10}$$
$$\forall i \in K,\ s_i, t_i \in \mathcal{S}_{\rangle} \tag{11}$$

Theorem 3.2. *The sensing game $G = (K, (\mathcal{S})_{i\in K}, (U_i)_{i\in K})$ is an exact potential game if and only if one of the two following conditions is satisfied.*

$$1)\ \forall i,j \in K \quad R_i g_i = R_j g_j$$
$$2)\ \forall i,j \in K,\ s_i, t_i \in S_i,\ \forall s_j, t_j \in S_j,\ \forall s_k \in S_{K\setminus\{i,j\}}$$
$$U^T(t_i, s_j, s_k) - U^S(s_i, s_j, s_k)$$
$$+U^S(s_i, t_j, s_k) - U^T(t_i, t_j, s_k) = 0$$

The Proof is given in the Appendix 4.

The potential functions of our game depends on which condition is satisfied in the above theorem. Suppose that the first condition is satisfied $\forall i,j \in K \quad R_i g_i = R_j g_j$. Then the Rosenthal's potential function writes :

$$\Phi(F,L) = \sum_{i=1}^{F} U^S(i, K-i) + \sum_{j=1}^{L} U^{NS}(K-j, j)$$

Theorem 3.3 (Potential Game [9]). *Every finite potential game is isomorphic to a congestion game.*

Definition 3.4 (Monderer and Shapley 1996 [9]). *The normal form game G is a weighted potential game if there is a vector $(w_i)_{i\in K}$ and a potential function $V : S \longrightarrow \mathbb{R}$ such that:*

$$U_i(s_i, s_{-i}) - U_i(t_i, s_{-i}) = w_i(V(s_i, s_{-i}) - V(t_i, s_{-i})),$$
$$\forall i \in K,\ s_i, t_i \in \mathcal{S}_i$$

Theorem 3.5. *The sensing game $G = (K, (\mathcal{S}_i)_{i\in K}, (U_i)_{i\in K})$ is a weighted potential game with the weight vector:*

$$\forall i \in K \quad w_i = \frac{R_i g_i}{\sigma^2} \tag{12}$$

The Proof is given in the Appendix 5.

3.3 Equilibrium Analysis

First of all, note that since the game is finite (i.e., both the number of players and the sets of actions are finite), the existence of at least one mixed Nash equilibrium is guaranteed [10]. Now, since we know that the game is weighted potential we know that there is at least one pure Nash equilibrium [9]. Indeed, the following theorem holds.

Theorem 3.6. *The equilibria of the above potential game is the set of maximizers of the Rosenthal potential function [12].*

$$\{S = (S_1, \ldots, S_K) | S \in NE\} = \arg\max_{(F,L)} \Phi(F, L)$$
$$= \arg\max_{(F,L)} \left[\sum_{i=1}^{F} U(S, i, K - i) + \sum_{j=1}^{L} U(NS, K - j, j) \right]$$

The proof follows directly the one of Rosenthal's theorem [12].

We may restrict our attention to pure and mixed Nash equilibria. However, as it will be clearly seen in the 2-player case study (Sec. 4.2), this may pose a problem of fairness. This is the main reason why we study the set of correlated equilibria of the sensing game. We introduce the concept of correlated equilibrium [1] in order to enlarge the set of equilibrium utilities. Every utility vector inside the convex hull of the equilibrium utilities is a correlated equilibrium. The convexification property of the correlated equilibrium allow the system to better chose an optimal sensing. The concept of correlated equilibrium is a generalization of the Nash equilibrium. It consist in the stage game G extended with a signalling structure Γ. A correlated equilibrium (CE) of a stage game correspond to a Nash equilibrium (NE) of the same game extended with an adequate signalling structure Γ. A canonical correlated equilibrium is a probability distribution $Q \in \Delta(A)$, $A = A_1 \times ... \times A_K$ over the action product of the players that satisfy some incentives conditions.

Definition 3.7. *A probability distribution $Q \in \Delta(A)$ is a canonical correlated equilibrium if for each player i, for each action $a_i \in A_i$ that satisfies $Q(a_i) > 0$ we have:*

$$\sum_{a_{-i} \in A_{-i}} Q(a_{-i} \mid a_i) u_i(a_i, a_{-i}) \geq \sum_{a-i \in A_{-i}} Q(a_{-i} \mid a_i) u_i(b_i, a_{-i}), \quad \forall b_i \in A_i$$

The result of Aumann 1987 [2] states that for any correlated equilibrium, it correspond a canonical correlated equilibrium.

Theorem 3.8 (Aumann 1987, prop. 2.3 [2]). *The utility vector u is a correlated equilibrium utility if and only if there exists a distribution $Q \in \Delta(A)$ satisfying the linear inequality contraint 13 with $u = E_Q U$.*

The convexification property of the correlated equilibrium allow the system to better chose an optimal sensing. Denote E the set of pure or mixed equilibrium utility vectors and Conv E the convex hull of the set E.

Theorem 3.9. *Every utility vector $u \in$ Conv E is a correlated equilibrium utility of the sensing game.*

Any convex combination of Nash equilibria is a correlated equilibrium. As example, let $(\underline{U}_j)_{j \in J}$ a family of equilibrium utilities and $(\lambda_j)_{j \in J}$ a family of positive parameters with $\sum_{j \in J} \lambda_j = 1$ such that:

$$\underline{U} = \sum_{j \in J} \lambda_j \underline{U}_j \tag{13}$$

Then $\underline{U}$ is a correlated equilibrium utility vector.

4 Detailed Analysis for the 2-Player Case

4.1 The 2-Player Hybrid Power Control Game

In the previous section, we consider the sensing game as if the players do not chose their own power control policy. Indeed, when a player chooses to sense, he cannot choose its own power control because, it would depend on whether the other transmitters sense or not. We investigate the case where the players are choosing their sensing and power control policy in a joint manner. It enlarges the set of actions of the sensing game and it turns that, as a Braess-type paradox, that the set of equilibria is dramatically reduced. The sensing game with power control has a stricly dominated strategy: the sensing strategy. It implies that the equilibria of such a game boils down to the Nash equilibrium without sensing.

We consider that the action set for player i consists in choosing to sense or not and the transmit power level. The action set of player i writes :

$$A_i = \{S_i, NS_i\} \times [0, \bar{P}_i] \tag{14}$$

Before to characterize the set of equilibria of such a game, remark that the two pure equilibria of the previous matrix game are no longer equilibria. Indeed, assume that player 2 will not sense its environment and transmit using the leading power p_2^L. Then player 1 best response would be to play the following transmit power p_1^F as for the classical Stackelberg equilibrium. Nevertheless in the above formulation, the player 1 has a sensing cost α that correspond to the fraction of time to sense its environment. In this context, player 1 is incited to play the following transition power without sensing. The strategy (S_1, p_1^F) and (NS_2, p_1^L) is not an equilibrium of the game with Discrete and Compact Action Set.

Theorem 4.1. *The unique Nash equilibrium of the Power Control and Sensing Game is the Nash equilibrium without sensing.*

Proof. This result comes from the cost of sensing activity. Indeed, the strategy (S_1, p_1) is always dominated by the strategy (NS_1, p_1). It turns out that the sensing is a dominated actions for both players 1 and 2. Thus every equilibria is of the form (NS_1, p_1), (NS_2, p_2) with the reduced action spaces $p_1 \in [0, \bar{P}_1]$ and $p_2 \in [0, \bar{P}_2]$. The previous analysis applies in that case, showing that the unique Nash equilibrium of the Power Control and Sensing Game is the Nash of the game without sensing (p_1^*, p_2^*). □

As a conclusion, we see that letting the choice to the transmitters to choose jointly their discrete and continuous actions lead to a performance which is less than the one obtained by choosing his discrete action first, and then choosing his continuous action. This is the reason why we assume, from now on, the existence of a mechanism imposing this order in the decision taking.

4.2 The 2-Player Sensing Game

We consider the following two players-two strategies matrix game where players 1 and 2 choose to sense the channel (action S) or not (action NS) before transmitting his data. We denote by x_i the mixed strategy of user i, that is the probability that user i takes action S (sense the channel). Sensing activity provide the possibility to play as a follower, knowing in advance the action of the leaders. Let α denote the sensing cost, we compare the strategic behavior of sensing by considering the equilibrium utilities at the Nash and at the Stackelberg equilibria as payoff functions.

	NS_2	S_2
NS_1	$\frac{R_1 g_1 f(\beta^*)(1-\beta^*)}{\sigma^2\beta^*}$, $\frac{R_2 g_2 f(\beta^*)(1-\beta^*)}{\sigma^2\beta^*}$	$\frac{R_1 g_1 f(\gamma^*)(1-\gamma^*\beta^*)}{\sigma^2\gamma^*(1+\beta^*)}$, $(1-\alpha)\frac{R_2 g_2 f(\beta^*)(1-\gamma^*\beta^*)}{\sigma^2\beta^*(1+\gamma^*)}$
S_1	$(1-\alpha)\frac{R_1 g_1 f(\beta^*)(1-\gamma^*\beta^*)}{\sigma^2\beta^*(1+\gamma^*)}$, $\frac{R_2 g_2 f(\gamma^*)(1-\gamma^*\beta^*)}{\sigma^2\gamma^*(1+\beta^*)}$	$(1-\alpha)\frac{R_1 g_1 f(\beta^*)(1-\beta^*)}{\sigma^2\beta^*}$, $(1-\alpha)\frac{R_2 g_2 f(\beta^*)(1-\beta^*)}{\sigma^2\beta^*}$

Fig. 1. The Utility Matrix of the Two-Player Sensing Game

The equilibria of this game are strongly related to the sensing parameter α.

Theorem 4.2. *The matrix game has three equilibria if and only if*

$$\alpha < \frac{\beta^* - \gamma^*}{1 - \beta^*\gamma^*} \tag{15}$$

Let us characterize the three equilibria. From Appendix 1, is it easy to see that:

$$\alpha < \frac{\beta^* - \gamma^*}{1 - \beta^*\gamma^*} \Longleftrightarrow$$
$$(1-\alpha)\frac{R_1 g_1 f(\beta^*)(1-\gamma^*\beta^*)}{\sigma^2\beta^*(1+\gamma^*)} > \frac{R_1 g_1 f(\beta^*)(1-\beta^*)}{\sigma^2\beta^*}$$

We conclude that the joint actions (NS_1, NS_2) and (S_1, S_2) are not Nash Equilibria:

$$U_1(NS_1, NS_2) < U_1(S_1, NS_2) \tag{16}$$
$$U_2(NS_1, NS_2) < U_2(NS_1, S_2) \tag{17}$$
$$U_1(S_1, S_2) < U_1(NS_1, S_2) \tag{18}$$
$$U_2(S_1, S_2) < U_2(S_1, NS_2) \tag{19}$$

The sensing parameter determines which one of the two options is optimal between leading and following.

Corollary 4.3. *Following is better than leading if and only if*

$$\alpha < \frac{f(\beta^*) - f(\gamma^*) + \frac{f(\beta^*)}{\beta^*} - \frac{f(\gamma^*)}{\gamma^*}}{f(\beta^*)\frac{1+\beta^*}{\beta^*}} \tag{20}$$

The proof is given in Appendix 3.

The above matrix game has two pure equilibria (NS_1, S_2) and (S_1, NS_2). There is also a completely mixed equilibrium we compute using the indifference principle. Let $(x, 1-x)$ a mixed strategy of player 1 and $(y, 1-y)$ a mixed strategy of player 2. We aim at characterize the optimal joint mixed strategy

$$x^* = y^* = \frac{(1-\alpha)\frac{f(\beta^*)}{\beta^*}(1-\beta^*) - \frac{f(\gamma^*)}{\gamma^*}\frac{1-\gamma^*\beta^*}{1+\beta^*}}{(1-\alpha)\frac{f(\beta^*)}{\beta^*}(1-\beta^*) - \frac{f(\gamma^*)}{\gamma^*}\frac{1-\gamma^*\beta^*}{1+\beta^*} + \frac{f(\beta^*)}{\beta^*}(1-\beta^*) - (1-\alpha)\frac{f(\beta^*)}{\beta^*}\frac{1-\gamma^*\beta^*}{1+\gamma^*}}$$

(x^*, y^*) satisfying the indifference principle (see Appendix 2 for more details). The above joint mixed strategy $(x^*, 1-x^*)$ and $(y^*, 1-y^*)$ is an equilibrium strategy. The corresponding utilities are computed in Appendix 2. and writes with Δ defined in(4.2).

$$U_1(x^*, y^*) = \frac{R_1 g_1}{\sigma^2}\Delta$$
$$U_2(x^*, y^*) = \frac{R_2 g_2}{\sigma^2}\Delta$$

$$\Delta = \frac{(1-\alpha)\frac{f(\beta^*)}{\beta^*}(1-\beta^*)\frac{f(\beta^*)}{\beta^*}(1-\beta^*) - \frac{f(\gamma^*)}{\gamma^*}\frac{1-\gamma^*\beta^*}{1+\beta^*}(1-\alpha)\frac{f(\beta^*)}{\beta^*}\frac{1-\gamma^*\beta^*}{1+\gamma^*}}{(1-\alpha)\frac{f(\beta^*)}{\beta^*}(1-\beta^*) - \frac{f(\gamma^*)}{\gamma^*}\frac{1-\gamma^*\beta^*}{1+\beta^*} + \frac{f(\beta^*)}{\beta^*}(1-\beta^*) - (1-\alpha)\frac{f(\beta^*)}{\beta^*}\frac{1-\gamma^*\beta^*}{1+\gamma^*}}$$

The equilibrium utilities are represented on the following figure. The two pure Nash equilibrium utilities are represented by a circle whereas the mixed Nash utility is represented by a square.

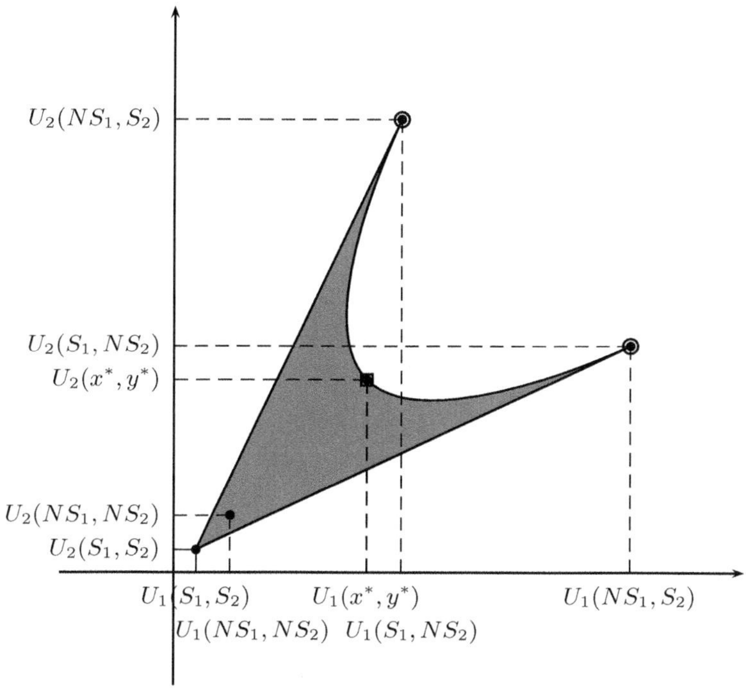

Fig. 2. The Equilibrium and Feasible Utilities

We also provide a characterization of the equilibria for the cases where α is greater or equal than $\frac{\beta^*-\gamma^*}{1-\beta^*\gamma^*}$.

Corollary 4.4. *The matrix game has a unique equilibrium if and only if*

$$\alpha > \frac{\beta^* - \gamma^*}{1 - \beta^*\gamma^*} \tag{21}$$

It has a infinity of equilibria if and only if

$$\alpha = \frac{\beta^* - \gamma^*}{1 - \beta^*\gamma^*} \tag{22}$$

First note that if the sensing cost is too high, the gain in terms of utility at Stackelberg instead of Nash equilibrium would be dominated by the loss of utility due to the sensing activity. In that case, the Nash equilibrium would be more efficient. Second remark that in case of equality, the action profiles (NS_1, NS_2), (NS_1, S_2), (S_1, NS_2) and every convex combination of the corresponding payoffs are all equilibrium payoffs.

Now that we have fully characterized the pure and mixed equilibria of the game, let us turn our attention to correlated equilibria.

Theorem (3.8) allows us to characterize the correlated equilibrium utility using the system of linear inequalities (13). We investigate the situation where the stage game has three Nash equilibria and following is better than leading. We suppose that the parameter α satisfies.

$$\alpha < \min\left(\frac{\beta^* - \gamma^*}{1 - \beta^*\gamma^*}, \frac{f(\beta^*) - f(\gamma^*) + \frac{f(\beta^*)}{\beta^*} - \frac{f(\gamma^*)}{\gamma^*}}{f(\beta^*)\frac{1+\beta^*}{\beta^*}}\right) \tag{23}$$

Note that the analysis is similar in the case where Leading is better than Following. However, if the parameter $\alpha > \frac{\beta^*-\gamma^*}{1-\beta^*\gamma^*}$ we have seen that the stage game has only one Nash equilibrium corresponding to play the Nash equilibrium power in the one-shot game. In such a case, no signalling device can increase the set of equilibria. The unique correlated equilibrium is the Nash equilibrium. We characterize an infinity of correlated equilibria.

Theorem 4.5. *Any convex combination of Nash equilibria is a correlated equilibrium. In particular if there exists a utility vector u and a parameter $\lambda \in [0, 1]$ such that:*

$$u_1 = \lambda U_1(S_1, NS_2) + (1 - \lambda)U_1(NS_1, S_2) \tag{24}$$
$$u_2 = \lambda U_2(S_1, NS_2) + (1 - \lambda)U_2(NS_1, S_2) \tag{25}$$

Then u is a correlated equilibrium.

The above result state that any distribution Q defined as follows with $\lambda \in [0, 1]$ is a correlated equilibrium. The canonical signalling device which should be added to the game consist in a lottery with parameter λ over the actions (S_1, NS_2)

	NS_2	S_2
NS_1	0	$1-\lambda$
S_1	λ	0

and (NS_1, S_2) and of signalling structure such that each player receives her component. For example, if (S_1, NS_2) is chosen the player 1 receives the signal "play S_1" whereas player 2 receives the signal "play NS_2".

The correlated equilibrium utilities are represented by the bold line. The signalling device increase the achievable utility region by adding the light gray area.

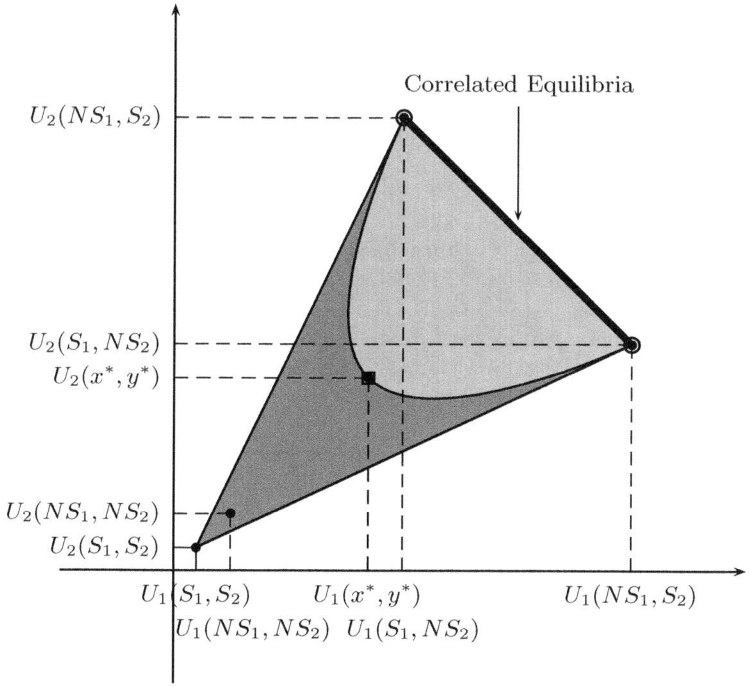

Fig. 3. The Correlated Equilibria

5 Conclusion

In this paper we have introduced a new power control game where the action of a player is hybrid, one component is discrete while the other is continuous. Whereas the general study of these games remains to be done, it turns out that in our case we can prove the existence of a Braess paradox which allows us to restrict our attention to two separate games played consecutively: a finite game where the players decide to sense or not and a compact game where the transmitter chooses his power level. We have studied in details the sensing game. In particular, it is proved it is weighted potential. Also, by characterizing the correlated equilibria of this game we show what is achievable in terms of fairness. Much work remains to be done to generalize all these results to games with arbitrary number of players and conduct simulations in relevant wireless scenarios.

References

1. Aumann, R.J.: Subjectivity and correlation in randomized strategies. Journal of Mathematics Economics 1(1), 67–96 (1974)
2. Aumann, R.J.: Correlated equilibrium as an expression of bayesian rationality. Econometrica 55(1), 1–18 (1987)
3. Belmega, E.V.: An information-theoretic look at mimo energy-efficient communications. In: ACM Proc. of the Intl. Conf. on Performance Evaluation Methodologies and Tools, VALUETOOLS (2009)
4. Schwartz, S.C., Meshkati, F., Poor, H.V., Mandayam, N.B.: An energy-efficient approach to power control and receiver design in wireless data networks. IEEE Trans. on Comm. 53(11) (2005)
5. Goodman, D.J., Mandayam, N.B.: Power control for wireless data. IEEE Person. Comm. 7, 48–54 (2000)
6. Goodman, D.J., Mandayam, N.: Power control for wireless data. IEEE Personal Communications 7(2), 45–54 (2000)
7. He, G., Lasaulce, S., Hayel, Y.: Stackelberg games for energy-efficient power control in wireless networks. In: Proc. INFOCOM 2011 (2011)
8. Lasaulce, S., Hayel, Y., El Azouzi, R., Debbah, M.: Introducing hierarchy in energy games. IEEE Trans. on Wireless Comm. 8(7), 3833–3843 (2009)
9. Monderer, D.: Potential games. Games and Economic Behavior 14, 124–143 (1996)
10. Nash, J.F.: Equilibrium points in n-points games. Proc. of the Nat. Academy of Science 36(1), 48–49 (1950)
11. Rodriguez, V.: An analytical foundation for resource management in wireless communication. In: IEEE Proc. of Globecom (2003)
12. Rosenthal, R.W.: A class of games possessing pure-strategy nash equilibria. International Journal of Game Theory 2, 65–67 (1973)
13. Stackelberg, V.H.: Marketform und Gleichgewicht. Oxford University Press (1934)

Appendix 1

$$
\begin{aligned}
& \alpha < \frac{\beta^* - \gamma^*}{1 - \beta^*\gamma^*} \\
\Longleftrightarrow\ & \frac{1 - \gamma^*\beta^* - \beta^* - \gamma^*}{(1 - \gamma^*\beta^*)} < 1 - \alpha \\
\Longleftrightarrow\ & (1 - \beta^*)(1 + \gamma^*) < (1 - \alpha)[(1 - \beta^*)(1 + \gamma^*) + \gamma^* + \beta^*] \\
\Longleftrightarrow\ & \frac{f(\beta^*)}{\beta^*}(1 - \beta^*) < (1 - \alpha)\frac{f(\beta^*)}{\beta^*}\frac{1 - \beta^*\gamma^*}{1 + \gamma^*} \\
\Longleftrightarrow\ & \frac{R_1 g_1 f(\beta^*)(1 - \beta^*)}{\sigma^2\beta^*} < (1 - \alpha)\frac{R_1 g_1 f(\beta^*)(1 - \gamma^*\beta^*)}{\sigma^2\beta^*(1 + \gamma^*)}
\end{aligned}
$$

Appendix 2

$$
\begin{aligned}
&\frac{R_1 g_1 f(\beta^*)(1-\beta^*)}{\sigma^2 \beta^*} \cdot y^* + \frac{R_1 g_1 f(\gamma^*)(1-\gamma^*\beta^*)}{\sigma^2 \gamma^*(1+\beta^*)} \cdot (1-y^*) \\
&= (1-\alpha)\frac{R_1 g_1 f(\beta^*)(1-\gamma^*\beta^*)}{\sigma^2 \beta^*(1+\gamma^*)} \cdot y^* + (1-\alpha)\frac{R_1 g_1 f(\beta^*)(1-\beta^*)}{\sigma^2 \beta^*} \cdot (1-y^*) \\
\Longleftrightarrow \quad & y^* \cdot [\frac{R_1 g_1 f(\beta^*)(1-\beta^*)}{\sigma^2 \beta^*} - (1-\alpha)\frac{R_1 g_1 f(\beta^*)(1-\gamma^*\beta^*)}{\sigma^2 \beta^*(1+\gamma^*)} \\
&+(1-\alpha)\frac{R_1 g_1 f(\beta^*)(1-\beta^*)}{\sigma^2 \beta^*} - \frac{R_1 g_1 f(\gamma^*)(1-\gamma^*\beta^*)}{\sigma^2 \gamma^*(1+\beta^*)}] \\
&= (1-\alpha)\frac{R_1 g_1 f(\beta^*)(1-\beta^*)}{\sigma^2 \beta^*} - \frac{R_1 g_1 f(\gamma^*)(1-\gamma^*\beta^*)}{\sigma^2 \gamma^*(1+\beta^*)} \\
\Longleftrightarrow \quad & y^* = \frac{(1-\alpha)\frac{f(\beta^*)}{\beta^*}(1-\beta^*) - \frac{f(\gamma^*)}{\gamma^*}\frac{1-\gamma^*\beta^*}{1+\beta^*}}{(1-\alpha)\frac{f(\beta^*)}{\beta^*}(1-\beta^*) - \frac{f(\gamma^*)}{\gamma^*}\frac{1-\gamma^*\beta^*}{1+\beta^*} + \frac{f(\beta^*)}{\beta^*}(1-\beta^*) - (1-\alpha)\frac{f(\beta^*)}{\beta^*}\frac{1-\gamma^*\beta^*}{1+\gamma^*}}
\end{aligned}
$$

Replacing the above y^* into the indifference equation, we obtain the utility of player 1 at the mixed equilibrium.

$$
\begin{aligned}
U_1(x^*, y^*) \;=\; & \frac{\frac{R_1 g_1 f(\beta^*)(1-\beta^*)}{\sigma^2\beta^*}\frac{R_1 g_1 f(\gamma^*)(1-\gamma^*\beta^*)}{\sigma^2\gamma^*(1+\beta^*)} - \frac{R_1 g_1}{\sigma^2}\frac{f(\gamma^*)}{\gamma^*}\frac{1-\gamma^*\beta^*}{1+\beta^*}\frac{R_1 g_1}{\sigma^2}(1-\alpha)\frac{f(\beta^*)}{\beta^*}\frac{1-\gamma^*\beta^*}{1+\gamma^*}}{\frac{R_1 g_1}{\sigma^2}(1-\alpha)\frac{f(\beta^*)}{\beta^*}(1-\beta^*) - \frac{R_1 g_1}{\sigma^2}\frac{f(\gamma^*)}{\gamma^*}\frac{1-\gamma^*\beta^*}{1+\beta^*} + \frac{R_1 g_1}{\sigma^2}\frac{f(\beta^*)}{\beta^*}(1-\beta^*) - \frac{R_1 g_1}{\sigma^2}(1-\alpha)\frac{f(\beta^*)}{\beta^*}\frac{1-\gamma^*\beta^*}{1+\gamma^*}} \\
+ \; & \frac{\frac{R_1 g_1}{\sigma^2}(1-\alpha)\frac{f(\beta^*)}{\beta^*}(1-\beta^*)\frac{R_1 g_1}{\sigma^2}\frac{f(\beta^*)}{\beta^*}(1-\beta^*) - \frac{R_1 g_1 f(\beta^*)(1-\beta^*)}{\sigma^2\beta^*}\frac{R_1 g_1 f(\gamma^*)(1-\gamma^*\beta^*)}{\sigma^2\gamma^*(1+\beta^*)}}{\frac{R_1 g_1}{\sigma^2}(1-\alpha)\frac{f(\beta^*)}{\beta^*}(1-\beta^*) - \frac{R_1 g_1}{\sigma^2}\frac{f(\gamma^*)}{\gamma^*}\frac{1-\gamma^*\beta^*}{1+\beta^*} + \frac{R_1 g_1}{\sigma^2}\frac{f(\beta^*)}{\beta^*}(1-\beta^*) - \frac{R_1 g_1}{\sigma^2}(1-\alpha)\frac{f(\beta^*)}{\beta^*}\frac{1-\gamma^*\beta^*}{1+\gamma^*}} \\
= \; & \frac{R_1 g_1}{\sigma^2}\frac{(1-\alpha)\frac{f(\beta^*)}{\beta^*}(1-\beta^*)\frac{f(\beta^*)}{\beta^*}(1-\beta^*) - \frac{f(\gamma^*)}{\gamma^*}\frac{1-\gamma^*\beta^*}{1+\beta^*}(1-\alpha)\frac{f(\beta^*)}{\beta^*}\frac{1-\gamma^*\beta^*}{1+\gamma^*}}{(1-\alpha)\frac{f(\beta^*)}{\beta^*}(1-\beta^*) - \frac{f(\gamma^*)}{\gamma^*}\frac{1-\gamma^*\beta^*}{1+\beta^*} + \frac{f(\beta^*)}{\beta^*}(1-\beta^*) - (1-\alpha)\frac{f(\beta^*)}{\beta^*}\frac{1-\gamma^*\beta^*}{1+\gamma^*}}
\end{aligned}
$$

The same argument applies:

$$
U_2(x^*, y^*) = \frac{R_2 g_2}{\sigma^2}\frac{(1-\alpha)\frac{f(\beta^*)}{\beta^*}(1-\beta^*)\frac{f(\beta^*)}{\beta^*}(1-\beta^*) - \frac{f(\gamma^*)}{\gamma^*}\frac{1-\gamma^*\beta^*}{1+\beta^*}(1-\alpha)\frac{f(\beta^*)}{\beta^*}\frac{1-\gamma^*\beta^*}{1+\gamma^*}}{(1-\alpha)\frac{f(\beta^*)}{\beta^*}(1-\beta^*) - \frac{f(\gamma^*)}{\gamma^*}\frac{1-\gamma^*\beta^*}{1+\beta^*} + \frac{f(\beta^*)}{\beta^*}(1-\beta^*) - (1-\alpha)\frac{f(\beta^*)}{\beta^*}\frac{1-\gamma^*\beta^*}{1+\gamma^*}}
$$

Appendix 3

$$
\begin{aligned}
&\alpha < \frac{f(\beta^*) - f(\gamma^*) + \frac{f(\beta^*)}{\beta^*} - \frac{f(\gamma^*)}{\gamma^*}}{f(\beta^*)\frac{1+\beta^*}{\beta^*}} \\
\Longleftrightarrow\; & 1-\alpha > \frac{f(\beta^*)\frac{1+\beta^*}{\beta^*} - f(\gamma^*)\frac{1+\gamma^*}{\gamma^*}}{f(\beta^*)\frac{1+\beta^*}{\beta^*}} \\
\Longleftrightarrow\; & (1-\alpha)\frac{f(\beta^*)}{\beta^*}\frac{1-\gamma^*\beta^*}{1+\beta^*} > \frac{f(\gamma^*)}{\gamma^*}\frac{1-\gamma^*\beta^*}{1+\gamma^*}
\end{aligned}
$$

Appendix 4

The proof comes from the theorem of Monderer and Shapley 1996 (see Sandholm "Decomposition of Potential" 2010)

Theorem 5.1. *The game G is a potential game if and only if for every players $i, j \in K$, every pair of actions $s_i, t_i \in S_i$ and $s_j, t_j \in S_j$ and every joint action $s_k \in S_{K \setminus \{i,j\}}$, we have that*

$$
\begin{aligned}
&U_i(t_i, s_j, s_k) - U_i(s_i, s_j, s_k) + U_i(s_i, t_j, s_k) - U_i(t_i, t_j, s_k) + \\
&U_j(t_i, t_j, s_k) - U_j(t_i, s_j, s_k) + U_j(s_i, s_j, s_k) - U_j(s_i, t_j, s_k) = 0
\end{aligned}
$$

Let us prove that the two conditions provided by our theorem are equivalent to the one of Monderer and Shapley's theorem. We introduce the following notation defined for each player $i \in K$ and each action $T \in \mathcal{S}$.

$$w_i = R_i g_i \tag{26}$$

$$U^T(t_i, t_j, s_k) = \frac{U_i^T(t_i, t_j, s_k)}{w_i} \tag{27}$$

For every players $i, j \in K$, every pair of actions $s_i, t_i \in S_i$ and $s_j, t_j \in S_j$ and every joint action $s_k \in S_{K \setminus \{i,j\}}$, we have the following equivalences:

$$
\begin{aligned}
& U_i(t_i, s_j, s_k) - U_i(s_i, s_j, s_k) \\
& +U_i(s_i, t_j, s_k) - U_i(t_i, t_j, s_k) \\
& + U_j(t_i, t_j, s_k) - U_j(t_i, s_j, s_k) \\
& +U_j(s_i, s_j, s_k) - U_j(s_i, t_j, s_k) = 0 \\
\Longleftrightarrow \quad & w_i(U^T(t_i, s_j, s_k) - U^S(s_i, s_j, s_k) \\
& +U^S(s_i, t_j, s_k) - U^T(t_i, t_j, s_k)) \\
& + w_j(U^T(t_i, t_j, s_k) - U^S(t_i, s_j, s_k) \\
& +U^S(s_i, s_j, s_k) - U^T(s_i, t_j, s_k)) = 0 \\
\Longleftrightarrow \quad & (w_i - w_j)(U^T(t_i, s_j, s_k) - U^S(s_i, s_j, s_k) \\
& +U^S(s_i, t_j, s_k) - U^T(t_i, t_j, s_k)) = 0 \\
& \Longleftrightarrow \begin{cases} w_i = w_j \\ U^T(t_i, s_j, s_k) - U^S(s_i, s_j, s_k) \\ +U^S(s_i, t_j, s_k) - U^T(t_i, t_j, s_k) = 0 \end{cases}
\end{aligned}
$$

Thus the sensing game is a potential game if and only if one of the two following condition is satisfied:

$$\forall i, j \in K \quad R_i g_i = R_j g_j \tag{28}$$

$$\forall i, j \in K,\ s_i, t_i \in S_i,\ \forall s_j, t_j \in S_j,\ \forall s_k \in S_{K \setminus \{i,j\}} \tag{29}$$

$$U^T(t_i, s_j, s_k) - U^S(s_i, s_j, s_k) \tag{30}$$

$$+U^S(s_i, t_j, s_k) - U^T(t_i, t_j, s_k) = 0 \tag{31}$$

Appendix 5

The proof of this theorem follows the same line of the previous theorem. It suffices to show that the auxiliary game defined as follows is a potential game.

$$\widetilde{G} = (K, (\mathcal{S})_{i \in K}, (\tilde{U}_i)_{i \in K}) \tag{32}$$

Where the utility are defined by the following equations with $w_i = \frac{R_i g_i}{\sigma^2}$.

$$\tilde{U}_i(s_i, s_{-i}) = \frac{U_i(s_i, s_{-i})}{w_i} \tag{33}$$

From the above demonstration, it is easy to show that, for every players $i, j \in K$, every pair of actions $s_i, t_i \in S_i$ and $s_j, t_j \in S_j$ and every joint action $s_k \in S_{K \setminus \{i,j\}}$:

$$\tilde{U}_i(t_i, s_j, s_k) - \tilde{U}_i(s_i, s_j, s_k) \tag{34}$$
$$+ \tilde{U}_i(s_i, t_j, s_k) - \tilde{U}_i(t_i, t_j, s_k) \tag{35}$$
$$+ \tilde{U}_j(t_i, t_j, s_k) - \tilde{U}_j(t_i, s_j, s_k) \tag{36}$$
$$+ \tilde{U}_j(s_i, s_j, s_k) - \tilde{U}_j(s_i, t_j, s_k) = 0 \tag{37}$$

We conclude that the sensing game is a weighted potential game.

Designing Social Norm Based Incentive Schemes to Sustain Cooperation in a Large Community

Yu Zhang, Jaeok Park, and Mihaela van der Schaar

Electrical Engineering Department, University of California, Los Angeles
yuzhang@ucla.edu, {jaeok,mihaela}@ee.ucla.edu

Abstract. Sustaining cooperation among self-interested agents is critical for the proliferation of emerging networked communities, such as the communities formed by social networking services. Providing incentives for cooperation in networked communities is particularly challenging because of their unique features: a large population of anonymous agents interacting infrequently, having asymmetric interests, and dynamically joining and leaving the network; network operation errors; and low-cost identity whitewashing. In this paper, taking these features into consideration, we propose a framework for the design and analysis of a class of incentive schemes based on social norms. We first define the concept of sustainable social norm under which no agent has an incentive to deviate. We then formulate the problem of designing an optimal social norm, which selects a social norm that maximizes overall social welfare among all sustainable social norms. Using the proposed framework, we study the structure of optimal social norms and the impacts of punishment lengths and whitewashing on optimal social norms. Our results show that optimal social norms are capable of sustaining cooperation, with the amount of cooperation varying depending on the community characteristics.

Keywords: Incentive Schemes, Networked Communities, Reputation Schemes, Social Norms.

1 Introduction

Recent developments in technology have expanded the boundaries of communities in which individuals interact with each other. However, a large population and the anonymity of individuals in network-based communities make it difficult to sustain cooperative behavior among self-interested individuals, with the so-called free-riding behavior prevailing [1]. Hence, incentive schemes are needed to provide individuals with incentives for cooperation.

The literature has proposed various incentive schemes. The popular forms of incentive devices used in many incentive schemes are payment and differential service. Pricing schemes use payments to reward and punish individuals for their behavior, which in principle can lead self-interested individuals to achieve social optimum by internalizing their external effects (see, for example, [8]). However, it is often claimed that pricing schemes are impractical because they require an accounting infrastructure [2]. Differential service schemes, on the other hand, reward and punish individuals by

R. Jain and R. Kannan (Eds.): GameNets 2011, LNICST 75, pp. 578–593, 2012.

providing differential services depending on their behavior. Such incentive schemes are based on the principle of reciprocity and can be classified into personal reciprocation and social reciprocation. In personal reciprocation schemes, individuals can identify each other, and behavior toward an individual is based on their personal experience with the individual. Personal reciprocation is effective in sustaining cooperation in a small community where individuals interact frequently and can identify each other, but it loses its force in a large community where anonymous individuals interact infrequently. In social reciprocation schemes, individuals obtain some information about other individuals (for example, rating) and decide their behavior toward an individual based on their information about that individual. Hence, an individual can be rewarded or punished by other individuals in the community who have not had direct interaction with it. Since social reciprocation requires neither observable identities nor frequent interactions, it has a potential to form a basis of successful incentive schemes for network-based communities. As such, this paper is devoted to the study of incentive schemes based on social reciprocation.

Sustaining cooperation using social reciprocation has been investigated in the economics literature using the framework of anonymous random matching games and social norm [7] and [10]. Each individual is attached a label indicating its reputation, status, etc. which contains information about its past behavior, and individuals with different labels are treated differently by other individuals they interact with. However, [7] and [10] have focused on obtaining the Folk Theorem by characterizing the set of equilibrium payoffs that can be achieved by using a social norm based strategy when the discount factor is sufficiently close to 1. Our work, on the contrary, addresses the problem of designing a social norm given a discount factor and other parameters arising from practical considerations. Specifically, our work takes into account the following features of network-based communities.

- *Asymmetry of interests.* We allow the possibility of asymmetric interests by modelling the interaction between a pair of individuals as a gift-giving game, instead of a prisoner's dilemma game, which assumes mutual interests between a pair of individuals.
- *Report errors.* In a social norm based incentive scheme, it is possible that the reputation of an individual is updated incorrectly because of errors in the report of individuals. Our model incorporates the possibility of report errors, which allows us to analyze its impact on the design and performance, whereas most existing works on reputation schemes [4][6] adopt an idealized assumption that reputations are always updated correctly.
- *Dynamic change in the population.* The members of a community change over time as individuals gain or lose interest in the services provided by community members. We model this feature by having a constant fraction of individuals leave and join the community in every period to study the impact of population turnover on the design and performance.

The remainder of this paper is organized as follows. In Section 2, we describe the repeated matching game and incentive schemes based on a social norm, and then formulate the problem of designing an optimal social norm. In Section 4, we provide analytical results about optimal social norms. We conclude the paper in Section 5.

2 Model

We consider an infinite-horizon discrete time model with a continuum of agents [4]. In a period, each agent generates a service request [9], which is sent to another agent that can provide the requested service. Each agent is equally likely to receive the request from a particular agent, and the matching in each period is independent. In a pair of matched agents, the agent that requests for a service is called a *client* while the agent that receives a service request is called a *server*. The interaction between a pair of matched agents is modelled as a gift-giving game [5], where the server has the binary choice from the set $\mathcal{A}=\{F,D\}$ of whether to fulfil, denoted as *F*, or decline the request, denoted as *D*, while the client has no choice. If the server fulfills the client's request, the client receives a service benefit of $b>0$ while the server suffers a service cost of $c>0$. If the server declines the request, both agents receive zero payoff. An agent plays the gift-giving game repeatedly with changing partners until it leaves the community. We assume that at the end of each period a fraction $\alpha \in [0,1]$ of agents in the current population leave and the same amount of new agents join the community. We refer to α as the *turnover rate* [4].

Social welfare in a time period is measured by the average payoff of the agents in that period. As we assume $b>c$, social welfare is maximized when all the servers choose action *F* in the gift-giving game they play, which yields payoff $b-c$ to every agent. On the contrary, action *D* is the dominant strategy for the server in the gift-giving game, which can be considered as the myopic equilibrium of the gift-giving game. When every server chooses its action to maximize its current payoff myopically, an inefficient outcome arises where every agent receives zero payoffs.

In order to improve the inefficiency of the myopic equilibrium, we use incentive schemes based on the idea of *social norms*. A social norm is defined as the rules that a group uses to regulate the behavior of members. We consider a social norm that consists of a *reputation scheme* and a *social strategy*, as in [7] and [10]. Formally, a reputation scheme determines the reputations of agents depending on their past actions as a server and is represented by two elements (Θ,τ). Θ is the set of reputations that an agent can hold, and τ is the reputation update rule. After a server takes an action, the client sends a report about the action of the server to the third-party device or infrastructure that manages the reputations of agents, but the report is subject to errors with a small probability ε. That is, with probability ε, *D* is reported when the server takes action *F*, and vice versa. Assuming a binary set of reports, it is without loss of generality to restrict ε in $[0,1/2]$. We consider a reputation update rule that updates the reputation of a server based only on the reputations of matched agents and the reported action of the server. Then, a reputation update rule can be represented by a mapping $\tau:\Theta\times\Theta\times\mathcal{A}\to\Theta$, where $\tau(\theta,\tilde{\theta},a_R)$ is the new reputation for a server with current reputation θ when it is matched with a client with reputation $\tilde{\theta}$ and its action is reported as $a_R \in \mathcal{A}$. A social strategy prescribes the actions that servers should take depending on the reputations of the matched agents and is represented by a mapping $\sigma : \Theta \times \Theta \rightarrow \mathcal{A}$, where $\sigma(\theta, \tilde{\theta})$ is the approved action for a server with reputation θ that is matched with a client with reputation $\tilde{\theta}$.

To simplify our analysis, we impose the following restrictions on reputation schemes we consider: (i) Θ is a nonempty finite set, i.e., $\Theta = \{0, 1, \ldots, L\}$ for some nonnegative integer L; (ii) τ is defined by

$$\tau(\theta, \tilde{\theta}, a_R) = \begin{cases} \min\{\theta + 1, L\} & \text{if } a_R = \sigma(\theta, \tilde{\theta}), \\ 0 & \text{if } a_R \neq \sigma(\theta, \tilde{\theta}). \end{cases} \tag{1}$$

Note that with the above restrictions a nonnegative integer L completely describes a reputation scheme, and thus a social norm can be represented by a pair $\kappa = (L, \sigma)$. We call the reputation scheme determined by L the *maximum punishment reputation scheme with punishment length* L. In the maximum punishment reputation scheme with punishment length L, there are total $L+1$ reputations, and the initial reputation for new peers entering the network is given as L. If the reported action of the server is the same as that specified by the social strategy σ, the server's reputation is increased by 1 while not exceeding L. Otherwise, the server's reputation is set as 0.

Below we summarize the sequence of events in a time period.

- Agents generate service requests and are matched.
- Each server observes the reputation of its client and then determines its action.
- Each client reports the action of its server.
- The reputations of agents are updated, and each agent observes its new reputation for the next period.
- A fraction of agents leave the community, and the same amount of new agents join the community.

As time passes, the reputations of agents are updated and agents leave and join the network. Thus, the distribution of reputations evolves over time. In this paper, we use the stationary distribution in our analysis, which will be written $\{\eta_L(\theta)\}$, where $\eta_L(\theta)$ be the fraction of θ-agents in the total population at the beginning of an arbitrary period t and a θ-agent means an agent with reputation θ. $\{\eta_L(\theta)\}$ satisfies the following equality as

$$\begin{aligned} &\eta_L(0) = (1-\alpha)\varepsilon, \\ &\eta_L(\theta) = (1-\alpha)(1-\varepsilon)\eta_L(\theta - 1) \quad \text{for } 1 \leq \theta \leq L-1, \\ &\eta_L(L) = (1-\alpha)(1-\varepsilon)\{\eta_L(L) + \eta_L(L-1)\} + \alpha. \end{aligned} \tag{2}$$

We now investigate the incentive of agents to follow a prescribed social strategy. Since we consider a non-cooperative scenario, we need to check whether an agent can improve its long-term payoff by a unilateral deviation.

Let $c_\sigma\left(\theta,\tilde{\theta}\right)$ be the cost suffered by a server with reputation θ that is matched with a client with reputation $\tilde{\theta}$ and follows a social strategy σ. Similarly, let $b_\sigma\left(\theta,\tilde{\theta}\right)$ be the benefit received by a client with reputation $\tilde{\theta}$ that is matched with a server with reputation θ following a social strategy σ. Since we consider uniform random matching, the expected period payoff of a θ-agent under social norm κ before it is matched is given by

$$v_\kappa(\theta)=\sum_{\tilde{\theta}\in\Theta}\eta_L\left(\tilde{\theta}\right)b_\sigma\left(\tilde{\theta},\theta\right)-\sum_{\tilde{\theta}\in\Theta}\eta_L\left(\tilde{\theta}\right)c_\sigma\left(\theta,\tilde{\theta}\right). \tag{3}$$

To evaluate the long-term payoff of an agent, we use the discounted sum criterion in which the long-term payoff of an agent is given by the expected value of the sum of discounted period payoffs from the current period. Let $p_\kappa\left(\theta' \mid \theta\right)$ be the transition probability that a θ-agent becomes a θ'-agent in the next period under social norm κ, the long-term payoff of an agent from the current period (before it is matched) can be solved by the following recursive equations

$$v_\kappa^\infty(\theta)=v_\kappa(\theta)+\delta\sum_{\theta'\in\Theta}p_\kappa\left(\theta' \mid \theta\right)v_\kappa^\infty\left(\theta'\right) \qquad \text{for } \theta\in\Theta, \tag{4}$$

where $\delta=\beta(1-\alpha)$ is the weight that an agent puts on its future payoff.

Now suppose that an agent deviates and uses a social strategy σ' under social norm κ. Since the deviation of a single agent does not affect the stationary distribution, the expected period payoff of a deviating θ-agent is given by

$$v_{\kappa,\sigma'}(\theta)=\sum_{\tilde{\theta}\in\Theta}\eta_L\left(\tilde{\theta}\right)b_\sigma\left(\tilde{\theta},\theta\right)+\sum_{\tilde{\theta}\in\Theta}\eta_L\left(\tilde{\theta}\right)c_{\sigma'}\left(\theta,\tilde{\theta}\right). \tag{5}$$

Let $p_{\kappa,\sigma'}\left(\theta' \mid \theta,\tilde{\theta}\right)$ be the transition probability that a θ-agent using social strategy σ' becomes a θ'-agent in the next period under social norm κ, when it is matched with a client with reputation $\tilde{\theta}$. The long-term payoff of a deviating agent from the current period (before it is matched) can be computed by solving

$$v_{\kappa,\sigma'}^\infty(\theta)=v_{\kappa,\sigma'}(\theta)+\delta\sum_{\theta'\in\Theta}p_{\kappa,\sigma'}\left(\theta' \mid \theta\right)v_{\kappa,\sigma'}^\infty\left(\theta'\right) \qquad \text{for } \theta\in\Theta. \tag{6}$$

In our model, a server decides whether to provide a service or not after it is matched with a client and observes the reputation of the client. Hence, we check the incentive for a server to follow a social strategy at the point when it knows the reputation of the client. Suppose that a server with reputation θ is matched with a client with reputation $\tilde{\theta}$. When the server follows the social strategy σ prescribed by social norm κ, it receives the long-term payoff $-c_\sigma\left(\theta,\tilde{\theta}\right)+\delta\sum_{\theta'}p_\kappa\left(\theta' \mid \theta\right)v_\kappa^\infty\left(\theta'\right)$, excluding the possible benefit as a client. On the contrary, when the server deviates to a social strategy σ', it receives the long-term payoff $-c_{\sigma'}\left(\theta,\tilde{\theta}\right)+\delta\sum_{\theta'}p_{\kappa,\sigma'}\left(\theta'\mid\theta,\tilde{\theta}\right)v_{\kappa,\sigma'}^{\infty}\left(\theta'\right)$, again excluding the possible benefit as a client. By comparing these two payoffs, we can check whether a θ-agent has an incentive to deviate to σ' when it is matched with a client with reputation $\tilde{\theta}$, and define a social norm κ is *sustainable* if no agent can gain from a unilateral deviation regardless of the reputation of the client it is matched with, i.e.

$$\begin{aligned}&-c_{\sigma}\left(\theta,\tilde{\theta}\right)+\delta\sum_{\theta'}p_{\kappa}\left(\theta'\mid\theta\right)v_{\kappa}^{\infty}\left(\theta'\right)\geq\\&-c_{\sigma'}\left(\theta,\tilde{\theta}\right)+\delta\sum_{\theta'}p_{\kappa,\sigma'}\left(\theta'\mid\theta,\tilde{\theta}\right)v_{\kappa,\sigma'}^{\infty}\left(\theta'\right)\end{aligned} \tag{7}$$

for all σ', for all $\left(\theta,\tilde{\theta}\right)$. Thus, under a sustainable social norm, agents follow the prescribed social strategy in their self-interest. Checking whether a social norm is sustainable using the above definition requires computing deviation gains from all possible social strategies, whose computation complexity can be high for moderate values of L. By employing the criterion of unimprovability in Markov decision theory [11], we establish the one-shot deviation principle for sustainable social norms. For notation, let c_a be the cost suffered by a server that takes action a, and let $p_{\kappa,a}\left(\theta'\mid\theta,\tilde{\theta}\right)$ be the transition probability that θ-agent becomes a θ'-agent in the next period under social norm κ when it takes action a to a client with reputation $\tilde{\theta}$. The values of $p_{\kappa,a}\left(\theta'\mid\theta,\tilde{\theta}\right)$ can be obtained in a similar way to obtain $p_{\kappa,\sigma'}\left(\theta'\mid\theta,\tilde{\theta}\right)$, by comparing a with $\sigma\left(\theta,\tilde{\theta}\right)$.

Lemma 1 (One-shot Deviation Principle). A social norm κ is sustainable if and only if

$$c_{\sigma}\left(\theta,\tilde{\theta}\right)-c_{a}\leq\delta\left[\sum_{\theta'}\left\{p_{\kappa}\left(\theta'\mid\theta\right)-p_{\kappa,a}\left(\theta'\mid\theta,\tilde{\theta}\right)\right\}v_{\kappa}^{\infty}\left(\theta'\right)\right] \tag{8}$$

for all $a\in\mathcal{A}$, for all $\left(\theta,\tilde{\theta}\right)$. ■

Lemma 1 shows that if an agent cannot gain by unilaterally deviating from σ only in the current period and following σ afterwards, it cannot gain by switching to any other social strategy σ' either, and vice versa. First, consider a pair of reputations $\left(\theta,\tilde{\theta}\right)$ such that $\sigma\left(\theta,\tilde{\theta}\right)=F$. If the server with reputation θ serves the client, it suffers the service cost of c in the current period while its reputation in the next period becomes $\min\{\theta+1,L\}$ with probability $(1-\varepsilon)$ and 0 with probability ε. Thus, the expected long-term payoff of a θ-agent when it provides a service is given by

$$V_{\theta}(F;F)=-c+\delta\left[(1-\varepsilon)v_{\kappa}^{\infty}\left(\min\{\theta+1,L\}\right)+\varepsilon v_{\kappa}^{\infty}(0)\right]. \tag{9}$$

Similarly, if a θ-agent deviates and declines the service request, the expected long-term payoff of a θ-agent when it does not provide a service is given by

$$V_\theta(D;F) = \delta\left[(1-\varepsilon)v_\kappa^\infty(0) + \varepsilon v_\kappa^\infty\left(\min\{\theta+1, L\}\right)\right]. \tag{10}$$

The incentive constraint that a θ-agent does not gain from a one-shot deviation is given by $V_\theta(F;F) \geq V_\theta(D;F)$, which can be expressed as,

$$\delta(1-2\varepsilon)\left[v_\kappa^\infty\left(\min\{\theta+1, L\}\right) - v_\kappa^\infty(0)\right] \geq c. \tag{11}$$

Now consider a pair of reputations $\left(\theta,\tilde{\theta}\right)$ such that $\sigma\left(\theta,\tilde{\theta}\right) = D$. Using a similar argument as above, we can show that the incentive constraint that a θ-agent does not gain from a one-shot deviation can be expressed as

$$\delta(1-2\varepsilon)\left[v_\kappa^\infty\left(\min\{\theta+1, L\}\right) - v_\kappa^\infty(0)\right] \geq -c. \tag{12}$$

Note that (11) implies (12), and thus for θ such that $\sigma\left(\theta,\tilde{\theta}\right) = F$ for some $\tilde{\theta}$, we can check only the first incentive constraint (11). Therefore, a social norm κ is sustainable if and only if (11) holds for all θ such that $\sigma\left(\theta,\tilde{\theta}\right) = F$ for some $\tilde{\theta}$ and (12) holds for all θ such that $\sigma\left(\theta,\tilde{\theta}\right) = D$ for all $\tilde{\theta}$. The left-hand side of the incentive constraints (11) and (12) can be interpreted as the loss from punishment that social norm κ applies to a θ-agent for not following the social strategy. Therefore, in order to induce a θ-agent to provide a service to some clients, the left-hand side should be at least as large as the service cost c, which can be interpreted as the deviation gain. We use $\min_{\theta\in\Theta}\left\{\delta(1-2\varepsilon)\left[v_\kappa^\infty\left(\min\{\theta+1, L\}\right) - v_\kappa^\infty(0)\right]\right\}$ to measure the strength of the *incentive for cooperation* under social norm κ, where cooperation means providing the requested service in our context.

Since we assume that the community operates at the stationary distribution of reputations, social welfare under social norm κ can be computed by

$$U_\kappa = \sum_\theta \eta_L(\theta) v_\kappa(\theta). \tag{13}$$

We assume that the community operator aims to choose a social norm that maximizes social welfare among sustainable social norms. Then the problem of designing a social norm can be formally expressed as

$$\begin{aligned}
\underset{(L,\sigma)}{\text{maximize}}\quad & U_\kappa = \sum_\theta \eta_L(\theta) v_\kappa(\theta)\\
\text{subject to}\quad & \delta(1-2\varepsilon)\left[v_\kappa^\infty\left(\min\{\theta+1, L\}\right) - v_\kappa^\infty(0)\right] \geq c,\ \forall\theta \text{ such that } \exists\tilde{\theta} \text{ such that } \sigma\left(\theta,\tilde{\theta}\right) = F,\\
& \delta(1-2\varepsilon)\left[v_\kappa^\infty\left(\min\{\theta+1, L\}\right) - v_\kappa^\infty(0)\right] \geq -c,\ \forall\theta \text{ such that } \sigma\left(\theta,\tilde{\theta}\right) = D\ \forall\tilde{\theta}.
\end{aligned} \tag{14}$$

A social norm that solves the problem (14) is called an *optimal social norm*.

3 Analysis of Optimal Social Norms

We first investigate whether there exists a sustainable social norm, i.e., whether the design problem (14) has a feasible solution. Fix the punishment length L and consider a social strategy σ_L^D where agents do not provide a service at all, i.e., for all $(\theta, \tilde{\theta})$. Since there is no service provided in the community when all the agents follow σ_L^D, we have $v_{(L,\sigma_L^D)}^{\infty}(\theta) = 0$ for all θ. Hence, the relevant incentive constraint (12) is satisfied for all θ, and the social norm (L, σ_L^D) is sustainable. This shows that the design problem (14) always has a feasible solution.

Assuming that an optimal social norm exists, let U^* be the optimal value of the design problem (14). In the following proposition, we study the properties of U^*.

Proposition 1. (i) $0 \leq U^* \leq b - c$; (ii) $U^* = 0$ if $\frac{c}{b} > \frac{\beta(1-\alpha)(1-2\varepsilon)}{1-\beta(1-\alpha)(2-3\varepsilon)}$; (iii) $U^* \geq [1-(1-\alpha)\varepsilon](b-c)$ if $\frac{c}{b} \leq \beta(1-\alpha)(1-2\varepsilon)$; (iv) $U^* < b - c$ if $\varepsilon > 0$; (v) $U^* = b - c$ if $\varepsilon = 0$ and $\frac{c}{b} \leq \beta(1-\alpha)$; (vi) $U^* = b - c$ only if $\varepsilon = 0$ and $\frac{c}{b} \leq \frac{\beta(1-\alpha)}{1-\beta(1-\alpha)}$.

Proof: See Appendix A. ∎

We obtain zero social welfare at myopic equilibrium, without using a social norm. Hence, we are interested in whether we can sustain a social norm in which agents cooperate in a positive proportion of matches. In other words, we look for conditions on the parameters $(b, c, \beta, \alpha, \varepsilon)$ that yield $U^* > 0$.

In order to obtain analytical results, we consider the design problem (14) with a fixed punishment length L, called *DP*$_L$. Note that *DP*$_L$ has a feasible solution, namely σ_L^D, for any L and that there are a finite number (total $2^{(L+1)^2}$) of possible social strategies given L. Therefore, *DP*$_L$ has an optimal solution for any L. We use and σ_L^* to denote the optimal value and the optimal social strategy of *DP*$_L$, respectively. We first show that increasing the punishment length cannot decrease the optimal value.

Proposition 2. $U_L^* \geq U_{L'}^*$ for all L and L' such that $L \geq L'$.

Proof: Choose an arbitrary L. To prove the result, we will construct a social strategy σ_{L+1} using punishment length $L+1$ that is feasible and achieves U_L^*. Define σ_{L+1} by

$$\sigma_{L+1}\left(\theta,\tilde{\theta}\right)=\begin{cases}\sigma_L^*\left(\theta,\tilde{\theta}\right) & \text{for } \theta\le L \text{ and } \tilde{\theta}\le L,\\ \sigma_L^*\left(L,\tilde{\theta}\right) & \text{for } \theta=L+1 \text{ and } \tilde{\theta}\le L,\\ \sigma_L^*\left(\theta,L\right) & \text{for } \theta\le L \text{ and } \tilde{\theta}=L+1,\\ \sigma_L^*\left(L,L\right) & \text{for } \theta=L+1 \text{ and } \tilde{\theta}=L+1.\end{cases} \tag{15}$$

Let $\kappa=\left(L,\sigma_L^*\right)$ and $\kappa'=\left(L+1,\sigma_{L+1}\right)$. From (2), we have $\eta_{L+1}(\theta)=\eta_L(\theta)$ for $\theta=0,\ldots,L-1$ and $\eta_{L+1}(L)+\eta_{L+1}(L+1)=\eta_L(L)$. Using this and (3), it is straightforward to see that $v_{\kappa'}(\theta)=v_\kappa(\theta)$ for all $\theta=0,\ldots,L$ and $v_{\kappa'}(L+1)=v_\kappa(L)$. Hence, we have that

$$\begin{aligned}U_{\kappa'}&=\sum_{\theta=0}^{L+1}\eta_{L+1}(\theta)v_{\kappa'}(\theta)=\sum_{\theta=0}^{L-1}\eta_{L+1}(\theta)v_{\kappa'}(\theta)+\sum_{\theta=L}^{L+1}\eta_{L+1}(\theta)v_{\kappa'}(\theta)\\ &=\sum_{\theta=0}^{L-1}\eta_L(\theta)v_\kappa(\theta)+\sum_{\theta=L}^{L+1}\eta_{L+1}(\theta)v_\kappa(L)\\ &=\sum_{\theta=0}^{L-1}\eta_L(\theta)v_\kappa(\theta)+\eta_L(L)v_\kappa(L)=U_\kappa=U_L^*.\end{aligned} \tag{16}$$

Using (25), we can show that $v_{\kappa'}^\infty(\theta)-v_{\kappa'}^\infty(0)=v_\kappa^\infty(\theta)-v_\kappa^\infty(0)$ for all $\theta=1,\ldots,L$ and $v_{\kappa'}^\infty(L+1)-v_{\kappa'}^\infty(0)=v_\kappa^\infty(L)-v_\kappa^\infty(0)$. By the definition of σ_{L+1}, the right-hand side of the relevant incentive constraint (i.e., c or $-c$) for each $\theta=0,\ldots,L$ is the same both under σ_L^* and under σ_{L+1}. Also, under σ_{L+1}, the right-hand side of the relevant incentive constraint for $\theta=L+1$ is the same as that for $\theta=L$. Therefore, σ_{L+1} satisfies the relevant incentive constraints for all $\theta=0,\ldots,L+1$. ■

Proposition 2 shows that U_L^* is non-decreasing in L. Since $U_L^*\le b-c$, we have $U^*=\lim_{L\to\infty}U_L^*=\sup_L U_L^*$. It may be the case that the incentive constraints eventually prevent the optimal value from increasing with L so that the supremum is attained by some finite L. If the supremum is not attained, the protocol designer can set an upper bound on L based on practical consideration. Now we analyze the structure of optimal social strategies given a punishment length.

Proposition 3. Suppose that $\varepsilon>0$ and $\alpha<1$. (i) If $\sigma_L^*\left(0,\hat{\theta}\right)=F$ for some $\hat{\theta}$, then $\sigma_L^*\left(0,\tilde{\theta}\right)=F$ for all $\tilde{\theta}\ge\min\left\{\ln\frac{c}{b}/\ln\beta,L\right\}$; (ii) If $\theta\in\{1,\ldots,L-1\}$ satisfies $\theta\ge L-\left(\ln\frac{c}{b}-\ln Y(\alpha,\varepsilon,L)\right)/\ln\beta$, where

$$Y(\alpha,\varepsilon,L)=\frac{(1-\alpha)^{L+1}(1-\varepsilon)^{L}\varepsilon-(1-\alpha)^{L+2}(1-\varepsilon)^{L+1}\varepsilon}{(1-\alpha)^{L+1}(1-\varepsilon)^{L}\varepsilon+\alpha} \tag{17}$$

and $\sigma_L^*\left(\theta,\hat{\theta}\right)=F$ for some $\hat{\theta}$, then $\sigma_L^*(\theta,L)=F$; (iii) If $\sigma_L^*\left(L,\hat{\theta}\right)=F$ for some $\hat{\theta}$, then $\sigma_L^*(L,L)=F$.

Proof: To facilitate the proof, we define $u_\kappa^\infty(\theta)$ by $u_\kappa^\infty(\theta)=\sum_{l=0}^{\infty}\gamma^l v_\kappa\left(\min\{\theta+l,L\}\right)$ for $\theta=0,\ldots,L$. Then, by (25), we have $v_\kappa^\infty(\theta)-v_\kappa^\infty(0)=u_\kappa^\infty(\theta)-u_\kappa^\infty(0)$ for all $\theta=1,\ldots,L$.

Suppose that $\sigma_L^*\left(0,\hat{\theta}\right)=F$ for some $\hat{\theta}$. Then the relevant incentive constraint for a 0-agent is $\delta(1-2\varepsilon)\left[u_\kappa^\infty(1)-u_\kappa^\infty(0)\right]\geq c$. Suppose that $\sigma_L^*\left(0,\overline{\theta}\right)=D$ for some $\overline{\theta}\in\{1,\ldots,L-1\}$ such that $\overline{\theta}\geq\ln\frac{c}{b}/\ln\beta$. Consider a social strategy $\sigma_L^{'}$ defined by

$$\sigma_L^{'}\left(\theta,\tilde{\theta}\right)=\begin{cases}\sigma_L^*\left(\theta,\tilde{\theta}\right) & \text{for } \left(\theta,\tilde{\theta}\right)\neq\left(0,\overline{\theta}\right),\\ F & \text{for } \left(\theta,\tilde{\theta}\right)=\left(0,\overline{\theta}\right).\end{cases} \tag{18}$$

Let $\kappa=\left(L,\sigma_L^*\right)$ and $\kappa^{'}=\left(L,\sigma_L^{'}\right)$. Note that $v_{\kappa^{'}}(0)-v_\kappa(0)=-\eta_\tau\left(\overline{\theta}\right)c<0$ and $v_{\kappa^{'}}\left(\overline{\theta}\right)-v_\kappa\left(\overline{\theta}\right)=\eta_\tau(0)b>0$ since $\varepsilon>0$ and $\alpha<1$. Thus, $U_{\kappa^{'}}-U_\kappa=\eta_L(0)\eta_L\left(\overline{\theta}\right)(b-c)>0$. Also,

$$u_{\kappa^{'}}^\infty(\theta)-u_\kappa^\infty(\theta)=\begin{cases}[v_{\kappa^{'}}(0)-v_\kappa(0)]+\gamma^{\overline{\theta}}[v_{\kappa^{'}}\left(\overline{\theta}\right)-v_\kappa\left(\overline{\theta}\right)] & \\ \quad=(1-\alpha)^{\overline{\theta}+1}(1-\varepsilon)^{\overline{\theta}}\varepsilon[\beta^{\overline{\theta}}b-c] & \text{for } \theta=0,\\ \gamma^{\overline{\theta}-\theta}[v_{\kappa^{'}}\left(\overline{\theta}\right)-v_\kappa\left(\overline{\theta}\right)] & \text{for } \theta=1,\ldots,\overline{\theta},\\ 0 & \text{for } \theta=\overline{\theta}+1,\ldots,L.\end{cases} \tag{19}$$

Since $\overline{\theta}\geq\ln\frac{c}{b}/\ln\beta$, we have $u_{\kappa^{'}}^\infty(0)-u_\kappa^\infty(0)\leq 0$. Thus, $u_{\kappa^{'}}^\infty(\theta)-u_{\kappa^{'}}^\infty(0)\geq u_\kappa^\infty(\theta)-u_\kappa^\infty(0)$ for all $\theta=1,\ldots,L$. Hence, $\sigma_L^{'}$ satisfies the incentive constraints of DP_L, which contradicts the optimality of σ_L^*. This proves that $\sigma_L^*(0,\tilde{\theta})=F$ for all $\tilde{\theta}\geq\ln\frac{c}{b}/\ln\beta$. Similar approaches can be used to prove $\sigma_L^*(0,L)=F$, (ii), and (iii). ■

We can represent a social strategy σ_L as an $(L+1)\times(L+1)$ matrix whose (i,j)-entry is given by $\sigma_L(i-1,j-1)$. Proposition 3 provides some structures of an optimal social strategy σ_L^* in the first row and the last column of the matrix representation, but it does not fully characterize the solution of DP_L. Here we aim to obtain the solution of DP_L for $L=1$ and 2 and analyze how it changes with the parameters. We first begin with the case of two reputations, i.e., $L=1$. In this case, if $\sigma_1\left(\theta,\tilde{\theta}\right)=F$ for some $\left(\theta,\tilde{\theta}\right)$, the relevant incentive constraint to sustain $\kappa=(1,\sigma_1)$ is $\delta(1-2\varepsilon)\left[v_\kappa^\infty(1)-v_\kappa^\infty(0)\right]\geq c$. By Proposition 3(i) and (iii), if $\sigma_1^*\left(\theta,\tilde{\theta}\right)=F$ for some $\left(\theta,\tilde{\theta}\right)$, then $\sigma_1^*(0,1)=\sigma_1^*(1,1)=F$, provided that $\varepsilon>0$ and $\alpha<1$. Hence, among the total of 16 possible social strategies, only four can be optimal social strategies. These four social strategies are

$$\sigma_1^1=\begin{bmatrix}D & F\\ F & F\end{bmatrix},\ \sigma_1^2=\begin{bmatrix}F & F\\ D & F\end{bmatrix},\ \sigma_1^3=\sigma_1^{D0}=\begin{bmatrix}D & F\\ D & F\end{bmatrix},\ \sigma_1^4=\sigma_1^D=\begin{bmatrix}D & D\\ D & D\end{bmatrix}. \quad (20)$$

The following proposition specifies the optimal social strategy given the parameters.

Proposition 4. Suppose that $0<(1-\alpha)\varepsilon<1/2$. Then

$$\sigma_1^*=\begin{cases}\sigma_1^1 & \text{if } 0<\frac{c}{b}\leq\frac{\beta(1-\alpha)^2(1-2\varepsilon)\varepsilon}{1+\beta(1-\alpha)^2(1-2\varepsilon)\varepsilon},\\ \sigma_1^2 & \text{if } \frac{\beta(1-\alpha)^2(1-2\varepsilon)\varepsilon}{1+\beta(1-\alpha)^2(1-2\varepsilon)\varepsilon}<\frac{c}{b}\leq\frac{\beta(1-\alpha)(1-2\varepsilon)[1-(1-\alpha)\varepsilon]}{1-\beta(1-\alpha)^2(1-2\varepsilon)\varepsilon},\\ \sigma_1^3 & \text{if } \frac{\beta(1-\alpha)(1-2\varepsilon)[1-(1-\alpha)\varepsilon]}{1-\beta(1-\alpha)^2(1-2\varepsilon)\varepsilon}<\frac{c}{b}\leq\beta(1-\alpha)(1-2\varepsilon),\\ \sigma_1^4 & \text{if } \beta(1-\alpha)(1-2\varepsilon)<\frac{c}{b}<1.\end{cases} \quad (21)$$

Proof: Let $\kappa^i=\left(1,\sigma_1^i\right)$, for $i=1,2,3,4$. We obtain that

$$\begin{aligned}&U_{\kappa^1}=\left(1-\eta_1(0)^2\right)(b-c),\quad U_{\kappa^2}=(1-\eta_1(0)\eta_1(1))(b-c),\\ &U_{\kappa^3}=(1-\eta_1(0))(b-c),\quad U_{\kappa^4}=0.\end{aligned} \quad (22)$$

Since $0<(1-\alpha)\varepsilon<1/2$, we have $\eta_1(0)<\eta_1(1)$. Thus, we have $U_{\kappa^1}>U_{\kappa^2}>U_{\kappa^3}>U_{\kappa^4}$. Also, we obtain that

$$\begin{aligned}&v_{\kappa^1}^\infty(1)-v_{\kappa^1}^\infty(0)=\eta_1(0)(b-c),\quad v_{\kappa^2}^\infty(1)-v_{\kappa^2}^\infty(0)=b-\eta_1(0)(b-c),\\ &v_{\kappa^3}^\infty(1)-v_{\kappa^3}^\infty(0)=b,\qquad\qquad\quad v_{\kappa^4}^\infty(1)-v_{\kappa^4}^\infty(0)=0.\end{aligned} \quad (23)$$

Thus, we have $v_{\kappa^3}^{\infty}(1) - v_{\kappa^3}^{\infty}(0) > v_{\kappa^2}^{\infty}(1) - v_{\kappa^2}^{\infty}(0) > v_{\kappa^1}^{\infty}(1) - v_{\kappa^1}^{\infty}(0) > v_{\kappa^4}^{\infty}(1) - v_{\kappa^4}^{\infty}(0)$. By choosing the social strategy that yields the highest social welfare among feasible ones, we obtain the result. ■

Proposition 4 shows that the optimal social strategy is determined by the ratio of the service cost and benefit, c/b. When c/b is sufficiently small, the social strategy σ_1^1 can be sustained, yielding the highest social welfare among the four candidate social strategies. As c/b increases, the optimal social strategy changes from σ_1^1 to σ_1^2 to σ_1^3 and eventually to σ_1^4. Figure 1 shows the optimal social strategies with $L=1$ as c varies. The parameters we use to obtain the results in the figures of this paper are set as follows unless otherwise stated: $\beta = 0.8$, $\alpha = 0.1$, $\varepsilon = 0.2$, and $b = 10$. Figure 1 (a) plots the incentive for cooperation of the four social strategies. We can find the region of c in which each strategy is sustained by comparing the incentive for cooperation with the service cost c for σ_1^1, σ_1^2, and σ_1^3, and with $-c$ for σ_1^4. The solid portion of the lines indicates that the strategy is sustained while the dashed portion indicates that the strategy is not sustained. Figure 1 (b) plots the social welfare of the four candidate strategies, with solid and dashed portions having the same meanings. The triangle-marked line represents the optimal value, which takes the maximum of the social welfare of all sustained strategies.

Next, we analyze the case of three reputations, i.e. $L=2$. In Figure 2, we show the optimal value and the optimal social strategy of DP_2 as we vary c. The optimal social strategy $\overset{*}{\sigma_2}$ changes in the following order before becoming σ_2^D as c increases:

$$\sigma_2^1 = \begin{bmatrix} F & F & F \\ D & F & F \\ F & F & F \end{bmatrix}, \sigma_2^2 = \begin{bmatrix} D & F & F \\ F & F & F \\ F & F & F \end{bmatrix}, \sigma_2^3 = \begin{bmatrix} D & F & F \\ D & F & F \\ F & F & F \end{bmatrix},$$
$$\sigma_2^4 = \begin{bmatrix} F & F & F \\ F & F & F \\ D & F & F \end{bmatrix}, \sigma_2^5 = \begin{bmatrix} F & F & F \\ D & F & F \\ D & F & F \end{bmatrix}, \sigma_2^6 = \begin{bmatrix} D & F & F \\ F & F & F \\ D & F & F \end{bmatrix}, \sigma_2^7 = \begin{bmatrix} D & F & F \\ D & F & F \\ D & F & F \end{bmatrix}. \tag{24}$$

Note that $\sigma_2^1 = \sigma_2^B$ for small c and $\sigma_2^7 = \sigma_2^{D0}$ for large c (but not too large to sustain cooperation), which are consistent with the discussion about Proposition 5. For the intermediate values of c, only the elements in the first column change in order to increase the incentive for cooperation. We find that the order of the optimal social strategies between $\sigma_2^1 = \sigma_2^B$ and $\sigma_2^7 = \sigma_2^{D0}$ depends on the community's parameters $(b, \beta, \alpha, \varepsilon)$.

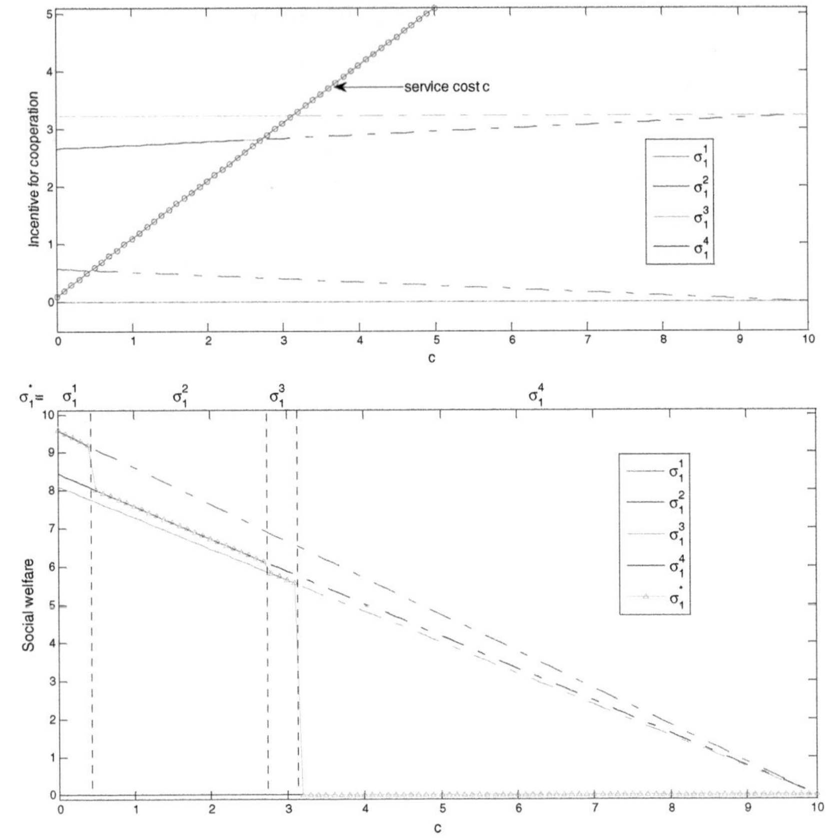

Fig. 1. Performance of the four candidate social strategies when $L = 1$: (a) incentive for cooperation, and (b) social welfare and the optimal social strategy

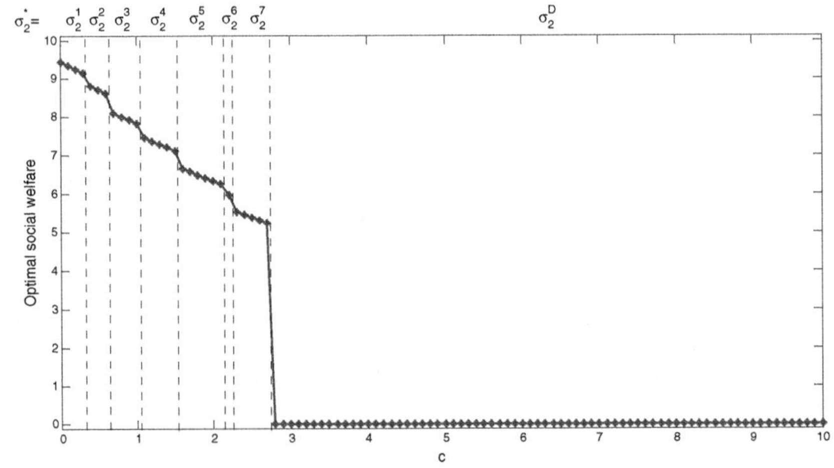

Fig. 2. Optimal social welfare and the optimal social strategy of DP_2

4 Conclusions

In this paper, we used the idea of "social norm" to establish a rigorous mathematical framework to analyze the incentive mechanisms based on indirect reciprocity in network-based communities. We proposed the idea of "sustainable social norm" in which no agent has the incentive to deliberately deviate. Based on this, we analyzed the optimal social norm problem and quantified the efficiency loss in the social norm as a trade-off to the incentive given to agents to follow it. We also showed that the optimal social strategy has some common properties which are independent from the number of reputations in the social norm. Our current analysis can be extended in different directions, among which we mention two. First, instead of implementing the maximum punishment reputation scheme, more complicated reputation schemes can be used where the agent's reputation does not fall to 0 directly upon deviation. Second, we grant the agents newly entering the network the highest reputation L in this paper. Nevertheless, this will provide agents with low reputations in the network the incentive to repeatedly leave and rejoin the network in the aim of obtaining a higher reputation and thus better long-term payoff, a phenomenon commonly known as "whitewash". Hence, more sophisticated rules of reputation assignment can be designed to discourage the "whitewash" behavior.

References

1. Adar, E., Huberman, B.A.: Free Riding on Gnutella. First Monday 5(10) (October 2000)
2. Buragohain, C., Agrawal, D., Suri, S.: A Game-theoretic Framework for Incentives in P2P Systems. In: Proc. Int. Conf. Agent-to-Agent Computing, pp. 48–56 (September 2003)
3. Cohen, B.: Incentives Building Robustness in BitTorrent. In: Proc. P2P Econ. Workshop, Berkeley, CA (2003)
4. Feldman, M., Lai, K., Stoica, I., Chuang, J.: Robust Incentive Techniques for Agent-to-Agent Networks. In: Proc. of the 5th ACM Conf. on Elec. Commerce, Session 4, pp. 102–111 (2004)
5. Johnson, P., Levine, D., Pesendorfer, W.: Evolution and Information in a Gift-Giving Game. J. Econ. Theory 100(1), 1–21 (2001)
6. Kamvar, S., Schlosser, M.T., Molina, H.G.: The Eigentrust Algorithm for Reputation Management in P2P Networks. In: Proc. 12th Int'l Conf. on World Wide Web, pp. 640–651 (2003)
7. Kandori, M.: Social Norms and Community Enforcement. Rev. Economic Studies 59(1), 63–80 (1992)
8. MacKie-Mason, J.K., Varian, H.R.: Pricing congestible network resources. IEEE J. Sel. Areas Commun. 13(7), 1141–1149 (1995)
9. Massoulie, L., Vojnovic, M.: Coupon Replication Systems. IEEE/ACM Trans. on Networking 16(3), 603–616 (2005)
10. Okuno-Fujiwara, M., Postlewaite, A.: Social norms and random matching games. Games Econ. Behavior 9(1), 79–109 (1995)
11. Whittle, P.: Optimization Over Time. Wiley, New York (1983)

5 Appendix A

Proof of Proposition 1:

(i) $U^* \geq 0$ follows by noting that (L, σ_L^D) is feasible. Note that the objective function can be rewritten as $U_\kappa = (b-c)\sum_{\theta,\tilde{\theta}} \eta_L(\theta)\eta_L(\tilde{\theta}) I(\sigma(\theta,\tilde{\theta}) = F)$, where I is an indicator function. Hence, $U_\kappa \leq b-c$ for all κ, which implies $U^* \leq b-c$.

(ii) By (4), we can express $v_\kappa^\infty(1) - v_\kappa^\infty(0)$ as

$$\begin{aligned} & v_\kappa^\infty(1) - v_\kappa^\infty(0) \\ &= v_\kappa(1) + \delta\left[(1-\varepsilon)v_\kappa^\infty(2) + \varepsilon v_\kappa^\infty(0)\right] - v_\kappa(0) - \delta\left[(1-\varepsilon)v_\kappa^\infty(1) + \varepsilon v_\kappa^\infty(0)\right] \\ &= v_\kappa(1) - v_\kappa(0) + \delta(1-\varepsilon)\left[v_\kappa^\infty(2) - v_\kappa^\infty(1)\right]. \end{aligned} \tag{25}$$

Similarly, we have

$$\begin{aligned} v_\kappa^\infty(2) - v_\kappa^\infty(1) &= v_\kappa(2) - v_\kappa(1) + \delta(1-\varepsilon)\left[v_\kappa^\infty(3) - v_\kappa^\infty(2)\right], \\ &\vdots \\ v_\kappa^\infty(L-1) - v_\kappa^\infty(L-2) &= v_\kappa(L-1) - v_\kappa(L-2) + \delta(1-\varepsilon)\left[v_\kappa^\infty(L) - v_\kappa^\infty(L-1)\right], \\ v_\kappa^\infty(L) - v_\kappa^\infty(L-1) &= v_\kappa(L) - v_\kappa(L-1). \end{aligned} \tag{26}$$

In general, for $\theta = 1, \ldots, L$,

$$v_\kappa^\infty(\theta) - v_\kappa^\infty(\theta-1) = \sum_{l=0}^{L-\theta} \gamma^l \left[v_\kappa(\theta+l) - v_\kappa(\theta+l-1)\right], \tag{27}$$

where we define $\gamma = \delta(1-\varepsilon)$. Thus, we obtain

$$\begin{aligned} & v_\kappa^\infty(\theta) - v_\kappa^\infty(0) \\ &= v_\kappa(\theta) - v_\kappa(0) + \gamma\left[v_\kappa(\theta+1) - v_\kappa(1)\right] + \cdots + \gamma^{L-\theta}\left[v_\kappa(L) - v_\kappa(L-\theta)\right] \\ &\quad + \gamma^{L-\theta+1}\left[v_\kappa(L) - v_\kappa(L-\theta+1)\right] + \cdots + \gamma^{L-1}\left[v_\kappa(L) - v_\kappa(L-1)\right] \\ &= \sum_{l=0}^{L-1} \gamma^l \left[v_\kappa(\min\{\theta+l, L\}) - v_\kappa(l)\right], \end{aligned} \tag{28}$$

for $\theta = 1, \ldots, L$. Since $-c \leq v_\kappa(\theta) \leq b$ for all θ, we have $v_\kappa(\theta) - v_\kappa(\tilde{\theta}) \leq b + c$ for all $(\theta, \tilde{\theta})$. Hence, by (25),

$$v_\kappa^\infty(\theta) - v_\kappa^\infty(0) \le \frac{1-\gamma^L}{1-\gamma}(b+c) \le \frac{b+c}{1-\gamma} \tag{29}$$

for all $\theta = 1,\dots,L$, for all $\kappa = (L,\sigma)$. Therefore, if $\delta(1-2\varepsilon)[(b+c)/(1-\gamma)] < c$, or equivalently, $c/b > [\beta(1-\alpha)(1-2\varepsilon)]/[1-\beta(1-\alpha)(2-3\varepsilon)]$, then the incentive constraint (11) cannot be satisfied for any θ, for any social norm (L,σ). This implies that any social strategy σ such that $\sigma\left(\theta,\tilde{\theta}\right) = F$ for some $\left(\theta,\tilde{\theta}\right)$ is not feasible, and thus $U^* = 0$.

(iii) For any L, define a social strategy σ_L^{D0} by $\sigma_L^{D0}\left(\theta,\tilde{\theta}\right) = D$ for $\tilde{\theta} = 0$ and $\sigma_L^{D0}\left(\theta,\tilde{\theta}\right) = F$ for all $\tilde{\theta} > 0$, for all θ. In other words, with σ_L^{D0} each agent declines the service request of 0-agents while providing a service to other agents. Consider a social norm $\kappa = (1,\sigma_1^{D0})$. Then $v_\kappa(0) = -\eta_1(1)c$ and $v_\kappa(1) = b - \eta_1(1)c$. Hence, $U_\kappa = [1-(1-\alpha)\varepsilon](b-c)$ and $v_\kappa^\infty(1) - v_\kappa^\infty(0) = b$, and thus the incentive constraint $\delta(1-2\varepsilon)\left(v_\kappa^\infty(1) - v_\kappa^\infty(0)\right) \ge c$ is satisfied by the hypothesis $c/b \le \beta(1-\alpha)(1-2\varepsilon)$. Since there exists a feasible solution that achieves $U_\kappa = \left[1-(1-\alpha)\varepsilon\right](b-c)$, we have $U^* \ge \left[1-(1-\alpha)\varepsilon\right](b-c)$.

(iv) Suppose, on the contrary to the conclusion, that $U^* = b-c$. If $\alpha = 1$, then (11) cannot be satisfied for any θ, for any κ, which implies $U^* = 0$. Hence, it must be the case that $\alpha < 1$. Let $\kappa^* = (L^*,\sigma^*)$ be an optimal social norm that achieves $U^* = b-c$. Since $U^* = U_{\kappa^*} = (b-c)\sum_{\theta,\tilde{\theta}} \eta_{L^*}(\theta)\eta_{L^*}\left(\tilde{\theta}\right) I(\sigma^*\left(\theta,\tilde{\theta}\right) = F)$, σ^* should have $\sigma^*\left(\theta,\tilde{\theta}\right) = F$ for all $\left(\theta,\tilde{\theta}\right)$. However, under this social strategy, all the agents are treated equally, and thus $v_{\kappa^*}^\infty(0) = \cdots = v_{\kappa^*}^\infty(L^*)$. Then σ^* cannot satisfy the relevant incentive constraint (11) for all θ since the left-hand side of (11) is zero, which contradicts the optimality of $\left(L^*,\sigma^*\right)$.

(v) The result can be obtained by combining (i) and (iii).

(vi) Suppose that $U^* = b-c$, and let (L,σ) be an optimal social norm that achieves $U^* = b-c$. By (iv), we obtain $\varepsilon = 0$. Then $\eta_L(\theta) = 0$ for all $0 \le \theta \le L-1$ and $\eta_L(L) = 1$. Hence, σ should have $\sigma(L,L) = F$ in order to attain $U^* = b-c$. Since $v_\kappa(L) = b-c$ and $v_\kappa(\theta) \ge -c$ for all $0 \le \theta \le L-1$, we have $v_\kappa^\infty(L) - v_\kappa^\infty(0) \le b/(1-\gamma)$ by (25). If $\delta b/(1-\delta) < c$, then the incentive constraint for L-agents, $\delta[v_\kappa^\infty(L) - v_\kappa^\infty(0)] \ge c$, cannot be satisfied. Therefore, we obtain $c/b \le \delta/(1-\delta)$.

Minimizing the Social Cost of an Epidemic

Elizabeth Bodine-Baron, Subhonmesh Bose,
Babak Hassibi, and Adam Wierman

California Institute of Technology,
Pasadena, CA 91125, USA
{eabodine,boses,hassibi,adamw}@caltech.edu

Abstract. In this paper we quantify the total cost of an epidemic spreading through a social network, accounting for both the immunization and disease costs. Previous research has typically focused on determining the optimal strategy to limit the lifetime of a disease, without considering the cost of such strategies. In the large graph limit, we calculate the exact expected disease cost for a general random graph, and we illustrate it for the specific example of an Erdös-Rényi network. We also give an upper bound on the expected disease cost for finite-size graphs, and show through simulation that the upper bound is tight for Erdös-Rényi networks and graphs with exponential degree distributions. Finally, we study how to optimally perform a one-shot immunization to minimize the social cost of a disease, including both the cost of the disease and the cost of immunization.

Keywords: Epidemic, immunization, information cascade, random matrix theory, generalized random graphs.

1 Introduction

The spread of epidemics over networks has been extensively studied since the Kermack-McKendrick SIR (Susceptible-Infectious-Recovered) model was proposed in 1927 [15]. Though initially the study of epidemics in networks was primarily motivated by the spread of disease, the results have had far reaching applications: the insights learned apply to many other settings where something is spreading over a network. For example, applications such as (i) network security, where the goal is to understand and limit the spread of computer viruses [7], [28], [10] (ii) viral advertising, where the goal is to create an epidemic to propagate interest in a product [24], [25], and (iii) information propagation, where the goal is to understand how quickly new ideas propagate through a network [12], [14], [5]. See [13] and [11] for comprehensive surveys of prior results. Note that in this paper we use the language of epidemiology, however the results and techniques can easily be seen to hold more broadly.

Typically, work in this area has focused on understanding two broad questions: First, what is the lifetime of an epidemic, and how does it depend on the structure of the network? Second, what is the optimal strategy to minimize/maximize the spread of the epidemic? The first type of research looks to determine the

R. Jain and R. Kannan (Eds.): GameNets 2011, LNICST 75, pp. 594–607, 2012.

"epidemic threshold" – the infection probability below which the disease will eventually die out. In [6] and [29], the authors show that this epidemic threshold depends on the largest eigenvalue of the adjacency matrix of the network graph, generalizing earlier results. In a series of papers, [20], [21], [2], [1], the authors show the absence of such an epidemic threshold in the presence of scale-free networks; the disease will not die out if it is even the smallest bit infectious. In the second branch of the research on modeling epidemics, the authors focus on ways to minimize or maximize the spread of an epidemic by targeting specific nodes. In [6], the authors use their eigenvalue model to predict which nodes are the best to immunize; in [23], the authors use similar methods to evaluate various immunization strategies and determine which is optimal. In contrast, [7] and [18] focus on real-world immunization strategies, where determining the optimal nodes to immunize might not be practical; they conclude that a good solution is to target various types of high-degree nodes. [22] comes to a similar conclusion in the case of scale-free networks. [17] proposes a static bond-percolation model for studying the classic SIR model and evaluates various immunization strategies within this model.

In this paper, we take a different, non-standard goal: instead of focusing on the spread of an epidemic, we focus on the "social cost" of an epidemic. This shift in focus is motivated by a growing trend in the medical community to quantify the cost of an epidemic by looking at the direct and indirect medical costs to both the hospitals and doctors treating and immunizing a population, as well as the cost to individuals in the population paying for medical care. See [4] and [26] for two examples of such studies. It is not just within the medical community that a focus on cost of epidemics is emerging. For example, within the computer science community [8] has recently studied the social cost of an epidemic, although in a very different model than studied in the current paper. In particular, their model assumes a single infected node and determines the maximum infection (reach) from that node. Note that looking at social cost is in general more difficult than studying the lifetime or spread of an epidemic, as it involves the transient, not just long-term, behavior of a disease.

To study the social cost of epidemics, we use a general random graph model, which can incorporate graphs with arbitrary degree distributions. The model generalizes the configuration model [19], and the details are provided in Section 2. To model the spread of an epidemic we use a standard model of, e.g., [29], [6], and [23], which is an approximation of the SIS model.

Within this context, the main contribution of the paper is the derivation of (i) the exact cost of an epidemic in the large graph limit (Theorem 1) and (ii) bounds on the cost of an epidemic for finite graphs (Theorem 3). To illustrate these results, we look at an Erdös-Rényi network and graphs with an exponential degree distribution.

To illustrate the usefulness of the cost calculations just described, we study immunization strategies for balancing the cost of disease with the cost of immunization in Section 4. In particular, we derive the optimal immunization strategy in a simple one-shot setting where immunization happens only at time zero. Our results allow the study of both random and degree-based immunization strategies. In this context, we derive the optimal immunization strategy in the large graph limit, and illustrate using simulations that it still provides a near-optimal choice for finite graphs.

2 Model Description

In this section we introduce the network model and epidemic spread model studied in this paper.

Network model. Let there be n nodes in the population, and let A be an $n \times n$ adjacency matrix corresponding to the network. We consider only undirected graphs. We assume that the network is drawn from a general class of random graphs, $\mathcal{G}$. For example, the network represented by A could be a realization of an Erdös-Rényi random graph, $G_{n,p}$, which would correspond to allowing each edge to exist independently with probability p.

We focus on a class of random graphs, which is related to the 'configuration' model [19] and the, so-called, 'general random graph' model [9], but which is slightly more general than each.

The construction of the graph works as follows. First, define a degree distribution $p(w)$, and obtain n i.i.d. samples $w = (w_1, \ldots, w_n)$. From this vector, generate a random graph given by the adjacency matrix:

$$A_{ij} = A_{ji} = \begin{cases} 1 & \text{w.p. } w_i w_j \rho \\ 0 & \text{w.p. } 1 - w_i w_j \rho \end{cases} \quad \text{where } \rho = \frac{1}{\sum_i w_i}. \tag{1}$$

Note that the expected degree of node i is $\sum_j w_i w_j \rho = w_i$. Since this model is fully determined by one degree distribution $p(w)$, for ease of reference, we call it $G_{n,p(w)}$.

To generate the Erdös-Rényi random graph, $G_{n,p}$, we would need $w = (np, np, \ldots, np)$. With this notation, we have $G_{n,p} = G_{n,\delta(w-np)}$, where $\delta(\cdot)$ is the Dirac δ-function.

The advantage of this model is that it can generate a random graph with a specified degree distribution, allowing us to examine very general random graph models. To relate this model to the configuration model [19] and the general random graph model [9], note that in the case of the configuration model the degree sequence is enforced deterministically and that in the general random graph model the expected degree sequence is fixed rather than the distribution.

Epidemic model. To model an epidemic, we consider an initial set of seed nodes that are infected at time $t = 0$ and then we consider a discrete-time evolution as follows. At $t = 0$, αn nodes are uniformly at random chosen to be infected. Then, the infection spreads according to the following, standard SIS-inspired model. Define, $P(t)$ a vector where the ith entry is the the probability that node i is infected at time step t. Then the evolution proceeds according to

$$P(t+1) = MP(t) \tag{2}$$

where $M = (1-\gamma)I + \beta A$ is termed the 'system matrix', γ is the probability that a node which is infected recovers in within the current time-step, and β is the infection rate, i.e., if a node has k infected neighbors then the probability it is infected in the current time-step is $\beta k + o(k)$.

This model is an approximation of the SIS (Susceptible-Infected-Susceptible) model where each node gets infected/infectious and then recovers to become susceptible to the disease again, using the approximation and independence assumptions made in [29], [6] and [23]. We give a more detailed analysis of the approximations inherent in this model in the extended version of this paper [3].

3 Cost of Disease

The primary goal of this paper is to derive an expression for the cost of an epidemic. Thus, we first need to model the cost of a disease. To do this, we assume a simple model where, c_d, is defined as the cost of an individual being infected during a time-step. Thus, c_d captures both the direct costs to the individual for medication, doctor visits, etc., as well as secondary costs such as missed work. We note that this model leaves open the question of how exactly to determine the parameter c_d. This is ongoing work in the medical community; see [4] and [26] for example studies in this area. In future work, we look to incorporate these results into determining the per-time cost of disease, but for the purposes of this paper, we leave it a general parameter of the model.

Given this model for the cost of disease to an individual, we can begin to calculate the cost of an epidemic. Define C_D, the 'disease cost', as the expected (averaged over the random spread) per node disease cost of an epidemic *during its entire course* when a fraction $\alpha < 1$ of the nodes are infected at time $t = 0$. Note that the expected per node disease cost in a *given time-step* t is simply $\frac{1^T P(t)}{n}$. Furthermore, since $P(0) = \alpha 1$ and $P(t) = M^{t-1}P(0)$, we can express the disease cost per node as

$$C_D = \frac{1}{n}\left[1^T\left(\sum_{t=0}^{\infty} M^t\right) 1\alpha c_d\right]. \tag{3}$$

When the infinite sum converges, i.e., when $||M|| < 1$, we have

$$C_D = \frac{1}{n}\alpha c_d \cdot \left[1^T (I - M)^{-1} 1\right]. \tag{4}$$

We emphasize that, when the underlying graph is drawn randomly from $G_{n,p(w)}$, say, the above C_D is a random variable.

The remainder of this section focuses on understanding (4). Though we cannot explicitly calculate it in all settings, it turns out that when $n \to \infty$ it converges almost surely to a deterministic constant which we can explicitly compute. Throughout, we use the Erdös-Rényi network as an illustrative example.

3.1 Exact Disease Cost for Large n

We first state a result for general random graphs.

Theorem 1. *For $G_{n,p(w)}$ where $p(w)$ has finite variance and M is almost surely stable*

$$\lim_{n\to\infty} C_D(n) = \alpha c_d \left(\frac{1+F^2}{\gamma} - \frac{F^2}{\gamma - \beta - \beta\sqrt{\langle w\rangle}/F} \right) \quad a.s.$$

$$F = \int_1^{\infty} \frac{wp(w)dw}{\gamma\sqrt{\langle w\rangle}/\beta - Fw} \quad and \ \langle w\rangle = \int_1^{\infty} wp(w)dw$$

We omit the proof for brevity; see the extended version of this paper, [3], for details. However, in order to demonstrate the idea of the above proof, we show the calculation of the expected per node disease cost for the case of an Erdös-Rényi network, where $p(w) = \delta(w - np)$. The proof for the general model is parallel but requires more detailed technical arguments.

Theorem 2. *For $G_{n,\delta(w-np)}$ with $\gamma > \beta np$, M is a.s. stable and*

$$\lim_{n\to\infty} C_D(n) = \frac{\alpha c_d}{\beta\sqrt{\langle w\rangle}} \left(\frac{F}{1-\sqrt{\langle w\rangle}F} \right) \quad a.s.$$

$$where\ F = \frac{\gamma - \sqrt{\gamma^2 - 4\beta^2\langle w\rangle}}{2\beta\sqrt{\langle w\rangle}} \quad and\ \langle w\rangle = np.$$

Proof. We can write the system matrix M as

$$M = (1-\gamma)I + \beta A = (1-\gamma-\beta p)I + \beta\underbrace{(A - p11^T + pI)}_{=G} + \beta p11^T$$
$$= (1-\gamma)I + \beta G + \beta p 11^T.$$

Note that G is a zero-mean matrix with i.i.d. lower triangular entries (taking the values $-p$ with probability $1-p$ and the value $1-p$ with probability p). The variance of the entries of G are

$$eG_{ij}^2 = p^2(1-p) + (1-p)^2 p = p - p^2 \tag{5}$$

Using the Matrix Inversion Lemma, we have

$$1^T(I-M)^{-1}1 = 1^T\left(\underbrace{(\gamma+\beta p)I - \beta G}_{=B} - \beta p 11^T\right)^{-1} 1 = 1^T B^{-1} 1 - \frac{1^T B^{-1} 1 1^T B^{-1} 1}{-\frac{1}{\beta p} + 1^T B^{-1} 1}.$$

Let

$$s = \frac{1}{n} 1^T B^{-1} 1 = \frac{1}{n} 1^T \left((\gamma+\beta p)I - \beta G\right)^{-1} 1,$$

then

$$C_D(n) = \frac{\alpha c_d}{n}\left[1^T(I-M)^{-1}1\right] = \left(\frac{\alpha c_d s}{1-\beta pns}\right)$$

and

$$\lim_{n\to\infty} C_D(n) = \lim_{n\to\infty}\left(\frac{\alpha c_d s}{1-\beta pns}\right). \tag{6}$$

Let $H = \frac{1}{\sqrt{(p-p^2)n}}G$. Then H is a symmetric matrix whose lower triangular entries are i.i.d. random variables with zero-mean and $\frac{1}{n}$ variance. Such a random matrix is called a *Wigner matrix.* We use the following Lemma to finish the calculation. □

Lemma 1. *Let H be a symmetric $n\times n$ random matrix whose lower triangular entries are i.i.d. random variable with zero-mean and $\frac{1}{n}$ variance. Then*

$$\begin{aligned}\lim_{n\to\infty}\frac{1}{n}1^T(xI-H)^{-1}1 &= \lim_{n\to\infty} e\frac{1}{n}\operatorname{tr}(xI-H)^{-1}\,a.s.\\ &= \begin{cases}\frac{x-\sqrt{x^2-4}}{2}, & |x|\geq 2\\ \infty, & otherwise\end{cases}\end{aligned}$$

We remark that the complex function $\lim_{n\to\infty} \mathrm{e}\frac{1}{n}\operatorname{tr}(zI-H)^{-1} = \mathrm{e}\frac{1}{z-\lambda}$is often called the *Stieltjes transform* of the random matrix H (or of its eigenvalue distribution). For a more detailed introduction to random matrices and Stieltjes transform, we refer the reader to [27]. It remains to calculate s.

$$\begin{aligned}\lim_{n\to\infty} s &= \lim_{n\to\infty}\frac{1}{n}1^T\left((\gamma+\beta p)I-\beta G\right)^{-1}1\\ &= \lim_{n\to\infty}\frac{1}{\beta\sqrt{(p-p^2)n^3}}\left(1^T\left(\frac{\gamma+\beta p}{\beta\sqrt{(p-p^2)n}}I-H\right)^{-1}1\right)\end{aligned}$$

Using Lemma 1, we get

$$\begin{aligned}\lim_{n\to\infty} s &= \lim_{n\to\infty}\frac{\frac{\gamma+\beta p}{\beta\sqrt{(p-p^2)n}}-\sqrt{\frac{(\gamma+\beta p)^2}{\beta^2(p-p^2)n}-4}}{2\beta\sqrt{(p-p^2)n}}\\ &= \lim_{n\to\infty}\frac{\gamma+\beta p-\sqrt{(\gamma+\beta p)^2-4\beta^2(p-p^2)n}}{2\beta^2(p-p^2)n}.\end{aligned} \tag{7}$$

Since M is stable *a.s.*, $\gamma > \beta np$. Continuing from (7) and removing $o(n)$ and $o(p)$ terms, we get

$$\lim_{n\to\infty} s = \frac{\gamma-\sqrt{\gamma^2-4\beta^2\langle w\rangle}}{2\beta^2\langle w\rangle} = \frac{F}{\beta\sqrt{\langle w\rangle}}\quad a.s. \tag{8}$$

where $\langle w\rangle = np$ and finally

$$\lim_{n\to\infty} C_D(n) = \lim_{n\to\infty}\left(\frac{\alpha c_d s}{1-\beta pns}\right) = \frac{\alpha c_d}{\beta\sqrt{\langle w\rangle}}\left(\frac{F}{1-\sqrt{\langle w\rangle}F}\right)\quad a.s. \tag{9}$$

□

3.2 Bounds for General n

The prior section provided exact calculation of the disease cost in the case of large n. We now provide bounds for finite n. Note that these bounds hold for all graphs, not only the random graphs described in Section 2.

Theorem 3. *For a graph with stable system matrix M, the cost of disease per node satisfies*

$$C_D \leq \frac{\alpha c_d}{1 - \lambda_{max}(M)} \tag{10}$$

Proof. From Perron-Frobenius theorem, it follows that

$$-1 < -\lambda_{max}(M) \leq \lambda_{min}(M) \leq \lambda_{max}(M) = |\lambda_{max}(M)| < 1 \tag{11}$$

Note that $(I - M)^{-1}$ is a positive definite matrix since all eigenvalues are positive. From (11), we have

$$\frac{1}{1 - \lambda(M)} \leq \frac{1}{1 - \lambda_{max}(M)}$$

and

$$1^T (I - M)^{-1} 1 \leq 1^T \left(\frac{I}{1 - \lambda_{max}(M)} \right) 1 \tag{12}$$

This follows from the fact that if A and B are positive definite matrices with eigenvalues $\lambda_A \leq \lambda_B$ for all eigenvalues λ_A and λ_B, then $x^T A x \leq x^T B x$. The bound follows. □

We note that the necessary and sufficient condition required for the disease to die out and the social cost to converge are the same, i.e., $\lambda_{max}(M) < 1$. Also, the bound only depends on $\lambda_{max}(M)$ from the disease propagation model, which is popularly known as the disease threshold. It is interesting that the same parameter of the disease plays the central role in both tapering off the disease and its total cost.

Comparison of Theorem 1 and Theorem 3

Theorem 1 involves the entire eigen-distribution of M, whereas Theorem 3 is dependent only on the maximum eigenvalue of M. We show here, through simulation, that in the case of an exponential $p(w)$ and for an Erdös-Rényi network, the upper bound in Theorem 3 is rather tight.

Define $p(w) = \lambda e^{-\lambda w}$; i.e., $p(w)$ is exponentially distributed with parameter λ. According to Theorem 1, the cost of the disease in the limit of large n will be

$$\lim_{n \to \infty} C_D(n) = \alpha c_d \left(\frac{1 + F^2}{\gamma} - \frac{F^2}{\gamma - \beta - \frac{\beta}{F}\sqrt{\frac{1}{\lambda}}} \right)$$

where

$$F = \int_1^\infty \frac{w \lambda e^{-\lambda w} dw}{\frac{\gamma}{\beta}\sqrt{\frac{1}{\lambda}} - F w},$$

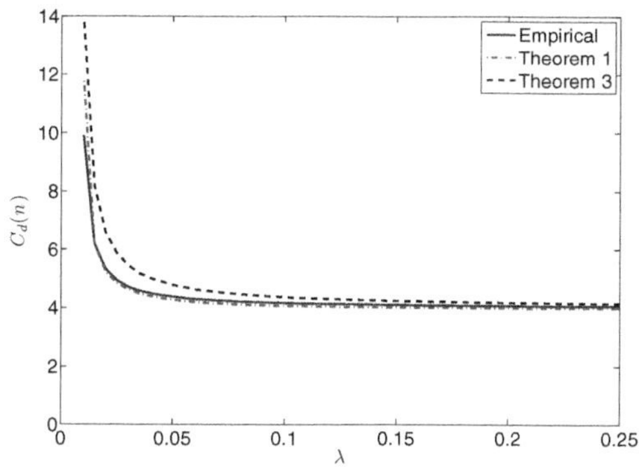

(a) Exponential $p(w)$

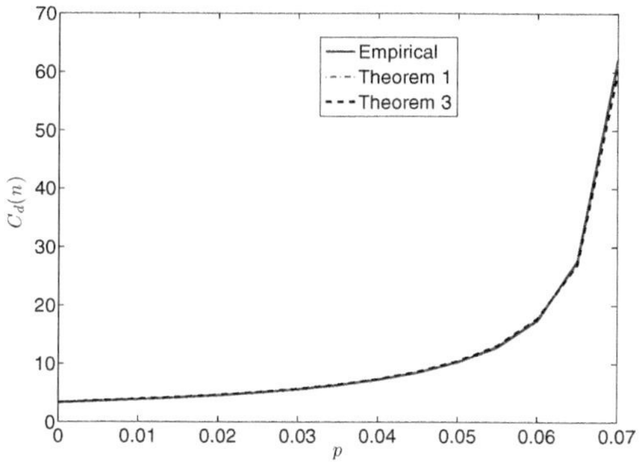

(b) Erdös-Rényi

Fig. 1. Comparison of Theorem 1 and Theorem 3

as $\langle w \rangle = 1/\lambda$. Using the parameters $n = 1000$, $\alpha = 0.2$, $\beta = 0.002$, $\gamma = 0.5$, and $c_d = 10$, we solve for F numerically and show the cost of the disease in the limit of large n as a function of the graph parameter λ in Figure 1a. We also plot the upper bound from Theorem 3 for comparison. The maximum eigenvalue for each λ was calculated as the average maximum eigenvalue of 10 realizations of M.

Now let $p(w) = \delta(w - np)$, for an Erdös-Rényi network. Using the parameters $n = 1000$, $\alpha = 0.2$, $\beta = 0.008$, $\gamma = 0.6$, and $c_d = 10$, we calculate F and $C_d(n)$ according to Theorem 2 and plot the result as a function of the graph parameter p in Figure 1b. Again, we also plot the upper bound from Theorem 3 for comparison. We note that for any probability $p(n)$, $\lambda_{max} = (1+O(1)) \max(\sqrt{\Delta}, np)$, where Δ denotes the maximum degree, almost surely [9], [16].

Note that the upper bound is tight, regardless of the parameter λ or p, indicating that only the maximum eigenvalue of M, rather than the entire eigendistribution, plays the central role in determining the cost of the disease.

4 Minimizing Social Cost with Immunization

The key impact of the cost calculations in Section 3 is to allow decisions that trade off immunization costs with disease cost. To illustrate this, in this section we study a simple example of such an optimization.

We define the "social cost" of an epidemic as the sum of the disease cost and the immunization cost. For simplicity, we assume that immunization or vaccination will occur at $t = 0$ and any immunized node will remain immune for all time, thus incurring a single immunization cost. This cost could represent the cost of a vaccine to an individual or the cost of quarantining. If the node is infected in the starting state and is selected to be immunized, it is quarantined. If it is susceptible, it is vaccinated. We assume that the cost of quarantining and the cost of vaccinating are the same, c_v.

We can incorporate the immunization process into the random graph model itself. Recall the original graph generation model, where the distribution $p(w)$ is sampled n times to create the degree vector w. Let us first consider a randomized immunization procedure where πn nodes are chosen uniformly at random for immunization. Then, the random graph is formed by sampling $p(w)$ only for the nodes that remain after the immunization. The nodes that are immunized are simply removed from the adjacency matrix A describing the network and the system matrix M, yielding $\tilde{M}$ and $\tilde{A}$. Using this procedure gives $\mathrm{e}[\dim \tilde{M}] = (1-\pi)n := \tilde{n}$. For a degree-based immunization scheme, we can simply truncate the degree distribution and sample as before.

In general, any one-shot immunization scheme simply results in a transformation of $p(w)$ and the social cost calculations in Section 3 still apply. For simplicity, in this paper, we focus on the case of random immunization of a fraction π of the nodes. The question is how to optimally choose π.

4.1 Social Cost Calculation

To determine the optimal immunization probability, we first need to incorporate the cost of vaccination into the calculations from Section 3. Define the per node social cost for a particular graph as:

$$S(M,\tilde{M}) = \frac{1}{n}\left[(\dim M - \dim \tilde{M})c_v + \left(1^T(I-\tilde{M})^{-1}1\right)\alpha c_d\right] \quad (13)$$

Considering a random immunization on a class of graphs $\mathcal{G}$, we define the expected social cost $S_{\mathcal{G}}(\pi)$ as a function of the fraction π of nodes removed as follows:

$$S_{\mathcal{G}}(\pi) = \pi c_v + (1-\pi)C_D\left(n(1-\pi)\right) \quad (14)$$

Thus, applying the results from Section 3, we obtain results for $S_{\mathcal{G}}(\pi)$. We illustrate this with the Erdös-Rényi case. Using Theorem 2 for $\mathcal{G} = G_{n,\delta(w-np)}$, we have that, as $n \to \infty$,

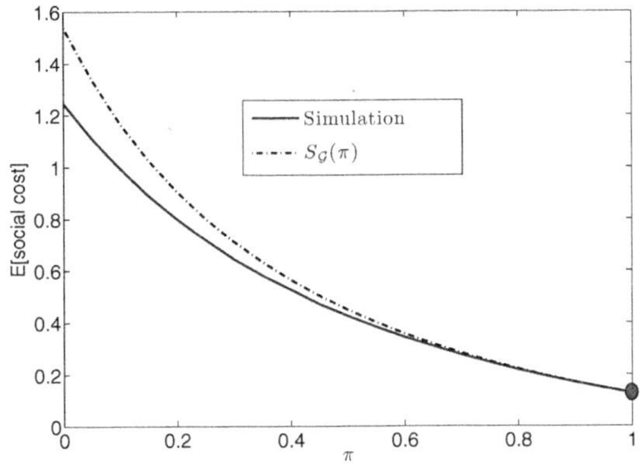

(a) Small C

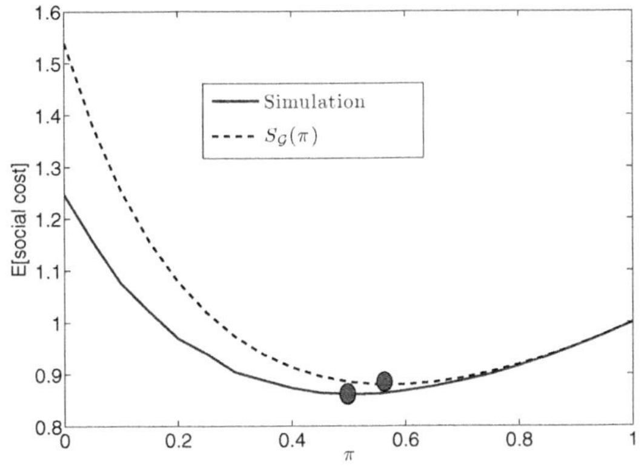

(b) Medium C

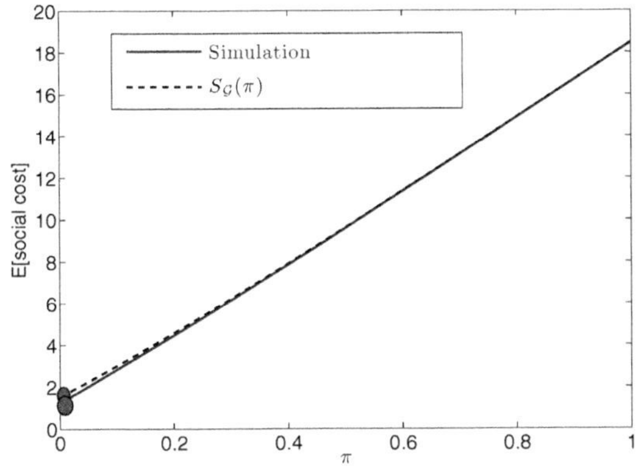

(c) Large C

Fig. 2. Social cost simulations for different values of C on Erdös-Rényi network as a function of the immunization probability π. The optimal immunization probability in each case is highlighted with a red circle.

$$S_{\mathcal{G}}(\pi) \to \pi c_v + \frac{\alpha c_d}{\beta\sqrt{(1-\pi)np}}\left(\frac{F}{1-\sqrt{(1-\pi)npF}}\right) \text{ a.s.} \tag{15}$$

$$\text{where } F = \frac{\gamma - \sqrt{\gamma^2 - 4\beta^2(1-\pi)np}}{2\beta\sqrt{(1-\pi)np}}.$$

Using $\sqrt{1-x} \approx 1 - \frac{x}{2}$, we get $F \approx \frac{\beta\sqrt{(1-\pi)np}}{\gamma}$. Thus

$$S(\pi) \approx \pi c_v + \frac{\alpha(1-\pi)c_d}{\gamma - (1-\pi)\beta np} \tag{16}$$

4.2 Optimal Immunization

To determine the optimal choice of π in the random immunization scheme for a large network, we use the social cost as calculated in Section 4.1. We illustrate this through the Erdös-Rényi case. For convenience, we normalize $c_d = 1$ and $c_v = C$, and consider the per node cost as given in equation ((16)).

Minimizing $S(\pi)$ over π, we obtain the approximate optimal random immunization, π^*. Define $a = \frac{\alpha}{\gamma}$ and $b = \frac{\alpha\gamma}{(\gamma-\beta np)^2}$, and note that $a < b$ always.

$$\pi^* = \begin{cases} 1 & C \leq a < b \\ 1 - \frac{\gamma - \sqrt{\gamma\alpha/C}}{\beta np} & a < C < b \\ 0 & a < b \leq C \end{cases} \tag{17}$$

To illustrate the above, we simulate a disease propagating on an Erdös-Rényi graph, with $n = 100000$, $p = 1.27 \times 10^{-4}$, $\alpha = 0.2$, $\beta = 0.02$, and $\gamma = 0.39$. We use a low $C = 0.1282$, medium $C = 1$, and high $C = 18.46$. The simulated cost as a function of π is shown in Figure 2, together with the approximate calculated cost as given in ((16)). Note that the simulation is of the disease propagating according to the SIS model, whereas $S_{\mathcal{G}}(\pi)$ as shown is an approximation on the per node social cost assuming infinite size graphs. In addition, the infection propagation model makes several assumptions explored further in [3]. However, note that the optimal π for both the bound and the simulation is very close, indicating the usefulness of the model for immunization decisions.

5 Conclusion and Future Work

In this paper, rather than taking the approach of previous work, which has typically focused on determining the optimal strategy to limit the lifetime of a disease, without considering the cost of such strategies, we look at the total "social cost" of an epidemic, including the cost of immunization and the cost to an individual being infected for a finite amount of time. Determining the social cost of an epidemic is a harder problem than estimating its lifetime, as it depends on the transient behavior of the epidemic, not just its steady state. This approach can help quantify the efficacy of immunization decisions from an economic standpoint for decision-making bodies like the CDC. It can also be applied to the dual problem, where the desire is to maximize the spread of information while keeping advertising costs low.

We model the physical interaction of individuals (and hence the transmission of disease) as an undirected random graph using a new random graph model and a dynamical system based on the popular SIS model. Within this framework, we derive the exact cost of an epidemic in the large graph limit (Theorem 1) and bounds on the cost of an epidemic for finite graphs (Theorem 3). The exact cost depends on the entire eigendistribution of the system matrix, whereas the upper bound depends only on the largest eigenvalue. Despite its simpler form, the upper bound appears to be tight, as shown by our simulations of Erdös-Rényi networks and graphs with exponential degree distributions. Our analysis makes use of ideas and techniques from random matrix theory, differentiating this work from previous work on the spread of epidemics. The key impact of our theorems calculating and bounding the disease cost is to allow decisions that trade off immunization costs with disease cost - as shown by our calculation of the social cost based on a one-shot random immunization strategy for an Erdös-Rényi network.

We note that the graph model proposed in Section 2 is a generalization of existing models and can capture any class of random graphs defined by an expected degree distribution. As such, it may be applicable to other areas outside of the scope of this paper and is worth further investigation.

To extend this work, we would like to investigate both theoretical and practical refinements of our model. We note that the linear model of disease propagation used here requires independence assumptions that do not hold in general and it would be worthwhile to explore other models that describe infection more accurately. We can also extend our model to include other classic disease models, such as the SIR model and more complicated models of infection. With regard to immunization, we would like to examine a more dynamic strategy where individuals vaccinate themselves at different points in time based on their currently available information (fraction of friends infected, etc.). Moreover, a case study with real data from communicable diseases like influenza or herpes would provide more insight into the accuracy and predictive power of these models in real-world scenarios, as well as help refine the cost of disease which we have assumed is a single parameter, c_d.

Acknowledgments. The authors would like to thank Professor K. Mani Chandy for his useful comments and suggestions for this paper. This work was supported in part by the National Science Foundation under grants NSF CNS-0846025, CCF-0729203, CNS-0932428 and CCF-1018927, by the Office of Naval Research under the MURI grant N00014-08-1-0747, and by Caltech's Lee Center for Advanced Networking.

References

1. Barthelemy, M., Barrat, A., Pastor-Satorras, R., Vespignani, A.: Velocity and Hierarchical Spread of Epidemic Outbreaks in Scale-Free Networks. Phys. Rev. Lett. 92(17) (2004)
2. Boguna, M., Pastor-Satorras, R., Vespignani, A.: Absence of Epidemic Threshold in Scale-Free Networks with Degree Correlations. Phys. Rev. Lett. 90(2) (2003)
3. Bose, S., Bodine-Baron, E., Wierman, A., Hassibi, B.: Cost of disease in large networks. Working paper (2010)
4. Bownds, L., Lindekugel, R., Stepak, P.: Economic impact of a Hepatitis A epidemic in a mid-sized urban community: The case of Spokane. Journal of Community Health 38(4) (2003)
5. Cha, M., Mislove, A., Gummadi, K.P.: A measurement-driven analysis of information propagation in the flickr social network. In: Proc. 18th WWW Conf., pp. 721–730 (2009)
6. Chakrabarti, D., Wang, Y., Wang, C., Leskovec, J., Faloutsos, C.: Epidemic thresholds in real networks. ACM Trans. on Info. and System Security (TISSEC) 10(4) (2008)
7. Chen, H., Jin, H., Sun, J., Han, Z.: Efficient Immunization Algorithm for Peer-to-Peer Networks. In: Bougé, L., Prasanna, V.K. (eds.) HiPC 2004. LNCS, vol. 3296, pp. 232–241. Springer, Heidelberg (2004)
8. Chen, P., David, M., Kempe, D.: Better vaccinations for better people. In: Proc. EC 2010 (2010)
9. Chung, F., Lu, L.: Complex Graphs and Networks. AMS (2006)
10. Cohen, R., Havlin, S., ben Avraham, D.: Efficient immunization strategies for computer networks and populations. Phys. Rev. Lett. 91(24) (2003)
11. Daley, D., Gani, J.: Epidemic Modeling: An Introduction. Cambridge University Press (2005)
12. Huang, L., Park, K., Lai, Y.C.: Information propagation on modular networks. Phys. Rev. E 73(3) (March 2006)
13. Jackson, M.O.: Social and Economic Networks. Princeton University Press (2008)
14. Jacquet, P., Mans, B., Rodolakis, G.: Information propagation speed in mobile and delay tolerant networks. IEEE Trans. on Info. Theory 56(10), 5001–5015 (2010)
15. Kermack, W.O., McKendrick, A.G.: A Contribution to the Mathematical Theory of Epidemics. Proc. Roy. Soc. Lond. A 115, 700–721 (1927)
16. Krivelevich, M., Sudakov, B.: The largest eigenvalue of sparse random graphs. Combinatorics, Probability, and Computing 12, 61–72 (2003)
17. Madar, N., Kalisky, T., Cohen, R., ben Avraham, D., Havlin, S.: Immunization and epidemic dynamics in complex networks. Euro. Phys. Journal B - Condensed Matter and Complex Systems 38, 269–276 (2004)
18. Miller, J., Hyman, J.: Effective vaccination strategies for realistic social networks. Physica A: Statistical Mechanics and its Applications 386(2), 780–785 (2007)
19. Newman, M.E.J.: Random graphs as models of networks. In: Handbook of Graphs and Networks (2005)
20. Pastor-Satorras, R., Vespignani, A.: Epidemic Spreading in Scale-Free Networks. Phys. Rev. Lett. 86, 3200–3203 (2001)
21. Pastor-Satorras, R., Vespignani, A.: Epidemic dynamics in finite size scale-free networks. Phys. Rev. E 65 (2002)
22. Pastor-Satorras, R., Vespignani, A.: Immunization of complex networks. Phys. Rev. E 65 (2002)

23. Peng, C., Jin, X., Shi, M.: Epidemic threshold and immunization on generalized networks. Physica A: Statistical Mechanics and its Applications 389(3), 549–560 (2010)
24. Phelps, J., Lewis, R., Mobilio, L., Perry, D., Raman, N.: Viral marketing or electronic word-of-mouth advertising: Examining consumer responses and motivations to pass along email. Journal of Advertising Research, 44 (2004)
25. Richardson, M., Domingos, P.: Mining knowledge-sharing sites for viral marketing. In: Proc. 8th ACM SIGKDD, pp. 61–70 (2002)
26. Rubin, R., Harrington, C., Poon, A., Dietrich, K., Greene, J., Moiduddin, A.: The Economic Impact of Staphylococcus aureus Infection in New York City Hospitals. Emerging Infectious Diseases 5(1) (1999)
27. Tulino, A.M., Verdú, S.: Random matrix theory and wireless communications. Comm. Inf. Theory 1, 1–182 (2004)
28. Wang, C., Knight, J.C., Elder, M.C.: On computer viral infection and the effect of immunization. In: Proc. ACSAC 2000, pp. 246–256 (2000)
29. Wang, Y., Chakrabarti, D., Wang, C., Faloutsos, C.: Epidemic spreading in real networks: An eigenvalue viewpoint. In: Proc. Symp. Reliable Dist. Systems (2003)

“Two Is a Crowd” - Optimal Trend Adoption in Social Networks

Lilin Zhang and Peter Marbach

University of Toronto, Toronto, ON M5S 3G4, Canada
{llzhang,marbach}@cs.toronto.edu

Abstract. In this paper, we study how an individual in a social network should decide whether or not to adopt a trend, based on how many people in his/her neighborhood in the social network adopted the trend. In particular we are interested in the question in what adoption policy leads to an optimal trend-adoption in the sense that an individual only adopts a trend if the majority of his/her social network will do so. We use a decision process on a Erdős-Rényi random graph model to model and study this situation. Using this model, we obtain the result that the optimal policy for an individual is to adopt a trend if two people in their neighborhood did. Interestingly, this result/behavior was experimentally observed in real social networks. We hope that this work will help towards building applications that is able to automatically push relevant content to users in online social networks.

1 Introduction

Online Social Networks have experienced an explosive growth over the recent a few years, and that growth has accompanied abundant interest in the study of the spread of trends in social networks. In this paper, we study how an individual in a social network should decide on whether or not to adopt a trend. We use the term “trend” quite broadly where a trend could for example be a fashion trend, a new technology, an idea, or a behavior.

Two paradigms for modeling trend adoption in social networks have been proposed in the literature: the threshold model and the cascade model. Both models are based on the natural assumption that individuals tend to be influenced by their acquaintances with whom they communicate. Yet the two are intrinsically modeling two different types of influence in social networks. In the threshold model, each individual observes the number of adoptions among his friends, and only adopts the trend if the number of adoptions among his/her friends exceeds a particular threshold value. On the other hand, in the cascade model, each individual can be convinced to adopt the trend through interactions with friends in the social network who have already adopted the trend.

However, two recent studies by Backstorm el. [4] on friendship links and community membership on LiveJournal, and the co-authorship network in the computer science bibliography hosted by the Digital Bibliography & Library Project

R. Jain and R. Kannan (Eds.): GameNets 2011, LNICST 75, pp. 608–627, 2012.

(DBLP), showed that neither the threshold model nor the cascade model is sufficiently realistic (for more details, see in Section 2). Hence in this paper, we re-investigate the problem of trend adoption in social networks, and propose a new model that better predicts the observed behavior in real-life social networks.

Our first contribution in this paper is thus a new hybrid model for studying how individuals are influenced by their acquaintances in social networks. The hybrid network model considers the social network population as two groups: one group in which individuals follow cascade process to decide whether or not to adopt; the other group in which users obey threshold strategy to make the decision. The basic idea behind the hybrid network is that, with respect to trend adoption, we can distinguish between two types of individuals: 1) individuals who make "informed decisions" and 2) individuals who "imitate". We refer to individuals who make informed decisions as *informed adopters.* Informed adopters have (insider) knowledge about the trend to adopt and decide whether or not to adopt based on information/discussion with other insiders who have already adopted. As a result, among the insider's group, the adoption process can be modeled by the cascade process. We refer to individuals who imitate as *followers.* Followers do not have enough knowledge about the trend so as to make informed decisions, and as a consequence, decide whether or not to adopt depending on how many other people already have adopted the item. Hence for the latter group, the adoption process can be captured by the threshold model. In addition, we consider an initial set of individuals who adopt the trend without being influenced by others. We refer to these individuals as *trendsetters (early adopters).* The trendsetters are important as they serve as "seeds" that enable the spread of the trend.

Note that followers have no special knowledge about the trend and therefore need to rely on their observation of what informed adopters do in order to decide on whether or not to adopt the trend. As such, the goal of a follower is to adopt the trend only if a large fraction of informed adopters does so, since this can be taken as an indication that the trend is indeed worth adopting. However, followers do not have a "global view" but only a "local view", i.e. they can only observe what the informed adopters who are neighbors in their social network do, but they can not observe the adoption behavior of all informed adopters. As such, the question we want to explore is whether it is possible to choose a threshold value that guarantees that followers make the correct decision and adopt the trend only if a majority of the informed adopters does so, even though they have only local information.

Our second contribution is to characterize a *optimal threshold strategy* for the followers. More precisely we show that is optimal for followers to adopt a trend if as soon as two friends in their social network adopted the trend. Interestingly, this behavior was observed in the Backstorm el. [4] on friendship links and community membership on LiveJournal, and the co-authorship network in DBLP.

To the best of our knowledge, it is the first time that a mathematical model is proposed to address and reassemble the real-life influence propagation pattern.

The rest of the paper is organized as follows. In Section 2, we discuss related work of empirical studies on real social networks. Section 3 presents our mathematical model, problem statement and numerical results. In Section 4, we propose a tractable model. In Section 5, we perform the analysis to address the choice of threshold value. Finally we conclude and discuss the future work in Section 6.

2 Related Work

In this section, we provide the related work of empirical studies concerning threshold strategy on real social networks. The subject of threshold is not new. In fact, it is well used in various fields of science [3]: as a method of segmentation in image processing, as a model in toxicology, as a phenomenon in particle physics, etc. However, the mathematical study of the threshold strategy and the optimum threshold value on social networks is the first time in the paper.

Backstorm el. [4] studied two large sources of data: friendship links and community membership on LiveJournal, and the co-authorship network in DBLP. They plotted the individual's probability p of joining a LiveJournal/DBLP community as a function of the number of friends, k, already in the community (Fig1, Fig2). Both plots exhibit qualitatively similar shapes in which p continues increasing, but more and more slowly as k increases. We take a closer look at Fig.1, since the error bars are smaller here. The probability of joining LiveJournal has a close fit to a logistic function for $k = 0, 1, 2$, with a rapid growth at $k = 2$. After this point, the probability of joining the community has a close fit to a logarithmic function. For the largest jump in p at $k = 2$, Backstorm el. claimed that the marginal benefit of having a second friend in a community is particularly strong. Such a curve suggests the individual's strategy of a threshold pattern: individuals have a low tendency to take an action if not many acquaintances did so; then the tendency increases very fast if a certain number of acquaintances have done so; beyond this point the tendency still increases, but does so slower and slower. The *action* herein is an abstract notation, which could be joining a community (Fig.1), or attending another conference (Fig.2). Though the empirical study of Backstorm el. reveals the individuals' threshold strategy, their research focus was on quite a different subject: the evolution of large social networks.

One variation of the above threshold strategy is in the context of voluntary vaccination [5]: once a sufficient proportion of the population is already immune, either naturally or by vaccination, people tend to skip vaccination under the voluntary policy due to morbidity risks of the vaccination itself. That is individuals tend to avoid taking an action once a threshold is reached. Bauch el. [5] performed a game theoretical analysis based on Nash equilibrium to explain the threshold strategy with respect to vaccination. However, they need to assume that all individuals are provided with the same information and use this information in the same way to assess risks.

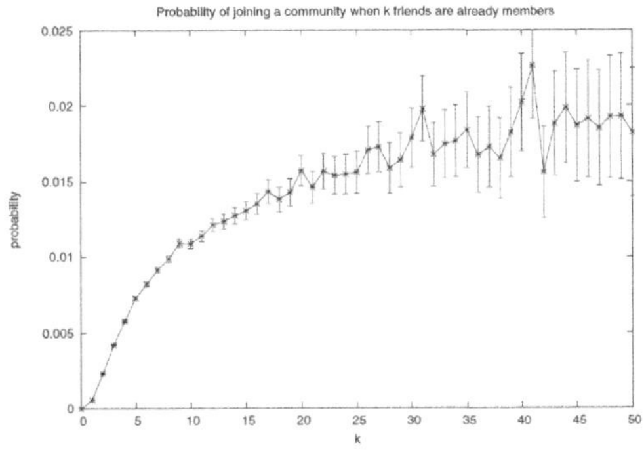

Fig. 1. The probability p of joining a LiveJournal community as a function of the number of friends k already in the community. Error bars represent two standard errors [4].

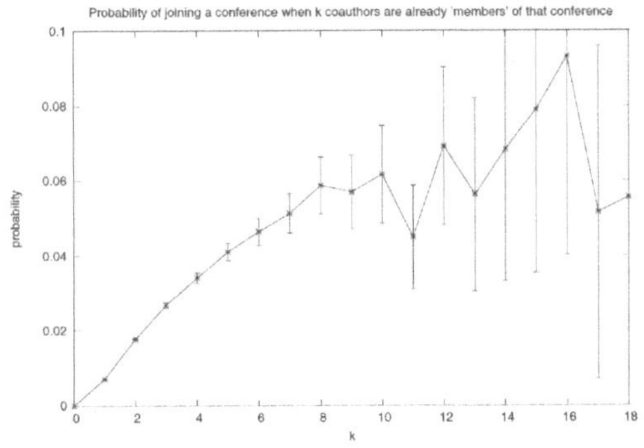

Fig. 2. The probability p of joining a DBLP community as a function of the number of friends k already in the community. Error bars represent two standard errors [4].

Here this paper lifts the assumption of synchronized global information, and considers the situations where individuals have social contacts with their friends, thus have merely local observations. And individuals make decisions based on the limited information. To the best of our knowledge, our work provides a unique approach to choose a threshold value so that individuals can make correct choices based on local observations.

3 Model and Problem Statement

In this section, we introduce our hybrid model for trend adoption in social networks, state the question of interests, and provide a numerical case study to illustrate our result.

Recall that our scenario of trend adoption in the social network: individuals can observe whether their friends have adopted the trend or not, and make the decision based on these observations. We refer to an individual who adopted the trend as *active*, and to an individual who has not adopted the trend as *non-active*, or inactive. Note that individuals only go from non-active to active status, but not from non-active to active.

3.1 Social Network Graph

Our model consider two main groups in the social network population. One is the "informed adopter" group consisting of individuals who have knowledge about the trend and make informed decision based on information/discussion with other insiders who have already adopted. For this group, we model their user behaviour as cascade process. The other is the "follower" group consisting of individuals who don't have enough knowledge about the trend to make informed decisions, and as a consequence, decide whether or not to adopt by "imitating" their acquaintances who already have adopted the trend. For the second group, we presume they obey threshold strategy.

We think of a *social network* as a collection of individuals who have connections with one another in social bonds such as friendship, kinship, or relationships of beliefs and knowledge [7]. We use a graph G to represent the entire social network. As part of the social network graph G, we define a undirected subgraph G_1 of G representing the "informed adopters" and directed subgraph G_2 representing the "followers".

In the rest of the paper, we will use the terms individuals and nodes/vertices interchangeably.

More precisely, we use the following Erdös-Rényi [11] random graph model. We denote with $V(G_1)$ the vertex set of G_1 given by

$$V(G_1) = \{1, 2, 3, \cdots, n_1\},$$

where n_1 denotes the total number of informed adopters in the social graph G. Each node in $V(G_1)$ represents an informed adopter in the social network.

Between any pair of nodes $i, j \in V(G_1)$, with probability $p_1 \in (0, 1]$ there exists an undirected edge between two informed adopters i and j in G_1, independent of everything else. Let $E(G_1)$ be the resulting edge set of G_1 given by

$$E(G_1) = \{e_{i,j} | i, j \in V(G_1)\}.$$

Each edge in $E(G_1)$ stands for the social connection within G_1.

The average node degree λ_1 in G_1 is then given by

$$\lambda_1 = p_1 n_1$$

and we have the relation

$$p_1 = \frac{\lambda_1}{n_1}. \tag{1}$$

Note that λ_1 represents the average number of informed adopters a node in G_1 communicates with.

Let $V(G_2)$ the vertex set of G_2 given by

$$V(G_2) = \{1, 2, 3, \cdots, n_2\},$$

where n_2 denotes the total number of followers in the social graph G. Each node in $V(G_2)$ represents a follower in the social network.

We assume that with probability $p_2 \in (0.1]$ there exists a directed edge $e(i,j)$ between a follower i in G_2 and an informed adopter j in G_1, independently of everything else.

Let $E(G_2)$ be the edge set of nodes in G_2. Each edge in $E(G_2)$ stands for the social connection between a follower in G_2 and an informed adopter in G_1, i.e. we have

$$E(G_2) = \{e_{i,j} | i \in V(G_1), j \in V(G_2)\}.$$

The average node degree λ_2 in G_2, i.e. the average number of informed adopters that a follower in G_2 knows, is then given by

$$\lambda_2 = p_2 n_1$$

and we have the relation

$$p_2 = \frac{\lambda_2}{n_1}.$$

Using the above notation, the graphs G_1 and G_2 are characterized by the parameters (n_1, λ_1) and (n_2, λ_2), respectively. Accordingly, we use the notation $G_1(n_1, \lambda_1)$ and $G_2(n_2, \lambda_2)$ to characterize the two graphs.

Aggregated, let $e(G)$ be the random variable denoting the number of edges in G.

$$e(G) = |E(G)|$$

And let $N(v)$ be the set of immediate *neighbours* of v, which stands for the set of friends, relatives or acquaintances of v on the social network.

$$N(v) = \{w | w \in V(G), \text{ and } e_{v,w} \in E(G)\}$$

3.2 Asymptotic Behavior

In the following we are interested in the asymptotic behavior as the social graph becomes large. As such, we consider an infinite sequence of social graphs denoted by $n = 1, 2, ...$, and let $G_1(n_1, \lambda_1)$ and $G_2(n_2, \lambda_2)$ be functions of n; i.e. the parameters (n_1, λ_1) and (n_2, λ_2) of G_1 and G_2, respectively, depend on n. We assume that n_1 and n_2 become large (approach infinity) as n increases.

Assumption 1. *We have*

$$\lim_{n \to \infty} n_1(n) = \infty$$

and

$$\lim_{n \to \infty} n_2(n) = \infty.$$

To keep the notation light, in the following we typically do not explicitly state the dependency of parameters such as (n_1, λ_1) and (n_2, λ_2) on n.

Note that the above implies that the edge probabilities p_1 and p_2 are also functions of n. For our analysis, we make the following assumption on $p_2(n)$.

Assumption 2. *We have*

$$\lim_{n to \infty} p_2(n) = \frac{\lambda_2(n)}{n_1(n)} = 0.$$

The above assumption states that the followers can only observe a vanishingly small fraction of the informed adopters.

3.3 Trendsetters

In addition to the informed adopters and followers, we consider a third group of individuals, namely the trendsetters or early adopters. Trendsetters adopt the trend without being influenced by others in the social network. Trendsetters serve as "seeds" that enable the spread of the trend.

In our model, the trendsetters are given by a subset A_1 of the informed adopters G_1, i.e. A_1 is the the initial active set within the insiders group, G_1. This size of the set A_1 can depend on n and we make the following technical assumption on the size of the set $A_1(n)$, i.e. the total number of trendsetters as a function of n.

Assumption 3. *There exist positive constants ϵ and k, and $c \in [0, \frac{1}{2})$ such that*

$$\lim_{n \to \infty} |A_1(n)| > \epsilon$$

and

$$\lim_{n \to \infty} |A_1(n)| \leq k * (n_1^c). \tag{2}$$

We restrict the size of $A_1(n)$ because we want to focus on the case where the numbers of trendsetters does not dominate the total number of informed adopters.

3.4 Trend Adoption

We consider the following policies for informed adopters and followers to adopt a trend.

Informed Adopters – Influence Cascade: Informed adopters in G_1 are influenced by the decisions made by other informed adopters who are within one hop friendship on $G_1(n_1, \lambda_1)$. We model this process using the *independent information cascade* model that was introduced to study the problem of the maximum influenced set in social networks [6,8,9].

More precisely, we divide the set $E(G_1)$ into two disjoint subsets of *open* and *blocked* edges. An edge $e(i, j) \in E(G_1)$ is open with probability $\rho_1 \in [0, 1]$, independent of everything else; otherwise it is blocked.

The trend adoption then only propagates over open edges, i.e. if node i has adopted the trend and node $e(i,j)$ is an open edge, then node j will also adopt the trend [6,8,9].

Followers – Threshold Strategy: Here we assume that followers use a *threshold policy* in order to adopt the trend, i.e. a follower will adopt the trend only if the number of informed adopters that adopted the trend exceeds a given threshold value. More precisely, let t_a be a threshold value used by the followers in $G_2(n_2, \lambda_2)$ in order to decide whether or not to adopt the trend. A follower then adopts the trend only if he observes t_a number of adoptions among his one-hop friends.

3.5 Adoption Process

The trend adoption process then proceeds as a discrete time. At time $t = 0$, only the trendsetters A_1 in G_1 have adopted the trend, i.e. are active in G_1. The trend then spreads from A_1 as follows. When a node $i \in V(G_1)$ first becomes active in step t, it is given a single chance to activate each currently non-active neighbour $j \in N(i)$, and it succeeds with probability ρ_1. If i succeeds, then j will become active in step $t+1$. The cascade terminates when no node has become newly active at a given time step. Followers $V(G_2)$ in G_2 observe whether their one-hop friends in G_1 and adopt the trend as soon as t_a of their friends have adopted the trend, i.e .have become active.

3.6 Problem Statement

In the above scenario of the individual adoptions on social networks, we are interested in the question of an optimal threshold decision policy, which allows followers to make the correct decision in the sense that they only adopts a trend if the majority of informed adopters in G_1 will adopt the trend. Thus we are interested in which threshold value t_a to choose such that followers will almost always make the correct decision to conform with the majority of individuals in G_1. Note that the threshold decision policy merely makes use of individuals' local observations.

It is well known [11] that (asymptotically) a majority of informed adopters in G_1 will adopt the trend only if

$$\lim_{n\to\infty} \rho_1\lambda_1 > 1$$

and only a small fraction of informed adopters in G_1 will adopt the trend if

$$\lim_{n\to\infty} \rho_1\lambda_1 < 1.$$

Therefore, the question we want to study is whether there exists a threshold value t_a that ensures that (asymptotically) no follower will adopt the trend in the case where $\lim_{n\to\infty} \rho_1\lambda_1 < 1$, and each follower will adopt the trend with positive probability if $\lim_{n\to\infty} \rho_1\lambda_1 > 1$.

3.7 Main Result

Our main result is to prove that choosing the threshold value $t_a = 2$ is optimal for followers in the sense that the threshold value $t_a = 2$ ensures that a follower never makes a mistake by adopting a trend that is adopted only by a small fraction of the informed adopters, while maximizing the probability that a follower adopts the trend when a majority of the informed adopters does so.

More precisely, we obtain the following result. Consider a follower i in G_2 and let the random variable K_i denote the number of active neighbours of i after the cascade process has finished, i.e. the numbers of active informed adopters in G_1 to whom follower i is connected in the social graph G. Conditional on $\lambda_1\rho_1$, let $F(k)$ be the probability there exists at least one node in G_2 who has at least k number of active neighbours in G_1, i.e. $F(k)$ is given by

$$F(k) = Pr(\exists i \in V(G_2), \text{ such that } K_i \geq k | \lambda_1\rho_1 < 1). \tag{3}$$

In our main result in Section 5 we show that

$$\lim_{n\to\infty} F(1) > 0$$

and

$$\lim_{n\to\infty} F(k) = 0, \quad \text{for } k = 2, 3, \cdots .$$

Note that this result implies that under a any threshold value $t_a \geq 2$, no follower will make the mistake to adopt a trend if only a small fraction of the informed adopters does so. This is not true for the threshold value $t_a = 1$, i.e. under the threshold value $t_a = 1$ some followers will make a the mistake and adopt the trend even though only a small fraction of the informed adopters does so. Hence, using a threshold $t_a \geq 2$ is "safe" as ensures that followers never make a mistake. Moreover, among all value $t_a \geq 2$ the value $t_2 = 2$ is optimal in the sense that it maximizes the probability of adopting the trend when a majority of the informed adopters does so. In particular, one can show that under the threshold $t_a = 2$ followers will adopt the trend with a strictly positive probability when a majority of the informed adopters adopt the trend.

3.8 Numerical Results

To illustrate our result, we provide a numerical case study.

First, we plot the fraction of nodes in G_2 that adopts the trend as the product of $\lambda_1\rho_1$ increases (Fig. 3). This leads to an important observation, known as *the double jump* [12].

Furthermore, in order to examine closely the relation between individual adoption behaviour and their neighbourhood adoption status, we plot Fig. 4 , in which we find that the fraction of nodes who adopt the trend has a sudden rise with two active neighbours presenting. Interestingly, this observation was experimentally observed in real social networks [4]. This is one important feature of our model. As before, no propagation models can reproduce this "second-adoption" phenomenon (Fig.1, 2) that appears in real life.

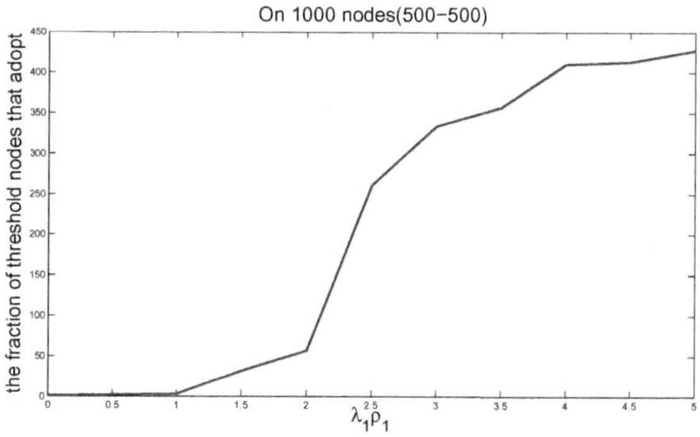

Fig. 3. Double jump at $\lambda_1\rho_1 = 1$

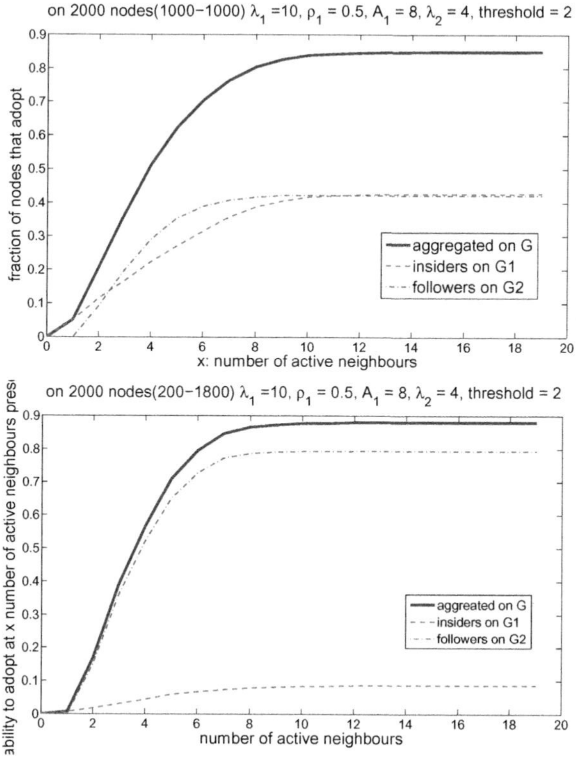

Fig. 4. Second-adoption phenomenon

4 Equivalent Model

In this section, we propose the equivalent model of G in Section 3, $G'(n_1, n_2, \lambda_1\rho_1, A_1, \lambda_2, t_a)$. In particular, we shall prove that the size of the influenced set produced by ρ_1-cascade process over $G_1(n_1, \lambda_1, A_1)$ has the same distribution as that of the influenced set on $G_1'(n_1, \lambda_1\rho_1, A_1)$.

Before proceeding, we need to define the *influenced set* I_1 on G_1. Let the influenced set be the set of vertices connected to A_1 via open edges in G_1, i.e.,

$$I_1 = \{v|v \in V(G_1), v \text{ connects to } A_1 \text{ via } \textit{open} \text{ edges}\}$$

Hence by the end of the influence cascade process, all the nodes in I_1 become active.

The equivalent model G_1' is a Erdös-Rényi graph. G_1' has the same vertex set as G_1, but with a different edge preserving probability

$$p_1' = p_1\rho_1,$$

where p_1 is defined in (3.1), the edge preserving probability of G_1, the success probability of the influence cascade process. G_1' have the same initial active set as G_1, A_1, and subjecting to the same size constraint.

Now we define the *influenced set* I_1' on G_1'. Let the influenced set on G_1' be the set of vertices connected to A_1 via edges in G_1', i.e.,

$$I_1' = \{v|v \in V(G_1'), v \text{ connects to } A_1 \text{ via edges } e \in E(G_1')\}$$

The following theorem guarantees that on both models the distribution in the size of the influenced set is the same. To prove that unfolding the influence cascade by determining the open/block status of edges gradually at each time step is the same as unfolding the influence cascade by determining the status of all the edges at the beginning during the construction of G_1', the idea is to argue that the distribution for a random node joining the influenced set is the same over time steps in both models.

Theorem 1. *Given an initial active set A_1, the distribution of the influenced sets obtained by cascade process on G_1 starting from A_1, is the same as the distribution of sets reachable from A_1 via edges on G_1'.*

Proof. First we argue the iterations of the influence cascade process over G_1. Define $A_1^{(t)}$ to be the set of active nodes at the end of iteration t, for $t = 0, 1, 2, 3, \cdots$. If a random node v has not become active by the end of iteration t, then under the cascade process, the probability that v will become active in iteration $t+1$ is equal to the chance that one of its active neighbours in $A_1^{(t)}$ but not in $A_1^{(t-1)}$ succeeds in activating v. This probability is $P(v \in A_1^{(t+1)}|v \notin A_1^{(t)}) = 1 - \prod_{w \in A_1^{(t)} \setminus A_1^{(t-1)}} (1 - \frac{\rho_1\lambda_1}{n_1})$.

Second we construct G'_1 gradually as follows. We start with the initial set A_1. For each node v with at least one edge stub, we determine whether it connects to A_1. If so, then v is reachable; if not, we keep the source of v's edge unknown subject to the condition that it comes from outside of A_1. Having now exposed a new set of reachable nodes $A_1^{(1)}$ in the first time step, we proceed to reveal further reachable nodes by performing the same process on edges from A_1, and in this way produce sets $A_1^{(2)}, A_1^{(3)}, A_1^{(4)}, \cdots$. If node v has not been determined to be reachable by the end of time step t, then the probability that it is determined to be reachable in time $t+1$ is equal to the chance that its edge comes from $A_1^{(t)}$ but not in $A_1^{(t-1)}$; this probability is $A_1^{(t+1)}|v \notin A_1^{(t)}) = 1 - \prod_{w \in A_1^{(t)} \setminus A_1^{(t-1)}} (1 - \frac{\rho_1 \lambda_1}{n_1})$.

Thus, by induction over the iterations, we see that the cascade process produces the same distribution over influenced sets as the construction of G'_1.

Since we have established the equivalence between the $G_1(n_1, \lambda_1, \rho_1, A_1)$ in Section 3 and $G'_1(n_1, \lambda_1 \rho_1, A_1)$ with respect to the distribution in the size of the influenced set, and G'_1 is essentially a Erdös-Rényi graph.

5 Analysis

In this section, we derive in detail the answer to the question of the optimal trend-adoption policy on $G'(n_1, n_2, \lambda_1 \rho_1, A_1, \lambda_2, t\_a)$. We shall specifically consider the adoption policy in subcritical phase, and obtain the optimal policy.

5.1 Decision Policy in Subcritical Phase

When $p'_1 = p_1 \rho_1 \in [0, \frac{1}{n_1})$, G'_1 is in subcritical phase. We first derive the probability, $u(k)$, for a random vertex in $G_2(n_2, \lambda_2)$ possessing k active neighbours. Then by making use of this formula, we derive the conditional probability, $F(k)$, for existing any node in $G_2(n_2, \lambda_2)$ with at least k active neighbours. And we thus prove that in the subcritical phase the correct decision policies are threshold strategies with threshold value: $t_a = 2, 3, \cdots$.

Let P_a be the probability for a random node in G'_1 being active after the influence cascade terminates. Then P_a is given by:

$$P_a = \frac{|I_1|}{n_1} \tag{4}$$

In order to quantify P_a, we make use of the expectation of $|I'_1|$, which can be expressed in closed form using percolation theory on random graphs. Percolation property of random graphs has been extensively studied in the field of epidemics spread [15,16,17]. It studies whether a disease will dominate the whole population. If the disease does not percolate the whole network, it quantifies the expected value of the size of the infected population. The main approach used is generating functions [18]. By the same approach, we are now deriving $E(|I_1'|)$ on Erdös-Rényi random graph, $G_{n,p}$, with percolation probability ρ.

Theorem 2. *On Erdös-Rényi random graph $G(n, p = \frac{\lambda}{n})$, the expectation of the size of the influenced set with one vertex percolated initially is* $\frac{1-\rho}{1-\rho-\lambda\rho}$.

Proof. Let p_k be the probability for a random vertex having degree k. p_k is given by:

$$p_k = \binom{n-1}{k} p^k (1-p)^{n-1-k} \approx \frac{\lambda^k e^{-\lambda}}{k!}. \tag{5}$$

Let $G_0(x)$ be the generating function of $\{p_k\}$, which is given by:

$$G_0(x) = \sum_{k \geq 0} p_k x^k$$

Follow a random edge and reach a vertex v. Let $\{q_k\}$ be the degree distribution of so-reached vertex v. Let $G_1(x)$ be the generating function of $\{q_k\}$, which is given by:

$$G_1(x) = \sum_{k \geq 0} q_k x^k$$

A random edge has probability $\frac{1}{e(G)}$ to be chosen by the uniform choosing policy, where $e(G)$ denotes the total number of edges in G. Conditional on v having degree k, the chosen edge has probability $\frac{k}{e(G)}$ being incident to v, i.e.,

$$P(\text{ the chosen edge incident to } v | deg(v) = k) = \frac{k}{e(G)} \tag{6}$$

By Total law of probability, from (5, (6) we have:

$$P(\text{the chosen edge incident to } v) = \frac{\sum_k p_k k}{e(G)}. \tag{7}$$

Thus by Bayes rule, from (5), (7) we have q_k which is given by:

$$\begin{aligned} q_k &= P(deg(v) = k | \text{ the chosen edge incident to } v) \\ &= \frac{p_k k}{\sum_k p_k k} = \frac{p_k k}{z} \end{aligned} \tag{8}$$

where z is the average vertex degree. And in the case of Erdös-Rényi random graph $z = \lambda$ on our model.

Let $\tilde{p}_m$ be the probability for a random vertex having m open edges. Let $G_0(x;\rho)$ be the generating function of $\{\tilde{p}_m\}$, which is given by:

$$
\begin{aligned}
&P(m \text{ edges of } v \text{ are open } | deg(v) = k)\\
&= \binom{k}{m}\rho^m(1-\rho)^{k-m} \text{ binomial distribution}\\
&\Rightarrow\\
&\tilde{p}_m = P(m \text{ edges of } v \text{ are open})\\
&= \sum_k p_k \binom{k}{m}\rho^m(1-\rho)^{k-m}\\
&\Rightarrow\\
&G_0(x;\rho) = \sum_{m\geq 0}\sum_{k\geq 0} p_k \binom{k}{m}\rho^m(1-\rho)^{k-m}x^m\\
&= \sum_{k\geq 0} p_k \sum_{m\geq 0}^{k} \binom{k}{m}(x\rho)^m(1-\rho)^{k-m}\\
&= \sum_{k\geq 0} p_k(1-\rho+x\rho)^k
\end{aligned}
$$

Thus

$$G_0(x;\rho) = G_0(1-\rho+x\rho) \tag{9}$$

Now follow a random edge and reach a vertex v. Let $\tilde{q}_m$ be the probability for so-reached v having m open edges. Let $G_1(x;\rho)$ be the generating function of $\{\tilde{q}_m\}$. Similar with the derivation of (9), $G_1(x;\rho)$ is given by:

$$G_1(x;\rho) = G_1(1-\rho+x\rho) \tag{10}$$

Let **Z** be a random variable denoting the size of influenced set connected to a random *vertex*. Define $H_0(x;\rho)$ be generating function for the distribution of **Z**. Let $\tilde{Z}$ be a random variable denoting the size of influenced set connected to a random *edge*. Define $H_1(x;\rho)$ be generating function for the distribution of $\tilde{Z}$. The influenced components generated by $H_0(x;\rho)$ consist of the initial infected node, plus any number of tree-like clusters, joined to it by single edges. Each tree-like cluster has the size distribution generated by $H_1(x;\rho)$. Thus we have:

$$
\begin{aligned}
H_0(x;\rho) = xq_0 + xq_1H_1(x;\rho)\\
+ xq_2[H_1(x;\rho)]^2 + xq_3[H_1(x;\rho)]^3 + \dots
\end{aligned}
$$

i.e.,

$$H_0(x;\rho) = xG_0(H_1(x;\rho);\rho)$$

and

$$H_1(x;\rho) = x\tilde{q_0} + x\tilde{q_1}H_1(x;\rho) + x\tilde{q_2}[H_1(x;\rho)]^2 + \dots$$

i.e.,

$$H_1(x;\rho) = xG_1(H_1(x;\rho);\rho) \tag{11}$$

The expected value of $\mathbf{Z}$ is given by differentiating $H_0(x;\rho)$, i.e.,

$$E[\mathbf{Z}] = H_0'(1;\rho) = 1 + G_0'(1;\rho)H_1'(1;\rho) \tag{12}$$

Differentiating equation (11), we have:

$$H_1'(1;\rho) = 1 + G_1'(1;\rho)H_1'(1;\rho) = \frac{1}{1 - G_1'(1;\rho)} \tag{13}$$

substitute (13) into (12), and make use of (8),(9), (10), we have:

$$E[\mathbf{Z}] = 1 + \frac{\rho G_0'(1)}{1 - \rho G_1'(1)} \tag{14}$$

We notice

$$G_0'(1) = \sum_{k\geq 0} kp_k x^{k-1}|_{x=1} = \sum_{k\geq 0} kp_k = \lambda \tag{15}$$

With (8), we also have

$$\begin{aligned} G_1'(1) &= \sum_{k\geq 0} kq_k x^{k-1}|_{x=1} \\ &= \sum_{k\geq 0} \frac{k^2 p_k}{z} = \frac{E(K^2)}{z} \\ &= \frac{Var(K) + E^2(K)}{z} = \frac{\lambda + \lambda^2}{\lambda} \\ &= 1 + \lambda \end{aligned} \tag{16}$$

where K in (16) is a r.v. denoting the vertex degree of a random node in Erdös-Rényi random graph $G(n, p = \frac{\lambda}{n})$. Thus, K has approximately Poisson distribution with $E(K) = \lambda$ and $Var(K) = \lambda$. Substituting (15) and (16) into (14), we have

$$E[\mathbf{Z}] = \frac{1-\rho}{1 - \rho(1+\lambda)}$$

By Theorem 2, we have the expectation of the influenced set given by:

$$E(|I_1'|) = \frac{1-\rho_1}{1-\rho_1(\lambda_1+1)} \text{ for } |A_1| = 1$$

Further for $|A_1| \leq k * n_1^c$ for some k, we have

$$E(|I_1'|) \leq \frac{1-\rho_1}{1-\rho_1(\lambda_1+1)} k * n_1^c = \leq k' * n_1^c, \text{ for some } k'.$$

If the influenced sets associated with each node in A_1 are independent, the sum of the influenced sets has the expectation of $\frac{1-\rho_1}{1-\rho_1(\lambda_1+1)} k * n_1^c$. The inequality comes from the possibility that those influenced sets overlap with one another.

By Markov inequality and Theorem 2, we have an upper bound of $|I_1'|$ given by:

$$Pr(|I_1'| \geq n^{2c}) \leq \frac{E(|I_1'|)}{n_1^{2c}} \leq \frac{kn_1^c}{n_1^{2c}} = \frac{k}{n_1^c} \rightarrow 0 \text{ for } n_1 \text{ large}$$

Since we have restricted the size of A_1 by assuming that $0 \leq c < \frac{1}{2}$ in (2), then $\lim_{n_1 \rightarrow \infty} P(|I'| \geq n_1^{c'}) = 0$, where $c' \in [0, 1)$. By Theorem 1, we have the same bound of $|I_1|$:

$$\lim_{n \rightarrow \infty} P(|I_1| \geq n_1^{c'}) = 0, \text{ where } c' \in [0, 1) \tag{17}$$

Alternatively, (17) can be written as:

$$\lim_{n_1 \rightarrow \infty} P(|I_1| < n_1^{c'}) = 1 \tag{18}$$

Now we bound P_a by substituting (18) into (4):

$$P_a = \frac{|I_1|}{n_1} < \frac{n_1^{c'}}{n_1} = 0 \text{ with probability 1 for } n_1 \text{ large} \tag{19}$$

Equation (19) has an intuitive interpretation. In the subcritical phase of $G_1'(n_1, \lambda_1\rho_1, A_1)$, we randomly pick a node when the influence cascade terminates. Such a node is non-active with probability 1 for large networks. The following theorem provides the distribution for a random node in $G_2(n_2, \lambda_2)$ having n_a number of active neighbours.

Theorem 3. *On our model $G'(n_1, n_2, \lambda 1\rho_1, A_1, \lambda_2, t_a)$, we have $P(K_v = n_a) = \frac{(\lambda_2 P_a)^{n_a}}{n_a! e^{\lambda_2 P_a}}$, where P_a is given by (4).*

Proof. First, we derive the degree distribution of a random node in $G_2(n_2, \lambda_2)$. Let p_k be the probability for a random node $v \in G_2$ having k neighbours. p_k satisfies binomial distribution and is given by:

$$p_k = \binom{n_1 - 1}{k} p_2^k (1 - p_2)^{n_1 - 1 - k}$$

where $p_2 = \frac{\lambda_2}{n_1}$.

We apply Poisson approximation as n_1 goes to infinity:

$$p_k = \binom{n_1 - 1}{k} p_2^k (1 - p_2)^{n_1 - 1 - k} \approx \frac{\lambda_2^k e^{-\lambda_2}}{k!} \tag{20}$$

Next we derive the distribution of the number of active neighbours a random node in G_2 possess. On the condition that a random node v has degree k, the probability for v having n_a active neighbours also satisfies binomial distribution:

$$P(K_v = n_a | deg(v) = k) = \binom{k}{n_a} P_a^{n_a} (1 - P_a)^{k - n_a} \tag{21}$$

where P_a is the probability for a random node being active, as defined in (4).

By Total law of probability, we remove the condition in (21), and have the probability for v having n_a active neighbours given by:

$$P(K_v = n_a) = \sum_k p_k \binom{k}{n_a} P_a^{n_a} (1 - P_a)^{k-n_a} \tag{22}$$

By substituting p_k in (20) into (22) and applying Taylor expansion [20], we have

$$\begin{aligned} P(K_v = n_a) &= \\ &= \sum_k \frac{\lambda_2^k e^{-\lambda_2}}{k!} \frac{k!}{(k-n_a)!n_a!} P_a^{n_a} (1-P_a)^{k-n_a} \\ &= \frac{e^{-\lambda_2} P_a^{n_a}}{n_a!(1-P_a)^{n_a}} \sum_k \frac{\lambda_2^k (1-P_a)^k}{(k-n_a)!} \\ &= \frac{e^{-\lambda_2} P_a^{n_a}}{n_a!(1-P_a)^{n_a}} [\lambda_2(1-P_a)]^{n_a} e^{\lambda_2(1-P_a)} \\ &= \frac{(\lambda_2 P_a)^{n_a}}{n_a! e^{\lambda_2 P_a}} \end{aligned}$$

Next, we use Theorem 3 to determine threshold decision policy. In particular, we define $F(k)$ to be a conditional probability. It denotes the probability that there exists one G_2 node having at least k number of active neighbours, conditional in subcritical phase. The next theorem quantifies $F(k)$, which in turn suggests that if choosing $t_a \geq 2$, with probability 1, no nodes in G_2 make a wrong decision.

Theorem 4. *We have*

$$\lim_{n \to \infty} F(1) > 0$$

and

$$\lim_{n \to \infty} F(k) = 0, \quad \textit{for } k = 2, 3, \cdots.$$

Proof. To see whether there exists a "follower" node in G_2 having at least one active insider-neighbour in subcritical phase, we compute $\lim_{n \to \infty} F(1)$:

$$\begin{aligned} \lim_{n \to \infty} F(1) &= \lim_{n \to \infty} 1 - [1 - Pr(K_v \geq 1)]^{n_2} \\ &= 1 - \lim_{n \to \infty} [Pr(K_v = 0)]^{n_2} \\ &= 1 - \lim_{n \to \infty} [\frac{(\lambda_2 P_a)^0}{0! e^{\lambda_2 P_a}}]^{n_2} \\ &= 1 - \lim_{n \to \infty} [\frac{(\lambda_2 \frac{|I_1|}{n_1})^0}{0! e^{\lambda_2 \frac{|I_1|}{n_1}}}]^{n_2} \\ &= 1 - \frac{1}{e^{\lambda_2 |I_1| \frac{n_2}{n_1}}} > 0 \end{aligned}$$

To see whether there exists a node in G_2 having at least two active neighbours in subcritical phase, we compute $\lim_{n\to\infty} F(2)$:

$$\begin{aligned}
\lim_{n\to\infty} F(2) &= \lim_{n\to\infty} 1 - [1 - Pr(K_v \geq 2)]^{n_2} \\
&= 1 - \lim_{n\to\infty} [Pr(K_v = 0) + Pr(K_v = 1)]^{n_2} \\
&= 1 - \lim_{n\to\infty} [\frac{(\lambda_2 P_a)^0}{0! e^{\lambda_2 P_a}} + \frac{(\lambda_2 P_a)^1}{1! e^{\lambda_2 P_a}}]^{n_2} \\
&= 1 - \lim_{n\to\infty} [\frac{(\lambda_2 \frac{|I_1|}{n_1})^0}{0! e^{\lambda_2 \frac{|I_1|}{n_1}}} + \frac{(\lambda_2 \frac{|I_1|}{n_1})^1}{1! e^{\lambda_2 \frac{|I_1|}{n_1}}}]^{n_2} \\
&= 1 - \lim_{n\to\infty} \frac{(1 + \frac{\lambda_2 |I_1|}{n_1})^n}{e^{\lambda_2 |I_1|}}
\end{aligned} \tag{23}$$

To further reduce (23), let $m = \frac{n_1}{\lambda_2 |I_1|}$. Note that $\lim_{n\to\infty} m = \infty$. Thus we reduce the nominator in (23) as follows:

$$\begin{aligned}
&\lim_{n\to\infty} (1 + \frac{\lambda_2 |I_1|}{n_1})^n_1 \\
&= \lim_{n\to\infty} \{(1 + \frac{\lambda_2 |I_1|}{n_1})^{\frac{n_1}{\lambda_2 |I_1|}}\}^{\lambda_2 |I_1|} \\
&= \lim_{m\to\infty} [(1 + \frac{1}{m})^m]^{\lambda_2 |I_1|} \\
&= e^{\lambda_2 |I_1|}
\end{aligned} \tag{24}$$

In the derivation of (24), we make use of the common limit [21]:

$$\lim_{m\to\infty} (1 + \frac{1}{m})^m = e$$

Substitute (24) into (23). We have:

$$\begin{aligned}
\lim_{n\to\infty} F(2) &= 1 - \lim_{n\to\infty} \frac{(1 + \frac{\lambda_2 |I_1|}{n_1})^{n_2}}{e^{\lambda_2 |I_1|}} \\
&= 1 - \frac{e^{\lambda_2 |I_1|}}{e^{\lambda_2 |I_1|}} \\
&= 0.
\end{aligned} \tag{25}$$

(25) says that there exists no node having 2 or more active neighbours in subcritical phase. It generalizes that there exists no node having 3 or more active neighbours in subcritical phase, and so forth. Thus we have $\lim_{n\to\infty} F(k) = 0$, for $k = 2, 3, \cdots$.

Theorem 4 suggests the decision policy for followers in subcritical phase. The first part of the theorem says that the threshold value should not be $t_a = 1$. Since $\lim_{n\to\infty} F(1) > 0$ suggests that the probability is positive for existing at least one node having 1 active neighbour by the end of influence cascade process. If we choose $t_a = 1$ as the threshold value, then by the strategy the nodes with one active neighbour shall adopt thus make the wrong decision in subcritical phase. The second part says that the threshold value could be $t_a = 2, 3, \cdots$. Since $\lim_{n\to\infty} F(k) = 0$ for $k = 2, 3, \cdots$ says the probability tends to zero for existing any nodes having k or more active neighbours. In words, with probability one there is no node with k or more active neighbours for $k \geq 2$. Hence these choices of k can guarantee all the "follower" nodes in G_2 not to adopt thus make the correct decision in the subcritical phase. Moreover, one can show that under the threshold $t_a = 2$ followers will adopt the trend with a strictly positive probability when a majority of the informed adopters adopt the trend (i.e. when $\lim_{n\to\infty} \lambda_1\rho_1 > 1$); and the larger the fraction of informed adopters who adopted the trend, the higher the probability that a follower will do so. In this sense, the threshold value $t_a = 2$ the optimal value.

6 Conclusions

In this paper, we study how an individual in a social network should decide whether or not to adopt a trend, based on how many people in their neighborhood in the social network adopted the trend. In particular we investigated the question in what adoption policy leads to an optimal trend-adoption in the sense that an individual only adopts a trend if the majority of their social network will do so. we obtain the result that the optimal policy for an individual is to adopt a trend if two people in their neighborhood did.

Our analysis is mainly done on the Erdös-Rényi random graph model. Though our work is based on an oversimplified model, it indeed shed some light on how individuals make choices in real life. this result/behavior was experimentally observed in real social networks. The work of Backstrom el. [4] noticed the special effect of the second friend and claimed that the marginal benefit of having a second friend in the community is particularly strong. Here in our paper, we prove the optimality of the threshold value of 2, which coincides with the discovery of the special second adoption found in [4].We hope that this work will help towards building applications that is able to automatically push relevant content to users in online social networks.

References

1. `http://en.wikipedia.org/wiki/Conformity` (access on March 15, 2010)
2. Aronson, E., Wilson, T.D., Akert, A.M.: Social Psychology, 6th edn. Pearson Prentice Hall, Upper Saddle River (2007)
3. `http://en.wikipedia.org/wiki/Threshold` (access on March 18, 2010)
4. Backstrom, L., Huttenlocher, D., Kleinberg, J., Lan, X.: Group Formation in Large Social Networks: Membership, Growth, and Evolution. In: Proc. 12th ACM SIGKDD Intl. Conf. on Knowledge Discovery and Data Mining (2006)
5. Bauch, C.T., Earn, D.J.D.: Vaccination and the theory of games. Proceedings of the National Academy of Sciences 101, 13391–13394 (2004)

6. Kempe, D., Kleinberg, J., Tardos, E.: Maximizing the Spread of Influence through a Social Network. In: Proc. 9th ACM SIGKDD Intl. Conf. on Knowledge Discovery and Data Mining (2003)
7. Easley, D., Kleinberg, J.: Networks, Crowds, and Markets: Reasoning About a Highly Connected World. Cambridge University Press (2010)
8. Kempe, D., Kleinberg, J.M., Tardos, É.: Influential Nodes in a Diffusion Model for Social Networks. In: Caires, L., Italiano, G.F., Monteiro, L., Palamidessi, C., Yung, M. (eds.) ICALP 2005. LNCS, vol. 3580, pp. 1127–1138. Springer, Heidelberg (2005)
9. Kleinberg, J.: Cascading behaviour in Networks: Algorithmic and Economic Issues. In: Nisan, N., Roughgarden, T., Tardos, E., Vazirani, V. (eds.) Algorithmic Game Theory. Cambridge University Press (2007)
10. Hacking, I.: Logic of statistical inference. Cambridge University Press (1965)
11. Bollobás, B.: Random graphs, 2nd edn. Cambridge University Press (2001)
12. Erdos, P., Renyi, A.: On the Evolution of Random Graphs. Publ. Math. Inst. Hungar. Acad. Sci. 5, 17–61 (1960)
13. Durrett, R.: Random graph dynamics. Cambridge University Press (2007)
14. Spencer, J.: The strange logic of random graphs. Springer (2001c)
15. Newman, M.E.J.: The spread of epidemic disease on networks. Phys. Rev. E 66, 16128 (2002)
16. Newman, M.E.J., Strogatz, S.H., Watts, D.J.: Random graphs with arbitrary degree distributions and their applications. Phys. Rev. E 64, 26118 (2001)
17. Newman, M.E.J.: Random graphs with clustering. Phys. Rev. Lett. 103, 58701 (2009)
18. Wilf, H.S.: Generating functionology, 3rd edn. A K Peters, Wellesley (2006)
19. NIST/SEMATECH, 6.3.3.1. Counts Control Charts, e-Handbook of Statistical Methods (March 2010), `http://www.itl.nist.gov/div898/handbook/pmc/section3/`
20. Greenberg, M.: Advanced Engineering Mathematics, 2nd edn. Prentice Hall (1998) ISBN 0-13-321431-1
21. Roberts, C., McLenaghan, R.: Continuous Mathematics in Standard Mathematical Tables and Formulae. In: Zwillinger, D. (ed.) Differential Calculus, ch. 5.1, p. 333. CRC Press, Boca Raton (1996)

Paris Metro Pricing for Internet Service Differentiation

Dongmyung Lee[1], Taehyun Kim[2], Jeonghoon Mo[2], and Jinwoo Park[1]

[1] Department of Industrial Engineering,
Seoul National University, Seoul, Korea
{leoleo333,autofact}@snu.ac.kr
[2] Department of Information and Industrial Engineering,
Yonsei University, Seoul, Korea
{taehyun84,j.mo}@yonsei.ac.kr

Abstract. This paper analyzes the Paris Metro Pricing (PMP) strategy for differentiating Internet service. PMP has several advantages over other pricing schemes that guarantee quality of service (QoS) such as simplicity and less bandwidth overhead. In this paper, we develop a simple analytical model for PMP. We first assume that there is only one network service provider (a monopolist) serving N users and model the user behavior and the provider's profit. Then we derive the optimal ratio of dividing a given network capacity in order to maximize the profit of the service provider. Our results show that, by maximizing providers profit, the subscription is also maximized which can be interpreted as a higher satisfaction of users compared to that of not using PMP. In addition, by taking into account various network types, we show that in a monopoly environment, it is always better to implement PMP regardless of user populations we considered. We then further extend our model to a duopoly setting. We found that there exist no Nash equilibrium even when both providers do not differentiate the network service.

Keywords: Internet services, price discrimination, Paris Metro Pricing, revenue maximization, user subscription, Nash equilibrium.

1 Introduction

The usage of the *Internet* was dominated by traditionally 'typical' data services (e.g., e-mail, web browsing, file transfer, etc.). These services do not require severe bandwidth overhead on the network since the traffic generated by those applications usually tolerate relatively large packet delays. However, as we clearly witness these days, new internet applications such as VoIP, IPTV, and many smart phone applications can be charaterized as delay-constrained (or delay-sensitive), and thus require higher requirements. In addition, by 2014, mobile data traffic will double every year through 2014, increasing 39 times between 2009 and 2014, according to Cisco forecast [1]. This increase of diverse quality of service (QoS) requirements gives the rationale to develop new methods capable of treating the delay-constrained and non-delay-constrained traffic differently. A

R. Jain and R. Kannan (Eds.): GameNets 2011, LNICST 75, pp. 628–641, 2012.

possible solution is to give priority to the delay-constrained traffic in the queues of the network [2] but without an appropriate *pricing* scheme, any prioritization is useless [3].

In this paper, we study the use of pricing mechanism called Paris Metro Pricing (PMP) proposed by Odlyzko [4]. Under the PMP scheme, the network is split into subnetworks. The tariff for each subnetwork is different, expecting a lower congestion for highly priced networks. This method does not offer any QoS guarantees, so that it is somewhat weak compared to several other approaches. However, due to its simplicity, it is, indeed, very attractive to many practitioners.

In this regard, a few papers have examined the PMP scheme for charging packet networks. In [5], the authors present a mathematical model in which all packets are generated by the same kind of application and all users have the same valuation of QoS. Even though they have showed the existence and uniqueness of the stability using queueing theory, the assumptions seem too strong to have practical implications. Paper [6] is the closest work to ours in that, based on their proposed model, they tried to show whether a network service provider (a single-constrained monopolist) can be profitable using PMP strategy. However, there are several differences with our study. First, although they modeled the user's satisfaction as the utility function they only showed the surplus from the providers perspective. In addition, the results they showed were merely for uniform distribution. Most of the papers in which PMP is analyzed take the uniform distribution of users as one of the assumptions for their model. However, in [7] the authors pointed out that it is required to assess the importance of those assumptions. In this spirit, we adopt a basic model from [8] and further extend it so as to analyze the economic aspects of this pricing scheme and to address the limitations of the existing papers.

The rest of the paper is organized as follows. In Section 2 we develop a simplified model to reflect both the provider revenue and user behavior. In section 3 we illustrate the impact of PMP on revenue and user subscription with several user distribution. In the subsequent section we analyze the PMP when two firms compete for the same set of users (duopoly). Section 5 concludes the paper.

2 System Model

2.1 Single Pricing

In this section, we develop a system model based on [8] with one network which will be used throughout the paper. Consider a communication system with a large population of N users each charaterized by a type θ that is an independent random variable distributed in $[0,1]$ with $f(\theta)$ and $F(\theta)$ as its probability density function and cumulative distribution function respectively. A user of type θ finds the network connection acceptable if the number of users X using the network and the price p are such that

$$\frac{X}{N} \leq 1-\theta \quad \text{and} \quad p \leq \theta$$

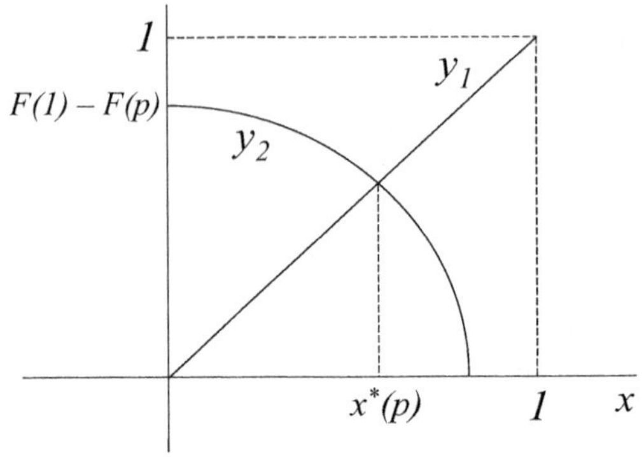

Fig. 1. The two functions y_1 and y_2 for the illustration of solution $x^*(p)$ of (2)

In this expression, N is the capacity of the network. Due to our analysis purpose, note that the capacity in the model is different from the conventional one that uses *bps*. Also, we will later divide the network into two subnetworks, each with capacity $N/2$. The expression interprets that a user with a large value of θ is willing to pay quite a lot for the connection but he expects a low utilization (or congestion) for a high quality of service. Conversely, a user with a small value of θ does not want to pay much for his connection but is willing to tolertate high delays. For example, we can view the users with large θ as users of VoIP and those with small θ as web browsers.

Assume that the network connection price is $p \in (0, 1)$. If the number of users in the network at a certain time period is X, then a user of type θ connects if the inequalities above are satisfied, i.e., if $\theta \in [p, 1 - X/N]$. Since θ is distributed in $[0, 1]$, the probability that a random user connects is $[F(1 - X/N) - F(p)]^+$. Accordingly, the number X of users that connect is binomial with mean $N \times [F(1 - X/N) - F(p)]^+$, so that

$$\frac{X}{N} \approx \left[F(1 - \frac{X}{N}) - F(p)\right]^+ \tag{1}$$

by the law of large numbers, since N is large. By letting $x = X/N$, (1) can be expressed as

$$x = [F(1 - x) - F(p)]^+ \tag{2}$$

Let the left-hand side and the right-hand side of (2) be $y_1(x)$ and $y_2(x)$ respectively. Then, for a fixed p, the solution of (2) can be illustrated as in Figure 1.

Since y_1 is increasing and y_2 is non-increasing, with both functions sharing the same domain of $[0, 1]$, a unique solution $x(p)$ exists. By taking the derivative of (2) with respect to p we have

$$\frac{dx}{dp} = -\frac{f(p)}{1 + f(1 - x)} < 0$$

which shows that x is a decreasing function of p.

Now, we solve the provider revenue ($R(p)$) maximization problem. In other words, we find the price p that maximizes the product of the number of users of the network and the price p, that is

$$p^* = \arg\max_p R(p) = p \cdot x(p) \tag{3}$$

Since $R(p)$ is a continuous function with $R(0) = 0$ and $R(1) = 0$, there exists at least one solution to the above maximization problem. Generally, it is nontrivial to find a closed form solution for $x(p)$, hence, for solving (3) a numerical method is performed. Nevertheless, to illustrate, we show a simple example when θ is uniformly distributed in $[0, 1]$. Equation (1), then, reduces to

$$\frac{X}{N} \approx \left(1 - \frac{X}{N} - p\right)^+$$

Solving this expression we find that $x := X/N = (1 - p)/2$. The operator can maximize his revenues by choosing the value of p that maximizes $px = p(1-p)/2$. The maximizing price is $p = 1/2$ and the corresponding value of px is $1/8$, which measures the revenue divided by N[1].

2.2 Differentiated Pricing

Consider now a situation where Paris Metro Pricing is applied. The network service provider divides the network into two subnetworks, each with different capacity. That is, there exist two networks: network 1 with price p_1 and capacity $N\alpha_1$ and network 2 with price p_2 and capacity $N\alpha_2$ (where $\alpha_1 + \alpha_2 = 1$). Without loss of generality, we assume $p_1 > p_2$. The users will select one of the two networks, based on the prices and utilizations. A user joins if there is an acceptable network and he chooses the cheapest network if both are acceptable. Moreover, if both networks are acceptable and, by any chance, have the same price, a user will join the one with the smallest utilization because it offers a marginally better QoS.

If the number of users in the two networks are X_1 and X_2 respectively, then a user of type θ chooses network 2 if $X_2/(N\alpha_2) \leq 1-\theta$ and $p_2 \leq \theta$. The probability that θ falls between p_2 and $1 - X_2/(N\alpha_2)$ is then

$$\left[F(1 - \frac{X_2}{N\alpha_2}) - F(p_2)\right]^+$$

Then, $x_2 := X_2/N$ is given by

$$x_2 = \left[F(1 - \frac{x_2}{\alpha_2}) - F(p_2)\right]^+$$

[1] For merely flat pricing, due to its simplicity, we were able to get results from curvilinear distributions (beta distribution). The pdf of the beta distribution is defined as $f(x, \alpha, \beta) = x^{\alpha-1}(1 - x)^{\beta-1} / \int_0^1 u^{\alpha-1}(1 - u)^{\beta-1}\, du$ and the results (maximum revenue) for some instances are as follows: 0.1045 ($\alpha = 0.5$, $\beta = 0.5$), 0.1176 ($\alpha = 2$, $\beta = 5$), 0.1620 ($\alpha = 5$, $\beta = 2$), 0.1488 ($\alpha = 2$, $\beta = 2$), and 0.1610 ($\alpha = 3$, $\beta = 3$).

A user will opt for network 1 if $X_1/(N\alpha_1) \leq 1-\theta$, $p_1 \leq \theta$ and $X_2/(N\alpha_2) > 1-\theta$. Thus, we find that $x_1 := X_1/N$ is such that

$$x_1 = \left[F(1 - \frac{x_1}{\alpha_1}) - F(\max\{p_1, 1 - \frac{x_2}{\alpha_2}\}) \right]^+$$

To determine the prices p_1 and p_2 that maximize the revenue of the operator, one needs to maximize the total revenue R obtained by both two subnetworks over p_1 and p_2, mathematically it is equivalent to solving the following problem

$$R = \max_{p_1, p_2} \; p_1 x_1 + p_2 x_2$$

By solving the problem when the network capacity is divided exactly half (again with uniform distribution of θ), one can show that the maximum occurs for $p_1 = 8/13$ and $p_2 = 11/26$ and that the maximum revenue equals to $25/156$. Eventually, we may conclude that the service differentiation with Paris Metro Pricing increases the revenue from $1/8$ to $25/156$, or by 28.2%. This can be one good rationale to use PMP instead of not using it. In this regard, we show, from another perspective, that it is recommended to use PMP. The provider who is willing to use PMP will want to know how much additional capacity is needed (without using PMP) in order to have as much as revenue obtained from using PMP (In this analysis we assume that the cost of increasing the network capacity is considerable).

Proposition 1. *PMP gives an increase in revenue by 28.2% which is equivalent to the amount when using only one network (and not using PMP) and increasing the network capacity by 78.6% given the original capacity N*

Proof. Let the increased capacity be $N(1+\beta)$. Considering the new increased capacity gives

$$\frac{X}{N} \approx \left(1 - \frac{X}{N(1+\beta)} - p\right)^+$$

solving for x, we have

$$x = \frac{1+\beta}{2+\beta}(1-p)$$

then solving the optimization problem (3), we have the optimal revenue with extended capacity as R_{ext}. By equating this with the optimal revenue obtained by PMP, R_{pmp} gives corresponding β as

$$R_{ext} = \frac{1}{4}\left(\frac{1+\beta}{2+\beta}\right) = \frac{25}{156} = R_{pmp}$$

$$\beta = \frac{44}{56}$$

3 Impact of PMP on Revenue and User Subscription

Since we have seen the price that maximizes the provider's revenue and its corresponding revenue with θ following uniform distribution, we will now see how these values as well as the fraction of users joining the network change with more general cdfs of θ. The rationale behind this is that uniform distribution alone cannot represent various type of user population. For instance, there might be a community of users in which the majorities use real-time services. On the other hand, one user group might consist of majority of people requiring non-delay-constrained services (e.g., email). Therefore, we need to represent these typical user distributions and incorporate them in the study of analyzing the impact of PMP.

In this regard, we define three additional probability distributions (f_1, f_3 and f_4) that capture three typical user distributions as follows (figure 2):

$$f_i(\theta) = \begin{cases} 2 - 2\theta & \text{if } i = 1 \\ 1 & \text{if } i = 2 \\ 2\theta & \text{if } i = 3 \end{cases}$$

where $\theta \in [0, 1]$ for all three distributions. Also,

$$f_4(\theta) = \begin{cases} 4\theta & \text{if } 0 \le \theta \le \frac{1}{2} \\ 4 - 4\theta & \text{if } \frac{1}{2} < \theta \le 1 \end{cases}$$

The first pdf (f_1) comes from a general concave cdf of θ and stands for a network consisting of high population of users generating non-delay-constrained traffic . The second function comes from a general linear cdf of θ and exactly corresponds to the uniform distribution and represents a well balanced network of users. The third one (f_3) is a pdf that represents a convex cumulative distribution function indicating the networks where the majority of users require high QoS levels. The last one (f_4) is a pdf of a 'S-shaped' cdf and designates a network with majority of users requiring intermediate QoS levels. We will denote the network environment for each user population type as NT1, NT2, NT3, and NT4 henceforth.

To see whether PMP actually gives benefits to both providers and users we, first, have to check whether it is always better irrespective of network environment. Secondly, even though this is true, we also have to check whether the average surplus of using PMP over every network environment is better than the value when it is not used. This comes from the fact that the type of user population will change over time since users consisting the network will require different service types at different time segment of interest. In light of this, we derive the maximum revenue and the corresponding user participation for each of the four user distributions. We will denote $\widehat{R}^1_{max}$, $\widehat{R}^2_{max}$, $\widehat{R}^3_{max}$ and $\widehat{R}^4_{max}$, and $\widehat{\alpha}^1_2$, $\widehat{\alpha}^2_2$, $\widehat{\alpha}^3_2$ and $\widehat{\alpha}^4_2$ as the maximum revenues and their corresponding α_2 values achieved by using PMP scheme for the distributions f_1, f_2, f_3 and f_4

3.1 Uniform Distribution

For the uniform distribution, we could express $\widehat{R}^2_{max}$, p_1, p_2, x_1, and x_2 with respect to just one variable α_2 after some manipulation. Each of these values

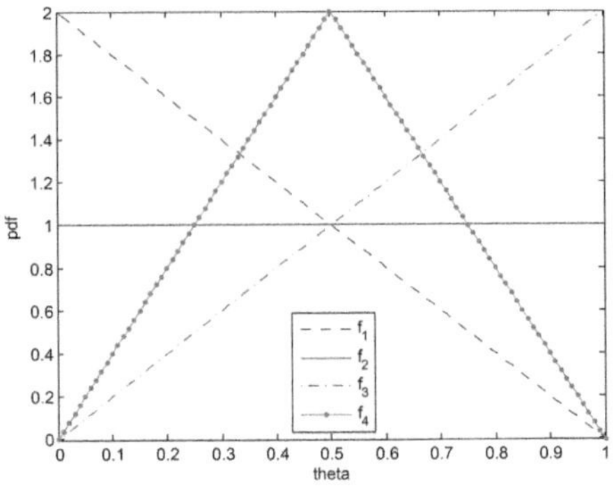

Fig. 2. The pdfs of θ representing each different network environment

are illustrated in Figure 3 and we find that the maximum revenue as well as the corresponding α_2 is as follows:

$$\widehat{R}^2_{max} = 0.1608 \ , \quad \widehat{\alpha}^2_2 = 0.43$$

In this example, the revenue increases by 28.6% compared to flat pricing. We could notice, from this analysis, that α_2 which maximizes revenue is in fact slightly less than 0.5. This tells us that when using two subnetworks to apply PMP, dividing it exactly half may not be the best choice for the network service provider. Also, since the range of α_2 that results in the revenue within 90% of the maximum value covers about 68% ($\geq$ 50%) of its whole interval, we could say that the revenue is not that sensitive to the ratio of dividing a network.

Figure 3c shows how the network utilization changes as α_2 increases. As you see from the figure, the total network utilization (denoted as subscription f2 in Fig. 3c) is at maximum when $\alpha_2 = 0.43$. This is the exact α_2 value that maximizes the total revenue. This implies that, when the network service provider maximizes his own profit, the network utilization is also maximized which is naturally true since the more users join the network the more revenue can be achieved (however, this does not imply that those two values have to be exactly the same, instead it means they will tend to have similar values). Moreover, it is easy to see from Figure 1c that whatever the value of α_2 is, PMP strategy always results in higher utilization than not performing it.

Non-uniform Distributions. In effect, if the user population is not represented by a uniform distribution, it is difficult to derive the maximum revenue of the service provider when PMP is used. This is due to the fact that it involves a maximization problem over three variables of α_2, p_1, and p_2. The results for the three non-uniform user distributions is illustrated in figure 3 and we find the maximum revenues and corresponding α_2 values as follows (for the derivation of all the values for uniform and non-uniform distributions see [9]):

$$\widehat{R}^1_{max} = 0.1512, \quad \widehat{\alpha}^1_2 = 0.45$$

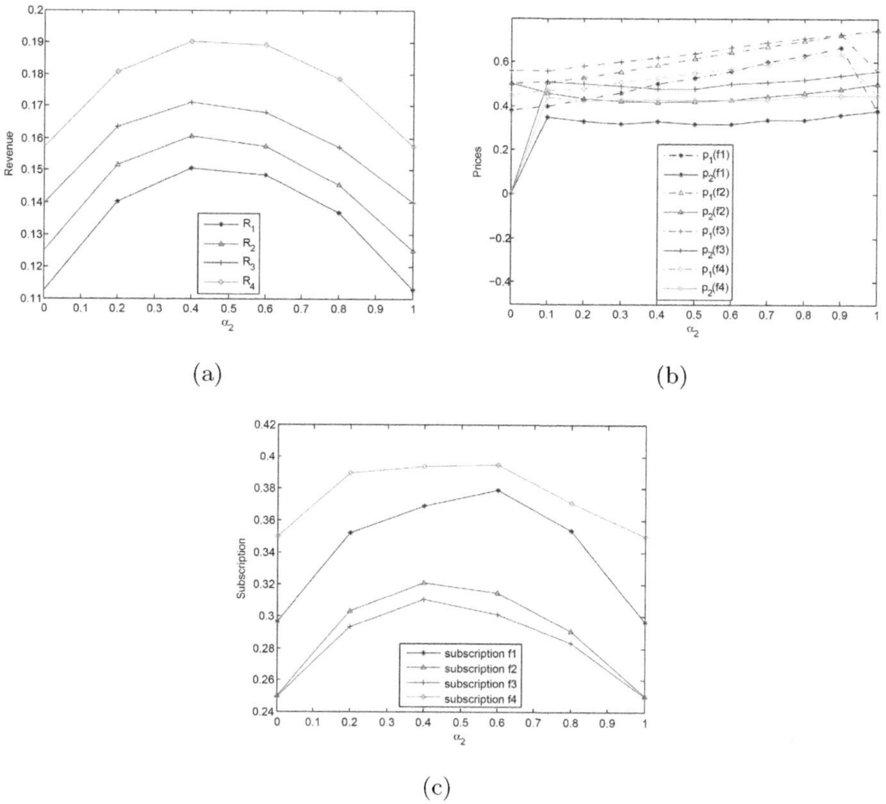

Fig. 3. Numerical results for the three non-uniform distribution cases (including that of uniform distribution) : (a) Revenues with respect to α_2 for each user distribution, (b) Prices with respect to α_2 for each user distribution, (c) User participation with respect to α_2 for each user distribution

$$\widehat{R}^3_{max} = 0.1713, \qquad \widehat{\alpha}^3_2 = 0.45$$

$$\widehat{R}^4_{max} = 0.1908, \qquad \widehat{\alpha}^4_2 = 0.50$$

From the previous section we have seen that when the network consists of a well-balanced users, the capacity ratio between two subnetworks is not that sensitive as long as it is close enough to the optimal ratio. We, now, try to see whether this is also true for the rest of the three network types. The sensitivity of the total revenue with respect to α_2 can be derived from the R-α_2 plots for each of the three non-uniform distributions (Figure 3a). That is, we would like to see how the revenue changes when the ratio of the capacities between two subnetwork is slightly altered from the optimal value (α_2^*). From Figure 3 we could see that the range of α_2 that produces the revenue within 90% of maximum value spans 65%, 74%, and 77% of its overall interval for NT1, NT3 and NT4 respectively. Thus, with the sensitivity of NT2 we could say that the revenue achieved in the

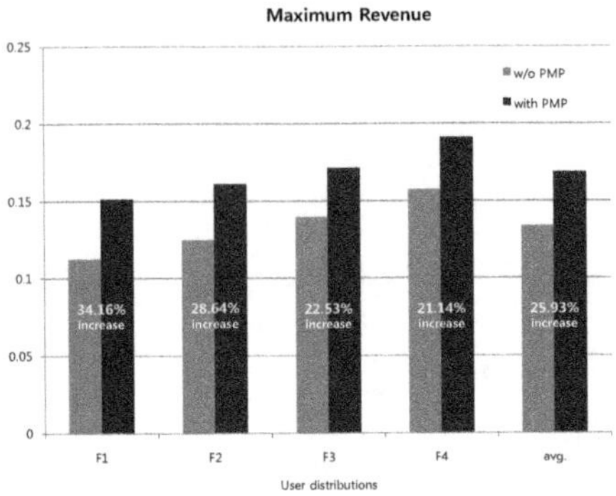

(a) Maximum revenue

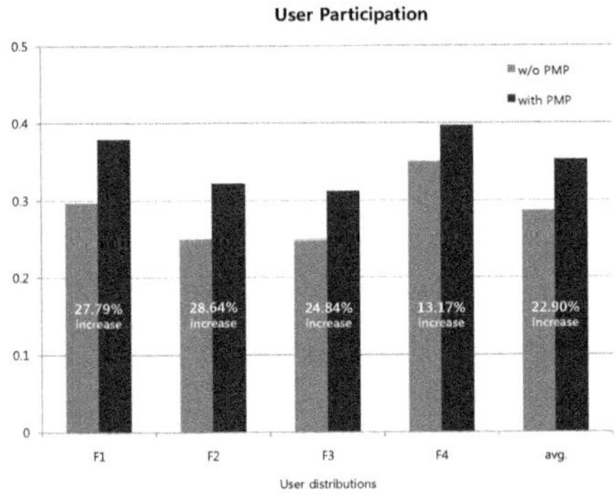

(b) User subscription

Fig. 4. Maximum revenues and user subscription for each user distribution

environment NT1 is relatively sensitive to the ratios of its subnetwork capacity which should be considered for a provider when its servicing environment is of NT1 most of the time. However, in general (and on average), determining the proportion of the subnetworks is not that critical.

Comparing with the maximum revenues when PMP is not used, we eventually see that no matter what type of user population may be, using PMP always produces a higher revenue for the network service providers (Figure 4a). Similarly, PMP scheme is always favored by network users, regardless of the characteristics of user groups as shown in Figure 4b. The users' satisfaction is measured by the number of users who join the network (user participation) because, based on our system model, only the satisfied users receive the network services. In summary, the average values of 4 network environment between using PMP and not using PMP clearly show that PMP is, indeed, a superior pricing scheme in terms of both the revenue and network utilization (user participation).

Observation 1. *In a network where only one network service provider (a monopolist) provides internet service, it is always better for the provider to use Paris Metro Pricing scheme than to just provide a general service with flat pricing. In addition, PMP scheme is always favored by the users consisting the network.*

Observation 2. *For a network service provider to achieve the maximum revenue, determining the fraction of the capacity of the subnetworks is less critical for all of the network environment.*

4 Competition: Duopoly

In previous sections, we have noticed that a single network service provider (a monopolist) will have an advantage for using PMP scheme. Then one might ask, "what happens when there are two network service providers? Does the Nash equilibrium even exist for the duopoly case?" In this section, we try to answer this question by modeling the situation using game theory.

4.1 Three Prices

We first analyze the situation where there are two service providers with only one using PMP. That is, one ISP (firm A) provides the service with a single price p_a and the other ISP (firm B) uses PMP with the lower price p_{b1} and the higher price p_{b2}. Both the competing firms have equal capacity where firm B splits the two subnetworks exactly half. Then, with this situation, there are 5 possible cases as follows:

1. $p_a < p_{b1} < p_{b2}$
2. $p_{b1} < p_a < p_{b2}$
3. $p_{b1} < p_{b2} < p_a$
4. $p_a = p_{b1} < p_{b2}$
5. $p_{b1} < p_a = p_{b2}$

Let R_A and R_B be the revenue (function) of the two firms respectively. From the similar derivation as in section 2 we can obtain the revenue functions of both firms for each of the 5 cases as follows:

1. $R_A = \frac{2}{3}(p_a - p_a^2)$,
   $R_B = p_{b1}\min(\frac{1-p_{b1}}{2}, \frac{1-p_a}{6}) + p_{b2}\min(\frac{1-p_{b2}}{2}, \frac{1}{2}\min(\frac{1-p_{b1}}{2}, \frac{1-p_a}{6}))$
2. $R_A = p_a\frac{2}{3}\min(1 - p_a, \frac{1-p_{b1}}{2})$,
   $R_B = \frac{p_{b1}(1-p_{b1})}{2} + p_{b2}\frac{1}{2}\min(1 - p_{b2}, \frac{2}{3}\min(1 - p_a, \frac{1-p_{b1}}{2}))$
3. $R_A = p_a\frac{2}{3}\min(1 - p_a, \frac{1}{2}\min(1 - p_{b2}, \frac{1-p_{b1}}{2}))$,
   $R_B = \frac{p_{b1}(1-p_{b1})}{2} + p_{b2}\frac{1}{2}\min(1 - p_{b2}, \frac{1-p_{b1}}{2})$
4. $R_A = \frac{1}{2}(p_a - p_a^2)$,
   $R_B = \frac{p_{b1}(1-p_{b1})}{4} + p_{b2}\frac{1}{2}\min(1 - p_{b2}, \frac{3(1-p_{b1})}{4})$

5. $R_A = p_a \frac{1}{2}\min(1-p_a, \frac{1-p_{b1}}{2})$,
   $R_B = \frac{p_{b1}(1-p_{b1})}{2} + p_{b2}\frac{1}{4}\min(1-p_{b2}, \frac{1-p_{b1}}{2})$

Let R_A^* and R_B^* be the optimal revenue for each firm and p_a^*, p_{b1}^*, and p_{b2}^* be the corresponding prices. Then the values for each of the 5 cases are summarized in table 1.

Based on the results, we try to find the Nash equilibrium price vector $(\bar{p}_a, \bar{p}_{b1}, \bar{p}_{b2})$ such that

$$R_A(\bar{p}_a, \bar{p}_{b1}, \bar{p}_{b2}) \geq R_A(p_a, \bar{p}_{b1}, \bar{p}_{b2}) \quad \forall p_a$$

and

$$R_B(\bar{p}_a, \bar{p}_{b1}, \bar{p}_{b2}) \geq R_B(\bar{p}_a, p_{b1}, p_{b2}) \quad \forall p_{b1}, p_{b2}$$

However, one can verify that there are no such vector and conclude that there exist no Nash equilibrium with three prices case.

Table 1. Optimal revenues and prices for firms A and B (three prices)

Case	p_a^*	p_{b1}^*	p_{b2}^*	R_A^*	R_B^*
1)	1/2	5/6	11/12	0.167	0.108
2)	3/4	1/2	5/6	0.125	0.194
3)	17/20	2/5	7/10	0.085	0.225
4)	1/2	1/2	5/8	0.125	0.180
5)	3/4	1/2	3/4	0.090	0.172

Table 2. User subscription for firms A and B (three prices)

Case	x_a^*	x_{b1}^*	x_{b2}^*
1)	1/3	1/12	1/24
2)	1/6	1/4	1/12
3)	1/10	3/10	3/20
4)	3/8	3/8	3/16
5)	3/16	1/4	3/16

4.2 Four Prices

We now consider the situation where the two firms both go for PMP scheme. In other words, firm A provides a differentiated service with two prices p_{a1} and $p_{a2}(> p_{a1})$ and, similarly, firm B divides the network into two subnetworks with prices p_{b1} and $p_{b2}(> p_{b1})$. Now, due to the symmetry, we need only consider 6 cases which are

1. $p_{b1} < p_{b2} < p_{a1} < p_{a2}$
2. $p_{b1} < p_{a1} < p_{b2} < p_{a2}$

3. $p_{a1} < p_{b1} < p_{b2} < p_{a2}$
4. $p_{b1} = p_{a1} < p_{b2} < p_{a2}$
5. $p_{b1} < p_{a1} < p_{b2} = p_{a2}$
6. $p_{b1} = p_{a1} < p_{b2} = p_{a2}$

And one can find the revenue functions of both firms for the three cases as follows:

1. $R_A = p_{a2}\frac{1}{2}\min(1-p_{a2}, x_{a1}) + p_{a1}\min(\frac{1-p_{a1}}{2}, \frac{1-p_{b2}}{4})$
   $R_B = p_{b2}\frac{1-p_{b2}}{2} + p_{b1}\frac{1-p_{b1}}{2}$
2. $R_A = p_{a2}\frac{1}{2}\min(1-p_{a2}, x_{b2}) + p_{a1}\min(\frac{1-p_{a1}}{2}, \frac{x_4}{2})$
   $R_B = p_{b2}\frac{1}{2}\min(1-p_{b2}, x_2) + p_{b1}\frac{1-p_{b1}}{2}$
3. $R_A = p_{a2}\frac{1}{2}\min(1-p_{a2}, x_{b2}) + p_{a1}\frac{1-p_{a1}}{2}$
   $R_B = p_{b2}\frac{1}{2}\min(1-p_{b2}, x_{b1}) + p_{b1}\frac{1}{2}\min(1-p_{b1}, x_{a1})$
4. $R_A = p_{a2}\frac{1}{2}\min(1-p_{a2}, x_{b2}) + p_{a1}(\frac{1-p_{a1}}{3})$
   $R_B = p_{b2}\frac{1}{2}\min(1-p_{b2}, x_{b1}) + p_{b1}\frac{1-p_{b1}}{3}$
5. $R_A = p_{a2}\frac{1}{3}\min(1-p_{a2}, x_{a1}) + p_{a1}\min(\frac{1-p_{a1}}{2}, \frac{x_4}{2})$
   $R_B = p_{b2}\frac{1}{3}\min(1-p_{b2}, x_2) + p_{b1}\frac{1-p_{b1}}{2}$
6. $R_A = p_{a2}\frac{1}{3}\min(1-p_{a2}, x_{a1}) + p_{a1}(\frac{1-p_{a1}}{3})$
   $R_B = p_{b2}\frac{1}{3}\min(1-p_{b2}, x_{b1}) + p_{b1}\frac{1-p_{b1}}{3}$

where x_{a2}, x_{a1}, x_{b2}, and x_{b1} are the number of users in the networks in which the prices are p_{a2}, p_{a1}, p_{b2}, and p_{b1} divided by the capacity respectively and are as follows:

$$x_{b1} = \frac{1}{2}(1-p_{b1}),\ x_{b2} = \min(\frac{1-p_{b2}}{2}, \frac{x_{b1}}{2})$$

$$x_{a1} = \min(\frac{1-p_{a1}}{2}, \frac{x_{b2}}{2}),\ x_{a2} = \min(\frac{1-p_{a2}}{2}, \frac{x_{a1}}{2})$$

For each of the three cases, the optimal revenues (R_A^*, R_B^*) and corresponding prices (p_{a2}^*, p_{a1}^*, p_{b2}^*, p_{b1}^*) are shown in table 3.

Table 3. Optimal revenues and prices for firms A and B (four prices)

Case	p_{a2}^*	p_{a1}^*	p_{b2}^*	p_{b1}^*	R_A^*	R_B^*
1)	0.93	0.86	0.7	0.4	0.0927	0.225
2)	0.93	0.75	0.88	0.5	0.1217	0.1778
3)	0.94	0.5	0.88	0.76	0.1532	0.144
4)	0.84	0.5	0.67	0.5	0.1505	0.1943
5)	0.86	0.7	0.86	0.4	0.1453	0.2007
6)	0.67	0.5	0.67	0.5	0.1578	0.1578

Similar to the three prices case, we observe that there are no Nash equilibrium price vector for this case. Thus, in the duopoly model, we conclude that there

Table 4. User subscription for firms A and B (four prices)

Case	x^*_{a2}	x^*_{a1}	x^*_{b2}	x^*_{b1}
1)	7/200	7/100	3/20	3/10
2)	3/100	1/8	3/50	1/4
3)	3/100	1/4	3/50	3/25
4)	1/12	1/3	1/6	1/3
5)	1/10	3/20	1/10	3/10
6)	2/9	1/3	2/9	1/3

are no equilibria based on the solution concept of Nash equilibrium when both firms use PMP[2]. Moreover, it is interesting to notice that even when both firms decide to provide services with single pricing there are no Nash equilibria [8].

4.3 Discussion

It is also interesting to notice that the result is model-dependent. In a related literature, Gibbens et al. [7] showed that if there are more than one firm (in the market) then they do not differentiate their networks in equilibrium. That is, in a duopoly setting, the result which states that there is no equilibrium with PMP is same as ours but the fact that there is indeed an equilibrium with single prices is different.

The nonexistence of equilibria does not mean that there is no incentive to differentiate the prices in the real-world duopoly setting[3]. For instance, assume there are two providers each dividing the market share equally with single prices. If firm A's price (p_A) is lower than firm B's price (p_B), clearly, firm A will increase its profit by introducing another price, say p_{A2}, around p_B to take away some of firm B's market share. On the other hand, firm B will introduce a lower price, say p_{B2}, around p_A to maximize its profit. Therefore, in reality, we might observe the situation where there are 2 providers each with 2 prices.

5 Conclusion

In this paper, we developed a model for Paris Metro Pricing strategy and demonstrated the profit incentive for a NSP to use PMP in a variety of scenarios. In particular, we analyzed the consumer behavior under PMP by allowing the model to define each users' condition which, when satisfied, they opt for joining the network. We evaluated the revenue of using PMP when there is a single provider (a monopolist) and determined the optimal fraction of the two subnetworks to be divided in order to maximize the profit. We have seen that using PMP, indeed,

[2] We have analyzed the cases with the concept of ϵ-equilibrium as well but the result did not change.

[3] Here we assume that, in the real world, ISPs cannot change its level of service price freely over time due to political and social reasons.

increases both the revenue and subscription. Also, we have looked at a competition setting where two NSPs provide PMP. As it turned out, we noticed that in a duopoly case there exist no (pure) Nash equilibrium even when the duopolists go for single pricing. As directions for future research, we wish to investigate more deeply the duopoly case where providers are capable to differentiate their services other than just PMP scheme.

Acknowledgments. This work was supported in part by BK21 Project of KRF, Basic Science Research Program through NRF, and ASRI at SNU, Korea. The authors would like to thank Prof. Jean Walrand and Dr. Nikhil Shetty for valuable comments and discussions.

References

1. Cisco: Cisco Visual Networking Index: Global Mobile Data Traffic Forecast Update, 2009-2014 (February 2010)
2. Shenker, S.: Fundamental design issues for the future internet. IEEE Journal on Selected Areas in Communications 13, 1176–1188 (1995)
3. Mandjes, M.: Pricing strategies under heterogeneous service requirements. Computer Networks 42, 231–249 (2003)
4. Odlyzko, A.: Paris metro pricing for the internet. In: Proceedings of the ACM Conference on Electronic Commerce, pp. 140–147 (1999)
5. Ros, D., Tuffin, B.: A mathematical model of the Paris Metro Pricing scheme for charging packet networks. Computer Networks 46, 73–85 (2004)
6. Jain, R., Mullen, T., Hausman, R.: Analysis of Paris Metro Pricing Strategy for QoS with a Single Service Provider. In: Wolf, L., Hutchinson, D.A., Steinmetz, R. (eds.) IWQoS 2001. LNCS, vol. 2092, pp. 44–58. Springer, Heidelberg (2001)
7. Gibbens, R., et al.: Internet service classes under competition. IEEE Journal on Selected Areas in Communications 18, 2490–2498 (2000)
8. Walrand, J.: Economic models of communication networks. In: Performance Modelling and Engineering. Springer, US (2008)
9. Lee, D., et al.: Analysis of Paris Metro Pricing for Wireless Internet Services. In: Proceedings of the International Conference on Information Networking (2011)

Joint Price and QoS Market Share Game with Adversarial Service Providers and Migrating Customers

Mohamed Baslam[1], Loubna Echabbi[2], Rachid El-Azouzi[3], and Essaid Sabir[1,3]

[1] LIMIARF, University of Mohammed V-Agdal, B.P. 1014 RP, Rabat, Morocco
[2] INPT, National Institute of Post and Telecommunication, 2, Av. Allal El Fassi, Madinat Al-Irfane, Rabat, Morocco
[3] LIA/CERI, University of Avignon, Agroparc BP 1228, Avignon, France

Abstract. In order to attract more and more customers, the price war between service providers (SPs) is becoming increasingly fierce. This kind of conflict situations has been analyzed and hugely studied by a plenty of works in the related literature. However unfortunately, almost all prior works neglect an important decision parameter, it is the promised quality of service which can be a serious advantage to register to an operator rather than the others. In this paper, we formulate the interaction between service providers as a non-cooperative game. First, each SP chooses the Quality of Service (QoS) to guarantee (it depends on the amount of requested bandwidth) and the corresponding price. Second, the customers decide to which SP to subscribe defining the market share for each SP. Then, each customers may migrate/churn to another SP or alternatively switch to "no subscription state" depending on the observed price/QoS. Furthermore, we build a Markovian model to derive the behaviour of customers depending on the strategic actions of the SPs. Finally, we provide extensive numerical examples to show the importance of taking price and QoS as a joint decision parameters and provide thereby some insights/heuristic on how to set them.

Keywords: Price, QoS, Behaviours of customers, Competition, Migration customers, Nash equilibrium.

1 Introduction

Compared to the earlier communication systems that are data centric, the current customers often transmit voice and video along with data. Definitively, voice and video (e.g., VoD, VoIP, streaming, mobile TV, $\cdots$) communications are more stringent in their quality requirements than raw data. These services are more sensitive to delay and require reliable transmission of the packets. Because the characteristics of traffic have changed, we need some reconsideration of the implications of architecture, service classes, and design principles on the pricing models of service providers. From customers point of view, a service provider

R. Jain and R. Kannan (Eds.): GameNets 2011, LNICST 75, pp. 642–657, 2012.

becomes more attractive as the quality it guarantee tends to 100% service quality (full delivery rate and reasonable delay). On one hand, when the network is over-provisioned, i.e., the service quality is 100%, more capacity does not lead to any performance improvement for the traffic. But on the other hand, and from service providers perspective this generate extra and useless fee which reduces their net revenue.

It is clear now that both price and promised quality encourage the migration/churn of customers from a service provider to another. Indeed, this phenomenon becomes relevant since the liberalization of telecommunications service and continuous proliferation of service providers. Churn is especially large in mobile networks, where yearly migration rates as high as 25% [1,9]. The migration of each customer, to the benefit of another service provider, implies both the loss of future revenues associated to that customer (this conflict situation can be modeled to a good extent by a zero-sum game). Service providers are therefore very keen on retaining their customers as well as on attracting new ones.

1.1 Related Works and Their Drawbacks

The interactions between the strategies of operators, seeking compromise between price and quality of service, are modeled by several works, most of these have been achieved by the use of non-cooperative game theory and assume a single characteristic through which an equilibrium is studied, such as price [1], [2], [8] and [11], QoS in terms of delay [5], or even loss probabilities [4]. However, to take in consideration the QoS on a network, and develop an real model, it is better to invest more than one parameter, an example is to incorporate jointly price and some measure of Quality of service, [3] assuming that the demand function for SPs is linear, authors establish conditions for existence and uniqueness of the equilibria for both measures of loss and delay of QoS. For more details on surveys techniques of competitive game in telecommunications, see [6,7].

This research examines the basic issue of designing a joint pricing and QoS model for service provider (SP) selection. By formulating a utility formula based on pricing schema drawn from related literature, see for example [1]. The novelty here is that we consider the quality of services (can be expressed in term of average throughput, expected delay, loss probability or any other performance metric) as an extra parameter that may potentially define the strategic decision of SPs. Nowadays, thanks to the democratization of telecommunications service, we observe that all SPs propose similar services at same price (e.g., in France, all SPs offer the triple-play service at 29,99Euros/month). Some SPs may consider this situation as unfair. Then and to attract more and more customers, they announce a better service quality compared to other competitors. This becomes a typical conflict situation that could be modeled as a non-cooperative game. This research provides new insights to understand how to fine-tune the pricing as well as the quality promised by service providers. This pricing-quality model is pretty flexible, however it bases the study on customers steady behaviour which is not easy to evaluate.

1.2 Organization of the Paper

The rest of this paper is organized as follows : in Section 2, we present the joint price-QoS problem and provide an analysis of the customers steady behaviour. Later, we study the non-cooperative game between the competitive service providers in Section 3. Finally, we provide an extensive numerical results to assist our analysis and conclude our paper in Section 4 and Section 5, respectively.

2 Problem Formulation : Customers Steady Behaviour

We consider a system with N customers and 2 service providers (SPs). Each customer seeks to find the operator which allows him to meet a QoS (this metric can be throughput, delay, loss probability, $\cdots$) sufficient to satisfy his needs, at suitable price. Let p_i and q_i be the pricing policy and the QoS guaranteed by service provider i[1], respectively. Based on this two parameters policy, each customer decides to register to one of the two operators or to stay at no subscription state, see Fig. 1.

Increasing market share is the most important objectives of each operator. Then, from operator point of view, the question is to define the best pricing strategy and the best amount of bandwidth to request from the network owner. The service providers are supposed to know the effect of their policy on the customers registration policy. Whereas from customers point of view, the question is to set the best probabilities vector to register to an operator. Conceptually, this situation is a typical Stackelberg game [13] where the operators are the leaders and the customers are the followers. This way, the providers play first, but using backward induction, they anticipate the resulting strategy of end users who actually make the last move. Sometimes customers may leave an operator (migration) in order to register under the services of another. We turn first to analyze the behavior of one customer, see figure 1. Here state "Service provider 1" means that the customer is with provider 1, state "Service provider 2" that he is with provider 2 and state "No subscription" that he does not use any service. We assume that customer behavior is represented by a continuous time Markov chain. The transition from state i to state j depends not only on parameters of current SP i that are p_i and q_i, but also on the price and QoS offered by its competitors, that is, π_i depends upon the entire price vector $\mathbf{p} = (p_1, p_2)$ and the entire QoS vector $\mathbf{q} = (q_1, q_2)$. To avoid negative profits, we consider here that both price and QoS indicator are bounded and positives, i.e., $0 < p_i^{min} \leq p_i \leq p_i^{max}$ and $0 < q_i^{min} \leq q_i \leq q_i^{max}$.

Let us denote by $\lambda_{i,j}$ the transition probability from state i to state j, the transition probabilities are depicted in figure 1. The resulting infinitesimal generator of Markov chain corresponding to the figure 1 is the following

[1] This QoS is strongly depending on the total bandwidth yield by service provider i and the number of customers registered to this latter.

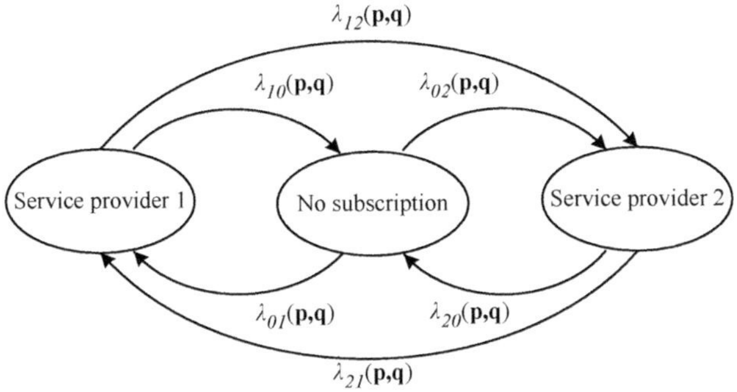

Fig. 1. Customer's behaviour as a continuous time Markov chain

$$\mathbf{Q} = \begin{pmatrix} -\left[\lambda_{01}(\mathbf{p},\mathbf{q}) + \lambda_{02}(\mathbf{p},\mathbf{q})\right] & \lambda_{01}(\mathbf{p},\mathbf{q}) & \lambda_{02}(\mathbf{p},\mathbf{q}) \\ \lambda_{10}(\mathbf{p},\mathbf{q}) & -\left[\lambda_{10}(\mathbf{p},\mathbf{q}) + \lambda_{12}(\mathbf{p},\mathbf{q})\right] & \lambda_{12}(\mathbf{p},\mathbf{q}) \\ \lambda_{20}(\mathbf{p},\mathbf{q}) & \lambda_{21}(\mathbf{p},\mathbf{q}) & -\left[\lambda_{20}(\mathbf{p},\mathbf{q}) + \lambda_{21}(\mathbf{p},\mathbf{q})\right] \end{pmatrix}$$

Let $\pi(\mathbf{p},\mathbf{q}) = \{\pi_i(\mathbf{p},\mathbf{q}),\ i \in \{0,1,2\}\}$ denotes the steady state of the Markovian system, where the probability that a given customer is with service provider i is π_i. From non-cooperative game theory perspective, π is the equilibrium mixed strategy of any given customer. Equivalently, π_i can also be seen as the market share of SP i, it follows that the average number of customers that are with SP i is $N\pi_i$. The customers problem is then a solution of system :

$$\begin{cases} \pi Q = 0, \\ \sum_{i=0}^{2} \pi_i = 1, \\ \pi_i \geq 0, \quad i = 0,1,2. \end{cases} \tag{1}$$

The solution of this system is quite easy to obtain for two SPs, it is given by :

$$\pi_0 = \left[\lambda_{10}(\mathbf{p},\mathbf{q})\lambda_{20}(\mathbf{p},\mathbf{q}) + \lambda_{10}(\mathbf{p},\mathbf{q})\lambda_{21}(\mathbf{p},\mathbf{q}) + \lambda_{12}(\mathbf{p},\mathbf{q})\lambda_{20}(\mathbf{p},\mathbf{q})\right] \div C,$$
$$\pi_1 = \left[\lambda_{01}(\mathbf{p},\mathbf{q})\lambda_{21}(\mathbf{p},\mathbf{q}) + \lambda_{21}(\mathbf{p},\mathbf{q})\lambda_{02}(\mathbf{p},\mathbf{q}) + \lambda_{20}(\mathbf{p},\mathbf{q})\lambda_{01}(\mathbf{p},\mathbf{q})\right] \div C,$$
$$\pi_2 = \left[\lambda_{02}(\mathbf{p},\mathbf{q})\lambda_{12}(\mathbf{p},\mathbf{q}) + \lambda_{10}(\mathbf{p},\mathbf{q})\lambda_{02}(\mathbf{p},\mathbf{q}) + \lambda_{01}(\mathbf{p},\mathbf{q})\lambda_{12}(\mathbf{p},\mathbf{q})\right] \div C,$$

where

$$\begin{aligned} C = \; & \lambda_{01}(\mathbf{p},\mathbf{q})\lambda_{12}(\mathbf{p},\mathbf{q}) + \lambda_{21}(\mathbf{p},\mathbf{q})\lambda_{02}(\mathbf{p},\mathbf{q}) + \lambda_{02}(\mathbf{p},\mathbf{q})\lambda_{10}(\mathbf{p},\mathbf{q}) \\ & + \lambda_{02}(\mathbf{p},\mathbf{q})\lambda_{12}(\mathbf{p},\mathbf{q}) + \lambda_{20}(\mathbf{p},\mathbf{q})\lambda_{01}(\mathbf{p},\mathbf{q}) + \lambda_{20}(\mathbf{p},\mathbf{q})\lambda_{10}(\mathbf{p},\mathbf{q}) \\ & + \lambda_{20}(\mathbf{p},\mathbf{q})\lambda_{12}(\mathbf{p},\mathbf{q}) + \lambda_{21}(\mathbf{p},\mathbf{q})\lambda_{01}(\mathbf{p},\mathbf{q}) + \lambda_{21}(\mathbf{p},\mathbf{q})\lambda_{10}(\mathbf{p},\mathbf{q}). \end{aligned}$$

Remark 1. It is important to note that the market share π_i is exactly the experienced demand by SP i.

3 Service Providers Strategic Decision

After we have defined the customers behavior, taking into consideration both the unit price and the guaranteed QoS, we turn now to define the best policy (price and QoS) for each SP. We first define the utility function and then analyze the equilibrium concept of interest. Each SP seeks to attract the maximum possible number of customers among a population of N customers. On one hand increasing the price (and/or decreasing the quality to guarantee) will increase the revenue per customer. On the other hand, this policy may potentially reduce the number of customers (equivalently the market share). Henceforth, there is a tradeoff to be analyzed. Having a market share π_i, the total revenue of SP i is then $N\pi_i p_i$. We assume that we have a single network owner, this latter charges each SP i a cost ϑ_i per unit of requested bandwidth. In order to insure the customers loyalty, the amount of bandwidth μ_i required by SP i should depend on π_i, N and on the QoS q_i it wishes to offer to its customers. Therefore, the net profit of SP i is simply the difference between the total revenue and the fee paid to the network owner :

$$U_i(\mathbf{p}, \mathbf{q}) = N\pi_i p_i - F_i(q_i, \pi_i), \quad \forall i \in \{1, 2\}. \tag{2}$$

The fee paid by SP i can be written as [3] : $F_i = \vartheta_i \mu_i (N, \pi_i, q_i)$, where $\mu_i (N, \pi_i, q_i)$ is the amount of bandwidth required by SP i to guarantee the announced QoS q_i, which has the following form :

$$\mu_i (N, \pi_i, q_i) = N\pi_i g_i (q_i) + h_i (q_i) \tag{3}$$

where $g_i (q_i)$ and $h_i (q_i)$ are positive functions, which mean that the profile function of SP i becomes :

$$U_i(\mathbf{p}, \mathbf{q}) = N\pi_i (p_i - \vartheta_i g_i (q_i)) - \vartheta_i h_i (q_i), \forall i \in \{1, 2\}. \tag{4}$$

Expected Delay as QoS Indicator: In the rest of this paper, we assume that the measure defining the QoS corresponds to some function of the expected delay. We consider the Kleinrock delay function which is a common delay function used in networking games, see [3] and [5]. This way, the maximization of QoS requires minimization of the delay. For this reason, and instead of minimizing delays, we consider the maximization of the reciprocal of its square root :

$$q_i := \frac{1}{\sqrt{Delay_i}} = \sqrt{\mu_i - N\pi_i}. \tag{5}$$

Therefore, $\mu_i (N, \pi_i, q_i) = N\pi_i g_i (q_i) + h_i (q_i)$. We will focus on the rest of this paper to the simple case where $g_i (q_i) = 1$ and$h_i (q_i) = q_i^2$. Thus, equation (4) becomes :

$$U_i(\mathbf{p}, \mathbf{q}) = N\pi_i (p_i - \vartheta_i) - \vartheta_i q_i^2, \quad i \in \{1, 2\}. \tag{6}$$

Each service provider strives to find its best strategy, i.e., its price and QoS to guarantee maximizing its revenue, which can be modified by the strategy of the competitor. The solution concept to adopt here is naturally that of a Nash equilibrium[2].

Definition 1 (Nash equilibrium). Since the customers steady state depends only on the fixed prices and the offered QoS by the two adversarial SP, then $(\boldsymbol{p}^*, \boldsymbol{q}^*) = (p_1^*, p_2^*, q_1^*, q_2^*)$ is a service providers Nash Equilibrium if it satisfies

1) $(p_i^*, q_i^*) \in \operatorname*{argmax}_{p_i, q_i} U_i\left(p_i, p_{-i}^*, q_i, q_{-i}^*\right), \quad i = 1, 2$ and
2) $(\mathbf{p}^*, \mathbf{q}^*)$ is a feasible strategy profile.

In order to solve this non-cooperative game we use a backward induction technique. We start with the customers registration and derive their steady behaviour as a function of the price set and the QoS offered by the service providers. Without extra assumptions, existence of a Nash equilibrium cannot be ensured, nor its uniqueness when existence is shown. In the case where utilities and rates functions are simple enough in terms of prices and QoS, we may find the form of the Nash equilibria analytically. Otherwise, the computations can be performed numerically using the following algorithm 1. We define the best response of each provider as a function of the strategy of its opponent by

$$BR_1(p_2, q_2) = \operatorname*{argmax}_{p_1, q_1} U_1(\mathbf{p}, \mathbf{q}), \tag{7}$$

$$BR_2(p_1, q_1) = \operatorname*{argmax}_{p_2, q_2} U_2(\mathbf{p}, \mathbf{q}). \tag{8}$$

Algorithm 1. Numerically finding the Nash equilibria of the joint Price-QoS game

1: Input : Transition rates of the Markov chain as function of $\mathbf{p}$ and $\mathbf{q}$,
2: For all feasible couple of values (p_2, q_2) of SP 2, find the set $BR_1(p_2, q_2)$,
3: For all feasible couple of values (p_1, q_1) of SP 1, find the set $BR_2(p_1, q_1)$,
4: Each 4-tuple $(p_1^*, p_2^*, q_1^*, q_2^*)$ such that $(p_1^*, q_1^*) \in BR_1(p_2^*, q_2^*)$ and $(p_2^*, q_2^*) \in BR_2(p_1^*, q_1^*)$, is a Nash equilibrium. Note that this game needs not have a Nash equilibrium.

3.1 Discussion : Special But Realistic Cases

Price Game with Fixed QoS: When the market price is high or similar for all service providers, it changes the customers perceptions about the quality of the product, see [12] . A different pricing strategy provides an "imaginary effect" on perceptions of quality and leads to a willingness to buy. However with presence

[2] A Nash equilibrium is a strategy profile such that no service provider can unilaterally increase its revenue.

of adversarial competitors, in particular when providing same quality of service, this turns to be false. This motivates us to study the pricing game with fixed QoS.

QoS Game with Fixed Price: In earlier days, quality of service meant delivering packets from a source to a destination without any transmission errors. As the Internet has become commercialized, however, QoS has started to become an important strategic tool for market competition [10]. Most network pricing researchers assume that when the price is high the arrival rate or demand for that service class is low. However, this assumption may not be always true. Some customers value high quality services more than lower-class services. Therefore, some people buy more expensive services because they believe that they can get better quality of service by paying more. Nowadays, all service providers propose services at same cost. To convince customers to join an SP i, this latter exposes its ability to provide a good QoS and invents many new services to attract more and more customers. This way, considering a game ruled only by the QoS is reasonable and realistic enough. This scheme is also used to analyze the interplay of quality of service on customers loyalty and service providers revenue.

4 Numerical Investigation

4.1 Migration Rate Model

The dependence of migration/transition rates on provider prices and qualities are too complicated. To solve the problem analytically is then a hard issue. Even with simple expressions of the transition rates, the problem remains too complicated to solve due to expression of steady market share expression. Obviously a service provider attracts more customers as it decreases the price and/or increase the promised QoS. Thus any function increasing with respect to p_i and decreasing with respect to q_i can be used to model the migration rate into SP i. Here, we consider a the following specific but realistic enough transition rate functions :

$$\begin{aligned}
\lambda_{01}(\mathbf{p},\mathbf{q}) &= \lambda_{02}(\mathbf{p},\mathbf{q}) = \alpha, \\
\lambda_{10}(\mathbf{p},\mathbf{q}) &= \alpha_1 \tfrac{p_1}{q_1}, \\
\lambda_{20}(\mathbf{p},\mathbf{q}) &= \alpha_2 \tfrac{p_2}{q_2}, \\
\lambda_{12}(\mathbf{p},\mathbf{q}) &= \beta_1 \tfrac{p_1 q_2}{p_2 q_1}, \\
\lambda_{21}(\mathbf{p},\mathbf{q}) &= \beta_2 \tfrac{p_2 q_1}{p_1 q_2}.
\end{aligned}$$

$\alpha_1, \alpha_2, \beta_1, \beta_2$ are strictly positive constants. The expressions of λ_{10} and λ_{20} shows that probability that customer leaves the SP $i \in \{1,2\}$ to state 0 (no subscription) if their prices become excessive, or alternatively their quality of service tends to become low. Incorporating these latter parameters helps us to introduce some asymmetry between service providers and then can be used to model the reputation of each service provider. Substituting the transition rates by their expressions, we can compute the steady state vector $\pi = (\pi_i)_{i\in\{0,1,2\}}$. The net revenues of each operator is then obtained using equation (6). Until contraindication, the parameter values considered for numerical examples are the following: $N = 1000$ (total number of customers), $p_i^{min} = 2, p_i^{max} = 100, q_i^{min} = 0, q_i^{max} = 120, \vartheta_1 = \vartheta_2 = 1$, $\lambda_{01} = \lambda_{02} = 3$, $\alpha = 3$ and $\beta_1 = \beta_2 = 20$. For sake of illustration, we perform an extensive numerical study with a focus on the two special but realistic schemes : Price game with fixed QoS and QoS game with fixed price. For the two-parameters game, and based on numerical analysis we noticed that an equilibrium point needs not exist. Moreover, the game may potentially has several equilibria.

Remark 2. We recall that the quality of service can be expressed as average throughput, expected delay or any other performance metric. Here, increasing the quality indicator q_i is equivalent to promise shorter delay.

4.2 Price Game with Fixed QoS

We consider here the classic scheme where the QoS is fixed and the price is the only action of the game. Later, we study the Nash equilibrium of this game and show how the net revenue and the market share evolves. Figure 2 plots the utility U_1 of provider 1 when its price p_1 varies, for different values of the adversary price p_2. We remark that the net revenue of provider 1 is first increasing, then decreasing in p_1. Also that the utility of provider 1 decreases when its price p_1 tends to infinity, which implies existence of a finite price p_1^o maximizing U_1. This price value constitutes the best response of provider 1 against the price set by provider 2. We remark also that the net revenue of SP_1 increase when adversary price p_2 increase, which is likely intuitive. Indeed, if the opponent chooses a higher pricing strategy, then it may loose a fraction of its customers. Moreover, service provider SP 1 may attract those latter and make incite them to churn. Consequently, its net revenue is increased. Surprisingly, we observed an interesting feature characterizing our proposed price game : Through hundreds of numerical example runs, we 1) checked existence of a unique best response, and 2) checked that as the price defined by SP 2 increases, the best response of SP 1 tends to increase also ! This motivated us to work about proving supermodularity of this game.

Figure 3 and 4 present respectively the influence of QoS q_1 and q_2 of both providers on the net revenue U_1 of SP_1. In terms of service provider's own quality of service, the net revenue behaves in two ways. When SP 1 sets a low price, it tends to be more beneficial to request relatively low amount of bandwidth providing low quality of service. This is a quite surprising behaviour, but can have a realistic good extent. Indeed, when an SP offers some given service at a low cost, its market share goes to 1. Absorbing the total number of customers, the tagged SP tends to no investing anymore and the offered service quality becomes poor. Another important result is that as the offered QOS increases as the best response increase respecting by the way "If you want more, you have to pay more". We depict in figure 4 the impact of QoS offered by the opponent. It shows that when SP 2 investment increases, i.e., its offered QoS, the best move for SP 1 is to decrease the cost of its services.

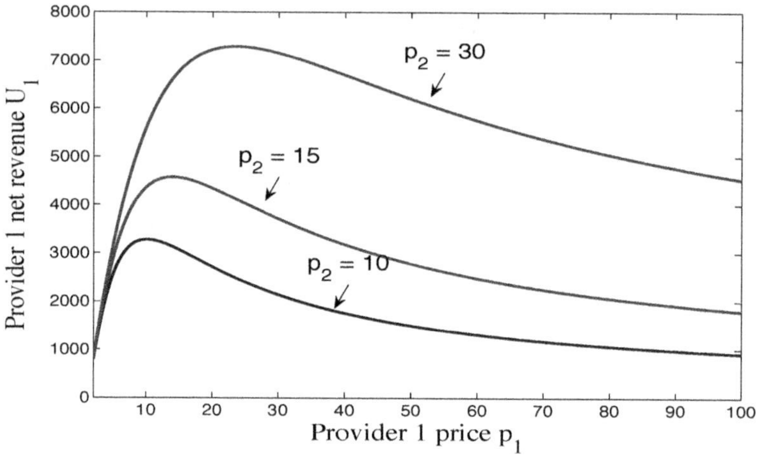

Fig. 2. Net revenue of provider 1 versus p_1 for different values of p_2, with $q_1 = 10$ and $q_2 = 10$

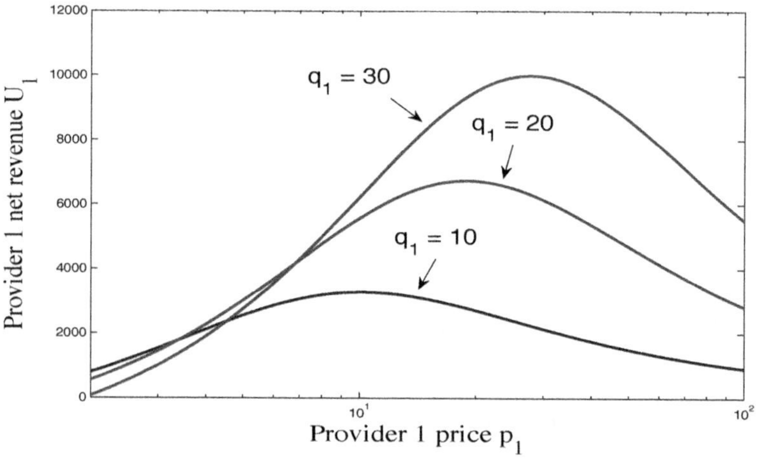

Fig. 3. Net revenue of provider 1 versus p_1 for different values of q_1, with $q_2 = 10$ and $p_2 = 10$

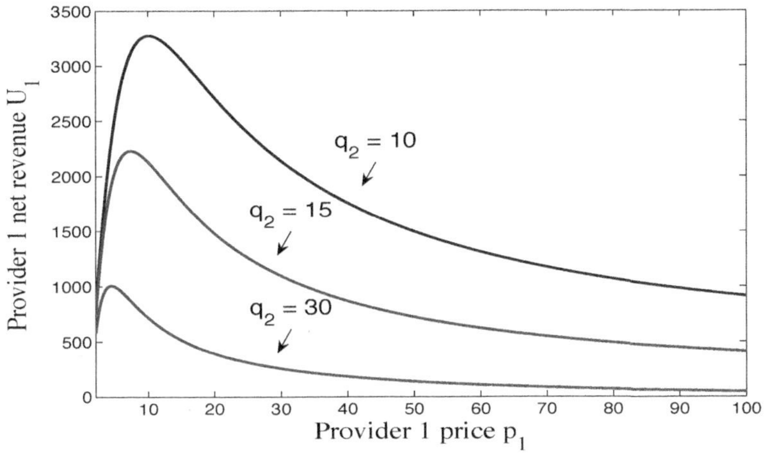

Fig. 4. Net revenue of provider 1 versus p_1 for different values of q_2, with $q_1 = 10$ and $p_2 = 10$

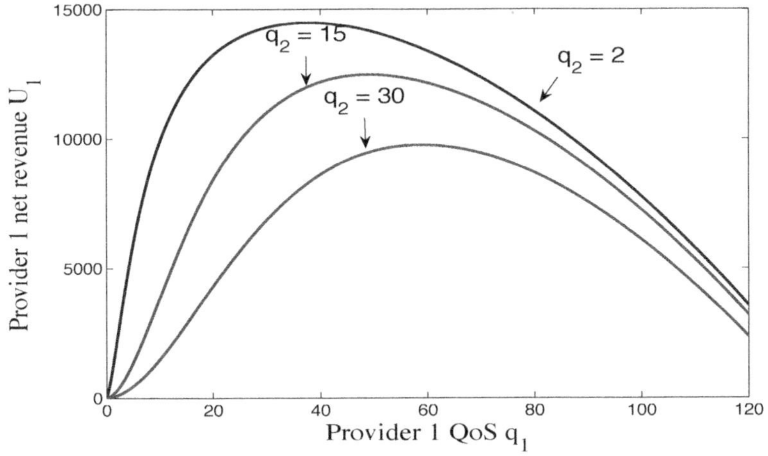

Fig. 5. Net revenue of provider 1 versus q_1 for several values of q_2, with $p_1 = 20$ and $p_2 = 20$

Next we depict the best response of the two adversarial service providers. We recall that this figure is obtained using algorithm 1 and the intersection between the two graphs represents the Nash equilibrium point of the game. Through several examples, we always obtained a unique Nash equilibrium. Another visible feature is that the fixed price should evolve in the same direction of the investment. In other words, the end price increases as the offered QoS increases.

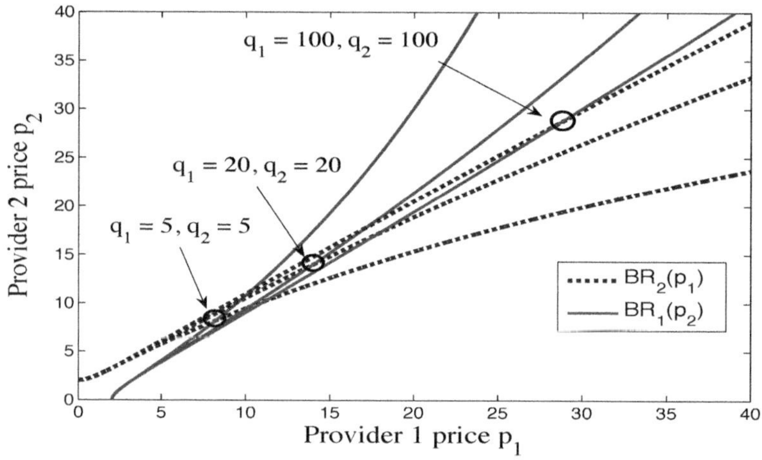

Fig. 6. Best response for both providers for different values of $q_1 = q_2$

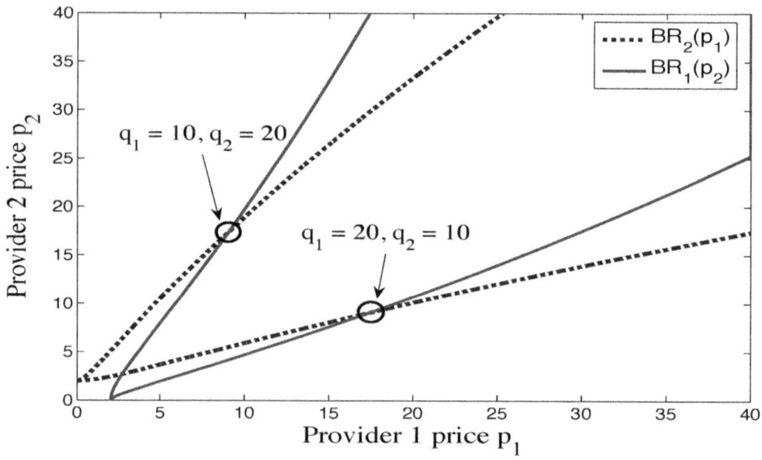

Fig. 7. Best responses for both providers for different values of $q_1 \neq q_2$

4.3 QoS Game with Fixed Price

Next we consider the QoS game while fixing the price for both SPs. We first notice that the net revenue is decreasing with the QoS. This is quite intuitive since it means for service provider to invest more and more while charging customers a constant price, see figure 5. It is clear from the same figure that, as the opponent SP 2 increases its quality of service, it is beneficial for SP 1 to increase its QoS. This latter result is well known in economics. Figures 8 and 9 shows the need to define a higher price as the own QoS improves and the need investment

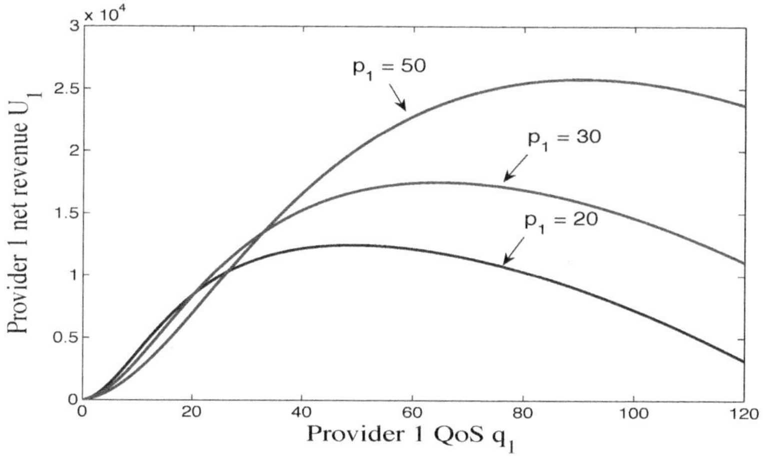

Fig. 8. Net revenue of provider 1 versus q_1 for several values of p_1, with $q_2 = 15$ and $p_2 = 20$

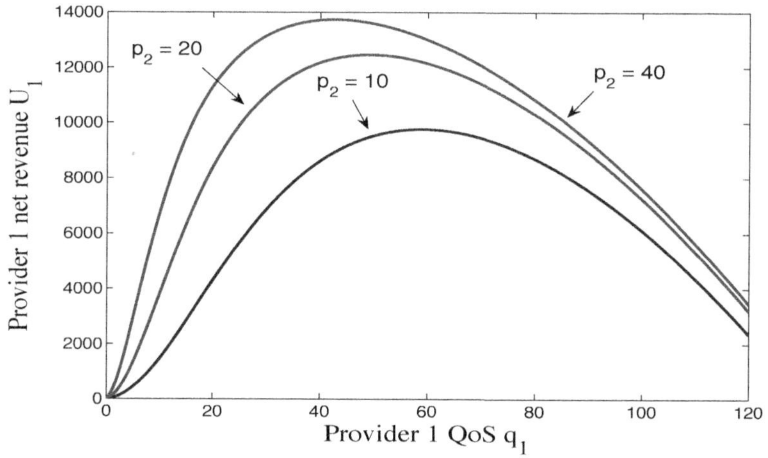

Fig. 9. Net revenue of provider 1 versus q_1 for several values of p_2, with $q_2 = 15$ and $p_1 = 20$

(improve promised QoS) as the competitor reduces its price. Later, we plot the best response in terms of QoS while the two service providers set the same price (figure 10) and where they set adopt different pricing strategies (figure 11).

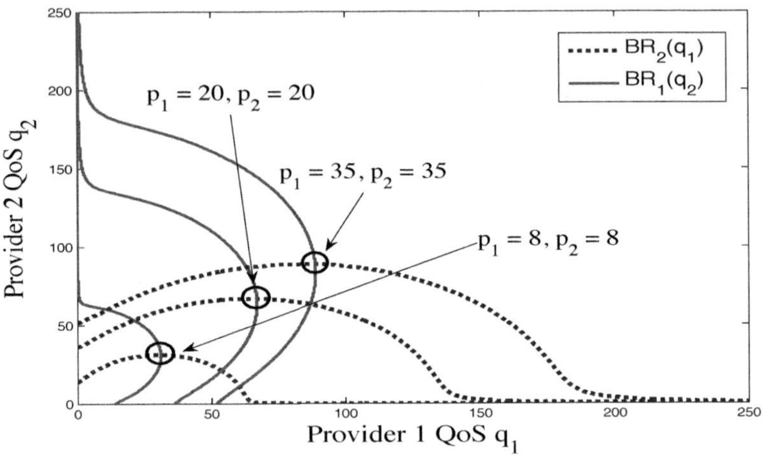

Fig. 10. Best responses QoSs of both provider for several values

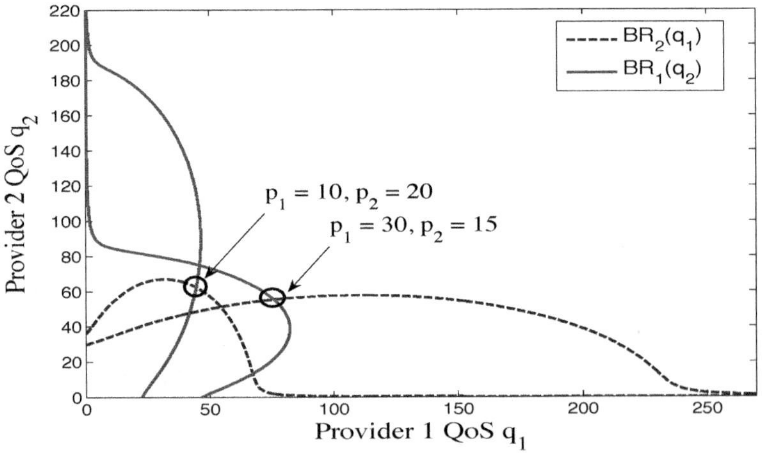

Fig. 11. Best responses Prices of both provider for several values

4.4 Convergence to Nash Equilibrium

If service providers allow free cost service with possibly null quality, i.e., price and quality may be 0, we notice that $(p_1, p_2) = (0, 0)$ is also a satisfying situation for both SPs, thus it represents also an equilibrium point. Since it brings a negative

Algorithm 2. Convergence to Nash equilibrium (when exists)

1: $z_i^{t+1} = BR_i(z_{-i}^t), \qquad i = 1, 2.$
where $BR_i(\cdot)$ is the best response function of SP i, and z_i is a generic variable representing p_i or q_i (while maintaining the other parameter constant).

revenue to the providers, and moreover it is not a stable Nash equilibrium : if any of the two providers slightly deviates from that situation by setting a strictly positive price, then an iterative best response-based algorithm leads to the other (stable) Nash equilibrium. By imposing non-null price and non-null

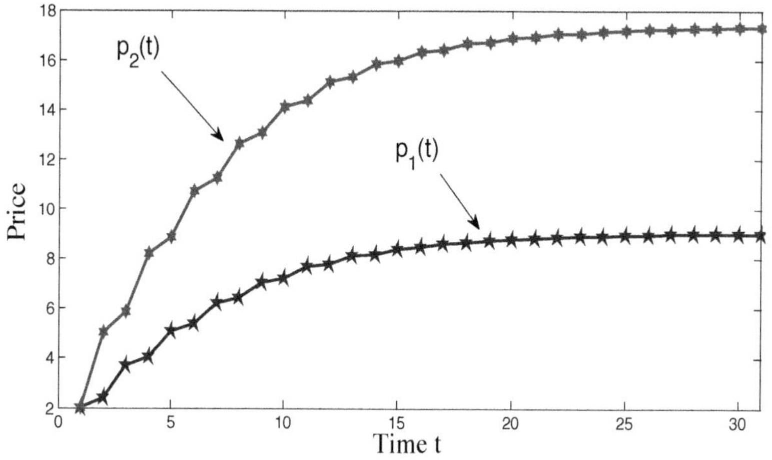

Fig. 12. Pricing game : Convergence to NE under $(q_1, q_2) = (10, 20)$

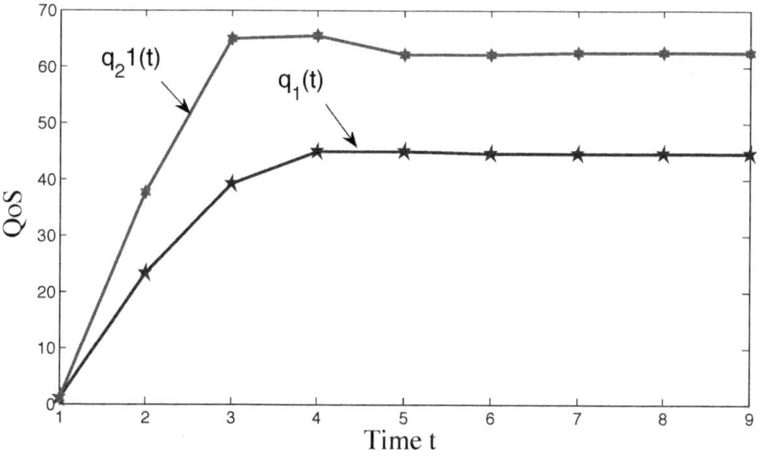

Fig. 13. QoS game : Convergence to NE under $(p_1, p_2) = (10, 20)$

quality, the Nash equilibrium would be a priori unique and the single-parameter game would be supermodular. We will consequently focus on that equilibrium in the following, when it exists. Algorithm 2 insures convergence of both SPs to the Nash equilibria points highlighted in previous figures, see figure 12 and figure 13. It is clear that the speed of convergence is relatively high (around 25 iterations to converge to price-baed equilibrium and around 8 iterations to converge to QoS-baed equilibrium).

5 Concluding Remarks

We have presented in this paper a non-cooperative game for pricing problem considering quality of service as an extra decision parameter. The service providers take into account the user migration/churn behavior to determine the best price and best quality of service to guarantee, so as to maximize their long term revenue. Through a numerical analysis, we remarked potential existence of many Nash equilibria or none for the joint price and QoS game. We remarked that the game my have a unique Nash equilibrium if minimum price p_{min} and minimum q_{min} are not zero. Furthermore, A special feature is obtained as the single game parameter seems to be supermodular. This way we checked that a simple successive best responses-based algorithm allows to learn the Nash equilibrium.

References

1. Maillé, P., Naldi, M., Tuffin, B.: Price war with migrating customers. In: 17th Annual Meeting of the IEEE/ACM International Symposium on Modelling, Analysis and Simulation of Computer and Telecommunication Systems (MASCOTS 2009). IEEE Computer Society, London (September 2009)
2. Maille, P., Tuffin, B.: Price War in Heterogeneous Wireless Networks. Computer Networks 54(13), 2281–2292 (2010)
3. El-Azouzi, R., Altman, E., Wynter, L.: Telecommunications Network Equilibrium with Price and Quality-of-Service Characteristics. In: Proc. of ITC, Berlin (September 2003)
4. Altman, E., El-Azouzi, R., Abramov, V.: Non-Cooperative Routing in Loss Networks. Performance Evaluation 49(1-4), 43–55 (2002)
5. Orda, A., Rom, R., Shimkin, N.: Competitive routing in multi-user environnements. IEEE/ACM Trans. on Netw., 510–521 (1993)
6. Altman, E., Boulogne, T., El-Azouzi, R., Jimenez, T., Wynter, L.: A survey on networking games in telecommunications. Computers and Operations Research 33(2), 286–311 (2006)
7. Altman, E., Wynter, L.: Equilibrium, games and pricing in transportation and telecommunications networks. Networks and Spacial Economics, special issue of on: Crossovers between Transportation Planning and Telecommunications 4(1), 7–21 (2004)
8. Danaher, P.: Optimal pricing of new subscription services: Analysis of a market experiment. Marketing Science 21(2), 119–138 (2002)
9. Wieland, K.: The customer retention challenge. Telecommunications 40(10), 14–17 (2006)

10. Ferguson, P., Huston, G.: Quality of Service in the Internet: Fact, Fiction, or Compromise? In: INET, Geneva, Switzerland (1998)
11. Rhee, S.H., Konstantopoulos, T.: Optimal flow control and capacity allocation in multi-service networks. In: 37th IEEE Conf. Decision and Control, Tampa, Florida, December 1998, McGraw-Hill, NY (2003)
12. Monroe, K.B.: Pricing: Making Profitable Decisions. McGraw-Hill, New York (2003)
13. Altman, E., Boulogne, T., El-Azouzi, R., Jimenez, T., Wynter, A.: A survey on networking games. In: Computers and Operations Research (2004)

Capacity Allocation Games for Network-Coded Multicast Streaming*

Elliot Anshelevich[1], Bugra Caskurlu[2], Koushik Kar[3], and Hang Zhang[3]

[1] Department of Computer Science,
Rensselaer Polytechnic Institute, Troy NY 12180, USA
eanshel@cs.rpi.edu

[2] Lane Department of Computer Science & Electrical Engineering,
West Virginia University, Morgantown WV 26506, USA
caskurlu@gmail.com

[3] Department of Electrical, Computer & Systems Engineering,
Rensselaer Polytechnic Institute, Troy NY 12180, USA
{kark,zhangh10}@rpi.edu

Abstract. In this paper we formulate and study a capacity allocation game between a set of receivers (players) that are interested in receiving multicast data (video/multimedia) being streamed from a server through a multihop network. We consider *fractional* multicast streaming, where the multicast stream from the source (origin-server) to any particular receiver (end-user) can be split over multiple paths. The receivers are selfish and non-cooperative, but must collaboratively purchase capacities of links in the network, as necessary for delivery of the multicast stream from the source to the individual receivers, assuming that the multicast stream is *network coded*. For this multicast capacity allocation (network formation) game, we show that the Nash equilibrium is guaranteed to exist in general. For a 2-tier network model where the receivers must obtain the multicast data from the source through a set of relay nodes, we show that the price-of-stability is at most 2, and provide a polynomial-time algorithm that computes a Nash equilibrium whose social cost is within a factor of 2 of the socially optimum solution. For more general network models, we give a polynomial time algorithm that computes a 2-approximate Nash equilibrium whose cost is at most 2 times the social optimum. Simulation studies show that our algorithms generate efficient Nash equilibrium allocation solutions for a vast majority of randomly generated network topologies.

Keywords: capacity allocation games, the price of stability, Nash equilibrium, multicast networks.

1 Introduction

The last decade has witnessed an explosive growth in the number of *streaming* video (multimedia) applications. Some of these involve live video streaming,

* This work is supported in part by NSF grants CCF-0914782, CNS-1017932, and CNS-1018398.

R. Jain and R. Kannan (Eds.): GameNets 2011, LNICST 75, pp. 658–673, 2012.

while others stream video that is already available in stored format but too large for download-and-play. Stored video streaming applications like YouTube [4] are already contributing to a large fraction of the Internet traffic today[1], and IPTV and similar other efforts are likely to boost live video streaming through the Internet in the coming years [3]. Streaming stored video may be unicast or multicast, depending on the application: while IPTV [2] video streaming may mostly be multicast (broadcast), receiver-driven video streaming (like video streaming from YouTube [4]) will typically be unicast. Streaming live video will typically be multicast to possibly many receivers.

For multicast data delivery, use of *network coding* allows individual receivers to simultaneously attain data rates that equal their maxflow capacities [6], which in general may not be achievable through a routing-only approach. Naturally, this makes network coding ideally suited for multicast data delivery over a multi-hop network.

In this paper, we consider a capacity allocation game that end-users will play in buying resources for multicast streaming data delivery. More specifically, receivers (users) buy capacities on the links of the distribution network at fixed (possibly different for different links) per-unit cost, so as to ensure delivery of a multicast stream (with a given source rate) from its source to the receiver. The receivers are selfish and non-cooperative and are only interested in minimizing their individual costs, but must collaboratively pay for capacities bought on network links, as necessary for network coded multicast data delivery from the data source to the individual receivers. We consider *fractional* multicast streaming, i.e., the multicast data between the source and any particular receiver can be split across multiple paths that exist between the source-receiver pair. The problem we consider is a *network formation* game where the amount of capacity collaboratively bought on the different links in the network must be such that the maxflow from the source to each receiver is no less than the desired multicast data rate. Using network coding [6], this ensures that all receivers are able to obtain their full data rate from the source. This paper focuses on the questions of the existence, efficiency, and computation of the equilibria of this game. Initially, we focus on the 2-tier network model where the receivers must obtain the multicast data from the source through a set of relay nodes, and derive some strong results by exploiting structural properties of such topologies. Later we consider arbitrary topology networks for multicast data distribution, and study the existence and efficiency of approximate equilibria for that case.

To measure efficiency of equilibrium, we use the common measures of the *price of anarchy* and the *price of stability* [22] — the supremum of the ratios between the costs of the worst and best pure Nash equilibrium, respectively, and that of the globally optimal solution over all instances of the game.

Our Contributions. The specific technical contributions of this paper are as follows. For our fractional multicast network formation game, we show that pure

[1] It was estimated that in 2007 YouTube consumed as much bandwidth as the entire Internet in 2000 [1].

Nash equilibrium is guaranteed to exist in general. For the 2-tier model, defined in Section 4, we show a tight bound of 2 on the price-of-stability, and provide a polynomial-time algorithm that returns a Nash equilibrium whose social cost is within a factor of 2 of the socially optimum solution. The 2-tier model is essentially equivalent to the case where all nodes in the network are receivers, which is itself an important special case (e.g. [17]). For more general network topologies, we give a polynomial time algorithm that computes a 2-approximate Nash equilibrium whose cost is at most 2 times the social optimum. Simulation studies show that our algorithms generate efficient Nash equilibrium allocation solutions for a vast majority of randomly generated networks. Due to lack of space, the proofs of Theorem 1, and Theorem 5 appear in the full version of this paper. (see *http://www.ecse.rpi.edu/~koushik/*)

While network formation games have been studied in other contexts (see Section 2), the questions we consider are new for the context of network-coded fractional multicast streaming. Unlike integral multicast, data distribution networks for socially optimal or Nash equilibrium solutions for network-coded multicast need not be trees (Figure 1), and techniques for integral multicast do not extend to this context (see Section 2). Interestingly, however, we show that there exist solutions based on tree topologies that are at Nash equilibrium (exactly or approximately) and attain a near-optimal social cost. For the 2-tier network model, the solution is based on the minimum spanning tree; for more general network models, it is based on the minimum Steiner tree or polynomial-time approximations of it. Despite the complexity of the problem, our results show that there exist easily-computable exact or approximate distribution networks where receivers have no motivation to deviate from it unilaterally, and yet results in the set of receivers paying near-minimal cost as a group for multicast data delivery.

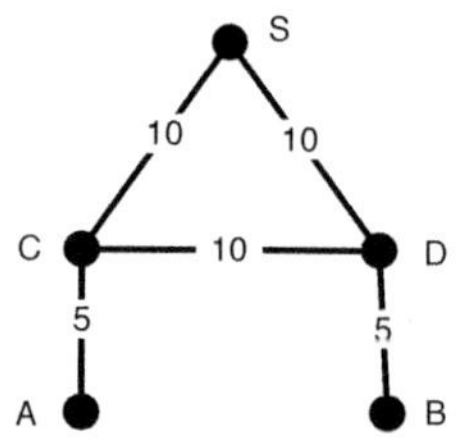

Fig. 1. Example: Fractional multicast streaming is better than integral multicast streaming. Source S sends multicast data at rate 1 to receivers A and B. All link capacities are 1 unit in each direction; numbers across links are per-unit capacity (bidirectional) purchase costs. Optimal fractional solution involves purchase of 1 unit capacities on links CA and DB, and 0.5 unit capacities on links SC, CD, SD; total cost = 25. Optimal integral solution involves purchase of 1 unit capacities along SC, SD, CA, DB; total cost = 30.

The paper is structured as follows. In Section 2, we outline related work on this topic. In Section 3 we describe the model and problem formulation. In Section 4 we state and prove our main results on the existence, efficiency, and properties of Nash equilibrium distribution topology solutions for the 2-tier network model. We extend these results to arbitrary topology networks in Section 5. In Section 6 we describe the results of experiments conducted on randomly generated network topologies.

2 Related Work

In contrast to the work presented in this paper, the models in most network formation game literature do not allow players to reserve an arbitrary amount of bandwidth on the links; rather, a link can either be constructed and be utilized to full extent or won't be constructed at all. Our game represents a more realistic scenario by allowing players to buy certain amount of capacity (bandwidth) on that link. This becomes particularly relevant for network coded multicast streaming [6,18,21,23] that we consider in this paper, where the capacity that needs to be bought on a link for successful multicast streaming is often less than the link capacity, as well as the source data rate.

There have been several variants of "integral" network formation games where instead of allocating capacity, an edge can be either fully present or non-existent. One of the most important decisions when modeling network design involving strategic agents is to determine how the total cost of the solution is going to be split among the players. Among various alternatives [13], the most popular one in the literature is the "fair sharing" mechanism [8,11,12,17]. In this cost sharing mechanism, the cost of each edge of the constructed network is shared equally by the players using that edge. Since in our model, each player is allowed to purchase any amount of capacity on an edge, the "arbitrary sharing" model of network formation [9,7,15,19,20] is closer to being an "integral" version of our game. In this model, players contribute to the cost of an edge, and an edge is present in the network if the player contributions are larger than its cost. This model has many differences from the "fair sharing" model: e.g., the game is not a congestion game, but the price of stability is much better than with fair sharing, etc.

While many interesting results have been proven for network formation games with arbitrary sharing, most do not extend to our "fractional" context where players (multicast data users/receivers in our case) are allowed to reserve an arbitrary amount of bandwidth on the links. For example, [9] proved that the minimum-cost Steiner tree is always a Nash equilibrium in this integral network formation game with arbitrary sharing, but the same does not hold for the fractional version. Consider, for example, a graph with n receiver nodes, node s, and an extra node v. All receivers wish to receive a rate of 1 from node s. For any receiver node u, the cost of allocating x capacity on an edge (u, v) is x, and the same is true for edge (v, s). The cost of allocating x capacity on an edge (u, s) is $x(2 - \epsilon)$. No other edges exist in the graph. Then, the minimum-cost Steiner tree has cost $n + 1$ (for this example, it is also the optimum fractional solution, although Figure 1 shows that this is not always the case). However, the min-cost Steiner tree in this example is not a Nash equilibrium: there must be some receiver u who is paying for 1 capacity of edge (u, v), and at least for $1/n$ capacity of edge (v, s). This player could reduce its payments to both edges by $1/n$, and instead pay for $1/n$ capacity on edge (u, s), costing it $(2 - \epsilon)/n$, and thus strictly decreasing its cost.

Our game assumes that there is no central authority that can dictate the network cost-sharing mechanism, and thus players simply purchase edge capacities directly. Other directions in multicast games include cooperative games and mechanism design (see, e.g. [16] and [22, Chapters 14-15] and references therein), where the goal is to come up with a cost-sharing mechanism with good properties that could be implemented by a central authority.

3 System Model and Formulation

We consider a network modeled by an undirected graph $G = (V, E, b)$, where vertices V denote the set of nodes, and (undirected) edges E denote the set of (bidirectional) links in the network. For each edge $e \in E$, $b(e) \geq 1$ denotes the capacity (bandwidth)[2] of the corresponding link in each direction. In other words, at full capacity $b(e)$ units of traffic can travel on link e in one direction, and simultaneously $b(e)$ units of traffic can travel on e in the opposite direction. This network is to be used for delivery of a given traffic stream of rate 1. One of the nodes, $s \in V$ is designated as the source of the traffic stream, and a subset of the nodes, $R \subseteq V$ are receivers (users) of the traffic stream. In our model, all receivers must collaboratively pay for capacities of the links that are used for carrying the traffic stream, for shared use by all receivers in R. We assume that an edge e is associated with a cost of $c(e)$ that buys the corresponding link, i.e., $c(e)$ is the cost of buying capacity of 1 of the link in each direction. If each receiver $i \in R$ pays $p_i(e)$ for edge e, then the total *purchased capacity* on the corresponding link is $\theta(e) = (\sum_{i \in R} p_i(e))/c(e)$. In other words, if the players R pay a fraction of edge e's cost in total, then the same fraction of e's desired capacity becomes available in both directions.

To formally define the game, we note that R is the set of players, and $\mathbf{p} = (p_i(e), i \in R, e \in E)$, or the prices paid by the receivers for the links, constitute the player strategies. (Formally, a strategy of player $i \in R$ is a function $p_i : E \to R_{\geq 0}$ that determines how much i is offering to pay for each edge.) Once the strategies are given, the network used for multicast traffic streaming, or the *distribution network*, is (V, E, θ), where the purchased capacities $\theta(\cdot)$ depend on the strategy vector $\mathbf{p}$. We assume that each node in the network is capable of network-coding. A necessary and sufficient condition for delivery of the given traffic stream to all receivers is that the maxflow from the source s to each receiver $i \in R$ in distribution network (V, E, θ) is no less than the stream rate of 1 [6]. In other words, a feasible strategy vector $\mathbf{p}$ is one that ensures that

$$f_i(\mathbf{p}) \geq 1, \tag{1}$$

holds for each receiver $i \in R$, where $f_i(\mathbf{p})$ denotes the maxflow from source node s to receiver node i in distribution network (V, E, θ). The goal of each receiver i is to receive the full rate of 1, but pay as little as possible. Formally, the cost of player i with strategy vector $\mathbf{p}$ is $\sum_e p_i(e)$ if $f_i(\mathbf{p}) \geq 1$, and is very large otherwise.

[2] We assume that the links are loss-free, and the terms 'capacity' and 'bandwidth' are used synonymously.

Among all strategies that form feasible networks, we are specifically interested in those that are Nash equilibria. A solution is a Nash equilibrium if each receiver does not have an incentive to unilaterally deviate from it. To state this formally, consider a strategy vector $\mathbf{p}^*$ that is feasible for all players, and let $\mathbf{p}^*{}_{-i} = (p_j^*(e), j \in R \setminus i, e \in E)$, denote the strategies of (purchase prices paid by) all receivers other than i. Then the strategy vector $\mathbf{p}^*$ is said to be a Nash equilibrium if for any $\mathbf{p}_i = (p_i(e), e \in E)$ such that $(\mathbf{p}_i, \mathbf{p}^*{}_{-i})$ satisfies the feasibilty condition (1) for receiver i, we have that $\sum_{e \in E} p_i^*(e) \leq \sum_{e \in E} p_i(e)$. In other words, given the strategies of (prices paid by) the other receivers, and subject to maintaining the feasibility condition $f_i(\mathbf{p}) \geq 1$ that is necessary for receiver i to receive the streamed data at the full rate, the total price paid by receiver i is minimized at Nash equilibrium. A solution that is not feasible for all players will never be a Nash equilibrium, since the infeasible players will have very large cost, and will have incentive to purchase more capacity in the network in order to become feasible.

To study the efficiency of the Nash equilibrium, we next define the *social optimum* against which the Nash equilibrium solution will be compared in terms of the total capacity purchase cost. A price vector $\mathbf{p}$ is said to be a social optimum if it minimizes $\sum_{i \in R, e \in E} p_i(e) = \sum_e \theta(e)c(e)$, subject to satisfying the feasibility constraints (1) for all receivers $i \in R$. We will denote this solution as OPT. The supremum of the ratio of the overall cost of the *worst* Nash equilibrium to that of the social optimum over all instances of a game is defined as the *price of anarchy* of the game. Similarly, the supremum of the ratio of the overall cost of the *best* Nash equilibrium to that of the social optimum over all instances of a game is defined as the *price of stability* of the game. In this paper, we use these two notions to characterize the efficiency of the worst and best Nash equlibria for the game.

4 Capacity Allocation Games on the 2-Tier Network Model

It has been envisioned that in the near future, streaming of video over the Internet will be done through the use of multiple, dedicated servers that would relay video from the source server to the end-users [5]. This network of video relay servers is expected to play a role similar to that of content distribution networks (CDN) for delivery of various kinds of non-streaming and non-real time data (content). This relay network will deliver video from the source to the receiver through multiple relay-hops, possibly performing network coding at the intermediate (relay) nodes/servers. This is illustrated in Figures 2-3.

In this section, we restrict our attention to the 2-tier network model. In this model, $V = \{s\} \cup L \cup R$, i.e., the vertex set of the graph is composed of the source vertex s, the set of receivers $R = \{r_1, r_2, \ldots, r_n\}$, and the set of relay nodes $L = \{l_1, l_2, \ldots, l_k\}$. Each receiver node r_i of G has *exactly* one incident edge and is adjacent to a relay node of G, i.e., for each $r_i \in R$ there exists $l_j \in L$ such that $(r_i, l_j) \in E$, and r_i has no other incident edges in G. Each relay node

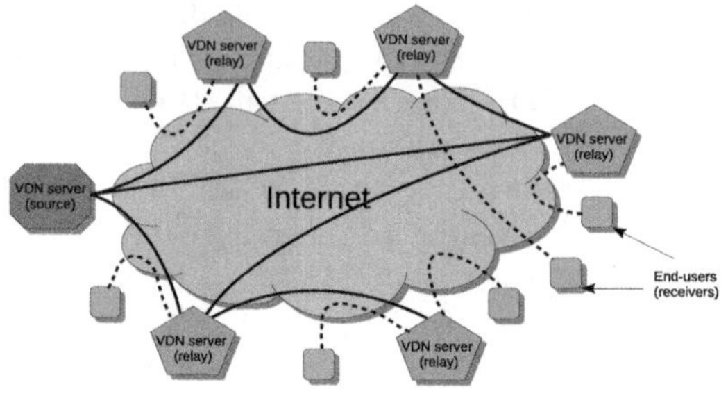

Fig. 2. Streaming video delivery in the 2-tier network model

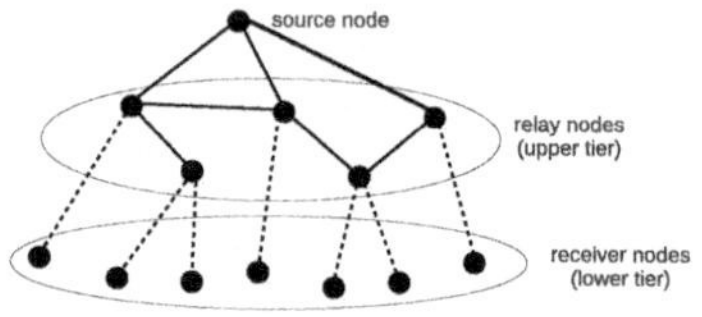

Fig. 3. Graph representation of the 2-tier network model in Figure 2

is adjacent to one or more receiver nodes, i.e., for each $l_j \in L$ there exists $r_i \in R$ such that $(r_i, l_j) \in E$. These assumptions imply that each receiver is directly connected to exactly one relay node, from which it must receive the data being streamed (this relay node can obtain the data from the source through other relay nodes, however). In addition, each relay node serves at least one receiver. This assumption can be interpreted in the following way: if a relay node has no receiver to serve, then it does not participate in the multicast data distribution.

Note that the 2-Tier model is equivalent to the model where all nodes are receivers, in the following sense. If G is an instance of our game in the 2-Tier model, let G' be a game obtained by contracting all the edges incident to a receiver node in G. Since every relay node in G must be adjacent to a receiver, then all nodes in G' except s are receivers, although there may be several receiver players located at the same node. Since every Nash equilibrium in G must have each receiver r_i purchasing 1 capacity on the edge (r_i, l_j), then it is easy to see that there is a one-to-one correspondence between the Nash equilibria in G and in G'. Below we argue about the price of stability for 2-Tier networks, but all the arguments can easily be extended to the model where the network has arbitrary topology, but every node except s is a receiver/player.

We prove below that if G has a 2-tier topology as described above, then there exists a Nash equilibrium solution that does not cost much more than the cost of the socially optimal solution. We first show, however, that there are examples where all Nash equilibria cost a factor of 2 more than the social optimum.

Theorem 1. *The price of stability of the capacity allocation game with the 2-Tier topology is at least 2.*

In the proof of Theorem 1, we show that for any $\epsilon > 0$ we can construct an instance of the capacity allocation game with the 2-Tier topology such that the social cost of all Nash equilibria are at least $(2 - \epsilon)$ times the social cost of OPT.

Consider the cheapest solution satisfying Condition (1) for our game, which we denoted by OPT. Notice that OPT is exactly the optimum fractional solution to the LP-relaxation of the classic Steiner tree problem (see, e.g. [24]), with the terminal nodes being $R \cup \{s\}$ and edge costs being $c(e)$. Since the integrality gap of this LP is at most 2 [24], this implies that a minimum-cost Steiner tree has cost at most twice the cost of OPT. More precisely, if we take a Steiner tree T of graph G with terminals $R \cup \{s\}$ and edge costs $c(e)$, and set the capacity of each edge in T to be 1, and each edge not in T to be 0, then the cost of this tree is at most twice the cost of OPT. We show in Section 5 that we can always form a 2-approximate Nash equilibrium on a Steiner tree, and there are simple examples where there is no Nash equilibrium that buys the minimum-cost Steiner tree. For the 2-Tier network topology, however, notice that the minimum-cost Steiner tree is simply the minimum spanning tree (MST) of G. This is because every node r_i of R has *exactly* one incident edge and is adjacent to a relay node of G, and each relay node is adjacent to one or more receiver nodes. We next prove that there always exists a Nash equilibrium solution that buys the minimum spanning tree, which gives the result stated by Theorem 2, since the cost of the minimum spanning tree is at most twice the cost of the socially optimal solution as argued above.

The proof of Theorem 2 gives a polynomial-time algorithm that returns a strategy profile $\mathbf{p} = (p_i(e), i \in R, e \in E)$ that is a Nash equilibrium, and reserves 1 unit of bandwidth on the edges of the MST. Since the cost of MST is at most 2 times the cost of OPT, the price of stability is at most 2. Theorem 1 and Theorem 2 together imply that the price of stability is 2.

Theorem 2. *There is a polynomial-time algorithm that returns a Nash equilibrium of the capacity allocation game for the 2-tier topology whose social cost is within a factor of 2 of the cost of OPT, and thus the price of stability is at most 2.*

Proof. We prove the result by showing that there always exists a Nash equilibrium solution that reserves 1 unit of bandwidth on the edges of the minimum spanning tree and 0 units of bandwidth on the remaining edges. Our proof is constructive, i.e., we explicitly form payments that purchase the minimum spanning tree. In our payment scheme, for each edge e of minimum-cost spanning tree T, there is a corresponding receiver r_i that reserves 1 unit of bandwidth on it. Notice that even though the 'arbitrary-sharing' cost-sharing scheme allows receivers to share the cost of reserving 1 unit of bandwidth on the edges of T, our payment scheme does not use this property. Therefore, in order to fully specify the payment scheme all we need to do is to assign one receiver for each edge e of T.

Without loss of generality, for an edge $e = (i, j)$, we will assume that j is the node that is closer to s on T than i. If $e = (i, j)$ is an edge incident to a receiver, i.e., $i \in R$, then the receiver i makes a payment on e that is sufficient to reserve 1 units of bandwidth on it. Let $e = (i, j)$ be an edge between two relay nodes, or a relay node and the source, i.e., $i \in L$. Let r_k be an arbitrary receiver that has a direct link to the relay node i, i.e., (r_k, i) is an edge of T. Then, r_k makes a payment of $c(e)$ on e.

Since we have fully specified the payment scheme on all the edges of T, let us now prove that this payment scheme is indeed a Nash equilibrium. Notice that in our payment scheme each receiver is reserving 1 units of bandwidth on either one or two edges of T. More precisely, each receiver r_i paying for the cost of the edge (r_i, l_j) incident to her node, and possibly she is also paying for the cost of the other incident edge of l_j.

Recall that any solution that is feasible for r_i must satisfy Inequality (1) for r_i. If a receiver r_i is only paying for the cost of her incident edge e in T, she trivially does not have an incentive of unilateral deviation since r_i does not have any other incident edges in G, and therefore any solution where r_i satisfies Equation 1 reserves 1 unit of bandwidth on e. Suppose r_i is paying for the cost of both $e = (r_i, l_j)$ incident to her node, and the cost of another incident edge f of l_j. Since the payments of the receivers $R - \{r_i\}$ reserve 1 unit of bandwidth on $T - \{e, f\}$, the best response of r_i must include enough capacity for a flow of size 1 between r_i and $T - \{e, f\}$. The cheapest way to obtain this capacity is to reserve 1 unit of bandwidth along a single path between r_i and $T - \{e, f\}$, and the cost of this path cannot be more than the total cost of e and f, since otherwise T would not be a minimum spanning tree. Thus, any deviation of r_i that results in a feasible solution for r_i is at least as expensive as $c(e) + c(f)$. Since no receiver has an incentive for unilateral deviation, the resulting payment scheme is a Nash equilibrium. ■

5 Generalizations for Arbitrary Network Models

In this section, we consider our general game, with the graph G having arbitrary topology, and an arbitrary subset of receiver nodes. The capacity allocation game is guaranteed to have a Nash equilibrium by Theorem 3, however, the cost of some Nash equilibria can be prohibitive by Theorem 5.

Theorem 3. *Nash equilibrium in pure strategies is guaranteed to exist in the capacity allocation game.*

Proof. For each player $i \in R$, let S_i denote the strategy space of player i. Player i selects a strategy $s_i \in S_i$ when she plays the game. Let $S = \prod_{i \in R} S_i$ denote the strategy space of the game. Notice that S is the product space of the strategy spaces of the players. A *strategy profile* $s \in S$ is an n-tuple $s = (s_1, \ldots, s_n)$ such that each entry s_i of s is a strategy of player i. We use the common notational convenience and write a strategy profile as $s = (s_i, s_{-i})$ where $s_{-i} \in \prod_{j \in R - \{i\}} S_j$. Notice that a strategy $s_i \in S_i$ of player i is a vector of size m (with m being the number of edges in the graph), since a strategy for a player consists of a nonnegative payment for each edge e of G. Without loss of generality, we will assume $s_i(e) \leq \max_e c(e)$. Notice that S is a nonempty, convex, and compact set since S is a cube in $R^{n \times m}$: specifically it is just the cross product of the closed interval $[0, \max_e c(e)]$ taken nm times.

In order to prove the result, we use the technique used in Nash's proof for showing existence of mixed Nash equilibrium in finite games, that uses Kakutani's fixed point theorem. Our proof uses standard techniques, except for the part showing that the graph $\Gamma(F)$ is closed, which requires somewhat different arguments due to the fact that our cost functions are not continuous.

Recall that Kakutani's fixed point theorem is defined as follows:

Theorem 4 (Kakutani's Fixed Point Theorem). *Let S be a non-empty, compact and convex subset of some Euclidean space R^n. Let $F : S \to 2^S$ be a set-valued function on S with a closed graph $\Gamma(F)$, and the property that $F(s)$ is nonempty and convex for all $s \in S$. Then F has a fixed point.*

A set-valued function $F : S \to 2^S$ is some rule that maps each element $s \in S$ to a subset of S, i.e., $F(s) \subset S$. Notice that each element of $F(s)$ is a strategy profile of the capacity allocation game. Since S is nonempty, compact, and convex, Kakutani's fixed point theorem states that if the function graph $\Gamma(F) = \{(s,t)|s \in S, t \in F(s)\}$ (which is a subset of the product space $S \times S$) is a closed set, and $F(s)$ is a nonempty and convex set for all $s \in S$, then there exists $s \in S$ such that $s \in F(s)$, i.e., a fixed point.

For a strategy profile $s = (s_i, s_{-i})$, let $\chi_i(s_{-i})$ denote the set of best responses of player i to the strategies s_{-i} of other players. Given the strategies s_{-i} of other players (which correspond to some capacity reservation on the edges of G), each element $s'_i \in \chi_i(s_{-i})$ is a minimum cost strategy of player i that will ensure that a flow of size 1 can be send from the source to i in the distribution network purchased by (s'_i, s_{-i}). It is easy to see that for each s_{-i} we can express $\chi_i(s_{-i})$ as the set of optimal solutions of a linear program, with only non-strict inequalities. Therefore, $\chi_i(s_{-i})$ is a closed and convex subset of S_i for all $s_{-i} \in \prod_{j \in R-\{i\}} S_j$. Moreover, $\chi_i(s_{-i})$ is non-empty, since player i always has at least one best response.

We define the mapping $F : S \to 2^S$ as follows. Given a strategy profile s, $t \in F(s)$ if t is a strategy profile that can be obtained if each player $i \in R$ deviates from her strategy s_i to one of her best responses, i.e., to an element of $\chi_i(s_{-i})$. Formally, we define F as $F(s_1, \ldots, s_n) = \{(t_1, \ldots, t_n)|t_i \in \chi_i(s_{-i})\}$. In other words, $F(s_1, \ldots, s_n) = \prod_{i \in R} \chi_i(s_{-i})$. Since $F(s)$ is the product space of nonempty, closed, and convex sets, then $F(s)$ is nonempty, closed, and convex for all $s \in S$. Therefore, if the graph $\Gamma(F) = \{(s,t)|s \in S, t \in F(s)\}$ is a closed set, then by Kakutani's fixed point theorem there exists $s \in S$ such that $s \in F(s)$. Notice that $s \in F(s)$ if and only if $s_i \in \chi_i(s_{-i})$ for all players i, i.e., the strategy of all the players is a best response of them to the strategies of the other players. Hence, a fixed point of F is a Nash equilibrium of the capacity allocation game. Therefore, in order to complete the proof all we need to show is that $\Gamma(F)$ is a closed set.

Let $(x^1, y^1), (x^2, y^2), \ldots$ be an arbitrary convergent sequence of points in $\Gamma(F)$, and denote its limit by (x^*, y^*). To show that $\Gamma(F)$ is closed, all we need to show is that $(x^*, y^*) \in \Gamma(F)$. Recall that $\Gamma(F) \subset S \times S$, and $S \times S$ is closed, so $(x^*, y^*) \in S \times S$. Therefore, all we need to show is that $y^*_i \in \chi_i(x^*_{-i})$ for all $i \in R$. In order to do this, fix some arbitrary player i.

Let $C_i(s)$ denote the cost of player $i \in R$ for strategy profile s, and let $\hat{C}_i(s)$ be the cost of i's best response to s, i.e., $\hat{C}_i(s) = \min_{s'_i \in S_i}\{C_i(s'_i, s_{-i})\}$. Since

$C_i(s'_i, s_{-i})$ is minimized if and only if $s'_i \in \chi_i(s_{-i})$, then we can equivalently define $\hat{C}_i$ as $\hat{C}_i(s) = C_i(s'_i, s_{-i})$ for some $s'_i \in \chi_i(s_{-i})$.

We next define a function $\overline{C_i} : S \to \mathcal{R}_{\geq 0}$ for each $i \in R$ as follows: $\overline{C_i}(s) = C_i(s) - \hat{C}_i(s)$. Notice that $\overline{C_i}(s) \geq 0$ for all $s \in S$ and $\overline{C_i}(s) = 0$ if and only if $s_i \in \chi_i(s_{-i})$. In other words, $\overline{C_i}(s) = 0$ if and only if s is a stable strategy profile for player i, i.e., player i does not have an incentive of unilateral deviation from s. Notice that for any point $(s,t) \in \Gamma(F)$, we have that $\overline{C_i}(t_i, s_{-i}) = 0$ since $t_i \in \chi_i(s_{-i})$ by definition of F. Therefore, $\overline{C_i}(y_i^k, x_{-i}^k) = 0$ for all $k > 0$. In the usual argument about the existence of Nash equilibrium, $\overline{C_i}$ is continuous over S and therefore this completes the proof, since this implies that $\overline{C_i}(y_i^*, x_{-i}^*) = 0$, and thus that $y_i^* \in \chi_i(x_{-i}^*)$. In our game, however, $\overline{C_i}$ is not continuous.

Recall that $\overline{C_i}$ is defined as the difference of two functions, i.e., $\overline{C_i}(s) = C_i(s) - \hat{C}_i(s)$. The function $\hat{C}_i(s)$ is continuous on S, since for any strategy profile s' such that $||s - s'|| \leq \epsilon$, we have that $|\hat{C}_i(s) - \hat{C}_i(s')| \leq \epsilon$. However, $C_i(s)$ is not necessarily continuous on S, since when the mincut between the source and i becomes less than 1 in the distribution network, then the cost for player i suddenly becomes unbounded. Let Δ_i be the set of strategy profiles where player i is feasible, i.e., $\Delta_i = \{s | f_i(s) \geq 1\}$. For any $s \in \Delta_i$, the cost $C_i(s)$ is simply equal to $|s_i| = \sum_e s_i(e)$. Thus C_i is clearly continuous on the domain Δ_i. Notice that Δ_i can be formulated as a set of linear constraints with non-strict inequalities and therefore, Δ_i is a closed (and convex) set. Since both $\hat{C}_i$ and C_i are continuous on Δ_i, then $\overline{C_i}$ is also continuous on Δ_i.

Notice that for any $k > 0$, we have that $(y_i^k, x_{-i}^k) \in \Delta_i$, since y_i^k is a best response of player i to x_{-i}^k, and thus results in a solution feasible for player i. Since Δ_i is closed, then (y_i^*, x_{-i}^*) is also in Δ_i. Thus, since $\overline{C_i}(y_i^k, x_{-i}^k) = 0$ for all k, and $\overline{C_i}$ is continuous on Δ_i, then $\overline{C_i}(y_i^*, x_{-i}^*) = 0$. This implies that $y_i^* \in \chi_i(x_{-i}^*)$. Since this is true for all i, then $(x^*, y^*) \in \Gamma(F)$, as desired. ■

Theorem 5. *The price of anarchy for the capacity allocation game is N and this bound is tight.*

In network formation games on undirected graphs, bounding the price of stability is usually a lot more challenging than bounding the price of anarchy [8,10]. The price of stability is known to be 1 for arbitrary sharing in the discrete model, i.e., there exists a Nash equilibrium that buys the Steiner Tree [9]. However, the analysis for the discrete model does not carry over for capacity allocation games (i.e., the fractional model), for which the price of stability is shown to be at least 2 in Section 4. In fact, the example in Section 2 shows that the Steiner tree is not necessarily a Nash equilibrium for our game, and so it is not the case that this factor of 2 arises simply because of the gap between integral and fractional solutions. Thus, even though there always exists a Nash equilibrium that buys the cheapest integral solution in any 2-tier topology (which is the minimum spanning tree in this case), this is not true for general undirected networks.

Though we do not have an upper bound for the price of stability in general undirected networks and therefore cannot guarantee the existence and efficient computation of cheap Nash equilibrium, we prove that there always exists a cheap *approximate* Nash equilibrium that can be efficiently computed. By an α-approximate Nash equilibrium, we mean a solution where no player can reduce her cost by a factor of α by unilateral deviation [9].

Similar to the 2-tier topology in spirit, we start with a cheap integral feasible solution T and form payments on the edges of T. The cheapest integral solution T is the Steiner tree that connects the source and the receiver nodes; however, the Steiner Tree is not efficiently computable. We therefore use an approximation to Steiner Tree that does not cost more than twice the cost of OPT. We obtain the integral solution by using the primal-dual approximation algorithm for minimum-cost Steiner Forest problems (see, e.g. [24]). We use T to denote the tree returned by the primal-dual algorithm with $R \cup \{s\}$ being terminal nodes, and root it at s. Since the primal-dual algorithm uses the fractional optimal solution OPT as its lower bound, the cost of the integral solution returned by the primal-dual algorithm is not only within a factor of 2 of the cost of the minimum-cost Steiner Tree, but also the cost of OPT. The proof of Theorem 6 gives a polynomial-time algorithm that returns a strategy profile $\mathbf{p} = (p_i(e), i \in R, e \in E)$ that is a 2-approximate Nash equilibrium, and reserves 1 unit of bandwidth on the edges of T.

The rest of this section is devoted to proving Theorem 6.

Theorem 6. *There is a polynomial-time algorithm that returns a 2-approximate Nash equilibrium whose social cost is at most twice the cost of OPT.*

We use the term *edge block* to refer to maximal length paths of T whose interior nodes are degree 2 nonreciever nodes. Notice that each edge of T is part of an edge block. It is easy to show that there are at most $2n - k$ edge blocks in T, where n is the number of receiver nodes, and k is the number of non-leaf receiver nodes. Notice that the removal of an edge block e from T will divide T into 2 connected components $T_1(e)$ and $T_2(e)$. Without loss of generality we will assume that e constitutes the cheapest path between these connected components in the rest of the text, since otherwise we can obtain a cheaper integral solution with this property by simply replacing the cheapest path between these 2 connected components for e.

Our payment algorithm assigns each edge block e to a receiver r_i and asks r_i to reserve 1 unit of bandwidth on *all* the edges of e. We use the notation $p_i = \{\mathsf{e}\}$ if edge block e is assigned to receiver r_i and all other edge blocks of T are assigned to the other receivers. Similarly, we say $p_i = \{\mathsf{e}, \mathsf{f}\}$ if edge blocks e and f are assigned to receiver r_i and all other edge blocks of T are assigned to the other receivers. Let $\chi_i\{\mathsf{e}\}$ and $\chi_i\{\mathsf{e}, \mathsf{f}\}$ denote the cheapest deviations of receiver r_i under the two possible strategies of her described above. Let $|\chi_i\{\mathsf{e}\}|$ and $|\chi_i\{\mathsf{e}, \mathsf{f}\}|$ denote the cost of these deviations to player i. We can now show the following lemmas.

Lemma 1. *For an edge block* e *between* s *and* r_i *on* T*, we have that* $\chi_i\{\mathsf{e}\} = \{\mathsf{e}\}$.

Proof. For an edge block e between s and r_i on T, the cheapest deviation $\chi_i\{\mathsf{e}\}$ of receiver r_i to strategy $p_i = \{\mathsf{e}\}$, where other players are reserving 1 unit of bandwidth on all other edge blocks of T, must reserve enough capacity to send 1 unit of traffic between $T_1(e)$ and $T_2(e)$. The cheapest way to do this is to reserve 1 unit of bandwidth along the cheapest path between $T_1(e)$ and $T_2(e)$, and thus $|\chi_i\{\mathsf{e}\}|$ is at least the cost of this path. Since e constitutes the cheapest path between $T_1(e)$ and $T_2(e)$, we have that $\chi_i\{\mathsf{e}\} = \{\mathsf{e}\}$. ■

Lemma 2. *For any two edge blocks* e *and* f *between* s *and* r_i *on* T*, we have* $|\chi_i\{\mathsf{e}, \mathsf{f}\}| \geq \max\{|\chi_i\{\mathsf{e}\}|, |\chi_i\{\mathsf{f}\}|\}$.

Proof. By Lemma 1, we know that $\chi_i\{\mathsf{e}\}$ is the cheapest path between s and r_i if all the edge blocks of T other than e are bought by other players, and similarly for $\chi_i\{\mathsf{f}\}$. By the same argument as in the previous lemma, we know that $\chi_i\{\mathsf{e}, \mathsf{f}\}$ costs at least as much as the cheapest path between s and r_i if all the edge blocks of T other than e and f are bought by other players. The lemma holds since r_i can find a cheaper path between r_i and s if other players buy more edges. ■

Lemma 1 and Lemma 2 imply that we will form a 2-approximate Nash equilibrium if for each receiver r_i, we assign at most 2 edge blocks between r_i and s to it. This is due to the fact that if r_i is assigned edge blocks e and f, then $|\chi_i\{\mathsf{e}, \mathsf{f}\}| \geq \max\{|\chi_i\{\mathsf{e}\}|, |\chi_i\{\mathsf{f}\}|\}$ (by Lemma 2), which is exactly the maximum of the cost of e and f by Lemma 1. Thus, the cost of r_i's best deviation $|\chi_i\{\mathsf{e}, \mathsf{f}\}|$ to its current strategy costs at least half of what r_i is currently paying. Since this is true for every player, this forms a 2-approximate Nash equilibrium. All that is left to show is that we can form such an assignment.

Recall that there are at most $2n$ edge blocks of T and we can make an assignment where each receiver is assigned at most 2 edge blocks. However, we cannot make an arbitrary assignment of edge blocks to receivers, since Lemma 1 and Lemma 2 hold only if a receiver r_i is assigned edge blocks that are between r_i and s. In order to make an assignment with the desired property, we will root T at s, and loop through the edge blocks of T in the reverse BFS order from s. For each edge block e in this order, we select an arbitrary receiver r_i under e that is not assigned 2 edge blocks yet, and assign e to r_i. It is easy to show by induction that at the time the algorithm decides the assignment of an edge block e, there is always a receiver r_i that is not assigned 2 edge blocks yet. By the above argument, this assignment creates a 2-approximate Nash equilibrium that purchases 1 unit of bandwidth on edges of T, and thus has cost at most twice that of OPT.

6 Experimental Results

In this section, we experimentally evaluate, using randomly generated networks, the performance of the algorithm used in the proof of Theorem 2 for the 2-tier network model, and the algorithm used in the proof of Theorem 6 for general undirected networks. Experimental studies indicate that both of the algorithms perform much better than the theoretical guarantees in the corresponding theorems.

We first present the experimental results for the 2-tier network model. Let β denotes the ratio of the cost of the Nash equilibrium (as computed by our MST-based algorithm used in the proof on Theorem 2) to the cost of the socially optimal solution. Recall that $\beta \leq 2$ by Theorem 2. In the representative results shown in Table 1, the total number of non-receiver nodes, i.e., the relay nodes plus the source, is varied from 5 to 25. For each value of the number of nodes, we compute the maximum and average values of β, namely β_{avg} and $\beta_{\max}$, over 200 random runs (network samples). Edges are drawn between the non-receiver nodes randomly, in the following manner: the nodes are picked one by one, and the node picked at any step is connected to each of the nodes already included (in previous steps) with probability $p = 0.5$. If the chosen node remains unconnected at the end of the step, to maintain connectivity, an edge is drawn between this node and a randomly chosen node that is already included. The cost of these edges follows a uniform distribution between 1 and 100. Finally receivers are assigned to the relay nodes randomly, such that each relay node is associated with at least one user: first we assign one receiver to each relay node, and then assign the remaining receiver nodes to the relay nodes randomly. The cost of edges between the receivers and their peers follows a uniform distribution between 1 and 5. Table 1(a) and (b) shows the results for two different numbers of (receiver nodes/non-receiver nodes) ratios. From the results, we observe that β value of the solution computed by our algorithm is very close to 1 on the average, and less than 1.5 in the worst case.

Table 1. Cost-approximation values for randomly generated 2-tier networks

#non-receivers	5	10	15	20	25
β_{avg}	1.093	1.199	1.201	1.195	1.194
$\beta_{\max}$	1.49	1.48	1.37	1.35	1.36

(a) #receiver nodes = 2×#non-receiver nodes

#non-receivers	5	10	15	20	25
β_{avg}	1.079	1.174	1.162	1.147	1.130
$\beta_{\max}$	1.37	1.42	1.34	1.29	1.28

(b) #receiver nodes = 4×#non-receiver nodes

Next we experimentally evaluate the performance of our algorithm used in the proof of Theorem 6 for general undirected networks (which buys a primal-dual approximation of the Steiner Tree), to compute α-approximate Nash equilibria that attains a cost that is within a factor of β of the cost of OPT. Recall that both α and β are at most 2 by Theorem 6. Our experimental study on randomly generated networks show that the observations on β are similar to those observed for the 2-tier model, so we only show the α values in Table 2. In these experiments, the total number of nodes (receivers, relays, and the source) is varied from 20 to 100. For each value of the number of nodes, we compute the maximum and average values of α, namely α_{avg} and $\alpha_{\max}$, over 500 random runs (network samples). Edges are drawn randomly as in the 2-tier model, but across all nodes instead of only the non-receiver nodes. The cost of these edges follows a uniform distribution between 1 and 100, as before. From Table 2, we observe that α value of the solution computed by our algorithm is very close to 1 on average, and equals 1 for a large fraction of the networks. Therefore, for general networks, our algorithm generates a solution that is an exact equilibrium or extremely close to an equilibrium, and has low cost (typically within 1.5 times the socially optimal cost). For (receiver/non-receiver) node number ratio of 2 (not shown in the table), α was observed to be 1 in all 500 runs (network samples). Therefore, from these results we conclude that a larger (receiver/non-receiver) node number ratio creates better (smaller) α values.

Table 2. Equilibrium-approximation values for randomly generated general networks

#nodes	20	40	60	80	100
α_{avg}	1.0003	1.0003	1.0002	1.0009	1.0007
α_{max}	1.1250	1.1429	1.0833	1.2000	1.3333

(a) #receiver nodes = 0.5×#non-receiver nodes

#nodes	20	40	60	80	100
α_{avg}	1.0000	1.0009	1.0000	1.0000	1.0000
α_{max}	1.0000	1.1667	1.0000	1.0000	1.0000

(b) #receiver nodes = #non-receiver nodes

References

1. Web could collapse as video demand soars. Daily Telegraph (April 7, 2008)
2. ITU IPTV Focus Group, `http://www.itu.int/ITU-T/IPTV/`
3. Future Looks Bright For IPTV. Satellite Today (May 2, 2005)
4. `http://www.youtube.com`
5. Cisco Virtual Video Infrastructure Managing Complexity and Scale in a Next-Generation Video Network. White Paper, Cisco Systems (2008)
6. Ahlswede, R., Cai, N., Li, S., Yeung, R.: Network information flow. IEEE Trans. Inform. Theory 46(4), 1204–1216 (2000)
7. Anshelevich, E., Caskurlu, B.: Price of Stability in Survivable Network Design. In: Mavronicolas, M., Papadopoulou, V.G. (eds.) SAGT 2009. LNCS, vol. 5814, pp. 208–219. Springer, Heidelberg (2009)
8. Anshelevich, E., Dasgupta, A., Kleinberg, J., Tardos, É., Wexler, T., Roughgarden, T.: The Price of Stability for Network Design with Fair Cost Allocation. SIAM Journal on Computing 38(4), 1602–1623 (2008)
9. Anshelevich, E., Dasgupta, A., Tardos, É., Wexler, T.: Near-Optimal Network Design with Selfish Agents. Theory of Computing 4, 77–109 (2008)
10. Bilò, V., Caragiannis, I., Fanelli, A., Monaco, G.: Improved Lower Bounds on the Price of Stability of Undirected Network Design Games. In: Kontogiannis, S., Koutsoupias, E., Spirakis, P.G. (eds.) SAGT 2010. LNCS, vol. 6386, pp. 90–101. Springer, Heidelberg (2010)
11. Chekuri, C., Chuzhoy, J., Lewin-Eytan, L., Naor, J., Orda, A.: Non-cooperative multicast and facility location games. In: Proceedings of the 7th ACM Conference on Electronic Commerce (EC), Ann Arbor, Michigan, pp. 72–81 (2006)
12. Chen, H., Roughgarden, T.: Network Design with Weighted Players. In: SPAA (2006)
13. Chen, H., Roughgarden, T., Valiant, G.: Designing Networks with Good Equilibria. In: SODA 2008 (2008)
14. Christodoulou, G., Koutsoupias, E.: On the Price of Anarchy and Stability of Correlated Equilibria of Linear Congestion Games. In: Brodal, G.S., Leonardi, S. (eds.) ESA 2005. LNCS, vol. 3669, pp. 59–70. Springer, Heidelberg (2005)

15. Epstein, A., Feldman, M., Mansour, Y.: Strong Equilibrium in Cost-Sharing Connection Games. In: EC 2007 (2007)
16. Feigenbaum, J., Papadimitriou, C., Shenker, S.: Sharing the Cost of Multicast Transmissions. Journal of Computer and System Sciences 63, 21–41 (2001)
17. Fiat, A., Kaplan, H., Levy, M., Olonetsky, S., Shabo, R.: On the Price of Stability for Designing Undirected Networks with Fair Cost Allocations. In: Bugliesi, M., Preneel, B., Sassone, V., Wegener, I. (eds.) ICALP 2006, Part I. LNCS, vol. 4051, pp. 608–618. Springer, Heidelberg (2006)
18. Ho, T., Medard, M., Shi, J., Effros, M., Karger, D.R.: On randomized network coding. In: Proc. 41st Annual Allerton Conf. Comm., Control, & Computing, Monticello, IL (October 2003)
19. Hoefer, M.: Non-cooperative Facility Location and Covering Games. In: Asano, T. (ed.) ISAAC 2006. LNCS, vol. 4288, pp. 369–378. Springer, Heidelberg (2006)
20. Hoefer, M., Krysta, P.: Geometric Network Design with Selfish Agents. In: Wang, L. (ed.) COCOON 2005. LNCS, vol. 3595, pp. 167–178. Springer, Heidelberg (2005)
21. Li, S.-Y.R., Yeung, R.W., Cai, N.: Linear network coding. IEEE Trans. Inform. Theory 49(2), 317–381 (2003)
22. Nisan, N., Roughgarden, T., Tardos, É., Vazirani, V.V.(eds.): Algorithmic Game Theory. Cambridge University Press
23. Sanders, P., Egner, S., Tolhuizen, L.: Polynomial time algorithms for network information flow. In: Proc.of the Fifteenth Annual ACM Symposium on Parallel Algorithms and Architectures, pp. 286–294 (2003)
24. Vazirani, V.V.: Approximation algorithms. Springer, Berlin (2001)

Author Index

GPSR Compliance

The European Union's (EU) General Product Safety Regulation (GPSR) is a set of rules that requires consumer products to be safe and our obligations to ensure this.

If you have any concerns about our products, you can contact us on ProductSafety@springernature.com

In case Publisher is established outside the EU, the EU authorized representative is:

Springer Nature Customer Service Center GmbH
Europaplatz 3
69115 Heidelberg, Germany

Batch number: 10844611

Printed by Printforce, the Netherlands